Handbook of Exact Solutions to the Nonlinear Schrödinger Equations (Second Edition)

Online at: https://doi.org/10.1088/978-0-7503-5954-2

Handbook of Exact Solutions to the Nonlinear Schrödinger Equations (Second Edition)

Usama Al Khawaja

Department of Physics, The University of Jordan, Amman, Jordan

Department of Physics, United Arab Emirates University, Al-Ain, UAE

Laila Al Sakkaf

College of Engineering, Abu Dhabi University, Al-Ain, UAE

IOP Publishing, Bristol, UK

ISBN   978-0-7503-5954-2 (ebook)
ISBN   978-0-7503-5952-8 (print)
ISBN   978-0-7503-5955-9 (myPrint)
ISBN   978-0-7503-5953-5 (mobi)

DOI   10.1088/978-0-7503-5954-2

Multimedia content is available for this book from https://doi.org/10.1088/978-0-7503-5954-2.

Version: 20241001

IOP ebooks

British Library Cataloguing-in-Publication Data: A catalogue record for this book is available from the British Library.

Published by IOP Publishing, wholly owned by The Institute of Physics, London

IOP Publishing, No.2 The Distillery, Glassfields, Avon Street, Bristol, BS2 0GR, UK

US Office: IOP Publishing, Inc., 190 North Independence Mall West, Suite 601, Philadelphia, PA 19106, USA

To our parents

Contents

Preface to First Edition

We have been involved for the past two decades with the nonlinear Schrödinger equation (NLSE) and its various variations including NLSE with higher-order terms, two- and three-dimensional NLSE, discrete NLSE, nonlocal NLSE, and coupled NLSEs. We noticed that, throughout the long history of NLSE, a large number of exact analytical solutions have been found and the number is still increasing as new solutions are being sought and discovered. As the basis for theoretical models of various research fields such as Bose–Einstein condensates of ultracold gases, nonlinear optics, and deep water waves, this equation is a subject of interest by the scientific communities in all three areas. Its known solutions are scattered in the literatures of the different fields. For a beginner as well as for an expert researcher, it is difficult to keep track of the large number of known solutions. It is important for a researcher or a reviewer to know if a certain solution is a new solution, belongs to an existing class of solutions, or can be trivially obtained from another solution by a transformation.

This book is the result of our effort towards serving the research communities involved with the NLSE in the different fields by collecting all known solutions in one document. In addition, the book organizes the solutions by classifying and grouping them based on the aspects and symmetries they have. Although most of the solutions presented in this book have been derived elsewhere using various methods, we attempt here to present a systematic derivation of many solutions and even have derived some new ones. We have also presented symmetries and reductions that connect different solutions through transformations and enable classifying new solutions into known classes.

For the user to verify that the presented solutions do satisfy the NLSE, we provide Mathematica notebooks containing all solutions in a one-to-one correspondence with the solutions in the text. The reader can run the Mathematica cell and see for themself that it indeed satisfies the NLSE. This is also an efficient method for detecting and avoiding possible typo mistakes in the text. A large number of figures and animations are included to help visualize solutions and their dynamics.

We have applied the following rules while collecting the solutions from the literature: (1) It has to be NLSE-related. As a result of this restriction, some interesting equations have been excluded such as the nonlinear Dirac equation, which we may consider in a future edition. (2) It has to satisfy the NLSE. We attempted to fix typo errors whenever it was obvious, but normally we did not invest much time in discovering what is wrong in the solution that does not satisfy the NLSE. (3) It has to be analytical. This excludes all numerical solutions. Some NLSEs do not admit analytical solutions but do have important stable numerical solutions such as in the case of two-dimensional and discrete NLSEs. Nonetheless, and in order to set a well-defined scope, we restrict our book to analytical solutions.

We made our best effort to cite the reference where the solution appeared first. Quite often, however, solutions were either rederived in a later reference and put in a different form, or were derived under different conditions. We cite the reference that we copied the solution from although a different variation of the solution may have appeared earlier. We do not claim any credit for the solutions collected, and we apologize for any citation mistake and for missing any solution. It should be mentioned that the solutions presented in this book have been the subject of study by a very large number of references. However, and as mentioned above, we cite only references that we copied the solutions from giving priority to any reference containing a large number of solutions. Due to the fact that we employed scaling transformations and symmetry reductions, many solutions published in the literature were not presented; the reader can reproduce those solutions using the transformations and symmetry reductions presented here. We gave numerous examples on such cases. We should be grateful, however, if readers would draw our attention to any missing solutions. We also welcome criticism and comments hoping that they will lead to an enhanced second edition.

Preface to Second Edition

The second edition contains the following three major additions: (i) updating the book with new solutions, (ii) adding an introductory section to each chapter, and (iii) adding a new chapter on the fractional nonlinear Schrödinger equation. In addition, we have corrected some typos in the first edition.

In this edition, we fulfill our promise to the readers in the first edition by updating the book with the new solutions that have appeared in the literature since the publication of the first edition. To the best of our knowledge, we have included all new solutions that have appeared since then.

We believe that an introductory section to each chapter will be useful. The reader will be briefed about the history, background, conservation laws, and physical systems described by the equation at hand. Some chapters will also contain a derivation of the NLSE. Such an overview will be particularly important to readers embarking on a new area of research. In addition, the introduction section will set a context to the solutions listed afterwards.

Noticing the increasing interest in the fractional nonlinear Schrödinger equation, we decided to add a new chapter that contains all known exact solutions of this equation. We remind readers by our criteria of adding new solutions, which are:

   (i) to be a solution to the NLSE or a version of it,
   (ii) to be an analytical solution
   (iii) to be free of mistakes or typos, i.e., the published solution should indeed satisfy the NLSE. In some cases, we were able to identify these typos and mistakes and to correct the solution. We included such corrected solutions and identified them by the term 'corrected'. These criteria were also applied on the solutions included in the new chapter.

We believe with these new additions, the book is now even closer towards achieving its goals which were set at the first edition. New contents added to this edition are summarized in the table below which highlights new sections, new solutions, and new figures in each chapter.

Here is a list of all new additions in the second edition:

Chapter	New chapters, sections, subsections	New solutions	New figures
1	–	–	–
2	2.1	(2.163), (2.176), (2.181), (2.182), (2.183), (2.184), (2.197), (2.198), (2.199), (2.200), (2.202), (2.203), (2.217), (2.218)	2.1, 2.2, 2.3, 2.19
3	3.1	(3.36), (3.37), (3.38)	3.1, 3.3
4	4.1, 4.7, 4.8	(4.28), (4.29), (4.41), (4.42), (4.43), (4.44), (4.45), (4.46), (4.47), (4.48), (4.49), (4.50)	
5	5.1, 5.8.2	(5.28), (5.120), (5.121), (5.122), (5.123), (5.124), (5.125)	5.6
6	6.1	–	
7	7.1, 7.10	(7.23), (7.24), (7.25), (7.26), (7.27), (7.28), (7.29), (7.30), (7.31), (7.32), (7.33), (7.34), (7.35), (7.36), (7.37), (7.127), (7.129), (7.131)	7.10, 7.11, 7.12
8	8.1		8.1
9	9.1, 9.2.2, 9.3	(9.21), (9.22), (9.23), (9.24), (9.25), (9.26), (9.27), (9.28), (9.29), (9.31), (9.32), (9.33), (9.34)	9.1
10	Chapter 10	All solutions	–

Notation

AL	Ablowitz–Ladik
CNLSE	Coupled nonlinear Schrödinger equation
CW	Continuous wave
DNLSE	Discrete nonlinear Schrödinger equation
DT	Darboux transformation
DW	Decaying wave
1D	One-dimensional/dimension
2D	Two-dimensional/dimensions
3D	Three-dimensional/dimensions
FNLSE	Fractional nonlinear Schrödinger equation
GN	General nonlinearity
HE	Hirota equation
HOE	Higher order effects
HONLSE	Nonlinear Schrödinger equation with higher order terms
IPS	Iterative power series
IST	Inverse scattering transform
KN	Kerr nonlinearity
LP	Lax-Pair
ND	N-dimensional/dimensions
$(N+1)$-D	N dimensions in space, 1 refers to time
NLSE	Nonlinear Schrödinger equation
PT	Parity-Time
SBS	Stimulated Brillouin scattering
SN	Saturable nonlinearity
sn, cn, dn, nd, cd, sd, cs, ds, dc, ns	Jacobi elliptic functions
SPM	Self-phase modulation
SRS	Stimulated Raman scattering
SSE	Sasa–Satsuma equation
SW	Solitary wave
WSC	Wide sense criterion
XPM	Cross-phase modulation

Acknowledgements

Fruitful discussions with our colleagues and collaborators, Lincoln Carr, Abdulaziz Alhaidari, Bakhtiyor Baizakov, Hocine Bahlouli, Saeed Al-Marzoug, Yuri Kivshar, Nail Akhmedeiv, Andrey Sokhorokov, Fathulla Abdullaev, Abdelaali Boudjumaa, Khelifa Elhadj, Houria Chachou, Mohamad Al-Refai, Majed Alotaiby, and Majid Taki are acknowledged. We thank authors whom replied to our enquiries about solutions which they have found and published. We acknowledge the support of our universities: The University of Jordan and United Arab Emirates University. We are mostly grateful to our families for their support and patience.

Author Biographies

Usama Al Khawaja

Usama Al Khawaja obtained his Bachelor's degree in Physics from the University of Jordan in 1992 and Master's degree in physics from the same university in 1996 with thesis research on *two-dimensional neutral Fermi systems*. He earned his PhD degree in theoretical physics with dissertation research on *Bose–Einstein condensation* from the University of Copenhagen in 1999. He spent afterwards three years of postdoctoral research at Utrecht University in the Netherlands before joining the United Arab Emirates University in 2002 as an assistant professor. In 2023 he moved to the University of Jordan as a full professor at the Department of Physics. His main research interests are integrability and exact solutions, nonlinear and quantum optics, quantum computation, and Bose–Einstein condensation. His main achievements in integrability and exact solutions include developing a systematic search method of finding Lax pairs of a given nonlinear partial differential equation. He also developed a highly-accurate convergent power series method for solving regular and fractional nonlinear partial differential equations. He has authored more than 100 papers and obtained two patents on applying discrete solitons in all-optical operations and converging light using acoustic waves.

Laila Al Sakkaf

Laila Al Sakkaf earned her Bachelor's degree in physics from the United Arab Emirates University in 2015. She then completed her Master's degree in physics from the same institution in 2018, with thesis research on the *Iterative Power Series Method for Solving Nonlinear Differential equations*. Subsequently, she earned her PhD degree in theoretical physics with a dissertation entitled *Lax Integrability and Solution Methods for the Nonlinear Schrödinger equation with External Potentials: Exact Solutions and Applications* from the United Arab Emirates University in 2023. She served as an instructor at the Department of Physics at the United Arab Emirates University in 2023. Laila is currently an assistant professor in the College of Engineering at Abu Dhabi University. Her research interests encompass integrability and exact solutions of differential equations, particularly those modeling nonlinear physical phenomena, including soliton scattering. She developed a numerical code that excels over existing methods in solving nonlinear differential equations. In 2023 Laila obtained a patent on focusing light using acoustic field.

IOP Publishing

Handbook of Exact Solutions to the Nonlinear Schrödinger Equations (Second Edition)

Usama Al Khawaja and Laila Al Sakkaf

Chapter 1

Introduction

The nonlinear Schrödinger equation is known in the literature, most commonly, with the following dimensionless form

$$i\,\psi_t + \frac{1}{2}\psi_{xx} + \sigma\,|\psi|^2\,\psi = 0, \tag{1.1}$$

where σ is a real constant, $\psi = \psi(x, t)$ is a complex function, and the subscripts are partial derivatives in terms of its two independent variables, x and t. It is the basis for theoretical models describing three major fields, namely: Bose–Einstein condensates of ultracold gases [1], nonlinear optics in fibers and waveguide arrays [2], and deep water waves [3].

In Bose–Einstein condensates, which is a quantum system, the nonlinear Schrödinger equation is the classical field limit of the analogous quantized field equation. The function $\psi(x, t)$ is the wave function of the macroscopic many-particle system. To realize Bose–Einstein condensation, a confining (trapping) magnetic and optical potential is needed. This is accounted for by adding a potential term, $V(x)\psi(x, t)$, to the NLSE which then becomes the *Gross–Pitaevskii equation* [4, 5]. The nonlinear term corresponds to the interatomic interaction known as the Hartree–Fock energy with σ being proportional to the s-wave scattering length. The sign of σ can be both positive and negative, corresponding to attractive or repulsive interatomic interactions, respectively. The dispersion term corresponds to the kinetic energy pressure [1].

In nonlinear optics, the NLSE describes the propagation of pulses in nonlinear media such as optical fibers, photonic crystals, or waveguide arrays. It can be derived from Maxwell's equations with $\psi(x, t)$ corresponding to the envelope of modulated electrical (or magnetic) field strength of the propagating pulse [6, 7]. The nonlinear term corresponds to the modulation of the refractive index of the medium as a response to the propagating light pulse, which is known in the nonlinear optics

doi:10.1088/978-0-7503-5954-2ch1

community as the Kerr nonlinearity. The constant σ represents, in this case, the strength of the Kerr nonlinearity, which can also be positive or negative, leading to the *focusing* or *defocusing* NLSE, respectively. Here, the term ψ_{xx} corresponds to the dispersion of the pulse [2].

The NLSE describes also surface water waves where $\psi(x, t)$ corresponds to the intensity and phase of the waves. This description is restricted to deep water waves with wavelength much smaller than the water depth. Shallow water waves are not described by the NLSE. The nonlinearity originates from the Bernoulli equation, its strength depends on the water depth, and it is always negative for deep water waves [3].

The above three examples suggest that the NLSE is a universal equation that describes the propagation of wave modulations in media with dispersion and nonlinearity.

The NLSE is integrable and admits, in principle, an infinite number of independent solutions [8]. It was first solved by Zakharov and Shabat using the Inverse Scattering Transform (IST) which relies on associating the NLSE with a linear system of differential equations [9]. The system was known since then as the Zakharov–Shabat system and the method was adopted to find other solutions of the NLSE and its variations. In general, the linear system is given in terms of a pair of matrices, known as the Lax pair, acting on an *auxiliary* field. The existence of a Lax pair establishes the integrability of a differential equation, at least within the Lax pair sense [10]. The IST is a powerful method for finding solutions of nonlinear differential equations [11]. It is distinguished among other methods, by generating classes of infinite hierarchy of solutions. It can also be used to exactly solve the nonlinear initial value problem for the given nonlinear differential equation. Many other methods of solving nonlinear differential equations have been applied to the NLSE. For the systematic derivations of solutions we present in this book, we use mainly the IST and separation of variables methods.

There are many variations of the NLSE including NLSE with higher-order terms, NLSE in higher dimensions, NLSE with function coefficients and potential terms, coupled NLSEs, discrete NLSE, nonlocal NLSE, and NLSE with fractional derivatives. It should be noted that we often refer to the NLSE and its variations simply by NLSE. Many of these variations turn out to be integrable and many others turn out to be related to the fundamental NLSE via some *scaling* transformations. All these variations will be considered in this book.

The book starts in chapter 2 with the fundamental NLSE. The Introduction section starts by presenting, discussing, and deriving some of the main features of the NLSE including its Lagrangian representation, conservation laws, integrability, analytical and numerical methods of solution. The NLSE is then derived for the three physical systems of Bose–Einstein condensates, optical pulses in fibers, and deep ocean waves. Solutions to the NLSE are then presented in this chapter with arbitrary constant coefficients a_1 and a_2 for dispersion and nonlinearity, respectively (see (2.160)). One may argue that this is not the 'fundamental' NLSE since a simple scaling transformation, as shown in chapter 5, transforms it into another NLSE with no coefficients ($i\,\psi_t + \psi_{xx} + |\psi|^2\psi = 0$) which may be more accurately denoted as the fundamental NLSE. This is indeed the case when a_2 is real, but the scaling transformation does not work when a_2 is complex. Therefore, by keeping the coefficients a_1 and a_2 explicitly in the NLSE, we will be able to consider solutions when the coefficients are complex. This chapter contains the largest number of solutions collected. The solutions of this chapter can be used as a *seed* for transformations generating many solutions of other NLSEs in the subsequent chapters. The solutions can be categorized as: (i) stationary solutions of the form $\psi(x, t) = u(x) \, e^{i \, \phi(x, t)}$, where $u(x)$ is a real function that can be localized or oscillatory and $\phi(x, t)$ is a real function, (ii) class of breathers family, (iii) class of N-bright solitons, and (iv) rational solutions that are fundamentally different from the breathers class.

In chapter 3, we consider the NLSE with power law and dual power law nonlinearities. Here, the cubic nonlinearity of the fundamental NLSE is replaced first by a nonlinearity with general power, n, that is not restricted to integers. Then we consider an NLSE with two nonlinear terms; one with power n and another with power m, where again n and m are arbitrary real constants that do not have to be integers. In the first case, we show, at the beginning of the chapter, that with a scaling transformation applied to stationary solutions, the NLSE with power law nonlinearity reduces to the fundamental NLSE, and thus all stationary solutions of chapter 2 lead to stationary solutions to the NLSE with power law nonlinearity. Many examples have been worked out explicitly. We could not find a similar transformation for the dual power law nonlinearity, but we have found 17 solutions for this case.

In chapter 4, we consider the NLSE with higher-order terms These include, following the nonlinear optics terminology: third-order dispersion, fourth-order dispersion, self-steepening, self-frequency shift, and power law nonlinearity. Finally, we consider an infinite hierarchy of integrable NLSEs. The first member of the hierarchy being the fundamental NLSE, the higher-order members turn out to comprise most of the known higher-order variations of the NLSE. Solutions to all member equations of the infinite hierarchy were presented.

In chapter 5, we present scaling transformations that reduce many variations of the NLSE to the fundamental one. These include, transforming the NLSE with arbitrary constant coefficients to the one with no coefficients, transforming the NLSE with focusing (defocusing) nonlinearity to the NLSE with defocusing (focusing) nonlinearity, Galilean transformation to obtain movable solutions from static ones, transforming the NLSE with function coefficients and complex potential to the fundamental NLSE, and introducing a solution-dependent transformation where a seed solution is used to construct the transformation operator. The latter allowed for more possibilities including transforming an NLSE with constant coefficients and PT-symmetric potential to the fundamental NLSE. A new sub-section (section 5.8.2) is now added to the second edition which contains solutions to the NLSE with the reflectionless Pöschl-Teller potential. These solutions are derived by the authors and we believe they are presented here for the first time in the literature.

In chapter 6, we consider the NLSE in higher dimensions. We start with a scaling transformation showing that an $(N + 1)$-dimensional NLSE, with N denoting the spatial dimensions and 1 denoting the temporal dimension, can be reduced to the one-dimensional NLSE in terms of a reduced spatial variable. A Galilean transformation is then shown to apply where movable solutions of the $(N + 1)$-dimensional NLSE are obtained from the static solutions of the one-dimensional NLSE. An NLSE with mixed derivatives is also considered. Then, solutions of the $(N + 1)$-dimensional NLSE with power law and dual power law nonlinearities were presented. A scaling transformation of the $(N + 1)$-dimensional NLSE in polar coordinates was also worked out allowing consideration of solutions with cylindrical and spherical symmetries in two- and three-dimensional geometries, respectively. Finally, we present our iterative power series method of obtaining convergent power series representations of the solutions. The method is applied here for nonintegrable cases such as some two- and three-dimensional NLSEs.

In chapter 7, we consider the coupled NLSE. First, we consider the fundamental coupled NLSE, known as the Manakov system. Being an integrable system, many solutions have been found and presented. Then, we show three simple symmetry reductions that transform the coupled NLSE to the scalar NLSE and symmetry reduction that transforms the vector NLSE (N-coupled NLSEs) to the Manakov system. Some interesting examples have been shown explicitly. We consider also a coupled system with additional linear coupling terms A scaling transformation is performed to reduce this system to the fundamental Manakov system. Here, we found that, as a special case, one may obtain the solutions of a Manakov system from those of another Manakov system that differs in the values of the constant coefficients. Furthermore, one may obtain a solution to the Manakov system from another solution to the same system, which invokes the superposition principle known for linear differential equations. It is interesting to find such a principle applying for nonlinear differential equations. We have worked out explicitly a nontrivial example on this case. We have also considered a coupled system with additional complex coupling terms, which again with a scaling transformation was reduced to the fundamental Manakov system. Then we considered the $(N + 1)$-dimensional coupled NLSEs. First, we show that, with a proper scaling transformation, this system reduces to the one-dimensional Manakov system. Then, we consider scaling transformations that reduce this system to the $(N + 1)$-dimensional scalar NLSE. Scaling transformations were then found for linear and nonlinear coupling, as mentioned above, but here generalized for $(N + 1)$ dimensions. A new section (section 7.10) is added to the second edition where composite solutions, which are linear superposition of other solutions, are presented.

In chapter 8, we consider the discrete NLSE. The chapter is started by the discrete NLSE with saturable nonlinearity, which is integrable. Here, the solutions are classified into staggered and nonstaggered solutions. We show the transformation that links the two kinds of solutions. Then, we consider other types of nonlinearity including a general form of nonlinearity. Then, we considered the Ablowitz–Ladik equation which is also integrable. Then, we consider an integrable discrete NLSE with cubic and quentic nonlinearites. A generalized discrete NLSE was then considered with many solutions satisfying the equation under some integrability conditions. Finally, we consider coupled discrete NLSEs including the coupled Salerno equations and the coupled Ablowitz–Ladik equations.

In chapter 9, we consider the nonlocal NLSE. We start with transformations that reduce the nonlocal NLSE to the fundamental NLSE. This is possible only for even or odd solutions in the variable x. Then we consider coupled nonlocal NLSEs and reduce them to the local Manakov system. Then, we consider nonlocal coupled NLSEs with linear, nonlinear, and complex coupling. With scaling transformation, they are reduced to their local counterparts. Finally, we consider the nonlocal discrete NLSE, nonlocal discrete coupled NLSEs, nonlocal Ablowitz–Ladik coupled system, and nonlocal NLSE with cubic and quentic nonlinearities.

Chapter 10 is a new addition to the second edition. Here, we collect solutions to the fractional NLSE including different versions with fractional time-derivative, fractional space-derivatives, or both.

Chapters 2 and 3 are supplemented by appendix A and appendix B, where we lay out detailed derivations of most of the solutions presented in these chapters. In appendix A, we follow the standard methods of solving differential equations in a systematic manner in order account for all possible solutions. Appendix B explains the Lax pair and Darboux transformation method and gives the detailed derivation of the bright soliton and breather solutions. Chapter 5 is supplemented with appendix C where we show the detailed derivations of the scaling transformation. All chapters start with a 'glance' that helps the reader to easily navigate through the chapter.

All solutions presented in the text are rewritten, together with the NLSE they satisfy, in a Mathematica notebook (available online at http://doi.org/10.1088/978-0-7503-5954-2). The reader can run the cells in the Mathematica notebook in order to verify that the solutions indeed satisfy the NLSE. This minimizes any possible typo errors in the text. It is also convenient to have the solution typed in for those readers who want to use the solutions in their calculations. Verification of the solutions was possible at different levels of accuracy. Some solutions satisfy the NLSE with all variables and parameters unspecified. For some other solutions, Mathematica could not verify the solution in a reasonable time, therefore we set arbitrary values for the variables and parameters. Here we used only integers or ratios of integers. The verification is then obtained with the integer '0' as a result of substituting the solution in the NLSE. This result is numerical but with infinite accuracy. For the rest of solutions, Mathematica could not verify the solutions with infinite accuracy. In this case, we set numerical values for the parameters and variables with a chosen number of digits. The verification leads to a numerical zero with accuracy increasing when the number of digits is increased. Some readers may still want to verify the last two cases with unspecified values of the variables. In this case they need to wait longer. For example, with a typical personal computer one may have to wait 15 minutes to verify the two-bright-solitons solution with unspecified variables.

There are many solutions which can be obtained from other solutions for some specific values of the parameters. For the reader to have a quick and easy access to the solutions of interest, we present such solutions explicitly.

This book can be considered as the nucleus of a growing collection of solutions to the NLSE and its various variations. The search for new solutions is an ongoing effort by many researchers. We believe that new solutions will continue to appear. Our book will be a suitable host and helpful to keep track of the new solutions and classify them into their proper classes, if any. The book will be also a useful reference to judge on what will be claimed as new solutions. In addition, the scaling transformations presented in this book will be an efficient tool to reveal whether a new solution can be obtained from an existing solution via a transformation. They will be helpful to derive new solutions for some specific setups. We aim at monitoring the literature for the appearance of new solutions of the NLSE and plan to update the book frequently with future editions. Researchers and users of this book are invited to contribute by suggesting and pointing out new or any missing solutions so that we incorporate them in future editions.

References

[1] Pethick C J and Smith H 2001 *Bose-Einstein Condensation in Dilute Gases* (Cambridge University Press)

[2] See for instance: Hasegawa A and Kodama Y 1995 *Solitons in Optical Communications* (New York: Oxford University Press); Mollenauer L F and Gordon J P 2006 *Solitons in Optical Fibers* (Boston, MA: Academic); Agrawal G P 2001 *Nonlinear Fiber Optics* 3rd edn (San Diego, CA: Academic); Akhmediev N N and Ankiexicz A 1997 *Solitons: Nonlinear Pulses and Beams* (London: Chapman and Hall); Akhmediev N N and Ankiewicz A 2008 *Dissipative Solitons: From Optics to Biology and Medicine* (Berlin: Springer); Taylor J (ed) 1992 *Optical Solitons: Theory and Experiment* Cambridge Studies in Modern Optics (Cambridge: Cambridge University Press) pp I–VI; Kivshar Y S and Agrawal G P 2003 *Optical Solitons* (Burlington, VT: Academic); Butcher P and Cotter D 1990 *The Elements of Nonlinear Optics* Cambridge Studies in Modern Optics (Cambridge: Cambridge University Press); Newell A C and Moloney J V 1992 *Nonlinear Optics* (Redwood City, CA: Addison-Wesley); Taylor J R (ed) 1992 *Optical Solitons—Theory and Experiment* (Cambridge: Cambridge University Press)

[3] Kharif C, Pelinovsky E and Slunyaev A 2009 Rogue waves in the ocean *Advances in Geophysical and Environmental Mechanics and Mathematics* (Berlin: Springer)

[4] Gross E P 1961 Structure of a quantized vortex in boson systems *Il Nuovo Cimento (1955–1965)* **20** 454–77

[5] Pitaevskii L P 1961 Vortex lines in an imperfect Bose gas *Sov. Phys.-JETP* **13** 451–4

[6] Hasegawa A and Tappert F 1973 Transmission of stationary nonlinear optical pulses in dispersive dielectric fibers. I. Anomalous dispersion *Appl. Phys. Lett.* **23** 142–4

[7] Hasegawa A and Tappert F 1973 Transmission of stationary nonlinear optical pulses in dispersive dielectric fibers. II. Normal dispersion *Normal dispersion Appl. Phys. Lett.* **23** 171–2

[8] Zakharov V E E and Manakov S V 1974 On the complete integrability of a nonlinear Schrodinger equation *Theor. Math. Phys.* **19** 332–43

[9] Shabat A and Zakharov V 1972 Exact theory of two-dimensional self-focusing and one-dimensional self-modulation of waves in nonlinear media *Sov. Phys.-JETP* **34** 62–9

[10] Al Khawaja U 2010 A comparative analysis of Painlevé, Lax pair, and similarity transformation methods in obtaining the integrability conditions of nonlinear Schrödinger equations *J. Math. Phys.* **51** 053506–11
[11] Ablowitz M J and Clarkson P A 1991 *Solitons, Nonlinear Evolution equations and Inverse Scattering* (Cambridge: Cambridge University Press) p 149

IOP Publishing

Handbook of Exact Solutions to the Nonlinear Schrödinger Equations (Second Edition)

Usama Al Khawaja and Laila Al Sakkaf

Chapter 2

Fundamental Nonlinear Schrödinger Equation

A Glance at Chapter 2

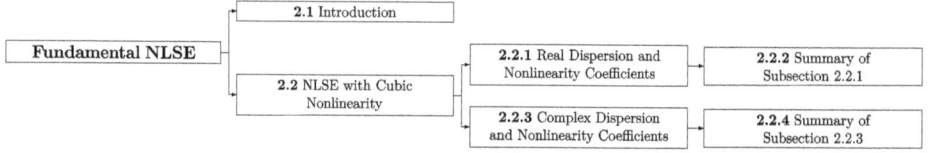

A Statistical View of Chapter 2

	Equation	Solutions
1	$i\,\psi_t + a_1\,\psi_{xx} + a_2\,|\psi|^2\,\psi = 0$	62
2	$i\,\psi_t + (a_{1r} + i\,a_{1i})\,\psi_{xx} + (a_{2r} + i\,a_{2i})\,|\psi|^2\,\psi = 0$	10
Total	2	72

2.1 Introduction

The nonlinear Schrödinger equation (NLSE) is, in a sense, a universal equation in that it describes the propagation of localized waves in any system with wave dispersion depending on the intensity of the wave. Intensity dependence of dispersion introduces nonlinearity in the equation. For instance, the fundamentally different physical systems of Bose–Einstein condensates [1], optical pulses in optical fibers, photonic crystals, and waveguide arrays [2–9], surface gravity waves in deep waters [10, 11, 12],

Langmuir waves in plasma physics [13, 14], and many others [15], are all described by the same NLSE with cubic nonlinearity. While they all share the same general feature of intensity-dependent wave dispersion, the nature of the field and the physical source of nonlinearity in each of these systems is different. In Bose–Einstein condensates, the NLSE describes mean-field intensity of matter-waves and the nonlinearity, which corresponds to the so-called Hartree–Fock energy, originates from the inter-atomic interactions. The NLSE can be derived from a mean-field theory of an interacting dilute gas of bosons, which is known in the condensed-matter community as the Gross–Pitaevskii equation [16, 17]. In optical fibers, the NLSE describes the propagation of the electric field envelope of an electromagnetic pulse [18]. The nonlinearity, which is in this case known as the Kerr nonlinearity, is caused by the interaction between the propagating electromagnetic wave and the medium. Specifically, for large pulse intensities, the interaction results in anharmonic electron oscillations generating nonlinear dependence of electric polarization on the field intensity, which in turn leads to the intensity-dependent dispersion. In water waves, the NLSE describes the propagation of gravity waves on the surface of deep waters where the field corresponds to the height of the wave and the nonlinearity arises from the kinetic energy of the water wave and the boundary conditions [12].

The fundamental nonlinear Schrödinger equation for a complex field $u(x, t)$ reads

$$iu_t + a_1 u_{xx} + a_2 |u|^2 u = 0, \tag{2.1}$$

where $u_t = \partial u/\partial t$, $u_{xx} = \partial^2 u/\partial x^2$, and a_1 and a_2 are arbitrary constants corresponding to the strengths of dispersion and nonlinearity, respectively. In Section 2.1.2, we present a generic derivation of this equation for a pulse with general intensity-dependent dispersion. In Section 2.1.3, the equation is derived for Bose–Einstein condensates. In Section 2.1.4, the equation is derived for optical pulses in fibers, and in Section 2.1.5, the equation is derived for surface water gravity waves. In each of the physical systems, the real dimensions of the field and parameters are specified and examples on their numerical values are given. Before delving into these derivations, we present in the following subsection some important general features and properties of the NLSE, including its Lagrangian representation, energy functional, conservation laws, integrability, and numerical solution.

2.1.1 Main Features of NLSE

2.1.1.1 Lagrangian and Energy Functional

It is useful to construct the Lagrangian that the NLSE corresponds to. Here, we show this for the case of one spatial dimension, namely $u = u(x, t)$. The NLSE is the solution for the action principle which requires a vanishing infinitesimal variation in the action with respect to the field, u, and its complex conjugate, u^*,

$$\delta S = \delta \int L dt = 0. \tag{2.2}$$

Being complex functions, u and u^* are considered as two independent functions. The Lagrangian

$$L = \int dx \mathcal{L} \tag{2.3}$$

is defined in terms of the lagrangian density

$$\mathcal{L} = \frac{i}{2}\left(u^* \frac{\partial u}{\partial t} - u\frac{\partial u^*}{\partial t}\right) - \mathcal{E} \tag{2.4}$$

and

$$\mathcal{E} = a_1 \left|\frac{\partial u}{\partial x}\right|^2 - \frac{1}{2}a_2|u|^4 \tag{2.5}$$

is the energy density, which defines the energy functional

$$E = \int \mathcal{E}dx. \tag{2.6}$$

The NLSE can now be obtained as the Euler–Lagrange equation of the Lagrangian density

$$\frac{d}{dt}\frac{\partial \mathcal{L}}{\partial u_t^*} + \frac{d}{dx}\frac{\partial \mathcal{L}}{\partial u_x^*} - \frac{\partial \mathcal{L}}{\partial u^*} = 0, \tag{2.7}$$

which gives the NLSE, Equation (2.1),

$$iu_t + a_1 u_{xx} + a_2|u|^2 u = 0.$$

The complex conjugate of the NLSE will be obtained as the Euler–Lagrange equation for u^*

$$\frac{d}{dt}\frac{\partial \mathcal{L}}{\partial u_t} + \frac{d}{dx}\frac{\partial \mathcal{L}}{\partial u_x} - \frac{\partial \mathcal{L}}{\partial u} = 0, \tag{2.8}$$

which gives

$$-iu_t^* + a_1 u_{xx}^* + a_2|u|^2 u^* = 0. \tag{2.9}$$

2.1.1.2 Conservation Laws

The NLSE supports an infinite number of conserved quantities. Most prominent among them are the norm:

$$N = \int_{-\infty}^{\infty} |u|^2 dx, \tag{2.10}$$

momentum:

$$J = \frac{i}{2}\int_{-\infty}^{\infty}\left(u_x^* u - u_x u^*\right)dx, \tag{2.11}$$

and energy:

$$E = \int_{-\infty}^{\infty}\left(a_1|u_x|^2 - \frac{1}{2}a_2|u|^4\right)dx. \tag{2.12}$$

According to Noether's theorem [19], each conservation law is associated with a transformation symmetry. The conservation of norm is a result of the Lagrangian invariance to a constant change in the phase of the field. The conservation of momentum results from the invariance to constant shift in space, and the conservation of energy results from the invariance to constant shift in time [20].

Conservation laws can be obtained from the NLSE and its complex conjugate as follows. Multiply the NLSE, Equation (2.1), by u^*, multiply the complex conjugate of the NLSE, Equation (2.9), by u, subtract the resulting two equations, and then integrate over x, to get

$$\int_{-\infty}^{\infty} \left[i\left(u^* u_t + uu_t^*\right) + a_1\left(u^* u_{xx} - uu_{xx}^*\right) \right] dx = 0. \tag{2.13}$$

Under the assumption that u and u^* are equal at the boundaries, but they do not necessarily have to vanish, and applying integration by parts to the x-derivative terms, gives $\int_{-\infty}^{\infty} u^* u_{xx} dx = -\int_{\infty}^{\infty} u_x u_x^* dx$ and $\int_{-\infty}^{\infty} uu_{xx}^* dx = -\int_{\infty}^{\infty} u_x u_x^* dx$, which results in $\int_{-\infty}^{\infty} (u^* u_{xx} - uu_{xx}^*)dx = 0$, and we get

$$\frac{\partial}{\partial t} \int_{-\infty}^{\infty} |u|^2 dx = 0, \tag{2.14}$$

showing that the norm, N, is a conserved quantity.

For the momentum conservation, multiply the NLSE, Equation (2.1), by u_x^*, multiply the complex conjugate of the NLSE, Equation (2.9), by u_x, multiply the x-derivative of (2.1) by $-u^*$, and multiply the x-derivative of (2.9) by $-u$, to get respectively

$$i\, u_t\, u_x^* + a_1\, u_{xx}\, u_x^* + a_2\, |u|^2\, u\, u_x^* = 0, \tag{2.15}$$

$$-i\, u_t^*\, u_x + a_1\, u_{xx}^*\, u_x + a_2\, |u|^2\, u^*\, u_x = 0, \tag{2.16}$$

$$-i\, u_{xt}\, u^* - a_1\, u_{xxx}\, u^* - a_2\, (|u|^2\, u)_x\, u^* = 0, \tag{2.17}$$

$$i\, u_{xt}^*\, u - a_1\, u_{xxx}^*\, u - a_2\, (|u|^2\, u^*)_x\, u = 0. \tag{2.18}$$

Adding these four equations and integrating over x, we obtain

$$\int_{-\infty}^{\infty} \left\{ \frac{\partial}{\partial t} i \left(u_x^*\, u - u_x\, u^*\right) \right.$$
$$+ a_1 \left(u_{xx}\, u_x^* + u_{xx}^*\, u_x - u_{xxx}\, u^* - u_{xxx}^*\, u\right) \tag{2.19}$$
$$\left. + a_2 \left[|u|^2\, u\, u_x^* + |u|^2\, u^*\, u_x - (|u|^2\, u)_x\, u^* - (|u|^2\, u^*)_x\, u \right] \right\} dx = 0.$$

The first two dispersion terms, namely $\int_{-\infty}^{\infty} u_{xx}\, u_x^*\, dx$ and $\int_{-\infty}^{\infty} u_{xx}^*\, u_x\, dx$, cancel out by applying integration by parts to any of these two terms. The last two terms, namely $\int_{-\infty}^{\infty} - u_{xxx}\, u^*\, dx$ and $\int_{-\infty}^{\infty} - u_{xxx}^*\, u\, dx$, also cancel out by applying the integration by parts two times to any of these terms and once to the other term. This requires the additional assumption that the x-derivative of u must be also the same for $x \to \infty$ or for $x \to -\infty$. Alternatively, one may apply integration by parts three times to one of the terms. However, this will require an additional condition on the second x-derivative of u to be equal at both boundaries. The nonlinear terms simplify to $2a_2 \int_{-\infty}^{\infty} |u|^2 (|u|^2)_x dx = a_2 |u|^4 |_{-\infty}^{\infty} = 0$. This leaves

$$\frac{\partial}{\partial t} \int_{-\infty}^{\infty} i \left(u_x^* u - u_x u^* \right) dx = 0, \tag{2.20}$$

which shows that the momentum $J = (i/2) \int_{-\infty}^{\infty} (u_x^* u - u_x u^*) dx$ is a conserved quantity. It is noted that the quantity $u_x^* u$ is also conserved. This can be seen by subtracting Equation (2.17) from (2.15), and then using integration by parts. This quantity can be used as an alternative definition for the momentum. However, it is in general complex while the former definition is always real.

For the conservation of energy, multiply (2.1) by u_t^*, multiply (2.9) by u_t, add the resulting equations, and then integrate over x, to get

$$\int_{-\infty}^{\infty} \left[-a_1 \left(u_t^* u_{xx} + u_t u_{xx}^* \right) - a_2 \left(|u|^2 u u_t^* + |u|^2 u^* u_t \right) \right] dx = 0, \tag{2.21}$$

where the pure temporal derivative terms cancel out. Applying the integration by parts to the dispersion terms, the last expression takes the form

$$\frac{\partial}{\partial t} \int_{-\infty}^{\infty} \left(-a_1 |u_x|^2 - \frac{1}{2} a_2 |u|^4 \right) dx = 0, \tag{2.22}$$

which is the conservation law for the energy functional. Higher-order invariants can be similarly derived.

The following procedure can, however, systematically produce conserved quantities of any order [23, 24]

$$b_{n+1} = a_1 u^* \frac{\partial}{\partial x} \left(\frac{b_n}{u^*} \right) - \frac{a_2}{2} \sum_{k=1}^{n-1} b_k b_{n-k}, \quad n \geqslant 1, \tag{2.23}$$

$$d_n = \frac{1}{a_1^2} \text{Re}[b_n] \tag{2.24}$$

with $b_1 = a_1 u u^*$. The conserved quantities are then expressed as

$$\frac{\partial}{\partial t} \int_{-\infty}^{\infty} d_n \, dx = 0. \tag{2.25}$$

For $n = 1, 2, 3$, the conserved quantities will be norm, momentum, and energy, respectively. While $\int_{-\infty}^{\infty} b_n \, dx$ is also conserved, the definition in (2.24) guarantees that the conserved quantities will be real. One can verify the conservation property (2.25) by inserting the partial t-derivative in the x-integration, using the NLSE and its complex conjugate to substitute for the u_t and u_t^*, and then using integration by parts.

2.1.1.3 Integrability

Possessing an infinite number of conserved quantities, as we have seen above, the NLSE is considered as an integrable differential equation. This was established by Zakharov and Shabat who employed the inverse scattering transform (IST) to exactly solve the NLSE by introducing what is now known as the Zakharov–Shabat system [25], and to show that the NLSE is completely integrable [26]. An alternative but equivalent approach was later developed by Ablowitz *et al*, which is known as the AKNS system [27–30]. In both systems, the NLSE is associated with a system of coupled linear differential equations through a *Lax pair*. The *Darboux transformation* is then applied to the linear system which ultimately results in a new solution to the NLSE. Finding a new solution using this transformation requires knowing one solution, denoted as the *seed* solution. A class of infinite chains of independent solutions can be obtained by repeated and successive action of the Darboux transformation. The procedure is explained in detail in appendix B. It should be noted however, that while the IST is a powerful method that generates systematically families of solutions to the NLSE, it is not the only analytical method available for solving the NLSE. There are many other methods that generate solutions that cannot be obtained by the IST. Specifically, there are many cases where a certain version of the NLSE does not have a Lax pair, nonetheless some of its exact analytical solutions exist.

2.1.1.4 Numerical Solution

Numerical methods of solving the NLSE are often used in cases where a version of the NLSE is not integrable, or a certain solution is not possible to generate in an analytical form. Many numerical solution methods have been developed, which are classified mainly as finite difference or spectral methods, such as the split-step Crank–Nicolson method and the Fourier spectral method, see for instance [31–33] and references therein. Based on power series expansions, we have recently developed a finite difference method with machine precision accuracy [34].

In the following section, we present a generic derivation of the NLSE for any system that supports the intensity-dependent dispersion. In the subsequent sections, we present the derivation for the above-mentioned three specific systems, namely Bose–Einstein condensates, optical pulses in fibers, and water waves.

2.1.2 Generic Derivation

Consider a complex field, $q(x, t)$, with a wave propagating in the form

$$q(x, t) = u(x, t)\, e^{i(k_0 x - \omega_0 t)}, \tag{2.26}$$

which corresponds to an envelope, $u(x, t)$, superposed on a propagating carrier wave with frequency ω_0 and wavenumber k_0. The envelope can be localized in space or time. The former is described by the so-called spacial NLSE and the latter is described by the temporal NLSE. The Fourier transform of this wave is given by

$$Q(k, \omega) \equiv \mathcal{F}[q(x, t)] = \int_0^\infty \int_{-\infty}^\infty q(x, t) \, e^{i(kx-\omega t)} dx \, dt, \qquad (2.27)$$

which is also localized in the frequency or wavenumber domain such that the localization is in general centered at $k = k_0$ and $\omega = \omega_0$. For simplicity, we take $k_0 = \omega_0 = 0$, which corresponds to a coordinate shift in the k- and ω-space. The wave dispersion is assumed to be intensity-dependent

$$\omega = \omega(k, |u|^2), \qquad (2.28)$$

which is expanded in a Taylor series around $k_0 = 0$ and $|u|^2 = 0$ as

$$\omega = \omega(0, 0) + k\frac{\partial \omega}{\partial k} + \frac{1}{2}k^2\frac{\partial^2 \omega}{\partial k^2} + |u|^2\frac{\partial \omega}{\partial |u|^2} + \cdots. \qquad (2.29)$$

Multiplying the last equation by $u(x, t) \, e^{i(kx-\omega t)}$ and then integrating with respect to x and t, we get

$$\int_0^\infty \int_{-\infty}^\infty \left(\omega - k\frac{\partial \omega}{\partial k} - \frac{1}{2}k^2\frac{\partial^2 \omega}{\partial k^2} - \frac{\partial \omega}{\partial |u|^2}|u|^2\right)u \, e^{i(kx-\omega t)} dx \, dt = 0, \qquad (2.30)$$

where we have ignored the higher order terms and used $\omega(0, 0) = \omega_0 = 0$. Using the Fourier transform properties for derivatives, $\mathcal{F}[u_x] = -ik\mathcal{F}[u]$, $\mathcal{F}[u_t] = i\omega\mathcal{F}[u]$, and $\mathcal{F}[u_{xx}] = -k^2\mathcal{F}[u]$, the last equation can be re-expressed as

$$\int_0^\infty \int_{-\infty}^\infty \left[\left(-i\frac{\partial}{\partial t} - i\omega'\frac{\partial}{\partial x} + \frac{1}{2}\omega''\frac{\partial^2}{\partial x^2} - \frac{\partial \omega}{\partial |u|^2}|u|^2\right)u\right] e^{i(kx-\omega t)} dx \, dt = 0, (2.31)$$

which vanishes identically for

$$\left(i\frac{\partial}{\partial t} + iv_g\frac{\partial}{\partial x} + a_1\frac{\partial^2}{\partial x^2} + a_2|u|^2\right)u = 0, \qquad (2.32)$$

with group velocity $v_g = \omega' = \partial\omega/\partial k$, dispersion coefficient $a_1 = -(1/2)\omega'' = -(1/2)\partial^2\omega/\partial k^2$, and nonlinearity strength $a_2 = \partial\omega/\partial|u|^2$. It is convenient to re-express this equation in terms of coordinates of a frame of reference moving along the x-axis with the group velocity via the transformation: $x \to x - v_g t$. In the moving frame, the single spacial derivative disappears, and we finally obtain the spacial NLSE (2.1), with cubic nonlinearity

$$\left(i\frac{\partial}{\partial t} + a_1\frac{\partial^2}{\partial x^2} + a_2|u|^2\right)u = 0.$$

The temporal NLSE can be similarly derived as follows. We start by expanding the inverse dispersion relation $k = k(\omega, |u|^2)$ around $\omega_0 = 0$ and $|u|^2 = 0$

$$k = k(0, 0) + \omega\frac{\partial k}{\partial \omega} + \frac{1}{2}\omega^2\frac{\partial^2 k}{\partial \omega^2} + |u|^2\frac{\partial k}{\partial |u|^2} + \cdots, \qquad (2.33)$$

which gives

$$\int_0^\infty \int_{-\infty}^\infty \left(k - \omega \frac{\partial k}{\partial \omega} - \frac{1}{2}\omega^2 \frac{\partial^2 k}{\partial \omega^2} - \frac{\partial k}{\partial |u|^2} |u|^2 \right) u \, e^{i(kx-\omega t)} dx \, dt = 0. \quad (2.34)$$

Using the Fourier transformation of derivatives, this equation is re-expressed as

$$\int_0^\infty \int_{-\infty}^\infty \left[\left(i\frac{\partial}{\partial x} + ik'\frac{\partial}{\partial t} + \frac{1}{2}k''\frac{\partial^2}{\partial t^2} - \frac{\partial k}{\partial |u|^2} |u|^2 \right) u \right] e^{i(kx-\omega t)} dx \, dt = 0, \quad (2.35)$$

leading to

$$\left(i\frac{\partial}{\partial x} + ik'\frac{\partial}{\partial t} + \frac{1}{2}k''\frac{\partial^2}{\partial t^2} - \frac{\partial k}{\partial |u|^2} |u|^2 \right) u = 0. \quad (2.36)$$

The transformation $t \to t - k'x$ removes the first t-derivative term and we obtain the temporal NLSE

$$\left(i\frac{\partial}{\partial x} + b_1 \frac{\partial^2}{\partial t^2} + b_2 |u|^2 \right) u = 0, \quad (2.37)$$

where $b_1 = 1/2k''$ and $b_2 = -\partial k/\partial |u|^2$.

Different versions of the NLSE may be derived following the above derivation but with some variations. For instance, the NLSE with power law and dual power law nonlinearities (Chapter 3), may be derived by extending the expansion of $\omega(k, |u|^2)$ to higher powers in $|u|$. The NLSE with higher order dispersion (Chapter 4), may also be derived by keeping higher order terms in the expansion of $\omega(k, |u|^2)$ in powers of k. Other higher order terms, as in Chapter 4, can also be accounted for, by taking the more general dependance of dispersion as $\omega = \omega(k, u, u_x, u^*, u_x^*, \ldots)$. Keeping all terms in the expansion of $\omega(k, |u|^2)$ in powers of k, may lead to the fractional NLSE (Chapter 10). Furthermore, the above derivation assumed $\omega(k, |u|^2)$ to be independent of x and t. Extending to x- and t-dependent dispersion, the NLSE with function coefficients (Chapter 5), may also be derived.

2.1.3 Bose–Einstein Condensates

Nonlinearity in the Schrödinger equation that describes Bose–Einstein condensates originates from two-body interatomic interactions. The equation can be derived as the Euler–Lagrange equation of motion for the action of weakly-interacting N bosons. The gas is described by the quantum N-particle state $\Psi(\mathbf{r}_1, \mathbf{r}_2, \ldots, \mathbf{r}_N, t)$. Since the gas is in a condensed phase, all particles occupy the same single particle state, $\phi(\mathbf{r}, t)$, and thus the N-body quantum state can be written as the product of N single particle states as follows

$$\Psi(\mathbf{r}_1, \mathbf{r}_2, \ldots, \mathbf{r}_N, t) = \prod_{i=1}^N \phi(\mathbf{r}_i, t). \quad (2.38)$$

The action of this system is written as

$$S = \int \Psi^*(\mathbf{r}_1, \mathbf{r}_2, \ldots, \mathbf{r}_N, t)\left(i\hbar\frac{\partial}{\partial t} - \hat{H}\right)\Psi(\mathbf{r}_1, \mathbf{r}_2, \ldots, \mathbf{r}_N, t)d\mathbf{r}_1 \ldots d\mathbf{r}_N \, dt, \quad (2.39)$$

with the Hamiltonian given by

$$\hat{H} = \sum_{i=1}^{N}\left[-\frac{\hbar^2}{2m}\nabla_i^2 + V_{\text{ext}}(\mathbf{r}_i)\right] + \frac{1}{2}\sum_{i=1}^{N}\sum_{j=1, j\neq i}^{N} V(\mathbf{r}_i - \mathbf{r}_j), \quad (2.40)$$

where V_{ext} is an external potential provided by the magnetic or optical trap, which is to a good approximation a parabolic potential. Here, $V(\mathbf{r}_i - \mathbf{r}_j)$ is the two-body interatomic interaction potential between two particles located at positions $\mathbf{r}_i$ and $\mathbf{r}_j$, $\hbar = h/2\pi$, where h is Planck's constant, and m is the mass of one particle. The interatomic interaction for the dilute gas can be approximated by a contact potential, which can be represented by a Dirac delta function

$$V(\mathbf{r}_i - \mathbf{r}_j) = g\delta(\mathbf{r}_i - \mathbf{r}_j), \quad (2.41)$$

where the constant g is determined by the normalization $\int V(\mathbf{r}_i - \mathbf{r}_j)d(\mathbf{r}_i - \mathbf{r}_j) = g$. Scattering theory relates this constant to the s-wave scattering length, a_s, as

$$g = \frac{4\pi\hbar^2 a_s}{m}. \quad (2.42)$$

Taking Equations (2.38), (2.40), (2.41), and (2.42) into consideration, the action (2.39) becomes

$$S = N\int \phi^*(\mathbf{r}, t)\left[i\hbar\frac{\partial}{\partial t} + \frac{\hbar^2}{2m}\nabla^2 - V_{\text{ext}}(\mathbf{r}) - \frac{N}{2}g|\phi(\mathbf{r}, t)|^2\right]\phi(r, t)\, d\mathbf{r}\, dt, \quad (2.43)$$

where we have approximated prefactor of the nonlinear term, $N - 1$, by N. The Lagrangian density per particle, thus reads

$$\mathcal{L} = i\hbar\phi^*(\mathbf{r}, t)\frac{\partial}{\partial t}\phi(r, t) - \frac{\hbar^2}{2m}|\nabla\phi(\mathbf{r}, t)|^2 - V_{\text{ext}}(\mathbf{r})|\phi(\mathbf{r}, t)|^2 - \frac{N}{2}g|\phi(\mathbf{r}, t)|^4, \quad (2.44)$$

where integration by parts was used to express the kinetic energy term in terms of $|\nabla\phi(r, t)|^2$. The Euler–Lagrange equation

$$\left[\frac{d}{dt}\frac{\partial}{\partial\phi_t^*(r, t)} + \frac{d}{dx}\frac{\partial}{\partial\phi_x^*(r, t)} + \frac{d}{dy}\frac{\partial}{\partial\phi_y^*(r, t)} + \frac{d}{dz}\frac{\partial}{\partial\phi_z^*(r, t)} - \frac{\partial}{\partial\phi^*(r, t)}\right]\mathcal{L} = 0 \quad (2.45)$$

gives

$$i\hbar\frac{\partial}{\partial t}\phi(\mathbf{r}, t) + \frac{\hbar^2}{2m}\nabla^2\phi(\mathbf{r}, t) - V_{\text{ext}}(r)\phi(\mathbf{r}, t) - Ng|\phi(\mathbf{r}, t)|^2\,\phi(\mathbf{r}, t) = 0. \quad (2.46)$$

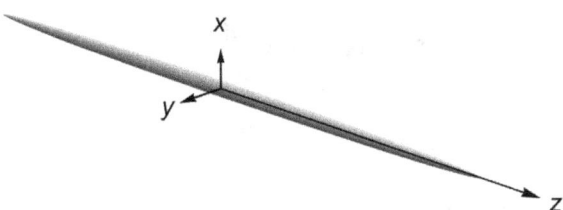

Figure 2.1. Elongated (cigar-shaped) Bose–Einstein condensate.

Alternatively, we may just differentiate the integrand of (2.43) with respect to ϕ^* to get the same equation. This is the Gross–Pitaevskii equation. For an elongated condensate, as shown in Figure 2.1, with strength of radial trapping potential being much larger than the axial one, the dynamics is quasi one-dimensional. The radial degrees of freedom can then be integrated out with the following separation of variables

$$\phi(\mathbf{r},\ t) = f(x,\ y)\psi(z,\ t). \tag{2.47}$$

Substituting back in Equation (2.46) and integrating over x and y results in the one-dimensional nonlinear Schrödinger equation

$$i\hbar\frac{\partial}{\partial t}\psi(z,\ t) + \frac{\hbar^2}{2m}\frac{\partial^2}{\partial z^2}\psi(z,\ t) - V_{\text{ext}}(z)\psi(z,\ t) - N\kappa|\psi(z,\ t)|^2\ \psi(z,\ t) = 0, \tag{2.48}$$

where κ is an effective interatomic interaction strength, which for a Gaussian (2.72) for the radial wave function of width q_r, will be given by $\kappa = a_s/2\pi q_r^2$. In experiments [35], the external trapping potential is given by

$$V_{\text{ext}}(z) = \frac{1}{2}m\omega_z^2 z^2. \tag{2.49}$$

Scaling time to $1/\omega_z$ and length to the characteristic harmonic trap length, $a_z = \sqrt{\hbar/m\omega_z}$, the nonlinear Schrödinger Equation (2.48), reads

$$i\frac{\partial}{\partial t}\psi(z,\ t) + \frac{1}{2}\frac{\partial^2}{\partial z^2}\psi(z,\ t) - \frac{1}{2}z^2\psi(z,\ t) - \gamma|\psi(z,\ t)|^2\ \psi(z,\ t) = 0, \tag{2.50}$$

where $\gamma = (1/2\pi)\ Na_s/a_z$ and z is the scaled length. For one of the early experiments on ^{87}Rb, $N = 10^4$, $\omega_z = 2\pi \times 100$ Hz, the nonlinearity strength is $\gamma \sim 100$.

2.1.4 Light Pulses in Optical Fibers

Propagation of electromagnetic waves in optical fibers is governed by Maxwell's equations

$$\nabla \cdot \mathbf{D} = \rho_f, \tag{2.51}$$

$$\nabla \cdot \mathbf{B} = 0, \tag{2.52}$$

$$\nabla \times \mathbf{H} = \mathbf{J} + \frac{\partial \mathbf{D}}{\partial t}, \tag{2.53}$$

$$\nabla \times \mathbf{E} = -\frac{\partial \mathbf{B}}{\partial t}, \tag{2.54}$$

where $\mathbf{E}$ is the electric field, $\mathbf{H}$ is the magnetic field, $\mathbf{D}$ is the electric displacement, $\mathbf{B}$ is the magnetic induction, ρ_f is the free charge density, and $\mathbf{J}$ is the current density. The electric displacement and magnetic induction are essentially the electric and magnetic fields dressed with polarization, $\mathbf{P}$, and magnetization, $\mathbf{M}$, respectively, as follows

$$\mathbf{D} = \epsilon_0 \mathbf{E} + \mathbf{P}, \tag{2.55}$$

$$\mathbf{B} = \mu_0 \mathbf{H} + \mathbf{M}, \tag{2.56}$$

where ϵ_0 and μ_0 are, respectively, the free space permittivity and permeability. Polarization, $\mathbf{P}$, and magnetization $\mathbf{M}$ result from the interaction between the electric and magnetic fields with the material of the medium. In free space, where $\mathbf{P} = \mathbf{M} = 0$, the electric field and electric displacement, as well as the magnetic field and magnetic induction, will be essentially the same, apart from the scaling constants ϵ_0 and μ_0.

Typically, optical fibers are characterized by the absence of free charges, $\rho_f = 0$, absence of electric currents, $\mathbf{J} = 0$, and absence of magnetization, $\mathbf{M} = 0$. The wave equation governing electromagnetic propagation can then be derived by taking the curl of Equation (2.54) and then using the rest of Maxwell's equations together with Equations (2.55) and (2.56), to get

$$\nabla^2 \mathbf{E} - \nabla(\nabla \cdot \mathbf{E}) = \frac{1}{c^2} \frac{\partial^2}{\partial t^2} E + \mu_0 \frac{\partial^2}{\partial t^2} \mathbf{P}, \tag{2.57}$$

where $c = 1/\sqrt{\epsilon_0 \mu_0}$ is the speed of light in free space (vacuum). Optical fibers are made of homogeneous materials where the polarization, $\mathbf{P}$, is parallel to the electric field $\mathbf{E}$. In this case, Equation (2.55) shows that the electric displacement, $\mathbf{D}$, will also be parallel to E, and thus, in the absence of free charges, Equation (2.51) gives $\nabla \cdot \mathbf{E} = 0$, which simplifies the wave equation to

$$\nabla^2 \mathbf{E} = \frac{1}{c^2} \frac{\partial^2}{\partial t^2} \mathbf{E} + \mu_0 \frac{\partial^2}{\partial t^2} \mathbf{P}. \tag{2.58}$$

Electric polarization, $\mathbf{P}$, is a response to the interaction between the electric field and the material of the medium. Polarization is expressed in terms of the electric field as

$$
\begin{aligned}
\mathbf{P}(\mathbf{r}, t) = {} & \epsilon_0 \int_{-\infty}^{\infty} \chi^{(1)}(\mathbf{r}, t - t') \cdot \mathbf{E}(r, t') dt' \\
& + \epsilon_0 \int_{-\infty}^{\infty} \int_{-\infty}^{\infty} \chi^{(2)}(r, t - t', t - t'') : \mathbf{E}(r, t') \mathbf{E}(r, t'') dt' dt'' \\
& + \epsilon_0 \int_{-\infty}^{\infty} \int_{-\infty}^{\infty} \int_{-\infty}^{\infty} \chi^{(3)}(r, t - t', t - t'', t - t''') \vdots \mathbf{E}(\mathbf{r}, t') \mathbf{E}(r, t'') \mathbf{E}(r, t''') dt' dt'' dt''' \\
& + \cdots,
\end{aligned}
\tag{2.59}
$$

where $\chi^{(j)}$ is the jth susceptibility tensor of rank $j + 1$. The ith component of the three-dimensional polarization vector is thus written explicitly as

$$
\begin{aligned}
P_i = \epsilon_0 \int_{-\infty}^{\infty} \sum_{j=1}^{3} \chi_{ij}^{(1)}(\mathbf{r}, t - t') E_j(r, t') dt' \\
+ \epsilon_0 \int_{-\infty}^{\infty} \int_{-\infty}^{\infty} \sum_{j,k=1}^{3} \chi_{ijk}^{(2)}(r, t - t', t - t'') E_j(\mathbf{r}, t') E_k(r, t'') dt' dt'' \\
+ \epsilon_0 \int_{-\infty}^{\infty} \int_{-\infty}^{\infty} \sum_{l,j,k=1}^{3} \chi_{ijkl}^{(3)}(r, t - t', t - t'', t - t''') E_j(\mathbf{r}, t') E_k(r, t'') E_l(\mathbf{r}, t''') dt' dt'' dt''' \\
+ \cdots,
\end{aligned}
\tag{2.60}
$$

where $i = 1, 2, 3$ correspond respectively to x, y, z, and similarly for the other integer subscripts. The first line in this equation corresponds to the linear response and the third line corresponds to the first nonvanishing nonlinear response, which is in cubic power of the electric field. For homogeneous materials, inversion symmetry requires $\chi_{ijk}^{(2)} = -\chi_{ijk}^{(2)}$, which can only be satisfied for $\chi_{ijk}^{(2)} = 0$ and thus the second order response vanishes. This is the case for most of materials used in optical fibers.

A light pulse is expressed by a localized envelope superposed on the carrier electromagnetic wave, as shown in Figure 2.2. It is assumed that the envelope profile is wide enough such that it changes slowly over a wavelength of the carrier wave. For a pulse with electric field polarized in the x-direction, $i = 1$, we decouple the fast moving carrier wave oscillation by writing

$$
\mathbf{E} = E_x \hat{x} = \mathrm{Re}\left[E(\mathbf{r}, t) e^{-i\omega_0 t}\right] \hat{x}, \qquad E_y = E_z = 0,
\tag{2.61}
$$

where $E(\mathbf{r}, t)$ is the envelope profile and ω_0 is the carrier wave frequency. Similarly, the polarization is written as

$$
\mathbf{P} = P_x \hat{x} = \mathrm{Re}\left[P(\mathbf{r}, t) e^{-i\omega_0 t}\right] \hat{x}, \qquad P_y = P_z = 0.
\tag{2.62}
$$

To express $P(\mathbf{r}, t)$ in terms of $E(\mathbf{r}, t)$, we substitute for $\mathbf{E}(\mathbf{r}, t)$ and $P(\mathbf{r}, t)$ from Equations (2.61) and (2.62) in Equation (2.59)

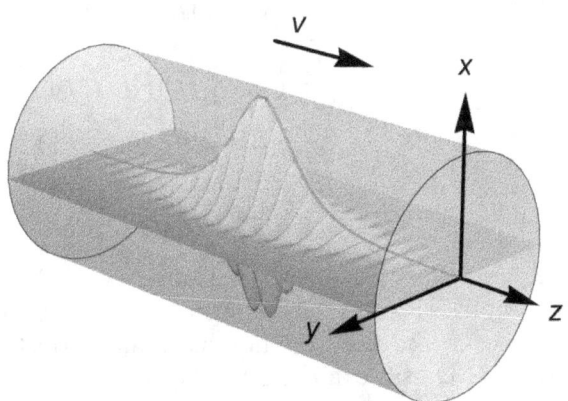

Figure 2.2. Light pulse propagating in an optical fiber.

$$\frac{1}{2}[P(\mathbf{r}, t)e^{-i\omega_0 t} + P^*(\mathbf{r}, t)e^{i\omega_0 t}] = \frac{1}{2}\epsilon_0 \int_{-\infty}^{\infty} \chi_{11}^{(1)}(t - t')[E(r, t')e^{-i\omega_0 t'} + E^*(\mathbf{r}, t')e^{i\omega_0 t'}]dt'$$

$$+ \frac{1}{8}\epsilon_0 \int_{-\infty}^{\infty} \int_{-\infty}^{\infty} \int_{-\infty}^{\infty} \chi_{1111}^{(3)}(r, t - t', t - t'', t - t''')$$

$$\times [E(\mathbf{r}, t')e^{-i\omega_0 t'} + E^*(\mathbf{r}, t')e^{i\omega_0 t'}]$$

$$\times [E(\mathbf{r}, t'')e^{-i\omega_0 t''} + E^*(\mathbf{r}, t'')e^{i\omega_0 t''}]$$

$$\times [E(\mathbf{r}, t''')e^{-i\omega_0 t'''} + E^*(\mathbf{r}, t''')e^{i\omega_0 t'''}]dt'dt''dt''', \tag{2.63}$$

where we have taken $\chi_{ijk}^{(2)} = 0$ into consideration. It is more convenient to proceed with frequency domain. To be able to calculate the Fourier transform of the nonlinear term, we assume instantaneous nonlinear response which is expressed mathematically by replacing the time dependence of $\chi^{(3)}$ with delta functions

$$\chi_{1111}^{(3)}(\mathbf{r}, t - t', t - t'', t - t''') = \chi_{1111}^{(3)}(r)\delta(t - t')\,\delta(t - t'')\,\delta(t - t'''), \tag{2.64}$$

where $\delta(\cdot)$ is the Dirac delta function. Taking the Fourier transform of Equation (2.63), we obtain

$$\frac{1}{2}\int_{-\infty}^{\infty} [P(\mathbf{r}, t)e^{-i\omega_0 t} + P^*(\mathbf{r}, t)e^{i\omega_0 t}]e^{i\omega t}dt$$

$$= \frac{\epsilon_0}{2}\int_{-\infty}^{\infty} \left\{ \int_{-\infty}^{\infty} \chi_{11}^{(1)}(t - t')[E(\mathbf{r}, t')e^{-i\omega_0 t'} + E^*(\mathbf{r}, t')e^{i\omega_0 t'}]dt' \right\}e^{i\omega t}dt$$

$$+ \frac{\epsilon_0}{8}\int_{-\infty}^{\infty} \left\{ \chi_{1111}^{(3)}(\mathbf{r})[E^3(r, t)e^{-3i\omega_0 t} + E^{*3}(\mathbf{r}, t)e^{3i\omega_0 t} \right.$$

$$\left. + 3|E(r, t)|^2 E(r, t)e^{-i\omega_0 t} + 3|E(r, t)|^2 E^*(r, t)e^{i\omega_0 t}]\right\}e^{i\omega t}dt. \tag{2.65}$$

The nonlinear term clearly requires another approximation in order to be calculated; the presence of the cubic terms, $|E(\mathbf{r}, t)|^2 E(\mathbf{r}, t)$, $E^3(\mathbf{r}, t)$, and their complex conjugates, hinders the calculation of their Fourier transform. Exploiting the assumption of slow envelope time evolution compared to that of the carrier wave, we may treat $|E(\mathbf{r}, t)|^2$ and $E^2(\mathbf{r}, t)$ as independent of time; leaving out $E(\mathbf{r}, t)$ from each term where the Fourier transform can be calculated for. The linear response term can be easily calculated using the convolution theorem, and thus we obtain

$$\frac{1}{2}[P(\mathbf{r}, \omega - \omega_0) + P^*(\mathbf{r}, \omega + \omega_0)] = \frac{1}{2}\epsilon_0\chi_{11}^{(1)}(\omega)[E(\omega - \omega_0) + E^*(\omega + \omega_0)]$$

$$+ \frac{1}{8}\epsilon_0\chi_{1111}^{(3)}(\mathbf{r})[3|E(\mathbf{r}, t)|^2 E(\omega - \omega_0)$$

$$+ 3|E(r, t)|^2 E^*(\omega + \omega_0)$$

$$+ E^2(\mathbf{r}, t)E(\omega - 3\omega_0)$$

$$+ E^2(r, t)E^*(\omega + 3\omega_0)]. \tag{2.66}$$

Phase matching requires equating terms oscillating with the same frequency. Nonlinear terms oscillating with $\omega \pm 3\omega_0$, which correspond to the third harmonic generation, are neglected since they do not have the correct phase matching. However, such terms are classified among the higher order nonlinearities which become important and thus need to be considered for very thin and highly intense pulses. Equating terms oscillating with frequency $\omega - \omega_0$, we obtain

$$P(\mathbf{r}, \omega - \omega_0) = \epsilon_0 \left[\chi_{11}^{(1)}(\omega) \, E(r, \omega - \omega_0) + \frac{3}{4}\chi_{1111}^{(3)}(\mathbf{r})|E(\mathbf{r}, t)|^2 E(r, \omega - \omega_0) \right]. \quad (2.67)$$

Coefficients of $\omega + \omega_0$ lead to the complex conjugate of this equation. Applying the Fourier transform on Equation (2.58) and taking into consideration Equation (2.67), we obtain

$$\nabla^2 E(r, \omega - \omega_0) + \frac{\omega^2}{c^2}[\epsilon(\omega)E(\mathbf{r}, \omega - \omega_0)] = 0. \quad (2.68)$$

Here, the dielectric constant is given by

$$\epsilon(\omega) = 1 + \text{Re}[\chi_{11}^{(1)}(\omega)]$$
$$+ i\left(\text{Im}[\chi_{11}^{(1)}(\omega)] + \frac{3}{4}\,\text{Im}[\chi_{1111}^{(3)}]|E(\mathbf{r}, t)|^2 \right) + \text{Re}[\chi_{1111}^{(3)}(\omega)]|E(\mathbf{r}, t)|^2 \quad (2.69)$$
$$= \epsilon_l(\omega) + \Delta\epsilon(\omega),$$

where

$$\epsilon_l(\omega) = 1 + \text{Re}[\chi_{11}^{(1)}(\omega)] \quad (2.70)$$

is the lossless linear dielectric constant, and

$$\Delta\epsilon(\omega) = i\left(\text{Im}[\chi_{11}^{(1)}(\omega)] + \frac{3}{4}\,\text{Im}[\chi_{1111}^{(3)}]|E(\mathbf{r}, t)|^2 \right) + \text{Re}[\chi_{1111}^{(3)}(\omega)]|E(\mathbf{r}, t)|^2 \quad (2.71)$$

is treated below as a perturbation to the dielectric constant as a result of the fiber loss and nonlinear susceptibility. Equation (2.69) describes the evolution of optical pulses in general. In the following, we solve this equation for pulses in optical fibers.

2.1.4.1 Linear Schrödinger Equation

In optical fibers, the pulse is guided by the fiber which confines the pulse in the radial direction and allows it to evolve only along its axis. Confinement introduces boundary conditions which result in radial eigen modes. The evolution rate of the pulse along the axis depends on the eigen mode. We consider first the lossless linear case, namely $\epsilon = \epsilon_l$, which will lead to the linear Schrödinger equation. For such confined pulses, we insert in Equation (2.68) the ansatz

$$E(\mathbf{r}, \omega - \omega_0) = F_l(x, y)A(z)e^{i\beta_l(\omega_0)z}, \quad (2.72)$$

where confinement is assumed to be along the x- and y-axes, and the pulse evolution is along the z-axis, which is also the axis of the fiber. The function $F(x, y)$ is the profile of the confined mode in the radial direction. The carrier wave $e^{i\beta_l(\omega_0)z}$ is modulated by the envelope $A(z)$, where $\beta_l(\omega_0)$ is an unknown constant which is essentially the carrier wavenumber but modulated by the radial confinement. Later, we will consider also nonlinearity and fiber losses, which will add to the modulation. Both functions $F(x, y)$ and $A(z)$ contain also $\omega - \omega_0$ dependence, which will be resurrected below in $A(z)$ when we take the inverse Fourier transform. Substituting the trial function (2.72) in the wave Equation (2.68), we get

$$\frac{1}{F_l(x, y)}\left(\frac{\partial^2}{\partial x^2} + \frac{\partial^2}{\partial y^2}\right)F_l(x, y) - \beta_l^2(\omega_0) + \frac{A''(z)}{A(z)} + 2i\beta_l(\omega_0)\frac{A'(z)}{A(z)} + \frac{\omega^2}{c^2}\epsilon_l(\omega) = 0. \quad (2.73)$$

The term $A''(z)/A(z)$ as compared to $A'(z)/A(z)$ can be neglected for the wide-pulse approximation. The separation of variables results in

$$\left(\frac{\partial^2}{\partial x^2} + \frac{\partial^2}{\partial y^2}\right)F_l(x, y) + \left[\frac{\omega^2}{c^2}\epsilon_l(\omega) - \beta_l^2(\omega)\right]F_l(x, y) = 0 \quad (2.74)$$

and

$$2i\beta_l(\omega_0)A_l'(z) + \left[\beta_l^2(\omega) - \beta_l^2(\omega_0)\right]A_l(z) = 0, \quad (2.75)$$

where $\beta_l(\omega)$ is an arbitrary constant. In the absence of confinement, i.e., $F(x, y) =$ constant, and very wide profile such that $A'(z)$ can also be neglected, and for $\omega = \omega_0$, the last two equations give $|\beta_l(\omega)| = |\beta_l(\omega_0)| = \omega_0 \epsilon_l(\omega_0)/c$. In the presence of confinement, Equation (2.74) is solved as an eigenvalue problem for specific eigenmode, F_l, and corresponding eigen value, $\beta_l(\omega)$. The eigenvalue, $\beta_l(\omega)$, is then used in Equation (2.75), which determines the propagation of the pulse. This is precisely how the fiber's confinement affects the propagation of the pulse. In most situations, however, an analytical expression to the eigenvalue, $\beta_l(\omega)$, is not available. Alternatively, we exploit the fact that pulses are localized in the frequency domain around ω_0 to expand $\beta_l(\omega)$ around $\beta_l(\omega_0)$, and Equation (2.75) gives

$$iA'(z) + [-a_0(\omega - \omega_0) - a_1(\omega - \omega_0)^2 + \cdots]A(z) = 0, \quad (2.76)$$

where we have defined

$$a_0 = -\left.\frac{d\beta_l}{d\omega}\right|_{\omega=\omega_0}, \quad (2.77)$$

$$a_1 = -\frac{1}{2}\left.\frac{d^2\beta_l}{d\omega^2}\right|_{\omega=\omega_0}. \quad (2.78)$$

We take the inverse Fourier transform of Equation (2.76), as

$$i\frac{\partial}{\partial z}\int_{-\infty}^{\infty} A_l(z, \omega - \omega_0)e^{i(\omega-\omega_0)t}d(\omega - \omega_0)$$

$$- a_0\int_{-\infty}^{\infty} (\omega - \omega_0)A_l(z, \omega - \omega_0)e^{i(\omega-\omega_0)t}d(\omega - \omega_0) \quad (2.79)$$

$$- a_1\int_{-\infty}^{\infty} (\omega - \omega_0)^2 A_l(z, \omega - \omega_0)e^{i(\omega-\omega_0)t}d(\omega - \omega_0) + \cdots = 0.$$

Using the property for the Fourier transformation of derivatives of a function, the last equation is rewritten as

$$i\frac{\partial}{\partial z}A(z,\,t) + ia_0\frac{\partial}{\partial t}A(z,\,t) + a_1\frac{\partial^2}{\partial t^2}A(z,\,t) = 0, \tag{2.80}$$

where we have neglected the higher order t-derivatives since in fibers they usually correspond to much weaker effects. Transforming to a frame co-moving with the pulse, the term with first derivative in t can be eliminated by substituting the transformation $A(z,\,t) \to A(z,\,t - a_0z)$, and the last equation becomes

$$i\frac{\partial}{\partial z}A(z,\,t) + a_1\frac{\partial^2}{\partial t^2}A(z,\,t) = 0, \tag{2.81}$$

where the coordinates in this equation are those of the moving frame, i.e., we have renamed $t - a_0z$ as t, for simplicity. This is the linear Schrödinger equation.

2.1.4.2 Nonlinear Schrödinger Equation

Since $\epsilon(\omega)$ depends also on E through the nonlinear contribution to susceptibility, Equation (2.74) becomes nonlinear and therefore will be solved using perturbation theory. The perturbative expansion is performed for all quantities around their values for the linear case, namely

$$\beta(\omega) = \beta_l(\omega) + \Delta\beta(\omega), \tag{2.82}$$

$$F(x,\,y) = F_l(x,\,y) + \Delta F(x,\,y), \tag{2.83}$$

and Equation (2.69) is invoked.

Substituting the perturbed quantities (2.69), (2.82) and (2.83) in Equation (2.73), and separating the variables, the zeroth order reproduces Equations (2.74) and (2.75), and the linear order gives

$$\left(\frac{\partial^2}{\partial x^2} + \frac{\partial^2}{\partial y^2}\right)\Delta F(x,\,y) + \left[\frac{\omega^2}{c^2}\epsilon_l(\omega) - \beta_l^2(\omega)\right]\Delta F(x,\,y) + \left[\frac{\omega^2}{c^2}\Delta\epsilon(\omega) - 2\beta_l(\omega)\Delta\beta(\omega)\right]F_l(x,\,y) = 0. \tag{2.84}$$

In this equation, the first two terms indicate that $\Delta F(x,\,y)$ satisfies the same equation for the zeroth order (2.74), and therefore $F(x,\,y)$ will not be affected by the perturbation. Equating these two terms to zero, then multiplying the remaining term by $|F_l(x,\,y)|^2$ and integrating over x and y, we obtain

$$\Delta\beta(\omega) = \frac{\omega^2}{2c^2\beta_l(\omega)} \frac{\displaystyle\int_{-\infty}^{\infty}\int_{-\infty}^{\infty}\Delta\epsilon(\omega)|F_l(x,\,y)|^2\,dx\,dy}{\displaystyle\int_{-\infty}^{\infty}\int_{-\infty}^{\infty}|F_l(x,\,y)|^2\,dx\,dy}. \tag{2.85}$$

For the nonlinear case and with fiber losses, the dielectric constant becomes complex and is expressed in terms of its real and imaginary parts, as

$$\epsilon(\omega) = \left(n + i\frac{\alpha}{2k_0}\right)^2, \tag{2.86}$$

where $k_0 = \omega/c$,

$$n = n_l + \Delta n \tag{2.87}$$

is the refractive index, with Δn being the change in the refractive index due to the nonlinear susceptibility, and similarly

$$\alpha = \alpha_l + \Delta\alpha \tag{2.88}$$

is the fiber absorption coefficient. Substituting Equations (2.87) and (2.88) in Equation (2.86), and expanding to first order in α, $\Delta\alpha$, and Δn, we get

$$\epsilon(\omega) = \epsilon_l + \Delta\epsilon = n_l^2 \left\{1 + \frac{2}{n_l}\left[\Delta n + \frac{i}{2k_0}(\alpha_l + \Delta\alpha)\right]\right\}. \tag{2.89}$$

This equation defines the linear dielectric constant, ϵ_l, in terms of the refractive index, n_l, through

$$\epsilon_l(\omega) = n_l^2, \tag{2.90}$$

and the perturbation in the dielectric constant as

$$\Delta\epsilon(\omega) = 2n_l\left[\Delta n + \frac{i}{2k_0}(\alpha_l + \Delta\alpha)\right]. \tag{2.91}$$

Substituting for this expression of $\Delta\epsilon$ in Equation (2.85), we get

$$\Delta\beta(\omega) = \frac{n_l\omega^2}{c^2\beta_l(\omega)}\frac{\int_{-\infty}^{\infty}\int_{-\infty}^{\infty}(\Delta n + i\alpha_l/2k_0 + i\Delta\alpha/2k_0)|F_l(x,y)|^2}{\int_{-\infty}^{\infty}\int_{-\infty}^{\infty}|F_l(x,y)|^2}. \tag{2.92}$$

Comparing Equations (2.69) and (2.86), we obtain

$$n_l + \Delta n + i\frac{\alpha_l + \Delta\alpha}{2k_0} = \sqrt{1 + \chi_{11}^{(1)} + \frac{3}{4}\chi_{1111}^{(3)}|E|^2}, \tag{2.93}$$

where the zeroth order, gives

$$n_l + i\frac{\alpha_l}{2k_0} = \sqrt{1 + \chi_{11}^{(1)}}, \tag{2.94}$$

and from the first order, we obtain

$$\Delta n = \frac{3}{8n_l}\,\text{Re}\left[\chi_{1111}^{(3)}\right]|E|^2, \tag{2.95}$$

$$\Delta\alpha = \frac{3\omega}{4n_l c} \, \text{Im} \left[\chi_{1111}^{(3)} \right] |E|^2. \tag{2.96}$$

Using Equations (2.72), (2.95), and (2.96), the expression for $\Delta\beta(\omega)$ in Equation (2.92), reduces to

$$\Delta\beta = i\frac{\alpha_l}{2} + (\eta + i\gamma)|A(z)|^2, \tag{2.97}$$

where

$$\eta = \frac{\omega}{c} n_2 \frac{\displaystyle\int_{-\infty}^{\infty} \int_{-\infty}^{\infty} |F_l(x, y)|^4 dx \, dy}{\displaystyle\int_{-\infty}^{\infty} \int_{-\infty}^{\infty} |F_l(x, y)|^2 dx \, dy}, \tag{2.98}$$

$$\gamma = \frac{\omega}{c} \frac{\alpha_2}{2k_0} \frac{\displaystyle\int_{-\infty}^{\infty} \int_{-\infty}^{\infty} |F_l(x, y)|^4 dx \, dy}{\displaystyle\int_{-\infty}^{\infty} \int_{-\infty}^{\infty} |F_l(x, y)|^2 dx \, dy}, \tag{2.99}$$

with

$$n_2 = \frac{3}{8n_l} \, \text{Re} \left[\chi_{1111}^{(3)} \right] \tag{2.100}$$

and

$$\alpha_2 = \frac{3\omega_l}{4n_l c} \, \text{Im} \left[\chi_{1111}^{(3)} \right]. \tag{2.101}$$

It should be noted that $\beta_l(\omega)$ is the propagation constant of the pulse guided by the fiber. As explained above, it is essentially the propagation constant of the pulse in an open medium modulated by the radial boundary conditions introduced by the fiber guidance. By considering a length scale, $1/\sigma$, for the radial part of the electric field, $F(x, y)$, one can show that the effect of the boundary conditions on β_l is of order β_l/σ, which for tight confinement, is a small quantity. Therefore, treating the effect of boundary conditions on β_l as a perturbation, β_l was approximated in Equations (2.98) and (2.99), by its value in the open medium, namely $\beta_l = n_l\omega/c$. This can be obtained from Equation (2.74) after removing the derivative terms since they are of order σ^2. The other terms are of order β_l^2, and thus can be treated as a different order.

The dispersion relation of the envelope can now be determined from Equation (2.75), which is written as

$$2\beta_l(\omega_0)iA'(z) + \left\{ [\beta_l(\omega) + \Delta\beta(\omega)]^2 - \beta_l^2(\omega_0) \right\} A(z) = 0. \tag{2.102}$$

Expanding to first order in $\Delta\beta(\omega)$, we get

$$iA'(z) + \left[\beta_l^2(\omega) - \beta_l^2(\omega_0) + 2\beta_l(\omega_0)\Delta\beta(\omega)\right]A(z) = 0. \tag{2.103}$$

This is essentially Equation (2.75), but with the additional term $2\beta_l(\omega_0)\Delta\beta(\omega)$, which introduces nonlinearity. Proceeding, as above, by expanding $\beta_l(\omega)$ around ω_0, and using (2.97), we obtain the nonlinear Schrödinger equation

$$i\frac{\partial}{\partial z}A(z,\,t) + a_1\frac{\partial^2}{\partial t^2}A(z,\,t) + (\eta + i\gamma)|A(z,\,t)|^2A(z,\,t) + i\frac{\alpha_l}{2}A(z,\,t) = 0, \tag{2.104}$$

where η and γ are calculated at $\omega = \omega_0$.

2.1.4.3 Scaling and Units
It is convenient to write the nonlinear Schrödinger equation in terms of scaled quantities using the parameters of the fiber and initial pulse profile. We neglect here the linear and nonlinear damping terms which will allow for scaling out all parameters and hence results in a universal dimensionless equation, often named as the fundamental nonlinear Schrödinger equation.

In practice, optical pulses are measured by their power, which is related to the electric field as

$$W = cn\epsilon_0 \int |E|^2 dS, \tag{2.105}$$

where dS is an element of the fiber's cross-sectional area, namely $dS = dx\,dy$, and thus

$$W = W_c|A|^2\int_{-\infty}^{\infty}\int_{-\infty}^{\infty}|F(x,\,y)|^2dx\,dy, \tag{2.106}$$

where $W_c = cn\epsilon_0$. Therefore, the pulse field, $A(z,\,t)$, will be expressed in terms of its power, $W(z,\,t)$, with the following transformation

$$A(z,\,t) = \frac{\sqrt{W(z,\,t)}}{\sqrt{W_c\int_{-\infty}^{\infty}\int_{-\infty}^{\infty}|F(x,\,y)|^2dx\,dy}}. \tag{2.107}$$

Equation (2.104) then takes the form

$$i\frac{\partial}{\partial z}q(z,\,t) + a_1\frac{\partial^2}{\partial t^2}q(z,\,t) + a_2|q(z,\,t)|^2q(z,\,t) = 0, \tag{2.108}$$

where

$$q(z,\,t) = \sqrt{W(z,\,t)}, \tag{2.109}$$

$$a_2 = \frac{\omega n_2}{cA_{eff}}, \tag{2.110}$$

and

$$A_{eff} = \frac{\left[\int_{-\infty}^{\infty}\int_{-\infty}^{\infty} |F(x, y)|^2 dx\, dy\right]^2}{\int_{-\infty}^{\infty}\int_{-\infty}^{\infty} |F(x, y)|^4 dx\, dy} \tag{2.111}$$

is the effective radial mode area. The factor W_c has been absorbed in n_2, as $n_2/W_c \to n_2$. The dimension of n_2 is 1/(electric field)^2, as can be seen from Equations (2.95) and (2.100). The dimension of W_c is power/((electric field)^2 × area). Therefore, the dimension of n_2/W_c is area/power and thus the dimension of a_2 is 1/(power × length).

The pulse can also be expressed in terms of its intensity, which is defined as the power per unit area, and thus will be proportional to the electric field amplitude, $I \propto |E|^2$

$$I = I_c |A|^2 \int_{-\infty}^{\infty}\int_{-\infty}^{\infty} |F(x, y)|^2 dx\, dy, \tag{2.112}$$

where I_c is essentially W_c but divided by an element of area which can be taken as the cross-section area of the fiber. It will not be important to specify which area to use since this constant will be absorbed in n_2; what matters is just its dimension, namely power/area. Scaling the field, $A(z, t)$, to intensity as

$$A(z, t) = \frac{\sqrt{I(z, t)}}{\sqrt{I_c \int_{-\infty}^{\infty}\int_{-\infty}^{\infty} |F(x, y)|^2 dx\, dy}}, \tag{2.113}$$

we obtain a similar equation to Equation (2.108), where the constant I_c is absorbed in n_2 and thus the dimension of n_2 will be area/power, which is the one used in practice.

For the relevant temporal and spacial scalings, we consider a traveling Gaussian pulse as

$$q(z, t) = q_0 e^{-\frac{2t^2}{T_0^2}} e^{-i\left(\frac{\sqrt{2}\, t}{T_0} + \frac{z}{L_D}\right)}, \tag{2.114}$$

where T_0 is essentially the width of the initial pulse and L_D is a length set by the dispersion parameter a_1. The time, T_0, is defined as the time interval over which the intensity of the pulse $I(z, t) \propto |q(z, t)|^2$ equals $1/e$ of its maximum, namely $|q(0, T_0)|^2 = (1/e)|q(0, 0)|^2$. In practice, the FWHM definition is used, which is twice the time at which the intensity of the pulse equals 1/2 of its maximum. The dispersion length is determined by the first two terms of Equation (2.108). In this case, the oscillatory part of (2.113), namely $q(z, t) \sim q_0 e^{-i(\frac{\sqrt{2}\, t}{T_0} + \frac{z}{L_D})}$, is a solution to (2.108) with only the first two terms, and we get

$$L_D = \frac{T_0^2}{2|a_1|}. \tag{2.115}$$

Another important length scale characterizes the nonlinear terms. It can be obtained by substituting the oscillatory part of

$$q(z, t) = q_0 e^{-\frac{2t^2}{T_0^2}} e^{-i\left(\frac{\sqrt{2}\,t}{T_0} + \frac{z}{L_{nl}}\right)}, \tag{2.116}$$

namely $q(z, t) \sim q_0 e^{-i\left(\frac{\sqrt{2}\,t}{T_0} + \frac{z}{L_{nl}}\right)}$, in Equation (2.108) after removing the dispersion term, which turns out to be

$$L_{nl} = \frac{1}{a_2 W_0}, \tag{2.117}$$

where $W_0 = q_0^2$ is the initial pulse power. Scaling in Equation (2.108) time to T_0, length to L_D, and field to $\sqrt{W_0}$, we get

$$i\frac{\partial}{\partial z}u(z, t) + \frac{1}{2}\frac{\partial^2}{\partial t^2}u(z, t) + g_0|u(z, t)|^2 u(z, t) = 0, \tag{2.118}$$

where $u(z, t) = q(z, t)/\sqrt{W_0}$, $g_0 = L_D/L_{nl}$, $L_{nl} = 1/a_2 W_0$, and we have retained the symbols t and z for the scaled quantities.

The parameter $g_0 = L_D/L_{nl}$ can also be scaled in u, and the $1/2$ factor of the dispersion term can be scaled in x, which result in the dimensionless fundamental nonlinear Schrödinger equation

$$i\frac{\partial}{\partial z}u(z, t) + \frac{\partial^2}{\partial t^2}u(z, t) + |u(z, t)|^2 u(z, t) = 0, \tag{2.119}$$

where u is scaled as $u \to u/\sqrt{|g_0|}$ and x is scaled as $x - >x/\sqrt{2}$.

As a realistic numerical example for a pulse in an SiO_2 fiber, $|a_1| \sim 10^{-26}$ s^2/m, $T_0 \sim 10^{-12}$ s, $n_2 \sim 3 \times 10^{-20}$ m^2/watt, $k_0 \sim 2\pi/(1.5 \times 10^{-6})$ m^{-1}, $A_{eff} \sim 40 \times 10^{-12}$ m^2, $W_0 \sim 1$ watt. These lead to $L_D \sim L_{nl} \sim 100$ m, and thus $g_0 = L_D/L_{nl} \sim 1$.

2.1.5 Surface Water Waves

The nonlinear Schrödinger equation describing, in general, hydrodynamic surface waves, can be derived from the mass and momentum conservation laws subject to boundary conditions at the surface and the bottom [15]. It turns out that, depending on the ratio between the wavelength and the water depth, surface waves can be classified into shallow or deep waves, which have fundamentally different dispersion relations. While, shallow water waves obey the linear Schrödinger equations, deep water waves are described by the nonlinear Schrödinger equation, where nonlinearity originates from the combination of momentum conservation and boundary conditions. We start by deriving the conservation laws which lead to a wave equation with a corresponding dispersion relation that is shown to depend on the intensity of the wave. The derivation assumes irrotational, incompressible, inviscid (zero viscosity), and homogeneous fluid. The precise mathematical definitions of these conditions will be presented below.

2.1.5.1 Conservation Laws

Consider an element of the fluid with mass M and mass density ρ. The mass is given by

$$M = \int \rho dV, \qquad (2.120)$$

where the integration is over the volume of the element. Mass conservation dictates that the rate at which this mass changes must be equal to the net mass crossing in and out of the surface. Taking, without loss of generality, the element of mass, dM, flowing through the element of surface, $d\mathbf{S}$, to be of cylindrical shape with length $|d\mathbf{l}|$, the amount of mass flowing through the element of surface will be equal to $dM = \rho \, dV = \rho \, d\mathbf{l} \cdot d\mathbf{S}$, which leads to

$$\frac{\partial M}{\partial t} = -\int \rho \mathbf{v} \cdot d\mathbf{S}, \qquad (2.121)$$

where $\mathbf{v} = d\mathbf{l}/dt$ is the velocity of the fluid, and the integration is over the surface of the element. Differentiating Equation (2.120) with respect to time and using Green's theorem: $\int A \cdot d\mathbf{S} = \int \nabla \cdot A \, dV$, the last equation becomes

$$\int \left[\frac{\partial}{\partial t} \rho + \nabla \cdot (\rho \mathbf{v}) \right] dV = 0, \qquad (2.122)$$

which vanishes for a vanishing integrand, and thus we obtain the continuity equation

$$\frac{\partial}{\partial t} \rho + \nabla \cdot (\rho \mathbf{v}) = 0. \qquad (2.123)$$

The momentum conservation equation is derived from applying Newton's second law on the fluid element, where the force is provided by the fluid pressure p. The net force on the element is given by

$$\mathbf{F} = -\int p \, d\mathbf{S}, \qquad (2.124)$$

where the negative sign is introduced since, for positive pressure, the force is opposite to the element of area. Using Gauss-divergence theorem: $\int \nabla \Phi \, dV = \int \Phi \, d\mathbf{S}$, the surface integral in the last equation turns to a volume integral

$$\mathbf{F} = -\int \nabla p \, dV, \qquad (2.125)$$

suggesting

$$d\mathbf{F} = -\nabla p \, dV. \qquad (2.126)$$

On the other hand, the force can be written as

$$d\mathbf{F} = \rho dV \frac{d\mathbf{v}}{dt}. \qquad (2.127)$$

Comparing the last two equations, we get

$$\rho \frac{d\mathbf{v}}{dt} = -\nabla p. \tag{2.128}$$

Since $\mathbf{v} = \mathbf{v}(x, y, z, t)$, we can write

$$\begin{aligned}
\frac{d\mathbf{v}}{dt} &= \frac{\partial \mathbf{v}}{\partial t} + \frac{\partial \mathbf{v}}{\partial x}\frac{\partial x}{\partial t} + \frac{\partial \mathbf{v}}{\partial y}\frac{\partial y}{\partial t} + \frac{\partial \mathbf{v}}{\partial z}\frac{\partial z}{\partial t} \\
&= \frac{\partial \mathbf{v}}{\partial t} + (\mathbf{v} \cdot \nabla)\mathbf{v}.
\end{aligned} \tag{2.129}$$

Substituting for $d\mathbf{v}/dt$ from the last equation in Equation (2.128), and using the identity: $(\mathbf{v} \cdot \nabla)\mathbf{v} = \frac{1}{2}\nabla v^2 - \mathbf{v} \times \nabla \times \mathbf{v}$, we get

$$\frac{\partial \mathbf{v}}{\partial t} - \mathbf{v} \times \nabla \times \mathbf{v} + \frac{1}{2}\nabla v^2 + \frac{1}{\rho}\nabla p = 0. \tag{2.130}$$

For irrotational flow, $\nabla \times \mathbf{v} = 0$, and in the presence of an external force, $d F_{\text{ext}}$, this equation takes the form

$$\frac{\partial \mathbf{v}}{\partial t} + \frac{1}{2}\nabla(\mathbf{v}^2) + \frac{1}{\rho}\nabla p - \frac{1}{\rho}\frac{d\mathbf{F}_{\text{ext}}}{dV} = 0. \tag{2.131}$$

This equation expresses the second conservation law, namely conservation of momentum. The external force is, in our case, the force of gravity. For a wave of height $\eta(x, y, t)$ along the z-direction, the force is given by

$$d\mathbf{F}_{\text{ext}} = -\rho dV g \hat{z}, \tag{2.132}$$

where $g = 9.8$ ms^{-2} is the acceleration due to gravity. At the wave surface, $\eta(x, y, t) = z$, the force can be written as

$$d\mathbf{F}_{\text{ext}} = -\rho dV g \nabla z. \tag{2.133}$$

For irrotational flow, the velocity can be expressed in terms of a scalar field

$$\mathbf{v} = \nabla \phi. \tag{2.134}$$

The momentum conservation Equation (2.131), can then be expressed in terms of φ as

$$\frac{\partial \nabla \phi}{\partial t} + \frac{1}{2}\nabla(\nabla \phi)^2 + \frac{1}{\rho}\nabla p + g\nabla \eta = 0. \tag{2.135}$$

For homogeneous incompressible fluid, $\rho = $ constant, the gradient operator may be pulled out to the left side as an overall operator, leaving

$$\frac{\partial \phi}{\partial t} + \frac{1}{2}(\nabla \phi)^2 + \frac{1}{\rho}p + g\eta = 0. \tag{2.136}$$

The constant P/ρ can be removed by the transformation $\phi \to \phi - Pt/\rho$, which leads to the momentum conservation law

$$\frac{\partial \phi}{\partial t} + \frac{1}{2}(\nabla \phi)^2 + g\eta = 0. \tag{2.137}$$

2.1.5.2 Boundary Conditions

Boundary conditions set the values of the wave at the surface and the bottom, see Figure 2.3, as:

Upper surface:

$$z = \eta(x, y, t), \tag{2.138}$$

which leads to

$$\frac{dz}{dt} = \frac{\partial \eta}{\partial x}\frac{\partial x}{\partial t} + \frac{\partial \eta}{\partial y}\frac{\partial y}{\partial t} + \frac{\partial \eta}{\partial t}. \tag{2.139}$$

In view of the relation $\mathbf{v} = \nabla \phi$, the last equation becomes

$$\frac{dz}{dt} = \frac{\partial \eta}{\partial x}\frac{\partial \phi}{\partial x} + \frac{\partial \eta}{\partial y}\frac{\partial \phi}{\partial y} + \frac{\partial \eta}{\partial t}, \tag{2.140}$$

which is known as the kinematic boundary condition on the wave surface.

Lower surface: The lower surface (sea bed, or tank bottom) is defined to be located at $z = -h$. For a solid horizontal boundary, the vertical component of the velocity vanishes

$$\left.\frac{\partial \phi}{\partial z}\right|_{z=-h} = 0. \tag{2.141}$$

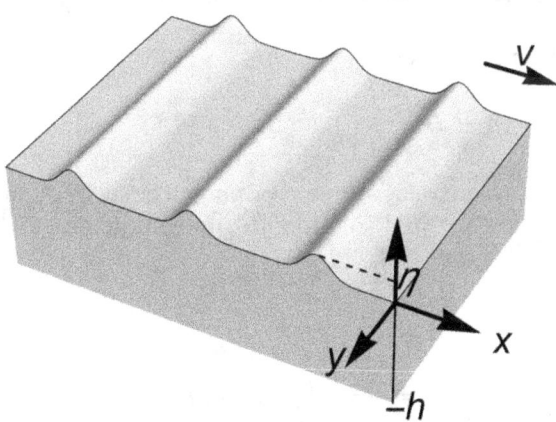

Figure 2.3. Surface water gravity waves.

2.1.5.3 Linear Surface Waves

For small wave amplitude such that the wave amplitude and velocity potential are much smaller than the wavelength and wave period, respectively, the nonlinear terms in Equation (2.137) and the boundary conditions can be neglected, which results in the linear system

$$\nabla^2 \phi = 0, \quad -h \leqslant z \leqslant 0 \tag{2.142}$$

and

$$\frac{\partial \phi}{\partial t} = -g\eta, \tag{2.143}$$

with boundary condition for the upper surface

$$\frac{\partial \eta}{\partial t} = \left. \frac{\partial \phi}{\partial z} \right|_{z=\eta} \tag{2.144}$$

and Equation (2.141) for the bottom. We may approximate $\partial \phi / \partial z$ at $z = \eta$ by its value at $z = 0$ since there is not much difference between the two values, given the small amplitude approximation. The last two equations, thus give

$$\frac{\partial^2 \phi}{\partial t^2} + g \left. \frac{\partial \phi}{\partial z} \right|_{z=\eta} = 0. \tag{2.145}$$

Specifying to one-dimensional waves, say propagating in the x-direction, and hence no y-dependence for the wave profile, and using the trial solution

$$\phi(x, z, t) = q(z)\sin(kx - \omega t) \tag{2.146}$$

in Equation (2.142), we get $q(z) = C_1 e^{kz} + C_2 e^{-kz}$, where C_1 and C_2 are arbitrary constants. Using the boundary condition (2.141), we get $C_2 = C_1 e^{-2kh}$, which gives

$$\phi(x, z, t) = 2C_1 e^{-kh} \cosh[k(z + h)]\sin(kx - \omega t). \tag{2.147}$$

The condition (2.144) gives then

$$\eta = \eta_0 + 2\frac{kC_1}{\omega} e^{-kh} \sinh(kh)\cos(kx - \omega t), \tag{2.148}$$

where $\eta_0 = \eta(t = 0)$ is an integration constant. The dispersion relation can be obtained by substituting the trial solution (2.146) in Equation (2.145)

$$\omega^2 = gk \tanh(kh), \tag{2.149}$$

with phase velocity

$$v_{ph} = \frac{\omega}{k} = \sqrt{\frac{g}{k} \tanh(kh)} \tag{2.150}$$

and group velocity

$$v_g = \frac{d\omega}{dk} = \frac{v_{ph}}{2}\left[1 + \frac{2kh}{\sinh(2kh)}\right].$$

(2.151)

The dimensionless parameter kh, which is essentially a ratio between the water depth and the wavelength, h/λ, defines two distinct regimes:

Shallow water waves: $h \ll \lambda$. In this case, the dispersion relation (2.149) is expanded in small kh to give, up to the second order,

$$\omega = \sqrt{gh}\, k.$$

(2.152)

Deep water waves: $h \gg \lambda$. In this case, the tanh function in (2.149) can be replaced by 1, and thus we get

$$\omega = \sqrt{gk}.$$

(2.153)

2.1.5.4 Nonlinear Surface Waves

In the nonlinear case, the effect of nonlinear terms is included to the lowest orders with a perturbative expansion known as the Stokes third order expansion [21, 22]. Both wave amplitude and velocity potential are expanded perturbatively as

$$\eta = \eta_0 + \epsilon\eta_1 + \epsilon^2\eta_2 + \cdots,$$

(2.154)

$$\phi = \phi_0 + \epsilon\phi_1 + \epsilon^2\phi_2 + \cdots,$$

(2.155)

where ϵ is the small perturbation parameter and η_i, ϕ_i, $i = 0, 1, 2, \ldots$ are the perturbative expansions of η and ϕ. The resulting dispersion relation takes the form

$$\omega = \sqrt{gk(1 + k|\eta|^2)},$$

(2.156)

where $|\eta|$ is the amplitude of the Stokes wave. As pointed out in Section 2.1.2, amplitude-dependence in the dispersion relation results in the nonlinear term in the Schrödinger equation. Comparing the dispersion relation (2.156) with that in the generic derivation (2.28), and then following the same procedure, we find

$$i\frac{\partial\eta}{\partial t} + a_1\frac{\partial^2\eta}{\partial x^2} + a_2|\eta|^2\eta = 0,$$

(2.157)

where

$$a_1 = -\frac{1}{2}\frac{\partial^2\omega}{\partial k^2}\bigg|_{k=k_0} = \frac{\sqrt{gk_0}}{8\,k_0^2}$$

(2.158)

and

$$a_2 = \frac{\partial\omega}{\partial|\eta|^2}\bigg|_{|\eta|^2=0} = \frac{1}{2}\sqrt{gk_0}\,k_0^2,$$

(2.159)

where $\sqrt{gk_0} = \omega_0$ and k_0 are the frequency and wave number of the carrier wave. Scaling time to $1/\omega_0$, length to $1/\sqrt{8}\, k_0$, and η to $\sqrt{2}/k_0$, the last equation becomes the dimensionless fundamental nonlinear Schrödinger equation

$$i\frac{\partial \eta}{\partial t} + \frac{\partial^2 \eta}{\partial x^2} + |\eta|^2 \eta = 0. \tag{2.160}$$

2.2 NLSE with Cubic Nonlinearity

Equation:

$$i\,\psi_t + a_1\,\psi_{xx} + a_2\,|\psi|^2\,\psi = 0, \tag{2.161}$$

where $\psi = \psi(x, t)$ is the complex function profile, x and t are its two independent variables, a_1 and a_2 are arbitrary constants.

Solutions:

2.2.1 Real Dispersion and Nonlinearity Coefficients

Solution 1. Constant amplitude I $\quad$ *continuous wave (CW), t-dependent phase*

$$\psi(x, t) = A_0\, e^{i\left[\, a_2\, A_0^2\, (t-t_0)+\phi_0 \right]}, \tag{2.162}$$

where A_0, t_0, and ϕ_0 are arbitrary real constants.
  • *Reference:* [36].

Solution 2. Constant amplitude II $\quad$ *CW, x-dependent phase*

$$\psi(x, t) = A_0\, e^{i\left[\pm A_0 \sqrt{\frac{a_2}{a_1}}\,(x-x_0)+\phi_0\right]}, \tag{2.163}$$

where $a_1 a_2 > 0$, A_0, x_0, and ϕ_0 are arbitrary real constants.
  • *Derived in appendix* A.

Solution 3. Constant amplitude III $\quad$ *CW, t- and x-dependent phase*

$$\psi(x, t) = A_0\, e^{i[A_1\,(x-x_0)+(|A_0|^2\, a_2 - A_1^2\, a_1)\,(t-t_0)+\phi_0]}, \tag{2.164}$$

where A_0 is an arbitrary complex constant, A_1, x_0, t_0, and ϕ_0 are arbitrary real constants.

Solution 4. Rational solution I $\quad$ *decaying wave (DW)*

$$\psi(x, t) = \frac{A_0}{\sqrt{A_1 + t - t_0}}\, e^{i\left[\frac{(a_1 A_2 + x - x_0)^2}{4\, a_1\, (A_1 + t - t_0)}+a_2\, A_0^2\,\ln(A_1 + t - t_0)+\phi_0\right]}, \tag{2.165}$$

where A_0, A_1, A_2, t_0, x_0, and ϕ_0 are arbitrary real constants.
  • *Reference:* [36].

Solution 5. **Rational Solution II**

$$\psi(x,\,t) = \sqrt{\frac{-2\,a_1}{a_2}}\,\frac{1}{x - x_0}\,e^{i\,\phi_0},\tag{2.166}$$

where $a_1\,a_2 < 0$, x_0, and ϕ_0 are arbitrary real constants.
- *Reference*: [37].

Solution 6. **Rational Solution III**   *higher order of* (2.166)

$$\psi(x,\,t) = \frac{1}{\sqrt{-a_2}}\,\frac{q_1(x,\,t)}{q_2(x,\,t)},\tag{2.167}$$

where

$$q_1(x,\,t) = 16\,x_1^2 + 2\left[2\,A_1\,x_0\left(\sqrt{\frac{2}{a_1}}\,x_1\,x + \lambda - \lambda^*\right) - A_0\,\lambda^2\,e^{2\,\lambda*\left(\frac{x}{\sqrt{2\,a_1}} - i\,\lambda*\,t\right)}\right]e^{2\,\lambda\left(\frac{x}{\sqrt{2\,a_1}} + i\,\lambda\,t\right)},$$

$$q_2(x,\,t) = 4\,x_0\,\frac{\lambda^2\,A_0}{A_1}\,e^{2\,\lambda*\left(\frac{x}{\sqrt{2\,a_1}} - i\,\lambda*\,t\right)}$$

$$+ e^{2\,\lambda\left(\frac{x}{\sqrt{2\,a_1}} + i\,\lambda\,t\right)}\left[2\,x_0\,A_1 + A_0\left(\sqrt{\frac{2}{a_1}}\,x_1\,x - x_0\right)e^{2\,\lambda*\left(\frac{x}{\sqrt{2\,a_1}} - i\,\lambda*\,t\right)}\right]$$

$$- 8\,\lambda^2\left(\sqrt{\frac{2}{a_1}}\,x_1\,x + x_0\right),$$

$\lambda = \lambda_r + i\,\lambda_i$, $c_1 = c_{1r} + i\,c_{1i}$, $c_2 = c_{2r} + i\,c_{2i}$, $x_0 = \lambda + \lambda^*$, $x_1 = |\lambda|^2$,
$A_0 = \frac{|c_2|^2}{2\,\lambda^{*2}\,|c_1|^2}$, $A_1 = \frac{c_2}{c_1}$, $a_1 > 0$, $a_2 < 0$, $\lambda_r,\,\lambda_i,\,c_{1r},\,c_{1i},\,c_{2r},\,c_{2i}$, and ϕ_0 are arbitrary
real constants. The * denotes the complex conjugate.
- *Reference*: [39].

Solution 7.

$$\psi(x,\,t) = A_0\,\sqrt{\frac{-2\,a_1}{a_2}}\,\sec[A_0\,(x - x_0)]\,e^{-i\left[a_1\,A_0^2\,(t - t_0) + \phi_0\right]},\tag{2.168}$$

where $a_1\,a_2 < 0$, A_0, x_0, t_0, and ϕ_0 are arbitrary real constants.
- *Reference*: [40] with $m = 1$ and $\alpha = \gamma = \lambda = \nu = 0$.

Solution 8.

$$\psi(x,\,t) = A_0\,\sqrt{\frac{-2\,a_1}{a_2}}\,\csc[A_0\,(x - x_0)]\,e^{-i\left[a_1\,A_0^2\,(t - t_0) + \phi_0\right]},\tag{2.169}$$

where $a_1 a_2 < 0$, A_0, x_0, t_0, and ϕ_0 are arbitrary real constants.
- *Reference*: [40] with $m = 1$ and $\alpha = \gamma = \lambda = \nu = 0$.

Solution 9.

$$\psi(x,\, t) = A_0 \sqrt{\frac{-2\, a_1}{a_2}}\, \tan[A_0\, (x - x_0)]\, e^{i\left[\, 2\, a_1\, A_0^2\, (t - t_0) + \phi_0\right]}, \qquad (2.170)$$

where $a_1 a_2 < 0$, A_0, x_0, t_0, and ϕ_0 are arbitrary real constants.
- *Reference*: [40] with $m = 1$ and $\alpha = \gamma = \lambda = \nu = 0$.

Solution 10.

$$\psi(x,\, t) = A_0 \sqrt{\frac{-2\, a_1}{a_2}}\, \cot[A_0\, (x - x_0)]\, e^{i\left[\, 2\, a_1\, A_0^2\, (t - t_0) + \phi_0\right]}, \qquad (2.171)$$

where $a_1 a_2 < 0$, A_0, x_0, t_0, and ϕ_0 are arbitrary real constants.
- *Reference*: [40] with $m = 1$ and $\alpha = \gamma = \lambda = \nu = 0$.

Solution 11.   *bright soliton*
(Figure 2.4)

$$\psi(x,\, t) = A_0 \sqrt{\frac{2\, a_1}{a_2}}\, \operatorname{sech}[A_0\, (x - x_0)]\, e^{i\left[\, a_1\, A_0^2\, (t - t_0) + \phi_0\right]}, \qquad (2.172)$$

where $a_1 a_2 > 0$, A_0, x_0, t_0, and ϕ_0 are arbitrary real constants.
- *Reference*: [36].

Solution 12.

$$\psi(x,\, t) = A_0 \sqrt{\frac{-2\, a_1}{a_2}}\, \operatorname{csch}[A_0\, (x - x_0)]\, e^{i\left[\, a_1\, A_0^2\, (t - t_0) + \phi_0\right]}, \qquad (2.173)$$

where $a_1 a_2 < 0$, A_0, x_0, t_0, and ϕ_0 are arbitrary real constants.
- *Reference*: [40] with $m = 1$ and $\alpha = \gamma = \lambda = \nu = 0$.

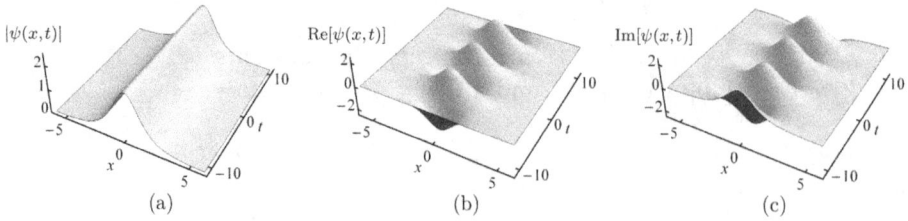

Figure 2.4. Bright soliton (2.172), with $a_1 = 1$, $a_2 = 1/2$, $A_0 = 1$, and $x_0 = t_0 = \phi_0 = 0$. (a) Absolute value of (2.172), (b) Real part of (2.172), and (c) Imaginary part of (2.172).

Solution 13. *dark soliton*
(Figure 2.5)

$$\psi(x,\, t) = A_0 \sqrt{\frac{-2\, a_1}{a_2}} \; \tanh[A_0\, (x - x_0)]\, e^{-i\left[2\, a_1\, A_0^2\, (t - t_0) + \phi_0\right]}, \qquad (2.174)$$

where $a_1\, a_2 < 0$, A_0, x_0, t_0, and ϕ_0 are arbitrary real constants.

- *Reference*: [40] with $m = 1$ and $\alpha = \gamma = \lambda = \nu = 0$.

Solution 14.

$$\psi(x,\, t) = A_0 \sqrt{\frac{-2\, a_1}{a_2}} \; \coth[A_0\, (x - x_0)]\, e^{-i\left[2\, a_1\, A_0^2\, (t - t_0) + \phi_0\right]}, \qquad (2.175)$$

where $a_1\, a_2 < 0$, A_0, x_0, t_0, and ϕ_0 are arbitrary real constants.

- *Reference*: [40] with $m = 1$ and $\alpha = \gamma = \lambda = \nu = 0$.

Solution 15.
(Figure 2.6)

$$\psi(x,\, t) = \left(A_0 + i\, A_1 \tan\left\{A_1\left[\sqrt{\frac{-a_2}{2\, a_1}}\, (x - x_0) + a_2\, A_0\, (t - t_0)\right]\right\}\right) e^{i\left[a_2\, (A_0^2 - A_1^2)\, (t - t_0) + \phi_0\right]}, \quad (2.176)$$

where $a_1 a_2 < 0$, A_0, A_1, x_0, t_0, and ϕ_0 are arbitrary real constants.

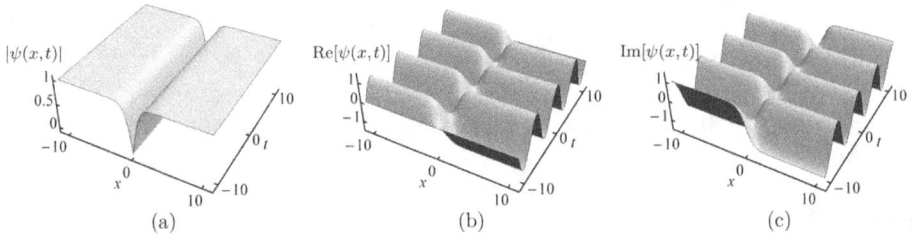

Figure 2.5. Dark soliton (2.174), with $a_1 = 1/2$, $a_2 = -1$, $A_0 = 1$, and $x_0 = t_0 = \phi_0 = 0$. (a) Absolute value of (2.174), (b) Real part of (2.174), and (c) Imaginary part of (2.174).

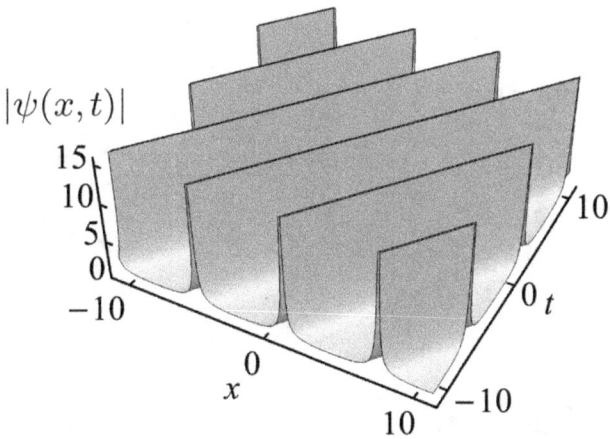

Figure 2.6. Plot of solution (2.176) with $a_1 = A_1 = 1/2$, $a_2 = -1$, $A_3 = 3/4$, and $x_0 = t_0 = \phi_0 = 0$.

- *Reference*: [41].
  Solution 16.

$$\psi(x, t) = A_0 \left[\sqrt{1 - A_1} + i \sqrt{A_1} \tanh\left(A_0 \sqrt{\frac{A_1}{2}} \{(x - x_0) + [A_0 \sqrt{2(1 - A_1)} + 2 A_3](t - t_0)\} \right) \right]$$
$$\times e^{-i [A_2 (t-t_0) + A_3 (x - x_0) + \phi_0]},$$ (2.177)

where $0 < A_1 < 1$, $A_2 = A_0^2 + A_3^2$, $a_1 = 0$, $a_2 = -1$, A_0, A_3, x_0, t_0, and ϕ_0 are arbitrary real constants.
- *Reference*: [42].

Solution 17.

$$\psi(x, t) = A_0 \sqrt{\frac{-a_1}{2 a_2}} \{\sec[A_0 (x - x_0)] + \tan[A_0 (x - x_0)]\} e^{i \left[\frac{a_1 A_0^2}{2} (t - t_0) + \phi_0 \right]},$$ (2.178)

where $a_1 a_2 < 0$, A_0, x_0, t_0, and ϕ_0 are arbitrary real constants.
- *Reference*: [40].

Solution 18.

$$\psi(x, t) = A_0 \sqrt{\frac{-2}{a_2}} \left(\cot\left\{ A_0 \left[\frac{x - x_0}{\sqrt{a_1}} - 2 A_1 (t - t_0) \right] \right\} - \tan\left\{ A_0 \left[\frac{x - x_0}{\sqrt{a_1}} - 2 A_1 (t - t_0) \right] \right\} \right)$$
$$\times e^{i \left[\frac{A_1}{\sqrt{a_1}} (x - x_0) - (A_1^2 - 8 A_0^2)(t - t_0) \right]},$$ (2.179)

where $a_1 > 0$, $a_2 < 0$, A_0, A_1, x_0, t_0, and ϕ_0 are arbitrary real constants.
- *Reference*: [41].

Solution 19.
  (Figure 2.7)

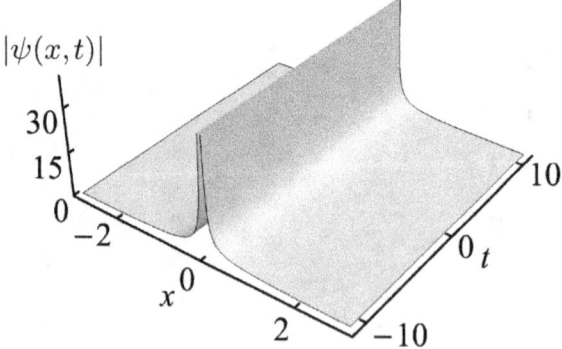

Figure 2.7. Plot of solution (2.180) with $a_1 = -1/2$, $a_2 = 1/2$, $A_0 = 1/20$, and $x_0 = t_0 = \phi_0 = 0$.

$$\psi(x,\,t) = A_0 \sqrt{\frac{-2\,a_1}{a_2}} \left\{ \frac{1 + \tanh^2[A_0\,(x - x_0)]}{\tanh[A_0\,(x - x_0)]} \right\} e^{-i\left[8\,a_1\,A_0^2\,(t - t_0) + \phi_0\right]}, \qquad (2.180)$$

where $a_1\,a_2 < 0$, A_0, x_0, t_0, and ϕ_0 are arbitrary real constants.
- *Reference*: [40].

Solution 20.

$$\psi(x,\,t) = A_0 \sqrt{\frac{-2}{a_2}} \left(\frac{b_0 \tan\left\{ A_0\left[\dfrac{x - x_0}{\sqrt{a_1}} - 2\,A_1\,(t - t_0) \right] \right\} - b_1}{b_0 + b_1 \tan\left\{ A_0\left[\dfrac{x - x_0}{\sqrt{a_1}} - 2\,A_1\,(t - t_0) \right] \right\}} \right) e^{\,i\left[\dfrac{A_1\,(x - x_0)}{\sqrt{a_1}} - (A_1^2 - 2\,A_0^2)\,(t - t_0) \right]}, \qquad (2.181)$$

where $a_1 > 0$, $a_2 < 0$, A_0, A_1, b_0, b_1, x_0, t_0, and ϕ_0 are arbitrary real constants.
- *Reference*: [41].

Solution 21.

$$\psi(x,\,t) = \pm\,A_0 \left(A_1 - \frac{4\,A_2}{A_1 + \sqrt{4\,A_2 - A_1^2}\,\tan\left\{ A_0\,\sqrt{4\,A_2 - A_1^2}\,[2\,A_3\,(t - t_0) - (x - x_0)] \right\}} \right) \qquad (2.182)$$
$$\times\, e^{i\left[A_3\,(x - x_0) + (-2\,A_0^2\,A_1^2 + 8\,A_0^2\,A_2 - A_3^2)\,(t - t_0) + \phi_0\right]},$$

where $a_1 = 1$, $a_2 = -2$, $4\,A_2 - A_1^2 > 0$, A_0, A_3, x_0, t_0, and ϕ_0 are arbitrary real constants.
- *Reference*: [38].

Solution 22.

$$\psi(x,\,t) = \pm\,A_0 \left(A_1 - \frac{4\,A_2}{A_1 + \sqrt{4\,A_2 - A_1^2}\,\cot\left\{ A_0\,\sqrt{4\,A_2 - A_1^2}\,[2\,A_3\,(t - t_0) - (x - x_0)] \right\}} \right) \qquad (2.183)$$
$$\times\, e^{i\left[A_3\,(x - x_0) + (-2\,A_0^2\,A_1^2 + 8\,A_0^2\,A_2 - A_3^2)\,(t - t_0) + \phi_0\right]},$$

where $a_1 = 1$, $a_2 = -2$, $4\,A_2 - A_1^2 > 0$, A_0, A_3, x_0, t_0, and ϕ_0 are arbitrary real constants.
- *Reference*: [38].

Solution 23.

$$\psi(x,\,t) = \pm\,A_0 \left(A_1 - \frac{4\,A_2}{A_1 - \sqrt{A_1^2 - 4\,A_2}\,\tanh\left\{ A_0\,\sqrt{A_1^2 - 4\,A_2}\,[2\,A_3\,(t - t_0) - (x - x_0)] \right\}} \right) \qquad (2.184)$$
$$\times\, e^{i\left[A_3\,(x - x_0) + (-2\,A_0^2\,A_1^2 + 8\,A_0^2\,A_2 - A_3^2)\,(t - t_0) + \phi_0\right]},$$

where $a_1 = 1$, $a_2 = -2$, $A_1^2 - 4\,A_2 > 0$, A_0, A_3, x_0, t_0, and ϕ_0 are arbitrary real constants.
- *Reference*: [38].

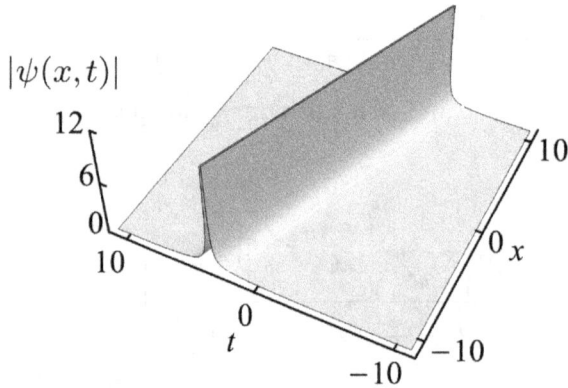

Figure 2.8. Plot of solution (2.186) with $a_1 = 1/2$, $a_2 = -1/4$, $A_0 = -2$, $b_0 = b_1 = -1$, and $x_0 = t_0 = \phi_0 = 0$.

Solution 24.

$$\psi(x,\, t) = \pm A_0 \left(A_1 - \frac{4\, A_2}{A_1 - \sqrt{A_1^2 - 4\, A_2}\, \coth\left\{ A_0\, \sqrt{A_1^2 - 4\, A_2}\, [2\, A_3\, (t - t_0) - (x - x_0)] \right\}} \right) \quad (2.185)$$

$$\times\, e^{i\left[A_3\, (x - x_0) + (-2\, A_0^2\, A_1^2 + 8\, A_0^2\, A_2 - A_3^2)\, (t - t_0) + \phi_0 \right]},$$

where $a_1 = 1$, $a_2 = -2$, $A_1^2 - 4\, A_2 > 0$, A_0, A_3, x_0, t_0, and ϕ_0 are arbitrary real constants.

- *Reference*: [38].

Solution 25.
(Figure 2.8)

$$\psi(x,\, t) = \frac{q_1(x,\, t)}{q_2(x,\, t)}\, e^{i\left[\frac{A_0}{\sqrt{a_1}}\, (x - x_0) - A_0^2\, (t - t_0) \right]}, \quad (2.186)$$

where

$$q_1(x,\, t) = -2\, b_0\, A_0\, \sqrt{-2\, a_2} + 2\, b_1 - \sqrt{\frac{-2\, a_2}{a_1}}\, (x - x_0) + 2\, A_0\, \sqrt{-2\, a_2}\, (t - t_0),$$

$$q_2(x,\, t) = 4\, b_0^2\, a_2\, A_0^2 + 2\, b_1^2 + \frac{4\, b_0\, a_2\, A_0}{\sqrt{a_1}}\, (x - x_0) - 8\, b_0\, a_2\, A_0^2\, (t - t_0)$$

$$+ \frac{a_2}{a_1}\, (x - x_0)^2 - \frac{4\, a_2\, A_0}{\sqrt{a_1}}\, (x - x_0)\, (t - t_0) + 4\, a_2\, A_0^2\, (t - t_0)^2, \quad \begin{array}{l} a_1 > 0, \end{array}$$

$a_2 < 0$, A_0, b_0, b_1, x_0, t_0, and ϕ_0 are arbitrary real constants.
- *Reference*: [41].

Solution 26.   *solitary wave (SW)*
(Figure 2.9)

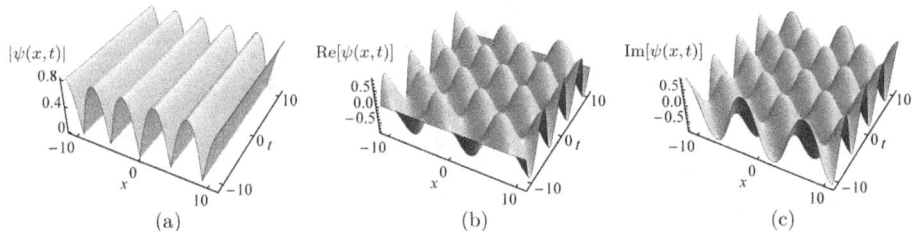

Figure 2.9. Solitary wave (2.187), with $a_1 = 1$, $a_2 = -1$, $A_0 = 1$, $x_0 = t_0 = \phi_0 = 0$, and $m = 1/2$. (a) Absolute value of (2.187), (b) Real part of (2.187), and (c) Imaginary part of (2.187).

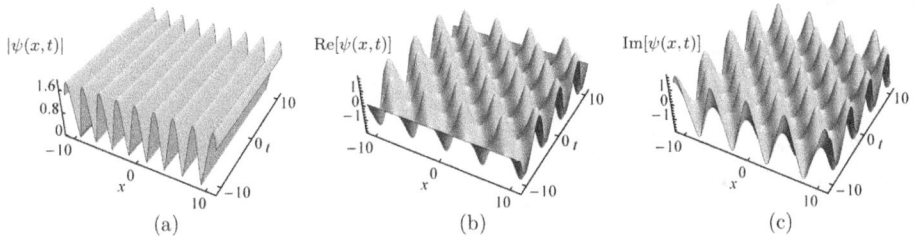

Figure 2.10. Solitary wave (2.189), with $a_1 = 1$, $a_2 = 1$, $A_0 = 1$, $x_0 = t_0 = \phi_0 = 0$, and $m = 7/10$. (a) Absolute value of (2.189), (b) Real part of (2.189), and (c) Imaginary part of (2.189).

$$\psi(x, t) = A_0 \sqrt{\frac{2\,m}{-a_2\,(1+m)}}\ \text{sn}\left[\frac{A_0}{\sqrt{a_1\,(1+m)}}\,(x-x_0),\,m\right] e^{-i\,[A_0^2\,(t-t_0)+\phi_0]}, \qquad (2.187)$$

where $a_1\,(1+m) > 0$, $a_2\,m\,(1+m) < 0$, $m \neq -1$, A_0, t_0, x_0, and ϕ_0 are arbitrary real constants.

- *Reference*: [43].

***Solution* 27.**   SW

$$\psi(x, t) = A_0 \sqrt{\frac{2\,a_1}{a_2}}\ \text{sn}[A_0\,(x-x_0),\,-1]\,e^{i\,\phi_0}, \qquad (2.188)$$

where $a_1\,a_2 > 0$, A_0, x_0, and ϕ_0 are arbitrary real constants.

- *Derived in appendix* A.

***Solution* 28.**   SW

(Figure 2.10)

$$\psi(x, t) = A_0 \sqrt{\frac{2\,m}{a_2\,(2\,m-1)}}\ \text{cn}\left[\frac{A_0}{\sqrt{a_1\,(2\,m-1)}}\,(x-x_0),\,m\right] e^{i\,[A_0^2\,(t-t_0)+\phi_0]}, \qquad (2.189)$$

where $a_1\,(2\,m-1) > 0$, $a_2\,m\,(2\,m-1) > 0$, $m \neq 1/2$, A_0, t_0, x_0, and ϕ_0 are arbitrary real constants.

- *Reference*: [43].

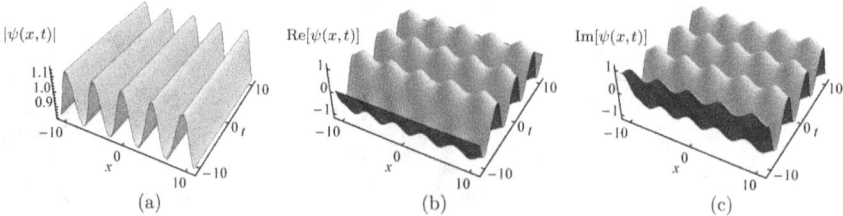

Figure 2.11. Solitary wave (2.191), with $a_1 = 1$, $a_2 = 1$, $A_0 = 1$, $x_0 = t_0 = \phi_0 = 0$, and $m = 1/2$. (a) Absolute value of (2.191), (b) Real part of (2.191), and (c) Imaginary part of (2.191).

Solution 29.   *SW*

$$\psi(x, t) = A_0 \sqrt{\frac{a_1}{a_2}} \, \mathrm{cn}\left[A_0 \, (x - x_0), \frac{1}{2} \right] e^{i\,\phi_0}, \tag{2.190}$$

where $a_1 a_2 > 0$, $m = 1/2$, A_0, x_0, and ϕ_0 are arbitrary real constants.
  • *Derived in appendix* A.

Solution 30.   *SW*
  (Figure 2.11)

$$\psi(x, t) = A_0 \sqrt{\frac{2}{a_2 \, (2 - m)}} \, \mathrm{dn}\left[\frac{A_0}{\sqrt{a_1 \, (2 - m)}} \, (x - x_0), m \right] e^{i\,[A_0^2\,(t-t_0)+\phi_0]}, \tag{2.191}$$

where $a_1 \, (2 - m) > 0$, $a_2 \, (2 - m) > 0$, $m \neq 2$, A_0, t_0, x_0, and ϕ_0 are arbitrary real constants.
  • *Reference*: [43].

Solution 31.   *SW*

$$\psi(x, t) = A_0 \sqrt{\frac{2\,a_1}{a_2}} \, \mathrm{dn}[A_0 \, (x - x_0), 2] \, e^{i\,\phi_0}, \tag{2.192}$$

where $a_1 a_2 > 0$, A_0, x_0, and ϕ_0 are arbitrary real constants.
  • *Derived in appendix* A.

Solution 32.   *SW*

$$\psi(x, t) = A_0 \sqrt{1 - m} \, \mathrm{nd}\left[\frac{A_1}{\sqrt{a_1}} \, (x - x_0), m \right] e^{-i\,[A_2\,(t-t_0)+\phi_0]}, \tag{2.193}$$

where $A_2 = (m - 2) \, A_1^2$, $a_1 > 0$, $a_2 = \dfrac{2\,A_1^2}{A_0^2}$, $0 < m \leqslant 1$, A_0, A_1, x_0, t_0, and ϕ_0 are arbitrary real constants.
  • *Reference*: [44], *taken from the nonlocal case.*

Solution 33.   *SW*

$$\psi(x, t) = A_0 \sqrt{m(1-m)} \, \mathrm{sd}\left[\frac{A_1}{\sqrt{a_1}} (x - x_0), m\right] e^{-i[A_2(t-t_0)+\phi_0]}, \qquad (2.194)$$

where $A_2 = (1 - 2m) A_1^2$, $a_1 > 0$, $a_2 = \frac{2A_1^2}{A_0^2}$, $0 < m \leqslant 1$, A_0, A_1, x_0, t_0, and ϕ_0 are arbitrary real constants.

- *Reference*: [44], *taken from the nonlocal case.*

Solution 34.   *SW*

$$\psi(x, t) = A_0 \sqrt{m} \, \mathrm{cd}\left[\frac{A_1}{\sqrt{a_1}} (x - x_0), m\right] e^{-i[A_2(t-t_0)+\phi_0]}, \qquad (2.195)$$

where $A_2 = (m + 1) A_1^2$, $a_1 > 0$, $a_2 = \frac{-2A_1^2}{A_0^2}$, $m > 0$, A_0, A_1, x_0, t_0, and ϕ_0 are arbitrary real constants.

- *Reference*: [44], *taken from the nonlocal case.*

Solution 35.   *solitary wave on a finite background*
(Figure 2.12)

$$\psi(x, t) = A(x) \, e^{i[\phi(x, t)+\phi_0]}, \qquad (2.196)$$

where $\quad A(x) = \sqrt{R_3 + (R_2 - R_3) \, \mathrm{sn}^2[A_0(x - x_0), m]}, \qquad \phi(x, t) =$

$$\frac{\sqrt{2} \, \lambda_0 \, \Pi\{\frac{R_3 - R_2}{R_3}, \, \mathrm{am}[A_0(x-x_0), m], m\} \, \mathrm{dn}[A_0(x-x_0), m]}{\sqrt{2} \, R_3 \, A_0 \sqrt{1 - m \, \mathrm{sn}^2[A_0(x-x_0), m]}} + \lambda_2(t - t_0), \qquad m = \frac{R_2 - R_3}{R_1 - R_3}, \qquad A_0 =$$

$\sqrt{\frac{a_2(R_3 - R_1)}{2a_1}}$, $a_1 a_2 (R_3 - R_1) > 0$, R_j, $j = 1, 2, 3$ are the three roots of $Y(x) = 2a_1 \lambda_0^2 - 2a_1 \lambda_1 x - 2\lambda_2 x^2 + a_2 x^3$, Π is the incomplete elliptic integral, am is the amplitude for Jacobi elliptic functions, x_0, t_0, ϕ_0, λ_0, λ_1, and λ_2 are arbitrary real constants,

- *Derived in appendix* A.
- *This solution can be written in terms of* cn^2 *or* dn^2 *by using:* $\mathrm{sn}^2(x, m) = 1 - \mathrm{cn}^2(x, m) = [1 - \mathrm{dn}^2(x, m)]/m.$

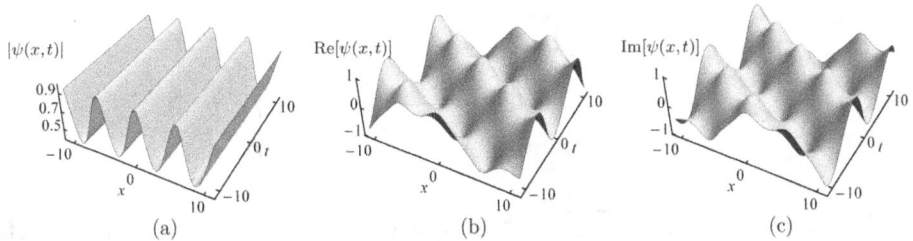

Figure 2.12. Solitary wave on a finite background (2.196), with $a_1 = 1$, $a_2 = 1$, $x_0 = t_0 = \phi_0 = 0$, $\lambda_0 = 1/10$, $\lambda_1 = 0$, and $\lambda_2 = 1/2$. (a) Absolute value of (2.196), (b) Real part of (2.196), and (c) Imaginary part of (2.196).

Solution 36.   *SW*

$$\psi(x, t) = \left\{ \frac{A_0}{2} \, dn\left[\frac{A_1}{\sqrt{a_1}} \, (x - x_0), m \right] + \frac{B_0 \sqrt{m}}{2} \, cn\left[\frac{A_1}{\sqrt{a_1}} \, (x - x_0), m \right] \right\} e^{-i \, [A_2 \, (t-t_0)+\phi_0]}, \quad (2.197)$$

where $A_2 = \dfrac{-(m + 1) A_1^2}{2}$, $B_0 = \pm A_0$, $a_1 > 0$, $a_2 = \dfrac{2 A_1^2}{A_0^2}$, $0 < m \leqslant 1$, A_0, A_1, x_0, t_0, and ϕ_0 are arbitrary real constants.

   • *Reference*: [44], *taken from the nonlocal case.*

Solution 37.   *SW*

$$\psi(x, t) = A_0 \sqrt{m} \, \frac{sn\left[\dfrac{A_1}{\sqrt{a_1}} \, (x - x_0), m \right]}{1 + dn\left[\dfrac{A_1}{\sqrt{a_1}} \, (x - x_0), m \right]} \, e^{-i \, [A_2 \, (t-t_0)+\phi_0]}, \quad (2.198)$$

where $A_2 = (2 - m) A_1^2/2$, $a_1 > 0$, $a_2 = -\dfrac{m A_1^2}{2 A_0^2}$, $m > 0$, A_0, A_1, x_0, t_0, and ϕ_0 are arbitrary real constants.

   • *Reference*: [42], *taken from the nonlocal case, corrected.*

Solution 38.   *SW*

$$\psi(x, t) = A_0 \sqrt{m} \, \frac{cn\left[\dfrac{A_1}{\sqrt{a_1}} \, (x - x_0), m \right]}{\sqrt{1 - m} + dn\left[\dfrac{A_1}{\sqrt{a_1}} \, (x - x_0), m \right]} \, e^{-i \, [A_2 \, (t-t_0)+\phi_0]}, \quad (2.199)$$

where $A_2 = (2 - m) A_1^2/2$, $a_1 > 0$, $a_2 = -\dfrac{m A_1^2}{2 A_0^2}$, $0 < m < 1$, A_0, A_1, x_0, t_0, and ϕ_0 are arbitrary real constants.

   • *Reference*: [42], *taken from the nonlocal case.*

Solution 39.   *SW*

$$\psi(x, t) = \left\{ \frac{A_0}{2} - \frac{A_0 \, dn\left[\dfrac{A_1}{\sqrt{a_1}} \, (x - x_0), m \right]}{(1 - m)^{1/4} + dn\left[\dfrac{A_1}{\sqrt{a_1}} \, (x - x_0), m \right]} \right\} e^{-i \, [A_2 \, (t-t_0)+\phi_0]}, \quad (2.200)$$

where $A_2 = (2 - m + 6 \sqrt{1 - m}\,) A_1^2/2$, $a_1 > 0$, $a_2 = -2 (1 - \sqrt{1 - m}\,)^2 A_1^2/A_0^2$, $0 < m < 1$, A_0, A_1, x_0, t_0, and ϕ_0 are arbitrary real constants.

   • *Reference*: [42], *taken from the nonlocal case.*

Solution 40.   *SW*

$$\psi(x, t) = A_0 \sqrt{m} \; \frac{\mathrm{dn}\left[\dfrac{A_1}{\sqrt{a_1}} (x - x_0), m\right]}{1 + \sqrt{m} \; \mathrm{sn}\left[\dfrac{A_1}{\sqrt{a_1}} (x - x_0), m\right]} \; e^{-i \, [A_2 \, (t - t_0) + \phi_0]}, \qquad (2.201)$$

where $A_2 = -(1 + m) \, A_1^2 / 2, a_1 > 0, a_2 = (1 - m) \, A_1^2 / (2 \, m \, A_0^2), m > 0, A_0, A_1, x_0,$ $t_0,$ and ϕ_0 are arbitrary real constants.
  • *Reference*: [42], *corrected.*

Solution 41.   *SW*

$$\psi(x, t) = A_0 \, m \; \frac{\mathrm{cn}\left[\dfrac{A_1}{\sqrt{a_1}} (x - x_0), m\right] \mathrm{sn}\left[\dfrac{A_1}{\sqrt{a_1}} (x - x_0), m\right]}{\mathrm{dn}\left[\dfrac{A_1}{\sqrt{a_1}} (x - x_0), m\right]} \; e^{-i \, [A_2 \, (t - t_0) + \phi_0]}, \qquad (2.202)$$

where $A_2 = 2 \, (2 - m) \, A_1^2, \, a_1 > 0, \, a_2 = -\dfrac{2 \, A_1^2}{A_0^2}, \, m > 0, \, A_0, \, A_1, \, x_0, \, t_0,$ and ϕ_0 are arbitrary real constants.
  • *Reference*: [44], *taken from the nonlocal case, we corrected the denominator.*

Solution 42.   *SW*

$$\psi(x, t) = A_0 \; \frac{\mathrm{dn}\left[\dfrac{A_1}{\sqrt{a_1}} (x - x_0), m\right] \mathrm{sn}\left[\dfrac{A_1}{\sqrt{a_1}} (x - x_0), m\right]}{\mathrm{cn}\left[\dfrac{A_1}{\sqrt{a_1}} (x - x_0), m\right]} \; e^{-i \, [A_2 \, (t - t_0) + \phi_0]}, \qquad (2.203)$$

where $A_2 = 2 \, (2 \, m - 1) \, A_1^2, \, a_1 > 0, \, a_2 = -\dfrac{2 \, A_1^2}{A_0^2}, \, m > 0, \, A_0, \, A_1, \, x_0, \, t_0,$ and ϕ_0 are arbitrary real constants.
  • *Reference*: [45], *corrected.*

Solution 43.   *SW*

$$\psi(x, t) = A_0 \; \frac{\mathrm{dn}\left[\dfrac{A_1}{\sqrt{a_1}} (x - x_0), m\right]}{\mathrm{sn}\left[\dfrac{A_1}{\sqrt{a_1}} (x - x_0), m\right] \mathrm{cn}\left[\dfrac{A_1}{\sqrt{a_1}} (x - x_0), m\right]} \; e^{-i \, [A_2 \, (t - t_0) + \phi_0]}, \qquad (2.204)$$

where $A_2 = 2 \, (2 - m) \, A_1^2, \, a_1 > 0, \, a_2 = -\dfrac{2 \, A_1^2}{A_0^2}, \, m > 0, \, A_0, \, A_1, \, x_0, \, t_0,$ and ϕ_0 are arbitrary real constants.
  • *Reference*: [45], *corrected.*

Solution 44.   *SW*

$$\psi(x, t) = A_0 \frac{\mathrm{cn}\left[\dfrac{A_1}{\sqrt{a_1}}(x - x_0), m\right]}{\mathrm{sn}\left[\dfrac{A_1}{\sqrt{a_1}}(x - x_0), m\right]\mathrm{dn}\left[\dfrac{A_1}{\sqrt{a_1}}(x - x_0), m\right]} e^{-i[A_2(t-t_0)+\phi_0]}, \qquad (2.205)$$

where $A_2 = 2(2m - 1)A_1^2$, $a_1 > 0$, $a_2 = -\dfrac{2A_1^2}{A_0^2}$, $m > 0$, A_0, A_1, x_0, t_0, and ϕ_0 are arbitrary real constants.

- *Reference*: [45], *corrected.*

Solution 45.   *N-bright solitons*

$$\psi(x, t) = \frac{1}{\sqrt{a_2}} \sum_{j=1}^{N} \psi_j(x, t), \qquad (2.206)$$

where $\psi_j(x, t)$ are solutions of

$$\sum_{k=1}^{N} M_{jk}\left[\gamma_j^{-1}(x, t) + \gamma_k^*(x, t)\right]\psi_k(x, t) = 1, \quad j = 0, 1, 2, \dots, N, \qquad (2.207)$$

$a_1 > 0$, $\qquad\qquad a_2 > 0$, $\qquad\qquad \lambda_j = \alpha_j + i\nu_j$, $\qquad\qquad M_{jk} = 1/(\lambda_j + \lambda_k^*)$,

$\gamma_j(x, t) = e^{\frac{\lambda_j}{\sqrt{2a_1}}(x-x_{0j})+i(\lambda_j^2(t-t_0)/2+\phi_{0j})}$, $\gamma_k^*(x, t)$ and $\gamma_j^{-1}(x, t)$ are the complex conjugate and the inverse of γ_k and γ_j, respectively,

λ_k^* is the complex conjugate of λ_k, α_j, ν_j, x_{0j}, t_0, and ϕ_{0j} are arbitrary real constants.

- *Reference*: [46].

Solution 46.   *two bright solitons*
    (Figure 2.13)

$$\psi(x, t) = \frac{1}{\sqrt{a_2}}[\psi_1(x, t) + \psi_2(x, t)], \qquad (2.208)$$

where

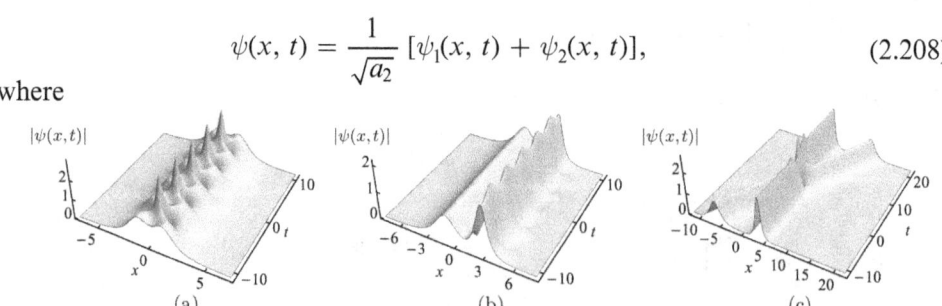

(a)   (b)   (c)

Figure   2.13. Two-bright-solitons   (2.208),   with $a_1 = 1/2$, $a_2 = 1$, $\alpha_1 = 1$, $\alpha_2 = 2$,   and $x_{01} = \nu_2 = \phi_{01} = \phi_{02} = 0$. (a) $x_{02} = \nu_1 = 0$, (b) $x_{02} = 2$ and $\nu_1 = 0$, and (c) $x_{02} = 3$ and $\nu_1 = 1/2$. Animation available online at http://doi.org/10.1088/978-0-7503-5954-2.

$$\psi_1(x, t) = \frac{M_{12}\left[\gamma_1^{-1}(x, t) + \gamma_2^{*}(x, t)\right] - M_{22}\left[\gamma_2^{-1}(x, t) + \gamma_2^{*}(x, t)\right]}{M_{12} M_{21}\left[\gamma_1^{*}(x, t) + \gamma_2^{-1}(x, t)\right]\left[\gamma_1^{-1}(x, t) + \gamma_2^{*}(x, t)\right] - M_{11} M_{22}\left[\gamma_1^{-1}(x, t) + \gamma_1^{*}(x, t)\right]\left[\gamma_2^{-1}(x, t) + \gamma_2^{*}(x, t)\right]},$$

$$\psi_2(x, t) = \frac{-M_{11}\left[\gamma_1^{-1}(x, t) + \gamma_1^{*}(x, t)\right] + M_{21}\left[\gamma_1^{*}(x, t) + \gamma_2^{-1}(x, t)\right]}{M_{12} M_{21}\left[\gamma_1^{*}(x, t) + \gamma_2^{-1}(x, t)\right]\left[\gamma_1^{-1}(x, t) + \gamma_2^{*}(x, t)\right] - M_{11} M_{22}\left[\gamma_1^{-1}(x, t) + \gamma_1^{*}(x, t)\right]\left[\gamma_2^{-1}(x, t) + \gamma_2^{*}(x, t)\right]},$$

$a_1 > 0, \qquad a_2 > 0, \qquad M_{jk} = 1/(\lambda_j + \lambda_k^{*}), \qquad \gamma_j(x, t) = e^{\frac{\lambda_j}{\sqrt{2 a_1}}(x - x_{0j}) + i\left[\lambda_j^2 (t - t_0)/2 + \phi_{0j}\right]},$

$\lambda_j = \alpha_j + i\,\nu_j,\ \alpha_j,\ \nu_j,\ x_{0j},\ t_0,$ and ϕ_{0j} are arbitrary real constants, $N = 2$ in (2.206)
  - *Reference*: [46].

***Solution* 47.**   *three bright solitons*
  (Figure 2.14)

$$\psi(x, t) = \frac{1}{\sqrt{a_2}}\left[\psi_1(x, t) + \psi_2(x, t) + \psi_3(x, t)\right], \tag{2.209}$$

where

$\psi_1(x, t) = -\left[(\{M_{13} M_{22}\left[\gamma_2^{-1}(x, t) + \gamma_2^{*}(x, t)\right]\left[\gamma_1^{-1}(x, t) + \gamma_3^{*}(x, t)\right]\right.$

$\qquad - M_{12} M_{23}\left[\gamma_1^{-1}(x, t) + \gamma_2^{*}(x, t)\right]\left[\gamma_2^{-1}(x, t) + \gamma_3^{*}(x, t)\right]\}\{-M_{13}\left[\gamma_1^{-1}(x, t)\right.$

$\qquad + \gamma_3^{*}(x, t)\right] + M_{33}\left[\gamma_3^{-1}(x, t) + \gamma_3^{*}(x, t)\right]\} - \{-M_{13}\left[\gamma_1^{-1}(x, t) + \gamma_3^{*}(x, t)\right]$

$\qquad + M_{23}\left[\gamma_2^{-1}(x, t) + \gamma_3^{*}(x, t)\right]\}\{M_{13} M_{32}\left[\gamma_3^{-1}(x, t) + \gamma_2^{*}(x, t)\right]\left[\gamma_1^{-1}(x, t)\right.$

$\qquad + \gamma_3^{*}(x, t)\right] - M_{12} M_{33}\left[\gamma_1^{-1}(x, t) + \gamma_2^{*}(x, t)\right]\left[\gamma_3^{-1}(x, t) + \gamma_3^{*}(x, t)\right]\})$

$\qquad \div (\{M_{13} M_{22}\left[\gamma_2^{-1}(x, t) + \gamma_2^{*}(x, t)\right]\left[\gamma_1^{-1}(x, t) + \gamma_3^{*}(x, t)\right]$

$\qquad - M_{12} M_{23}\left[\gamma_1^{-1}(x, t) + \gamma_2^{*}(x, t)\right]\left[\gamma_2^{-1}(x, t) + \gamma_3^{*}(x, t)\right]\}\{M_{13} M_{31}\left[\gamma_3^{-1}(x, t)\right.$

$\qquad + \gamma_1^{*}(x, t)\right]\left[\gamma_1^{-1}(x, t) + \gamma_3^{*}(x, t)\right] - M_{11} M_{33}\left[\gamma_1^{-1}(x, t) + \gamma_1^{*}(x, t)\right]\left[\gamma_3^{-1}(x, t)\right.$

$\qquad + \gamma_3^{*}(x, t)\right]\} - \{M_{13} M_{21}\left[\gamma_2^{-1}(x, t) + \gamma_1^{*}(x, t)\right]\left[\gamma_1^{-1}(x, t) + \gamma_3^{*}(x, t)\right]$

$\qquad - M_{11} M_{23}\left[\gamma_1^{-1}(x, t) + \gamma_1^{*}(x, t)\right]\left[\gamma_2^{-1}(x, t) + \gamma_3^{*}(x, t)\right]\}\{M_{13} M_{32}\left[\gamma_3^{-1}(x, t)\right.$

$\qquad + \gamma_2^{*}(x, t)\right]\left[\gamma_1^{-1}(x, t) + \gamma_3^{*}(x, t)\right] - M_{12} M_{33}\left[\gamma_1^{-1}(x, t) + \gamma_2^{*}(x, t)\right]\left[\gamma_3^{-1}(x, t)\right.$

$\qquad + \gamma_3^{*}(x, t)\right]\})],$

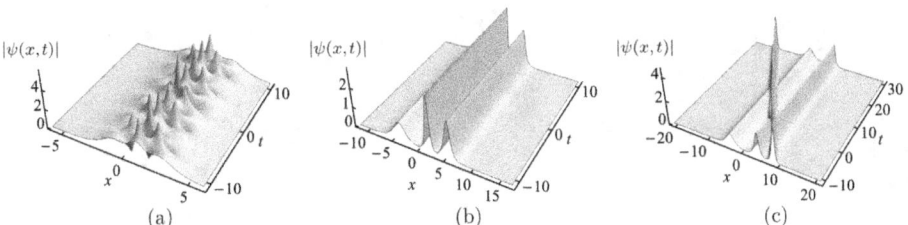

Figure 2.14. Three-bright-solitons (2.209), with $a_1 = 1/2$, $a_2 = 1$, $\alpha_1 = 2$, $\alpha_2 = 3$, $\alpha_3 = 1$, and $\phi_{01} = \phi_{02} = \phi_{03} = t_0 = 0$. (a) $x_{01} = x_{02} = x_{03} = 1$ and $\nu_1 = \nu_2 = \nu_3 = 0$, (b) $x_{01} = 4$, $x_{02} = 2$, $x_{03} = -2$, and $\nu_1 = \nu_2 = \nu_3 = 0$, and (c) $x_{01} = 4$, $x_{02} = 2$, $x_{03} = -2$, $\nu_1 = \nu_3 = 0$, and $\nu_2 = -1/2$. Animation available online at http://doi.org/10.1088/978-0-7503-5954-2.

$$\psi_2(x,\ t) = \{(M_{23}\ M_{31} - M_{21}\ M_{33})\ \gamma_2^{-1}(x,\ t)\ \gamma_3^{-1}(x,\ t) + [M_{21}\ (M_{13} - M_{33})\ \gamma_2^{-1}(x,\ t)$$
$$+ (-M_{13} + M_{23})\ M_{31}\ \gamma_3^{-1}(x,\ t)]\ \gamma_3^*(x,\ t) + \gamma_1^{-1}(x,\ t)\ [M_{13}\ (M_{21} - M_{31})$$
$$\times\ \gamma_1^*(x,\ t) + (M_{13}\ M_{21} - M_{11}\ M_{23})\ \gamma_2^{-1}(x,\ t) + (-M_{13}\ M_{31} + M_{11}\ M_{33})$$
$$\times\ \gamma_3^{-1}(x,\ t) + M_{11}\ (-M_{23} + M_{33})\ \gamma_3^*(x,\ t)] + \gamma_1^*(x,\ t)\ [M_{23}\ (-M_{11} + M_{31})$$
$$\times\ \gamma_2^{-1}(x,\ t) + (M_{11} - M_{21})\ M_{33}\ \gamma_3^{-1}(x,\ t) + (M_{13}\ M_{21} - M_{11}\ M_{23}$$
$$-\ M_{13}\ M_{31} + M_{23}\ M_{31} + M_{11}\ M_{33} - M_{21}\ M_{33})\ \gamma_3^*(x,\ t)]\}$$
$$\div\ (M_{12}\ (M_{23}\ M_{31} - M_{21}\ M_{33})\gamma_2^{-1}(x,\ t)\ \gamma_2^*(x,\ t)\ \gamma_3^{-1}(x,\ t) + \{(-M_{13}\ M_{22}$$
$$+\ M_{12}\ M_{23})\ M_{31}\ \gamma_2^*(x,\ t)\ \gamma_3^{-1}(x,\ t) + \gamma_2^{-1}(x,\ t)\ [M_{21}\ (M_{13}\ M_{32} - M_{12}\ M_{33})$$
$$\times\ \gamma_2^*(x,\ t) + M_{13}\ (-M_{22}\ M_{31} + M_{21}\ M_{32})\ \gamma_3^{-1}(x,\ t)]\}\ \gamma_3^*(x,\ t) + \gamma_1^{-1}(x,\ t)$$
$$\times\ \{M_{22}\ (-M_{13}\ M_{31} + M_{11}\ M_{33})\ \gamma_2^*(x,\ t)\ \gamma_3^{-1}(x,\ t) + [M_{11}\ (-M_{23}\ M_{32}$$
$$+\ M_{22}\ M_{33})\ \gamma_2^*(x,\ t) + M_{23}\ (M_{12}\ M_{31} - M_{11}\ M_{32})\ \gamma_3^{-1}(x,\ t)]\gamma_3^*(x,\ t)$$
$$+\ \gamma_2^{-1}(x,\ t)[(M_{13}\ M_{21} - M_{11}\ M_{23})\ M_{32}\ \gamma_2^*(x,\ t) + (-M_{13}\ M_{22}\ M_{31}$$
$$+\ M_{12}\ M_{23}\ M_{31} + M_{13}\ M_{21}\ M_{32} - M_{11}\ M_{23}\ M_{32} - M_{12}\ M_{21}\ M_{33}$$
$$+\ M_{11}\ M_{22}\ M_{33})\ \gamma_3^{-1}(x,\ t) + (-M_{12}\ M_{21} + M_{11}\ M_{22})\ M_{33}\ \gamma_3^*(x,\ t)]$$
$$+\ \gamma_1^*(x,\ t)\ [(-M_{13}\ M_{22} + M_{12}\ M_{23})\ M_{31}\ \gamma_2^{-1}(x,\ t) + M_{13}\ (-M_{22}\ M_{31}$$
$$+\ M_{21}\ M_{32})\ \gamma_2^*(x,\ t) + M_{21}\ (M_{13}\ M_{32} - M_{12}\ M_{33})\ \gamma_3^{-1}(x,\ t)$$
$$+\ M_{12}\ (M_{23}\ M_{31} - M_{21}\ M_{33})\ \gamma_3^*(x,\ t)]\}$$
$$\times\ \gamma_1^*(x,\ t)\ \{(M_{13}\ M_{21} - M_{11}\ M_{23})\ M_{32}\ \gamma_3^{-1}(x,\ t)\ \gamma_3^*(x,\ t)$$
$$+\ \gamma_2^{-1}(x,\ t)\ [M_{23}\ (M_{12}\ M_{31} - M_{11}\ M_{32})\ \gamma_2^*(x,\ t) + M_{11}\ (-M_{23}\ M_{32}$$
$$+\ M_{22}\ M_{33})\ \gamma_3^{-1}(x,\ t) + M_{22}\ (-M_{13}\ M_{31} + M_{11}\ M_{33})\ \gamma_3^*(x,\ t)]$$
$$+\ \gamma_2^*(x,\ t)\ [(-M_{12}\ M_{21} + M_{11}\ M_{22})\ M_{33}\ \gamma_3^{-1}(x,\ t) + (-M_{13}\ M_{22}\ M_{31}$$
$$+\ M_{12}\ M_{23}\ M_{31} + M_{13}\ M_{21}\ M_{32} - M_{11}\ M_{23}\ M_{32} - M_{12}\ M_{21}\ M_{33}$$
$$+\ M_{11}\ M_{22}\ M_{33})\ \gamma_3^*(x,\ t)]\}),$$

$$\psi_3(x, t) = \Big\{ M_{21} (M_{12} - M_{32}) \gamma_2^{-1}(x, t) \gamma_2^*(x, t) + [(M_{22} M_{31} - M_{21} M_{32}) \gamma_2^{-1}(x, t)$$

$$+ (-M_{12} + M_{22}) M_{31} \gamma_2^*(x, t)] \gamma_3^{-1}(x, t) + \gamma_1^*(x, t) [M_{22} (-M_{11} + M_{31})$$

$$\times \gamma_2^{-1}(x, t) + (M_{12} M_{21} - M_{11} M_{22} - M_{12} M_{31} + M_{22} M_{31} + M_{11} M_{32}$$

$$- M_{21} M_{32}) \gamma_2^*(x, t) + (M_{11} - M_{21}) M_{32} \gamma_3^{-1}(x, t)] + \gamma_1^{-1}(x, t) [M_{12} (M_{21}$$

$$- M_{31}) \gamma_1^*(x, t) + (M_{12} M_{21} - M_{11} M_{22}) \gamma_2^{-1}(x, t) + M_{11} (-M_{22} + M_{32})$$

$$\times \gamma_2^*(x, t) + (-M_{12} M_{31} + M_{11} M_{32}) \gamma_3^{-1}(x, t)] \Big\}$$

$$\div \Big(M_{12} (-M_{23} M_{31} + M_{21} M_{33}) \gamma_2^{-1}(x, t) \gamma_2^*(x, t) \gamma_3^{-1}(x, t) + \Big\{ (M_{13} M_{22}$$

$$- M_{12} M_{23}) M_{31} \gamma_2^*(x, t) \gamma_3^{-1}(x, t) + \gamma_2^{-1}(x, t) [M_{21} (-M_{13} M_{32} + M_{12} M_{33})$$

$$\times \gamma_2^*(x, t) + M_{13} (M_{22} M_{31} - M_{21} M_{32}) \gamma_3^{-1}(x, t)] \Big\} \gamma_3^*(x, t) + \gamma_1^{-1}(x, t)$$

$$\times \Big\{ M_{22} (M_{13} M_{31} - M_{11} M_{33}) \gamma_2^*(x, t) \gamma_3^{-1}(x, t) + [M_{11} (M_{23} M_{32} - M_{22} M_{33})$$

$$\times \gamma_2^*(x, t) + M_{23} (-M_{12} M_{31} + M_{11} M_{32}) \gamma_3^{-1}(x, t)] \gamma_3^*(x, t) + \gamma_2^{-1}(x, t)$$

$$\times [(-M_{13} M_{21} + M_{11} M_{23}) M_{32} \gamma_2^*(x, t) + (M_{13} M_{22} M_{31} - M_{12} M_{23} M_{31}$$

$$- M_{13} M_{21} M_{32} + M_{11} M_{23} M_{32} + M_{12} M_{21} M_{33} - M_{11} M_{22} M_{33}) \gamma_3^{-1}(x, t)$$

$$+ (M_{12} M_{21} - M_{11} M_{22}) M_{33} \gamma_3^*(x, t)] + \gamma_1^*(x, t) [(M_{13} M_{22} - M_{12} M_{23}) M_{31}$$

$$\times \gamma_2^{-1}(x, t) + M_{13} (M_{22} M_{31} - M_{21} M_{32}) \gamma_2^*(x, t) + M_{21} (-M_{13} M_{32} + M_{12} M_{33})$$

$$\times \gamma_3^{-1}(x, t) + M_{12} (-M_{23} M_{31} + M_{21} M_{33}) \gamma_3^*(x, t)] \Big\} + \gamma_1^*(x, t) \Big\{ (-M_{13} M_{21}$$

$$+ M_{11} M_{23}) M_{32} \gamma_3^{-1}(x, t) \gamma_3^*(x, t) + \gamma_2^{-1}(x, t) [M_{23} (-M_{12} M_{31} + M_{11} M_{32})$$

$$\times \gamma_2^*(x, t) + M_{11} (M_{23} M_{32} - M_{22} M_{33}) \gamma_3^{-1}(x, t) + M_{22} (M_{13} M_{31} - M_{11} M_{33})$$

$$\times \gamma_3^*(x, t)] + \gamma_2^*(x, t) [(M_{12} M_{21} - M_{11} M_{22}) M_{33} \gamma_3^{-1}(x, t) + (M_{13} M_{22} M_{31}$$

$$- M_{12} M_{23} M_{31} - M_{13} M_{21} M_{32} + M_{11} M_{23} M_{32} + M_{12} M_{21} M_{33}$$

$$- M_{11} M_{22} M_{33}) \gamma_3^*(x, t)] \Big\} \Big),$$

$a_1 > 0$, $a_2 > 0$, $M_{jk} = 1/(\lambda_j + \lambda_k^*)$, $\gamma_j(x, t) = e^{\frac{\lambda_j}{\sqrt{2 a_1}}(x - x_{0j}) + i [\lambda_j^2 (t - t_0)/2 + \phi_{0j}]}$, $\lambda_j = \alpha_j + i \nu_j$, α_j, ν_j, x_{0j}, t_0, and ϕ_{0j} are arbitrary real constants, $N = 3$ in (2.206).

- *Reference*: [46].

Solution 48. *N-dark solitons*

$$\psi(x, t) = \sqrt{\frac{-2}{a_2}} \left\{ 1 - 2 i \sum_n \mu_n(t) e^{\frac{-2 v_n}{\sqrt{a_1}} (x - x_0)} [-Q_{12}{}^n + (\lambda_n + i v_n) (Q_{11}{}^n - 1)] \right\} e^{-2 i [t - t_0 + \phi_0]}, (2.210)$$

where $v_j = \sqrt{1 - \lambda_j^2}$, $\mu_j(t) = e^{4 v_j \lambda_j (t - t_0)}$, Q_{11}^n and Q_{12}^n are obtained by solving the linear algebraic equations: $Q_{12}^j + \sum_n \frac{\mu_n(t)}{v_n + v_j} Q_{12}^n e^{\frac{-2 v_n (x - x_0)}{\sqrt{a_1}}} - \sum_n Q_{11}^n \frac{\mu_n(t) (\lambda_n + i v_n)}{v_n + v_j}$

$e^{\frac{-2 v_n (x - x_0)}{\sqrt{a_1}}} = \sum_n \frac{\mu_n(t) (\lambda_n + i v_n)}{v_n + v_j} e^{\frac{-2 v_n (x - x_0)}{\sqrt{a_1}}}$,

$$Q_{11}^j + \sum_n \frac{\mu_n(t)}{v_n + v_j} \, Q_{11}^n \, e^{\frac{-2\,v_n\,(x-x_0)}{\sqrt{a_1}}} + \sum_n Q_{12}^n \, \frac{\mu_n(t)\,(-\lambda_n + i\,v_n)}{v_n + v_j} \, e^{\frac{-2\,v_n\,(x-x_0)}{\sqrt{a_1}}} = \sum_n \frac{\mu_n(t)}{v_n + v_j} \, e^{\frac{-2\,v_n\,(x-x_0)}{\sqrt{a_1}}},$$

$a_1 > 0$, $a_2 < 0$, $-1 < \lambda_j < 1$, x_0, t_0, and ϕ_0 are arbitrary real constants.

- *Reference*: [47], *we corrected the exponential prefactor.*

***Solution* 49.**   *two dark solitons*
  (Figure 2.15)

$$\psi(x,\,t) = \left\{ 1 - \frac{2\,i}{p(x,\,t)} \left[\frac{2}{v_1 + v_2} \left(\frac{1}{\lambda_1 + i\,v_1} + \frac{1}{\lambda_2 + i\,v_2} \right) - (\lambda_1 - i\,v_1)\,q_1(x,\,t) \right.\right.$$
$$\left.\left. - (\lambda_2 - i\,v_2)\,q_2(x,\,t) \right] \right\} \sqrt{\frac{-2}{a_2}} \, e^{-2\,i\,[t-t_0 + \phi_0]}, \tag{2.211}$$

where $\quad v_1 = \sqrt{1 - \lambda_1^2}, \qquad v_2 = \sqrt{1 - \lambda_2^2}, \qquad q_1(x,\,t) = \frac{1}{v_1} + e^{\frac{2\,v_1}{\sqrt{a_1}}\,(x-x_0) - 4\,v_1\,\lambda_1\,(t-t_0)},$

$q_2(x,\,t) = \frac{1}{v_2} + e^{\frac{2\,v_2}{\sqrt{a_1}}\,(x-x_0) - 4\,v_2\,\lambda_2\,(t-t_0)},$

$p(x,\,t) = (\lambda_1 - i\,v_1)\,(\lambda_2 - i\,v_2)\,q_1(x,\,t)\,q_2(x,\,t) - \frac{1}{(v_1 + v_2)^2} \left(\frac{1}{\lambda_1 + i\,v_1} + \frac{1}{\lambda_2 + i\,v_2} \right)^2,$

$a_1 > 0$, $a_2 < 0$, $\lambda_1 = -\lambda_2$, $-1 < \lambda_2 < 1$, x_0, t_0, and ϕ_0 are arbitrary real constants.

- *Reference*: [47], *we corrected the exponential prefactor.*

***Solution* 50. Generalized first-order breather (form I*)**

$$\psi(x,\,t) = \frac{1}{\sqrt{a_2}} \left\{ \frac{\kappa^2 \cosh[\delta\,(t - t_0)] + 2\,i\,\kappa\,\nu\,\sinh[\delta\,(t - t_0)]}{2 \cosh[\delta\,(t - t_0)] - 2\,\nu\,\cos[\frac{\kappa}{\sqrt{2\,a_1}}\,(x - x_0)]} - 1 \right\} e^{i\,[t - t_0 + \phi_0]}, \tag{2.212}$$

where $a_1 > 0$, $a_2 > 0$, $\kappa = 2\sqrt{1 - \nu^2}$, $\delta = \kappa\,\nu$, ν, x_0, t_0, and ϕ_0 are arbitrary real constants.

- *Reference*: [48].

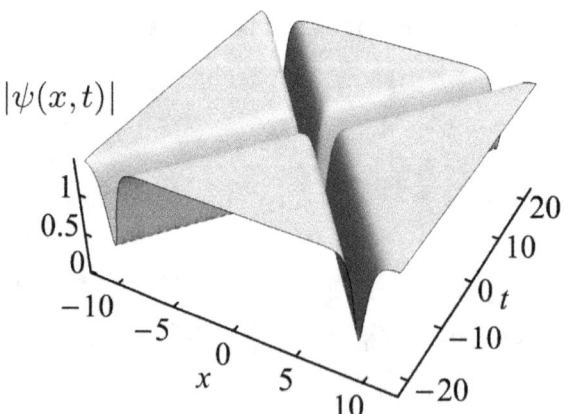

$|\psi(x,t)|$

Figure 2.15. Two dark solitons (2.211), with $a_1 = 1/2$, $a_2 = -1$, $\lambda_2 = 3/10$, and $x_0 = t_0 = \phi_0 = 0$.

Solution 51. Generalized first-order breather (form II*)

$$\psi(x,\,t) = \frac{\cos(A_0)\cos[q_1(x,\,t) + 2\,i\,A_1] - \cosh(A_1)\cosh[q_2(x,\,t) + 2\,i\,A_0]}{\cos(A_0)\cos[q_1(x,\,t)] - \cosh(A_1)\cosh[q_2(x,\,t)]}\,e^{i\,[a_2\,(t-t_0)+\phi_0]}, \quad (2.213)$$

where $\quad q_1(x,\,t) = v_3\,[\sqrt{\frac{2\,a_2}{a_1}}\,(x - x_0) - a_2\,v_0\,(t - t_0)], \qquad q_2(x,\,t) = v_2\,[\sqrt{\frac{2\,a_2}{a_1}}$

$(x - x_0) - a_2\,v_1\,(t - t_0)], \qquad v_0 = \frac{\sinh(2\,A_1)\cos(2\,A_0)}{\cosh(A_1)\sin(A_0)}, \qquad v_1 = -\frac{\cosh(2\,A_1)\sin(2\,A_0)}{\sinh(A_1)\cos(A_0)},$

$v_2 = -\sinh(A_1)\cos(A_0)$, $v_3 = \cosh(A_1)\sin(A_0)$, $a_1\,a_2 > 0$, x_0, t_0, A_0, A_1, and ϕ_0 are arbitrary real constants.

- *Reference*: [51].

Solution 52. Generalized first-order breather (form III*)

$$\psi(x,\,t) = \frac{A_0}{\sqrt{a_2}}\left(1 - \frac{\sqrt{8}\,\lambda_r}{A_0}\,p(x,\,t)\right)e^{i\,[A_0^2\,(t-t_0)+\phi_0]}, \quad (2.214)$$

where $\quad p(x,\,t) = \dfrac{(A_0^2 + \Gamma^2)\cos[q_1(x,\,t)] + i\,(A_0^2 - \Gamma^2)\sin[q_1(x,\,t)] + 2\,A_0\,\{\Gamma_r\cosh[q_2(x,\,t)] - i\,\Gamma_i\sinh[q_2(x,\,t)]\}}{2\,A_0\,\Gamma_r\cos[q_1(x,\,t)] + (\Gamma^2 + A_0^2)\cosh[q_2(x,\,t)]},$

$q_1(x,\,t) = \delta_i + \sqrt{2}\,[\frac{x - x_0}{\sqrt{a_1}}\,\Delta_i - 2\,(t - t_0)\,(\Delta_i\,\lambda_i + \Delta_r\,\lambda_r)],$

$q_2(x,\,t) = \delta_r + \sqrt{2}\,[\frac{x - x_0}{\sqrt{a_1}}\,\Delta_r - 2\,(t - t_0)\,\Delta_r\,\lambda_i + 2\,(t - t_0)\,\Delta_i\,\lambda_r],$

$\Delta_r = \text{Re}[\sqrt{2\,(\lambda_r - i\,\lambda_i)^2 - A_0^2}], \qquad\qquad \Delta_i = \text{Im}[\sqrt{2\,(\lambda_r - i\,\lambda_i)^2 - A_0^2}],$

$\Gamma_r = \Delta_r + \sqrt{2}\,\lambda_r$, $\Gamma_i = \Delta_i - \sqrt{2}\,\lambda_i$, $\Gamma = \sqrt{\Gamma_r^2 + \Gamma_i^2}$, $a_1 > 0$, $a_2 > 0$, A_0, λ_r, λ_i, x_0, t_0, and ϕ_0 are arbitrary real constants.

- *Reference*: [52].

*Remark: There are different forms of the breather in the literature, as given by Solutions 39, 40, 41. In appendix B.1.2 we derive a fourth expression valid for focusing and defocusing nonlinearities and the arbitrary constants are expressed in terms of physical parameters.

Solution 53. Periodicity in *t* and Localization in *x*   *Kuznetsov–Ma breather*
(Figure 2.16)

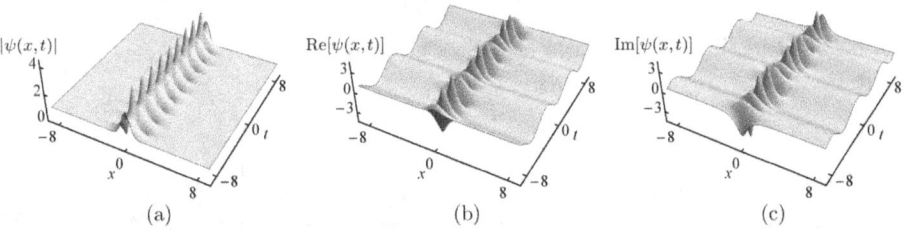

Figure 2.16. Kuznetsov–Ma breather (2.215), with $a_1 = a_2 = 1$, $\nu = 1.5$, and $x_0 = t_0 = \phi_0 = 0$. (a) Absolute value of (2.215), (b) Real part of (2.215), and (c) Imaginary part of (2.215). Animation available online at http://doi.org/10.1088/978-0-7503-5954-2.

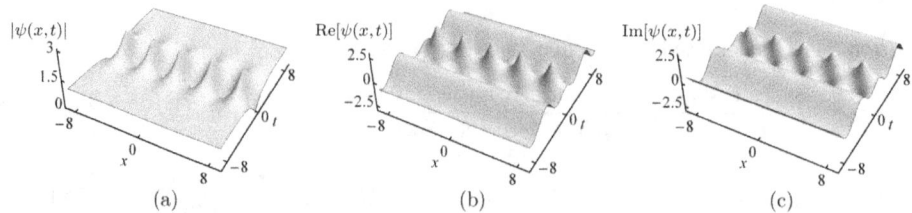

Figure 2.17. Akhmediev breather (2.216), with $a_1 = 1/2$, $a_2 = 1$, $\nu = 0.5$, and $x_0 = t_0 = \phi_0 = 0$. (a) Absolute value of (2.216), (b) Real part of (2.216), and (c) Imaginary part of (2.216). Animation available online at http://doi.org/10.1088/978-0-7503-5954-2.

$$\psi(x,\, t) = \frac{1}{\sqrt{a_2}} \left\{ \frac{-p^2 \cos[\omega\,(t - t_0)] - 2\,i\,p\,\nu\,\sin[\omega\,(t - t_0)]}{2 \cos[\omega\,(t - t_0)] - 2\,\nu\,\cosh\left[\dfrac{p}{\sqrt{2\,a_1}}\,(x - x_0)\right]} - 1 \right\} e^{i\,[t - t_0 + \phi_0]}, \qquad (2.215)$$

where $a_1 > 0$, $a_2 > 0$, $p = 2\,\sqrt{\nu^2 - 1}$, $\omega = p\,\nu$, $\nu > 1$, x_0, t_0, and ϕ_0 are arbitrary real constants.

- *Reference*: [48].

- This solution can be generated from (2.212) with $\nu > 1$.

***Solution* 54. Periodicity in x and Localization in t**   *Akhmediev breather*
(Figure 2.17)

$$\psi(x,\, t) = \frac{1}{\sqrt{a_2}} \left\{ \frac{\kappa^2 \cosh[\delta\,(t - t_0)] + 2\,i\,\kappa\,\nu\,\sinh[\delta\,(t - t_0)]}{2 \cosh[\delta\,(t - t_0)] - 2\,\nu\,\cos\left[\dfrac{\kappa}{\sqrt{2\,a_1}}\,(x - x_0)\right]} - 1 \right\} e^{i\,[t - t_0 + \phi_0]}, \qquad (2.216)$$

where $a_1 > 0$, $a_2 > 0$, $\kappa = 2\,\sqrt{1 - \nu^2}$, $\delta = \kappa\,\nu$, $\nu < 1$, x_0, t_0, and ϕ_0 are arbitrary real constants.

- *Reference*: [48].

***Solution* 55. Localization in x and t**   *Peregrine soliton*
(Figure 2.18)

$$\psi(x,\, t) = \frac{1}{\sqrt{a_2}} \left[\frac{4 + i\,8\,(t - t_0)}{1 + 4\,(t - t_0)^2 + \dfrac{2}{a_1}\,(x - x_0)^2} - 1 \right] e^{i\,[t - t_0 + \phi_0]}, \qquad (2.217)$$

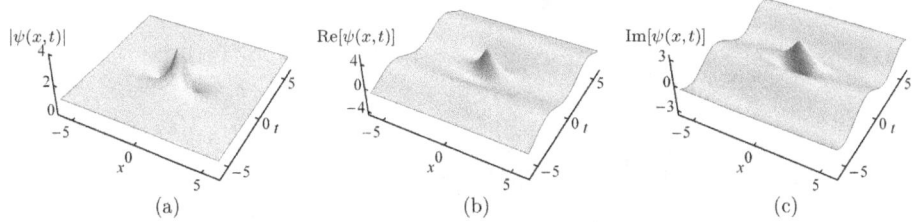

Figure 2.18. Peregrine soliton (2.217), with $a_1 = a_2 = 1$, and $x_0 = t_0 = \phi_0 = 0$. (a) Absolute value of (2.217), (b) Real part of (2.217), and (c) Imaginary part of (2.217). Animation available online at http://doi.org/10.1088/978-0-7503-5954-2.

where $a_2 > 0$, x_0, t_0, and ϕ_0 are arbitrary real constants.

- *Reference*: [48].

- This solution can be generated from (2.212) *in the limits* $\nu \to 1$ and $\kappa \to 0$.

Solution 56.   *doubly-periodic (type A)*

$$\psi(x, t) = [Q(x, t) + i\,\delta(t)]\,e^{i\,\phi(t)}, \tag{2.218}$$

where $\delta(t) = \sqrt{\frac{\alpha_3}{2}(1 - \nu)}\sqrt{\frac{1 + \mathrm{dn}(\mu\,t, k^2)}{1 + \nu\,\mathrm{cn}(\mu\,t, k^2)}}\,\mathrm{sn}(\frac{\mu\,t}{2}, k^2)$, $\phi(t) = (2\,\alpha_1 + \frac{\alpha_3}{\nu})\,t$

$-\frac{\alpha_3}{\nu\,\mu}\{\Pi[n; \mathrm{am}(\mu\,t, k^2)|k^2] - \nu\,\sigma\,\tan^{-1}[\frac{\mathrm{sd}(\mu\,t, k^2)}{\sigma}]\}$,

$Q(x, t) = s(t)\,z(t) - q_a(t)\frac{r(t) + \mathrm{cn}(p\,x, k_q)}{1 + r(t)\,\mathrm{cn}(p\,x, k_q)}$,

$q_a(t) = \sqrt{2\sqrt{[\delta^2(t) - \alpha_1]^2 + \alpha_2^2} + 2\,[\alpha_1 - \delta^2(t)]}$,

$q_b(t) = \sqrt{2\sqrt{[\delta^2(t) - \alpha_1]^2 + \alpha_2^2} - 2\,[\alpha_1 - \delta^2(t)]}$,

$q_c(t) = \sqrt{[2\,s(t)\,z(t) + q_a(t)]^2 + q_b(t)^2}$, $q_d(t) = \sqrt{[2\,s(t)\,z(t) - q_a(t)]^2 + q_b(t)^2}$,

$z(t) = \sqrt{\alpha_3 - \delta^2(t)}$, $r(t) = \frac{q_c(t) - q_d(t)}{q_c(t) + q_d(t)}$, $s(t) = 2\,\theta[\mathrm{cn}(\frac{\mu\,t}{2}, k^2)] - 1$,

$A = \sqrt{(\alpha_3 - \alpha_1)^2 + \alpha_2^2}$, $B = \sqrt{\alpha_1^2 + \alpha_2^2}$, $k = \sqrt{\frac{1}{2}[1 - \frac{\alpha_2^2 + \alpha_1(\alpha_1 - \alpha_3)}{A\,B}]}$,

$\nu = (A - B)/(A + B)$, $\sigma = \sqrt{\frac{1 - \nu^2}{k^2 + \nu^2(1 - k^2)}}$, $\mu = 4\sqrt{A\,B}$, $n = \nu^2/(\nu^2 - 1)$,

$p = 2\,[(\alpha_3 - \alpha_1)^2 + \alpha_2^2]^{1/4}$, $k_q = \frac{1}{2} + 2\,(\frac{\alpha_1 - \alpha_3}{p^2})$, $a_1 = 1/2$, $a_2 = 1$, $\alpha_3 > 0$, α_1 and α_3 are arbitrary real constants, $\mathrm{am}(\cdot)$ is the amplitude for Jacobi elliptic functions, $\Pi(\cdot)$ is the complete elliptic integral for the third kind, $\theta(\cdot)$ is the Heaviside theta function.

- *Reference*: [49, 50], *corrected*.

Solution 57.   *doubly-periodic (type B)*
    (Figure 2.19)

$$\psi(x, t) = [Q(x, t) + i\,\delta(t)]\,e^{i\,\phi(t)}, \tag{2.219}$$

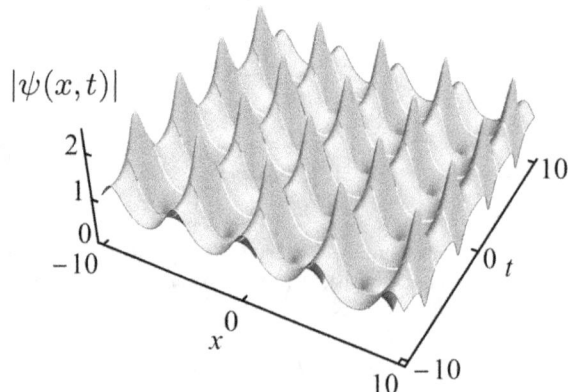

Figure 2.19. Doubly periodic solution I (2.219), with $\alpha_1 = 0.3$, $\alpha_2 = 0.7$, and $\alpha_3 = 1$.

where $\delta(t) = \dfrac{\sqrt{\alpha_1\,\alpha_3}\ \mathrm{sn}(\mu\,t, k^2)}{\sqrt{\alpha_3 - \alpha_1\,\mathrm{cn}^2(\mu\,t, k^2)}}$, $\phi(t) = (\alpha_1 + \alpha_2 - \alpha_3)\,t + (2\,\alpha_3/\mu)\,\Pi[n;\,\mathrm{am}(\mu\,t,\,k^2)$

$|k^2],\quad Q(x,\,t) = \dfrac{q_b(t)\,[q_a(t) - q_c(t)] + q_a(t)\,[q_c(t) - q_b(t)]\,\mathrm{sn}^2(p\,x, k_q^2)}{[q_a(t) - q_c(t)] + [q_c(t) - q_b(t)]\,\mathrm{sn}^2(p\,x, k_q^2)},\quad q_a(t) = s(t)\,\sqrt{\alpha_1 - \delta^2(t)}\,+$

$\sqrt{\alpha_2 - \delta^2(t)}\,+\,\sqrt{\alpha_3 - \delta^2(t)}$,

$q_b(t) = s(t)\,\sqrt{\alpha_1 - \delta^2(t)}\,-\,\sqrt{\alpha_2 - \delta^2(t)}\,-\,\sqrt{\alpha_3 - \delta^2(t)}$,

$q_c(t) = -s(t)\,\sqrt{\alpha_1 - \delta^2(t)}\,+\,\sqrt{\alpha_2 - \delta^2(t)}\,-\,\sqrt{\alpha_3 - \delta^2(t)}$,

$s(t) = 2\,\theta[\mathrm{cn}(\mu\,t, k^2)] - 1$, $\quad k = \sqrt{\dfrac{\alpha_1\,(\alpha_3 - \alpha_2)}{\alpha_2\,(\alpha_3 - \alpha_1)}}$, $\quad \mu = 2\,\sqrt{\alpha_2\,(\alpha_3 - \alpha_1)}$, $\quad n = \alpha_1/$

$(\alpha_1 - \alpha_3)$, $\quad p = \sqrt{\alpha_3 - \alpha_1}$, $\quad k_q = \sqrt{\dfrac{\alpha_2 - \alpha_1}{\alpha_3 - \alpha_1}}$, $\quad a_1 = 1/2$, $\quad a_2 = 1$, $\quad \alpha_3 - \alpha_1 > 0$,

$\alpha_2 - \alpha_1 > 0$, $\alpha_1\,(\alpha_3 - \alpha_2) > 0$, $\alpha_2 > 0$, $\mathrm{am}(\cdot)$ is the amplitude for Jacobi elliptic functions, $\Pi(\cdot)$ is the complete elliptic integral for the third kind, $\theta(\cdot)$ is the Heaviside theta function.

- *Reference*: [49, 50], *corrected*.

Solution 58. **Generalized two-breathers solution**

(Figures 2.20–2.24)

$$\psi(x,\,t) = \frac{1}{\sqrt{a_2}}\left[1 + \frac{\alpha(x,\,t) + i\,\beta(x,\,t)}{\gamma(x,\,t)}\right] e^{i\,[t - t_0 + \phi_0]}, \qquad (2.220)$$

where

$$\alpha(x,\,t) = (\kappa_2^2 - \kappa_1^2)\left\{ \frac{\delta_2\,\kappa_1^2\,\cos[\frac{\kappa_2}{\sqrt{2\,a_1}}\,(x - x_{02})]\,\cosh[\delta_1\,(t - t_{01})]}{\kappa_2}\right.$$

$$-\,\frac{\delta_1\,\kappa_2^2\,\cos\left[\frac{\kappa_1}{\sqrt{2\,a_1}}\,(x - x_{01})\right]\,\cosh\left[\delta_2\,(t - t_{02})\right]}{\kappa_1}$$

$$\left. -\,(\kappa_1^2 - \kappa_2^2)\,\cosh\left[\delta_1\,(t - t_{01})\right]\,\cosh\left[\delta_2(t - t_{02})\right]\right\},$$

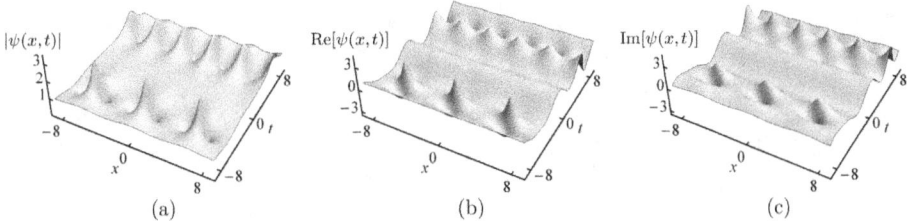

Figure 2.20. Plot of solution (2.220), with $a_1 = 1/2$, $a_2 = 1$, $\nu_1 = 0.5$, $\nu_2 = 0.85$, $x_{01} = x_{02} = t_0 = \phi_0 = 0$, $t_{01} = 5$, and $t_{02} = -5$. (a) Absolute value of (2.220), (b) Real part of (2.220), and (c) Imaginary part of (2.220).

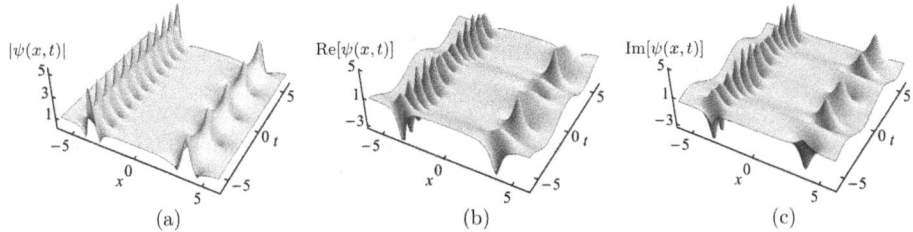

Figure 2.21. Plot of solution (2.220), with $a_1 = 1/2$, $a_2 = 1$, $\nu_1 = 1.3$, $\nu_2 = 1.85$, $x_{01} = 5$, $x_{02} = -5$, and $t_{01} = t_{02} = t_0 = \phi_0 = 0$. (a) Absolute value of (2.220), (b) Real part of (2.220), and (c) Imaginary part of (2.220).

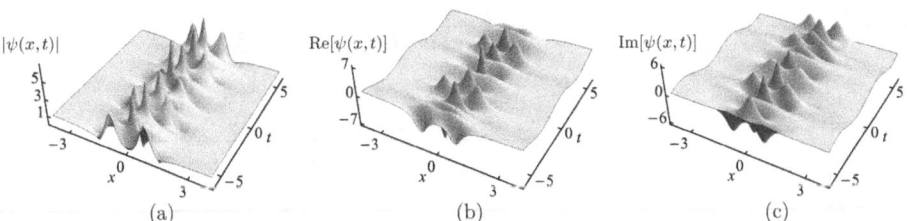

Figure 2.22. Plot of solution (2.220), with $a_1 = 1/2$, $a_2 = 1$, $\nu_1 = 1.3$, $\nu_2 = 1.85$, and $x_{01} = x_{02} = t_{01} = t_{02} = t_0 = \phi_0 = 0$. (a) Absolute value of (2.220), (b) Real part of (2.220), and (c) Imaginary part of (2.220).

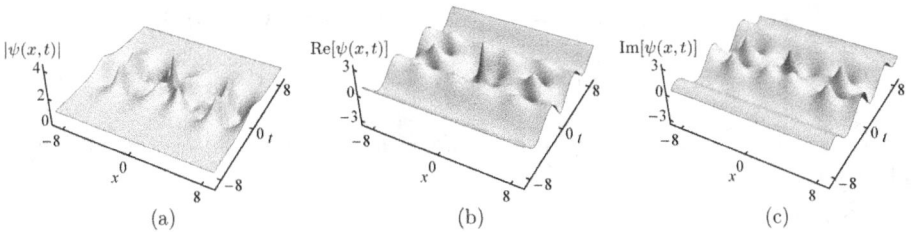

Figure 2.23. Plot of solution (2.220), with $a_1 = 1/2$, $a_2 = 1$, $\nu_1 = 0.5$, $\nu_2 = 0.85$, and $x_{01} = x_{02} = t_{01} = t_{02} = t_0 = \phi_0 = 0$. (a) Absolute value of (2.220), (b) Real part of (2.220), and (c) Imaginary part of (2.220).

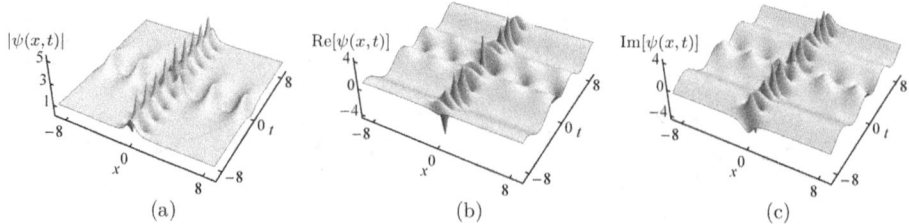

$|\psi(x,t)|$

$\mathrm{Re}[\psi(x,t)]$

$\mathrm{Im}[\psi(x,t)]$

(a)   (b)   (c)

Figure 2.24. Plot of solution (2.220), with $a_1 = 1/2$, $a_2 = 1$, $\nu_1 = 0.5$, $\nu_2 = 1.5$, and $x_{01} = x_{02} = t_{01} = t_{02} = t_0 = \phi_0 = 0$. (a) Absolute value of (2.220), (b) Real part of (2.220), and (c) Imaginary part of (2.220).

$$\beta(x,t) = -2(\kappa_1^2 - \kappa_2^2)\left\{ \frac{\delta_1\,\delta_2\cos[\frac{\kappa_2}{\sqrt{2\,a_1}}(x-x_{02})]\sinh[\delta_1(t-t_{01})]}{\kappa_2}\right.$$

$$- \delta_1\cosh[\delta_2(t-t_{02})]\sinh[\delta_1(t-t_{01})] + \delta_2\cosh[\delta_1(t-t_{01})]\sinh[\delta_2(t-t_{02})]$$

$$\left. - \frac{\delta_1\,\delta_2\cos[\frac{\kappa_1}{\sqrt{2\,a_1}}(x-x_{01})]\sinh[\delta_2(t-t_{02})]}{\kappa_1}\right\},$$

$$\gamma(x,t) = \frac{2\,\delta_1\,\delta_2(\kappa_1^2 + \kappa_2^2)\cos[\frac{\kappa_1}{\sqrt{2\,a_1}}(x-x_{01})]\cos[\frac{\kappa_2}{\sqrt{2\,a_1}}(x-x_{02})]}{\kappa_1\,\kappa_2}$$

$$- (2\,\kappa_1^2 + 2\,\kappa_2^2 - \kappa_1^2\,\kappa_2^2)\cosh[\delta_1(t-t_{01})]\cosh[\delta_2(t-t_{02})]$$

$$- 2(\kappa_1^2 - \kappa_2^2)\left\{ -\frac{\delta_2\cos[\frac{\kappa_2}{\sqrt{2\,a_1}}(x-x_{02})]\cosh[\delta_1(t-t_{01})]}{\kappa_2}\right.$$

$$\left. + \frac{\delta_1\cos[\frac{\kappa_1}{\sqrt{2\,a_1}}(x-x_{01})]\cosh[\delta_2(t-t_{02})]}{\kappa_1}\right\}$$

$$+ 4\,\delta_1\,\delta_2\left\{ \sin[\frac{\kappa_1}{\sqrt{2\,a_1}}(x-x_{01})]\sin[\frac{\kappa_2}{\sqrt{2\,a_1}}(x-x_{02})]\right.$$

$$\left. + \sinh[\delta_1(t-t_{01})]\sinh[\delta_2(t-t_{02})]\right\},$$

$a_1 > 0$, $a_2 > 0$, $\kappa_j = 2\sqrt{1-\nu_j^2}$, $\delta_j = \frac{\kappa_j}{2}\sqrt{4-\kappa_j^2}$, ν_j, and x_{0j}, t_{0j}, t_0, and ϕ_0 are arbitrary real constants, $j = 1, 2$.

- *Reference*: [48].

Solution 59. Specific two-breather solution I — *Nonlinear superposition of Kuznetsov–Ma or Akhmediev breather with a Peregrine soliton*

(Figures 2.25–2.28)

$$\psi(x, t) = \frac{1}{\sqrt{a_2}} \left[1 + \frac{\alpha(x, t) + i\,\beta(x, t)}{\gamma(x, t)} \right] e^{i\,[t - t_0 + \phi_0]}, \qquad (2.221)$$

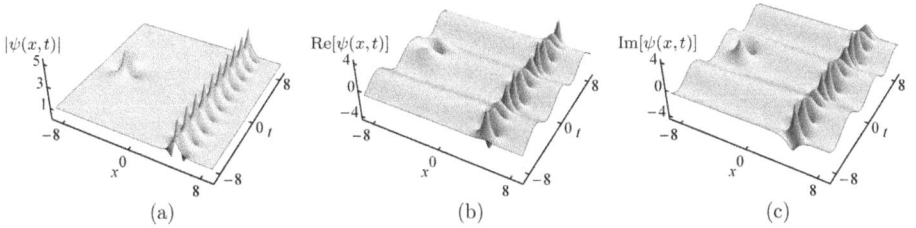

(a)   (b)   (c)

Figure 2.25. Plot of solution (2.221), with $a_1 = 1/2$, $a_2 = 1$, $\nu = 1.2$, $x_{01} = 5$, $x_{02} = -5$, and $t_{01} = t_{02} = t_0 = \phi_0 = 0$. (a) Absolute value of (2.221), (b) Real part of (2.221), and (c) Imaginary part of (2.221).

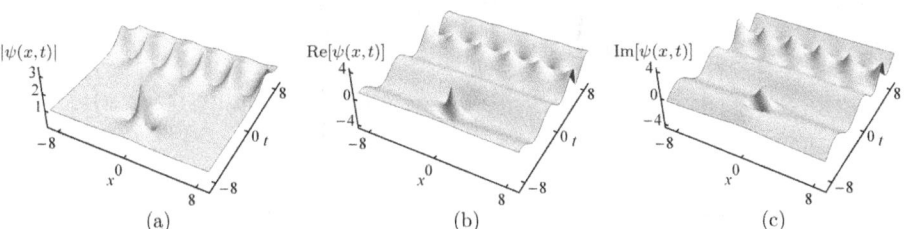

(a)   (b)   (c)

Figure 2.26. Plot of solution (2.221), with $a_1 = 1/2$, $a_2 = 1$, $\nu = 0.5$, $t_{01} = 5$, $t_{02} = -5$, and $x_{01} = x_{02} = t_0 = \phi_0 = 0$. (a) Absolute value of (2.221), (b) Real part of (2.221), and (c) Imaginary part of (2.221).

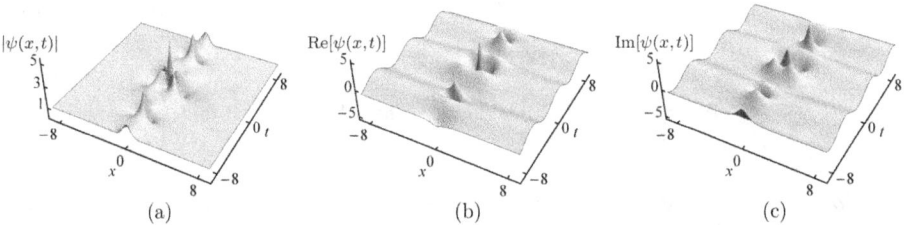

(a)   (b)   (c)

Figure 2.27. Plot of solution (2.221), with $a_1 = 1/2$, $a_2 = 1$, $\nu = 1.2$, and $x_{01} = x_{02} = t_{01} = t_{02} = t_0 = \phi_0 = 0$. (a) Absolute value of (2.221), (b) Real part of (2.221), and (c) Imaginary part of (2.221).

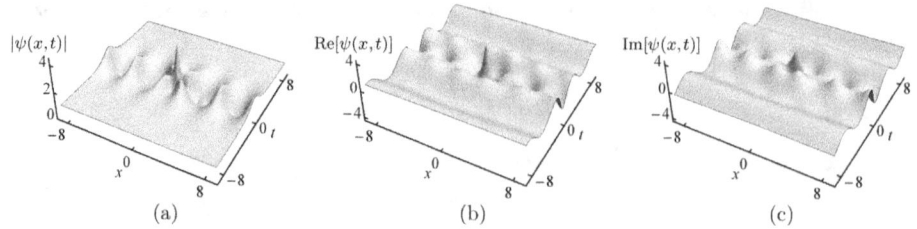

Figure 2.28. Plot of solution (2.221), with $a_1 = 1/2$, $a_2 = 1$, $\nu = 0.5$, $x_{01} = x_{02} = t_{01} = t_{02} = t_0 = \phi_0 = 0$. (a) Absolute value of (2.221), (b) Real part of (2.221), and (c) Imaginary part of (2.221).

where

$$\alpha(x,\,t) = \frac{\kappa}{8}\left(8\,\delta\,\cos[\frac{\kappa}{\sqrt{2\,a_1}}\,(x - x_{01})]\right.$$

$$\left. + \kappa\left\{-8 + [1 + 4\,(t - t_{02})^2 + \frac{2}{a_1}\,(x - x_{02})^2]\,\kappa^2\right\}\cosh[\delta\,(t - t_{01})]\right),$$

$$\beta(x,\,t) = \frac{\kappa}{4}\left(8\,(t - t_{02})\left\{\delta\,\cos[\frac{\kappa}{\sqrt{2\,a_1}}\,(x - x_{01})] - \kappa\,\cosh[\delta\,(t - t_{01})]\right\}\right.$$

$$\left. + \left[1 + 4\,(t - t_{02})^2 + \frac{2}{a_1}\,(x - x_{02})^2\right]\delta\,\kappa\,\sinh[\delta\,(t - t_{01})]\right),$$

$$\gamma(x,\,t) = -\frac{1}{4\,\kappa}\left[\delta\left\{-16 + [1 + 4\,(t - t_{02})^2 + \frac{2}{a_1}\,(x - x_{02})^2]\,\kappa^2\right\}\right.$$

$$\times \cos[\frac{\kappa}{\sqrt{2\,a_1}}\,(x - x_{01})] + \kappa\,(\{16 + [-3 + 4\,(t - t_{02})^2$$

$$+ \frac{2}{a_1}\,(x - x_{02})^2]\,\kappa^2\}\cosh[\delta\,(t - t_{01})] - 16\,\delta\left\{\frac{1}{\sqrt{2a_1}}\,(x - x_{02})\right.$$

$$\left.\left.\left.\times \sin[\frac{\kappa}{\sqrt{2\,a_1}}\,(x - x_{01})] + (t - t_{02})\,\sinh[\delta\,(t - t_{01})]\right\}\right)\right],$$

$a_1 > 0$, $a_2 > 0$, $\kappa = 2\sqrt{1 - \nu^2}$, $\delta = \frac{\kappa}{2}\sqrt{4 - \kappa^2}$, ν, x_{0j}, t_{0j}, t_0, and ϕ_0 are arbitrary real constants, $j = 1, 2$.

- *Reference*: [48].
- This solution can be generated from (2.220) in the limit of $\kappa_2 \to 0$ with $\kappa_1 \neq 0$.

Solution 60. Specific two-breather solution II

(Figures 2.29, 2.30)

$$\psi(x, t) = \frac{1}{\sqrt{a_2}} \left[1 + \frac{\alpha(x, t) + i\,\beta(x, t)}{\gamma(x, t)} \right] e^{i\,[t - t_0 + \phi_0]}, \tag{2.222}$$

where

$$\alpha(x, t) = -\frac{2\,\kappa}{\delta} \left\{ \cosh[\delta\,(t - t_0)] \left((\delta^2 + \kappa^2) \cos[\kappa\,\frac{1}{\sqrt{2a_1}}\,(x - x_0)] \right. \right.$$

$$\left. - 2\,\delta\,\kappa \cosh[\delta\,(t - t_0)] + \delta^2\,\kappa\,\frac{1}{\sqrt{2a_1}}\,(x - x_0)\sin[\kappa\,\frac{1}{\sqrt{2a_1}}\,(x - x_0)] \right)$$

$$\left. + \delta\,(2\,\delta^2 - \kappa^2)\,(t - t_0)\cos[\kappa\,\frac{1}{\sqrt{2a_1}}\,(x - x_0)]\,\sinh[\delta\,(t - t_0)] \right\},$$

$$\beta(x, t) = -\frac{1}{2\,\delta\,\kappa} \left\{ 8\,\delta\,(2\,\delta^2 - \kappa^2)\,(t - t_0) \left(-\kappa + \delta\,\cos[\kappa\,\frac{1}{\sqrt{2a_1}}\,(x - x_0)] \right. \right.$$

$$\times \cosh[\delta\,(t - t_0)]) + 8\,\delta^3 \left(\cos[\kappa\,\frac{1}{\sqrt{2a_1}}\,(x - x_0)] + \kappa\,\frac{1}{\sqrt{2a_1}}\,(x - x_0) \right.$$

$$\left. \times \sin[\kappa\,\frac{1}{\sqrt{2a_1}}\,(x - x_0)] \right) \sinh[\delta\,(t - t_0)] + \kappa\,(\kappa^4 - 4\,\delta^2)\,\sinh[2\,\delta\,(t - t_0)] \right\},$$

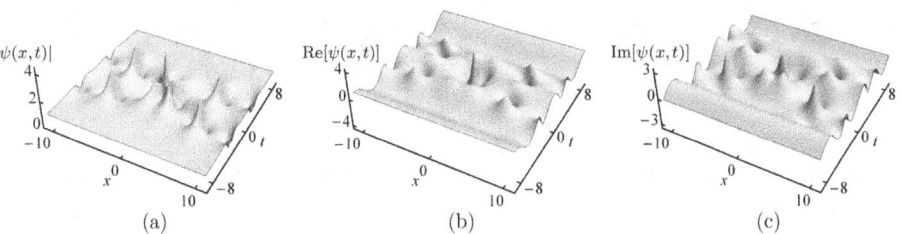

(a) (b) (c)

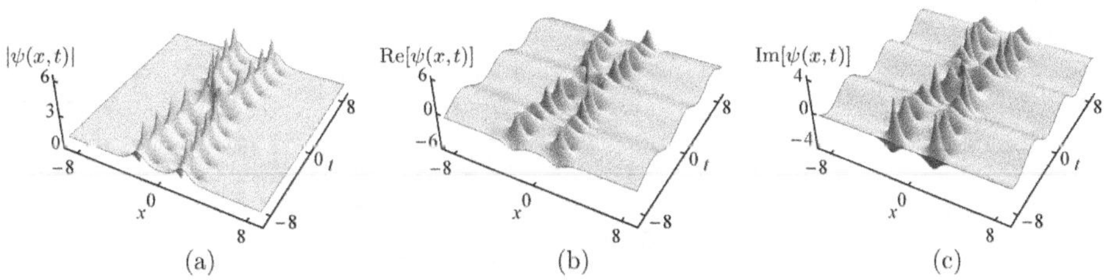

Figure 2.29. Plot of solution (2.222), with $a_1 = 1/2$, $a_2 = 1$, $\nu = 1.5$, and $x_0 = t_0 = \phi_0 = 0$. (a) Absolute value of (2.222), (b) Real part of (2.222), and (c) Imaginary part of (2.222).

(a) (b) (c)

Figure 2.30. Plot of solution (2.222), with $a_1 = 1/2$, $a_2 = 1$, $\nu = 0.8$, and $x_0 = t_0 = 0$. (a) Absolute value of (2.222), (b) Real part of (2.222), and (c) Imaginary part of (2.222).

$$\gamma(x,\ t) = -\frac{1}{4\ \delta^2\ \kappa^2}\ \{32\ \delta^4\ (\delta^2 - \kappa^2)\ (t - t_0)^2 + \kappa^4\ (\delta^2 + \kappa^2)$$

$$+\ 8\ \delta^2\ \kappa^2\ \left(\delta^2 \frac{1}{2a_1}\ (x - x_0)^2 + \kappa^2\ (t - t_0)^2\right)$$

$$+\ 4\left(\kappa^4\ \cosh[2\ \delta\ (t - t_0)] - \delta^4\ \cos[2\ \kappa\ \frac{1}{\sqrt{2a_1}}\ (x - x_0)]\right)$$

$$-\ 4\ \delta\ \kappa^2\ \cosh[\delta\ (t - t_0)]\left(\kappa^3\ \cos[\kappa\ \frac{1}{\sqrt{2a_1}}\ (x - x_0)]\right.$$

$$\left.+\ 4\ \delta^2\ \frac{1}{\sqrt{2a_1}}\ (x - x_0)\ \sin[\kappa\ \frac{1}{\sqrt{2a_1}}\ (x - x_0)]\right)$$

$$-\ 16\ \delta^2\ \kappa\ (2\ \delta^2 - \kappa^2)\ (t - t_0)\ \cos[\kappa\ \frac{1}{\sqrt{2a_1}}\ (x - x_0)]\ \sinh[\delta\ (t - t_0)]\},$$

$a_1 > 0$, $a_2 > 0$, $\kappa = 2\ \sqrt{1 - \nu^2}$, $\delta = \frac{\kappa}{2}\ \sqrt{4 - \kappa^2}$, ν, x_0, t_0, and ϕ_0 are arbitrary real constants.

- *Reference*: [48].
- This solution can be generated from (2.220) in the limit of $\kappa_2 \to \kappa_1$, $x_{01} = x_{02} = x_0$, and $t_{01} = t_{02} = t_0$.

Solution 61. Second-order Peregrine solution
(Figure 2.31)

$$\psi(x,\ t) = \frac{1}{\sqrt{a_2}}\left[1 + \frac{\alpha(x,\ t) + i\ \beta(x,\ t)}{\gamma(x,\ t)}\right] e^{i\ [t - t_0 + \phi_0]}, \tag{2.223}$$

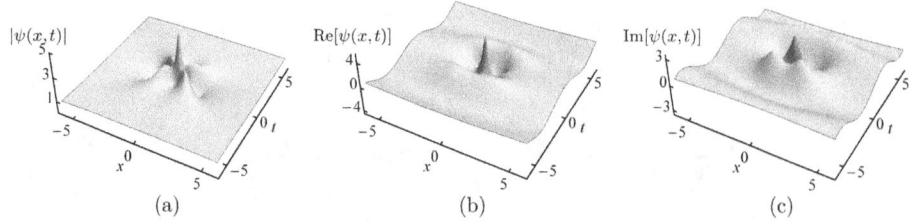

Figure 2.31. Second-order Peregrine soliton (2.223), with $a_1 = 1/2$, $a_2 = 1$, and $x_0 = t_0 = \phi_0 = 0$. (a) Absolute value of (2.223), (b) Real part of (2.223), and (c) Imaginary part of (2.223). Animation available online at http://doi.org/10.1088/978-0-7503-5954-2.

where

$$\alpha(x, t) = \frac{1}{96}\left[-3 + 72\,(t - t_0)^2 + 80\,(t - t_0)^4 + \frac{12}{a_1}\,(x - x_0)^2\right.$$
$$\left. + \frac{48}{a_1}\,(x - x_0)^2\,(t - t_0)^2 + \frac{4}{a_1^2}\,(x - x_0)^4\right],$$

$$\beta(x, t) = \frac{1}{48}\,(t - t_0)\left[-15 + 8\,(t - t_0)^2 + 16\,(t - t_0)^4 - \frac{12}{a_1}\,(x - x_0)^2\right.$$
$$\left. + \frac{16}{a_1}\,(x - x_0)^2\,(t - t_0)^2 + \frac{4}{a_1^2}\,(x - x_0)^4\right],$$

$$\gamma(x, t) = \frac{-1}{1152}\left[9 + 396\,(t - t_0)^2 + 432\,(t - t_0)^4 + 64\,(t - t_0)^6 + \frac{54}{a_1}\,(x - x_0)^2\right.$$
$$- \frac{144}{a_1}\,(x - x_0)^2\,(t - t_0)^2 + \frac{96}{a_1}\,(x - x_0)^2\,(t - t_0)^4 + \frac{12}{a_1^2}\,(x - x_0)^4$$
$$\left. + \frac{48}{a_1^2}\,(x - x_0)^4\,(t - t_0)^2 + \frac{8}{a_1^3}\,(x - x_0)^6\right],$$

$a_2 > 0$, x_0, t_0, and ϕ_0 are arbitrary real constants.
  • *Reference*: [48].
  • This solution can be generated from (2.222) in the limit of $\kappa \to 0$ and $\delta = 0$.

Solution **62. Rogue wave triplet**
  (Figure 2.32)

$$\psi(x, t) = \frac{1}{\sqrt{a_2}}\left[1 + \frac{\alpha(x, t) + i\,\beta(x, t)}{\gamma(x, t)}\right]e^{i\,[t - t_0 + \phi_0]}, \tag{2.224}$$

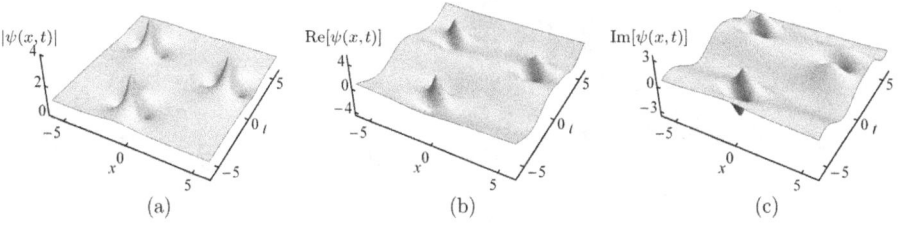

Figure 2.32. Rogue wave triplet (2.224), with $a_1 = 1/2$, $a_2 = 1$, $x_d = -15$, $t_d = -10$, and $x_0 = t_0 = \phi_0 = 0$. (a) Absolute value of (2.224), (b) Real part of (2.224), and (c) Imaginary part of (2.224).

where

$$\alpha(x, t) = -12 \left\{ -3 + \frac{4}{a_1^2} (x - x_0)^4 - 128 t_d \frac{1}{\sqrt{2a_1}} (x - x_0) + \frac{12}{a_1} (x - x_0)^2 [1 + 4 (t - t_0)^2] \right.$$
$$\left. + 8 [9 + 10 (t - t_0)^2] (t - t_0)^2 - 128 x_d (t - t_0) \right\},$$

$$\beta(x, t) = -24 \left(-128 \frac{1}{\sqrt{2 a_1}} t_d (x - x_0) (t - t_0) + (t - t_0) \left\{ -15 + \frac{4}{a_1^2} (x - x_0)^4 \right. \right.$$
$$\left. + 8 (t - t_0)^2 [1 + 2 (t - t_0)^2] + \frac{4}{a_1} (x - x_0)^2 [-3 + 4 (t - t_0)^2] \right\}$$
$$\left. + 16 [1 + \frac{2}{a_1} (x - x_0)^2 - 4 (t - t_0)^2] x_d \right),$$

$$\gamma(x, t) = 9 + \frac{8}{a_1^3} (x - x_0)^6 + 396 (t - t_0)^2 + 432 (t - t_0)^4 + 64 (t - t_0)^6$$
$$+ \frac{12}{a_1^2} (x - x_0)^4 [1 + 4 (t - t_0)^2] + \frac{6}{a_1} (x - x_0)^2 \left\{ 9 + 8 (t - t_0)^2 [-3 + 2 (t - t_0)^2] \right\}$$
$$+ 128 \left(8 t_d^2 + t_d \left\{ 4 (\frac{1}{2 a_1})^{3/2} (x - x_0)^3 + 3 \frac{1}{\sqrt{2 a_1}} (x - x_0) [-1 - 4 (t - t_0)^2] \right\} \right.$$
$$\left. + x_d \left\{ (t - t_0) [-9 + \frac{6}{a_1} (x - x_0)^2 - 4 (t - t_0)^2] + 8 x_d \right\} \right),$$

$a_1 > 0$, $a_2 > 0$, x_d, t_d, x_0, t_0, and ϕ_0 are arbitrary real constants.
- *Reference*: [53], solution (8) with $\delta = 0$.

2.2.2 Summary of Section 2.2.1

Note: For lengthy conditions, the reader is referred to the solutions in Section 2.2.1.

Equation

$$i\,\psi_t + a_1\,\psi_{xx} + a_2\,|\psi|^2\,\psi = 0$$

#	Solution	Conditions	Name	Equation #
1.	$\psi(x,t) = A_0\, e^{i[a_2 A_0^2 (t-t_0)+\phi_0]}$	$A_0, t_0,$ and ϕ_0 are arbitrary real constants	Continuous wave, t-dependent phase	(2.162)
2.	$\psi(x,t) = A_0\, e^{i[\pm A_0 \sqrt{\frac{a_2}{a_1}}\,(x-x_0)+\phi_0]}$	$a_1 a_2 > 0, A_0, x_0,$ and ϕ_0 are arbitrary real constants	Continuous wave, x-dependent phase	(2.163)
3.	$\psi(x,t) = A_0\, e^{i[A_1(x-x_0)+(|A_0|^2 a_2 - A_1^2 a_1)(t-t_0)+\phi_0]}$	A_0 is an arbitrary complex constant, $A_1, x_0, t_0,$ and ϕ_0 are arbitrary real constants	Continuous wave, t- and x-dependent phase	(2.164)
4.	$\psi(x,t) = \dfrac{A_0}{\sqrt{A_1+t-t_0}}\, e^{i\left\{\frac{[a_1 A_2+x-x_0]^2}{4 a_1(A_1+t-t_0)}+a_2 A_0^2 \ln[A_1+t-t_0]+\phi_0\right\}}$	$A_0, A_1, A_2, t_0, x_0,$ and ϕ_0 are arbitrary real constants	Decaying wave	(2.165)
5.	$\psi(x,t) = \sqrt{\dfrac{-2 a_1}{a_2}}\,\dfrac{1}{x-x_0}\,e^{i\phi_0}$	$a_1 a_2 < 0, x_0,$ and ϕ_0 are arbitrary real constants	—	(2.166)
6.	$\psi(x,t) = \dfrac{1}{\sqrt{-a_2}}\,\dfrac{q(x,t)}{q_2(x,t)}$	See text	Higher order of *Solution 5*	(2.167)
7.	$\psi(x,t) = A_0 \sqrt{\dfrac{-2 a_1}{a_2}}\,\sec[A_0 (x-x_0)]\, e^{-i[a_1 A_0^2 (t-t_0)+\phi_0]}$	$a_1 a_2 < 0, A_0, x_0, t_0,$ and ϕ_0 are arbitrary real constants	—	(2.168)
8.	$\psi(x,t) = A_0 \sqrt{\dfrac{-2 a_1}{a_2}}\,\csc[A_0 (x-x_0)]\, e^{-i[a_1 A_0^2 (t-t_0)+\phi_0]}$	$a_1 a_2 < 0, A_0, x_0, t_0,$ and ϕ_0 are arbitrary real constants	—	(2.169)
9.	$\psi(x,t) = A_0 \sqrt{\dfrac{-2 a_1}{a_2}}\,\tan[A_0 (x-x_0)]\, e^{i[2 a_1 A_0^2 (t-t_0)+\phi_0]}$	—	—	(2.170)

$a_1 a_2 < 0$, A_0, x_0, t_0, and ϕ_0 are arbitrary real constants

10. $\psi(x,t) = A_0 \sqrt{\dfrac{-2a_1}{a_2}}\, \cot[A_0(x-x_0)]\, e^{i[2a_1 A_0^2(t-t_0)+\phi_0]}$

$a_1 a_2 < 0$, A_0, x_0, t_0, and ϕ_0 are arbitrary real constants — (2.171)

11. $\psi(x,t) = A_0 \sqrt{\dfrac{2a_1}{a_2}}\, \mathrm{sech}[A_0(x-x_0)]\, e^{i[a_1 A_0^2(t-t_0)+\phi_0]}$

$a_1 a_2 > 0$, A_0, x_0, t_0, and ϕ_0 are arbitrary real constants. Bright soliton (2.172)

12. $\psi(x,t) = A_0 \sqrt{\dfrac{-2a_1}{a_2}}\, \mathrm{csch}[A_0(x-x_0)]\, e^{i[a_1 A_0^2(t-t_0)+\phi_0]}$

$a_1 a_2 < 0$, A_0, x_0, t_0, and ϕ_0 are arbitrary real constants — (2.173)

13. $\psi(x,t) = A_0 \sqrt{\dfrac{-2a_1}{a_2}}\, \tanh[A_0(x-x_0)]\, e^{-i[2a_1 A_0^2(t-t_0)+\phi_0]}$

$a_1 a_2 < 0$, A_0, x_0, t_0, and ϕ_0 are arbitrary real constants. Dark soliton (2.174)

14. $\psi(x,t) = A_0 \sqrt{\dfrac{-2a_1}{a_2}}\, \coth[A_0(x-x_0)]\, e^{-i[2a_1 A_0^2(t-t_0)+\phi_0]}$

$a_1 a_2 < 0$, A_0, x_0, t_0, and ϕ_0 are arbitrary real constants — (2.175)

15. $\psi(x,t) = (A_0 + i A_1 \tan\{A_1[\sqrt{\dfrac{a_2}{2a_1}}(x-x_0) + a_2 A_0(t-t_0)]\})$
$\times\, e^{i[a_2(A_0^2-A_1^2)(t-t_0)+\phi_0]}$

$a_1 a_2 < 0$, A_0, A_1, x_0, t_0, and ϕ_0 are arbitrary real constants — (2.176)

16. $\psi(x,t) = (A_0\sqrt{1-A_1}$
$+ i A_0\sqrt{A_1}\, \tanh\{A_0\sqrt{\dfrac{A_1}{2}}[(x-x_0) + (A_0\sqrt{2(1-A_1)} + 2A_3)(t-t_0)]\})$
$\times\, e^{-i[a_2(t-t_0)+A_3(x-x_0)+\phi_0]}$

$0 < A_1 < 1$, $A_2 = A_0^2 + A_3^2$, $a_1 = 0$, $a_2 = -1$, A_0, A_3, x_0, t_0, and ϕ_0 are arbitrary real constants — (2.177)

17. $\psi(x,t) = A_0 \sqrt{\dfrac{-a_1}{2a_2}}\, \{\sec[A_0(x-x_0)] + \tan[A_0(x-x_0)]\}$
$\times\, e^{i[\frac{a_1 A_0^2}{2}(t-t_0)+\phi_0]}$

$a_1 a_2 < 0$, A_0, x_0, t_0, and ϕ_0 are arbitrary real constants — (2.178)

18. $\psi(x,t) = A_0 \sqrt{\dfrac{-2a_1}{a_2}}\, (\cot\{A_0[\dfrac{x-x_0}{\sqrt{a_1}} - 2A_1(t-t_0)]\}$
$- \tan\{A_0[\dfrac{x-x_0}{\sqrt{a_1}} - 2A_1(t-t_0)]\})\, e^{i[\frac{A_1}{\sqrt{a_1}}(x-x_0)-(A_1^2-8A_0^2)(t-t_0)]}$

$a_1 > 0$, $a_2 < 0$, A_0, A_1, x_0, t_0, and ϕ_0 are arbitrary real constants — (2.179)

19. $\psi(x,t) = A_0 \sqrt{\dfrac{-2a_1}{a_2}}\, \{\dfrac{1+\tanh^2[A_0(x-x_0)]}{\tanh[A_0(x-x_0)]}\}\, e^{-i[8a_1 A_0^2(t-t_0)+\phi_0]}$

$a_1 a_2 < 0$, A_0, x_0, t_0, and ϕ_0 are arbitrary real constants — (2.180)

20. $\psi(x,t) = A_0 \sqrt{\dfrac{-2a_1}{a_2}}\, (\dfrac{b_0\tan\{A_0[\frac{x-x_0}{\sqrt{a_1}} - 2A_1(t-t_0)]\} - b_1}{b_0 + b_1\tan\{A_0[\frac{x-x_0}{\sqrt{a_1}} - 2A_1(t-t_0)]\}})$
$\times\, e^{i[\frac{A_1(x-x_0)}{\sqrt{a_1}}-(A_1^2-2A_0^2)(t-t_0)]}$

$a_1 > 0$, $a_2 < 0$, A_0, A_1, b_0, b_1, x_0, t_0, and ϕ_0 are arbitrary real constants — (2.181)

(Continued)

21.	$\psi(x,t) = \pm\left(A_1 - \dfrac{4A_2}{A_1 + \sqrt{4A_2 - A_1^2}\,\tan[A_0\sqrt{4A_2 - A_1^2}\,[2A_3(t-t_0)-(x-x_0)]]}\right)$ $\times A_0\, e^{i\,[A_3(x-x_0)+(-2A_0^2 A_2^2 + 8A_0^2 A_2 - A_3^2)(t-t_0)+\phi_0]}$	—	$a_1 = 1,\ a_2 = -2,\ 4A_2 - A_1^2 > 0,$ $A_0, A_3, x_0, t_0,$ and ϕ_0 are arbitrary real constants	(2.182)
22.	$\psi(x,t) = \pm\left(A_1 - \dfrac{4A_2}{A_1 + \sqrt{4A_2 - A_1^2}\,\cot[A_0\sqrt{4A_2 - A_1^2}\,[2A_3(t-t_0)-(x-x_0)]]}\right)$ $\times A_0\, e^{i\,[A_3(x-x_0)+(-2A_0^2 A_2^2 + 8A_0^2 A_2 - A_3^2)(t-t_0)+\phi_0]}$	—	$a_1 = 1,\ a_2 = -2,\ 4A_2 - A_1^2 > 0,$ $A_0, A_3, x_0, t_0,$ and ϕ_0 are arbitrary real constants	(2.183)
23.	$\psi(x,t) = \pm\left(A_1 - \dfrac{4A_2}{A_1 - \sqrt{A_1^2 - 4A_2}\,\tanh[A_0\sqrt{A_1^2 - 4A_2}\,[2A_3(t-t_0)-(x-x_0)]]}\right)$ $\times A_0\, e^{i\,[A_3(x-x_0)+(-2A_0^2 A_2^2 + 8A_0^2 A_2 - A_3^2)(t-t_0)+\phi_0]}$	—	$a_1 = 1,\ a_2 = -2,\ A_1^2 - 4A_2 > 0,$ $A_0, A_3, x_0, t_0,$ and ϕ_0 are arbitrary real constants	(2.184)
24.	$\psi(x,t) = \pm\left(A_1 - \dfrac{4A_2}{A_1 - \sqrt{A_1^2 - 4A_2}\,\coth[A_0\sqrt{A_1^2 - 4A_2}\,[2A_3(t-t_0)-(x-x_0)]]}\right)$ $\times A_0\, e^{i\,[A_3(x-x_0)+(-2A_0^2 A_2^2 + 8A_0^2 A_2 - A_3^2)(t-t_0)+\phi_0]}$	—	$a_1 = 1,\ a_2 = -2,\ A_1^2 - 4A_2 > 0,$ $A_0, A_3, x_0, t_0,$ and ϕ_0 are arbitrary real constants	(2.185)
25.	$\psi(x,t) = \dfrac{q_1(x,t)}{q_2(x,t)}\, e^{i\,[\frac{A_0}{\sqrt{a_1}}(x-x_0)-A_0^2(t-t_0)]}$	—	See text.	(2.186)
26.	$\psi(x,t) = A_0\sqrt{\dfrac{2m}{-a_2(1+m)}}\,\mathrm{sn}[\dfrac{A_0}{\sqrt{a_1(1+m)}}(x - x_0),\, m]e^{-i\,[A_0^2(t-t_0)+\phi_0]}$	Solitary wave	$a_1(1+m) > 0,\ a_2 m(1+m) < 0,$ $m \neq -1,\ A_0, t_0, x_0,$ and ϕ_0 are arbitrary real constants	(2.187)
27.	$\psi(x,t) = A_0\sqrt{\dfrac{2q}{a_2}}\,\mathrm{sn}[A_0(x - x_0),\, -1]\, e^{i\phi_0}$	Solitary wave	$a_2 > 0,\ A_0, x_0,$ and ϕ_0 are arbitrary real constants	(2.188)
28.	$\psi(x,t) = A_0\sqrt{\dfrac{2m}{a_2(2m-1)}}\,\mathrm{cn}[\dfrac{A_0}{\sqrt{a_1(2m-1)}}(x - x_0),\, m]e^{i\,[A_0^2(t-t_0)+\phi_0]}$	Solitary wave	$a_2(2m-1) > 0,$ $a_2 m(2m-1) > 0,\ m \neq 1/2,$ $A_0, t_0, x_0,$ and ϕ_0 are arbitrary real constants	(2.189)
29.	$\psi(x,t) = A_0\sqrt{\dfrac{q}{a_2}}\,\mathrm{cn}[A_0(x - x_0),\, \dfrac{1}{2}]\, e^{i\phi_0}$	Solitary wave	$a_2 > 0,\ m = 1/2,\ A_0, x_0,$ and ϕ_0 are arbitrary real constants	(2.190)
30.	$\psi(x,t) = A_0\sqrt{\dfrac{2}{a_2(2-m)}}\,\mathrm{dn}[\dfrac{A_0}{\sqrt{a_1(2-m)}}(x - x_0),\, m]e^{i\,[A_0^2(t-t_0)+\phi_0]}$	Solitary wave	$a_1(2-m) > 0,\ a_2(2-m) > 0,$ $m \neq 2,\ A_0, t_0, x_0,$ and ϕ_0 are arbitrary real constants	(2.191)
31.	$\psi(x,t) = A_0\sqrt{\dfrac{2q}{a_2}}\,\mathrm{dn}[A_0(x - x_0),\, 2]\, e^{i\phi_0}$	Solitary wave	$a_2 > 0,\ A_0, x_0,$ and ϕ_0 are arbitrary real constants	(2.192)
32.	$\psi(x,t) = A_0\sqrt{1 - m}\,\mathrm{nd}[\dfrac{A_1}{\sqrt{a_1}}(x - x_0),\, m]\, e^{-i\,[A_2^2(t-t_0)+\phi_0]}$	Solitary wave		(2.193)

No.	$\psi(x,t) = \cdots$	Conditions	Type	Eq.
33.	$\psi(x,t) = A_0\sqrt{m}\,(1-m)\,\text{sd}\left[\frac{A_1}{\sqrt{a_1}}(x-x_0),m\right] e^{-i[A_2(t-t_0)+\phi_0]}$	$A_2 = (m-2)A_1^2,\ a_1 > 0,\ a_2 = \frac{2A_1^2}{A_0^2}$, $0 < m \le 1,\ A_0, A_1, x_0, t_0$, and ϕ_0 are arbitrary real constants	Solitary wave	(2.194)
34.	$\psi(x,t) = A_0\sqrt{m}\,\text{cd}\left[\frac{A_1}{\sqrt{a_1}}(x-x_0),m\right] e^{-i[A_2(t-t_0)+\phi_0]}$	$A_2 = (1-2m)A_1^2,\ a_1 > 0$, $a_2 = \frac{2A_1^2}{A_0^2},\ 0 < m \le 1,\ A_0, A_1, x_0, t_0$, and ϕ_0 are arbitrary real constants	Solitary wave	(2.195)
35.	$\psi(x,t) = A(x)\,e^{i[\phi(x,t)+\phi_0]},\quad A(x) = \sqrt{R_3 + (R_2-R_3)\,\text{sn}^2[A_0(x-x_0),m]}$, $\phi(x,t) = \dfrac{\sqrt{2}\,A_0\,\Pi\left(\frac{R_3-R_2}{R_3},\ \text{am}[A_0(x-x_0),m],m\right)\text{dn}[A_0(x-x_0),m]}{\sqrt{2}\,R_3\,\sqrt{a_0}\sqrt{1-m\,\text{sn}^2[A_0(x-x_0),m]}} + \lambda_2(t-t_0)$	See text	Solitary wave on a finite background	(2.196)
36.	$\psi(x,t) = \left\{\frac{A_0}{2}\text{dn}\left[\frac{A_1}{\sqrt{a_1}}(x-x_0),m\right] + \frac{B_0\sqrt{m}}{2}\text{cn}\left[\frac{A_1}{\sqrt{a_1}}(x-x_0),m\right]\right\} e^{-i[A_2(t-t_0)+\phi_0]}$	$A_2 = \frac{-(m+1)A_1^2}{2},\ B_0 = \pm A_0,\ a_1 > 0$, $a_2 = -\frac{2A_1^2}{A_0^2},\ 0 < m \le 1,\ A_0, A_1, x_0, t_0$, and ϕ_0 are arbitrary real constants	Solitary wave	(2.197)
37.	$\psi(x,t) = A_0\sqrt{m}\ \dfrac{\text{sn}\left[\frac{A_1}{\sqrt{a_1}}(x-x_0),m\right]}{1+\text{dn}\left[\frac{A_1}{\sqrt{a_1}}(x-x_0),m\right]}\, e^{-i[A_2(t-t_0)+\phi_0]}$	$A_2 = (2-m)A_1^2/2,\ a_1 > 0$, $a_2 = -\frac{mA_1^2}{2A_0^2},\ m > 0,\ A_0, A_1, x_0, t_0$, and ϕ_0 are arbitrary real constants	Solitary wave	(2.198)
38.	$\psi(x,t) = A_0\sqrt{m}\ \dfrac{\text{cn}\left[\frac{A_1}{\sqrt{a_1}}(x-x_0),m\right]}{\sqrt{1-m}+\text{dn}\left[\frac{A_1}{\sqrt{a_1}}(x-x_0),m\right]}\, e^{-i[A_2(t-t_0)+\phi_0]}$	$A_2 = (2-m)A_1^2/2,\ a_1 > 0$, $a_2 = -\frac{mA_1^2}{2A_0^2},\ 0 < m < 1,\ A_0, A_1$, x_0, t_0, and ϕ_0 are arbitrary real constants	Solitary wave	(2.199)
39.	$\psi(x,t) = \left\{\frac{A_0}{2} - \dfrac{A_0\,\text{dn}\left[\frac{A_1}{\sqrt{a_1}}(x-x_0),m\right]}{(1-m)^{1/4}+\text{dn}\left[\frac{A_1}{\sqrt{a_1}}(x-x_0),m\right]}\right\} e^{-i[A_2(t-t_0)+\phi_0]}$	$A_2 = [2-m+6\sqrt{1-m}]A_1^2/2,\ a_1 > 0$, $a_2 = -2(1-\sqrt{1-m})^2 A_1^2/A_0^2$, $0 < m < 1,\ A_0, A_1, x_0, t_0$, and ϕ_0 are arbitrary real constants	Solitary wave	(2.200)

(Continued)

#	$\psi(x,t)$	Conditions	Type	Eq.
40.	$\psi(x,t) = A_0\sqrt{m}\,\dfrac{\text{dn}[\frac{A_1}{\sqrt{a_1}}(x-x_0),m]}{1+\sqrt{m}\,\text{sn}[\frac{A_1}{\sqrt{a_1}}(x-x_0),m]}\,e^{-i[A_2(t-t_0)+\phi_0]}$	$A_2 = -(1+m)A_1^2/2,\ a_1 > 0,$ $a_2 = (1-m)A_1^2/(2mA_0^2),$ $m > 0,\ A_0, A_1, x_0, t_0,$ and ϕ_0 are arbitrary real constants	Solitary wave	(2.201)
41.	$\psi(x,t) = A_0 m\,\dfrac{\text{cn}[\frac{A_1}{\sqrt{a_1}}(x-x_0),m]\,\text{sn}[\frac{A_1}{\sqrt{a_1}}(x-x_0),m]}{\text{dn}[\frac{A_1}{\sqrt{a_1}}(x-x_0),m]}\,e^{-i[A_2(t-t_0)+\phi_0]}$	$A_2 = 2(2-m)A_1^2,\ a_1 > 0,$ $a_2 = -\frac{2A_1^2}{A_0^2},\ m > 0,\ A_0, A_1, x_0, t_0,$ and ϕ_0 are arbitrary real constants	Solitary wave	(2.202)
42.	$\psi(x,t) = A_0\,\dfrac{\text{dn}[\frac{A_1}{\sqrt{a_1}}(x-x_0),m]\,\text{sn}[\frac{A_1}{\sqrt{a_1}}(x-x_0),m]}{\text{cn}[\frac{A_1}{\sqrt{a_1}}(x-x_0),m]}\,e^{-i[A_2(t-t_0)+\phi_0]}$	$A_2 = -(1+m)A_1^2/2,\ a_1 > 0,$ $a_2 = (1-m)A_1^2/(2mA_0^2),$ $m > 0,\ A_0, A_1, x_0, t_0,$ and ϕ_0 are arbitrary real constants	Solitary wave	(2.203)
43.	$\psi(x,t) = A_0\,\dfrac{\text{dn}[\frac{A_1}{\sqrt{a_1}}(x-x_0),m]}{\text{sn}[\frac{A_1}{\sqrt{a_1}}(x-x_0),m]}\,e^{-i[A_2(t-t_0)+\phi_0]}$	$A_2 = 2(2-m)A_1^2,\ a_1 > 0,$ $a_2 = -\frac{2A_1^2}{A_0^2},\ m > 0,\ A_0, A_1, x_0, t_0,$ and ϕ_0 are arbitrary real constants	Solitary wave	(2.204)
44.	$\psi(x,t) = A_0\,\dfrac{\text{cn}[\frac{A_1}{\sqrt{a_1}}(x-x_0),m]}{\text{sn}[\frac{A_1}{\sqrt{a_1}}(x-x_0),m]\,\text{dn}[\frac{A_1}{\sqrt{a_1}}(x-x_0),m]}\,e^{-i[A_2(-t_0)+\phi_0]}$	$A_2 = 2(2m-1)A_1^2,\ a_1 > 0,$ $a_2 = -\frac{2A_1^2}{A_0^2},\ m > 0,\ A_0, A_1, x_0, t_0,$ and ϕ_0 are arbitrary real constants	Solitary wave	(2.205)
45.	$\psi(x,t) = \frac{1}{\sqrt{a_2}}\sum_{j=1}^{N}\psi_j(x,t),$ $\sum_{k=1}^{N}M_{jk}[\Gamma_j^{-1}(x,t)+\gamma_k^*(x,t)]\psi_k(x,t)=1,\quad j=0,1,2,\ldots,N$	See text	N-bright solitons	(2.206)
46.	$\psi(x,t) = \frac{1}{\sqrt{a_2}}[\psi_1(x,t)+\psi_2(x,t)]$	See text	Two bright solitons	(2.208)
47.	$\psi(x,t) = \frac{1}{\sqrt{a_2}}[\psi_1(x,t)+\psi_2(x,t)+\psi_3(x,t)]$	See text	Three bright solitons	(2.209)
48.	$\psi(x,t) = \sqrt{\frac{-2}{a_2}}\{1-2i\sum_m\mu_n(t)\,e^{\frac{-2v_n}{\sqrt{a_1}}(x-x_0)}[-Q_{12}^n+(\lambda_n+iv_n)(Q_{11}^n-1)]\}$ $\times e^{-2i[t-t_0+\phi_0]}$	See text	N-dark solitons	(2.210)
49.	$\psi(x,t) = \sqrt{\frac{-2}{a_2}}\{1-\frac{2i}{p(x,t)}[\frac{2}{v_1+v_2}(\frac{1}{\lambda_1+iv_1}+\frac{1}{\lambda_2+iv_2})$ $-(\lambda_1-iv_1)q_1(x,t)-(\lambda_2-iv_2)q_2(x,t)]\}e^{-2i[t-t_0+\phi_0]}$	See text	Two dark solitons	(2.211)

#	$\psi(x,t)$	Conditions	Name	Eq.
50.	$\psi(x,t) = \dfrac{1}{\sqrt{a_2}}\left\{ \dfrac{\kappa^2\cosh[\delta(t-t_0)]+2i\nu\sinh[\delta(t-t_0)]}{2\cosh[\delta(t-t_0)]-2\nu\cos[\frac{\kappa}{\sqrt{2}q}(x-x_0)]} - 1 \right\} e^{i[t-t_0+\phi_0]}$	$a_1 > 0,\ a_2 > 0,\ \kappa = 2\sqrt{1-\nu^2},$ $\delta = \kappa\nu,\ \nu,\ x_0,\ t_0,$ and ϕ_0 are arbitrary real constants	Generalized first-order breather I	(2.212)
51.	$\psi(x,t) = \dfrac{\cos(A_0)\cos[q_1(x,t)+2iA_1]-\cosh(A_1)\cosh[q_2(x,t)+2iA_0]}{\cos(A_0)\cos[q_1(x,t)]-\cosh(A_1)\cosh[q_2(x,t)]}$ $\times\, e^{i[a_2(t-t_0)+\phi_0]}$	See text	Generalized first-order breather II	(2.213)
52.	$\psi(x,t) = \dfrac{A_0}{\sqrt{a_2}}\left(1 - \dfrac{\sqrt{8}}{A_0}\lambda_r\, p(x,t)\right) e^{i[A_0^2(t-t_0)+\phi_0]}$	See text	Generalized first-order breather III	(2.214)
53.	$\psi(x,t) = \dfrac{1}{\sqrt{a_2}}\left\{ \dfrac{-p^2\cos[\omega(t-t_0)]-2ip\nu\sin[\omega(t-t_0)]}{2\cos[\omega(t-t_0)]-2\nu\cosh[\frac{p}{\sqrt{2}q}(x-x_0)]} - 1 \right\} e^{i[t-t_0+\phi_0]}$	$a_1 > 0,\ a_2 > 0,\ p = 2\sqrt{\nu^2-1},$ $\omega = p\nu,\ \nu > 1,\ x_0,\ t_0,$ and ϕ_0 are arbitrary real constants	Kuznetsov–Ma breather	(2.215)
54.	$\psi(x,t) = \dfrac{1}{\sqrt{a_2}}\left\{ \dfrac{\kappa^2\cosh[\delta(t-t_0)]+2i\nu\sinh[\delta(t-t_0)]}{2\cosh[\delta(t-t_0)]-2\nu\cos[\frac{\kappa}{\sqrt{2}q}(x-x_0)]} - 1 \right\} e^{i[t-t_0+\phi_0]}$	$a_1 > 0,\ a_2 > 0,\ \kappa = 2\sqrt{1-\nu^2},$ $\delta = \kappa\nu,\ \nu < 1,\ x_0,\ t_0,$ and ϕ_0 are arbitrary real constants	Akhmediev breather	(2.216)
55.	$\psi(x,t) = \dfrac{1}{\sqrt{a_2}}\left[\dfrac{4+8i(t-t_0)}{1+4(t-t_0)^2+\frac{2}{q}(x-x_0)^2} - 1 \right] e^{i[t-t_0+\phi_0]}$	$a_2 > 0,\ x_0,\ t_0,$ and ϕ_0 are arbitrary real constants	Peregrine soliton	(2.217)
56.	$\psi(x,t) = [Q(x,t)+i\,\delta(t)]\, e^{i\,\phi(t)}$	See text	Doubly-periodic (type A)	(2.218)
57.	$\psi(x,t) = [Q(x,t)+i\,\delta(t)]\, e^{i\,\phi(t)}$	See text	Doubly-periodic (type B)	(2.219)
58.	$\psi(x,t) = \dfrac{1}{\sqrt{a_2}}\left[1 + \dfrac{\alpha(x,t)+i\,\beta(x,t)}{\gamma(x,t)}\right] e^{i[t-t_0+\phi_0]}$	See text	Generalized two-breathers	(2.220)
59.	$\psi(x,t) = \dfrac{1}{\sqrt{a_2}}\left[1 + \dfrac{\alpha(x,t)+i\,\beta(x,t)}{\gamma(x,t)}\right] e^{i[t-t_0+\phi_0]}$	See text	Specific two-breathers I	(2.221)
60.	$\psi(x,t) = \dfrac{1}{\sqrt{a_2}}\left[1 + \dfrac{\alpha(x,t)+i\,\beta(x,t)}{\gamma(x,t)}\right] e^{i[t-t_0+\phi_0]}$	See text	Specific two-breathers II	(2.222)
61.	$\psi(x,t) = \dfrac{1}{\sqrt{a_2}}\left[1 + \dfrac{\alpha(x,t)+i\,\beta(x,t)}{\gamma(x,t)}\right] e^{i[t-t_0+\phi_0]}$	See text	Second-order Peregrine soliton	(2.223)
62.	$\psi(x,t) = \dfrac{1}{\sqrt{a_2}}\left[1 + \dfrac{\alpha(x,t)+i\,\beta(x,t)}{\gamma(x,t)}\right] e^{i[t-t_0+\phi_0]}$	See text	Rogue wave triplet	(2.224)

2.2.3 Complex Dispersion and Nonlinearity Coefficients

Here, $a_1 = a_{1r} + i\, a_{1i}$, $a_2 = a_{2r} + i\, a_{2i}$, where a_{1r}, a_{1i}, a_{2r}, and a_{2i} are real constants.

***Solution* 1. Constant amplitude I**   *CW, t-dependent phase*

$$\psi(x,\, t) = A_0\, e^{i\left[\, a_{2r}\, A_0^2\, (t-t_0) + \phi_0\right]}, \tag{2.225}$$

where $a_{2i} = 0$, A_0, t_0, and ϕ_0 are arbitrary real constants.
  • *Derived in appendix* A.

***Solution* 2. Constant amplitude II**   *CW, x-dependent phase*

$$\psi(x,\, t) = A_0\, e^{i\left[\pm A_0\, \sqrt{\frac{a_{2r}}{a_{1r}}}\, (x-x_0) + \phi_0\right]}, \tag{2.226}$$

where $a_{1r}\, a_{2r} > 0$, $a_{1i} = \frac{a_{1r}\, a_{2i}}{a_{2r}}$, A_0, x_0, and ϕ_0 are arbitrary real constants.
  • *Derived in appendix* A.

***Solution* 3. Rational solution I**   *DW*

$$\psi(x,\, t) = \frac{1}{\sqrt{2\, a_{2i}\, (t - t_0)}}\, e^{i\, \phi_0}, \tag{2.227}$$

where $a_{2i} > 0$,
  $a_{2r} = 0$,
  t_0 and ϕ_0 are arbitrary real constants.
  • *Derived in appendix* A.

***Solution* 4. Rational solution II**   *DW*

$$\psi(x,\, t) = \frac{1}{\sqrt{2\, a_{2i}\, (t - t_0)}}\, e^{i\left\{A_0\, (x-x_0) - a_{1r}\, A_0^2\, (t-t_0) + \frac{a_{2r}}{2\, a_{2i}}\, \ln[2\, a_{2i}\, (t-t_0)] + \phi_0\right\}}, \tag{2.228}$$

where $a_{2i} > 0$, $a_{1i} = 0$, A_0, t_0, x_0, and ϕ_0 are arbitrary real constants.
  • *Derived in appendix* A.

***Solution* 5.**

$$\psi(x,\, t) = A_0\, \sqrt{\frac{a_{1r}}{e^{2\, a_{1r}\, A_0^2\, (t-t_0)} - a_{2i}}}$$
$$\times\, e^{i\left\{A_0\, (x-x_0) - \frac{a_{1r}\, (a_{2i} + a_{2r})\, A_0^2}{a_{2i}}\, (t-t_0) + \frac{a_{2r}}{2\, a_{2i}}\, \ln\left[-a_{2i} + e^{2\, a_{1r}\, A_0^2\, (t-t_0)}\right] + \phi_0\right\}}, \tag{2.229}$$

where $a_{1i} = -a_{1r}$, A_0, t_0, x_0, and ϕ_0 are arbitrary real constants.
  • *Derived in appendix* A.

Solution 6.

$$\psi(x,\ t) = A_0\ \frac{\sqrt{a_{1i}}\ e^{a_{1i}\ A_0^2\ (t-t_0)}}{\sqrt{a_{2i}\ e^{2\ a_{1i}\ A_0^2\ (t-t_0)} - 1}}$$

$$\times\ e^{i\left(A_0\ (x-x_0) - a_{1r}\ A_0^2\ (t-t_0) + \frac{a_{2r}}{2\ a_{2i}}\ \ln\left\{e^{2\ a_{1i}\ A_0^2\ A_1}\left[a_{2i}\ e^{2\ a_{1i}\ A_0^2\ (t-t_0)} - 1\right]\right\} + \phi_0\right)}, \tag{2.230}$$

where A_0, A_1, t_0, x_0, and ϕ_0 are arbitrary real constants.
- *Derived in appendix* A.

Solution 7.

$$\psi(x,\ t) = A_0\ e^{-a_{1r}\ A_1^2\ (t-t_0)}\ e^{i\left[A_1\ (x-x_0) - a_{1r}\ A_1^2\ (t-t_0) - \frac{a_{2r}\ A_0^2}{2\ a_{1r}\ A_1^2}\ e^{-2\ a_{1r}\ A_1^2\ (t-t_0)} + \phi_0\right]}, \tag{2.231}$$

where $a_{1i} = -a_{1r}$, $a_{2i} = 0$, A_0, A_1, t_0, x_0, and ϕ_0 are arbitrary real constants.
- *Derived in appendix* A.

Solution 8.   *SW*

$$\psi(x,\ t) = A_0\ \sqrt{\frac{2\ a_{1r}}{a_{2r}}}\ \mathrm{sn}[A_0\ (x - x_0),\ -1]\ e^{i\ \phi_0}, \tag{2.232}$$

where $a_{1r}\ a_{2r} > 0$, $a_{1i} = \frac{a_{1r}\ a_{2i}}{a_{2r}}$, A_0, x_0, and ϕ_0 are arbitrary real constants.
- *Derived in appendix* A.

Solution 9.   *SW*

$$\psi(x,\ t) = A_0\ \sqrt{\frac{a_{1r}}{a_{2r}}}\ \mathrm{cn}\left[A_0\ (x - x_0),\ \frac{1}{2}\right]\ e^{i\ \phi_0}, \tag{2.233}$$

where $a_{1r}\ a_{2r} > 0$, $a_{1i} = \frac{a_{1r}\ a_{2i}}{a_{2r}}$, A_0, x_0, and ϕ_0 are arbitrary real constants.
- *Derived in appendix* A.

Solution 10.   *SW*

$$\psi(x,\ t) = A_0\ \sqrt{\frac{2\ a_{1r}}{a_{2r}}}\ \mathrm{dn}[A_0\ (x - x_0),\ 2]\ e^{i\ \phi_0}, \tag{2.234}$$

where $a_{1r}\ a_{2r} > 0$, $a_{1i} = \frac{a_{1r}\ a_{2i}}{a_{2r}}$, A_0, x_0, and ϕ_0 are arbitrary real constants.
- *Derived in appendix* A.

2.2.4 Summary of Section 2.2.3

Equation

$$i\psi_t + (a_{1r} + i a_{1i})\psi_{xx} + (a_{2r} + i a_{2i})|\psi|^2\psi = 0$$

#	Solution	Conditions	Name	Equation #
1.	$\psi(x,t) = A_0\, e^{i[a_{2r} A_0^2 (t-t_0)+\phi_0]}$	$a_{2i} = 0$, A_0, t_0, and ϕ_0 are arbitrary real constants	Continuous wave, t-dependent phase	(2.225)
2.	$\psi(x,t) = A_0\, e^{i[\pm A_0 \sqrt{\frac{a_{2r}}{a_{1r}}}(x-x_0)+\phi_0]}$	$a_{1r} a_{2r} > 0$, $a_{1i} = \frac{a_{1r} a_{2i}}{a_{2r}}$, A_0, x_0, and ϕ_0 are arbitrary real constants	Continuous wave, x-dependent phase	(2.226)
3.	$\psi(x,t) = \dfrac{1}{\sqrt{2 a_{2i}(t-t_0)}}\, e^{i\phi_0}$	$a_{2i} > 0$, $a_{2r} = 0$, t_0 and ϕ_0 are arbitrary real constants	Decaying wave	(2.227)
4.	$\psi(x,t) = \dfrac{1}{\sqrt{2 a_{2i}(t-t_0)}}\, e^{i\{A_0(x-x_0)-a_{1r}A_0^2(t-t_0)+\frac{a_{2r}}{2 a_{2i}}\ln[2 a_{2i}(t-t_0)]+\phi_0\}}$	$a_{2i} > 0$, $a_{1i} = 0$, A_0, t_0, x_0, and ϕ_0 are arbitrary real constants	Decaying wave	(2.228)
5.	$\psi(x,t) = A_0 \sqrt{\dfrac{a_{1r}}{e^{2 a_{1r} A_0^2 (t-t_0)}-a_{2i}}}$ $\times e^{i\{A_0(x-x_0)-\frac{a_{1r}(a_{2i}-a_{2r})A_0^2}{a_{2i}}(t-t_0)+\frac{a_{2r}}{2 a_{2i}}\ln[-a_{2i}+e^2 a_{1r} A_0^2 (t-t_0)]+\phi_0\}}$	$a_{1i} = -a_{1r}$, A_0, t_0, x_0, and ϕ_0 are arbitrary real constants	—	(2.229)
6.	$\psi(x,t) = A_0 \sqrt{\dfrac{a_{1i}\, e^{a_{1i} A_0^2 (t-t_0)}}{a_{2i}\, e^{2 a_{1i} A_0^2 (t-t_0)}-1}}$ $\times e^{i\{A_0(x-x_0)-a_{1r}A_0^2(t-t_0)+\frac{a_{2r}}{2 a_{2i}}\ln[e^{2 a_{1i} A_0^2 (t-t_0)} A_1 \{a_{2i} e^{2 a_{1i} A_0^2 (t-t_0)}-1\}]+\phi_0\}}$	A_0, A_1, t_0, x_0, and ϕ_0 are arbitrary real constants	—	(2.230)
7.	$\psi(x,t) = A_0\, e^{-a_{1r} A_1^2 (t-t_0)}\, e^{i[A_1(x-x_0)-a_{1r}A_1^2(t-t_0)-\frac{a_2 A_0^2}{2 a_{1r} A_1^2}e^{-2 a_{1r} A_1^2 (t-t_0)}+\phi_0]}$	$a_{1i} = -a_{1r}$, $a_{2i} = 0$, A_0, A_1, t_0, x_0, and ϕ_0 are arbitrary real constants	—	(2.231)
8.	$\psi(x,t) = A_0\sqrt{\dfrac{2 a_{1r}}{a_{2r}}}\, \mathrm{sn}[A_0(x-x_0),\,-1]\, e^{i\phi_0}$	$a_{1r} a_{2r} > 0$, $a_{1i} = \frac{a_{1r} a_{2i}}{a_{2r}}$, A_0, x_0, and ϕ_0 are arbitrary real constants	Solitary wave	(2.232)
9.	$\psi(x,t) = A_0\sqrt{\dfrac{a_{1r}}{a_{2r}}}\, \mathrm{cn}[A_0(x-x_0),\,\tfrac{1}{\sqrt{2}}]\, e^{i\phi_0}$	$a_{1r} a_{2r} > 0$, $a_{1i} = \frac{a_{1r} a_{2i}}{a_{2r}}$, A_0, x_0, and ϕ_0 are arbitrary real constants	Solitary wave	(2.233)
10.	$\psi(x,t) = A_0\sqrt{\dfrac{2 a_{1r}}{a_{2r}}}\, \mathrm{dn}[A_0(x-x_0),\,2]\, e^{i\phi_0}$	$a_{1r} a_{2r} > 0$, $a_{1i} = \frac{a_{1r} a_{2i}}{a_{2r}}$, A_0, x_0, and ϕ_0 are arbitrary real constants	Solitary wave	(2.234)

References

[1] Pethick C J and Smith H 2008 *Bose–Einstein Condensation in Dilute Gases* (Cambridge: Cambridge University Press)

[2] Hasegawa A and Kodama Y 1995 *Solitons in Optical Communications* (Oxford: Oxford University Press)

[3] Sulem C and Sulem P-L 1999 The nonlinear Schrödinger equation: self-focusing and wave collapse *Appl. Math. Sci.* **139** 123–44

[4] Akhmediev N and Ankiewicz A 1997 *Solitons: Nonlinear Pulses and Beams* (London: Chapman and Hall)

[5] Agrawal G 2001 *Nonlinear fiber optics* 3rd ed (San Diego, CA: Academic)

[6] Kivshar Y S and Agrawal G P 2003 *Optical Solitons* (San Diego, CA: Academic)

[7] Mollenauer L and Gordon J 2006 *Solitons in optical fibers* (Boston, MA: Academic)

[8] Eisenberg H, Silberberg Y, Morandotti R, Boyd A R and Aitchison J S 1998 Discrete spatial optical solitons in waveguide arrays *Phys. Rev. Lett.* **81** 3383

[9] Garanovich I L, Longhi S, Sukhorukova A A and Kivshar Y S 2012 Light propagation and localization in modulated photonic lattices, and waveguides *Phys. Rep.* **518** 1–79

[10] Kharif C, Pelinovsky E and Slunyaev A 2009 *Rogue Waves in the Ocean* Advances in Geophysical and Environmental Mechanics and Mathematics (Berlin: Springer)

[11] Osborne A R 2010 *Nonlinear Ocean Waves and the Inverse Scattering Transform* (San Diego, CA: Academic)

[12] Zakharov V E 1968 Stability of periodic waves of finite amplitude on the surface of a deep fluid *J. Appl. Mech. Tech. Phys.* **9** 190–4 Originally in: *Zhurnal Prikdadnoi Mekhaniki i Tekhnicheskoi Fiziki* **9** 86–94 (1968)

[13] Zakharov V E 1983 Collapse and self-focusing of Langmuir waves *Basic Plasma Physics: Selected Chapters, Handbook of Plasma Physics* (Amsterdam: North-Holland) vol 1

[14] Zakharov V E 1972 Collapse of Langmuir waves *Sov. Phys. JETP* **35** 908–14

[15] Remoissenet M 1999 *Waves Called Solitons: Concepts and Experiments* 3rd edn (Berlin, Heidelberg: Springer)

[16] Gross E P 1961 Structure of a quantized vortex in boson systems Il *Nuovo Cimento* **20** 454–7

[17] Pitaevskii L P 1961 Vortex lines in an imperfect Bose gas *Sov. Phys.-JETP* **13** 451–4

[18] Hasegawa A and Tappert F 1973 Transmission of stationary nonlinear optical pulses in dispersive dielectric fibers. I. Anomalous dispersion *Appl. Phys. Lett.* **23** 142–4
Hasegawa A and Tappert F 1973 Transmission of stationary nonlinear optical pulses in dispersive dielectric fibers. II. Normal dispersion *Appl. Phys. Lett.* **23** 171–2

[19] Noether E 1918 Invariante Variationsprobleme *Nachrichten von der Gesellschaft der Wissenschaften zu Göttingen. Mathematisch-Physikalische Klasse* pp 235–57

[20] See for instance, Dauxois T and Peyrard M 2006 *Physics of Solitons* (Cambridge: Cambridge University Press)

[21] Stokes G G 1847 On the theory of oscillatory waves *Trans. Camb. Phil. Soc.* **8** 441–73

[22] Stokes G G 1966 *Mathematical and Physical Papers* **vol 1** (Johnson Reprint Corp)

[23] Gutkin E 1985 Conservation laws for the nonlinear Schrödinger equation *Ann. Inst. Henri Poincaré* **C2** 67–74

[24] De Nicola S 1993 Conservation laws for the non-linear Schrödinger equation *Pure Appl. Opt.* **2** 5–7

[25] Zakharov V E and Shabat A B 1972 Exact theory of two-dimensional self-focusing and one-dimensional self-modulation of waves in nonlinear media *Sov. Phys. JETP* **34** 62–9

[26] Zakharov V E and Manakov S V 1974 On the complete integrability of a nonlinear Schrödinger equation *Theor. Math. Phys.* **19** 332–43

[27] Ablowitz M J, Kaup D J, Newell A C and Segur H 1974 The inverse scattering transform. Fourier analysis for nonlinear problems *Stud. Appl. Math.* **53** 249–315

[28] Ablowitz M J and Segur H 1981 *Solitons and the Inverse Scattering Transform* **vol 4** (Philadelphia, PA: SIAM)

[29] Ablowitz M J and Herbst B M 1990 On homoclinic structure and numerically induced chaos for the nonlinear Schrödinger equation *SIAM J. Appl. Math.* **50** 339–51

[30] Ablowitz M J, Kaup D J, Newell A C and Segur H 1974 The inverse scattering transform-Fourier analysis for nonlinear problems *Stud. Appl. Math.* **53** 249–315

[31] Muruganandam P and Adhikari S K 2003 Bose–Einstein condensation dynamics in three dimensions by the pseudo-spectral and finite-difference methods *J. Phys.* **B36** 2501–14

[32] Muruganandam P and Adhikari S K 2009 Fortran programs for the time-dependent Gross-Pitaevskii equation in a fully anisotropic trap *Comput. Phys. Commun.* **180** 1888–912

[33] Antoine X, Bao W and Besse C 2013 Computational methods for the dynamics of the nonlinear Schrd"inger/Gross-Pitaevskii equations *Comput. Phys. Commun.* **84** 2621–33

[34] Al Sakkaf L and Al Khawaja U 2022 High accuracy power series method for solving scalar, vector, and inhomogeneous nonlinear Schrödinger equations *Alexandria Eng. J.* **61** 11803–24

[35] Anderson M H, Ensher J R, Matthews M R, Wieman C E and Cornell E A 1995 Observation of Bose–Einstein condensation in a dilute atomic vapor *Science* **269** 198–201
Davis K B, Mewes M-O, Andrews M R, van Druten N J, Durfee D S, Kurn D M and Ketterle W 1995 Bose–Einstein condensation in a gas of sodium atoms *Phys. Rev. Lett.* **75** 3969–73
Bradley C C, Sackett C A, Tollett J J and Hulet R G 1995 Evidence of Bose–Einstein condensation in an atomic gas with attractive interactions *Phys. Rev. Lett.* **75** 1687–90
erratum Erratum 1997 *Phys. Rev. Lett.* **79** 1170

[36] Zaitsev V F and Polyanin A D 2003 *Handbook of Nonlinear Partial Differential equations* (Chapman and Hall)

[37] He B and Meng Q 2016 Qualitative analysis and explicit exact solitary, kink and anti-kink wave solutions of the generalized nonlinear Schrödinger equation with parabolic law nonlinearity *Commun. Theor. Phys.* **65** 1–10

[38] Abdelrahman M A, Ammar S I, Abualnaja K M and Inc M 2020 New solutions for the unstable nonlinear Schrödinger equation arising in natural science *Aims Math.* **5** 1893–912

[39] Elhadj K M, Al Sakkaf L, Al Khawaja U and Boudjemâa A 2020 Singular soliton molecules of the nonlinear Schrödinger equation *Phys. Rev.* **E101** 042221

[40] Zayed E M and Al-Nowehy A G 2017 Exact solutions for the perturbed nonlinear Schrödinger equation with power law nonlinearity and Hamiltonian perturbed terms *Optik Int. J. Light Electron Opt.* **139** 123–44

[41] Ma W X and Chen M 2009 Direct search for exact solutions to the nonlinear Schrödinger equation *Appl. Math. Comput.* **215** 2835–42

[42] Saxena A and Khare A 2023 New solutions of nonlocal Nls, Mkdv and Hirota equations *Ann. Phys.* **460** 169561

[43] Bronski J C, Carr L D, Deconinck B and Kutz J N 2001 Bose-Einstein condensates in standing waves: the cubic nonlinear Schrödinger equation with a periodic potential *Phys. Rev. Lett.* **86** 1402–5

[44] Khare A and Saxena A 2015 Periodic and hyperbolic soliton solutions of a number of nonlocal nonlinear equations *J. Math. Phys.* **56** 032104–27

[45] Ali K K, Tarla S, Ali M R, Yusuf A and Yilmazer R 2023 Physical wave propagation and dynamics of the Ivancevic option pricing model *Results Phys.* **52** 106751

[46] Gordon J P 1983 Interaction forces among solitons in optical fibers *Opt. lett.* **8** 596–8

[47] Blow K J and Doran N J 1985 Multiple dark soliton solutions of the nonlinear Schrödinger equation *Phys. Lett.* **A107** 55–8

[48] Kedziora D J, Ankiewicz A and Akhmediev N 2012 Second-order nonlinear Schrödinger equation breather solutions in the degenerate and rogue wave limits *Phys. Rev.* **E85** 066601–9

[49] Akhmediev N N, Eleonskii V M and Kulagin N E 1987 Exact first-order solutions of the nonlinear Schrödinger equation *Theor. Math. Phys.* **72** 809–18

[50] Conforti M, Mussot A, Kudlinski A, Trillo S and Akhmediev N 2020 Doubly periodic solutions of the focusing nonlinear Schrödinger equation: recurrence, period doubling, and amplification outside the conventional modulation-instability band *Phys. Rev.* **A101** 023843

[51] Kharif C, Pelinovsky E and Slunyaev A 2010 Rogue waves in the ocean *Advances in Geophysical and Environmental Mechanics and Mathematics* (Berlin: Springer)

[52] Al Khawaja U and Taki M 2013 Rogue waves management by external potentials *Phys. Lett.* **A37** 2944–9

[53] Chowdury A, Kedziora D J, Ankiewicz A and Akhmediev N 2015 Breather solutions of the integrable quintic nonlinear Schrödinger equation and their interactions *Phys. Rev.* **E91** 022919–11

Chapter 3

Nonlinear Schrödinger Equation with Power Law and Dual Power law Nonlinearities

A Glance at Chapter 3

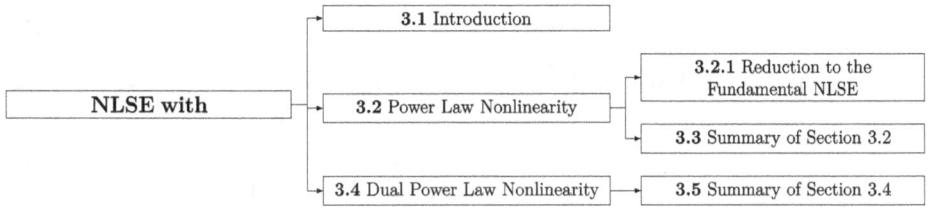

A Statistical View of Chapter 3

	Equation	Solutions
1	$i\,\psi_t + a_1\,\psi_{xx} + a_2\,|\psi|^n\,\psi = 0$	13
2	$i\,\psi_t + a_1\,\psi_{xx} + a_2\,|\psi|^n\,\psi + a_3\,|\psi|^m\,\psi = 0$	17
Total:	2	30

3.1 Introduction

Nonlinear Schrödinger equation with a generalized nonlinearity function, $F[|u(x,\,t)|^2]$, takes the form

$$iu_t + a_1 u_{xx} + F[|u|^2]u = 0. \tag{3.1}$$

Quadratic nonlinearity function, $F[|u|^2] = |u|^2$, corresponds to the Kerr nonlinearity. The equation is often denoted in this case as the NLSE with cubic nonlinearity, where cubic refers to $F[|u|^2]u = |u|^2 u$. Other forms of nonlinearity are referred to as non-Kerr nonlinearities [1–3]. While the NLSE with cubic nonlinearity is completely integrable, NLSE with non-Kerr nonlinearity is not integrable [1]. Nonetheless, some exact solutions are known in compact analytical form for specific forms of non-Kerr nonlinearity. A manifestation of the non-integrability of non-Kerr NLSEs is the finite number of conserved quantities, in contrast with an infinite number of conserved quantities for the NLSE with cubic nonlinearity. There are also fundamental differences between the solitonic solutions of the NLSE with Kerr nonlinearity and those of the NLSE with non-Kerr nonlinearity [1, 2]. While Kerr solitons collide elastically, non-Kerr solitons tend to coalesce and in general their collisions are inelastic [4]. Another fundamental difference is in the stability of solutions. While solitons of the NLSE with Kerr nonlinearity are always stable, solitons of the NLSE with non-Kerr nonlinearity are not necessarily stable; their stability depends on the parameters of the NLSE [5]. Stability can be investigated most simply for stationary solutions of the form [3]

$$u(x, t) = U(x) e^{-i\omega t}, \tag{3.2}$$

which renders the NLSE (3.1), to an ordinary differential equation

$$\omega U + a_1 U'' + a_2 F[U^2]U = 0. \tag{3.3}$$

Here, $U(x)$ and ω are assumed to be real. The last equation can be rewritten in the form

$$a_1 U'' = -\frac{d}{dU}\left[\int (a_2 F[U^2]U + \omega U)dU\right], \tag{3.4}$$

where the left hand side can be interpreted as an effective 'force' and the quantity between square brackets on the right hand side is interpreted as an effective 'potential'. Stability can then be investigated by seeking minima of the effective potential. This of course requires specifying the functional form of nonlinearity. While we have started in (3.1) with spacial NLSE, a similar result can be obtained had we started with temporal NLSE. The only difference will be in the interpretation of the effective force, which in this case will be $a_1 \ddot{U}(t)$.

Many specific forms of the nonlinearity are relevant to various physical systems, such as plasmas, semiconducting waveguides, and doped glass. They can be categorized in four main classes: (1) power law nonlinearity, (2) dual power law or competing nonlinearity, (3) saturable nonlinearity, and (4) transitive or threshold nonlinearity. Power law nonlinearity,

$$F[|u|^2] = a_3|u|^n, \tag{3.5}$$

where n is an integer and a_3 is a constant, appears in plasmas, semiconducting materials, and in the Effimov limit of Bose–Einstein condensates (see below). Dual nonlinearity

$$F[|u|^2] = a_3|u|^n + a_4|u|^m, \tag{3.6}$$

where n and m are different integers and a_3 and a_4 are constants, appears in nonlinear optics when the power of the optical pulse is highly intense such that the Kerr nonlinearity is insufficient to describe the dynamics of the pulse. In nonlinear optics, the cubic-quintic case, $n = 2$ and $m = 4$, is relevant [6]. Triple power law is also considered in the literature. Saturable nonlinearity

$$F[|u|^2] = \frac{a_3|u|^n}{1 + a_4|u|^m}, \tag{3.7}$$

is common in many nonlinear materials [7]. There is no system that is modeled by the transitive nonlinearity

$$F[|u|^2] = \begin{cases} a_3|u|^2, & |u|^2 < |u_{th}|^2 \\ a_4|u|^2, & |u|^2 > |u_{th}|^2, \end{cases} \tag{3.8}$$

where $|u_{th}|^2$ is a threshold pulse intensity. However, it is considered for the interesting properties of the relevant solitonic solutions [8] and their potential applications in the all-optical data processing [2].

In Bose–Einstein condensates, two-body collisions are represented in the NLSE by the cubic nonlinearity and three-body collisions correspond to the quintic nonlinearity [9]. Two-body collisions are typically dominant over three-body collisions. However, Feshbach resonance can be exploited to weaken the effect of the two-body collisions in favor of three-body collisions. In the extreme case where three-body collisions dominate, the condensate reaches the Effimov limit [10]. The Lee–Huang–Yang quantum correction [11] leads to a quadratic nonlinearity, which may, under certain circumstances, compete with the cubic nonlinearity, leading to the so-called *quantum droplets* [12], as was later observed in dipolar condensates [13] and in Bose–Bose mixtures [14, 15].

3.2 NLSE with Power Law Nonlinearity

Equation:

$$i\,\psi_t + a_1\,\psi_{xx} + a_2\,|\psi|^n\,\psi = 0, \tag{3.9}$$

where $\psi = \psi(x, t)$ is the complex function profile, x and t are its two independent variables, n, a_1 and a_2 are arbitrary real constants.

3.2.1 Reduction to the Fundamental NLSE

Case I: CW Profile:

If

$$\psi(x, t) = A_0\, e^{i\left[A_1(x-x_0) + \left(A_0^2 a_2 - A_1^2 a_1\right)(t-t_0) + \phi_0\right]} \tag{3.10}$$

is a stationary solution to the fundamental NLSE with cubic nonlinearity (2.160), then

$$\psi(x, t) = A_0^{2/n}\, e^{i\left[A_1(x-x_0) + \left(A_0^2 a_2 - A_1^2 a_1\right)(t-t_0) + \phi_0\right]} \tag{3.11}$$

is a stationary solution to the NLSE with power law nonlinearity (3.9).

Case II: x-dependent profile:

If

$$\phi(x, t) = u(x) \, e^{i \lambda t} \tag{3.12}$$

is a stationary solution to the fundamental NLSE with cubic nonlinearity (2.160), then

$$\psi(x, t) = \left(\frac{n + 2}{n^2}\right)^{1/n} u(x)^{2/n} \, e^{i \frac{4\lambda}{n^2} t} \tag{3.13}$$

is a stationary solution to the NLSE with power law nonlinearity (3.9).

Solutions:

***Solution* 1. Constant Amplitude I**   *continuous wave (CW), t-dependent phase*

$$\psi(x, t) = A_0^{2/n} \, e^{i \left[A_0^2 \, a_2 \, (t - t_0) + \phi_0 \right]}, \tag{3.14}$$

where A_0, t_0, and ϕ_0 are arbitrary real constants.
- *Reference*: [16].

***Solution* 2. Constant Amplitude II**   *CW, x-dependent phase*

$$\psi(x, t) = A_0^{2/n} \, e^{i \left[A_0 \sqrt{\frac{a_2}{a_1}} \, (x - x_0) + \phi_0 \right]}, \tag{3.15}$$

where $a_1 a_2 > 0$, A_0, x_0, and ϕ_0 are arbitrary real constants.
- *Derived in appendix* A.

***Solution* 3. Constant Amplitude III**   *CW, t- and x-dependent phase*

$$\psi(x, t) = A_0^{2/n} \, e^{i \left[A_1 \, (x - x_0) + \left(A_0^2 \, a_2 - A_1^2 \, a_1 \right) (t - t_0) + \phi_0 \right]}, \tag{3.16}$$

where A_0, A_1, x_0, t_0, and ϕ_0 are arbitrary real constants.
- *Reference*: [16].

***Solution* 4. Rational Solution I**   *decaying wave (DW)*

$$\psi_0(x, t) = \frac{A_0}{\sqrt{t - t_0}} \, e^{i \left[\frac{2 a_2 \, |A_0|^n \, (t - t_0)^{\frac{2-n}{2}}}{2 - n} + \frac{(x - x_0)^2}{4 a_1 \, (t - t_0)} + \phi_0 \right]}, \tag{3.17}$$

where A_0, x_0, t_0, and ϕ_0 are arbitrary real constants.
- *Reference*: [16].

***Solution* 5. Rational Solution II**   *DW*

$$\psi(x, t) = \left[\frac{\pm 1}{n \sqrt{\frac{-a_2}{2 a_1 (2 + n)}} \, (x - x_0)} \right]^{\frac{2}{n}} e^{i \phi_0}, \tag{3.18}$$

where $a_1 a_2 (2 + n) < 0$, n, A_0, x_0, and ϕ_0 are arbitrary real constants.
  • *Reference*: [17].

Solution 6.

$$\psi(x, t) = \left\{ \frac{-2 A_0^2 a_1 (n + 2)}{a_2 n^2} \sec^2[A_0 (x - x_0)] \right\}^{\frac{1}{n}} e^{-i\left[\frac{4 a_1 A_0^2}{n^2} (t-t_0)+\phi_0 \right]}, \quad (3.19)$$

where $a_1 a_2 (n + 2) < 0$, A_0, x_0, t_0, and ϕ_0 are arbitrary real constants.
  • *Reference*: [23] with $\alpha = \gamma = \lambda = \nu = 0$.

Solution 7.

$$\psi(x, t) = \left\{ \frac{-2 A_0^2 a_1 (n + 2)}{a_2 n^2} \csc^2[A_0 (x - x_0)] \right\}^{\frac{1}{n}} e^{-i\left[\frac{4 a_1 A_0^2}{n^2} (t-t_0)+\phi_0 \right]}, \quad (3.20)$$

where $a_1 a_2 (n + 2) < 0$, A_0, x_0, t_0, and ϕ_0 are arbitrary real constants.
  • *Reference*: [23] with $\alpha = \gamma = \lambda = \nu = 0$.

Solution 8.   *bright soliton*

$$\psi(x, t) = \left\{ \frac{2 A_0^2 a_1 (n + 2)}{a_2 n^2} \operatorname{sech}^2[A_0 (x - x_0)] \right\}^{\frac{1}{n}} e^{i\left[\frac{4 a_1 A_0^2}{n^2} (t-t_0)+\phi_0 \right]}, \quad (3.21)$$

where $a_1 a_2 (n + 2) > 0$, A_0, x_0, t_0, and ϕ_0 are arbitrary real constants.
  • *Reference*: [23] with $\alpha = \gamma = \lambda = \nu = 0$.

Solution 9.

$$\psi(x, t) = \left\{ \frac{-2 A_0^2 u_1 (n + 2)}{a_2 n^2} \operatorname{csch}^2[A_0 (x - x_0)] \right\}^{\frac{1}{n}} e^{i\left[\frac{4 a_1 A_0^2}{n^2} (t-t_0)+\phi_0 \right]}, \quad (3.22)$$

where $a_1 a_2 (n + 2) < 0$, A_0, x_0, t_0, and ϕ_0 are arbitrary real constants.
  • *Reference*: [23] with $\alpha = \gamma = \lambda = \nu = 0$.

Solution 10. Generalized Oscillatory Solution

$$\psi(x, t) = A(x) e^{i \phi_0}, \quad (3.23)$$

where $A(x) = Y^{-1}(x - x_0)$, $Y[A(x)] = \frac{A(x)}{\sqrt{A_0}} {}_2F_1\left[\frac{1}{2}, \frac{1}{n+2}, \frac{n+3}{n+2}, \frac{2 a_2 A^{n+2}(x)}{a_1 A_0 (n+2)} \right]$, ${}_2F_1$ is the hypergeometric function and Y^{-1} is the inverse operator of the function $Y[A(x)]$, $A_0 > 0$, x_0 and ϕ_0 are arbitrary real constants.
  • *Derived in appendix* A.

Solution 11.   *solitary wave (SW)*

$$\psi(x, t) = \{R_3 + (R_2 - R_3)\, \text{sn}^2[A_0\,(x - x_0),\, m]\}\, e^{i\,[\lambda_1\,(t - t_0) + \phi_0]}, \tag{3.24}$$

where $m = \frac{R_2 - R_3}{R_1 - R_3}$, R_j, $j = 1, 2, 3$ are the three roots of $Y(x) = 3\,\lambda_0 + \frac{3\,\lambda_1}{a_1}\,x^2 - \frac{a_2}{a_1}\,x^3$, $A_0 = \sqrt{\frac{a_2\,(R_3 - R_1)}{6\,a_1}}$, sn is the Jacobi elliptic function of the modulus m, $a_1\,a_2\,(R_3 - R_1) > 0$, $n = 1$, λ_0, λ_1, x_0, t_0, and ϕ_0 are arbitrary real constants.

- *Derived in appendix* A.

Solution 12.   SW

$$\psi(x, t) = \frac{\sqrt{R_1}\,\text{sn}[A_0\,(x - x_0),\, m]}{\sqrt{\dfrac{R_1 - R_3}{R_3} + \text{sn}^2[A_0\,(x - x_0),\, m]}}\, e^{i\,[\lambda_1\,(t - t_0) + \phi_0]}, \tag{3.25}$$

where $m = \frac{R_3\,(R_1 - R_2)}{R_2\,(R_1 - R_3)}$, R_j, $j = 1, 2, 3$ are the three roots of $Y(x) = 3\,\lambda_0 + \frac{3\,\lambda_1}{a_1}\,x - \frac{a_2}{a_1}\,x^3$, $A_0 = \sqrt{\frac{a_2\,R_2\,(R_1 - R_3)}{3\,a_1}}$, $R_1 > 0$, $a_1\,a_2\,R_2\,(R_1 - R_3) > 0$, $n = 4$, λ_0, λ_1, x_0, t_0, and ϕ_0 are arbitrary real constants.

- *Derived in appendix* A.

Solution 13.   *SW on a finite background*

$$\psi(x, t) = \frac{\sqrt{R_1\,(R_2 - R_4) + R_2\,(R_4 - R_1)\,\text{sn}^2\,[A_0\,(x - x_0),\, m]}}{\sqrt{-R_2 + R_4 + (R_1 - R_4)\,\text{sn}^2\,[A_0\,(x - x_0),\, m]}}\, e^{i\,\phi(x,\, t)}, \tag{3.26}$$

where

$$\phi(x, t) = \frac{\lambda_0\left(\dfrac{(R_2 - R_1)}{A_0}\,\Pi\left\{\dfrac{R_2\,(R_1 - R_4)}{R_1\,(R_2 - R_4)},\, \text{am}[A_0\,(x - x_0),\, m],\, m\right\}\,\text{dn}[A_0\,(x - x_0),\, m]\right)}{R_1\,R_2\,\sqrt{1 - \dfrac{(R_2 - R_3)\,(R_1 - R_4)\,\text{sn}^2[A_0\,(x - x_0),\, m]}{(R_1 - R_3)\,(R_2 - R_4)}}}$$

$$+ \frac{\lambda_0}{R_2}\,(x - x_0) + \lambda_2\,(t - t_0) + \phi_0,$$

$m = \frac{(R_2 - R_3)\,(R_1 - R_4)}{(R_1 - R_3)\,(R_2 - R_4)}$, R_j, $j = 1, 2, 3, 4$ are the four roots of $Y(x) = 3\,a_1\,\lambda_0^2 - 3\,a_1\,\lambda_1\,x - 3\,\lambda_2\,x^2 + a_2\,x^4$, $A_0 = \sqrt{\frac{-a_2\,(R_1 - R_3)\,(R_2 - R_4)}{3\,a_1}}$, Π is the incomplete elliptic integral, am is the amplitude for Jacobi elliptic functions, $a_1\,a_2\,(R_1 - R_3)\,(R_2 - R_4) < 0$, $n = 4$, λ_0, λ_1, λ_2, x_0, t_0, and ϕ_0 are arbitrary real constants.

- *Derived in appendix* A.

3.3 Summary of Section 3.2

Equation:

$$i\psi_t + a_1\psi_{xx} + a_2|\psi|^n\psi = 0$$

#	Solution	Conditions	Name	Equation #
1.	$\psi(x,t) = A_0^{2/n} e^{i[A_0^2 a_2 (t-t_0)+\phi_0]}$	$A_0, t_0,$ and ϕ_0 are arbitrary real constants	Continuous wave, t-dependent phase	(3.14)
2.	$\psi(x,t) = A_0^{2/n} e^{i[A_0\sqrt{\frac{a_2}{a_1}}(x-x_0)+\phi_0]}$	$a_1 a_2 > 0, A_0, x_0,$ and ϕ_0 are arbitrary real constants	Continuous wave, x-dependent phase	(3.15)
3.	$\psi(x,t) = A_0^{2/n} e^{i[A_1(x-x_0)+(A_0^2 a_2-A_1^2 a_1)(t-t_0)+\phi_0]}$	$A_0, A_1, x_0, t_0,$ and ϕ_0 are arbitrary real constants	Continuous wave, t- and x- dependent phase	(3.16)
4.	$\psi(x,t) = \frac{A_0}{\sqrt{t-t_0}}\, e^{i[\frac{2 a_2 |A_0|^n (t-t_0)^{\frac{2-n}{2}}}{2-n} + \frac{(x-x_0)^2}{4 a_1 (t-t_0)}+\phi_0]}$	$A_0, x_0, t_0,$ and ϕ_0 are arbitrary real constants	Decaying wave	(3.17)
5.	$\psi(x,t) = \left[\dfrac{\pm 1}{n\sqrt{\frac{-a_2}{2 a_1 (2+n)}}\,(x-x_0)}\right]^{\frac{1}{n}} e^{i\phi}$	$a_1 a_2 (2+n) < 0,\ n,\ A_0,\ x_0,$ and ϕ_0 are arbitrary real constants	—	(3.18)
6.	$\psi(x,t) = \left\{\dfrac{-2 A_0^2 a_1 (n+2)}{a_2\, n^2}\sec^2[A_0 (x-x_0)]\right\}^{\frac{1}{n}} e^{-i[\frac{4 a_1 A_0^2}{n^2}(t-t_0)+\phi_0]}$	$a_1 a_2 (n+2) < 0, A_0, x_0, t_0,$ and ϕ_0 are arbitrary real constants	—	(3.19)
7.	$\psi(x,t) = \left\{\dfrac{-2 A_0^2 a_1 (n+2)}{a_2\, n^2}\csc^2[A_0 (x-x_0)]\right\}^{\frac{1}{n}} e^{-i[\frac{4 a_1 A_0^2}{n^2}(t-t_0)+\phi_0]}$	$a_1 a_2 (n+2) < 0, A_0, x_0, t_0,$ and ϕ_0 are arbitrary real constants	—	(3.20)
8.	$\psi(x,t) = \left\{\dfrac{2 A_0^2 a_1 (n+2)}{a_2\, n^2}\operatorname{sech}^2[A_0 (x-x_0)]\right\}^{\frac{1}{n}} e^{i[\frac{4 a_1 A_0^2}{n^2}(t-t_0)+\phi_0]}$	$a_1 a_2 (n+2) > 0, A_0, x_0, t_0,$ and ϕ_0 are arbitrary real constants	Bright soliton	(3.21)
9.	$\psi(x,t) = \left\{\dfrac{-2 A_0^2 a_1 (n+2)}{a_2\, n^2}\operatorname{csch}^2[A_0 (x-x_0)]\right\}^{\frac{1}{n}} e^{i[\frac{4 a_1 A_0^2}{n^2}(t-t_0)+\phi_0]}$	$a_1 a_2 (n+2) < 0, A_0, x_0, t_0,$ and ϕ_0 are arbitrary real constants	—	(3.22)

(Continued)

(Continued)

Equation:

$$i\,\psi_t + a_1\,\psi_{xx} + a_2\,|\psi|^n\,\psi = 0$$

#	Solution	Name	Conditions	Equation #
10.	$\psi(x,t) = A(x)\,e^{j\phi_0}$, $A(x) = Y^{-1}(x - x_0)$	Generalized oscillatory solution	$Y[A(x)] = \dfrac{A(x)}{\sqrt{A_0}}\,2F_1\left[\dfrac{1}{2},\ \dfrac{1}{n+2},\ \dfrac{n+3}{n+2},\ \dfrac{2a_2\,A^{n+2}(x)}{q\,A_0(n+2)}\right]$, $2F_1$ is the hypergeometric function and Y^{-1} is the inverse operator of the function $Y[A(x)]$, $A_0 > 0$, x_0 and ϕ_0 are arbitrary real constants	(3.23)
11.	$\psi(x,t) = \{R_3 + (R_2 - R_3)\,\mathrm{sn}^2[A_0(x - x_0), m]\}\,e^{j[\lambda_1(t-t_0)+\phi_0]}$	Solitary wave	$m = \dfrac{R_2 - R_3}{R_1 - R_3}$, R_j, $j = 1, 2, 3$ are the three roots of $Y(x) = 3\lambda_0 + \dfrac{3\lambda_1}{q}x^2 - \dfrac{a_2}{q}x^3$, $A_0 = \sqrt{\dfrac{a_2(R_3 - R_1)}{6q}}$, $a_1\,a_2\,(R_3 - R_1) > 0$, $n = 1$, $\lambda_0, \lambda_1, x_0, t_0$, and ϕ_0 are arbitrary real constants	(3.24)
12.	$\psi(x,t) = \dfrac{\sqrt{R_1}\,\mathrm{sn}[A_0(x-x_0), m]}{\sqrt{\dfrac{R_1 - R_3}{R_3} + \mathrm{sn}^2[A_0(x-x_0), m]}}\,e^{j[\lambda_1(t-t_0)+\phi_0]}$	Solitary wave	$m = \dfrac{R_3(R_1 - R_2)}{R_2(R_1 - R_3)}$, R_j, $j = 1, 2, 3$ are the three roots of $Y(x) = 3\lambda_0 + \dfrac{3\lambda_1}{q}x - \dfrac{a_2}{q}x^3$, $A_0 = \sqrt{\dfrac{a_2\,R_2(R_1 - R_3)}{3q}}$, $n = 4$, $R_1 > 0$, $a_1\,a_2\,R_2\,(R_1 - R_3) > 0$, $\lambda_0, \lambda_1, x_0, t_0$, and ϕ_0 are arbitrary real constants	(3.25)
13.	$\psi(x,t) = \sqrt{\dfrac{R_1(R_2 - R_4) + R_2(R_4 - R_1)\,\mathrm{sn}^2[A_0(x-x_0),m]}{-R_2 + R_4 + (R_1 - R_4)\,\mathrm{sn}^2[A_0(x-x_0),m]}}\,e^{j\,\phi(x,t)}$, $\phi(x,t) = \dfrac{\lambda_0\,\dfrac{(R_2 - R_1)}{A_0}\,\Pi\!\left[\dfrac{R_2(R_1 - R_4)}{R_1(R_2 - R_4)}, \mathrm{am}[A_0(x - x_0),m],m\right]\mathrm{dn}[A_0(x - x_0),m]}{R_1 R_2\sqrt{1 - \dfrac{(R_2 - R_3)(R_1 - R_4)}{(R_1 - R_3)(R_2 - R_4)}\,\mathrm{sn}^2[A_0(x-x_0),m]}}$ $+ \dfrac{\lambda_0}{R_2}(x - x_0) + \lambda_2(t - t_0) + \phi_0$	Solitary wave	$m = \dfrac{(R_2 - R_3)(R_1 - R_4)}{(R_1 - R_3)(R_2 - R_4)}$, R_j, $j = 1, 2, 3, 4$ are the four roots of $Y(x) = \dfrac{3a_1\lambda_0^2 - 3a_1\lambda_1 x - 3\lambda_2 x^2 + a_2 x^4}{3q}$, $A_0 = \sqrt{\dfrac{a_2(R_1 - R_3)(R_2 - R_4)}{3q}}$, Π is the incomplete elliptic integral, am is the amplitude for Jacobi elliptic functions, $a_1\,a_2\,(R_1 - R_3)(R_2 - R_4) < 0$, $n = 4$, $\lambda_0, \lambda_1, \lambda_2, x_0, t_0$, and ϕ_0 are arbitrary real constants	(3.26)

3.4 NLSE with Dual Power Law Nonlinearity

Equation:

$$i\,\psi_t + a_1\,\psi_{xx} + a_2\,|\psi|^n\,\psi + a_3\,|\psi|^m\,\psi = 0, \tag{3.27}$$

where $\psi = \psi(x, t)$ is the complex function profile, x and t are its two independent variables, n, m, a_1, a_2, and a_3 are arbitrary real constants.

Solutions:

***Solution* 1. Constant Amplitude I**   *CW, t-dependent phase*

$$\psi(x,\,t) = A_0\,e^{i\,[(|A_0|^n\,a_2 + |A_0|^m\,a_3)\,(t-t_0)+\phi_0]}, \tag{3.28}$$

where A_0, t_0, and ϕ_0 are arbitrary real constants.
  • *Derived in appendix* A.

***Solution* 2. Constant Amplitude II**   *CW, x-dependent phase*

$$\psi(x,\,t) = A_0\,e^{i\left[\sqrt{\frac{|A_0|^n\,a_2 + |A_0|^m\,a_3}{a_1}}\,(x-x_0)+\phi_0\right]}, \tag{3.29}$$

where A_0, x_0, and ϕ_0 are arbitrary real constants.
  • *Derived in appendix* A.

***Solution* 3. Constant Amplitude III**   *CW, t- and x-dependent phase*

$$\psi(x,\,t) = A_0\,e^{i\left[A_1\,(x-x_0)+(|A_0|^n\,a_2 + |A_0|^m\,a_3 - a_1\,A_1^2)\,(t-t_0)+\phi_0\right]}, \tag{3.30}$$

where A_0, A_1, x_0, t_0, and ϕ_0 are arbitrary real constants.
  • *Derived in appendix* A.

***Solution* 4. Rational Solution I**   *DW*

$$\psi(x,\,t) = \frac{A_0}{\sqrt{t-t_0}}\,e^{i\left\{-\frac{2\,|A_0|^m\,a_3\,(t-t_0)^{\frac{2-m}{2}}}{m-2} - \frac{2\,|A_0|^n\,a_2\,(t-t_0)^{\frac{2-n}{2}}}{n-2} + \frac{[2\,a_2\,A_1 + (x-x_0)]^2}{4\,a_1\,(t-t_0)}+\phi_0\right\}}, \tag{3.31}$$

where $n \neq 2$, $m \neq 2$, A_0, A_1, x_0, t_0, and ϕ_0 are arbitrary real constants.
  • *Derived in appendix* A.

***Solution* 5. Rational Solution II**

$$\psi(x,\,t) = \left[\frac{-2\,a_1\,a_2\,(n+1)\,(n+2)}{a_1\,a_3\,(n+2)^2 + a_2^2\,n^2\,(1+n)\,(x-x_0)^2}\right]^{\frac{1}{n}}\,e^{i\,\phi_0}, \tag{3.32}$$

where $a_1\,a_2 < 0$, $a_1\,a_3 > 0$, $m = 2\,n$, x_0 and ϕ_0 are arbitrary real constants.
  • *Reference*: [17].

Solution 6.

$$\psi(x, t) = \sqrt{\frac{-a_2^3}{2 a_3}} \; \frac{x - x_0}{\sqrt{12 a_1 a_3 + a_2^2 (x - x_0)^2}} \; e^{-i \left[\frac{a_2^2}{4 a_3} (t - t_0) + \phi_0 \right]}, \qquad (3.33)$$

where $n = 2$, $m = 4$, $a_2 a_3 < 0$, $a_1 a_3 > 0$, x_0, t_0, and ϕ_0 are arbitrary real constants.
  • *Reference*: [17] with $m = 1$.

Solution 7.

$$\psi(x, t) = \left\{ \frac{n + 1}{(n + 2) [1 + 2 e^{\sqrt{A_0} (x - x_0)}]} \right\}^{\frac{1}{n}} e^{i [A_1 (t - t_0) + \phi_0]}, \qquad (3.34)$$

where $A_0 = \frac{a_2 n^2 (n + 1)}{a_1 (n + 2)^2}$, $A_1 = \frac{a_1 A_0}{n^2}$, $m = 2 n$, $a_3 = -a_2$, $a_1 a_2 (n + 1) > 0$, x_0, t_0, and ϕ_0 are arbitrary real constants.
  • *Reference*: [17].

Solution 8. Generalized Soliton Solution

$$\psi(x, t) = \left(\frac{A_0 (n + 2)}{a_2 + a_2 \sqrt{1 + \gamma} \cosh\left[n \sqrt{\frac{A_0}{a_1}} (x - x_0) \right]} \right)^{\frac{1}{n}} e^{i [A_0 (t - t_0) + \phi_0]}, \qquad (3.35)$$

where $a_3 = \gamma a_{03}$, $a_{03} = a_2^2 (n + 1)/[A_0 (n + 2)^2]$, $m = 2 n$, $a_1 A_0 > 0$, γ, x_0, t_0, A_0, and ϕ_0 are arbitrary real constants.
  • *Reference*: [18], *generalized for n*.

Solution 9.   *Kink Soliton*
    (Figure 3.1)

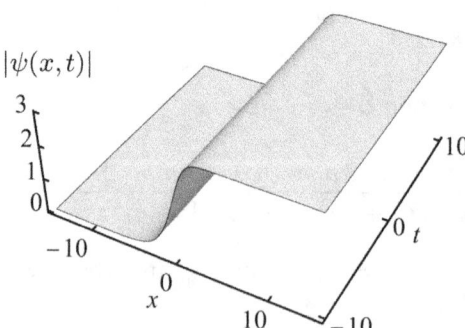

Figure 3.1. Kink soliton (3.36) with $a_1 = 3$, $a_2 = 1$, $A_0 = 2.2$, $n = 2$, and $x_0 = t_0 = \phi_0 = 0$.

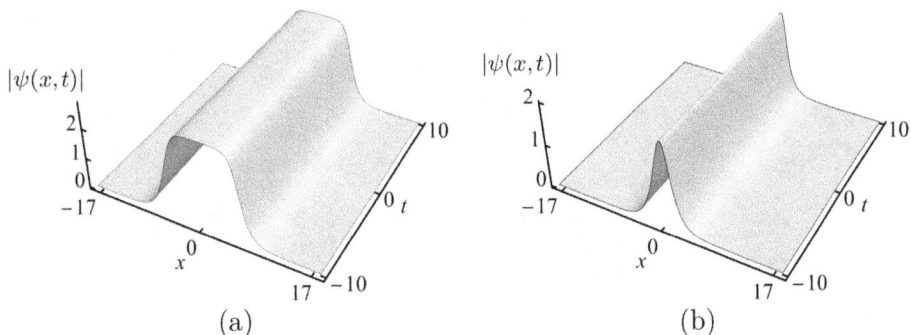

Figure 3.2. Plot of solution (3.37). (a) flat-top soliton with $\theta = \pi/9998$, (b) bright soliton with $\theta = \pi/2$. Values of the other parameters are: $a_1 = 3$, $a_2 = 1$, $A_0 = 2$, $n = 2$, and $x_0 = t_0 = \phi_0 = 0$.

$$\psi(x, t) = \left[\frac{A_0 (n + 2)}{a_2 + a_2 \sqrt{\dfrac{a_1}{4 a_2^2}} \, e^{-n \sqrt{\frac{A_0}{a_1}} (x - x_0)}} \right]^{\frac{1}{n}} e^{i\,[A_0\,(t - t_0) + \phi_0]}, \qquad (3.36)$$

where $a_3 = \gamma\, a_{03}$, $\gamma = -1$, $a_{03} = a_2^2 (n + 1)/[A_0 (n + 2)^2]$, $m = 2\,n$, $a_1 A_0 > 0$, x_0, t_0, A_0, and ϕ_0 are arbitrary real constants, $\gamma \to -1$ in (3.35).
- *Reference*: [18], *generalized for n.*

***Solution* 10.**　　*flat-top soliton*
(Figure 3.2)

$$\psi(x, t) = \left\{ \frac{A_0 (n + 2)}{a_2 + a_2 \sin(\theta) \cosh\!\left[n \sqrt{\dfrac{A_0}{a_1}} (x - x_0) \right]} \right\}^{\frac{1}{n}} e^{i\,[A_0\,(t - t_0) + \phi_0]}, \qquad (3.37)$$

where $a_3 = \gamma\, a_{03}$, $\gamma = -\cos^2(\theta)$, $0 < \theta < \pi/2$, $a_{03} = a_2^2 (n + 1)/[A_0 (n + 2)^2]$, $m = 2\,n$, $a_1 A_0 > 0$, x_0, t_0, A_0, and ϕ_0 are arbitrary real constants, $0 < \gamma < -1$ in (3.35).
- *Reference*: [18], *generalized for n.*

***Solution* 11.**　　*thin-top soliton*
(Figure 3.3)

$$\psi(x, t) = \left\{ \frac{A_0 (n + 2)}{a_2 + a_2 \cosh(\theta) \cosh\!\left[n \sqrt{\dfrac{A_0}{a_1}} (x - x_0) \right]} \right\}^{\frac{1}{n}} e^{i\,[A_0\,(t - t_0) + \phi_0]}, \qquad (3.38)$$

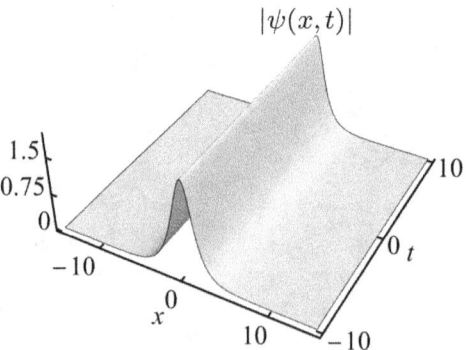

Figure 3.3. Thin-top soliton (3.38) with $a_1 = 3$, $a_2 = 1$, $A_0 = 2.2$, $\theta = \pi/4$, $n = 2$, and $x_0 = t_0 = \phi_0 = 0$.

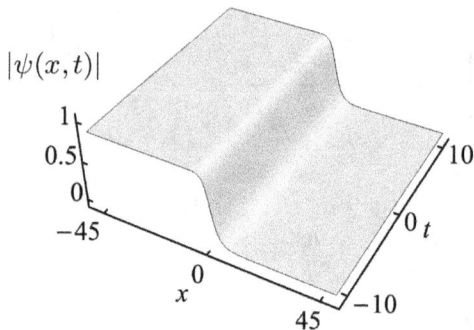

Figure 3.4. Dark soliton (3.39) with $a_1 = a_2 = 1$, $a_3 = -1$, $n = 2$, and $x_0 = t_0 = \phi_0 = 0$.

where $a_3 = \gamma\, a_{03}$, $\gamma = \sinh^2(\theta)$, $\theta \neq 0$, $a_{03} = a_2^2\,(n + 1)/[A_0\,(n + 2)^2]$, $m = 2\,n$, $a_1\,A_0 > 0$, x_0, t_0, A_0, and ϕ_0 are arbitrary real constants, $\gamma > 0$ in (3.35).
  • *Reference*: [18], *generalized for n.*

***Solution* 12.**   *Dark Soliton*
  (Figure 3.4)

$$\psi(x,\,t) = \left(\frac{2\,a_1\,A_0^2\,(n + 2)}{a_2\,n^2}\,\{1 - \tanh[A_0\,(x - x_0)]\} \right)^{\frac{1}{n}} e^{i\left[\frac{4\,a_1\,A_0^2}{n^2}\,(t - t_0) + \phi_0 \right]}, \quad (3.39)$$

where $A_0 = \sqrt{\dfrac{-a_2^2\,n^2\,(1 + n)}{4\,a_1\,a_3\,(2 + n)^2}}$, $m = 2\,n$, $a_1\,a_2\,(n + 2) > 0$, $a_1\,a_3\,(n + 1) < 0$, x_0, t_0, and ϕ_0 are arbitrary real constants.
  • *Reference*: [20].

Solution 13.

$$\psi(x, t) = \left(\frac{2\, a_1\, A_0^2\, (n + 2)}{a_2\, n^2} \left\{ 1 - \coth[A_0\, (x - x_0)] \right\} \right)^{\frac{1}{n}} e^{i \left[\frac{4\, a_1\, A_0^2}{n^2} (t - t_0) + \phi_0 \right]}, \quad (3.40)$$

where $A_0 = \sqrt{\dfrac{-a_2^2\, n^2\, (1 + n)}{4\, a_1\, a_3\, (2 + n)^2}}$, $m = 2\, n$, $a_1\, a_2\, (n + 2) > 0$, $a_1\, a_3\, (n + 1) < 0$, x_0, t_0, and ϕ_0 are arbitrary real constants.

* *Reference*: [21].

Solution 14.

$$\psi(x, t) = \pm \frac{\mu_1\, \mu_2}{\sqrt{\mu_1^2 + (\mu_1^2 + 4\, \mu_2^2)\, \sinh^2[A_0\, (x - x_0)]}}\; e^{i \left[\frac{A_0^2\, A_1\, a_1}{a_3\, \mu_1^2\, \mu_2^2} (t - t_0) + \phi_0 \right]}, \quad (3.41)$$

where $\mu_1 = \sqrt{\dfrac{3\, a_2}{a_3} + \sqrt{\dfrac{9\, a_2^2}{a_3^2} + \dfrac{4\, A_1}{a_3}}}$, $\mu_2 = \dfrac{1}{2} \sqrt{\dfrac{-3\, a_2}{a_3} + \sqrt{\dfrac{9\, a_2^2}{a_3^2} + \dfrac{4\, A_1}{a_3}}}$, $A_0 = \sqrt{\dfrac{a_3\, \mu_1^2\, \mu_2^2}{12\, a_1}}$,

$\dfrac{3\, a_2}{a_3} < \sqrt{\dfrac{9\, a_2^2}{a_3^2} + \dfrac{4\, A_1}{a_3}}$, $a_1\, a_3 > 0$, $n = 2$, $m = 4$, x_0, t_0, A_1, and ϕ_0 are arbitrary real constants.

* *Reference*: [17] with $m = 1$.

Solution 15.

$$\psi(x, t) = \left(\frac{n + 1}{2\, a_2^n\, a_3\, n^2} \right)^{\frac{1}{2n}} \left\{ \frac{A_0^2}{\sqrt{\dfrac{2\, a_2\, n^2\, (n + 1)}{a_3\, (n + 2)^2} + A_0^2}\; \cosh\left[\dfrac{A_0}{\sqrt{2\, a_1}}\, (x - x_0) \right] + \sqrt{\dfrac{2\, a_2\, n^2\, (n + 1)}{a_3\, (n + 2)^2}}} \right\}^{\frac{1}{n}} \quad (3.42)$$

$$\times\, e^{i \left[\frac{A_0^2}{2\, n^2} (t - t_0) + \phi_0 \right]},$$

where $a_1 > 0$, $a_2 = 1$, $a_3\, (n + 1) > 0$, $m = 2\, n$, x_0, t_0, A_0, and ϕ_0 are arbitrary real constants.

* *Reference*: [22] with $A_2 = B_2 = 0$.

Solution 16.

$$\psi(x,\,t) = \left(\frac{-1}{2}\right)^{\frac{1}{n}} \left[\frac{2\,(n+1)}{a_2^n\,a_3\,n^2}\right]^{\frac{1}{2n}} \left\{\frac{A_0^2}{\sqrt{\dfrac{2\,a_2\,n^2\,(n+1)}{a_3\,(n+2)^2} - A_0^2}\,\sin\left[\dfrac{A_0}{\sqrt{2\,a_1}}\,(x-x_0)\right] + \sqrt{\dfrac{2\,a_2\,n^2\,(n+1)}{a_3\,(n+2)^2}}}\right\}^{\frac{1}{n}}$$

$$\times e^{i\left[\frac{-A_0^2}{2\,n^2}\,(t-t_0)+\phi_0\right]},$$

(3.43)

where $a_1 > 0$, $a_2 = 1$, $a_3\,(n+1) > 0$, $m = 2\,n$, x_0, t_0, A_0, and ϕ_0 are arbitrary real constants.

 • *Reference*: [22] with $A_2 = B_2 = 0$.

Solution 17.

$$\psi(x,\,t) = \left(\frac{-1}{2}\right)^{\frac{1}{n}} \left[\frac{2\,(n+1)}{a_2^n\,a_3\,n^2}\right]^{\frac{1}{2n}} \left\{\frac{A_0^2}{\sqrt{\dfrac{2\,a_2\,n^2\,(n+1)}{a_3\,(n+2)^2} - A_0^2}\,\cos\left[\dfrac{A_0}{\sqrt{2\,a_1}}\,(x-x_0)\right] + \sqrt{\dfrac{2\,a_2\,n^2\,(n+1)}{a_3\,(n+2)^2}}}\right\}^{\frac{1}{n}}$$

$$\times e^{i\left[\frac{-A_0^2}{2\,n^2}\,(t-t_0)+\phi_0\right]},$$

(3.44)

where $a_1 > 0$, $a_2 = 1$, $a_3\,(n+1) > 0$, $m = 2\,n$, x_0, t_0, A_0, and ϕ_0 are arbitrary real constants.

 • *Reference*: [22] with $A_2 = B_2 = 0$.

3.5 Summary of Section 3.4

Equation

$$i\psi_t + a_1\psi_{xx} + a_2|\psi|^n\psi + a_3|\psi|^m\psi = 0$$

#	Solution	Conditions	Name	Equation #
1.	$\psi(x,t) = A_0\,e^{i[(|A_0|^n a_2 + |A_0|^m a_3)(t-t_0)+\phi_0]}$	$A_0,\ t_0,$ and ϕ_0 are arbitrary real constants	Continuous wave, t-dependent phase	(3.28)
2.	$\psi(x,t) = A_0\,e^{i\left[\sqrt{\frac{|A_0|^n a_2 + |A_0|^m a_3}{a_1}}\,(x-x_0)+\phi_0\right]}$	$A_0,\ x_0,$ and ϕ_0 are arbitrary real constants	Continuous wave, x-dependent phase	(3.29)
3.	$\psi(x,t) = A_0\,e^{i\left[A_1(x-x_0)+(|A_0|^n a_2 + |A_0|^m a_3 - a_1 A_1^2)(t-t_0)+\phi_0\right]}$	$A_0,\ A_1,\ x_0,\ t_0,$ and ϕ_0 are arbitrary real constants	Continuous wave, t- and x-dependent phase	(3.30)
4.	$\psi(x,t) = \dfrac{A_0}{\sqrt{t-t_0}}\,e^{i\left(\frac{-2A_0^m a_3(t-t_0)^{\frac{2-m}{2}}}{m-2} + \frac{2A_0^n a_2(t-t_0)^{\frac{2-n}{2}}}{n-2} + \frac{[2a_2 A_1+(x-x_0)]^2}{4a_1(t-t_0)}+\phi_0\right)}$	$n\neq 2,\ m\neq 2,\ A_0,\ A_1,\ x_0,\ t_0,$ and ϕ_0 are arbitrary real constants	Decaying wave	(3.31)
5.	$\psi(x,t) = \left[\dfrac{-2q\,a_2(n+1)(n+2)}{q\,a_3(n+2)^2 + a_2^2\,n^2(1+n)(x-x_0)^2}\right]^{\frac{1}{n}} e^{i\phi_0}$	$a_1 a_2 < 0,\ a_1 a_3 > 0,\ m = 2n,\ x_0$ and ϕ_0 are arbitrary real constants	—	(3.32)
6.	$\psi(x,t) = \sqrt{\dfrac{-a_2^3}{2a_3}}\,\sqrt{\dfrac{x-x_0}{12q\,a_3 + a_2^2(x-x_0)^2}}\,e^{-i\left[\frac{a_2^2}{4a_3}(t-t_0)+\phi_C\right]}$	$n=2,\ m=4,\ a_2 a_3 < 0,\ a_1 a_3 > 0,\ x_0,\ t_0,$ and ϕ_0 are arbitrary real constants	—	(3.33)
7.	$\psi(x,t) = \left\{\dfrac{n+1}{(n+2)\left[1+2e^{\sqrt{\frac{A_1}{A_0}}(x-x_0)}\right]}\right\}^{\frac{1}{n}} e^{i[A_1(t-t_0)+\phi_0]}$	$A_0 = \dfrac{a_2\,n^2(n+1)}{q(n+2)^2},\ A_1 = \dfrac{q\,A_0}{n^2},\ m = 2n,\ a_3 = -a_2,$ $a_1 a_2(n+1) > 0,\ x_0,\ t_0,$ and ϕ_0 are arbitrary real constants	—	(3.34)
8.	$\psi(x,t) = \left(\dfrac{A_0(n+2)}{a_2 + a_2\sqrt{1+\gamma}\,\cosh\!\left[a\sqrt{\frac{A_0}{q}}(x-x_0)\right]}\right)^{\frac{1}{n}} e^{i[A_0(t-t_0)+\phi_0]}$	$a_3 = \gamma\,a_{03},\ a_{03} = a_2^2(n+1)/[A_0(n+2)^2],$ $m = 2n,\ a_1 A_0 > 0,\ \gamma,\ x_0,\ t_0,\ A_0,$ and ϕ_0 are arbitrary real constants	Generalized soliton	(3.35)

(Continued)

(Continued)

Equation

$$i\psi_t + a_1\psi_{xx} + a_2|\psi|^n\psi + a_3|\psi|^m\psi = 0$$

#	Solution	Conditions	Name	Equation #
9.	$\psi(x,t) = \left[\dfrac{A_0(n+2)}{a_2 + a_2\sqrt{\frac{q}{4a_2^2}}\,e^{-n\sqrt{\frac{A_0}{a_1}}(x-x_0)}}\right]^{\frac{1}{n}} e^{i[A_0(t-t_0)+\phi_0]}$	$a_{03} = a_2^2(n+1)/[A_0(n+2)^2]$, $m=2n$, $a_3 = \gamma a_{03}$, $\gamma = -1$, $a_1 A_0 > 0$, $x_0, t_0, A_0,$ and ϕ_0 are arbitrary real constants	Kink soliton	(3.36)
10.	$\psi(x,t) = \left\{\dfrac{A_0(n+2)}{a_2 + a_2\sin(\theta)\cosh\left[n\sqrt{\frac{A_0}{q}}(x-x_0)\right]}\right\}^{\frac{1}{n}} e^{i[A_0(t-t_0)+\phi_0]}$	$a_3 = \gamma a_{03}$, $\gamma = -\cos^2(\theta)$, $0 < \theta < \pi/2$, $a_{03} = a_2^2(n+1)/[A_0(n+2)^2]$, $m=2n$, $a_1 A_0 > 0$, $x_0, t_0, A_0,$ and ϕ_0 are arbitrary real constants	Flat-top soliton	(3.37)
11.	$\psi(x,t) = \left\{\dfrac{A_0(n+2)}{a_2 + a_2\cosh(\theta)\cosh\left[n\sqrt{\frac{A_0}{q}}(x-x_0)\right]}\right\}^{\frac{1}{n}} e^{i[A_0(t-t_0)+\phi_0]}$	$a_3 = \gamma a_{03}$, $\gamma = \sinh^2(\theta)$, $\theta \neq 0$, $a_{03} = a_2^2(n+1)/[A_0(n+2)^2]$, $m=2n$, $a_1 A_0 > 0$, $x_0, t_0, A_0,$ and ϕ_0 are arbitrary real constants	Thin-top soliton	(3.38)
12.	$\psi(x,t) = \left(\dfrac{2q A_0^2(n+2)}{a_2 n^2}\{1 - \tanh[A_0(x-x_0)]\}\right)^{\frac{1}{n}} e^{i\left[\frac{4a_1 A_0^2}{n^2}(t-t_0)+\phi_0\right]}$	$A_0 = \sqrt{\dfrac{-a_2^2 n^2(1+n)}{4a_1 a_3(2+n)^2}}$, $m=2n$, $a_1 a_2(n+2) > 0$, $a_1 a_3(n+1) < 0$, x_0, t_0 are arbitrary real constants	Dark soliton	(3.39)
13.	$\psi(x,t) = \dfrac{2q A_0^2(n+2)}{a_2 n^2}\{1 - \coth[A_0(x-x_0)]\}]^{\frac{1}{n}} e^{i\left[\frac{4a_1 A_0^2}{n^2}(t-t_0)+\phi_0\right]}$	$A_0 = \sqrt{\dfrac{-a_2^2 n^2(1+n)}{4a_1 a_3(2+n)^2}}$, $m=2n$, $a_1 a_2(n+2) > 0$, $a_1 a_3(n+1) < 0$, $x_0, t_0,$ and ϕ_0 are arbitrary real constants	—	(3.40)
14.	$\psi(x,t) = \pm\dfrac{\mu_1\mu_2}{\sqrt{\mu_1^2 + (\mu_1^2 + 4\mu_2^2)\sinh^2[4\mu_2(x-x_0)]}}\,e^{i\left[\frac{A_1^2 A_1 a_1}{a_3\mu_1^2\mu_2^2}(t-t_0)+\phi_1\right]}$	$\mu_1 = \sqrt{\dfrac{3a_2}{a_3} + \sqrt{\dfrac{9a_2^2}{a_3^2} + \dfrac{4A_1}{a_3}}}$, $\mu_2 = \dfrac{1}{2}\sqrt{\dfrac{-3a_2}{a_3} + \sqrt{\dfrac{9a_2^2}{a_3^2} + \dfrac{4A_1}{a_3}}}$, $A_0 = \sqrt{\dfrac{a_3\mu_1^2\mu_2^2}{12q}}$, $\dfrac{3a_2}{a_3} < \sqrt{\dfrac{9a_2^2}{a_3^2} + \dfrac{4A_1}{a_3}}$, $a_1 a_3 > 0$, $n=2$, $m=4$, $x_0,$ $t_0, A_1,$ and ϕ_0 are arbitrary real constants	—	(3.41)

15. $\psi(x,t) = \left(\dfrac{n+1}{2\,a_2^n\,a_3\,n^2}\right)^{\frac{1}{2n}}$

$\times \left\{ \dfrac{A_0^2}{\sqrt{\dfrac{2\,a_2\,n^2(n+1)}{a_3(n+2)^2}} + A_0^2\,\cosh[\dfrac{A_0}{\sqrt{2a_2}}(x-x_0)] + \sqrt{\dfrac{2\,a_2\,n^2(n+1)}{a_3(n+2)^2}}} \right\}^{\frac{1}{n}}$

$\times\, e^{i\,[\frac{A_0^2}{2n^2}(t-t_0)+\phi_0]}$

$a_1 > 0,\ a_2 = 1,\ a_3\,(n+1) > 0,\ m = 2\,n, x_0, t_0, A_0,$ and ϕ_0 are arbitrary real constants $\quad\text{—}$ (3.42)

16. $\psi(x,t) = \left(\dfrac{-1}{2}\right)^{\frac{1}{n}} \left[\dfrac{2(n+1)}{a_2^n\,a_3\,n^2}\right]^{\frac{1}{2n}}$

$\times \left\{ \dfrac{A_0^2}{\sqrt{\dfrac{2\,a_2\,n^2(n+1)}{a_3(n+2)^2}} - A_0^2\,\sin[\dfrac{A_0}{\sqrt{2a_2}}(x-x_0)] + \sqrt{\dfrac{2\,a_2\,n^2(n+1)}{a_3(n+2)^2}}} \right\}^{\frac{1}{n}}$

$\times\, e^{i\,[\frac{-A_0^2}{2n^2}(t-t_0)+\phi_0]}$

$a_1 > 0,\ a_2 = 1,\ a_3\,(n+1) > 0,\ m = 2\,n, x_0, t_0, A_0,$ and ϕ_0 are arbitrary real constants $\quad\text{—}$ (3.43)

17. $\psi(x,t) = \left(\dfrac{-1}{2}\right)^{\frac{1}{n}} \left[\dfrac{2(n+1)}{a_2^n\,a_3\,n^2}\right]^{\frac{1}{2n}}$

$\times \left\{ \dfrac{A_0^2}{\sqrt{\dfrac{2\,a_2\,n^2(n+1)}{a_3(n+2)^2}} - A_0^2\,\cos[\dfrac{A_0}{\sqrt{2a_2}}(x-x_0)] + \sqrt{\dfrac{2\,a_2\,n^2(n+1)}{a_3(n+2)^2}}} \right\}^{\frac{1}{n}}$

$\times\, e^{i\,[\frac{-A_0^2}{2n^2}(t-t_0)+\phi_0]}$

$a_1 > 0,\ a_2 = 1,\ a_3\,(n+1) > 0,\ m = 2\,n, x_0, t_0, A_0,$ and ϕ_0 are arbitrary real constants $\quad\text{—}$ (3.44)

References

[1] Akhmediev N and Ankiewicz A 1997 *Solitons: Nonlinear Pulses and Beams* (London: Chapman and Hall)

[2] Kivshar Y S and Agrawal G P 2003 *Optical Solitons* (San Diego, CA: Academic)

[3] Biswas A and Konar S 2007 *Introduction to non-Kerr Law Optical Solitons* (New York: Chapman and Hall/CRC Taylor and Francis Group)

[4] Abdulloev K O, Bogolubsky I L and Makhankov V G 1976 One more example of inelastic soliton interaction *Phys. Lett.* **A56** 427–8

[5] Kaplan A E 1985 Bistable solitons *Phys. Rev. Lett.* **55** 1291

[6] Lawrence B, Torruellas W E, Cha M, Sundheimer M L, Stegeman G I, Meth J, Etemad S and Baker G 1994 Identification and role of two-photon excited states in a π-conjugated polymer *Phys. Rev. Lett.* **73** 597

[7] Coutaz J and Kull M 1991 Saturation of the nonlinear index of refraction in semiconductor-doped glass *J. Opt. Soc. Am.* **B8** 95–8

[8] Kaplan A E 1985 Multistable self-trapping of light and multistable soliton pulse propagation *IEEE J. Quantum Electron.* **21** 1538–43

[9] Pethick C J and Smith H 2008 *Bose–Einstein Condensation in Dilute Gases* (Cambridge: Cambridge University Press)

[10] Efimov V 1990 Energy levels arising from resonant two-body forces in a three-body system *Phys. Lett.* **B33** 563
(Comment) 1990 Is a qualitative approach to the three-body problem useful? *Nucl. Part. Phys.* **19** 271

[11] Lee T D, Huang K and Yang C N 1957 Eigenvalues and eigenfunctions of a Bose system of hard spheres and its low-temperature properties *Phys. Rev.* **106** 1135

[12] Petrov D S 2015 Quantum mechanical stabilization of a collapsing Bose-Bose mixture *Phys. Rev. Lett.* **115** 155302

[13] Ferrier-Barbut I, Kadau H, Schmitt M, Wenzel M and Pfau T 2016 Observation of quantum droplets in a strongly dipolar Bose gas *Phys. Rev. Lett.* **116** 215301

[14] Cabrera C R, Tanzi L, Sanz J, Naylor B, Thomas P, Cheiney P and Tarruell L 2018 Quantum liquid droplets in a mixture of Bose–Einstein condensates *Science* **359** 301–4

[15] Cheiney P, Cabrera C R, Sanz J, Naylor B, Tanzi L and Tarruell L 2018 Bright soliton to quantum droplet transition in a mixture of Bose-Einstein condensates *Phys. Rev. Lett.* **120** 135301

[16] Zaitsev V F and Polyanin A D 2003 *Handbook of Nonlinear Partial Differential equations* (Amsterdam: North-Holland)

[17] He B and Meng Q 2016 Qualitative analysis and explicit exact solitary, kink and anti-kink wave solutions of the generalized nonlinear Schrödinger equation with parabolic law nonlinearity *Commun. Theor. Phys.* 1–10

[18] Al Sakkaf L and Al Khawaja U 2023 Reflectionless potentials and resonant scattering of flat-top and thin-top solitons *Phys. Rev. E* **107** 014202

[19] Al Khawaja U and Bahlouli H 2019 Integrability conditions and solitonic solutions of the nonlinear Schrödinger equation with generalized dual-power nonlinearities, PT-symmetric potentials, and space-and time-dependent coefficients *Commun. Nonlinear Sci. Numer. Simul.* **69** 248–60

[20] Triki H and Biswas A 2011 Dark solitons for a generalized nonlinear Schrödinger equation with parabolic law and dual-power law nonlinearities *Math. Methods Appl. Sci.* **34** 958–62

[21] Mirzazadeh M, Eslami M, Milovic D and Biswas A 2014 Topological solitons of resonant nonlinear Schödinger's equation with dual-power law nonlinearity by GG-expansion technique *Optik Int. J. Light Electron Opt.* **125** 5480–9

[22] Zhang L H and Si J G 2010 New soliton and periodic solutions of (1+2)-dimensional nonlinear Schrödinger equation with dual-power law nonlinearity *Commun. Nonlinear Sci. Numer. Simul.* **15** 2747–54

[23] Zayed E M E and Al-Nowehy A-G 2017 Exact solutions for the perturbed nonlinear Schrödinger equation with power law nonlinearity and Hamiltonian perturbed terms *Optik* **139** 123–44

IOP Publishing

Handbook of Exact Solutions to the Nonlinear Schrödinger
Equations (Second Edition)

Usama Al Khawaja and Laila Al Sakkaf

Chapter 4

Nonlinear Schrödinger Equation with Higher Order Terms

A Glance at Chapter 4

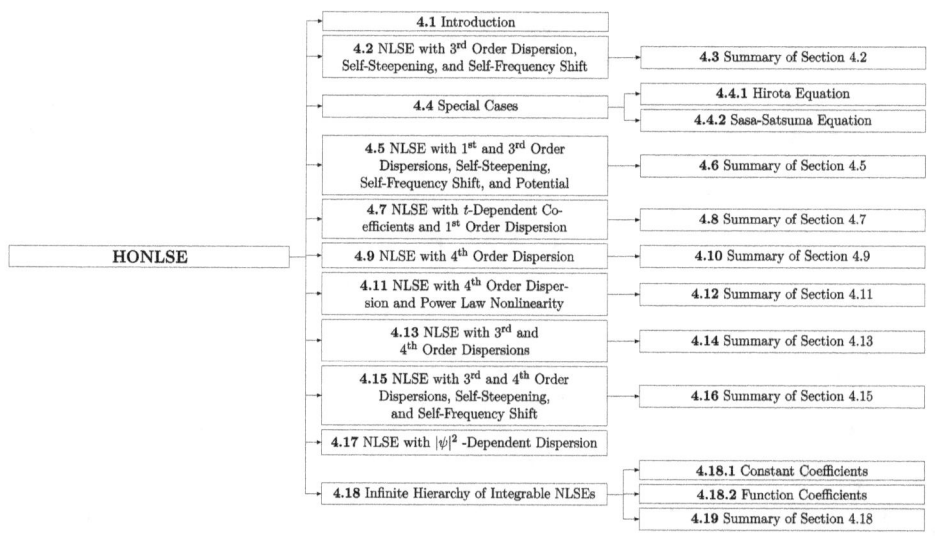

A Statistical View of Chapter 4

	Equation	Solutions
1	$i\,\psi_t + a_1\,\psi_{xx} + a_2\,|\psi|^2\,\psi + i\,a_3\,\psi_{xxx} + i\,a_4\,(|\psi|^2\,\psi)_x + i\,a_5\,(|\psi|^2)_x\,\psi = 0$	15
2	$i\,\psi_t + a_1\,\psi_{xx} + a_2\,|\psi|^2\,\psi + i\,a_3\,\psi_{xxx} + i\,a_4\,|\psi|^2\,\psi_x = 0$	0
3	$i\,\psi_t + a_1\,\psi_{xx} + a_2\,|\psi|^2\,\psi + i\,a_4\,[\psi_{xxx} + (|\psi|^2)_x\,\psi + |\psi|^2\,\psi_x] = 0$	0
4	$i\,\psi_t + i\,a_1\,\psi_x + a_2\,\psi_{xx} - i\,a_3\,\psi_{xxx} + a_4\,|\psi|^2\,\psi - i\,a_5\,|\psi|^2\,\psi_x - i\,a_6\,\psi^2\,\psi^*_{\ x} - a_7\,\psi = 0$	4
5	$i\,\psi_t + i\,a_1(t)\,\psi_x + a_2(t)\,\psi_{xx} + a_3(t)\,|\psi|^2\,\psi = 0$	10
6	$i\,\psi_t + a_1\,\psi_{xx} + a_2\,|\psi|^2\,\psi + a_3\,\psi_{xxxx} = 0$	5
7	$i\,\psi_t + a_1\,\psi_{xx} + a_2\,|\psi|^{2n}\,\psi + a_3\,\psi_{xxxx} = 0$	5
8	$i\,\psi_t + a_1\,\psi_{xx} + a_2\,|\psi|^2\,\psi + a_3\,|\psi|^4\,\psi + i\,a_4\,\psi_{xxx} + a_5\,\psi_{xxxx} = 0$	9
9	$i\,\psi_t + a_1\,\psi_{xx} + a_2\,|\psi|^2\,\psi + a_3\,|\psi|^4\,\psi + i\,a_4\,\psi_{xxx} + a_5\,\psi_{xxxx} + i\,a_6\,(|\psi|^2\,\psi)_x + i\,a_7$ $(|\psi|^2)_x\,\psi = 0$	9
10	$i\,\psi_t + (1 - |\mu|\,|\psi|^2)\,\psi_{xx} + 2\,(1 - |\mu|)\,|\psi|^2\,\psi = 0$	2
11	$i\,\psi_t + a_2\,k_2 - i\,a_3\,k_3 + a_4\,k_4 - i\,a_5\,k_5 + \cdots = 0$	8
12	$i\,\psi_t + a_2(t)\,k_2 - i\,a_3(t)\,k_3 + a_4(t)\,k_4 - i\,a_5(t)\,k_5 + \cdots = 0$	2
Total	12	69

4.1 Introduction

Cubic nonlinearity is the dominant nonlinear effect in physical systems described by the NLSE. However, other nonlinear effects may also exist. Usually these are much weaker than the cubic nonlinearity, hence they are denoted as higher order effects (HOEs). In certain circumstances, the HOEs become important and need to be taken into account. In general, all HOEs can be mathematically accounted for by expanding the intensity-dependent dispersion relation $k(\omega, A, A^*, ...)$, in powers of the carrier wave frequency, ω_0, the wave field, A, its complex conjugate, A^*, and their powers. Then proceeding with the wide-envelope approximation, an NLSE with higher order terms can be derived. Physical processes determine which higher order terms are relevant and with what strength, as explained below for the three systems of optical pulses, Bose–Einstein condensates, and water surface waves.

4.1.1 Optical Pulses

One of the physical systems that exhibits HOEs most clearly is optical pulses in fibers [1, 2]. Cubic, or Kerr, nonlinearity in this system originates from the nonlinear electric polarization response to the electric field of the optical pulse. Higher order effects result from elastic and inelastic scattering of light by the material of the medium [3, 4]. Elastic scattering processes include nonlinear effects such as self-phase modulation (SPM), cross-phase modulation (XPM), four-wave mixing, higher order dispersion, and self-steepening. Physically, these originate from the same physical mechanism as the Kerr nonlinearity, namely the anharmonic vibrational modes excited by the electric field of the light pulse. Inelastic light scattering includes mainly the stimulated Raman scattering (SRS) and stimulated Brillouin scattering (SBS). In this case, the optical pulse excites vibrational modes of the molecules of the medium. A photon with a certain frequency is absorbed by the medium and then is re-emitted but with a smaller frequency (frequency downshift). The difference in energy is carried by an emerging phonon. The emitted photon may propagate along the same direction as the incident one, or opposite to it. This type of inelastic photon scattering is known as the SRS or SBS. The difference between the two types is that in the Raman scattering, the emitted photon may be along or opposite to the direction of the incident photon and the created phonon is an *optical phonon*, while in Brillouin scattering, the emitted photon is mainly opposite to the incident one and the created phonon is an *acoustic phonon*.

In optical fibers, Raman scattering is the effect to be considered, as the pulse propagates forward. It is also the dominant HOE. In situations when SRS is not small, the two HOEs: self-steepening and third-order dispersion, need also to be included, since they will not be significantly smaller than SRS.

Raman scattering is a delayed type of response. Therefore, it may be accounted for by a generalized nonlinear susceptibility, $\chi^{(3)}$, as [1]

$$\chi^{(3)}_{1111}(\mathbf{r},\, t - t',\, t - t'',\, t - t''') = \chi^{(3)}_{1111}(r)R(t - t')\, \delta(t - t'')\, \delta(t - t'''), \qquad (4.1)$$

where $R(t - t')$ is a response function replacing a Dirac delta function in order to account for the non-instantaneous property of Raman scattering. An approximate analytical form of the function $R(t - t')$, that fits accurately the experimental measurements, is then used to proceed with the derivation of the NLSE, as shown in Chapter 2. The consistency of the perturbative expansion, requires the inclusion of one additional term in the Taylor expansions of β and $\Delta\beta$. Apart from these changes, following the same procedure as in Chapter 2, the NLSE with HOEs, is shown to be

$$
i\frac{\partial}{\partial z}A(z, t) + a_1\frac{\partial^2}{\partial t^2}A(z, t) + (\eta + i\gamma)|A(z, t)|^2 A(z, t) + i\frac{\alpha_l}{2}A(z, t)
$$
$$
= ia_3\frac{\partial^3}{\partial z^3}A(z, t) - ia_{ss}\frac{\partial}{\partial z}[|A(z, t)|^2 A(z, t)] + a_R |A(z, t)|^2\frac{\partial}{\partial z}A(z, t), \tag{4.2}
$$

where

$$
a_3 = -\frac{1}{6}\frac{\partial\omega_l}{\partial z}\bigg|_{\omega=\omega_0}, \tag{4.3}
$$

$$
a_{ss} = \frac{1}{\omega_0}, \tag{4.4}
$$

and

$$
a_R = \int_0^\infty t\, R(t)\, dt, \tag{4.5}
$$

are the coefficients of HOEs: third-order dispersion, self-steepening, and Raman scattering, respectively.

To estimate the strengths of the HOEs in real units, we employ the same scaling as in Chapter 2, namely scaling time to the initial pulse width, T_0, length to the dispersion length, $L_D - T_0^2/2|a_1|$, and field to $1/\sqrt{L_D|a_2|}$. Neglecting, for simplicity, the linear and nonlinear damping terms, the NLSE with higher order terms, takes the form

$$
i\frac{\partial}{\partial z}u(z, t) + \frac{1}{2}\frac{\partial^2}{\partial t^2}u(z, t) + |u(z, t)|^2 u(z, t)
$$
$$
= -ia_3\frac{\partial^3}{\partial z^3}u(z, t) - ia_{ss}\frac{\partial}{\partial z}[|u(z, t)|^2 u(z, t)] + a_R |u(z, t)|^2\frac{\partial}{\partial z}u(z, t), \tag{4.6}
$$

where the parameters are now dimensionless, and read

$$
a_3 = -\frac{1}{6|a_1|T_0}\frac{\partial^3\omega_l}{\partial z^3}\bigg|_{\omega=\omega_0}, \tag{4.7}
$$

$$
a_{ss} = \frac{1}{\omega_0 T_0}, \tag{4.8}
$$

and

$$a_R = \frac{1}{T_0} \int_0^\infty tR(t)dt. \tag{4.9}$$

Higher order effects become important for pulse widths on the order of the femto second scale. For example, $a_3 \sim a_{ss} \approx 10^{-2}$ and $a_R \approx 10^{-1}$ for a pulse with $T_0 \approx 10$ fs and Silica fiber with $\int_0^\infty tR(t)dt = 1$ fs [2].

Before we conclude this part, three remarks are in order.

(i) While HOEs become increasingly significant for very short pulses, there is a lower limit on the width of short pulses. For pulses with width on the order of the width of the carrier wave, the wide envelope approximation breaks down and the NLSE will not be applicable. In such a case, one needs to solve Maxwell's equations directly.

(ii) For cases with extreme spectral broadening, referred to as *supercontinuum*, a large number of the higher order terms, usually more than 10 terms, need to be taken into account in the perturbative expansion.

(iii) The Raman coefficient, a_R, is a real constant. However, the NLSE is integrable only for imaginary a_R, which of course does not correspond to realistic situations. Nonetheless, we present in this chapter the exact analytical solutions for the case with imaginary a_R.

4.1.2 Bose–Einstein Condensates

In Bose–Einstein condensates, taking into account the energy dependence of the two-body scattering amplitude, leads to the following Gross–Pitaevskii equation with higher order terms [5]

$$i\hbar\frac{\partial}{\partial t}\phi(\mathbf{r}, t) + \frac{\hbar^2}{2m}\nabla^2\phi(\mathbf{r}, t) - V_{ext}(r)\phi(\mathbf{r}, t) - g_0|\phi(\mathbf{r}, t)|^2 - g_1|\phi(\mathbf{r}, t)|^3 - \frac{g_2}{2}\nabla^2[|\phi(\mathbf{r}, t)|^2]\phi(\mathbf{r}, t) = 0, \tag{4.10}$$

where the $g_1|\phi(\mathbf{r}, t)|^3$ term corresponds to the Lee–Huang–Yang quantum correction [6, 7], and the term $\frac{g_2}{2}\nabla^2[|\phi(\mathbf{r}, t)|^2]$ corresponds to the energy dependence of the two-body scattering correction [5]. Here, V_{ext} is an external potential, which in Bose–Einstein condensate experiments is a magnetic or optical confining potential, $g_{0, 1, 2}$ are real parameters, $\hbar = h/2\pi$ with h being Planck's constant, and m is the mass of an atom.

4.1.3 Water Surface Waves

Surface gravity waves on deep waters are described by the NLSE with cubic nonlinearity. Including corrections for pulses with narrow band width results in an NLSE with HOEs, known as the Dysthe Equation [8].

$$i\frac{\partial\eta}{\partial t} + a_1\frac{\partial^2\eta}{\partial x^2} + a_2|\eta|^2\eta = \sum_{i=3}^{5}a_i\frac{\partial^i\eta}{\partial x^i} + a_6\eta^2\frac{\partial\eta^*}{\partial x} + a_7\frac{\partial\phi}{\partial x}, \tag{4.11}$$

where a_{1-7} are constants [8], $\eta(x, t)$ is the wave amplitude, and $\phi(x, t)$ is the velocity field, defined by $v_x = \partial\phi/\partial x$. This equation is coupled to the equation of motion for the velocity field subject to boundary conditions

$$\frac{\partial^2 \phi}{\partial x^2} = 0, \ -h < z < 0, \tag{4.12}$$

$$\frac{\partial \phi}{\partial z} = \frac{1}{2} \frac{\partial |\eta|^2}{\partial x}, \ z = 0, \tag{4.13}$$

$$\frac{\partial \phi}{\partial z} = 0, \ z = -h, \tag{4.14}$$

where the wave amplitude is assumed to be independent of y, the water bottom is at $-h$, and the origin of the coordinates is taken on the surface $z = 0$.

Another relevant equation describing surface gravity waves in two spacial dimensions is the Davey–Stewartson Equation [9]

$$i\frac{\partial u}{\partial t} + a_1 \frac{\partial^2 u}{\partial x^2} + \frac{\partial^2 u}{\partial y^2} + a_2 |u|^2 u + a_3 u \frac{\partial \phi}{\partial x} = 0, \tag{4.15}$$

$$\frac{\partial^2 \phi}{\partial x^2} + a_4 \frac{\partial^2 \phi}{\partial y^2} = \frac{\partial(|u|^2)}{\partial x}, \tag{4.16}$$

which reduces to the NLSE for one spacial dimension.

4.2 NLSE with Third Order Dispersion, Self-Steepening, and Self-Frequency Shift

Equation:

$$i\,\psi_t + a_1\,\psi_{xx} + a_2\,|\psi|^2\,\psi + i\,a_3\,\psi_{xxx} + i\,a_4\,(|\psi|^2\,\psi)_x + i\,a_5\,(|\psi|^2)_x\,\psi = 0, \tag{4.17}$$

where $\psi = \psi(x, t)$ is the complex function profile, x and t are its two independent variables, a_j are arbitrary real constants, $j = 1, 2, 3, 4, 5$.

Solutions:

***Solution* 1. Constant Amplitude**   *continuous wave (CW), t- and x-dependent phase*

$$\psi(x, t) = c\ e^{i\,[c\,(x-x_0)+c^2\,(-a_1+c\,a_3+a_2-c\,a_4)\,(t-t_0)+\phi_0]}, \tag{4.18}$$

where x_0, t_0, c, and ϕ_0 are arbitrary real constants.

***Solution* 2.**

$$\psi(x, t) = \pm\sqrt{-\mu_1}\ \mathrm{sec}\{\sqrt{\mu_2}\ [x - x_0 + c_1\,(t - t_0)]\}e^{i\,[c_2\,(x-x_0)+c_3\,(t-t_0)+\phi_0]}, \tag{4.19}$$

where $\mu_1 = \frac{6\,(c_1 + 2\,a_1\,c_2 - 3\,a_3\,c_2^2)}{3\,a_4 + 2\,a_5} < 0$, $\mu_2 = \frac{c_1 + 2\,a_1\,c_2 - 3\,a_3\,c_2^2}{a_3} > 0$, $c_2 = \frac{-3\,a_2\,a_3 + a_1\,(3\,a_4 + 2\,a_5)}{6\,a_3\,(a_4 + a_5)}$,

$c_3 = 8\,a_1\,c_2^2 - 8\,a_3\,c_2^3 - \frac{a_1\,c_1 + 2\,a_1^2\,c_2}{a_3} + 3\,c_1\,c_2$, x_0, t_0, c_1, and ϕ_0 are arbitrary real constants.

- *Reference*: [10].

Solution 3.

$$\psi(x,\, t) = \pm\sqrt{-\mu_1}\, \csc\{\sqrt{\mu_2}\, [x - x_0 + c_1\, (t - t_0)]\}e^{i\, [c_2\, (x - x_0) + c_3\, (t - t_0) + \phi_0]}, \quad (4.20)$$

where $\mu_1 = \frac{6\, (c_1 + 2\, a_1\, c_2 - 3\, a_3\, c_2^2)}{(3\, a_4 + 2\, a_5)} < 0$, $\mu_2 = \frac{c_1 + 2\, a_1\, c_2 - 3\, a_3\, c_2^2}{a_3} > 0$, $c_2 = \frac{-3\, a_2\, a_3 + a_1\, (3\, a_4 + 2\, a_5)}{6\, a_3\, (a_4 + a_5)}$,

$c_3 = 8\, a_1\, c_2^2 - 8\, a_3\, c_2^3 - \frac{(a_1\, c_1 + 2\, a_1^2\, c_2)}{a_3} + 3\, c_1\, c_2$, x_0, t_0, c_1, and ϕ_0 are arbitrary real constants.

- *Reference*: [10].

Solution 4.

$$\psi(x,\, t) = \pm\sqrt{\mu_1}\, \tan\{\sqrt{-\mu_2}\, [x - x_0 + c_1\, (t - t_0)]\}e^{i\, [c_2\, (x - x_0) + c_3\, (t - t_0) + \phi_0]}, \quad (4.21)$$

where $\mu_1 = \frac{3\, (c_1 + 2\, a_1\, c_2 - 3\, a_3\, c_2^2)}{(3\, a_4 + 2\, a_5)} > 0$, $\mu_2 = \frac{c_1 + 2\, a_1\, c_2 - 3\, a_3\, c_2^2}{2\, a_3} < 0$, $c_2 = \frac{-3\, a_2\, a_3 + a_1\, (3\, a_4 + 2\, a_5)}{6\, a_3\, (a_4 + a_5)}$,

$c_3 = 8\, a_1\, c_2^2 - 8\, a_3\, c_2^3 - \frac{(a_1\, c_1 + 2\, a_1^2\, c_2)}{a_3} + 3\, c_1\, c_2$, x_0, t_0, c_1, and ϕ_0 are arbitrary real constants.

- *Reference*: [10].

Solution 5.

$$\psi(x,\, t) = \pm\sqrt{\mu_1}\, \cot\{\sqrt{-\mu_2}\, [x - x_0 + c_1\, (t - t_0)]\}e^{i\, [c_2\, (x - x_0) + c_3\, (t - t_0) + \phi_0]}, \quad (4.22)$$

where $\mu_1 = \frac{3\, (c_1 + 2\, a_1\, c_2 - 3\, a_3\, c_2^2)}{(3\, a_4 + 2\, a_5)} > 0$, $\mu_2 = \frac{c_1 + 2\, a_1\, c_2 - 3\, a_3\, c_2^2}{2\, a_3} < 0$, $c_2 = \frac{-3\, a_2\, a_3 + a_1\, (3\, a_4 + 2\, a_5)}{6\, a_3\, (a_4 + a_5)}$,

$c_3 = 8\, a_1\, c_2^2 - 8\, a_3\, c_2^3 - \frac{(a_1\, c_1 + 2\, a_1^2\, c_2)}{a_3} + 3\, c_1\, c_2$, x_0, t_0, c_1, and ϕ_0 are arbitrary real constants.

- *Reference*: [10].

Solution 6.   *Bright Soliton*
(Figure 4.1)

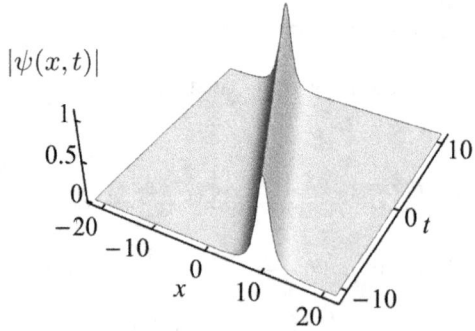

Figure 4.1. Bright soliton (4.23) with $a_1 = 1/2$, $a_2 = 1$, $a_3 = -1$, $a_4 = -3$, $a_5 = 2$, $c_1 = 9/10$, and $x_0 = t_0 = \phi_0 = 0$.

$$\psi(x, t) = \pm\sqrt{-\mu_1} \operatorname{sech}\{\sqrt{-\mu_2} \ [x - x_0 + c_1 (t - t_0)]\}e^{i \ [c_2 (x-x_0)+c_3 (t-t_0)+\phi_0]}, \qquad (4.23)$$

where $\mu_1 = \frac{6 (c_1 + 2 a_1 c_2 - 3 a_3 c_2^2)}{(3 a_4 + 2 a_5)} < 0$, $\mu_2 = \frac{c_1 + 2 a_1 c_2 - 3 a_3 c_2^2}{a_3} < 0$, $c_2 = \frac{-3 a_2 a_3 + a_1 (3 a_4 + 2 a_5)}{6 a_3 (a_4 + a_5)}$,

$c_3 = 8 a_1 c_2^2 - 8 a_3 c_2^3 - \frac{(a_1 c_1 + 2 a_1^2 c_2)}{a_3} + 3 c_1 c_2$, x_0, t_0, c_1, and ϕ_0 are arbitrary real constants.

• *Reference*: [10].

Solution 7.

$$\psi(x, t) = \pm\sqrt{\mu_1} \operatorname{csch}\{\sqrt{-\mu_2} \ [x - x_0 + c_1 (t - t_0)]\}e^{i \ [c_2 (x-x_0)+c_3 (t-t_0)+\phi_0]}, \qquad (4.24)$$

where $\mu_1 = \frac{6 (c_1 + 2 a_1 c_2 - 3 a_3 c_2^2)}{(3 a_4 + 2 a_5)} > 0$, $\mu_2 = \frac{c_1 + 2 a_1 c_2 - 3 a_3 c_2^2}{a_3} < 0$, $c_2 = \frac{-3 a_2 a_3 + a_1 (3 a_4 + 2 a_5)}{6 a_3 (a_4 + a_5)}$,

$c_3 = 8 a_1 c_2^2 - 8 a_3 c_2^3 - \frac{(a_1 c_1 + 2 a_1^2 c_2)}{a_3} + 3 c_1 c_2$, x_0, t_0, c_1, and ϕ_0 are arbitrary real constants.

• *Reference*: [10].

Solution 8. *dark soliton*
(Figure 4.2)

$$\psi(x, t) = \pm\sqrt{-\mu_1} \tanh\{\sqrt{\mu_2} \ [x - x_0 + c_1 (t - t_0)]\}e^{i \ [c_2 (x-x_0)+c_3 (t-t_0)+\phi_0]}, \qquad (4.25)$$

where $\mu_1 = \frac{3 (c_1 + 2 a_1 c_2 - 3 a_3 c_2^2)}{(3 a_4 + 2 a_5)} < 0$, $\mu_2 = \frac{c_1 + 2 a_1 c_2 - 3 a_3 c_2^2}{2 a_3} > 0$, $c_2 = \frac{-3 a_2 a_3 + a_1 (3 a_4 + 2 a_5)}{6 a_3 (a_4 + a_5)}$,

$c_3 = 8 a_1 c_2^2 - 8 a_3 c_2^3 - \frac{(a_1 c_1 + 2 a_1^2 c_2)}{a_3} + 3 c_1 c_2$, x_0, t_0, c_1, and ϕ_0 are arbitrary real constants.

• *Reference*: [10].

Solution 9.

$$\psi(x, t) - \pm\sqrt{-\mu_1} \coth\{\sqrt{\mu_2} \ [x - x_0 + c_1 (t - t_0)]\}e^{i \ [c_2 (x-x_0)+c_3 (t-t_0)+\phi_0]}, \qquad (4.26)$$

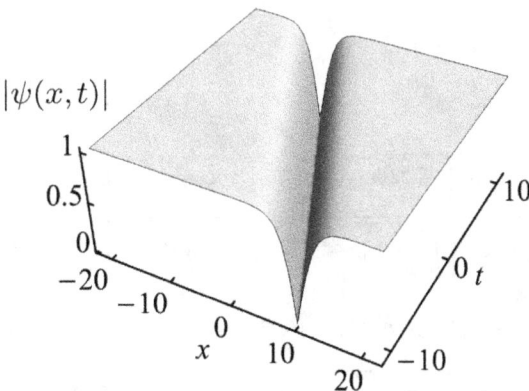

Figure 4.2. Dark soliton (4.25) with $a_1 = -4$, $a_2 = 1$, $a_3 = -1$, $a_4 = -1$, $a_5 = 2$, $c_1 = 9/10$, and $x_0 = t_0 = \phi_0 = 0$.

where $\mu_1 = \frac{3\,(c_1 + 2\,a_1\,c_2 - 3\,a_3\,c_2^2)}{(3\,a_4 + 2\,a_5)} < 0$, $\mu_2 = \frac{c_1 + 2\,a_1\,c_2 - 3\,a_3\,c_2^2}{2\,a_3} > 0$, $c_2 = \frac{-3\,a_2\,a_3 + a_1\,(3\,a_4 + 2\,a_5)}{6\,a_3\,(a_4 + a_5)}$,

$c_3 = 8\,a_1\,c_2^2 - 8\,a_3\,c_2^3 - \frac{(a_1\,c_1 + 2\,a_1^2\,c_2)}{a_3} + 3\,c_1\,c_2$. x_0, t_0, c_1, and ϕ_0 are arbitrary real constants.

- *Reference*: [10].

Solution 10. **Rational Solution**   *decaying wave (DW)*

$$\psi(x, t) = \pm \left\{ \frac{\sqrt{-6\,a_3}}{\sqrt{3\,a_4 + 2\,a_5}\,[x - x_0 + (3\,a_3\,c_2^2 - 2\,a_1\,c_2)\,(t - t_0) + c_1]} \right\} e^{i\,[c_2\,(x - x_0) + c_3\,(t - t_0) + \phi_0]}, \quad (4.27)$$

where $c_2 = \frac{a_1(3\,a_4 + 2\,a_5) - 3\,a_2\,a_3}{6\,a_3\,(a_4 + a_5)}$, $c_3 = a_3\,c_2^3 - a_1\,c_2^2$, $a_3 < 0$, x_0, t_0, c_1, and ϕ_0 are arbitrary real constants.

- *Reference*: [10].

Solution 11.

$$\psi(x, t) = p_1 \sqrt{\frac{3\,\mu_1}{\mu_2}} \left(\frac{\cos\left\{ \sqrt{\frac{-2\,\mu_1}{a_3}}\,[x - x_0 - c_1\,(t - t_0)] \right\}}{p_2 \sin\left\{ \sqrt{\frac{-2\,\mu_1}{a_3}}\,[x - x_0 - c_1\,(t - t_0)] \right\} + 1} \right) e^{i\,[c_2\,(x - x_0) + c_3\,(t - t_0) + \phi_0]}, \quad (4.28)$$

where $\mu_1 = 2\,a_1\,c_2 - c_1 - 3\,a_3\,c_2^2$, $\mu_2 = 3\,a_4 + 2\,a_5$, $c_2 = \frac{-3\,a_2\,a_3 + a_1\,(3\,a_4 + 2\,a_5)}{6\,a_3\,(a_4 + a_5)}$,

$c_3 = 8\,a_1\,c_2^2 - 8\,a_3\,c_2^3 + \frac{(a_1\,c_1 - 2\,a_1^2\,c_2)}{a_3} - 3\,c_1\,c_2$, $(\mu_1\,a_3) < 0$, $(\mu_1\,\mu_2) > 0$, $p_1 = \pm 1$, $p_2 = \pm 1$, independent from p_1, x_0, t_0, c_1, and ϕ_0 are arbitrary real constants.

This solution can be written in terms of sec *and* tan *as:*

$$\psi(x, t) = p_1 \sqrt{\frac{3\,\mu_1}{\mu_2}} \left(\sec\left\{ \sqrt{\frac{-2\,\mu_1}{a_3}}\,[x - x_0 - c_1\,(t - t_0)] \right\} - \tan\left\{ \sqrt{\frac{-2\,\mu_1}{a_3}}\,[x - x_0 - c_1\,(t - t_0)] \right\} \right)$$
$$\times\, e^{i\,[c_2\,(x - x_0) + c_3\,(t - t_0) + \phi_0]}.$$

- *Reference*: [11], *corrected.*

Solution 12.

$$\psi(x, t) = p_1 \sqrt{\frac{3\,\mu_1}{\mu_2}} \left(\frac{\sin\left\{ \sqrt{\frac{-2\,\mu_1}{a_3}}\,[x - x_0 - c_1\,(t - t_0)] \right\}}{p_2 \cos\left\{ \sqrt{\frac{-2\,\mu_1}{a_3}}\,[x - x_0 - c_1\,(t - t_0)] \right\} + 1} \right) e^{i\,[c_2\,(x - x_0) + c_3\,(t - t_0) + \phi_0]}, \quad (4.29)$$

where $\mu_1 = 2\,a_1\,c_2 - c_1 - 3\,a_3\,c_2^2$, $\mu_2 = 3\,a_4 + 2\,a_5$, $c_2 = \frac{-3\,a_2\,a_3 + a_1\,(3\,a_4 + 2\,a_5)}{6\,a_3\,(a_4 + a_5)}$,

$c_3 = 8\,a_1\,c_2^2 - 8\,a_3\,c_2^3 + \frac{(a_1\,c_1 - 2\,a_1^2\,c_2)}{a_3} - 3\,c_1\,c_2$, $(\mu_1\,a_3) < 0$, $(\mu_1\,\mu_2) > 0$, $p_1 = \pm 1$, $p_2 = \pm 1$, independent from p_1, x_0, t_0, c_1, and ϕ_0 are arbitrary real constants.

This solution can be written in terms of csc and cot as:

$$\psi(x, t) = p_1 \sqrt{\frac{3\,\mu_1}{\mu_2}} \left(\csc\left\{ \sqrt{\frac{-2\,\mu_1}{a_3}}\,[x - x_0 - c_1\,(t - t_0)] \right\} - \cot\left\{ \sqrt{\frac{-2\,\mu_1}{a_3}}\,[x - x_0 - c_1\,(t - t_0)] \right\} \right)$$

$$\times\, e^{i\,[c_2\,(x - x_0) + c_3\,(t - t_0) + \phi_0]}.$$

- *Reference*: [11], *corrected*.

Solution 13.

$$\psi(x, t) = \pm \sqrt{\frac{\mu_1}{\mu_2}} \left(\frac{\cot\{\sqrt{-2\,\mu_1}\,[x - x_0 - c_1\,(t - t_0)]\}}{\csc\{\sqrt{-2\,\mu_1}\,[x - x_0 - c_1\,(t - t_0)]\} + 1} \right) e^{i\,[c_2\,(x - x_0) + c_3\,(t - t_0) + \phi_0]}, \qquad (4.30)$$

where $\mu_1 = \dfrac{-c_1 + 2\,a_1\,c_2 - 3\,a_3\,c_2^2}{a_3} < 0$, $\mu_2 = \dfrac{3\,a_4 + 2\,a_5}{3\,a_3} < 0$, $c_2 = \dfrac{-3\,a_2\,a_3 + a_1\,(3\,a_4 + 2\,a_5)}{6\,a_3\,(a_4 + a_5)}$,

$c_3 = 8\,a_1\,c_2^2 - 8\,a_3\,c_2^3 + \dfrac{(a_1\,c_1 - 2\,a_1^2\,c_2)}{a_3} - 3\,c_1\,c_2$, x_0, t_0, c_1, and ϕ_0 are arbitrary real constants.

- *Reference*: [10].

Solution 14.

$$\psi(x, t) = \pm \left(\frac{\sqrt{-3}\,\sec\{\sqrt{2\,\mu_1}\,[x - x_0 - c_1\,(t - t_0)]\} + \tan\{\sqrt{2\,\mu_1}\,[x - x_0 - c_1\,(t - t_0)]\}}{2\,\sec\{\sqrt{2\,\mu_1}\,[x - x_0 - c_1\,(t - t_0)]\} + 1} \right) \qquad (4.31)$$

$$\times\, \sqrt{\frac{\mu_1}{\mu_2}}\, e^{i\,[c_2\,(x - x_0) + c_3\,(t - t_0) + \phi_0]},$$

where $\mu_1 = \dfrac{c_1 - 2\,a_1\,c_2 + 3\,a_3\,c_2^2}{a_3} > 0$, $\mu_2 = \dfrac{-3\,a_4 - 2\,a_5}{3\,a_3} > 0$, $c_2 = -\dfrac{3\,a_2\,a_3 - a_1\,(3\,a_4 + 2\,a_5)}{6\,a_3\,(a_4 + a_5)}$,

$c_3 = 8\,a_1\,c_2^2 - 8\,a_3\,c_2^3 + \dfrac{(a_1\,c_1 - 2\,a_1^2\,c_2)}{a_3} - 3\,c_1\,c_2$, x_0, t_0, c_1, and ϕ_0 are arbitrary real constants.

- *Reference*: [10].

Solution 15.

$$\psi(x, t) = \pm \left(\frac{\sqrt{5}\,\operatorname{csch}\{\sqrt{2\,\mu_1}\,[x - x_0 - c_1\,(t - t_0)]\} + \coth\{\sqrt{2\,\mu_1}\,[x - x_0 - c_1\,(t - t_0)]\}}{2\,\operatorname{csch}\{\sqrt{2\,\mu_1}\,[x - x_0 - c_1\,(t - t_0)]\} + 1} \right) \qquad (4.32)$$

$$\times\, \sqrt{-\frac{\mu_1}{\mu_2}}\, e^{i\,[c_2\,(x - x_0) + c_3\,(t - t_0) + \phi_0]},$$

where $\mu_1 = \dfrac{-c_1 + 2\,a_1\,c_2 - 3\,a_3\,c_2^2}{a_3} > 0$, $\mu_2 = \dfrac{3\,a_4 + 2\,a_5}{3\,a_3} < 0$, $c_2 = \dfrac{a_1\,(3\,a_4 + 2\,a_5) - 3\,a_2\,a_3}{6\,a_3\,(a_4 + a_5)}$,

$c_3 = 8\,a_1\,c_2^2 - 8\,a_3\,c_2^3 + \dfrac{(a_1\,c_1 - 2\,a_1^2\,c_2)}{a_3} - 3\,c_1\,c_2$, x_0, t_0, c_1, and ϕ_0 are arbitrary real constants.

- *Reference*: [10].

4.3 Summary of Section 4.2

Equation

$$i \psi_t + a_1 \psi_{xx} + a_2 |\psi|^2 \psi + i a_3 \psi_{xxx} + i a_4 (|\psi|^2 \psi)_x + i a_5 (|\psi|^2)_x \psi = 0$$

#	Solution	Conditions	Name	Equation #
1.	$\psi(x,t) = c\, e^{i[c(x-x_0)+c^2(-a_1+c a_3+a_2-c a_4)(t-t_0)+\phi_0]}$	$x_0, t_0, c,$ and ϕ_0 are arbitrary real constants	Continuous wave, t- and x-dependent phase	(4.18)
2.	$\psi(x,t) = \pm\sqrt{-\mu_1}\,\sec\{\sqrt{\mu_2}\,[x-x_0+c_1(t-t_0)]\}e^{i[c_2(x-x_0)+c_3(t-t_0)+\phi_0]}$	$\mu_1 = \dfrac{6(c_1+2c_2 c_3 - 3 a_3 c_2^2)}{(3a_4+2a_5)} < 0,$ $\mu_2 = \dfrac{c_1+2c_2 c_3 - 3 a_3 c_2^2}{-3 a_2 a_3 + a_1(3 a_4+2 a_5)} > 0,$ $c_2 = \dfrac{-3a_2 a_3 + a_1(3 a_4+2 a_5)}{6 a_3(a_4+a_5)},$ $c_3 = 8 a_1 c_2^2 - 8 a_3 c_2^3$ $- \dfrac{(c_1+2a_1^2 c_2)}{a_3} + 3 c_1 c_2,$ $x_0, t_0, c_1,$ and ϕ_0 are arbitrary real constants	—	(4.19)
3.	$\psi(x,t) = \pm\sqrt{-\mu_1}\,\csc\{\sqrt{\mu_2}\,[x-x_0+c_1(t-t_0)]\}e^{i[c_2(x-x_0)+c_3(t-t_0)+\phi_0]}$	$\mu_1 = \dfrac{6(c_1+2c_2 c_3 - 3 a_3 c_2^2)}{(3a_4+2a_5)} < 0,$ $\mu_2 = \dfrac{c_1+2c_2 c_3 - 3 a_3 c_2^2}{-3 a_2 a_3 + a_1(3 a_4+2 a_5)} > 0,$ $c_2 = \dfrac{-3a_2 a_3 + a_1(3 a_4+2 a_5)}{6 a_3(a_4+a_5)},$ $c_3 = 8 a_1 c_2^2 - 8 a_3 c_2^3$ $- \dfrac{(c_1+2a_1^2 c_2)}{a_3} + 3 c_1 c_2,$ $x_0, t_0, c_1,$ and ϕ_0 are arbitrary real constants	—	(4.20)
4.	$\psi(x,t) = \pm\sqrt{\mu_1}\,\tan\{\sqrt{-\mu_2}\,[x-x_0+c_1(t-t_0)]\}e^{i[c_2(x-x_0)+c_3(t-t_0)+\phi_0]}$	$\mu_1 = \dfrac{3(c_1+2c_2 c_3 - 3 a_3 c_2^2)}{(3a_4+2a_5)} > 0,$ $\mu_2 = \dfrac{c_1+2c_2 c_3 - 3 a_3 c_2^2}{-3 a_2 a_3 + a_1(3 a_4+2 a_5)} < 0,$ $c_2 = \dfrac{-3a_2 a_3 + a_1(3 a_4+2 a_5)}{6 a_3(a_4+a_5)},$ $c_3 = 8 a_1 c_2^2 - 8 a_3 c_2^3$ $- \dfrac{(c_1+2a_1^2 c_2)}{a_3} + 3 c_1 c_2,$ $x_0, t_0, c_1,$ and ϕ_0 are arbitrary real constants	—	(4.21)

(Continued)

(Continued)

Equation

$$i\psi_t + a_1\psi_{xx} + a_2|\psi|^2\psi + i a_3\psi_{xxx} + i a_4(|\psi|^2\psi)_x + i a_5(|\psi|^2)_x\psi = 0$$

#	Solution	Conditions	Name	Equation #
5.	$\psi(x,t) = \pm\sqrt{\mu_1}\cot\{\sqrt{-\mu_2}\,[x - x_0 + c_1(t - t_0)]\}e^{i[c_2(x-x_0)+c_3(t-t_0)+\phi_0]}$	$\mu_1 = \dfrac{3(q+2qc_2-3a_3c_2^2)}{(3a_4+2a_5)} > 0,$ $\mu_2 = \dfrac{q+2qc_2-3a_3c_2^2}{-3a_2a_3+q(3a_4+2a_5)} < 0,$ $c_2 = \dfrac{2a_3}{6a_3(a_4+a_5)},$ $c_3 = 8a_1c_2^2 - 8a_3c_2^3$ $\quad -\dfrac{a_3}{a_1}(q+2a_1^2c_2) + 3c_1c_2,$ $x_0, t_0, c_1, \text{ and } \phi_0$ are arbitrary real constants	—	(4.22)
6.	$\psi(x,t) = \pm\sqrt{-\mu_1}\,\mathrm{sech}\{\sqrt{-\mu_2}\,[x - x_0 + c_1(t - t_0)]\}e^{i[c_2(x-x_0)+c_3(t-t_0)+\phi_0]}$	$\mu_1 = \dfrac{6(q+2qc_2-3a_3c_2^2)}{(3a_4+2a_5)} < 0,$ $\mu_2 = \dfrac{q+2qc_2-3a_3c_2^2}{-3a_2a_3+q(3a_4+2a_5)} < 0,$ $c_2 = \dfrac{a_3}{6a_3(a_4+a_5)},$ $c_3 = 8a_1c_2^2 - 8a_3c_2^3$ $\quad -\dfrac{a_3}{a_1}(q+2a_1^2c_2) + 3c_1c_2,$ $x_0, t_0, c_1, \text{ and } \phi_0$ are arbitrary real constants	Bright soliton	(4.23)
7.	$\psi(x,t) = \pm\sqrt{\mu_1}\,\mathrm{csch}\{\sqrt{-\mu_2}\,[x - x_0 + c_1(t - t_0)]\}e^{i[c_2(x-x_0)+c_3(t-t_0)+\phi_0]}$	$\mu_1 = \dfrac{6(q+2qc_2-3a_3c_2^2)}{(3a_4+2a_5)} > 0,$ $\mu_2 = \dfrac{q+2qc_2-3a_3c_2^2}{-3a_2a_3+q(3a_4+2a_5)} < 0,$ $c_2 = \dfrac{a_3}{6a_3(a_4+a_5)},$ $c_3 = 8a_1c_2^2 - 8a_3c_2^3$ $\quad -\dfrac{a_3}{a_1}(q+2a_1^2c_2) + 3c_1c_2,$ $x_0, t_0, c_1, \text{ and } \phi_0$ are arbitrary real constants	—	(4.24)
8.	$\psi(x,t) = \pm\sqrt{-\mu_1}\tanh\{\sqrt{\mu_2}\,[x - x_0 + c_1(t - t_0)]\}e^{i[c_2(x-x_0)+c_3(t-t_0)+\phi_0]}$	$\mu_1 = \dfrac{3(q+2qc_2-3a_3c_2^2)}{(3a_4+2a_5)} < 0,$ $\mu_2 = \dfrac{q+2qc_2-3a_3c_2^2}{-3a_2a_3+q(3a_4+2a_5)} > 0,$ $c_2 = \dfrac{2a_3}{6a_3(a_4+a_5)},$	Dark soliton	(4.25)

$$c_3 = 8 a_1 c_2^2 - 8 a_3 c_2^3$$
$$- \frac{(q_1 q + 2 a_1^2 c_2)}{a_3} + 3 c_1 c_2,$$

$x_0, t_0, c_1,$ and ϕ_0 are arbitrary real constants

9. $\psi(x, t) = \pm \sqrt{-\mu_1} \coth\{\sqrt{\mu_2} [x - x_0 + c_1 (t - t_0)]\} e^{i [c_2 (x-x_0)+c_3 (t-t_0)+\phi_0]}$

$\mu_1 = \dfrac{3 (q + 2 q c_2 - 3 a_3 c_2^2)}{(3 a_4 + 2 a_5)} < 0,$

$\mu_2 = \dfrac{q + 2 q c_2 - 3 a_3 c_2^2}{2 a_3} > 0,$

$c_2 = \dfrac{-3 a_2 a_3 + q (3 a_4 + 2 a_5)}{6 a_3 (a_4 + a_5)},$

$c_3 = 8 a_1 c_2^2 - 8 a_3 c_2^3$
$- \dfrac{(q_1 q + 2 a_1^2 c_2)}{a_3} + 3 c_1 c_2,$

$x_0, t_0, c_1,$ and ϕ_0 are arbitrary real constants

(4.26)

—

10. $\psi(x, t) = \pm \left\{ \dfrac{\sqrt{-6 a_3}}{\sqrt{3 a_4 + 2 a_5}\, [x - x_0 + (3 a_3 c_2^2 - 2 q c_2)(t-t_0) + q]} \right\} e^{i [c_2 (x-x_0)+c_3 (t-t_0)+\phi_0]}$

$c_2 = \dfrac{q(3 a_4 + 2 a_5) - 3 a_2 a_3}{6 a_3 (a_4 + a_5)},$ $c_3 = a_3 c_2^3 - a_1 c_2^2,$

$a_3 < 0,$ $x_0, t_0, c_1,$ and ϕ_0 are arbitrary real constants

(4.27)

Decaying wave

11. $\psi(x, t) = p_1 \sqrt{\dfrac{3 \mu_1}{\mu_2}} \left(\dfrac{\cos[\sqrt{\frac{-2\mu_1}{c_3}}\, [x - x_0 - q(t-t_0)]]}{p_2 \sin[\sqrt{\frac{-2\mu_1}{a_3}}\, [x - x_0 - q(t-t_0)]] + 1} \right) e^{i [c_2 (x-x_0)+c_3 (t-t_0)+\phi_0]}$

$\mu_1 = \dfrac{2 a_1 c_2 - c_1 - 3 a_3 c_2^2}{-3 a_2 a_3 + q (3 a_4 + 2 a_5)}, \mu_2 = 3 a_4 + 2 a_5,$

$c_2 = \dfrac{(q_1 q - 2 a_1^2 c_2)}{6 a_3 (a_4 + a_5)} +$

$c_3 = 8 a_1 c_2^2 - 8 a_3 c_2^3 + \dfrac{(q_1 q - 2 a_1^2 c_2)}{a_3} - 3 c_1 c_2,$

$(\mu_1 a_3) < 0, (\mu_1 \mu_2) > 0, p_1 = \pm 1, p_2 = \pm 1,$
independent from $p_1, x_0, t_0, c_1,$ and ϕ_0 are arbitrary real constants

(4.28)

—

12. $\psi(x, t) = p_1 \sqrt{\dfrac{3 \mu_1}{\mu_2}} \left(\dfrac{\sin[\sqrt{\frac{-2\mu_1}{c_3}}\, [x - x_0 - q(t-t_0)]]}{p_2 \cos[\sqrt{\frac{-2\mu_1}{a_3}}\, [x - x_0 - q(t-t_0)]] + 1} \right) e^{i [c_2 (x-x_0)+c_3 (t-t_0)+\phi_0]}$

$\mu_1 = \dfrac{2 a_1 c_2 - c_1 - 3 a_3 c_2^2}{-3 a_2 a_3 + q (3 a_4 + 2 a_5)}, \mu_2 = 3 a_4 + 2 a_5,$

$c_2 = \dfrac{(q_1 q - 2 a_1^2 c_2)}{6 a_3 (a_4 + a_5)} +$

$c_3 = 8 a_1 c_2^2 - 8 a_3 c_2^3 + \dfrac{(q_1 q - 2 a_1^2 c_2)}{a_3} - 3 c_1 c_2,$

$(\mu_1 a_3) < 0, (\mu_1 \mu_2) > 0, p_1 = \pm 1, p_2 = \pm 1,$
independent from $p_1, x_0, t_0, c_1,$ and ϕ_0 are arbitrary real constants

(4.29)

—

(*Continued*)

(Continued)

Equation

$$i\,\psi_t + a_1\,\psi_{xx} + a_2\,|\psi|^2\,\psi + i\,a_3\,\psi_{xxx} + i\,a_4\,(|\psi|^2\,\psi)_x + i\,a_5\,(|\psi|^2)_x\,\psi = 0$$

#	Solution	Conditions	Name	Equation #
13.	$\psi(x,t) = \pm\sqrt{\dfrac{\mu_1}{\mu_2}}\left(\dfrac{\cot\{\sqrt{-2\mu_1}\,[x-x_0-q(t-t_0)]\}}{\sec\{\sqrt{-2\mu_1}\,[x-x_0-q(t-t_0)]\}+1}\right) e^{i\,[c_2(x-x_0)+c_3(t-t_0)+\phi_0]}$	$\mu_1 = \dfrac{-q+2q\,c_2-3a_3\,c_2^2}{-3a_2\,a_3+q\,(3a_4+2a_5)} < 0,\ \mu_2 = \dfrac{3a_4+2a_5}{3a_3} < 0,$ $c_2 = \dfrac{a_3}{6a_3(a_4+a_5)},$ $c_3 = 8a_1\,c_2^2 - 8a_3\,c_2^3$ $+\dfrac{(q\,q-2a_1^2\,c_2)}{a_3} - 3c_1\,c_2,$ $x_0,\ t_0,\ c_1,$ and ϕ_0 are arbitrary real constants	—	(4.30)
14.	$\psi(x,t) = \pm\left(\dfrac{\sqrt{-3}\,\sec\{\sqrt{2\mu_1}\,[x-x_0-q(t-t_0)]\}+\tan\{\sqrt{2\mu_1}\,[x-x_0-q(t-t_0)]\}}{2\sec\{\sqrt{2\mu_1}\,[x-x_0-q(t-t_0)]\}+1}\right)$ $\times \sqrt{\dfrac{\mu_1}{\mu_2}}\, e^{i\,[c_2(x-x_0)+c_3(t-t_0)+\phi_0]}$	$\mu_1 = \dfrac{q-2q\,c_2+3a_3\,c_2^2}{3a_2\,a_3-q\,(3a_4+2a_5)} > 0,\ \mu_2 = \dfrac{-3a_4-2a_5}{3a_3} > 0,$ $c_2 = -\dfrac{a_3}{6a_3(a_4+a_5)},$ $c_3 = 8a_1\,c_2^2 - 8a_3\,c_2^3$ $+\dfrac{(q\,q-2a_1^2\,c_2)}{a_3} - 3c_1\,c_2,$ $x_0,\ t_0,\ c_1,$ and ϕ_0 are arbitrary real constants	—	(4.31)
15.	$\psi(x,t) = \pm\sqrt{-\dfrac{\mu_1}{\mu_2}}\, e^{i\,[c_2(x-x_0)+c_3(t-t_0)+\phi_0]}$ $\times\left(\dfrac{\sqrt{5}\,\operatorname{csch}\{\sqrt{2\mu_1}\,[x-x_0-q(t-t_0)]\}+\coth\{\sqrt{2\mu_1}\,[x-x_0-q(t-t_0)]\}}{2\operatorname{csch}\{\sqrt{2\mu_1}\,[x-x_0-q(t-t_0)]\}+1}\right)$	$\mu_1 = \dfrac{-q+2q\,c_2-3a_3\,c_2^2}{q\,(3a_4+2a_5)-3a_2\,a_3} > 0,\ \mu_2 = \dfrac{3a_4+2a_5}{3a_3} < 0,$ $c_2 = \dfrac{q\,(3a_4+2a_5)-3a_2\,a_3}{6a_3(a_4+a_5)},$ $c_3 = 8a_1\,c_2^2 - 8a_3\,c_2^3$ $+\dfrac{(q\,q-2a_1^2\,c_2)}{a_3} - 3c_1\,c_2,$ $x_0,\ t_0,\ c_1,$ and ϕ_0 are arbitrary real constants	—	(4.32)

4.4 Special Cases of Equation (4.17)

4.4.1 Case I: Hirota Equation (HE)

$$i\,\psi_t + a_1\,\psi_{xx} + a_2\,|\psi|^2\,\psi + i\,a_3\,\psi_{xxx} + i\,a_4\,|\psi|^2\,\psi_x = 0. \qquad (4.33)$$

Solutions to (4.33) can be obtained from solutions to (4.2) for $a_5 = -a_4$.

4.4.2 Case II: Sasa–Satsuma Equation (SSE)

$$i\,\psi_t + a_1\,\psi_{xx} + a_2\,|\psi|^2\,\psi + i\,a_4\,[\psi_{xxx} + (|\psi|^2)_x\,\psi + |\psi|^2\,\psi_x] = 0. \qquad (4.34)$$

Solutions to (4.34) can be obtained from solutions to (4.2) for $a_4 = a_3$ and $a_5 = 0$.

4.5 NLSE with First and Third Order Dispersions, Self-Steepening, Self-Frequency Shift, and Potential

Equation:

$$i\,\psi_t + i\,a_1\,\psi_x + a_2\,\psi_{xx} - i\,a_3\,\psi_{xxx} + a_4\,|\psi|^2\,\psi - i\,a_5\,|\psi|^2\,\psi_x - i\,a_6\,\psi^2\,\psi^*_x - a_7\,\psi = 0, \quad (4.35)$$

where $\psi = \psi(x,\,t)$ is the complex function profile, x and t are its two independent variables, a_j are real constants, $j = 1, 2, \ldots, 7$.
Solutions:

Solution 1. **Constant Amplitude**   *CW, t- and x-dependent phase*

$$\psi(x,\,t) = A_0\,e^{i\,[A_1\,(x-x_0)+A_2\,(t-t_0)+\phi_0]}, \qquad (4.36)$$

where $A_2 = A_0^2\,[a_4 + A_1\,(a_5 - a_6)] - A_1\,[a_1 + A_1\,(a_2 + a_3\,A_1)] - a_7$, x_0, t_0, A_0, A_1, ϕ_0, and a_j are arbitrary real constants, $j = 1, 2, \ldots, 7$.

Solution 2.

$$\psi(x,\,t) = \lambda\,\tanh\{\eta\,[x - x_0 - \chi\,(t - t_0)]\} + i\,\rho\,\text{sech}\{\eta\,[x - x_0 - \chi\,(t - t_0)]\}, \qquad (4.37)$$

where $a_1 = -2\,\alpha_1\,\Omega + 3\,a_3\,\Omega^2$, $a_2 = \alpha_1 - 3\,a_3\,\Omega$, $a_3 = \frac{\alpha_1\,\alpha_4}{3\,\alpha_2}$, $a_4 = \alpha_2 - \alpha_3\,\Omega$,

$a_5 = 2\,\alpha_3 + \alpha_4$, $a_6 = \alpha_3 + \alpha_4$, $a_7 = \kappa + \alpha_1\,\Omega^2 - a_3\,\Omega^3$, $\alpha_4 = -\frac{\alpha_3}{2}$, $\kappa = -\frac{2\,\alpha_1\,\alpha_2^2}{3\,\alpha_3^2}$,

$\Omega = \frac{\alpha_2}{\alpha_3}$, $\chi = -(\alpha_1\,\Omega + \alpha_3\,\lambda^2) - a_3\,\eta^2$, $\eta = \sqrt{\frac{\alpha_3}{3\,a_3}\,(\rho^2 - \lambda^2)}$, $\frac{\alpha_2}{\alpha_1}\,(\rho^2 - \lambda^2) > 0$, x_0,

t_0, α_1, α_2, α_3, ρ, and λ are arbitrary real constants.
   • *Reference*: [12].

Solution 3.

$$\psi(x,\,t) = i\,\beta + \lambda\,\tanh\{\eta\,[x - x_0 - \chi\,(t - t_0)]\} + i\,\lambda\,\text{sech}\{\eta\,[x - x_0 - \chi\,(t - t_0)]\}, \qquad (4.38)$$

where $a_1 = -2\,\alpha_1\,\Omega + 3\,a_3\,\Omega^2$, $a_2 = \alpha_1 - 3\,a_3\,\Omega$, $a_3 = 0$, $a_4 = \alpha_2 - \alpha_3\,\Omega$,
$a_5 = 2\,\alpha_3 + \alpha_4$, $a_6 = \alpha_3 + \alpha_4$, $a_7 = \kappa + \alpha_1\,\Omega^2 - a_3\,\Omega^3$, $\alpha_4 = -\alpha_3$,

$$\kappa = (\alpha_2 - \alpha_3 \, \Omega)(\lambda^2 + \beta^2) - \alpha_1 \, \Omega^2, \qquad \chi = -(2 \, \alpha_1 \, \Omega + \alpha_3 \, \beta^2), \qquad \eta = -\frac{\alpha_3 \, \beta \, \lambda}{\alpha_1},$$

$\lambda = \frac{\sqrt{2 \, \alpha_1 \, (\alpha_3 \, \Omega - \alpha_2)}}{\alpha_3}$, $\alpha_1 \, (\alpha_3 \, \Omega - \alpha_2) > 0$, x_0, t_0, α_1, α_2, α_3, β, ρ, and Ω are arbitrary real constants.

- *Reference*: [12].

Solution 4.

(Figure 4.3)

$$\psi(x, \, t) = i \, \beta + \lambda \, \tanh[\eta \, (x - x_0)] + i \, \rho \, \text{sech}[\eta \, (x - x_0)], \qquad (4.39)$$

where $a_1 = -2 \, \alpha_1 \, \Omega + 3 \, a_3 \, \Omega^2$, $a_2 = \alpha_1 - 3 \, a_3 \, \Omega$, $a_3 = 0$, $a_4 = \alpha_2 - \alpha_3 \, \Omega$, $a_5 = 2 \, \alpha_3 + \alpha_4$, $a_6 = \alpha_3 + \alpha_4$, $a_7 = \kappa + \alpha_1 \, \Omega^2 - a_3 \, \Omega^3$, $\alpha_1 = 0$, $\alpha_4 = -\frac{3 \, \alpha_3}{2}$, $\kappa = (\alpha_2 - \alpha_3 \, \Omega)(\lambda^2 + \beta^2)$, $\eta = -\frac{\beta \, (\alpha_2 - \alpha_3 \, \Omega)}{\lambda \, (\alpha_3 + \alpha_4)}$, $\rho = \sqrt{\lambda^2 + 2 \, \beta^2}$, x_0, t_0, α_2, α_3, β, λ, and Ω are arbitrary real constants.

- *Reference*: [12].

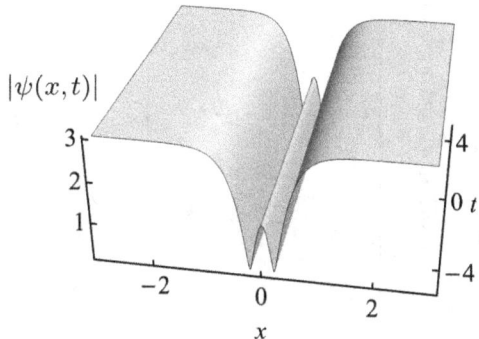

Figure 4.3. Plot of solution (4.39) with $\alpha_2 = 1$, $\alpha_4 = 3$, $\Omega = \lambda = 1/5$, $\beta = -3$, and $x_0 = t_0 = 0$.

4.6 Summary of Section 4.5

Equation

$$i\psi_t + i a_1 \psi_x + a_2 \psi_{xx} - i a_3 \psi_{xxx} + a_4 |\psi|^2 \psi - i a_5 |\psi|^2 \psi_x - i a_6 \psi^2 \psi_x^* - a_7 \psi = 0$$

#	Solution	Conditions	Name	Equation #
1.	$\psi(x,t) = A_0\, e^{i[A_1(x-x_0)+A_2(t-t_0)+\phi_0]}$	$A_2 = A_0^2[a_4 + A_1(a_5-a_6)] - A_1[a_1 + A_1(a_2+a_3 A_1)] - a_7$, $x_0, t_0, A_0, A_1, \phi_0$, and a_j are arbitrary real constants, $j = 1, 2, \ldots, 7$	Continuous wave, t- and x-dependent phase	(4.36)
2.	$\psi(x,t) = \lambda \tanh\{\eta[x - x_0 - \chi(t - t_0)]\}$ $+ i\rho\, \mathrm{sech}\{\eta[x - x_0 - \chi(t - t_0)]\}$	$a_1 = -2\alpha_1\Omega + 3 a_3\Omega^2$, $a_2 = \alpha_1 - 3 a_3\Omega$, $a_3 = \dfrac{q\alpha_4}{3\alpha_2}$, $\dfrac{\alpha_2}{q}(\rho^2-\lambda^2) > 0$ $\quad a_4 = \alpha_2 - \alpha_3\Omega$, $a_5 = 2\alpha_3 + \alpha_4$, $a_6 = \alpha_3 + \alpha_4$, $\alpha_4 = -\dfrac{\alpha_3}{2}$, $a_7 = \kappa + \alpha_1\Omega^2 - \alpha_3\Omega^3$, $\kappa = -\dfrac{2 q \alpha_2^2}{3\alpha_3^2}$, $\Omega = \dfrac{\alpha_2}{\alpha_3}$, $\chi = -(\alpha_1 + \alpha_3\lambda^2) - \alpha_3\eta^2$, $\eta = \sqrt{\dfrac{\alpha_3}{3\alpha_3}(\rho^2 - \lambda^2)}$, $x_0, t_0, \alpha_1, \alpha_2, \alpha_3, \rho$, and λ are arbitrary real constants	—	(4.37)
3.	$\psi(x,t) = i\beta + \lambda \tanh\{\eta[x - x_0 - \chi(t - t_0)]\}$ $+ i\lambda\, \mathrm{sech}\{\eta[x - x_0 - \chi(t - t_0)]\}$	$a_1 = -2\alpha_1\Omega + 3 a_3\Omega^2$, $a_2 = \alpha_1 - 3 a_3\Omega$, $a_3 = 0$, $a_4 = \alpha_2 - \alpha_3\Omega$, $a_5 = 2\alpha_3 + \alpha_4$, $a_6 = \alpha_3 + \alpha_4$, $\alpha_4 = -\alpha_3$, $a_7 = \kappa + \alpha_1\Omega^2 - \alpha_3\Omega^3$, $\kappa = (\alpha_2 - \alpha_3\Omega)(\lambda^2 + \beta^2) - \alpha_1\Omega^2$, $\chi = -(2\alpha_1\Omega + \alpha_3\beta^2)$, $\eta = -\dfrac{\alpha_3\lambda}{q}$, $\lambda = \dfrac{\sqrt{2 q(\alpha_3\Omega - \alpha_2)}}{\alpha_3}$, $\alpha_1(\alpha_3\Omega - \alpha_2) > 0$, $x_0, t_0, \alpha_1, \alpha_2, \alpha_3, \beta, \rho$, and Ω are arbitrary real constants	—	(4.38)
4.	$\psi(x,t) = i\beta + \lambda \tanh[\eta(x - x_0)] + i\rho\, \mathrm{sech}[\eta(x - x_0)]$	$a_1 = -2\alpha_1\Omega + 3 a_3\Omega^2$, $a_2 = \alpha_1 - 3 a_3\Omega$, $a_3 = 0$, $a_4 = \alpha_2 - \alpha_3\Omega$, $a_5 = 2\alpha_3 + \alpha_4$, $a_6 = \alpha_3 + \alpha_4$, $\alpha_1 = 0$, $a_4 = -\dfrac{\alpha_3}{3\alpha_3}$, $a_7 = \kappa + \alpha_1\Omega^2 - \alpha_3\Omega^3$, $\kappa = (\alpha_2 - \alpha_3\Omega)(\lambda^2 + \beta^2)$, $\eta = -\dfrac{\beta(\alpha_2 - \alpha_3\Omega)}{\lambda(\alpha_3 + \alpha_4)}$, $\rho = \sqrt{\lambda^2 + 2\beta^2}$, $x_0, t_0, \alpha_2, \alpha_3, \beta, \lambda$, and Ω are arbitrary real constants	—	(4.39)

4.7 NLSE with t-dependent Coefficients and First Order Dispersion

Equation:

$$i\,\psi_t + i\,a_1(t)\,\psi_x + a_2(t)\,\psi_{xx} + a_3(t)\,|\psi|^2\,\psi = 0, \qquad (4.40)$$

where $\psi = \psi(x, t)$ is the complex function profile, x and t are its two independent variables, $a_j(t)$ are real functions of t, $j = 1, 2, 3$.

Solutions:

***Solution* 1. Constant Amplitude**   *CW, t- and x-dependent phase*

$$\psi(x,\,t) = A_0\,e^{i\,[A_1\,(x-x_0)+A_2\,(t-t_0)+\phi_0]}, \qquad (4.41)$$

where $A_0 = \pm A_1$, $a_1(t) = -A_2/A_1$, $a_2(t) = a_3(t)$, $a_3(t)$ is an arbitrary real function, x_0, t_0, A_1, A_2, and ϕ_0 are arbitrary constants.

***Solution* 2.**   *bright soliton*

$$\psi(x,\,t) = A_0\,\text{sech}[\omega(x,\,t)]\,e^{i\,\phi(x,\,t)}, \qquad (4.42)$$

where $\omega(x,\,t) = A_0\,x - A_0\int [a_1(t) + 2\,A_1\,a_2(t)]\,dt$,

$\phi(x,\,t) = A_1\,x + \int [(A_0^2 - A_1^2)\,a_2(t) - A_1\,a_1(t)]\,dt$, $a_3(t) = 2\,a_2(t)$, $a_1(t)$ and $a_2(t)$ are arbitrary real functions, A_0 and A_1 are arbitrary real constants.

- *Reference*: [13].

***Solution* 3.**   *dark soliton*

$$\psi(x,\,t) = A_0\,\tanh[\omega(x,\,t)]\,e^{i\,\phi(x,\,t)}, \qquad (4.43)$$

where $\omega(x,\,t) = A_0\,x - A_0\int [a_1(t) + 2\,A_1\,a_2(t)]\,dt$, $\phi(x,\,t) = A_1\,x -$
$\int [(2\,A_0^2 + A_1^2)\,a_2(t) + A_1\,a_1(t)]\,dt$, $a_3(t) = -2\,a_2(t)$, $a_1(t)$ and $a_2(t)$ are arbitrary real functions, A_0 and A_1 are arbitrary real constants.

- *Reference*: [13].

***Solution* 4.**

$$\psi(x,\,t) = A_0\,\text{csch}[\omega(x,\,t)]\,e^{i\,\phi(x,\,t)}, \qquad (4.44)$$

where $\omega(x,\,t) = A_0\,x - A_0\int [a_1(t) + 2\,A_1\,a_2(t)]\,dt$, $\phi(x,\,t) = A_1\,x +$
$\int [(A_0^2 - A_1^2)\,a_2(t) - A_1\,a_1(t)]\,dt$, $a_3(t) = -2\,a_2(t)$, $a_1(t)$ and $a_2(t)$ are arbitrary real functions, A_0 and A_1 are arbitrary real constants.

- *Reference*: [13].

***Solution* 5.**

$$\psi(x,\,t) = A_0\,\coth[\omega(x,\,t)]\,e^{i\,\phi(x,\,t)}, \qquad (4.45)$$

where $\omega(x,\,t) = A_0\,x - A_0\int [a_1(t) + 2\,A_1\,a_2(t)]\,dt$, $\phi(x,\,t) = A_1\,x -$
$\int [(2\,A_0^2 + A_1^2)\,a_2(t) + A_1\,a_1(t)]\,dt$, $a_3(t) = -2\,a_2(t)$, $a_1(t)$ and $a_2(t)$ are arbitrary real functions, A_0 and A_1 are arbitrary real constants.

- *Reference*: [13].

Solution 6.

$$\psi(x, t) = A_0 \sec[\omega(x, t)] \, e^{i\,\phi(x,\,t)}, \qquad (4.46)$$

where $\qquad \omega(x, t) = A_0\, x - A_0 \int [a_1(t) + 2\, A_1\, a_2(t)]\, dt, \qquad \phi(x, t) =$
$A_1\, x - \int [(A_0^2 + A_1^2)\, a_2(t) + A_1\, a_1(t)]\, dt$, $a_3(t) = -2\, a_2(t)$, $a_1(t)$ and $a_2(t)$ are arbitrary real functions, A_0 and A_1 are arbitrary real constants.

• *Reference*: [13].

Solution 7.

$$\psi(x, t) = A_0 \csc[\omega(x, t)] \, e^{i\,\phi(x,\,t)}, \qquad (4.47)$$

where $\qquad\qquad\qquad \omega(x, t) = A_0\, x - A_0 \int [a_1(t) + 2\, A_1\, a_2(t)]\, dt,$
$\phi(x, t) = A_1\, x - \int [(A_0^2 + A_1^2)\, a_2(t) + A_1\, a_1(t)]\, dt$, $a_3(t) = -2\, a_2(t)$, $a_1(t)$ and $a_2(t)$ are arbitrary real functions, A_0 and A_1 are arbitrary real constants.

• *Reference*: [13].

Solution 8.

$$\psi(x, t) = A_0 \tan[\omega(x, t)] \, e^{i\,\phi(x,\,t)}, \qquad (4.48)$$

where $\qquad\qquad\qquad \omega(x, t) = A_0\, x - A_0 \int [a_1(t) + 2\, A_1\, a_2(t)]\, dt,$
$\phi(x, t) = A_1\, x + \int [(2\, A_0^2 - A_1^2)\, a_2(t) - A_1\, a_1(t)]\, dt$, $a_3(t) = -2\, a_2(t)$, $a_1(t)$ and $a_2(t)$ are arbitrary real functions, A_0 and A_1 are arbitrary real constants.

• *Reference*: [13].

Solution 9.

$$\psi(x, t) = A_0 \cot[\omega(x, t)] \, e^{i\,\phi(x,\,t)}, \qquad (4.49)$$

where $\qquad\qquad\qquad \omega(x, t) = A_0\, x - A_0 \int [a_1(t) + 2\, A_1\, a_2(t)]\, dt,$
$\phi(x, t) = A_1\, x + \int [(2\, A_0^2 - A_1^2)\, a_2(t) - A_1\, a_1(t)]\, dt$, $a_3(t) = -2\, a_2(t)$, $a_1(t)$ and $a_2(t)$ are arbitrary real functions, A_0 and A_1 are arbitrary real constants.

• *Obtained by the authors.*

Solution 10.

$$\psi(x, t) = A_0 \left\{ \sec[\omega(x, t)] \pm \tan[\omega(x, t)] \right\} e^{i\,\phi(x,\,t)}, \qquad (4.50)$$

where $\qquad\qquad\qquad \omega(x, t) = A_0\, x - A_0 \int [a_1(t) + 2\, A_1\, a_2(t)]\, dt,$
$\phi(x, t) = A_1\, x + \int [(\frac{1}{2}\, A_0^2 - A_1^2)\, a_2(t) - A_1\, a_1(t)]\, dt$, $a_3(t) = -(1/2)\, a_2(t)$, $a_1(t)$ and $a_2(t)$ are arbitrary real functions, A_0 and A_1 are arbitrary real constants.

• *Reference*: [13].

4.8 Summary of Section 4.7

Equation

$$i\psi_t + i\,a_1(t)\,\psi_x + a_2(t)\,\psi_{xx} + a_3(t)\,|\psi|^2\,\psi = 0$$

#	Solution	Name	Conditions	Equation #
1.	$\psi(x,t) = A_0\, e^{i[A_1(x-x_0)+A_2(t-t_0)+\phi_0]}$	Continuous wave	$A_0 = \pm A_1$, $a_1(t) = -A_2/A_1$, $a_2(t) = a_3(t)$, $a_3(t)$ is an arbitrary real function, x_0, t_0, A_1, A_2, and ϕ_0 are arbitrary constants	(4.41)
2.	$\psi(x,t) = A_0\,\mathrm{sech}[\omega(x,t)]\,e^{i\,\phi(x,t)}$	Bright soliton	$\omega(x,t) = A_0\,x - A_0\int[a_1(t) + 2\,A_1\,a_2(t)]\,dt$, $\phi(x,t) = A_1\,x + \int[(A_0^2 - A_1^2)\,a_2(t) - A_1\,a_1(t)]\,dt$, $a_3(t) = 2\,a_2(t)$, $a_1(t)$ and $a_2(t)$ are arbitrary real functions, A_0 and A_1 are arbitrary real constants	(4.42)
3.	$\psi(x,t) = A_0\,\tanh[\omega(x,t)]\,e^{i\,\phi(x,t)}$	Dark soliton	$\omega(x,t) = A_0\,x - A_0\int[a_1(t) + 2\,A_1\,a_2(t)]\,dt$, $\phi(x,t) = A_1\,x - \int[(2\,A_0^2 + A_1^2)\,a_2(t) + A_1\,a_1(t)]\,dt$, $a_3(t) = -2\,a_2(t)$, $a_1(t)$ and $a_2(t)$ are arbitrary real functions, A_0 and A_1 are arbitrary real constants	(4.43)
4.	$\psi(x,t) = A_0\,\mathrm{csch}[\omega(x,t)]\,e^{i\,\phi(x,t)}$	—	$\omega(x,t) = A_0\,x - A_0\int[a_1(t) + 2\,A_1\,a_2(t)]\,dt$, $\phi(x,t) = A_1\,x + \int[(A_0^2 - A_1^2)\,a_2(t) - A_1\,a_1(t)]\,dt$, $a_3(t) = -2\,a_2(t)$, $a_1(t)$ and $a_2(t)$ are arbitrary real functions, A_0 and A_1 are arbitrary real constants	(4.44)
5.	$\psi(x,t) = A_0\,\coth[\omega(x,t)]\,e^{i\,\phi(x,t)}$	—	$\omega(x,t) = A_0\,x - A_0\int[a_1(t) + 2\,A_1\,a_2(t)]\,dt$, $\phi(x,t) = A_1\,x - \int[(2\,A_0^2 + A_1^2)\,a_2(t) + A_1\,a_1(t)]\,dt$, $a_3(t) = -2\,a_2(t)$, $a_1(t)$ and $a_2(t)$ are arbitrary real functions, A_0 and A_1 are arbitrary real constants	(4.45)
6.	$\psi(x,t) = A_0\,\sec[\omega(x,t)]\,e^{i\,\phi(x,t)}$	—	$\omega(x,t) = A_0\,x - A_0\int[a_1(t) + 2\,A_1\,a_2(t)]\,dt$, $\phi(x,t) = A_1\,x - \int[(A_0^2 + A_1^2)\,a_2(t) + A_1\,a_1(t)]\,dt$, $a_3(t) = -2\,a_2(t)$, $a_1(t)$ and $a_2(t)$ are arbitrary real functions, A_0 and A_1 are arbitrary real constants	(4.46)

7. $\psi(x, t) = A_0 \csc[\omega(x, t)] \, e^{i\,\phi(x, t)}$ —

$\omega(x, t) = A_0\, x - A_0 \int [a_1(t) + 2\, A_1\, a_2(t)]\, dt,$
$\phi(x, t) = A_1\, x - \int [(A_0^2 + A_1^2)\, a_2(t) + A_1\, a_1(t)]\, dt,\ a_3(t) = -2\, a_2(t),$
$a_1(t)$ and $a_2(t)$ are arbitrary real functions, A_0 and A_1 are arbitrary real constants

(4.47)

8. $\psi(x, t) = A_0 \tan[\omega(x, t)] \, e^{i\,\phi(x, t)}$ —

$\omega(x, t) = A_0\, x - A_0 \int [a_1(t) + 2\, A_1\, a_2(t)]\, dt,$
$\phi(x, t) = A_1\, x + \int [(2\, A_0^2 - A_1^2)\, a_2(t) - A_1\, a_1(t)]\, dt,$
$a_3(t) = -2\, a_2(t), a_1(t)$ and $a_2(t)$ are arbitrary real functions, A_0 and A_1 are arbitrary real constants

(4.48)

9. $\psi(x, t) = A_0 \cot[\omega(x, t)] \, e^{i\,\phi(x, t)}$ —

$\omega(x, t) = A_0\, x - A_0 \int [a_1(t) + 2\, A_1\, a_2(t)]\, dt,$
$\phi(x, t) = A_1\, x + \int [(2\, A_0^2 - A_1^2)\, a_2(t) - A_1\, a_1(t)]\, dt,$
$a_3(t) = -2\, a_2(t), a_1(t)$ and $a_2(t)$ are arbitrary real functions, A_0 and A_1 are arbitrary real constants

(4.49)

10. $\psi(x, t) = A_0 \{\sec[\omega(x, t)] \pm \tan[\omega(x, t)]\} \, e^{i\,\phi(x, t)}$ —

$\omega(x, t) = A_0\, x - A_0 \int [a_1(t) + 2\, A_1\, a_2(t)]\, dt,$
$\phi(x, t) = A_1\, x + \int [(\tfrac{1}{2}\, A_0^2 - A_1^2)\, a_2(t) - A_1\, a_1(t)]\, dt,$
$a_3(t) = -(1/2)\, a_2(t), a_1(t)$ and $a_2(t)$ are arbitrary real functions, A_0 and A_1 are arbitrary real constants

(4.50)

4.9 NLSE with Fourth Order Dispersion

Equation:

$$i\,\psi_t + a_1\,\psi_{xx} + a_2\,|\psi|^2\,\psi + a_3\,\psi_{xxxx} = 0, \tag{4.51}$$

where $\psi = \psi(x,\,t)$ is the complex function profile, x and t are its two independent variables, a_j are arbitrary real constants, $j = 1,\,2,\,3$.

Solutions:

***Solution* 1. Constant Amplitude**   *CW, t- and x-dependent*

$$\psi(x,\,t) = c\,e^{i\left[\sqrt{\frac{a_1}{a_3}}\,(x-x_0)+c^2\,a_2\,(t-t_0)+\phi_0\right]}, \tag{4.52}$$

where $a_1\,a_3 > 0$, x_0, t_0, c and ϕ_0 are arbitrary real constants.

***Solution* 2.**

$$\psi(x,\,t) = a_1\sqrt{\frac{-3}{10\,a_2\,a_3}}\,\sec^2\left[\sqrt{\frac{a_1}{20\,a_3}}\,(x-x_0)\right]e^{i\left[\frac{-4\,a_1^2}{25\,a_3}(t-t_0)+\phi_0\right]}, \tag{4.53}$$

where $a_1\,a_3 > 0$, $a_2\,a_3 < 0$, x_0, t_0, and ϕ_0 are arbitrary real constants.
  • *Reference*: [14].

***Solution* 3.**

$$\psi(x,\,t) = a_1\sqrt{\frac{-3}{10\,a_2\,a_3}}\,\csc^2\left[\sqrt{\frac{a_1}{20\,a_3}}\,(x-x_0)\right]e^{i\left[\frac{-4\,a_1^2}{25\,a_3}(t-t_0)+\phi_0\right]}, \tag{4.54}$$

where $a_1\,a_3 > 0$, $a_2\,a_3 < 0$, x_0, t_0, and ϕ_0 are arbitrary real constants.
  • *Reference*: [14].

***Solution* 4.**   *bright soliton*

$$\psi(x,\,t) = a_1\sqrt{\frac{-3}{10\,a_2\,a_3}}\,\text{sech}^2\left[\sqrt{\frac{-a_1}{20\,a_3}}\,(x-x_0)\right]e^{i\left[\frac{-4\,a_1^2}{25\,a_3}(t-t_0)+\phi_0\right]}, \tag{4.55}$$

where $a_1\,a_3 < 0$, $a_2\,a_3 < 0$, x_0, t_0, and ϕ_0 are arbitrary real constants.
  • *Reference*: [14].

***Solution* 5.**

$$\psi(x,\,t) = a_1\sqrt{\frac{-3}{10\,a_2\,a_3}}\,\text{csch}^2\left[\sqrt{\frac{-a_1}{20\,a_3}}\,(x-x_0)\right]e^{i\left[\frac{-4\,a_1^2}{25\,a_3}(t-t_0)+\phi_0\right]}, \tag{4.56}$$

where $a_1\,a_3 < 0$, $a_2\,a_3 < 0$, x_0, t_0, and ϕ_0 are arbitrary real constants.
  • *Reference*: [14].

4.10 Summary of Section 4.9

Equation

$$i\,\psi_t + a_1\,\psi_{xx} + a_2\,|\psi|^2\,\psi + a_3\,\psi_{xxxx} = 0$$

#	Solution	Conditions	Name	Equation #
1.	$\psi(x,t) = c\,e^{i\,[\sqrt{\frac{a_1}{a_3}}\,(x-x_0)+c^2\,a_2\,(t-t_0)+\phi_0]}$	$a_1\,a_3 > 0$, x_0, t_0, c and ϕ_0 are arbitrary real constants	Continuous wave, t- and x-dependent phase	(4.52)
2.	$\psi(x,t) = a_1\,\sqrt{\frac{-3}{10\,a_2\,a_3}}\,\sec^2[\,\sqrt{\frac{a_1}{20\,a_3}}\,(x-x_0)]e^{i\,[\frac{-4\,a_1^2}{25\,a_3}\,(t-t_0)+\phi_0]}$	$a_1\,a_3 > 0$, $a_2\,a_3 < 0$, x_0, t_0, and ϕ_0 are arbitrary real constants	—	(4.53)
3.	$\psi(x,t) = a_1\,\sqrt{\frac{-3}{10\,a_2\,a_3}}\,\csc^2[\,\sqrt{\frac{a_1}{20\,a_3}}\,(x-x_0)]e^{i\,[\frac{-4\,a_1^2}{25\,a_3}\,(t-t_0)+\phi_0]}$	$a_1\,a_3 > 0$, $a_2\,a_3 < 0$, x_0, t_0, and ϕ_0 are arbitrary real constants	—	(4.54)
4.	$\psi(x,t) = a_1\,\sqrt{\frac{-3}{10\,a_2\,a_3}}\,\mathrm{sech}^2[\,\sqrt{\frac{-a_1}{20\,a_3}}\,(x-x_0)]e^{i\,[\frac{-4\,a_1^2}{25\,a_3}\,(t-t_0)+\phi_0]}$	$a_1\,a_3 < 0$, $a_2\,a_3 < 0$, x_0, t_0, and ϕ_0 are arbitrary real constants	Bright soliton	(4.55)
5.	$\psi(x,t) = a_1\,\sqrt{\frac{-3}{10\,a_2\,a_3}}\,\mathrm{csch}^2[\,\sqrt{\frac{-a_1}{20\,a_3}}\,(x-x_0)]e^{i\,[\frac{-4\,a_1^2}{25\,a_3}\,(t-t_0)+\phi_0]}$	$a_1\,a_3 < 0$, $a_2\,a_3 < 0$, x_0, t_0, and ϕ_0 are arbitrary real constants	—	(4.56)

4.11 NLSE with Fourth Order Dispersion and Power Law Nonlinearity

Equation:

$$i\,\psi_t + a_1\,\psi_{xx} + a_2\,|\psi|^{2n}\,\psi + a_3\,\psi_{xxxx} = 0, \tag{4.57}$$

where $\psi = \psi(x, t)$ is the complex function profile, x and t are its two independent variables, n and a_j are arbitrary real constants, $j = 1, 2, 3$.

Solutions:

***Solution* 1. Constant Amplitude**   *CW, t- and x-dependent*

$$\psi(x, t) = c\, e^{i\left[\sqrt{\frac{a_1}{a_3}}\,(x-x_0)+c^{2n}\,a_2\,(t-t_0)+\phi_0\right]}, \tag{4.58}$$

where $a_1 a_3 > 0$, x_0, t_0, c, and ϕ_0 are arbitrary real constants.

***Solution* 2.**

$$\psi(x, t) = \{\sqrt{-\mu_1}\,\sec^2\,[\sqrt{\mu_2}\,(x - x_0)]\}^{\frac{1}{n}}\, e^{-i\,[\mu_3\,(t-t_0)+\phi_0]}, \tag{4.59}$$

where $\mu_1 = \dfrac{a_1^2\,(n+1)\,(n+2)\,(3\,n+2)}{4\,a_3\,a_2\,(n^2+2\,n+2)^2} < 0$, $\mu_2 = \dfrac{a_1\,n^2}{4\,a_3\,(n^2+2\,n+2)} > 0$, $\mu_3 = \dfrac{a_1^2\,(n+1)^2}{a_3\,(n^2+2\,n+2)^2}$, x_0, t_0, and ϕ_0 are arbitrary real constants.

- *Reference*: [14].

***Solution* 3.**

$$\psi(x, t) = \{\sqrt{-\mu_1}\,\csc^2[\sqrt{\mu_2}\,(x - x_0)]\}^{\frac{1}{n}} e^{-i\,[\mu_3\,(t-t_0)+\phi_0]}, \tag{4.60}$$

where $\mu_1 = \dfrac{a_1^2\,(n+1)\,(n+2)\,(3\,n+2)}{4\,a_3\,a_2\,(n^2+2\,n+2)^2} < 0$, $\mu_2 = \dfrac{a_1\,n^2}{4\,a_3\,(n^2+2\,n+2)} > 0$, $\mu_3 = \dfrac{a_1^2\,(n+1)^2}{a_3\,(n^2+2\,n+2)^2}$, x_0, t_0, and ϕ_0 are arbitrary real constants.

- *Reference*: [14].

***Solution* 4.**   *bright soliton*

$$\psi(x, t) = \{\sqrt{-\mu_1}\,\mathrm{sech}^2\,[\sqrt{-\mu_2}\,(x - x_0)]\}^{\frac{1}{n}}\, e^{-i\,[\mu_3\,(t-t_0)+\phi_0]}, \tag{4.61}$$

where $\mu_1 = \dfrac{a_1^2\,(n+1)\,(n+2)\,(3\,n+2)}{4\,a_3\,a_2\,(n^2+2\,n+2)^2} < 0$, $\mu_2 = \dfrac{a_1\,n^2}{4\,a_3\,(n^2+2\,n+2)} < 0$, $\mu_3 = \dfrac{a_1^2\,(n+1)^2}{a_3\,(n^2+2\,n+2)^2}$, x_0, t_0, and ϕ_0 are arbitrary real constants.

- *Reference*: [14].

***Solution* 5.**

$$\psi(x, t) = \{\sqrt{-\mu_1}\,\mathrm{csch}^2[\sqrt{-\mu_2}\,(x - x_0)]\}^{\frac{1}{n}} e^{-i\,[\mu_3\,(t-t_0)+\phi_0]}, \tag{4.62}$$

where $\mu_1 = \dfrac{a_1^2\,(n+1)\,(n+2)\,(3\,n+2)}{4\,a_3\,a_2\,(n^2+2\,n+2)^2} < 0$, $\mu_2 = \dfrac{a_1\,n^2}{4\,a_3\,(n^2+2\,n+2)} < 0$, $\mu_3 = \dfrac{a_1^2\,(n+1)^2}{a_3\,(n^2+2\,n+2)^2}$, x_0, t_0, and ϕ_0 are arbitrary real constants.

- *Reference*: [14].

4.12 Summary of Section 4.11

Equation

$$i\psi_t + a_1\psi_{xx} + a_2|\psi|^{2n}\psi + a_3\psi_{xxxx} = 0$$

#	Solution	Conditions	Name	Equation #
1.	$\psi(x,t) = c\,e^{i[\sqrt{\frac{a_1}{a_3}}(x-x_0)+c^{2n}a_2(t-t_0)+\phi_0]}$	$a_1 a_3 > 0$, x_0, t_0, c, and ϕ_0 are arbitrary real constants	Continuous wave, t- and x-dependent phase	(4.58)
2.	$\psi(x,t) = \{\sqrt{-\mu_1}\,\sec^2[\sqrt{\mu_2}\,(x-x_0)]\}^{\frac{1}{n}}e^{-i[\mu_3(t-t_0)+\phi_0]}$	$\mu_1 = \dfrac{a_1^2(n+1)(n+2)(3n+2)}{4a_3 a_2(n^2+2n+2)^2} < 0,\ \mu_2 = \dfrac{a_1 n^2}{4a_3(n^2+2n+2)} > 0,$ $\mu_3 = \dfrac{\frac{a_1^2 a_2(n^2+2n+2)^2}{a_1^2(n+1)^2}}{a_3(n^2+2n+2)^2}$, x_0, t_0, and ϕ_0 are arbitrary real constants	—	(4.59)
3.	$\psi(x,t) = \{\sqrt{-\mu_1}\,\csc^2[\sqrt{\mu_2}\,(x-x_0)]\}^{\frac{1}{n}}e^{-i[\mu_3(t-t_0)+\phi_0]}$	$\mu_1 = \dfrac{a_1^2(n+1)(n+2)(3n+2)}{4a_3 a_2(n^2+2n+2)^2} < 0,\ \mu_2 = \dfrac{a_1 n^2}{4a_3(n^2+2n+2)} > 0,$ $\mu_3 = \dfrac{\frac{a_1^2 a_2(n^2+2n+2)^2}{a_1^2(n+1)^2}}{a_3(n^2+2n+2)^2}$, x_0, t_0, and ϕ_0 are arbitrary real constants	—	(4.60)
4.	$\psi(x,t) = \{\sqrt{-\mu_1}\,\text{sech}^2[\sqrt{-\mu_2}\,(x-x_0)]\}^{\frac{1}{n}}e^{-i[\mu_3(t-t_0)+\phi_0]}$	$\mu_1 = \dfrac{a_1^2(n+1)(n+2)(3n+2)}{4a_3 a_2(n^2+2n+2)^2} < 0,\ \mu_2 = \dfrac{a_1 n^2}{4a_3(n^2+2n+2)} < 0,$ $\mu_3 = \dfrac{\frac{a_1^2 a_2(n^2+2n+2)^2}{a_1^2(n+1)^2}}{a_3(n^2+2n+2)^2}$, x_0, t_0, and ϕ_0 are arbitrary real constants	Bright soliton	(4.61)
5.	$\psi(x,t) = \{\sqrt{-\mu_1}\,\text{csch}^2[\sqrt{-\mu_2}\,(x-x_0)]\}^{\frac{1}{n}}e^{-i[\mu_3(t-t_0)+\phi_0]}$	$\mu_1 = \dfrac{a_1^2(n+1)(n+2)(3n+2)}{4a_3 a_2(n^2+2n+2)^2} < 0,\ \mu_2 = \dfrac{a_1 n^2}{4a_3(n^2+2n+2)} < 0,$ $\mu_3 = \dfrac{\frac{a_1^2 a_2(n^2+2n+2)^2}{a_1^2(n+1)^2}}{a_3(n^2+2n+2)^2}$, x_0, t_0, and ϕ_0 are arbitrary real constants	—	(4.62)

4.13 NLSE with Third and Fourth Order Dispersions and Cubic and Quintic Nonlinearities

Equation:

$$i\,\psi_t + a_1\,\psi_{xx} + a_2\,|\psi|^2\,\psi + a_3\,|\psi|^4\,\psi + i\,a_4\,\psi_{xxx} + a_5\,\psi_{xxxx} = 0, \qquad (4.63)$$

where $\psi = \psi(x, t)$ is the complex function profile, x and t are its two independent variables, a_j are arbitrary real constants, $j = 1, 2, 3, 4, 5$.

Solutions:

Solution 1. **Constant Amplitude**   *CW, t- and x-dependent*

$$\psi(x, t) = c_1\,e^{i\,\{c_2\,(x-x_0)+[c_1^2\,(a_2+c_1^2\,a_3)+c_2^2\,(-a_1+c_2\,a_4+c_2^2\,a_5)]\,(t-t_0)+\phi_0\}}, \qquad (4.64)$$

where x_0, t_0, c_1, c_2, and ϕ_0 are arbitrary real constants.

Solution 2.

$$\psi(x, t) = \pm 2\,c_1\,\sqrt{\frac{-6\,a_5}{a_3}}\,\sec[c_1\,(x - x_0) + c_2\,(t - t_0)]\,e^{i\,[c_3\,(x-x_0)+c_4\,(t-t_0)+\phi_0]}, (4.65)$$

where $\qquad c_1 = -\sqrt{\dfrac{3\,a_4^2 + 8\,a_5\,(a_1 + a_2\,\sqrt{\frac{-6\,a_5}{a_3}})}{80\,a_5^2}}, \qquad c_2 = -c_1\,(\dfrac{a_4^3 + 4\,a_1\,a_4\,a_5}{192\,a_5^3}), \qquad c_3 = \dfrac{-a_4}{4\,a_5},$

$c_4 = a_5\,c_1^4 - \dfrac{(8\,a_5\,a_1 + 3\,a_4^2)}{8\,a_5}\,c_1^2 - \dfrac{3\,a_4^4 + 16\,a_1\,a_4^2\,a_5}{256\,a_5^3}, \qquad\qquad a_3\,a_5 < 0,$

$3\,a_4^2 + 8\,a_5\,(a_1 + a_2\,\sqrt{\frac{-6\,a_5}{a_3}}) > 0$, x_0, t_0, and ϕ_0 are arbitrary real constants.

- *Reference*: [15].

Solution 3.

$$\psi(x, t) = \pm 2\,c_1\,\sqrt{\frac{-6\,a_5}{a_3}}\,\csc[c_1\,(x - x_0) + c_2\,(t - t_0)]\,e^{i\,[c_3\,(x-x_0)+c_4\,(t-t_0)+\phi_0]}, (4.66)$$

where $\qquad c_1 = -\sqrt{\dfrac{3\,a_4^2 + 8\,a_5\left(a_1 + a_2\,\sqrt{\frac{-6\,a_5}{a_3}}\right)}{80\,a_5^2}}, \qquad c_2 = -c_1\,(\dfrac{a_4^3 + 4\,a_1\,a_4\,a_5}{192\,a_5^3}), \qquad c_3 = \dfrac{-a_4}{4\,a_5},$

$c_4 = a_5\,c_1^4 - \dfrac{(8\,a_5\,a_1 + 3\,a_4^2)}{8\,a_5}\,c_1^2 - \dfrac{3\,a_4^4 + 16\,a_1\,a_4^2\,a_5}{256\,a_5^3}, \qquad\qquad a_3\,a_5 < 0,$

$3\,a_4^2 + 8\,a_5\,(a_1 + a_2\,\sqrt{\frac{-6\,a_5}{a_3}}) > 0$, x_0, t_0, and ϕ_0 are arbitrary real constants.

- *Reference*: [15].

Solution 4.

$$\psi(x, t) = \pm 2\,c_1\,\sqrt{\frac{-6\,a_5}{a_3}}\,\tan[c_1\,(x - x_0) + c_2\,(t - t_0)]\,e^{i\,[c_3\,(x-x_0)+c_4\,(t-t_0)+\phi_0]}, \qquad (4.67)$$

where $\quad c_1 = -\sqrt{\dfrac{-3\,a_4^2 - 8\,a_5\left(a_1 + a_2\,\sqrt{\tfrac{-6\,a_5}{a_3}}\right)}{160\,a_5^2}}, \qquad c_2 = -c_1\,(\dfrac{a_4^3 + 4\,a_1\,a_4\,a_5}{192\,a_5^3}), \qquad c_3 = \dfrac{-a_4}{4\,a_5},$

$c_4 = 16\,a_5\,c_1^4 + \dfrac{(8\,a_5\,a_1 + 3\,a_4^2)}{4\,a_5}\,c_1^2 - \dfrac{3\,a_4^4 + 16\,a_1\,a_4^2\,a_5}{256\,a_5^3}, \qquad\qquad\qquad a_3\,a_5 < 0,$

$3\,a_4^2 + 8\,a_5\left(a_1 + a_2\,\sqrt{\tfrac{-6\,a_5}{a_3}}\right) < 0$, x_0, t_0, and ϕ_0 are arbitrary real constants.

- *Reference: [15].*

Solution 5.

$$\psi(x, t) = \pm 2\,c_1\,\sqrt{\dfrac{-6\,a_5}{a_3}}\,\cot[c_1\,(x - x_0) + c_2\,(t - t_0)]\,e^{i\,[c_3\,(x-x_0)+c_4\,(t-t_0)+\phi_0]}, \qquad (4.68)$$

where $\quad c_1 = -\sqrt{\dfrac{-3\,a_4^2 - 8\,a_5\left(a_1 + a_2\,\sqrt{\tfrac{-6\,a_5}{a_3}}\right)}{160\,a_5^2}}, \qquad c_2 = -c_1\left(\dfrac{a_4^3 + 4\,a_1\,a_4\,a_5}{192\,a_5^3}\right), \qquad c_3 = \dfrac{-a_4}{4\,a_5},$

$c_4 = 16\,a_5\,c_1^4 + \dfrac{(8\,a_5\,a_1 + 3\,a_4^2)}{4\,a_5}\,c_1^2 - \dfrac{3\,a_4^4 + 16\,a_1\,a_4^2\,a_5}{256\,a_5^3}, \qquad\qquad\qquad a_3\,a_5 < 0,$

$3\,a_4^2 + 8\,a_5\left(a_1 + a_2\,\sqrt{\tfrac{-6\,a_5}{a_3}}\right) < 0$, x_0, t_0, and ϕ_0 are arbitrary real constants.

- *Reference: [15].*

Solution 6.

$$\psi(x, t) = \pm 2\,c_1\,\sqrt{\dfrac{-6\,a_5}{a_3}}\,\operatorname{csch}[c_1\,(x - x_0) + c_2\,(t - t_0)]\,e^{i\,[c_3\,(x-x_0)+c_4\,(t-t_0)+\phi_0]}, \qquad (4.69)$$

where $\quad c_1 = -\sqrt{\dfrac{-3\,a_4^2 - 4\,a_5\left(2\,a_1 + a_2\,\sqrt{\tfrac{-24\,a_5}{a_3}}\right)}{80\,a_5^2}}, \qquad c_2 = -c_1\left(\dfrac{a_4^3 + 4\,a_1\,a_4\,a_5}{192\,a_5^3}\right), \qquad c_3 = \dfrac{-a_4}{4\,a_5},$

$c_4 = a_5\,c_1^4 + \dfrac{(8\,a_5\,a_1 + 3\,a_4^2)}{8\,a_5}\,c_1^2 - \dfrac{3\,a_4^4 + 16\,a_1\,a_4^2\,a_5}{256\,a_5^3}, \qquad\qquad\qquad a_3\,a_5 < 0,$

$-3\,a_4^2 - 4\,a_5\left(2\,a_1 + a_2\,\sqrt{\tfrac{-24\,a_5}{a_3}}\right) > 0$, x_0, t_0, and ϕ_0 are arbitrary real constants.

- *Reference: [15], we corrected the constant prefactor.*

Solution 7.   *bright soliton*

$$\psi(x, t) = \pm 2\,c_1\,\sqrt{\dfrac{-6\,a_5}{a_3}}\,\operatorname{sech}[c_1\,(x - x_0) + c_2\,(t - t_0)]\,e^{i\,[c_3\,(x-x_0)+c_4\,(t-t_0)+\phi_0]}, \qquad (4.70)$$

where $\quad c_1 = -\sqrt{\dfrac{-3\,a_4^2 - 8\,a_5\left(a_1 - a_2\,\sqrt{\tfrac{-6\,a_5}{a_3}}\right)}{80\,a_5^2}}, \qquad c_2 = -c_1\,(\dfrac{a_4^3 + 4\,a_1\,a_4\,a_5}{192\,a_5^3}), \qquad c_3 = \dfrac{-a_4}{4\,a_5},$

$c_4 = a_5\,c_1^4 + \dfrac{(8\,a_5\,a_1 + 3\,a_4^2)}{8\,a_5}\,c_1^2 - \dfrac{3\,a_4^4 + 16\,a_1\,a_4^2\,a_5}{256\,a_5^3}, \qquad\qquad\qquad a_3\,a_5 < 0,$

$-3\,a_4^2 - 8\,a_5\,(a_1 - a_2\,\sqrt{\tfrac{-6\,a_5}{a_3}}) > 0$, x_0, t_0, and ϕ_0 are arbitrary real constants.

- *Reference*: [15].

Solution 8.   *dark soliton*

$$\psi(x,\, t) = \pm 2\, c_1 \sqrt{\frac{-6\, a_5}{a_3}}\ \tanh[c_1\, (x \,-\, x_0) + c_2\, (t \,-\, t_0)]\ e^{i\, [c_3\, (x - x_0) + c_4\, (t - t_0) + \phi_0]}, \quad (4.71)$$

where $\quad c_1 = -\sqrt{\dfrac{3\, a_4^2 + 8\, a_5 \left(a_1 + a_2\, \sqrt{\frac{-6\, a_5}{a_3}}\right)}{160\, a_5^2}}, \qquad c_2 = -c_1\, (\dfrac{a_4^3 + 4\, a_1\, a_4\, a_5}{192\, a_5^3}), \qquad c_3 = \dfrac{-a_4}{4\, a_5},$

$c_4 = 16\, a_5\, c_1^4 - \dfrac{(8\, a_5\, a_1 + 3\, a_4^2)}{4\, a_5}\, c_1^2 - \dfrac{3\, a_4^4 + 16\, a_1\, a_4^2\, a_5}{256\, a_5^3}, \qquad\qquad\qquad a_3\, a_5 < 0,$

$3\, a_4^2 + 8\, a_5 \left(a_1 + a_2\, \sqrt{\dfrac{-6\, a_5}{a_3}}\right) > 0,\ x_0,\ t_0,$ and ϕ_0 are arbitrary real constants.
- *Reference*: [15].

Solution 9.

$$\psi(x,\, t) = \pm 2\, c_1 \sqrt{\frac{-6\, a_5}{a_3}}\ \coth[c_1\, (x \,-\, x_0) + c_2\, (t \,-\, t_0)]\ e^{i\, [c_3\, (x - x_0) + c_4\, (t - t_0) + \phi_0]}, \qquad (4.72)$$

where $\quad c_1 = -\sqrt{\dfrac{3\, a_4^2 + 8\, a_5 \left(a_1 + a_2\, \sqrt{\frac{-6\, a_5}{a_3}}\right)}{160\, a_5^2}}, \qquad c_2 = -c_1\, (\dfrac{a_4^3 + 4\, a_1\, a_4\, a_5}{192\, a_5^3}), \qquad c_3 = \dfrac{-a_4}{4\, a_5},$

$c_4 = 16\, a_5\, c_1^4 - \dfrac{(8\, a_5\, a_1 + 3\, a_4^2)}{4\, a_5}\, c_1^2 - \dfrac{3\, a_4^4 + 16\, a_1\, a_4^2\, a_5}{256\, a_5^3}, \qquad\qquad\qquad a_3\, a_5 < 0,$

$3\, a_4^2 + 8\, a_5 \left(a_1 + a_2\, \sqrt{\dfrac{-6\, a_5}{a_3}}\right) > 0,\ x_0,\ t_0,$ and ϕ_0 are arbitrary real constants.
- *Reference*: [15].

4.14 Summary of Section 4.13

Equation

$$i\psi_t + a_1\psi_{xx} + a_2|\psi|^2\psi + a_3|\psi|^4\psi + i a_4\psi_{xxx} + a_5\psi_{xxxx} = 0$$

#	Solution	Conditions	Name	Equation #
1.	$\psi(x,t) = c_1 e^{i\{c_2(x-x_0)+[c_1^2(a_2+c_1^2 a_3)+c_2^2(-a_1+c_2 a_4+c_2^2 a_5)](t-t_0)+\phi_0\}}$	$x_0, t_0, c_1, c_2,$ and ϕ_0 are arbitrary real constants	Continuous wave, t- and x-dependent phase	(4.64)
2.	$\psi(x,t) = \pm 2c_1\sqrt{\dfrac{-6a_5}{a_3}}\ \sec[c_1(x-x_0)+c_2(t-t_0)]$ $\times e^{i[c_3(x-x_0)+c_4(t-t_0)+\phi_0]}$	$c_1 = -\sqrt{\dfrac{3a_4^2+8a_5\left(a_1+a_2\sqrt{\frac{-6a_5}{a_3}}\right)}{80a_5^2}}$, $c_2 = -c_1\left(\dfrac{a_4^3+4a_1a_4a_5}{192a_5^3}\right)$, $c_3 = \dfrac{-a_4}{4a_5}$, $c_4 = a_5 c_1^4 - \dfrac{(8a_5 a_1 + 3a_4^2)}{8a_5}c_1^2 - \dfrac{3a_4^4+16a_1 a_4^2 a_5}{256 a_5^3}$, $a_3 a_5 < 0$, $3a_4^2 + 8a_5\left(a_1+a_2\sqrt{\frac{-6a_5}{a_3}}\right) > 0$, $x_0, t_0,$ and ϕ_0 are arbitrary real constants	—	(4.65)
3.	$\psi(x,t) = \pm 2c_1\sqrt{\dfrac{-6a_5}{a_3}}\ \csc[c_1(x-x_0)+c_2(t-t_0)]$ $\times e^{i[c_3(x-x_0)+c_4(t-t_0)+\phi_0]}$	$c_1 = -\sqrt{\dfrac{3a_4^2+8a_5\left(a_1+a_2\sqrt{\frac{-6a_5}{a_3}}\right)}{80a_5^2}}$, $c_2 = -c_1\left(\dfrac{a_4^3+4a_1a_4a_5}{192a_5^3}\right)$, $c_3 = \dfrac{-a_4}{4a_5}$, $c_4 = a_5 c_1^4 - \dfrac{(8a_5 a_1 + 3a_4^2)}{8a_5}c_1^2 - \dfrac{3a_4^4+16a_1 a_4^2 a_5}{256 a_5^3}$, $a_3 a_5 < 0$, $3a_4^2 + 8a_5\left(a_1+a_2\sqrt{\frac{-6a_5}{a_3}}\right) > 0$, $x_0, t_0,$ and ϕ_0 are arbitrary real constants	—	(4.66)
4.	$\psi(x,t) = \pm 2c_1\sqrt{\dfrac{-6a_5}{a_3}}\ \tan[c_1(x-x_0)+c_2(t-t_0)]$ $\times e^{i[c_3(x-x_0)+c_4(t-t_0)+\phi_0]}$	$c_1 = -\sqrt{\dfrac{-3a_4^2-8a_5\left(a_1+a_2\sqrt{\frac{-6a_5}{a_3}}\right)}{160a_5^2}}$, $c_2 = -c_1\left(\dfrac{a_4^3+4a_1a_4a_5}{192a_5^3}\right)$, $c_3 = \dfrac{-a_4}{4a_5}$, $c_4 = 16a_5 c_1^4 + \dfrac{(8a_5 a_1 + 3a_4^2)}{4a_5}c_1^2 - \dfrac{3a_4^4+16a_1 a_4^2 a_5}{256 a_5^3}$, $a_3 a_5 < 0, 3 a_4^2 + 8 a_5\left(a_1+a_2\sqrt{\frac{-6a_5}{a_3}}\right) < 0$, $x_0, t_0,$ and ϕ_0 are arbitrary real constants	—	(4.67)
5.	$\psi(x,t) = \pm 2c_1\sqrt{\dfrac{-6a_5}{a_3}}\ \cot[c_1(x-x_0)+c_2(t-t_0)]$ $\times e^{i[c_3(x-x_0)+c_4(t-t_0)+\phi_0]}$	$c_1 = -\sqrt{\dfrac{-3a_4^2-8a_5\left(a_1+a_2\sqrt{\frac{-6a_5}{a_3}}\right)}{160a_5^2}}$, $c_2 = -c_1\left(\dfrac{a_4^3+4a_1a_4a_5}{192a_5^3}\right)$, $c_3 = \dfrac{-a_4}{4a_5}$, $c_4 = 16a_5 c_1^4 + \dfrac{(8a_5 a_1 + 3a_4^2)}{4a_5}c_1^2 - \dfrac{3a_4^4+16a_1 a_4^2 a_5}{256 a_5^3}$, $a_3 a_5 < 0, 3 a_4^2 + 8 a_5\left(a_1+a_2\sqrt{\frac{-6a_5}{a_3}}\right) < 0$, $x_0, t_0,$ and ϕ_0 are arbitrary real constants	—	(4.68)

(Continued)

Equation

$$i\psi_t + a_1\psi_{xx} + a_2|\psi|^2\psi + a_3|\psi|^4\psi + i a_4\psi_{xxx} + a_5\psi_{xxxx} = 0$$

#	Solution	Conditions	Name	Equation #
6.	$\psi(x,t) = \pm 2c_1\sqrt{\dfrac{-6a_5}{a_3}}\,\text{csch}[c_1(x-x_0)+c_2(t-t_0)]$ $\times e^{i[c_3(x-x_0)+c_4(t-t_0)+\phi_0]}$	$c_1 = -\sqrt{\dfrac{-3a_4^2 - 4a_5(2q+a_2\sqrt{\frac{-24a_5}{a_3}})}{80\,a_5^2}}$, $c_2 = -c_1\left(\dfrac{a_4^3 + 4q\,a_4\,a_5}{192\,a_5^3}\right)$, $c_3 = \dfrac{-a_4}{4a_5}$, $c_4 = d_5 c_1^4 + \dfrac{(8a_5 q + 3a_4^2)}{8a_5}c_1^4 - \dfrac{3a_4^4 + 16q\,a_4^2\,a_5}{256\,a_5^3}$, $a_3 a_5 < 0$, $-3a_4^2 - 4a_5(2a_1 + a_2\sqrt{\frac{-24a_5}{a_3}}) > 0$, x_0, t_0, and ϕ_0 are arbitrary real constants	—	(4.69)
7.	$\psi(x,t) = \pm 2c_1\sqrt{\dfrac{-6a_5}{a_3}}\,\text{sech}[c_1(x-x_0)+c_2(t-t_0)]$ $\times e^{i[c_3(x-x_0)+c_4(t-t_0)+\phi_0]}$	$c_1 = -\sqrt{\dfrac{-3a_4^2 - 8a_5(q-a_2\sqrt{\frac{-6a_5}{a_3}})}{80\,a_5^2}}$, $c_2 = -c_1\left(\dfrac{a_4^3 + 4q\,a_4\,a_5}{192\,a_5^3}\right)$, $c_3 = \dfrac{-a_4}{4a_5}$, $c_4 = d_5 c_1^4 + \dfrac{(8a_5 q + 3a_4^2)}{8a_5}c_1^4 - \dfrac{3a_4^4 + 16q\,a_4^2\,a_5}{256\,a_5^3}$, $a_3 a_5 < 0$, $-3a_4^2 - 8a_5(2a_1 + a_2\sqrt{\frac{-6a_5}{a_3}}) > 0$, x_0, t_0, and ϕ_0 are arbitrary real constants	Bright soliton	(4.70)
8.	$\psi(x,t) = \pm 2c_1\sqrt{\dfrac{-6a_5}{a_3}}\,\tanh[c_1(x-x_0)+c_2(t-t_0)]$ $\times e^{i[c_3(x-x_0)+c_4(t-t_0)+\phi_0]}$	$c_1 = -\sqrt{\dfrac{3a_4^2 + 8a_5(q+a_2\sqrt{\frac{-6a_5}{a_3}})}{160\,a_5^2}}$, $c_2 = -c_1\left(\dfrac{a_4^3 + 4q\,a_4\,a_5}{192\,a_5^3}\right)$, $c_3 = \dfrac{-a_4}{4a_5}$, $c_4 = 16 a_5 c_1^4 - \dfrac{(8a_5 q + 3a_4^2)}{4a_5}c_1^2 - \dfrac{3a_4^4 + 16q\,a_4^2\,a_5}{256\,a_5^3}$, $a_3 a_5 < 0$, $3a_4^2 + 8a_5(a_1 + a_2\sqrt{\frac{-6a_5}{a_3}}) > 0$, x_0, t_0, and ϕ_0 are arbitrary real constants	Dark soliton	(4.71)
9.	$\psi(x,t) = \pm 2c_1\sqrt{\dfrac{-6a_5}{a_3}}\,\coth[c_1(x-x_0)+c_2(t-t_0)]$ $\times e^{i[c_3(x-x_0)+c_4(t-t_0)+\phi_0]}$	$c_1 = -\sqrt{\dfrac{3a_4^2 + 8a_5(q+a_2\sqrt{\frac{-6a_5}{a_3}})}{160\,a_5^2}}$, $c_2 = -c_1\left(\dfrac{a_4^3 + 4q\,a_4\,a_5}{192\,a_5^3}\right)$, $c_3 = \dfrac{-a_4}{4a_5}$, $c_4 = 16 a_5 c_1^4 - \dfrac{(8a_5 q + 3a_4^2)}{4a_5}c_1^2 - \dfrac{3a_4^4 + 16q\,a_4^2\,a_5}{256\,a_5^3}$, $a_3 a_5 < 0$, $3a_4^2 + 8a_5(a_1 + a_2\sqrt{\frac{-6a_5}{a_3}}) > 0$, x_0, t_0, and ϕ_0 are arbitrary real constants	—	(4.72)

4.15 NLSE with Third and Fourth Order Dispersions, Self-Steepening, Self-Frequency Shift, and Cubic and Quintic Nonlinearities

Equation:

$$i\,\psi_t + a_1\,\psi_{xx} + a_2\,|\psi|^2\,\psi + a_3\,|\psi|^4\,\psi + i\,a_4\,\psi_{xxx} + a_5\,\psi_{xxxx} + i\,a_6\,(|\psi|^2\,\psi)_x + i\,a_7\,(|\psi|^2)_x\,\psi = 0, \quad (4.73)$$

where $\psi = \psi(x, t)$ is the complex function profile, x and t are its two independent variables, a_j are arbitrary real constants, $j = 1, 2, 3, 4, 5, 6, 7$.

Solutions:

Solution 1. Constant Amplitude *CW, t- and x-dependent phase*

$$\psi(x, t) = c_1\,e^{i\,[c_2\,(x-x_0)+(c_1^2\,a_2+c_1^4\,a_3)\,(t-t_0)+\phi_0]}, \quad (4.74)$$

where $a_4 = \dfrac{a_6\,c_1^2 + a_1\,c_2 - a_5\,c_2^3}{c_2^2}$, x_0, t_0, c_1, c_2, and ϕ_0 are arbitrary real constants.

Solution 2.

$$\psi(x, t) = \sqrt{\frac{-6\,(4\,a_5\,c_3 + a_4)}{3\,a_6 + 2\,a_7}}\,\sec[x - x_0 + c_1\,(t - t_0)]\,e^{i\,[c_3\,(x-x_0)+c_2\,(t-t_0)]}, \quad (4.75)$$

where $c_1 = -2\,a_1\,c_3 + a_4\,(3\,c_3^2 + 1) + 4\,a_5\,(c_3^3 + c_3)$, $c_2 = -a_1\,(c_3^2 + 1) + a_4\,(c_3^3 + 3\,c_3) + a_5\,(c_3^4 + 6\,c_3^2 + 1)$, $a_1 = \dfrac{(4\,a_5\,c_3+a_4)\,(3\,a2 + 2\,c_3\,a7)}{3\,a_6 + 2\,a_7} + 2\,a_4\,c_3 + 2\,a_5\,(c_3^2 + 5)$,

$a_3 = \dfrac{-2\,a_5\,(3\,a_6 + 2\,a_7)^2}{3\,(a_4 + 4\,a_5\,c_3)^2}$, $(4\,a_5\,c_3 + a_4)\,(3\,a_6 + 2\,a_7) < 0$, x_0, t_0, c_3, and ϕ_0 are arbitrary real constants.

- *Reference:* [16].

Solution 3.

$$\psi(x, t) = \sqrt{\frac{-6\,(4\,a_5\,c_3 + a_4)}{3\,a_6 + 2\,a_7}}\,\csc[x - x_0 + c_1\,(t - t_0)]\,e^{i\,[c_3\,(x-x_0)+c_2\,(t-t_0)]}, \quad (4.76)$$

where $c_1 = -2\,a_1\,c_3 + a_4\,(3\,c_3^2 + 1) + 4\,a_5\,(c_3^3 + c_3)$, $c_2 = -a_1\,(c_3^2 + 1) + a_4\,(c_3^3 + 3\,c_3) + a_5\,(c_3^4 + 6\,c_3^2 + 1)$, $a_1 = \dfrac{(4\,a_5\,c_3+a_4)\,(3\,a2 + 2\,c_3\,a7)}{3\,a_6 + 2\,a_7} + 2\,a_4\,c_3 + 2\,a_5\,(c_3^2 + 5)$,

$a_3 = \dfrac{-2\,a_5\,(3\,a_6 + 2\,a_7)^2}{3\,(a_4 + 4\,a_5\,c_3)^2}$, $(4\,a_5\,c_3 + a_4)\,(3\,a_6 + 2\,a_7) < 0$, x_0, t_0, c_3, and ϕ_0 are arbitrary real constants.

- *Reference:* [16].

Solution 4.

$$\psi(x, t) = \sqrt{\frac{-6\,(4\,a_5\,c_3 + a_4)}{3\,a_6 + 2\,a_7}}\,\tan[x - x_0 + c_1\,(t - t_0)]\,e^{i\,[c_3\,(x-x_0)+c_2\,(t-t_0)]}, \quad (4.77)$$

where $c_1 = -2\,a_1\,c_3 + a_4\,(3\,c_3^2 - 2) + 4\,a_5\,(c_3^3 - 2\,c_3)$, $c_2 = a_1\,(2 - c_3^2) + a_4\,(c_3^3 - 6\,c_3) + a_5\,(c_3^4 - 12\,c_3^2 + 16)$,

$$a_1 = \frac{(4\,a_5\,c_3 + a_4)\,(3\,a2 + 2\,c_3\,a7)}{3\,a_6 + 2\,a_7} + 2\,a_4\,c_3 + 2\,a_5\,(c_3^2 - 10), \qquad a_3 = \frac{-2\,a_5\,(3\,a_6 + 2\,a_7)^2}{3\,(a_4 + 4\,a_5\,c_3)^2},$$

$(4\,a_5\,c_3 + a_4)\,(3\,a_6 + 2\,a_7) < 0$, x_0, t_0, c_3, and ϕ_0 are arbitrary real constants.

- *Reference*: [16].

Solution 5.

$$\psi(x, t) = \sqrt{\frac{-6\,(4\,a_5\,c_3 + a_4)}{3\,a_6 + 2\,a_7}}\, \cot[x - x_0 + c_1\,(t - t_0)]\, e^{i\,[c_3\,(x - x_0) + c_2\,(t - t_0)]}, \quad (4.78)$$

where
$$c_1 = -2\,a_1\,c_3 + a_4\,(3\,c_3^2 - 2) + 4\,a_5\,(c_3^3 - 2\,c_3),$$
$$c_2 = a_1\,(2 - c_3^2) + a_4\,(c_3^3 - 6\,c_3) + a_5\,(c_3^4 - 12\,c_3^2 + 16),$$
$$a_1 = \frac{(4\,a_5\,c_3 + a_4)\,(3\,a2 + 2\,c_3\,a7)}{3\,a_6 + 2\,a_7} + 2\,a_4\,c_3 + 2\,a_5\,(c_3^2 - 10), \qquad a_3 = \frac{-2\,a_5\,(3\,a_6 + 2\,a_7)^2}{3\,(a_4 + 4\,a_5\,c_3)^2},$$

$(4\,a_5\,c_3 + a_4)\,(3\,a_6 + 2\,a_7) < 0$, x_0, t_0, c_3, and ϕ_0 are arbitrary real constants.

- *Reference*: [16].

Solution 6.   *bright soliton*

$$\psi(x, t) = \sqrt{\frac{6\,(4\,a_5\,c_3 + a_4)}{3\,a_6 + 2\,a_7}}\, \mathrm{sech}[x - x_0 + c_1\,(t - t_0)]\, e^{i\,[c_3\,(x - x_0) + c_2\,(t - t_0)]}, \quad (4.79)$$

where
$$c_1 = -2\,a_1\,c_3 + a_4\,(3\,c_3^2 - 1) + 4\,a_5\,(c_3^3 - c_3),$$
$$c_2 = a_1\,(1 - c_3^2) + a_4\,(c_3^3 - 3\,c_3) + a_5\,(c_3^4 - 6\,c_3^2 + 1),$$
$$a_1 = \frac{(4\,a_5\,c_3 + a_4)\,(3\,a2 + 2\,c_3\,a7)}{3\,a_6 + 2\,a_7} + 2\,a_4\,c_3 + 2\,a_5\,(c_3^2 - 5), \qquad a_3 = \frac{-2\,a_5\,(3\,a_6 + 2\,a_7)^2}{3\,(a_4 + 4\,a_5\,c_3)^2},$$

$(4\,a_5\,c_3 + a_4)\,(3\,a_6 + 2\,a_7) > 0$, x_0, t_0, c_3, and ϕ_0 are arbitrary real constants.

- *Reference*: [16].

Solution 7.

$$\psi(x, t) = \sqrt{\frac{-6\,(4\,a_5\,c_3 + a_4)}{3\,a_6 + 2\,a_7}}\, \mathrm{csch}[x - x_0 + c_1\,(t - t_0)]\, e^{i\,[c_3\,(x - x_0) + c_2\,(t - t_0)]}, \quad (4.80)$$

where
$$c_1 = -2\,a_1\,c_3 + a_4\,(3\,c_3^2 - 1) + 4\,a_5\,(c_3^3 - c_3),$$
$$c_2 = a_1\,(1 - c_3^2) + a_4\,(c_3^3 - 3\,c_3) + a_5\,(c_3^4 - 6\,c_3^2 + 1),$$
$$a_1 = \frac{(4\,a_5\,c_3 + a_4)\,(3\,a2 + 2\,c_3\,a7)}{3\,a_6 + 2\,a_7} + 2\,a_4\,c_3 + 2\,a_5\,(c_3^2 - 5), \qquad a_3 = \frac{-2\,a_5\,(3\,a_6 + 2\,a_7)^2}{3\,(a_4 + 4\,a_5\,c_3)^2},$$

$(4\,a_5\,c_3 + a_4)\,(3\,a_6 + 2\,a_7) < 0$, x_0, t_0, c_3, and ϕ_0 are arbitrary real constants.

- *Reference*: [16].

Solution 8.   *dark soliton*

$$\psi(x, t) = \sqrt{\frac{-6\,(4\,a_5\,c_3 + a_4)}{3\,a_6 + 2\,a_7}}\, \tanh[x - x_0 + c_1\,(t - t_0)]\, e^{i\,[c_3\,(x - x_0) + c_2\,(t - t_0)]}, \quad (4.81)$$

where
$$c_1 = -2\,a_1\,c_3 + a_4\,(3\,c_3^2 + 2) + 4\,a_5\,(c_3^3 + 2\,c_3),$$

$$c_2 = -a_1\,(c_3^2 + 2) + a_4\,(6\,c_3 + c_3^3) + a_5\,(c_3^4 + 12\,c_3^2 + 16),$$

$$a_1 = \frac{(4\,a_5\,c_3 + a_4)\,(3\,a2 + 2\,c3\,a7)}{3\,a_6 + 2\,a_7} + 2\,a_4\,c_3 + 2\,a_5\,(c_3^2 + 10), \qquad a_3 = \frac{-2\,a_5\,(3\,a_6 + 2\,a_7)^2}{3\,(a_4 + 4\,a_5\,c_3)^2},$$

$(4\,a_5\,c_3 + a_4)\,(3\,a_6 + 2\,a_7) < 0$, x_0, t_0, c_3, and ϕ_0 are arbitrary real constants.

  • *Reference*: [16].

Solution 9.

$$\psi(x,\,t) = \sqrt{\frac{-6\,(4\,a_5\,c_3 + a_4)}{3\,a_6 + 2\,a_7}}\ \coth[x - x_0 + c_1\,(t - t_0)]\ e^{i\,[c_3\,(x-x_0)+c_2\,(t-t_0)]}, \quad (4.82)$$

where
$$c_1 = -2\,a_1\,c_3 + a_4\,(3\,c_3^2 + 2) + 4\,a_5\,(c_3^3 + 2\,c_3), \qquad c_2 = -a_1\,(c_3^2 + 2) +$$
$$a_4\,(6\,c_3 + c_3^3) + a_5\,(c_3^4 + 12\,c_3^2 + 16),$$

$$a_1 = \frac{(4\,a_5\,c_3 + a_4)\,(3\,a2 + 2\,c3\,a7)}{3\,a_6 + 2\,a_7} + 2\,a_4\,c_3 + 2\,a_5\,(c_3^2 + 10), \qquad a_3 = \frac{-2\,a_5\,(3\,a_6 + 2\,a_7)^2}{3\,(a_4 + 4\,a_5\,c_3)^2},$$

$(4\,a_5\,c_3 + a_4)\,(3\,a_6 + 2\,a_7) < 0$, x_0, t_0, c_3, and ϕ_0 are arbitrary real constants.

  • *Reference*: [16].

4.16 Summary of Section 4.15

Equation

$$i\psi_t + a_1\psi_{xx} + a_2|\psi|^2\psi + a_3|\psi|^4\psi + i\,a_4\,\psi_{xxx} + a_5\,\psi_{xxxx} + i\,a_6\,(|\psi|^2\,\psi)_x + i\,a_7\,(|\psi|^2)_x\,\psi = 0$$

#	Solution	Conditions	Name	Equation #
1.	$\psi(x,t) = c_1\,e^{i[c_2(x-x_0)+(c_1^2 a_2+c_1^4 a_3)(t-t_0)+\phi_0]}$	$a_4 = \dfrac{a_6 c_1^2 + a_1 c_2 - a_5 c_2^3}{c_2^2}$, x_0, t_0, c_1, c_2, and ϕ_0 are arbitrary real constants	Continuous wave, t- and x-dependent phase	(4.74)
2.	$\psi(x,t) = \sqrt{\dfrac{-6(4a_5 c_3+a_4)}{3a_6+2a_7}}\,\sec[x - x_0 + c_1(t - t_0)]$ $\times\, e^{i[c_3(x-x_0)+c_2(t-t_0)]}$	$c_1 = -2\,a_1 c_3 + a_4\,(3\,c_3^2 + 1) + 4\,a_5\,(c_3^3 + c_3)$, $c_2 = -a_1\,(c_3^2 + 1) + a_4\,(c_3^3 + 3\,c_3) + a_5\,(c_3^4 + 6\,c_3^2 + 1)$, $a_1 = \dfrac{(4\,a_5 c_3 + a_4)(3\,a2 + 2\,c_3\,a7)}{3\,a_6 + 2\,a_7} + 2\,a_4 c_3 + 2\,a_5\,(c_3^2 + 5)$, $a_3 = \dfrac{-2\,a_5\,(3\,a_6 + 2\,a_7)^2}{3\,(a_4 + 4\,a_5 c_3)^2}$, $(4\,a_5 c_3 + a_4)\,(3\,a_6 + 2\,a_7) < 0$, x_0, t_0, c_3, and ϕ_0 are arbitrary real constants	—	(4.75)
3.	$\psi(x,t) = \sqrt{\dfrac{-6(4a_5 c_3+a_4)}{3a_6+2a_7}}\,\csc[x - x_0 + c_1(t - t_0)]$ $\times\, e^{i[c_3(x-x_0)+c_2(t-t_0)]}$	$c_1 = -2\,a_1 c_3 + a_4\,(3\,c_3^2 + 1) + 4\,a_5\,(c_3^3 + c_3)$, $c_2 = -a_1\,(c_3^2 + 1) + a_4\,(c_3^3 + 3\,c_3) + a_5\,(c_3^4 + 6\,c_3^2 + 1)$, $a_1 = \dfrac{(4\,a_5 c_3 + a_4)(3\,a2 + 2\,c_3\,a7)}{3\,a_6 + 2\,a_7} + 2\,a_4 c_3 + 2\,a_5\,(c_3^2 + 5)$, $a_3 = \dfrac{-2\,a_5\,(3\,a_6 + 2\,a_7)^2}{3\,(a_4 + 4\,a_5 c_3)^2}$, $(4\,a_5 c_3 + a_4)\,(3\,a_6 + 2\,a_7) < 0$, x_0, t_0, c_3, and ϕ_0 are arbitrary real constants	—	(4.76)
4.	$\psi(x,t) = \sqrt{\dfrac{-6(4a_5 c_3+a_4)}{3a_6+2a_7}}\,\tan[x - x_0 + c_1(t - t_0)]$ $\times\, e^{i[c_3(x-x_0)+c_2(t-t_0)]}$	$c_1 = -2\,a_1 c_3 + a_4\,(3\,c_3^2 - 2) + 4\,a_5\,(c_3^3 - 2\,c_3)$, $c_2 = a_1\,(2 - c_3^2) + a_4\,(c_3^3 - 6\,c_3) + a_5\,(c_3^4 - 12\,c_3^2 + 16)$, $a_1 = \dfrac{(4\,a_5 c_3 + a_4)(3\,a2 + 2\,c_3\,a7)}{3\,a_6 + 2\,a_7} + 2\,a_4 c_3 + 2\,a_5\,(c_3^2 - 10)$, $a_3 = \dfrac{-2\,a_5\,(3\,a_6 + 2\,a_7)^2}{3\,(a_4 + 4\,a_5 c_3)^2}$, $(4\,a_5 c_3 + a_4)\,(3\,a_6 + 2\,a_7) < 0$, x_0, t_0, c_3, and ϕ_0 are arbitrary real constants	—	(4.77)

5.
$$\psi(x,t) = \sqrt{\frac{-6(4a_5c_3+a_4)}{3a_6+2a_7}}\;\cot[x - x_0 + c_1(t-t_0)]$$
$$\times\, e^{i[c_3(x-x_0)+c_2(t-t_0)]}$$

$c_1 = -2a_1c_3 + a_4(3c_3^2-2) + 4a_5(c_3^3-2c_3)$,

$c_2 = \dfrac{a_1(2-c_3^2) + a_4(c_3^3-6c_3) + a_5(c_3^4-12c_3^2+16)}{(4a_5c_3+a_4)(3a_2+2c_3a_7)} + 2a_4c_3 + 2a_5(c_3^2-10)$,

$a_1 = \dfrac{3a_6+2a_7}{}$

$a_3 = \dfrac{-2a_5(3a_6+2a_7)^2}{3(a_4+4a_5c_3)^2}$, $\;(4a_5c_3+a_4)(3a_6+2a_7) < 0$, x_0, t_0, c_3, and

ϕ_0 are arbitrary real constants

(4.78)   —

6.
$$\psi(x,t) = \sqrt{\frac{6(4a_5c_3+a_4)}{3a_6+2a_7}}\;\mathrm{sech}[x - x_0 + c_1(t-t_0)]$$
$$\times\, e^{i[c_3(x-x_0)+c_2(t-t_0)]}$$

$c_1 = -2a_1c_3 + a_4(3c_3^2-1) + 4a_5(c_3^3-c_3)$,

$c_2 = \dfrac{a_1(1-c_3^2) + a_4(c_3^3-3c_3) + a_5(c_3^4-6c_3^2+1)}{(4a_5c_3+a_4)(3a_2+2c_3a_7)} + 2a_4c_3 + 2a_5(c_3^2-5)$,

$a_1 = \dfrac{3a_6+2a_7}{}$

$a_3 = \dfrac{-2a_5(3a_6+2a_7)^2}{3(a_4+4a_5c_3)^2}$, $\;(4a_5c_3+a_4)(3a_6+2a_7) > 0$, x_0, t_0, c_3, and

ϕ_0 are arbitrary real constants

(4.79)   Bright soliton

7.
$$\psi(x,t) = \sqrt{\frac{-6(4a_5c_3+a_4)}{3a_6+2a_7}}\;\mathrm{csch}[x - x_0 + c_1(t-t_0)]$$
$$\times\, e^{i[c_3(x-x_0)+c_2(t-t_0)]}$$

$c_1 = -2a_1c_3 + a_4(3c_3^2-1) + 4a_5(c_3^3-c_3)$,

$c_2 = \dfrac{a_1(1-c_3^2) + a_4(c_3^3-3c_3) + a_5(c_3^4-6c_3^2+1)}{(4a_5c_3+a_4)(3a_2+2c_3a_7)} + 2a_4c_3 + 2a_5(c_3^2-5)$,

$a_1 = \dfrac{3a_6+2a_7}{}$

$a_3 = \dfrac{-2a_5(3a_6+2a_7)^2}{3(a_4+4a_5c_3)^2}$, $\;(4a_5c_3+a_4)(3a_6+2a_7) < 0$, x_0, t_0, c_3, and

ϕ_0 are arbitrary real constants

(4.80)   —

8.
$$\psi(x,t) = \sqrt{\frac{-6(4a_5c_3+a_4)}{3a_6+2a_7}}\;\tanh[x - x_0 + c_1(t-t_0)]$$
$$\times\, e^{i[c_3(x-x_0)+c_2(t-t_0)]}$$

$c_1 = -2a_1c_3 + a_4(3c_3^2+2) + 4a_5(c_3^3+2c_3)$,

$c_2 = \dfrac{-a_1(c_3^2+2) + a_4(6c_3+c_3^3) + a_5(c_3^4+12c_3^2+16)}{(4a_5c_3+a_4)(3a_2+2c_3a_7)} + 2a_4c_3 + 2a_5(c_3^2+10)$,

$a_1 = \dfrac{3a_6+2a_7}{}$

$a_3 = \dfrac{-2a_5(3a_6+2a_7)^2}{3(a_4+4a_5c_3)^2}$, $\;(4a_5c_3+a_4)(3a_6+2a_7) < 0$, x_0, t_0, c_3, and

ϕ_0 are arbitrary real constants

(4.81)   Dark soliton

9.
$$\psi(x,t) = \sqrt{\frac{-6(4a_5c_3+a_4)}{3a_6+2a_7}}\;\coth[x - x_0 + c_1(t-t_0)]$$
$$\times\, e^{i[c_3(x-x_0)+c_2(t-t_0)]}$$

$c_1 = -2a_1c_3 + a_4(3c_3^2+2) + 4a_5(c_3^3+2c_3)$,

$c_2 = \dfrac{-a_1(c_3^2+2) + a_4(6c_3+c_3^3) + a_5(c_3^4+12c_3^2+16)}{(4a_5c_3+a_4)(3a_2+2c_3a_7)} + 2a_4c_3 + 2a_5(c_3^2+10)$,

$a_1 = \dfrac{3a_6+2a_7}{}$

$a_3 = \dfrac{-2a_5(3a_6+2a_7)^2}{3(a_4+4a_5c_3)^2}$, $\;(4a_5c_3+a_4)(3a_6+2a_7) < 0$, x_0, t_0, c_3, and

ϕ_0 are arbitrary real constants

(4.82)   —

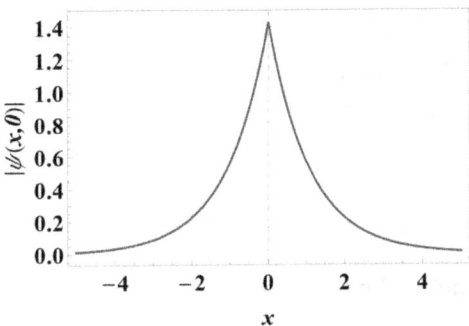

Figure 4.4. Peakon solution (4.85) at $t = 0$, with $A_0 = 1$, $a_2 = -1/2$, $\mu = 0.7$, and $\phi_0 = 0$.

4.17 NLSE with $|\psi|^2$-Dependent Dispersion

Equation:

$$i\,\psi_t + (1 - |\mu|\,|\psi|^2)\,\psi_{xx} + 2\,(1 - |\mu|)\,|\psi|^2\,\psi = 0, \tag{4.83}$$

where $\psi = \psi(x, t)$ is the complex function profile, x and t are its two independent variables, μ is an arbitrary real constant.

Solutions:

Solution 1. **Constant Amplitude** *CW, t- and x-dependent*

$$\psi(x, t) = A_0\,e^{i\,[A_1\,(x-x_0)+A_2\,(t-t_0)+\phi_0]}, \tag{4.84}$$

where $A_2 = 2\,A_0^2 + A_0^2\,|\mu|\,(A_1^2 - 2) - A_1^2$, x_0, t_0, A_0, A_1, μ, and ϕ_0 are arbitrary real constants.

Solution 2. *peakon*
  (Figure 4.4)

$$\psi(x, t) = A_0\,e^{-\sqrt{\frac{2-2\,|\mu|}{|\mu|}}\,\left|x-x_0\right|}\,e^{-i\,[A_1\,(t-t_0)+\phi_0]}, \tag{4.85}$$

where $2 - 2\,|\mu| > 0, |\mu| < 1$, *(for peakon solution)* $A_1 = 2 - \dfrac{2}{|\mu|}$, x_0, t_0, A_0, and ϕ_0 are arbitrary real constants.

  •*Reference*: [18], *we corrected the prefactors of x and t.*

4.18 Infinite Hierarchy of Integrable NLSEs with Higher Order Terms

4.18.1 Constant Coefficients

Equation:

$$i\,\psi_t + a_2\,k_2 - i\,a_3\,k_3 + a_4\,k_4 - i\,a_5\,k_5 + \cdots = 0, \tag{4.86}$$

where $\psi = \psi(x, t)$ is the complex function profile, x and t are its two independent variables, a_j are in general arbitrary real constants, $j = 2, 3, 4, 5, ...,$

$k_j = (-1)^j \frac{\delta}{\delta \psi^*} \int p_{j+1} \, dx, \quad p_j = \psi \frac{\partial}{\partial x}(\frac{p_{j-1}}{\psi}) + \sum_{j_1+j_2=j-1} p_{j_1} p_{j_2}.$ Starting with $p_1 = |\psi|^2$, the next three p_j terms are then $p_2 = \psi \psi_x^*$, $p_3 = |\psi|^4 + \psi \psi_{xx}^*$, $p_4 = \psi(\psi_x \psi^{*2} + 4|\psi|^2 \psi_x^* + \psi_{xxx}^*)$, while the first four k_j terms are $k_2 = \psi_{xx} + 2|\psi|^2 \psi$, $\quad k_3 = \psi_{xxx} + 6|\psi|^2 \psi_x$, $\quad k_4 = \psi_{xxxx} + 8|\psi|^2 \psi_{xx} + 6|\psi|^4 \psi + 4|\psi_x|^2 \psi + 6\psi_x^2 \psi^* + 2\psi^2 \psi_{xx}^*$, $k_5 = \psi_{xxxxx} + 10|\psi|^2 \psi_{xxx} + 10(|\psi_x|^2 \psi)_x + 20\psi^* \psi_x \psi_{xx} + 30|\psi|^4 \psi_x.$

Hierarchy of Integrable NLSEs:
$j = 2$,

$$i\psi_t + a_2 (\psi_{xx} + 2|\psi|^2 \psi) = 0, \tag{4.87}$$

$j = 3$,

$$i\psi_t + a_2 (\psi_{xx} + 2|\psi|^2 \psi) - i a_3 (\psi_{xxx} + 6|\psi|^2 \psi_x) = 0, \tag{4.88}$$

...

Solutions to All Equations in the Hierarchy:

***Solution* 1. Constant Amplitude**   *CW, t-dependent phase*

$$\psi(x, t) = c \, e^{i\left[c^2 \sum_{n=1}^{\infty} \frac{(2n)!}{(n!)^2} a_{2n} c^{2n-2} (t-t_0) + \phi_0\right]}, \tag{4.89}$$

where x_0, t_0, c, and ϕ_0 are arbitrary real constants.
  • *Reference*: [17].

***Solution* 2.**   *bright soliton*

$$\psi(x, t) = c \, \text{sech}[c(x - x_0) + v(t - t_0)] \, e^{i[\phi_1 (t-t_0) + \phi_0]}, \tag{4.90}$$

where $\phi_1 = \sum_{n=1}^{\infty} a_{2n} c^{2n}$, $v = \sum_{n=1}^{\infty} a_{2n+1} c^{2n+1}$, x_0, t_0, c, and ϕ_0 are arbitrary real constants.
  • *Reference*: [17].

Solution* 3. Localization in *x* and *t   *Peregrine soliton*

$$\psi(x, t) = c \left\{ 4 \left[\frac{1 + 2ib(t - t_0)}{d(x, t)} \right] - 1 \right\} e^{i[\phi_1 (t-t_0) + \phi_0]}, \tag{4.91}$$

where $\quad \phi_1 = c^2 \sum_{n=1}^{\infty} \frac{(2n)!}{(n!)^2} a_{2n} c^{2n-2}$, $\quad\quad d(x, t) = 1 + 4b^2 (t - t_0)^2 + 4[c(x - x_0) + v(t - t_0)]^2$, $b = \sum_{n=1}^{\infty} \frac{n(2n)!}{(n!)^2} a_{2n} c^{2n}$, $v = \sum_{n=1}^{\infty} \frac{(2n+1)!}{(n!)^2} a_{2n+1} c^{2n+1}$, x_0, t_0, c, and ϕ_0 are arbitrary real constants.
  • *Reference*: [17].

Solution **4. Rational Solution** $\qquad$ *DW*

$$\psi(x,\,t) = c \left\{ \frac{4}{1 + 4[c\,(x - x_0) + v\,(t - t_0)]^2} - 1 \right\} e^{i\,\phi_0}, \qquad (4.92)$$

where $v = \sum_{n=1}^{\infty} \frac{(2\,n + 1)!}{(n\,!)^2} a_{2\,n+1}\,c^{2\,n+1}$, x_0, t_0, c, and ϕ_0 are arbitrary real constants.

- *Reference*: [17].

Solution **5. Periodicity in x and Localization in t** $\qquad$ *Akhmediev breather*

$$\psi(x,\,t) = c \left\{ 1 + \frac{\kappa^2 \cosh\left[b\,\kappa\,\sqrt{1 - \frac{\kappa^2}{4}}\,(t - t_0)\right] + i\,\kappa\,\sqrt{4 - \kappa^2}\,\sinh\left[b\,\kappa\,\sqrt{1 - \frac{\kappa^2}{4}}\,(t - t_0)\right]}{\sqrt{4 - \kappa^2}\,\cos[\kappa\,c\,(x - x_0) + \kappa\,v\,(t - t_0)] - 2\cosh[b\,\kappa\,\sqrt{1 - \frac{\kappa^2}{4}}\,(t - t_0)]} \right\} \qquad (4.93)$$

$$\times\, e^{i\,[\phi_1\,(t - t_0) + \phi_0]},$$

where $\quad v = \sum_{n=1}^{\infty} \frac{(2\,n + 1)!}{n\,!} a_{2\,n+1}\,c^{2\,n+1}[\sum_{r=0}^{n} \frac{(-1)^r \kappa^{2\,r}\,r\,!}{(n - r)!\,(2\,r + 1)!}], \quad \phi_1 = \sum_{n=1}^{\infty} \frac{(2\,n)!}{(n\,!)^2} a_{2\,n}\,c^{2\,n},$

$b = 2\sum_{n=0}^{\infty} \frac{(2\,n + 1)!}{n\,!} a_{2\,n+2}\,c^{2\,n+2}[\sum_{r=0}^{n} \frac{(-1)^r \kappa^{2\,r}\,r\,!}{(n - r)!\,(2\,r + 1)!}]$, $0 < \kappa < 2$, x_0, t_0, c, and ϕ_0 are arbitrary real constants.

Solution **6. Periodicity in t and Localization in x** $\qquad$ *Kuznetsov–Ma breather*

$$\psi(x,\,t) = c\,\sqrt{2} \left[\frac{2(1 - \kappa)\,d_1(t) - \sqrt{2\,\kappa}\,d_3(x,\,t) + 2\,i\,\sqrt{1 - 2\,\kappa}\,d_2(t)}{\sqrt{2}\,d_3(x,\,t) - 2\,\sqrt{\kappa}\,d_1(t)} \right] e^{i\,[\phi_1\,(t - t_0) + \phi_0]}, \qquad (4.94)$$

where $\quad d_1(t) = \cos[2\,\sqrt{1 - 2\,\kappa}\,b\,(t - t_0)], \quad d_2(t) = \sin[2\,\sqrt{1 - 2\,\kappa}\,b\,(t - t_0)],$

$d_3(x,\,t) = \cosh[2\,\sqrt{1 - 2\,\kappa}\,[c\,(x - x_0) + v\,(t - t_0)]$,

$v = \sum_{n=1}^{\infty} 4^n\,a_{2\,n+1}\,c^{2\,n+1}[1 + \sum_{r=1}^{n} \frac{(2\,r - 1)!!\,\kappa^r}{r\,!}], \qquad \phi_1 = \sum_{n=1}^{\infty} \frac{(2\,n)!}{(n\,!)^2} \alpha_{2\,n}\,c^{2\,n}\,(2\,\kappa)^n,$

$b = 2\sum_{n=0}^{\infty} 4^n\,a_{2\,n+2}\,c^{2\,n+2}[1 + \sum_{r=1}^{n} \frac{(2\,r - 1)!!\,\kappa^r}{r\,!}]$, $0 < \kappa < 1/2$, x_0, t_0, c, and ϕ_0 are arbitrary real constants.

- *Reference*: [17].

Solution **7.** $\qquad$ *solitary wane (SW)*

$$\psi(x,\,t) = c\,\mathrm{dn}[c\,(x - x_0) + v\,(t - t_0),\,m]\,e^{i\,[\phi_1\,(t - t_0) + \phi_0]}, \qquad (4.95)$$

where $v = \sum_{n=1}^{\infty} a_{2\,n+1}\,c^{2\,n+1}\,m^n\,P_n(\frac{2}{m} - 1)$, $\phi_1 = \sum_{n=1}^{\infty} a_{2\,n}\,c^{2\,n}\,m^n\,P_n(\frac{2}{m} - 1)$, P_n is the set of orthogonal Legendre polynomials of the first kind, $0 < m < 1$, x_0, t_0, c, and ϕ_0 are arbitrary real constants.

- *Reference*: [17].

Solution **8.** $\qquad$ *SW*

$$\psi(x,\,t) = \frac{c\,\coth(\kappa)}{\sqrt{2}}\,\mathrm{cn}\left[\frac{c\,(x - x_0) + v\,(t - t_0)}{\sinh(\kappa)},\,m \right] e^{i\,[\phi_1\,(t - t_0) + \phi_0]}, \qquad (4.96)$$

where $v = \sum_{n=1}^{\infty} a_{2n+1} c^{2n+1} \sinh^{-2n}(\kappa) P_n[\sinh^2(\kappa)]$, $\phi_1 = \sum_{n=1}^{\infty} a_{2n} c^{2n} \sinh^{-2n}(\kappa) P_n[\sinh^2(\kappa)]$, P_n is the set of orthogonal Legendre polynomials of the first kind, $0 < [m = \frac{1}{2}\cosh^2(\kappa)] < 1$ with κ real, example: $\kappa = 1/2$.

• *Reference*: [17].

4.18.2 Function Coefficients

If $\psi(x, t)$ is a solution to an infinite hierarchy with constant coefficients

$$i\,\psi_t + a_{02}\,k_2 - i\,a_{03}\,k_3 + a_{04}\,k_4 - i\,a_{05}\,k_5 + \cdots = 0, \tag{4.97}$$

then $\psi(x, t)$ will be also a solution to the same infinite hierarchy with t-dependent coefficients

$$i\,\psi_t + a_2(t)\,k_2 - i\,a_3(t)\,k_3 + a_4(t)\,k_4 - i\,a_5(t)\,k_5 + \cdots = 0, \tag{4.98}$$

after making the following replacements in ψ:

1. $a_{02j}\,(t - t_0) \rightarrow \int a_{2j}(t)\,d\,t$,
2. $a_{02j+1}\,(t - t_0) \rightarrow \int a_{2j+1}(t)\,d\,t$, $j = 1, 2, 3, 4, \ldots$,

where a_{0n} are in general arbitrary real constants, $a_n(t)$ are arbitrary real functions, $n = 2, 3, 4, 5, \ldots$.

Example 1.   *bright soliton*

Given $\psi(x, t) = c\,\text{sech}[c\,(x - x_0) + v\,(t - t_0)]\,e^{i\,[\phi_1\,(t-t_0)+\phi_0]}$ is a solution to (4.97), where $\phi_1 = \sum_{n=1}^{\infty} a_{02n}\,c^{2n}$, $v = \sum_{n=1}^{\infty} a_{02n+1}\,c^{2n+1}$, then

$$\psi(x, t) = c\,\text{sech}\left[c\,(x - x_0) + \sum_{n=1}^{\infty} \int a_{2n+1}(t)\,d\,t\,c^{2n+1}\right] e^{i\left[\sum_{n=1}^{\infty}\int a_{2n}(t)\,d\,t\,c^{2n}+\phi_0\right]} \tag{4.99}$$

is a solution to (4.98), where x_0, t_0, c, and ϕ_0 are arbitrary real constants.

• *Reference*: [17].

Example 2. **Localization in x and t**   *Peregrine soliton*

Given $\psi(x, t) = c\,\{4\,[\frac{1 + 2\,i\,b\,(t - t_0)}{d(x,t)}] - 1\}\,e^{i\,[\phi_1\,(t-t_0)+\phi_0]}$ is a solution to (4.97), where $\phi_1 = c^2\sum_{n=1}^{\infty} \frac{(2n)!}{(n!)^2}\,a_{02n}\,c^{2n-2}$, $d(x, t) = 1 + 4\,b^2\,(t - t_0)^2 + 4\,[c\,(x - x_0) + v\,(t - t_0)]^2$, $b = \sum_{n=1}^{\infty} \frac{n(2n)!}{(n!)^2}\,a_{02n}\,c^{2n}$, $v = \sum_{n=1}^{\infty} \frac{(2n+1)!}{(n!)^2}\,a_{02n+1}\,c^{2n+1}$, then

$$\psi(x, t) = c\,\left\{4\,\left[\frac{1 + 2\,i\,\sum_{n=1}^{\infty}\frac{n(2n)!}{(n!)^2}\int a_{2n}(t)\,d\,t\,c^{2n}}{d(x,t)}\right] - 1\right\}\,e^{i\left[c^2\sum_{n=1}^{\infty}\frac{(2n)!}{(n!)^2}\int a_{2n}(t)\,d\,t\,c^{2n-2}+\phi_0\right]}, \tag{4.100}$$

is a solution to (4.98), where $d(x, t) = 1 + 4\left[\sum_{n=1}^{\infty}\frac{n(2n)!}{(n!)^2}\int a_{2n}(t)\,d\,t\,c^{2n}\right]^2 + 4\left[c\,(x - x_0) + \sum_{n=1}^{\infty}\frac{(2n+1)!}{(n!)^2}\int a_{2n+1}(t)\,d\,t\,c^{2n+1}\right]^2$, x_0, t_0, c, and ϕ_0 are arbitrary real constants.

• *Reference*: [17].

4.19 Summary of Section 4.18

Constant Coefficients

Equation: $i\,\psi_t + a_2\,k_2 - i\,a_3\,k_3 + a_4\,k_4 - i\,a_5\,k_5 + \cdots := 0$

#	Solution	Conditions	Name	Equation #
1.	$\psi(x,t) = c\,e^{i[c^2\sum_{n=1}^{\infty}\frac{(2n)!}{(n!)^2}a_{2n}c^{2n-2}(t-t_0)+\phi_0]}$	x_0, t_0, c, and ϕ_0 are arbitrary real constants	Continuous wave, t-dependent phase	(4.89)
2.	$\psi(x,t) = c\,\mathrm{sech}[c(x-x_0)+v(t-t_0)]\,e^{i[\phi_1(t-t_0)+\phi_0]}$	$\phi_1 = \sum_{n=1}^{\infty}a_{2n}c^{2n}$, $v = \sum_{n=1}^{\infty}a_{2n+1}c^{2n+1}$, x_0, t_0, c, and ϕ_0 are arbitrary real constants	Bright soliton	(4.90)
3.	$\psi(x,t) = c\left\{4\left[\frac{1+2\,i\,b\,(t-t_0)}{d(x,t)}\right]-1\right\}e^{i[\phi_1(t-t_0)+\phi_0]}$	$\phi_1 = c^2\sum_{n=1}^{\infty}\frac{(2n)!}{(n!)^2}a_{2n}c^{2n-2}$, $d(x,t) = 1 + 4b^2(t-t_0)^2 + 4[c(x-x_0)+v(t-t_0)]^2$, $b = \sum_{n=1}^{\infty}\frac{m(2m)!}{(n!)^2}a_{2n}c^{2n}$, $v = \sum_{n=1}^{\infty}\frac{(2n+1)!}{(n!)^2}a_{2n+1}c^{2n+1}$, x_0, t_0, c, and ϕ_0 are arbitrary real constants	Peregrine soliton	(4.91)
4.	$\psi(x,t) = c\left\{\frac{4}{1+4[c(x-x_0)+v(t-t_0)]^2}-1\right\}e^{i\,\phi_0}$	$v = \sum_{n=1}^{\infty}\frac{(2n+1)!}{(n!)^2}a_{2n+1}c^{2n+1}$, x_0, t_0, c, and ϕ_0 are arbitrary real constants	Decaying wave	(4.92)
5.	$\psi(x,t) = c\left\{1+\frac{q(t)}{q_2(x,t)}\right\}e^{i[\phi_1(t-t_0)+\phi_0]}$, $\quad q_1(t) = \kappa^2\cosh[b\,\kappa\sqrt{1-\frac{\kappa^2}{4}}\,(t-t_0)]$ $\quad + i\,\kappa\sqrt{4-\kappa^2}\sinh[b\,\kappa\sqrt{1-\frac{\kappa^2}{4}}\,(t-t_0)]$, $\quad q_2(x,t) = \sqrt{4-\kappa^2}\cos[\kappa\,[c(x-x_0)+v(t-t_0)]]$ $\quad - 2\cosh[b\,\kappa\sqrt{1-\frac{\kappa^2}{4}}\,(t-t_0)]]$	$v = \sum_{n=1}^{\infty}\frac{(2n+1)!}{(n!)^2}a_{2n+1}c^{2n+1}\left(\sum_{r=0}^{n}\frac{(-1)^r\kappa^{2r}r!}{(n-r)!(2r+1)!}\right)$, $\phi_1 = \sum_{n=1}^{\infty}\frac{(2n+1)!}{(n!)^2}a_{2n}c^{2n}$, $b = 2\sum_{n=0}^{\infty}\frac{(2n+1)!}{n!}a_{2n+2}c^{2n+2}\left(\sum_{r=0}^{n}\frac{(-1)^r\kappa^{2r}r!}{(n-r)!(2r+1)!}\right)$, $0<\kappa<2$, x_0, t_0, c, and ϕ_0 are arbitrary real constants	Akhmediev breather	(4.93)
6.	$\psi(x,t) = c\sqrt{2}\left[\frac{2(1-\kappa)d(t)-\sqrt{2\kappa}\,d_3(x,t)+2\,i\,\sqrt{1-2\kappa}\,d_2(t)}{\sqrt{2}\,d_3(x,t)-2\sqrt{\kappa}\,d(t)}\right]$ $\times e^{i[\phi_1(t-t_0)+\phi_0]}$	$d(t) = \cos[2\sqrt{1-2\kappa}\,b\,(t-t_0)]$, $d_2(t) = \sin[2\sqrt{1-2\kappa}\,b\,(t-t_0)]$, $d_3(x,t) = \cosh[2\sqrt{1-2\kappa}\,[c(x-x_0)+v(t-t_0)]]$, $v = \sum_{n=1}^{\infty}4^n a_{2n+1}c^{2n+1}(1+\sum_{r=1}^{n}\frac{(-1)^{r-1}!}{r!}\kappa^r)$, $\phi_1 = \sum_{n=1}^{\infty}\frac{(2n)!}{(n!)^2}a_{2n}c^{2n}(2\kappa)^n$, $0<\kappa<1/2$,	Kuznetsov–Ma breather	(4.94)

#	Solution	Function coefficients	Name	equation #
7.	$\psi(x,t) = c\,\mathrm{dn}[c(x-x_0) + v(t-t_0), m]\, e^{i[\phi_1(t-t_0)+\phi_0]}$	$b = 2\sum_{n=0}^\infty 4^n a_{2n+2} c^{2n+2}\left(1 + \sum_{r=1}^n \frac{(2r-1)!!\,\kappa^r}{r!}\right)$, x_0, t_0, c, and ϕ_0 are arbitrary real constants $v = \sum_{n=1}^\infty a_{2n+1} c^{2n+1} m^n P_n^m\!\left(\frac{2}{m} - 1\right)$, $\phi_1 = \sum_{n=1}^\infty a_{2n} c^{2n} m^n P_n\!\left(\frac{2}{m} - 1\right)$, P_n is the set of orthogonal Legendre polynomials of the first kind, $0 < m < 1$, x_0, t_0, c, and ϕ_0 are arbitrary real constants	Solitary wave	(4.95)
8.	$\psi(x,t) = \dfrac{c\coth(\kappa)}{\sqrt{2}}\,\mathrm{cn}\!\left[\dfrac{c(x-x_0)+v(t-t_0)}{\sinh(\kappa)}, m\right] e^{i[\phi_1(t-t_0)+\phi_0]}$	$v = \sum_{n=1}^\infty a_{2n+1} c^{2n+1} \sinh^{-2n}(\kappa)\, P_n[\sinh^2(\kappa)]$, $\phi_1 = \sum_{n=1}^\infty a_{2n} c^{2n} \sinh^{-2n}(\kappa)\, P_n[\sinh^2(\kappa)]$, P_n is the set of orthogonal Legendre polynomials of the first kind, $0 < [m = \frac{1}{2}\cosh^2(\kappa)] < 1$ with κ real, example: $\kappa = 1/2$	Solitary wave	(4.96)

Function coefficients

Equation: $i\psi_t + a_2(t)k_2 - i a_3(t)k_3 + a_4(t)k_4 - i a_5(t)k_5 + \cdots = 0$

#	Solution	Conditions	Name	equation #
1.	$\psi(x,t) = c\,\mathrm{sech}\left\{c(x-x_0) + \sum_{n=1}^\infty \int a_{2n+1}(t)\,dt\,c^{2n+1}\right\} \times e^{i\left(\sum_{n=1}^\infty \int a_{2n}(t)\,dt\,c^{2n}+\phi_0\right)}$	x_0, c, and ϕ_0 are arbitrary real constants	Bright soliton	(4.99)
2.	$\psi(x,t) = c\left\{\left\{4\left[\dfrac{1+2i\sum_{n=1}^\infty \frac{m(2n)!}{(n!)^2}\int a_{2n}(t)\,dt\,c^{2n}}{d(x,t)}\right] - 1\right\} \times e^{i\left(c^2\sum_{n=1}^\infty \frac{(2n)!}{(n!)^2}\int a_{2n}(t)\,dt\,c^{2n-2}+\phi_0\right)} - \right.$	$d(x,t) = 1 + 4\left[\sum_{n=1}^\infty \frac{m(2n)!}{(n!)^2}\int a_{2n}(t)\,dt\,c^{2n}\right]^2 + 4[c(x-x_0) + \sum_{n=1}^\infty \frac{(2n+1)!}{(n!)^2}\int a_{2n+1}(t)\,dt\,c^{2n+1}]^2$, x_0, c, and ϕ_0 are arbitrary real constants	Peregrine soliton	(4.100)

References

[1] Agrawal G 2001 *Nonlinear Fiber Optics* 3rd ed (San Diego, CA: Academic)

[2] Kivshar Y S and Agrawal G P 2003 *Optical Solitons* (San Diego, CA: Academic)

[3] Hellwarth R W 1977 Third-order optical susceptibilities of liquids and solids *Progr. Quant. Electron.* **5** 2–68

[4] Butcher P N and Cotter D 1990 *The Elements of Nonlinear Optics* (Cambridge: Cambridge University Press)

[5] Fu H, Wang Y and Gao B 2003 Beyond Fermi pseudopotential: a modified GP equation *Phys. Rev.* A **67** 053612

[6] Lee T D and Yang C N 1957 Many-body problem in quantum mechanics and quantum statistical mechanics *Phys. Rev.* **105** 1119–20

[7] Lee T D, Huang K and Yang C N 1957 Eigenvalues and eigenfunctions of a Bose system of hard spheres and its low-temperature properties *Phys. Rev.* **106** 1135

[8] Dysthe K B 1979 Note on a modification to the nonlinear Schrödinger equation for application to deep water waves *Proc. R. Soc.* A **369** 105–14

[9] Davey A and Stewartson K 1974 On three-dimensional packets of surface waves *Proc. R. Soc.* A **338** 101–10

[10] Zayed E M and Al-Nowehy A G 2017 Exact solutions for the perturbed nonlinear Schrödinger equation with power law nonlinearity and Hamiltonian perturbed terms *Optik Int. J. Light Electron Opt.* **139** 123–44

[11] Mahak N and Akram G 2019 Extension of rational sine-cosine and rational sinh-cosh techniques to extract solutions for the perturbed NLSE with Kerr law nonlinearity *Eur. Phys. J. Plus* **134** 1–10

[12] Li Z, Li L, Tian H and Zhou G 2000 New types of solitary wave solutions for the higher order nonlinear Schrödinger equation *Phys. Rev. Lett.* **84** 4096–9

[13] El-Shiekh R M 2019 Classes of new exact solutions for nonlinear Schrödinger equations with variable coefficients arising in optical fiber *Results Phys.* **13** 102214

[14] Wazwaz A M 2006 Exact solutions for the fourth order nonlinear Schrödinger equations with cubic and power law nonlinearities *Math. Comput. Modelling* **43** 802–8

[15] Huang Y and Liu P 2014 New exact solutions for a class of high-order dispersive cubic-quintic nonlinear Schrödinger equation *J. Math. Res.* **6** 104–8

[16] Zayed E M and Al-Nowehy A G 2017 Exact solutions and optical soliton solutions for the nonlinear Schrödinger equation with fourth-order dispersion and cubic-quintic nonlinearity *Ricerche di Matematica* **66** 531–52

[17] Ankiewicz A, Kedziora D J, Chowdury A, Bandelow U and Akhmediev N 2016 Infinite hierarchy of nonlinear Schrödinger equations and their solutions *Phys. Rev.* E **93** 012206–10

[18] Kevrekidis P G 2009 *The Discrete Nonlinear Schrödinger equation: Mathematical Analysis, Numerical Computations and Physical Perspectives* (Springer) p 232

IOP Publishing

Handbook of Exact Solutions to the Nonlinear Schrödinger
Equations (Second Edition)

Usama Al Khawaja and Laila Al Sakkaf

Chapter 5

Scaling Transformations

Known also as similarity transformations

A Glance at Chapter 5

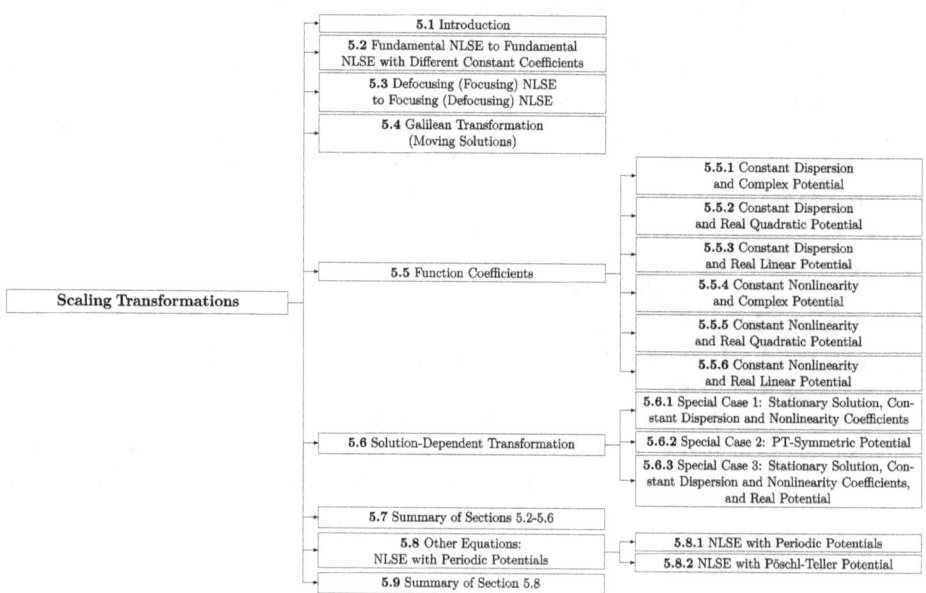

A Statistical View of Chapter 5

	Equation	Solution
1	$i\,\Phi_t + a_{11}\,\Phi_{xx} + a_{22}\,|\Phi|^2\,\Phi = 0$	2
2	$i\,\Phi_t + a_1\,\Phi_{xx} - a_2\,|\Phi|^2\,\Phi = 0$	2
3	$i\,\Phi_t + a_1\,\Phi_{xx} + a_2\,|\Phi|^2\,\Phi = 0$	3
4	$i\,\Phi_t + a_1\,\Phi_{xx} + a_2\,|\Phi|^n\,\Phi = 0$	1
5	$i\,\Phi_t + a_1\,\Phi_{xx} + a_2\,|\Phi|^n\,\Phi + a_3\,|\Phi|^m\,\Phi = 0$	1
6	$i\,\Phi_t + b_1(x,t)\,\Phi_{xx} + b_2(x,t)\,|\Phi|^2\,\Phi + [b_{3r}(x,t) + i\,b_{3i}(x,t)]\,\Phi = 0$	0
7	$i\,\Phi_t + b_{10}\,\Phi_{xx} + b_2(x,t)\,|\Phi|^2\,\Phi + [b_{3r}(x,t) + i\,b_{3i}(x,t)]\,\Phi = 0$	0
8	$i\,\Phi_t + b_{10}\,\Phi_{xx} + \dfrac{a_2\,b_{10}\,g_5(t)}{a_1\,c_2^2}\,|\Phi|^2\,\Phi - \dfrac{g_5(t)\,g_5''(t) - 2\,g_5'^2(t)}{4\,b_{10}\,g_5^2(t)}\,x^2\,\Phi = 0$	3
9	$i\,\Phi_t + b_{10}\,\Phi_{xx} + \dfrac{a_2\,b_{10}\,[\alpha+\beta\sin(\gamma\,t)]}{a_1\,c_2^2}\,|\Phi|^2\,\Phi + \dfrac{\beta\,\gamma^2\,[3\,\beta + \beta\cos(2\,\gamma\,t) + 2\,\alpha\sin(\gamma\,t)]}{8\,b_{10}\,[\alpha + \beta\sin(\gamma\,t)]^2}\,x^2\,\Phi = 0$	3
10	$i\,\Phi_t + b_{10}\,\Phi_{xx} + \dfrac{a_2\,b_{10}\,e^{\gamma\,t}}{a_1\,c_2^2}\,|\Phi|^2\,\Phi + \dfrac{\gamma^2}{4\,b_{10}}\,x^2\,\Phi = 0$	3
11	$i\,\Phi_t + b_{10}\,\Phi_{xx} + \dfrac{a_2\,b_{10}\,c_5}{a_1\,c_2^2\,c_7}\,|\Phi|^2\,\Phi - \dfrac{c_7\,g_4''(t)}{2\,b_{10}\,c_5}\,x\,\Phi = 0$	3
12	$i\,\Phi_t + b_1(x,t)\,\Phi_{xx} + b_{20}\,|\Phi|^2\,\Phi + [b_{3r}(x,t) + i\,b_{3i}(x,t)]\,\Phi = 0$	0
13	$i\,\Phi_t + \dfrac{a_1\,b_{20}\,c_2^2}{a_2\,g_5(t)}\,\Phi_{xx} + b_{20}\,|\Phi|^2\,\Phi + \dfrac{a_2\,[g_5'^2(t) - g_5(t)\,g_5''(t)]}{4\,a_1\,b_{20}\,c_2^2\,g_5(t)}\,x^2\,\Phi = 0$	0
14	$i\,\Phi_t + \dfrac{a_1\,b_{20}\,c_2^2\,e^{-c_6\,t}}{a_2\,c_5}\,\Phi_{xx} + b_{20}\,|\Phi|^2\,\Phi + \dfrac{a_2\,[c_6\,g_4'(t) - g_4''(t)]}{2\,a_1\,b_{20}\,c_2^2}\,x\,\Phi = 0$	0
15	$i\,\Phi_t + b_{10}\,\Phi_{xx} + b_{20}\,|\Phi|^2$ $\Phi + \left[B_t(x,t) + b_{10}\,B_x^2(x,t) - i\,b_{10}\left(\dfrac{\sqrt{\frac{4\,a_1}{b_{10}}}\,f'\!\left[\sqrt{\frac{a_1}{b_{10}}}\,(x-x_0)\right]\,B_x(x,t)}{f\!\left[\sqrt{\frac{a_1}{b_{10}}}\,(x-x_0)\right]} + B_{xx}(x,t) \right) \right]\Phi = 0$	0
16	$i\,\Phi_t + b_{10}\,\Phi_{xx} + b_{20}\,|\Phi|^2\,\Phi + [V_{even}(x) + i\,V_{odd}(x)]\Phi = 0$	0
17	$i\,\Phi_t + b_{10}\,\Phi_{xx} + b_{20}\,|\Phi|^2\,\Phi + b_{10}\,[\cos^2(x - x_0) + i\sin(x - x_0)]\,\Phi = 0$	0
18	$i\Phi_t + b_{10}\,\Phi_{xx} + b_{20}\,|\Phi|^2\,\Phi + \left\{ \dfrac{b_{10}\,g_1^2(t)}{f^4[\sqrt{\frac{a_1}{b_{10}}}(x-x_0)]} + g_1'(t)\int \dfrac{d\,x}{f^2[\sqrt{\frac{a_1}{b_{10}}}(x-x_0)]} + g_2'(t) \right\}\Phi = 0$	0
19	$i\,\psi_t + \tfrac{1}{2}\,\psi_{xx} - |\psi|^2\,\psi + V_0\,\mathrm{sn}^2(x,m)\,\psi = 0$	3
20	$i\,\psi_t + \tfrac{1}{2}\,\psi_{xx} - |\psi|^2\,\psi + V_0\,\sin^2(x)\,\psi = 0$	2
21	$i\,\psi_t + a_1\,\psi_{xx} + a_2\,|\psi|^2\,\psi + V_0\,\mathrm{sech}^2(V_1\,x)\,\psi = 0$	6
Total	21	32

5.1 Introduction

Many of the different versions of the NLSE turn out to be related to each other through some functional transformations. The most significant among these are the ones that can be reduced to the fundamental NLSE or any other integrable version of it. Functional transformations may include the amplitude, phase, or independent variables of the solution.

Assuming $u(x, t)$ is a solution to a certain version of the NLSE, then the transformation

$$u(x, t) \rightarrow A(x, t)U[X(x, t), T(x, t)], \qquad (5.1)$$

where $A(x, t)$, $X(x, t)$, and $T(x, t)$ are in general complex functions, reduces the equation to the fundamental NLSE and maps its solution, $u(x, t)$, to a solution to the fundamental NLSE, namely $U(X, T)$. The reduction determines the functions $A(x, t)$, $X(x, t)$, and $T(x, t)$, and thus maps all solutions of the new equation to those of the fundamental NLSE. The reverse process is of course possible. For instance, one may start with the following form of general NLSE

$$i\frac{\partial}{\partial t}u(x, t) = f(x, t)\frac{\partial^2}{\partial x^2}u(x, t) + g(x, t)\frac{\partial}{\partial x}u(x, t) + h(x, t)u(x, t) + k(x, t)|u(x, t)|^2 u(x, t), \quad (5.2)$$

which contains first derivative and potential terms as additional terms compared with the fundamental NLSE, in addition to the space- and time-dependent coefficients. This equation can be reduced to the fundamental NLSE with constant coefficients, using the transformation (5.1), which upon substituting for $u(x, t)$ from (5.1) in (5.2) gives

$$\begin{aligned}
&S_0(x, t)U_T(X, T) + S_1(x, t)U_{XX}(X, T) + S_2|U(X, T)|^2 U(X, T) \\
&+ S_3(x, t)U_X(X, T) + S_4(x, t)U_{TT}(X, T) + S_5(x, t)U_{XT}(X, T) + S_6 U(X, T) = 0,
\end{aligned} \qquad (5.3)$$

where

$$S_0(x, t) = -2fA_x T_x + A(iT_t - gT_x - fT_{xx}), \qquad (5.4)$$

$$S_1(x, t) = -AfX_x^2, \qquad (5.5)$$

$$S_2(x, t) = -A^3 k, \qquad (5.6)$$

$$S_3(x, t) = -2fA_x X_x + A(iX_t - gX_x - fX_{xx}), \qquad (5.7)$$

$$S_4(x, t) = -AfT_x^2, \qquad (5.8)$$

$$S_5(x, t) = -2AfT_x X_x, \qquad (5.9)$$

$$S_6(x, t) = -Ah + iA_t - gA_x - fA_{xx}. \qquad (5.10)$$

Clearly, for Equation (5.3) to reduce to the fundamental NLSE (Equation (2.1)), we must have

$$S_0 = i, \ S_1 = a_1, \ S_2 = a_2, \ S_3 = S_4 = S_5 = S_6 = 0. \tag{5.11}$$

The six unknown functions, f, g, h, k, X, T should satisfy the seven Equations (5.11). This is an over-determined system which may not be satisfied. Nonetheless, certain solutions do exist, but with restrictions on the coefficients of the NLSE, which are known as the integrability conditions, and lead to versions of the NLSE of particular importance [1]. Many such examples are included in this chapter.

The transformation (5.1) is not the most general transformation. Other transformations such as $u(x, t) \rightarrow u^n(x, t)$ are useful for the NLSE with power-law nonlinearity [2]. One may also consider integral or derivative transformations that may lead to wider range of NLSE versions.

In conclusion, this chapter does not include fundamentally new NLSEs; it only reveals the links between the different versions of the NLSE by reducing the more complicated ones to simpler NLSEs. It also maps the solutions of many versions of NLSE to those of more fundamental and integrable NLSEs.

5.2 Fundamental NLSE to Fundamental NLSE with Different Constant Coefficients

If $\psi(x, t)$ is a solution to the fundamental NLSE (2.160),

$$i \, \psi_t + a_1 \, \psi_{xx} + a_2 \, |\psi|^2 \, \psi = 0,$$

then

$$\Phi(x, t) = \sqrt{\frac{a_2}{a_{22}}} \, \psi\left(\sqrt{\frac{a_1}{a_{11}}} \, x, \, t \right) \tag{5.12}$$

is a solution to

$$i \, \Phi_t + a_{11} \, \Phi_{xx} + a_{22} \, |\Phi|^2 \, \Phi = 0, \tag{5.13}$$

with arbitrary real constants a_1, a_2, a_{11} and a_{22}.

***Example* 1:**   *Bright soliton*

Given $\psi(x, t) = A_0 \sqrt{\frac{2 \, a_1}{a_2}} \, \text{sech}[A_0 \, (x - x_0)] \, e^{i \, [a_1 \, A_0^2 \, (t - t_0) + \phi_0]}$ is a solution to (2.160), then

$$\Phi(x, t) = A_0 \sqrt{\frac{2 \, a_1}{a_{22}}} \, \text{sech}\left[A_0 \sqrt{\frac{a_1}{a_{11}}} \, (x - x_0) \right] e^{i \, [a_1 \, A_0^2 \, (t - t_0) + \phi_0]} \tag{5.14}$$

is a solution to (5.13), where $a_1 \, a_{22} > 0$, $a_1 \, a_{11} > 0$, A_0, x_0, t_0, and ϕ_0 are arbitrary real constants.

***Example* 2:**   *Dark soliton*

Given $\psi(x, t) = A_0 \sqrt{\frac{-2 \, a_1}{a_2}} \, \tanh[A_0 \, (x - x_0)] \, e^{-i \, [2a_1 \, A_0^2 \, (t - t_0) + \phi_0]}$ is a solution to (2.160), then

$$\Phi(x,\ t) = A_0 \sqrt{\frac{-2\,a_1}{a_{22}}}\ \tanh\left[A_0 \sqrt{\frac{a_1}{a_{11}}}\ (x - x_0)\right] e^{-i\,[2\,a_1\,A_0^2\,(t - t_0) + \phi_0]} \qquad (5.15)$$

is a solution to (5.13), where $a_1\,a_{22} < 0$, $a_1\,a_{11} > 0$, A_0, x_0, t_0, and ϕ_0 are arbitrary real constants.

5.3 Defocusing (Focusing) NLSE to Focusing (Defocusing) NLSE

If $\psi(x,\ t)$ is a solution to the fundamental NLSE (2.160),

$$i\,\psi_t + a_1\,\psi_{xx} + a_2\,|\psi|^2\,\psi = 0,$$

then

$$\Phi(x,\ t) = \psi(i\,x,\ -t) \qquad (5.16)$$

is a solution to

$$i\,\Phi_t + a_1\,\Phi_{xx} - a_2\,|\Phi|^2\,\Phi = 0, \qquad (5.17)$$

with arbitrary real constants a_1 and a_2. Here $\psi(x,\ t)$ should be an even function in x.

Example **1:**
(Figure 5.1)

Given $\psi(x,\ t) = A_0 \sqrt{\frac{2\,a_1}{a_2}}\ \mathrm{sech}[A_0\,(x - x_0)]\ e^{i\,[a_1\,A_0^2\,(t - t_0) + \phi_0]}$ is a solution to (2.160), then

$$\Phi(x,\ t) = A_0 \sqrt{\frac{2\,a_1}{a_2}}\ \mathrm{sech}[i\,A_0\,(x - x_0)]\ e^{i\,[-a_1\,A_0^2\,(t - t_0) + \phi_0]} \qquad (5.18)$$

is a solution to (5.17), where $a_1\,a_2 > 0$, A_0, x_0, t_0, and ϕ_0 are arbitrary real constants.

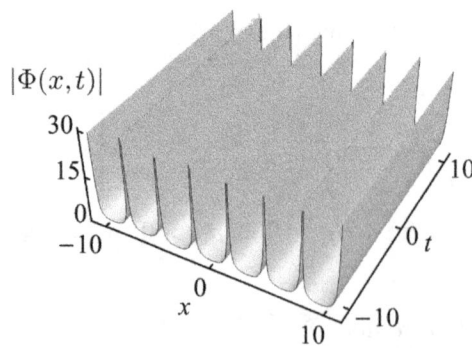

Figure 5.1. Plot of solution (5.18) with $a_1 = a_2 = 1$, $A_0 = 1$, and $x_0 = t_0 = \phi_0 = 0$.

Example 2:
(Figure 5.2)

Given $\psi(x, t) = \frac{1}{\sqrt{a_2}} \left[\frac{4 + i\, 8\, (t - t_0)}{1 + 4\, (t - t_0)^2 + \frac{2}{a_1} ((x - x_0)^2} - 1 \right] e^{i\, [t - t_0 + \phi_0]}$ is a solution to

(2.160), then

$$\Phi(x, t) = \frac{1}{\sqrt{a_2}} \left[\frac{4 - i\, 8\, (t - t_0)}{1 + 4\, ((t - t_0)^2 - \frac{2}{a_1}\, ((x - x_0)^2} - 1 \right] e^{i\, [-(t - t_0) + \phi_0]} \quad (5.19)$$

is a solution to (5.17), where $a_2 > 0$, x_0, t_0, and ϕ_0 are arbitrary real constants.

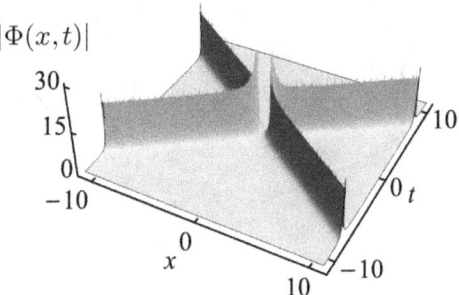

Figure 5.2. Plot of solution (5.19) with $a_1 = a_2 = 1$, and $x_0 = t_0 = \phi_0 = 0$. Animation available online at http://doi.org/10.1088/978-0-7503-5954-2.

5.4 Galilean Transformation (Moving Solutions)

If $\psi(x, t)$ is a solution to one of the three equations, fundamental NLSE (2.160), NLSE with power law nonlinearity (3.9), and NLSE with dual power law nonlinearity (3.27), $i\, \psi_t + a_1\, \psi_{xx} + a_2\, |\psi|^2\, \psi = 0$, $i\, \psi_t + a_1\, \psi_{xx} + a_2\, |\psi|^n\, \psi = 0$, $i\, \psi_t + a_1\, \psi_{xx} + a_2\, |\psi|^n\, \psi + a_3\, |\psi|^m\, \psi = 0$, then

$$\Phi(x, t) = \psi(x - v\, t, t)\, e^{i\left[\frac{v}{2\, a_1}\, (x - x_0) - \frac{v^2}{4\, a_1}\, (t - t_0) \right]} \quad (5.20)$$

is a moving solution to the same equation

$$i\, \Phi_t + a_1\, \Phi_{xx} + a_2\, |\Phi|^2\, \Phi = 0, \quad (5.21)$$

$$i\, \Phi_t + a_1\, \Phi_{xx} + a_2\, |\Phi|^n\, \Phi = 0, \quad (5.22)$$

$$i\, \Phi_t + a_1\, \Phi_{xx} + a_2\, |\Phi|^n\, \Phi + a_3\, |\Phi|^m\, \Phi = 0, \quad (5.23)$$

respectively, with real constants x_0, t_0, v, a_1, a_2, a_3, n, and m.

Example 1:   *Moving bright soliton*
(Figure 5.3)

Given $\psi(x,\, t) = A_0 \sqrt{\frac{2a_1}{a_2}} \text{sech}[A_0\,(x - x_0)]\, e^{i\,[a_1\, A_0^2\,(t-t_0)+\phi_0]}$ is a static solution to (2.160), then

$$\Phi(x,\, t) = A_0 \sqrt{\frac{2\,a_1}{a_2}}\ \text{sech}\{A_0\,[x - (x_0 + v\,t)]\}\, e^{i\left[\frac{v}{2\,a_1}\,(x-x_0)+\frac{4a_1^2\,A_0^2-v^2}{4\,a_1}\,(t-t_0)+\phi_0\right]} \qquad (5.24)$$

is a moving solution to (5.21), where $a_1\, a_2 > 0$, A_0, x_0, t_0, v, and ϕ_0 are arbitrary real constants.

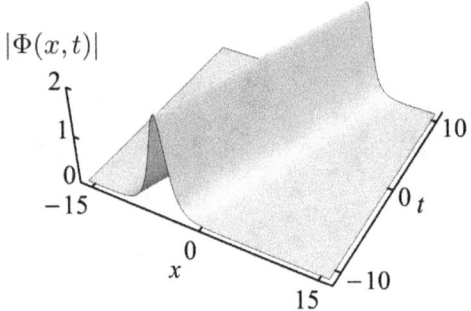

Figure 5.3. Moving bright soliton (5.24) with $a_1 = 1$, $a_2 = 1/2$, $A_0 = 1$, $v = 1/2$, and $x_0 = t_0 = \phi_0 = 0$. Animation available online at http://doi.org/10.1088/978-0-7503-5954-2.

Example 2:   *Moving dark soliton*
(Figure 5.4)

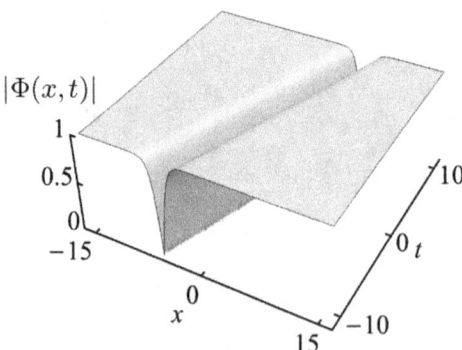

Figure 5.4. Moving dark soliton (5.25) with $a_1 = 1/2$, $a_2 = -1$, $A_0 = 1$, $v = 1/2$, and $x_0 = t_0 = \phi_0 = 0$. Animation available online at http://doi.org/10.1088/978-0-7503-5954-2.

Given $\psi(x, t) = A_0 \sqrt{\frac{-2\,a_1}{a_2}} \tanh[A_0\,(x - x_0)]\, e^{-i\,[2\,a_1\,A_0^2\,(t-t_0)+\phi_0]}$ is a static solution to (2.160), then

$$\Phi(x, t) = A_0 \sqrt{\frac{-2\,a_1}{a_2}} \tanh\{A_0\,[x - (x_0 + v\,t)]\}\, e^{-i\left[-\frac{v}{2\,a_1}(x-x_0)+\frac{8\,a_1^2\,A_0^2+v^2}{4\,a_1}(t-t_0)+\phi_0\right]} \quad (5.25)$$

is a moving solution to (5.21), where $a_1\,a_2 < 0$, A_0, x_0, t_0, v and ϕ_0 are arbitrary real constants.

Example 3: Localization in x and t *Moving Peregrine soliton* (Figure 5.5)

Given $\psi(x, t) = \frac{1}{\sqrt{a_2}}\,[\frac{4 + i\,8\,(t - t_0)}{1 + 4\,((t - t_0)^2 + \frac{2}{a_1}\,((x - x_0)^2}} - 1]\,e^{i\,[t - t_0 + \phi_0]}$ is a static solution to (2.160), then

$$\Phi(x, t) = \frac{1}{\sqrt{a_2}}\left[\frac{4 + i\,8\,(t - t_0)}{1 + 4\,((t - t_0)^2 + \frac{2}{a_1}\,(x - x_0 - v\,t)^2} - 1\right]\,e^{i\left[\frac{v}{2a_1}(x-x_0)-\frac{v^2}{4a_1}(t-t_0)+(t-t_0)+\phi_0\right]} \quad (5.26)$$

is a moving solution to (5.21), where $a_2 > 0$, x_0, t_0, v, and ϕ_0 are arbitrary real constants.

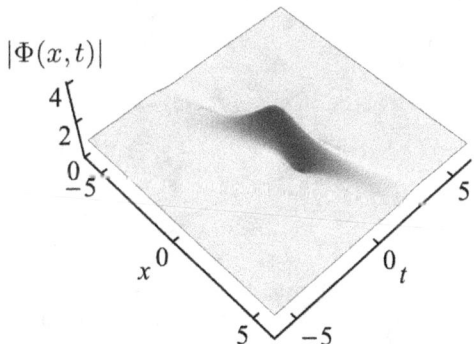

Figure 5.5. Moving Peregrine soliton (5.26) with $a_1 = a_2 = 1$, $A_0 = 1$, $v = 2$, and $x_0 = t_0 = \phi_0 = 0$. Animation available online at http://doi.org/10.1088/978-0-7503-5954-2.

Example 4:   *Moving bright soliton*

Given $\psi(x, t) = \{\frac{2\,A_0^2\,a_1\,(n + 2)}{a_2\,n^2}\,\mathrm{sech}^2[A_0\,(x - x_0)]\}^{\frac{1}{n}}\,e^{i\,[\frac{4\,a_1\,A_0^2}{n^2}(t-t_0)+\phi_0]}$ is a static solution to (3.9), then

$$\Phi(x, t) = \left(\frac{2\,A_0^2\,a_1\,(n + 2)}{a_2\,n^2}\,\mathrm{sech}^2\,\{A_0\,[x - (x_0 + v\,t)]\}\right)^{\frac{1}{n}}$$

$$\times\, e^{i\left[\frac{v}{2a_1}(x-x_0)-\frac{v^2}{4a_1}(t-t_0)+\frac{4\,a_1\,A_0^2}{n^2}(t-t_0)+\phi_0\right]} \quad (5.27)$$

is a moving solution to (5.22), where $a_1 a_2 (n + 2) > 0$, A_0, x_0, t_0, v, and ϕ_0 are arbitrary real constants.

Example 5. sech(x,t) *Moving flat-top soliton*
(Figure 5.6)

Given $\psi(x,\, t) = \left\{ \dfrac{A_0\,(n+2)}{a_2 + a_2 \sin(\theta)\, \cosh[n\, \sqrt{\tfrac{A_0}{a_1}}\,(x-x_0)]} \right\}^{\frac{1}{n}} e^{i\,[A_0\,(t-t_0)+\phi_0]}$, is a static solution to (3.27), then

$$\Phi(x,\, t) = \left\{ \dfrac{A_0\,(n+2)}{a_2 + a_2 \sin(\theta)\, \cosh\left[n\, \sqrt{\dfrac{A_0}{a_1}}\,(x - x_0)\right]} \right\}^{\frac{1}{n}} e^{i\left[A_0\,(t-t_0)+\frac{v}{2a_1}(x-x_0)-\frac{v^2}{4a_1}(t-t_0)+\phi_0\right]} \quad (5.28)$$

is a moving solution to (5.23), where $a_3 = \gamma\, a_{03}$, $\gamma = -\cos^2(\theta)$, $0 < \theta < \pi/2$, $a_{03} = a_2^2 (n + 1)/[A_0 (n + 2)^2]$, $m = 2n$, $a_1 A_0 > 0$, x_0, t_0, A_0, and ϕ_0 are arbitrary real constants.

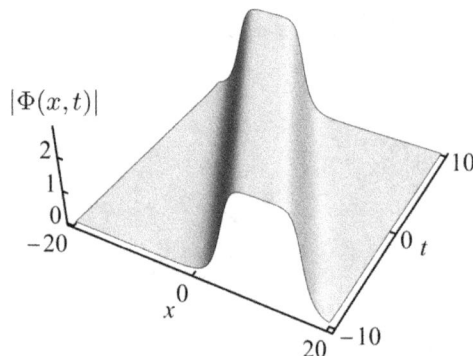

Figure 5.6. Moving flat-top soliton (5.28) with $a_1 = 3$, $a_2 = 1$, $\theta = \pi/9998$, $n = 2$, $v = -1$, and $x_0 = t_0 = \phi_0 = 0$. Animation available online at http://doi.org/10.1088/978-0-7503-5954-2.

5.5 Function Coefficients

If $\psi(x,\, t)$ is a solution to the fundamental NLSE (2.160),

$$i\,\psi_t + a_1\,\psi_{xx} + a_2\,|\psi|^2\,\psi = 0,$$

then

$$\Phi(x,\, t) = A(x,\, t)\, e^{i\,B(x,\,t)}\, \psi[X(x,\, t),\, T(x,\, t)] \quad (5.29)$$

is a solution to

$$i\,\Phi_t + b_1(x,\, t)\,\Phi_{xx} + b_2(x,\, t)\,|\Phi|^2\,\Phi + [b_{3r}(x,\, t) + i\,b_{3i}(x,\, t)]\,\Phi = 0, \quad (5.30)$$

where

$$T(x, t) = g_1(t), \tag{5.31}$$

$$A(x, t) = \frac{g_3(t)}{\sqrt{X_x(x, t)}}, \tag{5.32}$$

$$B(x, t) = g_2(t) - \int \frac{X_t(x, t)\, d\,x}{2\, b_1(x, t)\, X_x(x, t)}, \tag{5.33}$$

$$b_1(x, t) = \frac{a_1\, g_1{}'(t)}{X_x^2(x, t)}, \tag{5.34}$$

$$b_2(x, t) = \frac{a_2\, g_1{}'(t)}{A^2(x, t)}, \tag{5.35}$$

$$b_{3r}(x, t) = \frac{\int \left\{ g_1{}''(t)\, X_t(x, t)\, X_x(x, t) - g_1{}'(t)\, [X_{tt}(x, t)\, X_x(x, t) + X_t(x, t)\, X_{xt}(x, t)] \right\} d\,x}{2\, a_1\, g_1{}'^2(t)}$$
$$+ g_2{}'(t) + \frac{X_t^2(x, t)}{4\, a_1\, g_1{}'(t)^2} + \frac{a_1\, g_1{}'(t)\, [2\, X_x(x, t)\, X_{xxx}(x, t) - 3\, X_{xx}^2(x, t)]}{4\, X_x^4(x, t)}, \tag{5.36}$$

$$b_{3i}(x, t) = \frac{X_{xt}(x, t)}{X_x(x, t)} - \frac{g_3{}'(t)}{g_3(t)}, \tag{5.37}$$

$g_1(t)$, $g_2(t)$, $g_3(t)$, and $X(x, t)$ are arbitrary real functions.

5.5.1 Constant Dispersion and Complex Potential

If $\psi(x, t)$ is a solution to the fundamental NLSE (2.160), $i\, \psi_t + a_1\, \psi_{xx} + a_2\, |\psi|^2\, \psi = 0$, then

$$\Phi(x, t) = A(x, t)\, e^{i\, B(x, t)}\, \psi[X(x, t), T(x, t)] \tag{5.38}$$

is a solution to

$$i\, \Phi_t + b_{10}\, \Phi_{xx} + b_2(x, t)\, |\Phi|^2\, \Phi + [b_{3r}(x, t) + i\, b_{3i}(x, t)]\, \Phi = 0, \tag{5.39}$$

where

$$X(x, t) = g_4(t) + g_5(t)\, x, \tag{5.40}$$

$$T(x, t) = g_1(t) = c_0 + \frac{b_{10}}{a_1} \int g_5^2(t)\, d\,t, \tag{5.41}$$

$$A(x, t) = \frac{g_3(t)}{\sqrt{g_5(t)}}, \tag{5.42}$$

$$B(x, t) = g_2(t) - \frac{[2 \, g_4'(t) + x \, g_5'(t)] \, x}{4 \, b_{10} \, g_5(t)}, \tag{5.43}$$

$$b_2(x, t) = \frac{a_2 \, b_{10} \, g_5^3(t)}{a_1 \, g_3^2(t)}, \tag{5.44}$$

$$b_{3i}(x, t) = \frac{g_5'(t)}{g_5(t)} - \frac{g_3'(t)}{g_3(t)}, \tag{5.45}$$

$$b_{3r}(x, t) = g_2'(t) + \frac{g_4'^2(t)}{4 \, b_{10} \, g_5^2(t)} + \frac{2 \, g_4'(t) \, g_5'(t) - g_5(t) \, g_4''(t)}{2 \, b_{10} \, g_5^2(t)} \, x$$
$$+ \frac{2 \, g_5'^2(t) - g_5(t) \, g_5''(t)}{4 \, b_{10} \, g_5^2(t)} \, x^2, \tag{5.46}$$

$g_2(t)$, $g_3(t)$, $g_4(t)$ and $g_5(t)$ are arbitrary real functions and b_{10}, a_1, a_2 and c_0 are arbitrary real constants.

5.5.2 Constant Dispersion and Real Quadratic Potential

If $\psi(x, t)$ is a solution to the fundamental NLSE (2.160),
$$i \, \psi_t + a_1 \, \psi_{xx} + a_2 \, |\psi|^2 \, \psi = 0,$$
then

$$\Phi(x, t) = c_2 \, \sqrt{g_5(t)} \, e^{i \left[c_1 - \frac{c_4^2}{4 \, b_{10}} \int g_5^2(t) \, d \, t - \frac{2 \, c_4 \, g_5^2(t) + g_5'(t) \, x}{4 \, b_{10} \, g_5(t)} \, x \right]}$$
$$\times \psi \left[c_3 + g_5(t) \, x + c_4 \int g_5^2(t) \, d \, t, \; c_0 + \frac{b_{10}}{a_1} \int g_5^2(t) \, d \, t \right] \tag{5.47}$$

is a solution to

$$i \, \Phi_t + b_{10} \, \Phi_{xx} + \frac{a_2 \, b_{10} \, g_5(t)}{a_1 \, c_2^2} \, |\Phi|^2 \, \Phi - \frac{g_5(t) \, g_5''(t) - 2 \, g_5'^2(t)}{4 \, b_{10} \, g_5^2(t)} \, x^2 \, \Phi = 0, \quad (5.48)$$

where

$$g_2(t) = c_1 - \frac{c_4^2}{4 \, b_{10}} \int g_5^2(t) \, d \, t, \tag{5.49}$$

$$g_3(t) = c_2 \, g_5(t), \tag{5.50}$$

$$g_4(t) = c_3 + c_4 \int g_5^2(t) \, d \, t, \tag{5.51}$$

c_1, c_2, c_3, and c_4 are arbitrary real constants, $g_5(t)$ is an arbitrary real function of t.

Example 1:   Bright soliton

Given $\psi(x, t) = A_0 \sqrt{\frac{2 \, a_1}{a_2}} \, \mathrm{sech}[A_0 \, (x - x_0)] \, e^{i \, [a_1 \, A_0^2 \, (t - t_0) + \phi_0]}$ is a solution to (2.160), then

$$\Phi(x,\,t) = \sqrt{\frac{2\,A_0^2\,a_1\,c_2^2\,g_5(t)}{a_2}}\ \mathrm{sech}\left\{A_0\left[c_3 - x_0 + g_5(t)\,x + c_4\int g_5^2(t)\,d\,t\right]\right\} \tag{5.52}$$
$$\times\ e^{i\,\phi(x,\,t)}$$

is a solution to (5.48), where

$$\phi(x,\,t) = A_0^2\,[a_1\,(c_0 - t_0) + b_{10}\int g_5^2(t)\,d\,t]$$
$$-\frac{c_4^2}{4\,b_{10}}\int g_5^2(t)\,d\,t - \frac{\left[2\,c_4\,g_5^2(t) + x\,g_5'(t)\right]x}{4\,b_{10}\,g_5(t)} + c_1 + \phi_0,$$

$a_1\,a_2 > 0$, A_0, x_0, t_0, and ϕ_0 are arbitrary real constants.

Case I: $g_5(t) = \alpha + \beta\,\sin(\gamma\,t)$
(Figure 5.7)

$$\Phi(x,\,t) = \sqrt{\frac{2\,A_0^2\,a_1\,c_2^2\,[\alpha + \beta\,\sin(\gamma\,t)]}{a_2}}\ e^{i\,\phi(x,\,t)}$$
$$\times\ \mathrm{sech}(A_0\,\{c_3 - x_0 + [\alpha + \beta\,\sin(\gamma\,t)]\,x \tag{5.53}$$
$$+\ \frac{c_4\,[2\,\gamma\,(2\,\alpha^2 + \beta^2)\,t - 8\,\alpha\,\beta\,\cos(\gamma\,t) - \beta^2\,\sin(2\,\gamma\,t)]}{4\,\gamma}\})$$

is a solution to

$$i\,\Phi_t + b_{10}\,\Phi_{xx} + \frac{a_2\,b_{10}\,[\alpha + \beta\,\sin(\gamma\,t)]}{a_1\,c_2^2}\,|\Phi|^2\,\Phi + \frac{\beta\,\gamma^2\,[3\,\beta + \beta\,\cos(2\,\gamma\,t) + 2\,\alpha\,\sin(\gamma\,t)]}{8\,b_{10}\,[\alpha + \beta\,\sin(\gamma\,t)]^2}\,x^2\,\Phi = 0, \tag{5.54}$$

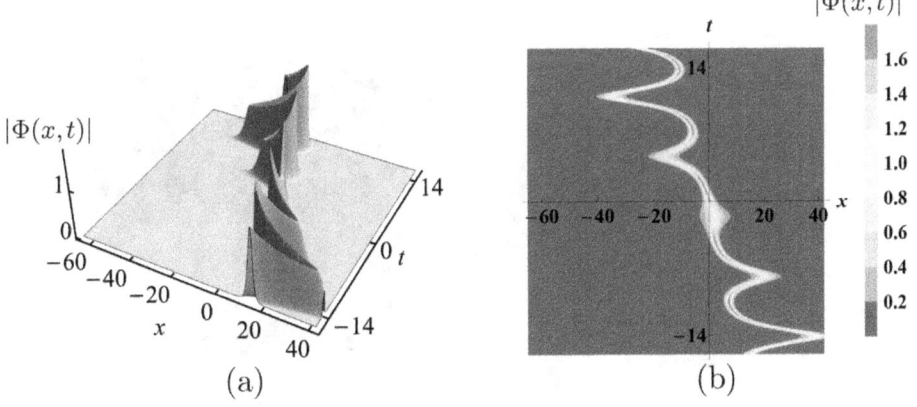

(a)   (b)

Figure 5.7. Plot of solution (5.53). (a) 3D plot, (b) contour plot, with $a_1 = a_2 = b_{10} = A_0 = \alpha = \gamma = c_0 = c_1 = c_2 = c_3 = c_4 = 1$, $\beta = 6/10$, and $x_0 = t_0 = \phi_0 = 0$. Animation available online at http://doi.org/10.1088/978-0-7503-5954-2.

where

$$\phi(x,\ t) = -\ \frac{\beta\ \gamma\ \cos(\gamma\ t)\ x^2 + 2\ c_4\ [\alpha + \beta\ \sin(\gamma\ t)]^2\ x}{4\ b_{10}\ [\alpha + \beta\ \sin(\gamma\ t)]}$$

$$-\ \frac{c_4^2\ [2\ \gamma\ (2\ \alpha^2 + \beta^2)\ t - 8\ \alpha\ \beta\ \cos(\gamma\ t) - \beta^2\ \sin(2\ \gamma\ t)]}{16\ b_{10}\ \gamma}$$

$$+\ \frac{A_0^2}{4\ \gamma}\ \{4\ \gamma\ a_1\ (c_0 - t_0) + b_{10}\ [2\ \gamma\ (2\ \alpha^2 + \beta^2)\ t - 8\ \alpha\ \beta\ \cos(\gamma\ t) - \beta^2\ \sin(2\ \gamma\ t)]\} + c_1 + \phi_0,$$

α, β, and γ are arbitrary real constants.

Case II: $g_5(t) = e^{\gamma\ t}$
(Figure 5.8)

$$\Phi(x,\ t) = c_2\ A_0\ \sqrt{\frac{2\ a_1}{a_2}\ e^{\gamma\ t}}\ \text{sech}\left[A_0\left(\frac{c_4\ e^{2\gamma\ t}}{2\ \gamma} + e^{\gamma\ t}\ x - x_0 + c_3\right)\right]\ e^{i\ \phi(x,\ t)} \quad (5.55)$$

is a solution to

$$i\ \Phi_t + b_{10}\ \Phi_{xx} + \frac{a_2\ b_{10}\ e^{\gamma\ t}}{a_1\ c_2^2}\ |\Phi|^2\ \Phi + \frac{\gamma^2}{4\ b_{10}}\ x^2\ \Phi = 0, \quad (5.56)$$

where

$$\phi(x,\ t) = \frac{1}{8\ b_{10}\ \gamma}\ \left\{-c_4\ e^{\gamma\ t}\ (c_4\ e^{\gamma\ t} + 4\ \gamma\ x) + 2\ \gamma\ [4\ b_{10}\ (c_1 + \phi_0) - \gamma\ x^2]\right.$$

$$\left. +\ 4\ A_0^2\ b_{10}\ [b_{10}\ e^{2\gamma\ t} + 2\ a_1\ \gamma\ (c_0 - t_0)]\right\} + \phi_0,$$

γ is an arbitrary real constant.

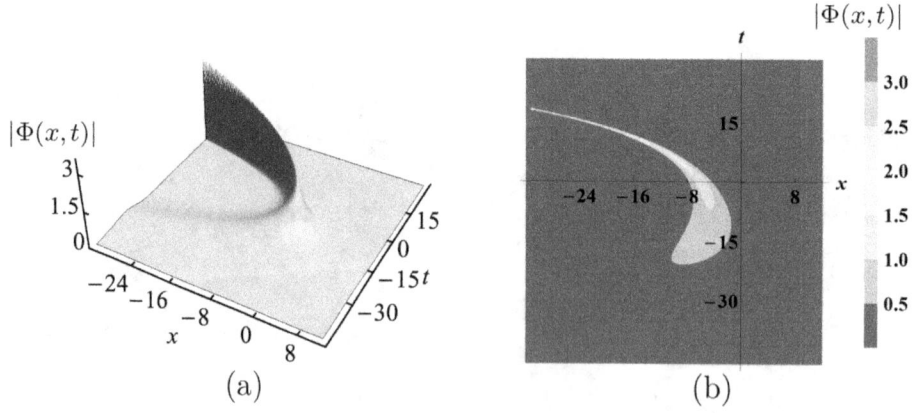

Figure 5.8. Plot of solution (5.55). (a) 3D plot, (b) contour plot, with $a_1 = a_2 = b_{10} = A_0 = c_0 = c_1 = c_2 = c_3 = c_4 = 1$, $\gamma = 1/10$, and $x_0 = t_0 = \phi_0 = 0$. Animation available online at http://doi.org/10.1088/978-0-7503-5954-2.

Example 2: *Dark soliton*

Given $\psi(x, t) = A_0 \sqrt{\frac{-2\,a_1}{a_2}} \tanh[A_0\,(x - x_0)]\, e^{-i\,[2\,a_1\,A_0^2\,(t-t_0)+\phi_0]}$ is a solution to (2.160), then

$$\Phi(x, t) = \sqrt{\frac{-2\,A_0^2\,a_1\,c_2^2\,g_5(t)}{a_2}}\;\tanh\left\{A_0\left[c_3 - x_0 + g_5(t)\,x + c_4 \int g_5^2(t)\,d\,t\right]\right\} \qquad (5.57)$$
$$\times\, e^{-i\,\phi(x, t)}$$

is a solution to (5.48), where

$$\phi(x, t) = 2\,A_0^2\left[a_1\,(c_0 - t_0) + b_{10} \int g_5^2(t)\,d\,t\right] + \frac{c_4^2}{4\,b_{10}} \int g_5^2(t)\,d\,t\;a_1\,a_2 < 0,\; A_0,\; x_0,$$

$$+\, \frac{[2\,c_4\,g_5^2(t) + x\,g_5{}'(t)]\,x}{4\,b_{10}\,g_5(t)} - c_1 + \phi_0,$$

t_0, and ϕ_0 are arbitrary real constants.

Case I: $g_5(t) = \alpha + \beta \sin(\gamma\,t)$
(Figure 5.9)

$$\Phi(x, t) = \sqrt{\frac{-2\,A_0^2\,a_1\,c_2^2\,[\alpha + \beta \sin(\gamma\,t)]}{a_2}}\; e^{-i\,\phi(x, t)}$$
$$\times \tanh[A_0\,(c_3 - x_0 + [\alpha + \beta \sin(\gamma\,t)]\,x \qquad (5.58)$$
$$+\, \frac{c_4\,[2\,\gamma\,(2\,\alpha^2 + \beta^2)\,t - 8\,\alpha\,\beta\,\cos(\gamma\,t) - \beta^2 \sin(2\,\gamma\,t)]}{4\,\gamma})]$$

is a solution to (5.54), where

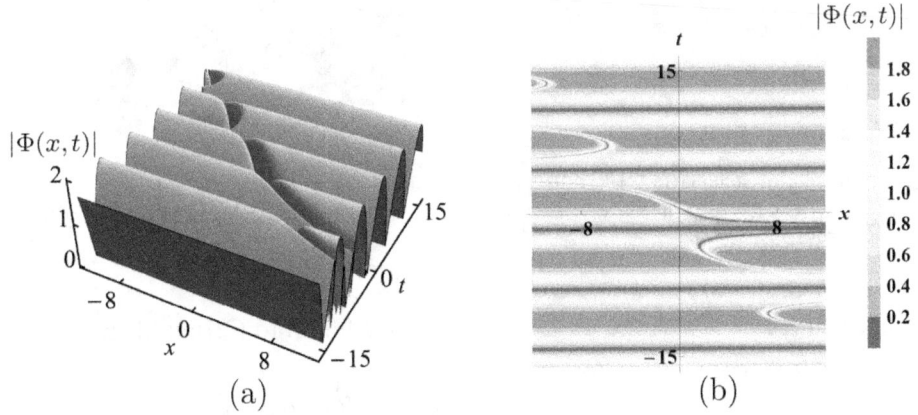

(a)   (b)

Figure 5.9. Plot of solution (5.58). (a) 3D plot, (b) contour plot, with $a_1 = b_{10} = A_0 = c_0 = c_1 = c_2 = c_3 = c_4 = \alpha = \beta = \gamma = 1$, $a_2 = -1$, and $x_0 = t_0 = \phi_0 = 0$. Animation available online at http://doi.org/10.1088/978-0-7503-5954-2.

$$\phi(x, t) = \frac{\beta\,\gamma\,\cos(\gamma\,t)\,x^2 + 2\,c_4\,[\alpha + \beta\,\sin(\gamma\,t)]^2\,x}{4\,b_{10}\,[\alpha + \beta\,\sin(\gamma\,t)]}$$

$$+ \frac{c_4^2\,[2\,\gamma\,(2\,\alpha^2 + \beta^2)\,t - 8\,\alpha\,\beta\,\cos(\gamma\,t) - \beta^2\,\sin(2\,\gamma\,t)]}{16\,b_{10}\,\gamma}$$

$$+ \frac{A_0^2}{2\,\gamma}\,\{4\,\gamma\,a_1\,(c_0 - t_0) + b_{10}\,[2\,\gamma\,(2\,\alpha^2 + \beta^2)\,t - 8\,\alpha\,\beta\,\cos(\gamma\,t) - \beta^2\,\sin(2\,\gamma\,t)]\} - c_1 + \phi_0,$$

$2\,A_0^2\,a_1\,c_2^2\,[\alpha + \beta\,\sin(\gamma\,t)]/a_2 < 0$, α, β, and γ are arbitrary real constants.

Case II: $g_5(t) = e^{\gamma\,t}$
(Figure 5.10)

$$\Phi(x, t) = c_2\,A_0\,\sqrt{\frac{-2\,a_1}{a_2}\,e^{\gamma\,t}}\,\tanh\left[A_0\left(\frac{c_4\,e^{2\,\gamma\,t}}{2\,\gamma} + e^{\gamma\,t}\,x - x_0 + c_3\right)\right]e^{i\,\phi(x,\,t)} \qquad (5.59)$$

is a solution to (5.56), where

$$\phi(x, t) = \frac{-1}{8\,b_{10}\,\gamma}\,\left\{c_4\,e^{\gamma\,t}\,(c_4\,e^{\gamma\,t} + 4\,\gamma\,x) + 2\,\gamma\,[4\,b_{10}\,(\phi_0 - c_1) + \gamma\,x^2]\right.$$

$$\left. + 8\,A_0^2\,b_{10}\,[b_{10}\,e^{2\,\gamma\,t} + 2\,a_1\,\gamma\,(c_0 - t_0)]\right\} + \phi_0,$$

γ is an arbitrary real constant.

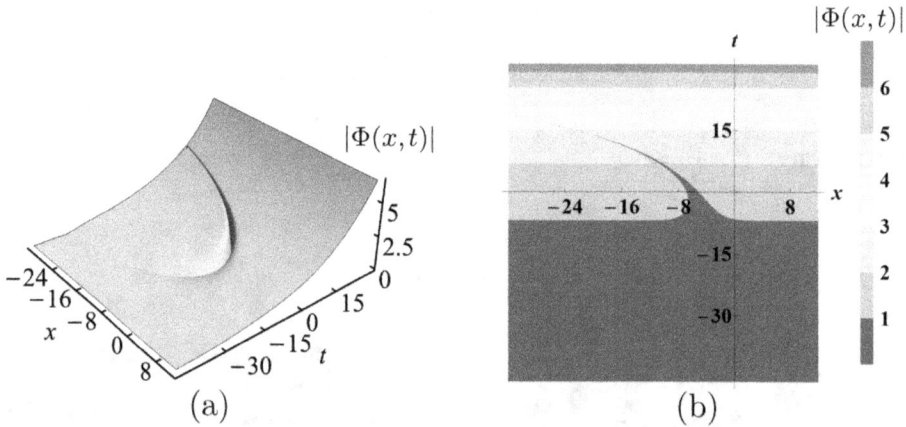

Figure 5.10. Plot of solution (5.59). (a) 3D plot, (b) contour plot, with $a_1 = b_{10} = A_0 = c_0 = c_1 = c_2 = c_3 = c_4 = 1$, $a_2 = -1$, $\gamma = 1/10$, and $x_0 = t_0 = \phi_0 = 0$. Animation available online at http://doi.org/10.1088/978-0-7503-5954-2.

Example 3: _Two bright solitons_
Given $\psi(x, t) = \frac{1}{\sqrt{a_2}}\,[\psi_1(x, t) + \psi_2(x, t)]$ is a solution to (2.160), then

$$\Phi(x, t) = c_2\,\sqrt{\frac{g_5(t)}{a_2}}\,[\psi_1(x, t) + \psi_2(x, t)]\,e^{i\left[c_1 - \frac{2\,c_4\,g_5^2(t)\,x + c_4^2\,g_5(t)\int g_5^2(t)\,d\,t + g_5'(t)x^2}{4\,b_{10}\,g_5(t)}\right]}, \qquad (5.60)$$

is a solution to (5.48), where

$$\psi_1(x, t) = \frac{M_{12}\left[\gamma_1^{-1}(x, t) + \gamma_2^*(x, t)\right] - M_{22}\left[\gamma_2^{-1}(x, t) + \gamma_2^*(x, t)\right]}{M_{12} M_{21}\left[\gamma_1^*(x, t) + \gamma_2^{-1}(x, t)\right]\left[\gamma_1^{-1}(x, t) + \gamma_2^*(x, t)\right] - M_{11} M_{22}\left[\gamma_1^{-1}(x, t) + \gamma_1^*(x, t)\right]\left[\gamma_2^{-1}(x, t) + \gamma_2^*(x, t)\right]},$$

$$\psi_2(x, t) = \frac{-M_{11}\left[\gamma_1^{-1}(x, t) + \gamma_1^*(x, t)\right] + M_{21}\left[\gamma_1^*(x, t) + \gamma_2^{-1}(x, t)\right]}{M_{12} M_{21}\left[\gamma_1^*(x, t) + \gamma_2^{-1}(x, t)\right]\left[\gamma_1^{-1}(x, t) + \gamma_2^*(x, t)\right] - M_{11} M_{22}\left[\gamma_1^{-1}(x, t) + \gamma_1^*(x, t)\right]\left[\gamma_2^{-1}(x, t) + \gamma_2^*(x, t)\right]},$$

$M_{11} = 1/(\lambda_1 + \lambda_1^*)$, $M_{12} = 1/(\lambda_1 + \lambda_2^*)$, $M_{21} = 1/(\lambda_2 + \lambda_1^*)$, $M_{22} = 1/(\lambda_2 + \lambda_2^*)$,

$\lambda_1 = \alpha_1 + i\, \nu_1$,

$\lambda_2 = \alpha_2 + i\, \nu_2$,

$$\gamma_1(x, t) = e^{i\left\{\frac{\lambda_1^2}{2}\left[c_0 - t_0 + \frac{b_{10}}{a_1}\int g_5^2(t)\,d\,t\right] + \phi_{01}\right\} + \frac{\lambda_1\left[c_3 + g_5(t)\,x + c_4\int g_5^2(t)\,d\,t\right]}{\sqrt{2\,a_1}} - x_{01}\,\lambda_1},$$

$$\gamma_2(x, t) = e^{i\left\{\frac{\lambda_2^2}{2}\left[c_0 - t_0 + \frac{b_{10}}{a_1}\int g_5^2(t)\,d\,t\right] + \phi_{02}\right\} + \frac{\lambda_2\left[c_3 + g_5(t)\,x + c_4\int g_5^2(t)\,d\,t\right]}{\sqrt{2\,a_1}} - x_{02}\,\lambda_2}.$$

Case I: $g_5(t) = \alpha + \beta \sin(\gamma\, t)$
(Figure 5.11)

$$\Phi(x, t) = c_2 \sqrt{\frac{\alpha + \beta \sin(\gamma\, t)}{a_2}}\ [\psi_1(x, t) + \psi_2(x, t)]\, e^{i\,\phi(x,\, t)} \qquad (5.61)$$

is a solution to (5.54), where

$$\phi(x, t) = -\frac{\beta\,\gamma \cos(\gamma\, t)\, x^2 + 2\, c_4\, [\alpha + \beta \sin(\gamma\, t)]^2\, x}{4\, b_{10}\, [\alpha + \beta \sin(\gamma\, t)]}$$
$$-\frac{c_4^2\, [2\,\gamma\, (2\,\alpha^2 + \beta^2)\, t - 8\,\alpha\,\beta \cos(\gamma\, t) - \beta^2 \sin(2\,\gamma\, t)]}{16\, b_{10}\,\gamma} + c_1,$$

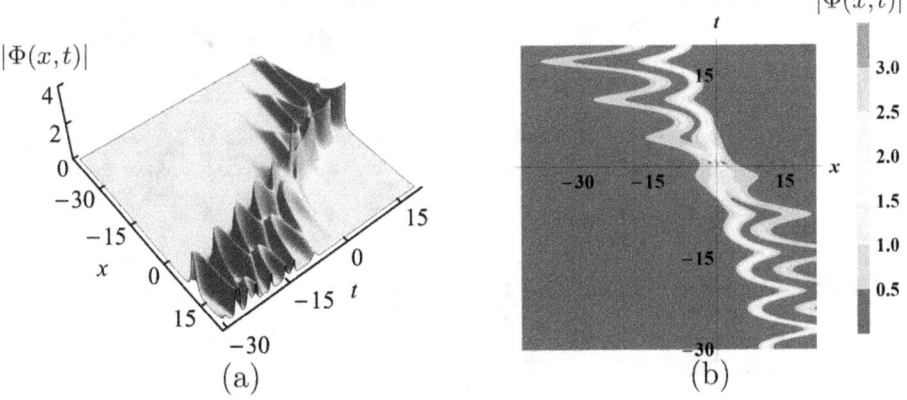

Figure 5.11. Plot of solution (5.61). (a) 3D plot, (b) contour plot, with $a_1 = A_0 = 2$, $a_2 = b_{10} = c_2 = c_4 = \alpha = \gamma = 1$, $\beta = 4/10$, $c_0 = c_1 = c_3 = 0$, $\alpha_1 = 1$, $\alpha_2 = 2$, $\nu_1 = 0$, $\nu_2 = 1/2$, and $x_{01} = x_{02} = \phi_{01} = \phi_{02} = 0$.

$$\gamma_1(x,\ t) = e^{i\left\{\frac{\lambda_1^2}{2}\left[c_0-t_0+\frac{b_{10}}{4\gamma a_1}p(t)\right]+\phi_{01}\right\}+\frac{\lambda_1}{\sqrt{2 a_1}}\left\{c_3+[\alpha+\beta\sin(\gamma t)]x-\frac{c_4}{4\gamma}p(t)\right\}-x_{01}\lambda_1},$$

$$\gamma_2(x,\ t) = e^{i\left\{\frac{\lambda_2^2}{2}\left[c_0-t_0+\frac{b_{10}}{4\gamma a_1}p(t)\right]+\phi_{02}\right\}+\frac{\lambda_2}{\sqrt{2 a_1}}\left\{c_3+[\alpha+\beta\sin(\gamma t)]x-\frac{c_4}{4\gamma}p(t)\right\}-x_{02}\lambda_2},$$

$p(t) = 8\,\alpha\,\beta\,\cos(\gamma\ t) + \beta^2\,\sin(2\,\gamma\ t) - 2\,(2\,\alpha^2 + \beta^2)\,t,$
$2\,A_0^2\,a_1\,c_2^2\,[\alpha+\beta\sin(\gamma\ t)]/a_2 < 0,\ \alpha,\ \beta,$ and γ are arbitrary real constants.

Case II: $g_5(t) = e^{\gamma\ t}$
(Figure 5.12)

$$\Phi(x,\ t) = \sqrt{\frac{e^{\frac{t}{2}}}{a_2}}\ [\psi_1(x,\ t) + \psi_2(x,\ t)]\,e^{-i\frac{1}{8}\left(2\,e^t+4\,e^{\frac{t}{2}}\,x+x^2\right)} \qquad (5.62)$$

is a solution to (5.54), where $\gamma_1(x,\ t) = e^{i\,[\frac{\lambda_1^2}{2}(c_0-t_0+\frac{b_{10}\,e^{2\gamma t}}{2 a_1\gamma})+\phi_{01}]+\frac{\lambda_1}{\sqrt{2 a_1}}[c_3+e^{\gamma\ t}\,x+\frac{c_4\,e^{2\gamma t}}{2\gamma}]-x_{01}\lambda_1},$

$\gamma_2(x,\ t) = e^{i\,[\frac{\lambda_2^2}{2}(c_0-t_0+\frac{b_{10}\,e^{2\gamma t}}{2 a_1\gamma})+\phi_{02}]+\frac{\lambda_2}{\sqrt{2 a_1}}[c_3+e^{\gamma\ t}\,x+\frac{c_4\,e^{2\gamma t}}{2\gamma}]-x_{02}\lambda_2},$ γ is an arbitrary real constant.

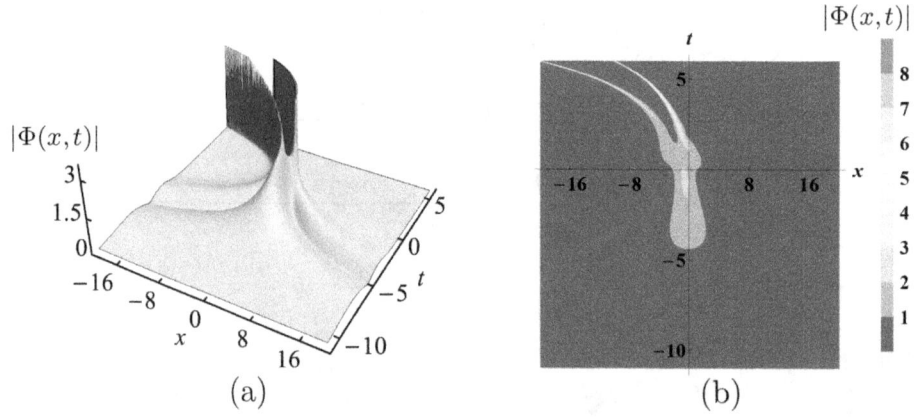

Figure 5.12. Plot of solution (5.62). (a) 3D plot, (b) contour plot, with $a_1 = A_0 = 2$, $a_2 = b_{10} = c_2 = c_4 = 1$, $\gamma = 1/2$, $c_0 = c_1 = c_3 = 0$, $\alpha_1 = 1$, $\alpha_2 = 2$, $\nu_1 = 0$, $\nu_2 = 1/2$, and $x_{01} = x_{02} = \phi_{01} = \phi_{02} = 0$.

5.5.3 Constant Dispersion and Real Linear Potential

If $\psi(x,\ t)$ is a solution to the fundamental NLSE (2.160), $i\,\psi_t + a_1\,\psi_{xx} + a_2\,|\psi|^2\,\psi = 0$, then

$$\Phi(x,\ t) = \sqrt{\frac{c_2^2\,c_5}{c_7}}\ \psi\left[\frac{c_5\,x}{c_7} + g_4(t),\ c_0 + \frac{b_{10}\,c_5^2\,t}{a_1\,c_7^2}\right]e^{i\left[c_1-\frac{c_7^2}{4 b_{10}\,c_5^2}\int g_4'^2(t)\,d\,t-\frac{c_7\,g_4'(t)\,x}{2 b_{10}\,c_5}\right]} \qquad (5.63)$$

is a solution to

$$i\,\Phi_t + b_{10}\,\Phi_{xx} + \frac{a_2\,b_{10}\,c_5}{a_1\,c_2^2\,c_7}\,|\Phi|^2\,\Phi - \frac{c_7\,g_4''(t)}{2\,b_{10}\,c_5}\,x\,\Phi = 0, \qquad (5.64)$$

where

$$g_2(t) = c_1 - \frac{1}{4\,b_{10}\,c_5^2}\int \left[c_7^2\,g_4'^2(t) + 2\,c_6\,c_7\,t\,g_4'^2(t) + c_6^2\,t^2\,g_4'^2(t)\right]d\,t, \quad (5.65)$$

$$g_5(t) = \frac{c_5}{c_6\,t + c_7}, \qquad (5.66)$$

$g_4(t)$ is an arbitrary real function of t, c_1, c_2, c_5, c_6, and c_7 are arbitrary real constants. In (5.63) and (5.64), c_6 is taken to be zero to obtain a solution with a norm proportional to that of $\psi(x,\,t)$.

***Example* 1:**   *Bright soliton*

Given $\psi(x,\,t) = A_0\sqrt{\dfrac{2\,a_1}{a_2}}\,\text{sech}[A_0\,(x - x_0)]\,e^{i\,[a_1\,A_0^2\,(t-t_0)+\phi_0]}$ is   a   solution   to (2.160), then

$$\Phi(x,\,t) = c_2\,A_0\sqrt{\frac{2\,a_1\,c_5}{a_2\,c_7}}\,\text{sech}\left\{A_0\left[\frac{c_5}{c_7}\,x - x_0 + g_4(t)\right]\right\}$$

$$\times\,e^{i\left\{A_0^2\,a_1\left[c_0+\frac{b_{10}\,c_5^2}{a_1\,c_7^2}\,(t-t_0)\right]-\frac{c_7^2}{4\,b_{10}\,c_5^2}\int g_4'^2(t)\,d\,t-\frac{c_7\,g_4'(t)}{2\,b_{10}\,c_5}\,x+c_1+\phi_0\right\}} \qquad (5.67)$$

is a solution to (5.64), where $a_1\,a_2\,c_5\,c_7 > 0$, A_0, x_0, t_0, and ϕ_0 are arbitrary real constants.

Case 1: $g_4(t) = \alpha\,t^2$
(Figure 5.13)

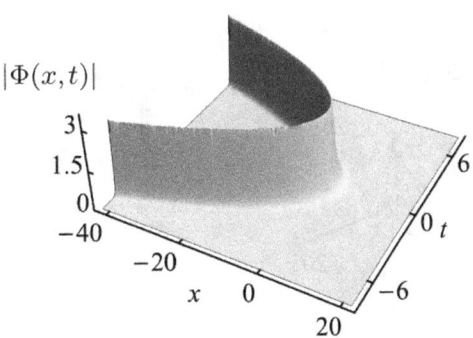

$|\Phi(x,t)|$

Figure   **5.13.** Plot   of   (5.68)   with $a_1 = a_2 = A_0 = \alpha = c_5 = c_7 = 1$, $c_2 = b_{10} = 2$,   and $c_0 = c_1 = x_0 = t_0 = \phi_0 = 0$. Animation available online at http://doi.org/10.1088/978-0-7503-5954-2.

$$\Phi(x,\ t) = c_2\ A_0\ \sqrt{\frac{2\ a_1\ c_5}{a_2\ c_7}}\ \mathrm{sech}\left[A_0\left(\frac{c_5}{c_7}\ x\ -\ x_0\ +\ \alpha\ t^2\right)\right]$$
$$\times\ e^{i\left\{A_0^2\ a_1\left[c_0 + \frac{b_{10}\ c_5^2}{a_1\ c_7^2}\ (t-t_0)\right] - \frac{c_7^2\ \alpha^2\ t^3}{3\ b_{10}\ c_5^2} - \frac{c_7\ \alpha\ t}{b_{10}\ c_5}\ x + c_1 + \phi_0\right\}}$$
(5.68)

is a solution to

$$i\ \Phi_t\ +\ b_{10}\ \Phi_{xx}\ +\ \frac{a_2\ b_{10}\ c_5}{a_1\ c_2^2\ c_7}\ |\Phi|^2\ \Phi\ -\ \frac{c_7\ \alpha}{b_{10}\ c_5}\ x\ \Phi\ =\ 0,$$
(5.69)

where α is an arbitrary real constant.

Example 2:   *Dark soliton*

Given $\psi(x,\ t) = A_0\ \sqrt{\frac{-2\ a_1}{a_2}}\ \tanh[A_0\ (x\ -\ x_0)]\ e^{-i\ [2\ a_1\ A_0^2\ (t-t_0) + \phi_0]}$ is a solution to (2.160), then

$$\Phi(x,\ t) = c_2\ A_0\ \sqrt{\frac{-2\ a_1\ c_5}{a_2\ c_7}}\ \tanh\left\{A_0\left[\frac{c_5}{c_7}\ x\ -\ x_0\ +\ g_4(t)\right]\right\}$$
$$\times\ e^{-i\left\{2\ A_0^2\ a_1\left[c_0 + \frac{b_{10}\ c_5^2}{a_1\ c_7^2}\ (t-t_0)\right] + \frac{c_7^2}{4\ b_{10}\ c_5^2}\int g_4'^2(t)\ d\ t + \frac{c_7\ g_4'(t)}{2\ b_{10}\ c_5}\ x - c_1 + \phi_0\right\}}$$
(5.70)

is a solution to (5.64), where $a_1\ a_2\ c_5\ c_7 < 0$, A_0, x_0, t_0, and ϕ_0 are arbitrary real constants.

Case I: $g_4(t) = \alpha\ t^2$
(Figure 5.14)

$$\Phi(x,\ t) = c_2\ A_0\ \sqrt{\frac{-2\ a_1\ c_5}{a_2\ c_7}}\ \tanh\left[A_0\left(\frac{c_5}{c_7}\ x\ -\ x_0\ +\ \alpha\ t^2\right)\right]$$
$$\times\ e^{-i\left\{2\ A_0^2\ a_1\left[c_0 + \frac{b_{10}\ c_5^2}{a_1\ c_7^2}\ (t-t_0)\right] + \frac{c_7^2\ \alpha^2\ t^3}{3\ b_{10}\ c_5^2} + \frac{c_7\ \alpha\ t}{b_{10}\ c_5}\ x - c_1 + \phi_0\right\}}$$
(5.71)

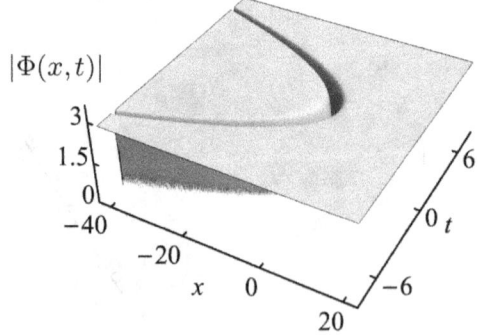

Figure 5.14. Plot of (5.71) with $a_1 = A_0 = \alpha = c_5 = c_7 = 1$, $a_2 = -1$, $c_2 = b_{10} = 2$, and $c_0 = c_1 = x_0 = t_0 = \phi_0 = 0$. Animation available online at http://doi.org/10.1088/978-0-7503-5954-2.

is a solution to (5.69) where α ia an arbitrary real constant.

Example 3: Generalized First-Order Breather

Given $\psi(x, t) = \frac{A_0}{\sqrt{a_2}} (1 - \frac{\sqrt{8}\,\lambda_r}{A_0} p(x, t))\, e^{i\,[A_0^2\,(t-t_0)+\phi_0]}$ is a solution to (2.160), then

$$\Phi(x, t) = A_0 \sqrt{\frac{c_2^2\, c5}{a_2\, c_7}} \left[1 - \frac{\sqrt{8}\,\lambda_r}{A_0} p(x, t) \right]$$

$$\times\, e^{i\left[A_0^2 \left(c_0 + \frac{b_{10}\, c_5^2\, t}{a_1\, c_7^2} - t_0 \right) - \frac{c_7^2}{4\, b_{10}\, c_5} \int g_4'^2(t)\, d\,t - \frac{c_7\, g_4'(t)\, x}{2\, b_{10}\, c_5} + c_1 + \phi_0 \right]}$$

(5.72)

is a solution to (5.64), where

$$p(x, t) = \frac{(A_0^2 + \Gamma^2)\cos[q_1(x, t)] + i\,(A_0^2 - \Gamma^2)\sin[q_1(x, t)] + 2\,A_0\,\{\Gamma_r\cosh[q_2(x, t)] - i\,\Gamma_i\sinh[q_2(x, t)]\}}{2\,A_0\,\Gamma_r\cos[q_1(x, t)] + (\Gamma^2 + A_0^2)\cosh[q_2(x, t)]},$$

$$q_1(x, t) = \delta_i + \sqrt{2}\,[\frac{\frac{c_5\,x}{c_7} + g_4(t) - x_0}{\sqrt{a_1}}\,\Delta_i - 2\,(c_0 + \frac{b_{10}\,c_5^2\,t}{a_1\,c_7^2} - t_0)\,(\Delta_i\,\lambda_i + \Delta_r\,\lambda_r)],$$

$q_2(x, t)$

$$= \delta_r + \sqrt{2}\,[\frac{\frac{c_5\,x}{c_7} + g_4(t) - x_0}{\sqrt{a_1}}\,\Delta_r - 2\,(c_0 + \frac{b_{10}\,c_5^2\,t}{a_1\,c_7^2} - t_0)\,\Delta_r\,\lambda_i + 2\,(c_0 + \frac{b_{10}\,c_5^2\,t}{a_1\,c_7^2} - t_0)$$

$$\Delta_i\,\lambda_r]$$

$\Delta_r = \text{Re}\,[\sqrt{2\,(\lambda_r - i\,\lambda_i)^2 - A_0^2}], \qquad \Delta_i = \text{Im}\,[\sqrt{2\,(\lambda_r - i\,\lambda_i)^2 - A_0^2}], \qquad \Gamma_r = \Delta_r + \sqrt{2}\,\lambda_r,$

$\Gamma_i = \Delta_r + \sqrt{2}\,\lambda_i, \Gamma = \sqrt{\Gamma_r^2 + \Gamma_i^2}, a_1 > 0, a_2 > 0, A_0, \lambda_r, \lambda_i, x_0, t_0,$ and ϕ_0 are arbitrary real constants.

Case I: $g_4(t) = \alpha\, t^2$
(Figure 5.15)

$$\Phi(x, t) = A_0 \sqrt{\frac{c_2^2\, c_5}{a_2\, c_7}} \left[1 - \frac{\sqrt{8}\,\lambda_r}{A_0} p(x, t) \right]$$

$$\times\, e^{i\left[A_0^2 \left(c_0 + \frac{b_{10}\, c_5^2\, t}{a_1\, c_7^2} - t_0 \right) - \frac{c_7^2}{4\, b_{10}\, c_5} \int g_4'^2(t)\, d\,t - \frac{c_7\, g_4'(t)\, x}{2\, b_{10}\, c_5} + c_1 + \phi_0 \right]}$$

(5.73)

is a solution to (5.69),

where $\qquad p(x, t) = \frac{(A_0^2 + \Gamma^2)\cos[q_1(x, t)] + i\,(A_0^2 - \Gamma^2)\sin[q_1(x, t)] + 2\,A_0\,\{\Gamma_r\cosh[q_2(x, t)] - i\,\Gamma_i\sinh[q_2(x, t)]\}}{2\,A_0\,\Gamma_r\cos[q_1(x, t)] + (\Gamma^2 + A_0^2)\cosh[q_2(x, t)]},$

$$q_1(x, t) = \delta_i + \sqrt{2}\,[\frac{\frac{c_5\,x}{c_7} + \alpha\,t^2 - x_0}{\sqrt{a_1}}\,\Delta_i - 2\,(c_0 + \frac{b_{10}\,c_5^2\,t}{a_1\,c_7^2} - t_0)\,(\Delta_i\,\lambda_i + \Delta_r\,\lambda_r)],$$

$q_2(x, t)$

$$= \delta_r + \sqrt{2}\,[\frac{\frac{c_5\,x}{c_7} + \alpha\,t^2 - x_0}{\sqrt{a_1}}\,\Delta_r - 2\,(c_0 + \frac{b_{10}\,c_5^2\,t}{a_1\,c_7^2} - t_0)\,\Delta_r\,\lambda_i + 2\,(c_0 + \frac{b_{10}\,c_5^2\,t}{a_1\,c_7^2} - t_0)$$

$$\Delta_i\,\lambda_r]$$

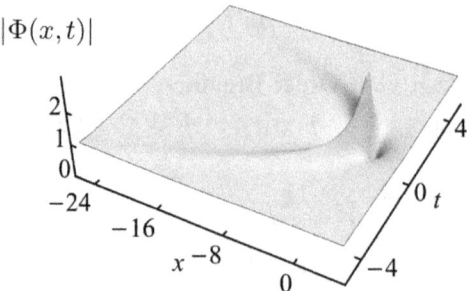

Figure 5.15. Plot of (5.73) with $a_1 = a_2 = A_0 = \alpha = c_2 = c_5 = c_7 = b_{10} = 1$, $\lambda_r = -1/\sqrt{2}$, $\lambda_i = 1/1000$, and $c_0 = c_1 = \delta_r = \delta_i = x_0 = t_0 = \phi_0 = 0$.

$$\Delta_r = \text{Re}\left[\sqrt{2(\lambda_r - i\,\lambda_i)^2 - A_0^2}\right], \qquad\qquad \Delta_i = \text{Im}\left[\sqrt{2(\lambda_r - i\,\lambda_i)^2 - A_0^2}\right],$$

$\Gamma_r = \Delta_r + \sqrt{2}\,\lambda_r$, $\Gamma_i = \Delta_r + \sqrt{2}\,\lambda_i$, $\Gamma = \sqrt{\Gamma_r^2 + \Gamma_i^2}$, $a_1 > 0$, $a_2 > 0$, A_0, α, λ_r, λ_i, x_0, t_0, and ϕ_0 are arbitrary real constants.

Case II: $g_4(t) = -\alpha\cos(\beta\,t)$
(Figure 5.16)

$$\Phi(x,\,t) = A_0 \sqrt{\frac{c_2^2\,c5}{a_2\,c_7}}\left[1 - \frac{\sqrt{8}\,\lambda_r}{A_0}\,p(x,\,t)\right]$$

$$\times\, e^{i\left\{A_0^2\left(c_0 + \frac{b_{10}\,c_5^2\,t}{a_1\,c_7^2} - t_0\right) - \frac{c_7^2\,\alpha^2\,\beta\,[2\,\beta\,t - \sin(2\,\beta\,t)]}{16\,b_{10}\,c_5^2} - \frac{c_7\,\alpha\,\beta\,\sin(\beta\,t)\,x}{2\,b_{10}\,c_5} + c_1 + \phi_0\right\}} \tag{5.74}$$

is a solution to

$$i\,\Phi_t + b_{10}\,\Phi_{xx} + \frac{a_2\,b_{10}\,c_5}{a_1\,c_2^2\,c_7}\,|\Phi|^2\,\Phi - \frac{c_7\,\alpha\,\beta^2\,\cos(\beta\,t)}{2\,b_{10}\,c_5}\,x\,\Phi = 0, \tag{5.75}$$

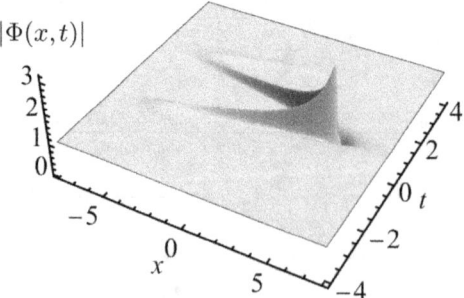

Figure 5.16. Plot of (5.74) with $a_1 = b_{10} = 1/2$, $a_2 = 2$, $A_0 = 1.5$, $\alpha = 18/4$, $\beta = 2.25$, $c_2 = c_5 = c_7 = 1$, $\lambda_r = -A_0/\sqrt{2}$, $\lambda_i = 1/100$, and $c_0 = c_1 = \delta_r = \delta_i = x_0 = t_0 = \phi_0 = 0$. (Used in the front cover page of the 1st edition).

where $\quad p(x,\,t) = \dfrac{(A_0^2 + \Gamma^2)\cos[q_1(x,t)] + i\,(A_0^2 - \Gamma^2)\sin[q_1(x,t)] + 2\,A_0\,\{\Gamma_r\cosh[q_2(x,t)] - i\,\Gamma_i\sinh[q_2(x,t)]\}}{2\,A_0\,\Gamma_r\cos[q_1(x,t)] + (\Gamma^2 + A_0^2)\cosh[q_2(x,t)]},$

$$q_1(x,\,t) = \delta_i + \sqrt{2}\,[\frac{\frac{c_5\,x}{c_7} - \alpha\cos(\beta\,t) - x_0}{\sqrt{a_1}}\,\Delta_i - 2\,(c_0 + \frac{b_{10}\,c_5^2\,t}{a_1\,c_7^2} - t_0)\,(\Delta_i\,\lambda_i + \Delta_r\,\lambda_r)],$$

$$q_2(x,\,t)$$

$$= \delta_r + \sqrt{2}\,[\frac{\frac{c_5\,x}{c_7} - \alpha\cos(\beta\,t) - x_0}{\sqrt{a_1}}\,\Delta_r - 2\,(c_0 + \frac{b_{10}\,c_5^2\,t}{a_1\,c_7^2} - t_0)\,\Delta_r\,\lambda_i$$

$$+\,2\,(c_0 + \frac{b_{10}\,c_5^2\,t}{a_1\,c_7^2} - t_0)\,\Delta_i\,\lambda_r]$$

$\Delta_r = \mathrm{Re}\,[\sqrt{2\,(\lambda_r - i\,\lambda_i)^2 - A_0^2}], \qquad\qquad \Delta_i = \mathrm{Im}\,[\sqrt{2\,(\lambda_r - i\,\lambda_i)^2 - A_0^2}],$

$\Gamma_r = \Delta_r + \sqrt{2}\,\lambda_r,\ \Gamma_i = \Delta_r + \sqrt{2}\,\lambda_i,\ \Gamma = \sqrt{\Gamma_r^2 + \Gamma_i^2},\ a_1 > 0,\ a_2 > 0,\ A_0,\ \alpha,\ \beta,\ \lambda_r,$ $\lambda_i,\ x_0,\ t_0,$ and ϕ_0 are arbitrary real constants.

5.5.4 Constant Nonlinearity and Complex Potential

If $\quad\psi(x,\,t)\quad$ is $\quad$ a $\quad$ solution $\quad$ to $\quad$ the $\quad$ fundamental $\quad$ NLSE $\quad$ (2.160), $i\,\psi_t + a_1\,\psi_{xx} + a_2\,|\psi|^2\,\psi = 0$, then

$$\Phi(x,\,t) = A(x,\,t)\,e^{i\,B(x,\,t)}\,\psi[X(x,\,t),\,T(x,\,t)] \tag{5.76}$$

is a solution to

$$i\,\Phi_t + b_1(x,\,t)\,\Phi_{xx} + b_{20}\,|\Phi|^2\,\Phi + [b_{3r}(x,\,t) + i\,b_{3i}(x,\,t)]\,\Phi = 0, \tag{5.77}$$

where

$$X(x,\,t) = g_4(t) + g_5(t)\,x, \tag{5.78}$$

$$T(x,\,t) = g_1(t) = c_0 + \frac{b_{20}}{a_2}\,\int \frac{g_3^2(t)}{g_5(t)}\,d\,t, \tag{5.79}$$

$$A(x,\,t) = \frac{g_3(t)}{\sqrt{g_5(t)}}, \tag{5.80}$$

$$B(x,\,t) = g_2(t) - \frac{a_2\,[2\,g_4{}'(t) + x\,g_5{}'(t)]\,x\,g_5^2(t)}{4\,a_1\,b_{20}\,g_3^2(t)}, \tag{5.81}$$

$$b_1(x,\,t) = \frac{a_1\,b_{20}\,g_3^2(t)}{a_2\,g_5^3(t)}, \tag{5.82}$$

$$b_{3i}(x,\,t) = \frac{g_5{}'(t)}{g_5(t)} - \frac{g_3{}'(t)}{g_3(t)}, \tag{5.83}$$

$$b_{3r}(x, t) = g_2'(t) + \frac{a_2 \, g_4'^2(t) \, g_5(t)}{4 \, a_1 \, b_{20} \, g_3^2(t)}$$

$$+ \frac{a_2 \, g_5(t) \, [2 \, g_5(t) \, g_3'(t) \, g_4'(t) - g_3(t) \, g_4'(t) \, g_5'(t) - g_3(t) \, g_4''(t) \, g_5(t)]}{2 \, a_1 \, b_{20} \, g_3^3(t)} \, x \qquad (5.84)$$

$$+ \frac{a_2 \, g_5(t) \, [2 \, g_3'(t) \, g_5(t) \, g_5'(t) - g_3(t) \, g_5'^2(t) - g_3(t) \, g_5(t) \, g_5''(t)]}{4 \, a_1 \, b_{20} \, g_3^3(t)} \, x^2,$$

$g_4(t)$ and $g_5(t)$ are arbitrary real functions, b_{20}, and c_0 are arbitrary real constants.

5.5.5 Constant Nonlinearity and Real Quadratic Potential

If $\psi(x, t)$ is a solution to the fundamental NLSE (2.160), $i \, \psi_t + a_1 \, \psi_{xx} + a_2 \, |\psi|^2 \, \psi = 0$, then

$$\Phi(x, t) = c_2 \, \sqrt{g_5(t)} \; e^{i \left\{ c_1 - \frac{a_2 \left[2 \, c_4 \, g_5(t) \, x + g_5'(t) \, x^2 + c_4^2 \int g_5(t) \, d \, t \right]}{4 \, a_1 \, b_{20} \, c_2^2} \right\}}$$

$$\times \psi \left[c_3 + g_5(t) \, x + c_4 \int g_5(t) \, d \, t, \; c_0 + \frac{b_{20} \, c_2^2}{a_2} \int g_5(t) \, d \, t \right] \qquad (5.85)$$

is a solution to

$$i \, \Phi_t + \frac{a_1 \, b_{20} \, c_2^2}{a_2 \, g_5(t)} \, \Phi_{xx} + b_{20} \, |\Phi|^2 \, \Phi + \frac{a_2 \, [g_5'^2(t) - g_5(t) \, g_5''(t)]}{4 \, a_1 \, b_{20} \, c_2^2 \, g_5(t)} \, x^2 \, \Phi = 0, \quad (5.86)$$

where $g_5(t)$ is an arbitrary real function of t, c_1, c_2, c_3, and c_4 are arbitrary real constants.

5.5.6 Constant Nonlinearity and Real Linear Potential

If $\psi(x, t)$ is a solution to the fundamental NLSE (2.160), $i \, \psi_t + a_1 \, \psi_{xx} + a_2 \, |\psi|^2 \, \psi = 0$, then

$$\Phi(x, t) = c_2 \, \sqrt{c_5 \, e^{c_6 \, t}} \; e^{i \left\{ c_1 - \frac{a_2 \left[c_5^2 \, c_6 \, e^{c_6 \, t} \, x^2 + 2 \, c_5 \, g_4'(t) \, x + \int g_4'^2(t) \, e^{-c_6 \, t} \, d \, t \right]}{4 \, a_1 \, b_{20} \, c_2^2 \, c_5} \right\}}$$

$$\times \psi \left[c_5 \, e^{c_6 \, t} \, x + g_4(t), \; c_0 + \frac{b_{20} \, c_2^2 \, c_5 \, (e^{c_6 \, t} - e^{c_6})}{a_2 \, c_6} \right] \qquad (5.87)$$

is a solution to

$$i \, \Phi_t + \frac{a_1 \, b_{20} \, c_2^2 \, e^{-c_6 \, t}}{a_2 \, c_5} \, \Phi_{xx} + b_{20} \, |\Phi|^2 \, \Phi + \frac{a_2 \, [c_6 \, g_4'(t) - g_4''(t)]}{2 \, a_1 \, b_{20} \, c_2^2} \, x \, \Phi = 0, \quad (5.88)$$

where

$$g_2(t) = c_1 - \frac{a_2}{4\, a_1\, b_{20}\, c_2^2\, c_5} \int g_4'(t)^2\, e^{-c_6\, t}\, d\, t, \tag{5.89}$$

$$g_5(t) = c_5\, e^{c_6\, t}, \tag{5.90}$$

$g_4(t)$ is an arbitrary real function of t, c_1, c_2, c_5, and c_6 are arbitrary real constants.

5.6 Solution-Dependent Transformation

If $\psi(x, t)$ is a solution to the fundamental NLSE (2.160), $i\, \psi_t + a_1\, \psi_{xx} + a_2\, |\psi|^2\, \psi = 0$, then

$$\Phi(x, t) = A(x, t)\, e^{i\, B(x, t)}\, \psi[X(x, t), T(x, t)] \tag{5.91}$$

is a solution to

$$i\, \Phi_t + b_1(x, t)\, \Phi_{xx} + b_2(x, t)\, |\Phi|^2\, \Phi + [b_{3r}(x, t) + i\, b_{3i}(x, t)]\, \Phi = 0, \tag{5.92}$$

where

$$T(x, t) = g_1(t), \tag{5.93}$$

$$b_1(x, t) = \frac{a_1\, g_1'(t)}{X_x^2(x, t)}, \tag{5.94}$$

$$b_2(x, t) = \frac{a_2\, g_1'(t)}{A^2(x, t)}, \tag{5.95}$$

$$
\begin{aligned}
b_{3r}(x, t) = {}& \frac{a_1\, g_1'(t)\left[A(x, t)\, B_x^2(x, t) - A_{xx}(x, t) \right]}{A(x, t)\, X_x^2(x, t)} + B_t(x, t) \\
& - \frac{a_1\, g1'(t)[2\, A_x(x, t)\, X_x(x, t) + A(x, t)\, X_{xx}(x, t)]}{A(x, t) X_x^2(x, t)}\, \mathrm{Re}\left[\frac{\psi_x(x, t)}{\psi(x, t)} \right] \\
& + \left[\frac{2\, a_1\, g_1'(t) B_x(x, t)}{X_x(x, t)} + X_t(x, t) \right] \mathrm{Im}\left[\frac{\psi_x(x, t)}{\psi(x, t)} \right],
\end{aligned}
\tag{5.96}
$$

$$
\begin{aligned}
b_{3i}(x, t) = {}& - \frac{a_1\, g_1'(t)[2\, A_x(x, t)\, B_x(x, t) + A(x, t)\, B_{xx}(x, t)]}{X_x^2(x, t)\, A(x, t)} - \frac{A_t(x, t)}{A(x, t)} \\
& - \left[\frac{2\, a_1\, g_1'(t)\, B_x(x, t)}{X_x(x, t)} + X_t(x, t) \right] \mathrm{Re}\left[\frac{\psi_x(x, t)}{\psi(x, t)} \right] \\
& - \frac{a_1\, g_1'(t)\, [2\, A_x(x, t)\, X_x(x, t) + A(x, t)\, X_{xx}(x, t)]}{X_x^2(x, t)\, A(x, t)}\, \mathrm{Im}\left[\frac{\psi_x(x, t)}{\psi(x, t)} \right],
\end{aligned}
\tag{5.97}
$$

$g_1(t)$, $X(x, t)$, and $B(x, t)$ are arbitrary real functions.

5.6.1 Special Case 1: Stationary Solution, Constant Dispersion and Nonlinearity Coefficients

If $\psi(x, t) = A_0 f(x) e^{i A_0^2 (t - t_0)}$ is a stationary solution to the fundamental NLSE (2.160), $i \psi_t + a_1 \psi_{xx} + a_2 |\psi|^2 \psi = 0$, then

$$\Phi(x, t) = \sqrt{\frac{a_2}{b_{20}}} \, e^{i B(x, t)} \psi \left[\sqrt{\frac{a_1}{b_{10}}} (x - x_0), (t - t_0) \right] \tag{5.98}$$

$$= A_0 \sqrt{\frac{a_2}{b_{20}}} \, f \left[\sqrt{\frac{a_1}{b_{10}}} (x - x_0) \right] e^{i \left[B(x, t) + A_0^2 (t - t_0) \right]} \tag{5.99}$$

is a solution to

$$
\begin{aligned}
i \, \Phi_t + b_{10} \, \Phi_{xx} + b_{20} \, |\Phi|^2 \, \Phi & \\
+ \, (B_t(x, t) + b_{10} \, B_x^2(x, t) & \\
- \, i \, b_{10} \left\{ \frac{\sqrt{\frac{4 a_1}{b_{10}}} \, f' \left[\sqrt{\frac{a_1}{b_{10}}} (x - x_0) \right] B_x(x, t)}{f \left[\sqrt{\frac{a_1}{b_{10}}} (x - x_0) \right]} + B_{xx}(x, t) \right\} \Bigg) \Phi = 0,
\end{aligned}
\tag{5.100}
$$

where $B(x, t)$ is an arbitrary real function, b_{10}, b_{20}, A_0 and t_0 are arbitrary real constants.

5.6.2 Special Case 2: PT-Symmetric Potential

If $\psi(x, t) = A_0 f(x) e^{i A_0^2 (t - t_0)}$ is a stationary solution to the fundamental NLSE (2.160), $i \psi_t + a_1 \psi_{xx} + a_2 |\psi|^2 \psi = 0$, then

$$\Phi(x, t) = \sqrt{\frac{a_2}{b_{20}}} \, e^{i B(x)} \psi \left[\sqrt{\frac{a_1}{b_{10}}} (x - x_0), (t - t_0) \right] \tag{5.101}$$

$$= A_0 \sqrt{\frac{a_2}{b_{20}}} \, f \left[\sqrt{\frac{a_1}{b_{10}}} (x - x_0) \right] e^{i [B(x) + A_0^2 (t - t_0)]} \tag{5.102}$$

is a solution to

$$i \, \Phi_t + b_{10} \, \Phi_{xx} + b_{20} \, |\Phi|^2 \, \Phi + [V_{\text{even}}(x) + i \, V_{\text{odd}}(x)] \Phi = 0, \tag{5.103}$$

where

$$V_{\text{even}} = b_{10} \, B_x^2(x), \tag{5.104}$$

and

$$V_{\text{odd}} = - b_{10} \left\{ \frac{\sqrt{\dfrac{4 a_1}{b_{10}}} \, f'\left[\sqrt{\dfrac{a_1}{b_{10}}} \, (x - x_0) \right] B_x(x)}{f\left[\sqrt{\dfrac{a_1}{b_{10}}} \, (x - x_0) \right]} + B_{xx}(x) \right\},$$ (5.105)

form a PT-symmetric potential for some choices of $B(x)$.

Case I: $B(x, t) = \sin(x - x_0)$

Given $\psi(x, t) = A_0 \, e^{i \, [a_2 \, A_0^2 \, (t - t_0) + \phi_0]}$ is a stationary solution to the fundamental NLSE (2.160), $i \, \psi_t + a_1 \, \psi_{xx} + a_2 \, |\psi|^2 \, \psi = 0$, then

$$\Phi(x, t) = \sqrt{\frac{a_2}{b_{20}}} \, e^{i \, \sin(x - x_0)} \, \psi\left[\sqrt{\frac{a_1}{b_{10}}} \, (x - x_0), (t - t_0) \right]$$ (5.106)

is a solution to

$$i \, \Phi_t + b_{10} \, \Phi_{xx} + b_{20} \, |\Phi|^2 \, \Phi + b_{10} \, [\cos^2(x - x_0) + i \, \sin(x - x_0)] \, \Phi = 0, \quad (5.107)$$

where $B(x)$ is an arbitrary real function, b_{10}, b_{20}, A_0, x_0, and t_0 are real constants.

5.6.3 Special Case 3: Stationary Solution, Constant Dispersion and Nonlinearity Coefficients, and Real Potential

If $\quad \psi(x, t) = A_0 f(x) \, e^{i \, [a_2 \, A_0^2 \, (t - t_0) + \phi_0]} \quad$ is a stationary solution to $i \, \psi_t + a_1 \, \psi_{xx} + a_2 \, |\psi|^2 \, \psi = 0$, then

$$\Phi(x, t) = \sqrt{\frac{a_2}{b_{20}}} \, e^{i \, B(x, t)} \, \psi\left[\sqrt{\frac{a_1}{b_{10}}} \, (x - x_0), (t - t_0) \right]$$ (5.108)

$$= A_0 \sqrt{\frac{a_2}{b_{20}}} \, f\left[\sqrt{\frac{a_1}{b_{10}}} \, (x - x_0) \right] e^{i \, \left[B(x, t) + A_0^2 \, (t - t_0) \right]}$$ (5.109)

is a solution to

$$i\Phi_t + b_{10} \, \Phi_{xx} + b_{20} \, |\Phi|^2 \, \Phi$$

$$+ \left\{ \frac{b_{10} \, g_1^2(t)}{f^4\left[\sqrt{\dfrac{a_1}{b_{10}}} \, (x - x_0) \right]} + g_1'(t) \int \frac{d x}{f^2\left[\sqrt{\dfrac{a_1}{b_{10}}} \, (x - x_0) \right]} + g_2'(t) \right\} \Phi = 0, \quad (5.110)$$

where

$$B(x, t) = g_1(t) \int \frac{d x}{f^2\left[\sqrt{\dfrac{a_1}{b_{10}}} \, (x - x_0) \right]} + g_2(t),$$ (5.111)

$g_1(t)$ and $g_2(t)$ are arbitrary real functions of t, b_{10}, b_{20}, A_0, x_0, and t_0 are real constants.

5.7 Summary of Sections 5.2–5.6

Note: For lengthy conditions, the reader is referred to the solutions in sections 5.2–5.6.

Fundamental NLSE to Fundamental NLSE with Different Constant Coefficients

Transformation: $\Phi(x,t) = \sqrt{\frac{a_2}{a_{22}}}\,\psi\left(\sqrt{\frac{a_1}{a_{11}}}\,x,\,t\right)$, ψ is a solution to the fundamental NLSE (2.160).

Equation: $i\,\Phi_t + a_{11}\,\Phi_{xx} + a_{22}\,|\Phi|^2\,\Phi = 0$

#	Example	Conditions	Name	Equation #
1.	$\Phi(x,t) = A_0\sqrt{\frac{2a_1}{a_{22}}}\,\operatorname{sech}[A_0\sqrt{\frac{a_1}{a_{11}}}(x-x_0)]\,e^{i[a_1 A_0^2(t-t_0)+\phi_0]}$	$a_1\,a_{22} > 0$, $a_1\,a_{11} > 0$, A_0, x_0, t_0, and ϕ_0 are arbitrary real constants	Bright soliton	(5.14)
2.	$\Phi(x,t) = A_0\sqrt{\frac{-2a_1}{a_{22}}}\,\tanh[A_0\sqrt{\frac{a_1}{a_{11}}}(x-x_0)]$ $\times\, e^{-i[2a_1 A_0^2(t-t_0)+\phi_0]}$	$a_1\,a_{22} < 0$, $a_1\,a_{11} > 0$, A_0, x_0, t_0, and ϕ_0 are arbitrary real constants	Drak soliton	(5.15)

Defocusing (Focusing) NLSE to Focusing (Defocusing) NLSE

Transformation: $\Phi(x,t) = \psi(i\,x,\,-t)$

Equation: $i\,\Phi_t + a_1\,\Phi_{xx} - a_2\,|\Phi|^2\,\Phi = 0$

#	Example	Conditions	Name	Equation #
1.	$\Phi(x,t) = A_0\sqrt{\frac{2a_1}{a_2}}\,\operatorname{sech}[i\,A_0(x-x_0)]\,e^{i[-a_1 A_0^2(t-t_0)+\phi_0]}$	$a_1\,a_2 > 0$, A_0, x_0, t_0, and ϕ_0 are arbitrary real constants	–	(5.18)
2.	$\Phi(x,t) = \frac{1}{\sqrt{a_2}}\left[\frac{4-i8(t-t_0)}{1+4((t-t_0)^2-\frac{2}{a_1}(x-x_0)^2)} - 1\right]e^{i[-(t-t_0)+\phi_0]}$	$a_2 > 0$, x_0, t_0, and ϕ_0 are arbitrary real constants	Peregrine soliton, two-solitons	(5.19)

(Continued)

Galilean Transformation (Moving Solutions)

Transformation: $\Phi(x, t) = \psi(x - v t, t)\, e^{i[\frac{v}{2a_1}(x-x_0) - \frac{v^2}{4a_1}(t-t_0)]}$

Equation: $i\,\Phi_t + a_1\,\Phi_{xx} + a_2\,|\Phi|^2\,\Phi = 0$

#	Example	Conditions	Name	Equation #
1.	$\Phi(x, t) = A_0 \sqrt{\frac{2 a_1}{a_2}}\, \operatorname{sech}\{A_0 [x - (x_0 + v t)]\}$ $\times e^{i[\frac{v}{2a_1}(x-x_0) + \frac{4a_1^2 A_0^2 - v^2}{4a_1}(t-t_0) + \phi_0]}$	$a_1 a_2 > 0, A_0, x_0, t_0, v,$ and ϕ_0 are arbitrary real constants	Moving bright soliton	(5.24)
2.	$\Phi(x, t) = A_0 \sqrt{\frac{-2 a_1}{a_2}}\, \tanh\{A_0 [x - (x_0 + v t)]\}$ $\times e^{-i[\frac{v}{2a_1}(x-x_0) + \frac{8a_1^2 A_0^2 + v^2}{4a_1}(t-t_0) + \phi_0]}$	$a_1 a_2 < 0, A_0, x_0, t_0, v$ and ϕ_0 are arbitrary real constants	Moving dark soliton	(5.25)
3.	$\Phi(x, t) = \frac{1}{\sqrt{a_2}}\left[\frac{4 + i\,8(t-t_0)}{1 + 4[(t-t_0)^2 + \frac{2}{a_1}(x - x_0 - v\,t)^2]} - 1\right]$ $\times e^{i[\frac{v}{2a_1}(x-x_0) - \frac{v^2}{4a_1}(t-t_0)]}\, e^{i[t - t_0 + \phi_0]}$	$a_2 > 0, x_0, t_0, v,$ and ϕ_0 are arbitrary real constants	Moving Peregrine soliton	(5.26)

Equation: $i\,\Phi_t + a_1\,\Phi_{xx} + a_2\,|\Phi|^n\,\Phi = 0$

#	Example	Conditions	Name	Equation #
1.	$\Phi(x, t) = \left(\frac{2 A_0^2 a_1 (n+2)}{a_2\, n^2}\, \operatorname{sech}^2\{A_0 [x - (x_0 + v t)]\}\right)^{\frac{1}{n}}$ $\times e^{i[\frac{v}{2a_1}(x-x_0) - \frac{v^2}{4a_1}(t-t_0)]}\, e^{i[\frac{4 a_1 A_0^2}{n^2}(t-t_0) + \phi_0]}$	$a_1 a_2 (n + 2) > 0, A_0, x_0, t_0, v,$ and ϕ_0 are arbitrary real constants	Moving bright soliton	(5.27)

Equation: $i\,\Phi_t + a_1\,\Phi_{xx} + a_2\,|\Phi|^n\,\Phi + a_3\,|\Phi|^{2n}\,\Phi = 0$

#	Example	Conditions	Name	Equation #
1.	$\Phi(x, t) = \left\{\frac{A_0 (n+2)}{a_2 + a_2 \sin(\theta) \cosh[n\sqrt{\frac{A_0}{a_1}}(x - x_0)]}\right\}^{\frac{1}{n}}$ $\times e^{i[A_0 (t-t_0)+\phi_0]}\, e^{i[\frac{v}{2a_1}(x-x_0) - \frac{v^2}{4a_1}(t-t_0)]}$	$a_3 = \gamma\, a_{03}, \gamma = -\cos^2(\theta), 0 < \theta < \pi/2,$ $a_{03} = a_2^2 (n + 1)/[A_0 (n + 2)^2], m = 2\,n, a_1\, A_0 > 0, x_0, t_0, A_0,$ and ϕ_0 are arbitrary real constants	Moving flat-top soliton	(5.28)

Function Coefficients

Transformation: $\Phi(x,t) = A(x,t)\, e^{j\,B(x,t)}\, \psi[X(x,t),\, T(x,t)]$

Equation: $i\,\Phi_t + b_1(x,t)\,\Phi_{xx} + b_2(x,t)\,|\Phi|^2\,\Phi + [b_{3r}(x,t) + i\,b_{3i}(x,t)]\,\Phi = 0$

Constant Dispersion and Complex Potential

Equation: $i\,\Phi_t + b_{10}\,\Phi_{xx} + b_2(x,t)\,|\Phi|^2\,\Phi + [b_{3r}(x,t) + i\,b_{3i}(x,t)]\,\Phi = 0$

Constant Dispersion and Real Quadratic Potential

Transformation:
$$\Phi(x,t) = c_2\,\sqrt{g_5(t)}\; e^{j\left[c_1 - \frac{c_4^2}{4\,b_{10}}\int g_5^2(t)\,dt - \frac{2\,a_4\,g_5^2(t) - g_5'(t)\,x}{4\,b_{10}\,g_5(t)}\,x\right]}$$
$$\times\,\psi\left[\,c_3 + g_5(t)\,x + c_4\int g_5^2(t)\,dt,\; c_0 + \frac{b_{10}}{q}\int g_5^2(t)\,dt\,\right]$$

Equation (I): $i\,\Phi_t + b_{10}\,\Phi_{xx} + \frac{a_2\,b_{10}\,g_5(t)}{q\,c_2^2}\,|\Phi|^2\,\Phi - \frac{g_5(t)\,g_5''(t) - 2\,g_5'^2(t)}{4\,b_{10}\,g_5^2(t)}\,x^2\,\Phi = 0$

#	Example	Conditions	Name	Equation #
1.	$\Phi(x,t) = \sqrt{\dfrac{2\,A_0^2\,q\,c_2^2\,g_5(t)}{a_2}}\; e^{j\,\phi(x,t)}$ $\times\,\mathrm{sech}\{A_0\,[c_3 - x_0 + g_5(t)\,x + c_4\int g_5^2(t)\,dt]\}$	$\phi(x,t) = A_0^2\left[a_1\,(c_0 - t_0) + b_{10}\int g_5^2(t)\,dt\right] - \dfrac{c_4^2}{4\,b_{10}}\int g_5^2(t)\,dt$ $-\dfrac{[2\,c_4\,g_5^2(t) + x\,g_5'(t)]\,x}{4\,b_{10}\,g_5(t)} + c_1 + \phi_0$ $a_1\,a_2 > 0,$ $A_0,\,x_0,\,t_0,\text{ and }\phi_0\text{ are arbitrary}$ real constants	Bright soliton	(5.52)
2.	$\Phi(x,t) = \sqrt{\dfrac{-2\,A_0^2\,q\,c_2^2\,g_5(t)}{a_2}}\; e^{-i\,\phi(x,t)}$ $\times\,\tanh\{A_0\,[c_3 - x_0 + g_5(t)\,x + c_4\int g_5^2(t)\,dt]\}$	$\phi(x,t) = 2\,A_0^2\left[a_1\,(c_0 - t_0) + b_{10}\int g_5^2(t)\,dt\right] + \dfrac{c_4^2}{4\,b_{10}}\int g_5^2(t)\,dt$ $+\dfrac{[2\,c_4\,g_5^2(t) + x\,g_5'(t)]\,x}{4\,b_{10}\,g_5(t)} - c_1 + \phi_0$ $a_1\,a_2 < 0,$ $A_0,\,x_0,\,t_0,\text{ and }\phi_0$ are arbitrary real constants	Dark soliton	(5.57)

(Continued)

3. $\Phi(x,t) = c_2 \sqrt{\frac{g_5(t)}{a_2}} \left[\psi_1(x,t) + \psi_2(x,t)\right]$

$\times e^{i\left[c_1 - \frac{2 c_4 g_5^2(t) x - c_4^2 g_5(t) \int g_5^2(t) \, dt + t_5'(t) x^2}{4 b_{10} g_5(t)}\right]}$

Two-bright-solitons $\qquad$ See text $\qquad$ (5.60)

Equation (2): $i \, \Phi_t + b_{10} \, \Phi_{xx} + \frac{a_2 b_{10} [\alpha + \beta \sin(\gamma \, t)]}{\eta \, c_2^4} \left|\Phi\right|^2 \Phi + \frac{\beta \gamma^2 [3\beta - \beta \cos(2\gamma \, t) + 2\alpha \sin(\gamma \, t)]}{8 b_{10} [\alpha + \beta \sin(\gamma \, t)]^2} x^2 \, \Phi = 0$

#	Example	Name	Conditions	Equation #
1.	$\Phi(x,t) = \sqrt{\frac{2 A_0^2 \eta \, c_2^4 [\alpha + \beta \sin(\gamma t)]}{a_2}}$ $\times \text{sech}(A_0 \{c_3 - x_0 + [\alpha + \beta \sin(\gamma \, t)] \, x$ $+ \frac{c_4^2 [2\gamma(2\alpha^2 + \beta^2) t - 8\alpha\beta\cos(\gamma t) - \beta^2 \sin(2\gamma t)]}{4\gamma}\})$ $\times e^{i \, \phi(x,t)}$	Bright soliton	$\phi(x,t) = -\frac{\beta\gamma\cos(\gamma t) x^2 + 2 c_4 [\alpha + \beta\sin(\gamma t)]^2 x}{4 b_{10}[\alpha + \beta\sin(\gamma t)]}$ $-\frac{c_4^2 [2\gamma(2\alpha^2 + \beta^2) t - 8\alpha\beta\cos(\gamma t) - \beta^2 \sin(2\gamma t)]}{16 b_{10}\gamma}$ $+ \frac{A_0^2}{4\gamma}[4\gamma a_1 (c_0 - t_0) + b_{10}[2\gamma(2\alpha^2 + \beta^2) t$ $- 8\alpha\beta\cos(\gamma \, t) - \beta^2 \sin(2\gamma \, t)]] + c_1 + \phi_0$	(5.53)
2.	$\Phi(x,t) = \sqrt{\frac{-2 A_0^2 \eta \, c_2^4 [\alpha + \beta \sin(\gamma t)]}{a_2}}$ $\times \tanh[A_0 \{c_3 - x_0 + [\alpha + \beta \sin(\gamma \, t)] \, x$ $+ \frac{c_4^2 [2\gamma(2\alpha^2 + \beta^2) t - 8\alpha\beta\cos(\gamma t) - \beta^2 \sin(2\gamma t)]}{4\gamma}\}]$ $\times e^{-i \, \phi(x,t)}$	Dark soliton	$\phi(x,t) = \frac{\beta\gamma\cos(\gamma t) x^2 + 2 c_4 [\alpha + \beta\sin(\gamma t)]^2 x}{4 b_{10}[\alpha + \beta\sin(\gamma t)]}$ $+ \frac{c_4^2 [2\gamma(2\alpha^2 + \beta^2) t - 8\alpha\beta\cos(\gamma t) - \beta^2 \sin(2\gamma t)]}{16 b_{10}\gamma}$ $+ \frac{A_0^2}{2\gamma}[4\gamma a_1 (c_0 - t_0) + b_{10}[2\gamma(2\alpha^2 + \beta^2) t$ $- 8\alpha\beta\cos(\gamma \, t) - \beta^2 \sin(2\gamma \, t)]] - c_1 + \phi_0$	(5.58)
3.	$\Phi(x,t) = c_2 \sqrt{\frac{\alpha + \beta\sin(\gamma t)}{a_2}} \left[\psi_1(x,t) + \psi_2(x,t)\right] e^{i \, \phi(x,t)}$	Two-bright-soliton	See text	(5.61)

Equation (3): $i \, \Phi_t + b_{10} \, \Phi_{xx} + \frac{a_2 h_0 e^{\gamma t}}{\eta \, c_2^4} \left|\Phi\right|^2 \Phi + \frac{\gamma^2}{4 b_{10}} x^2 \, \Phi = 0$

#	Example	Name	Conditions	Equation #
1.	$\Phi(x,t) = c_2 A_0 \sqrt{\frac{2\eta}{a_2}} \, e^{\gamma t} \, \text{sech}[A_0 (\frac{4 c_4 e^{2\gamma t}}{2\gamma} + e^{\gamma t} x - x_0 + c_3)]$ $\times e^{i \, \phi(x,t)}$	Bright soliton	$\phi(x,t) = \frac{1}{8 h_0 \gamma} \{-c_4 e^{\gamma t} (c_4 e^{\gamma t} + 4\gamma x) + 2\gamma [4 b_{10} (c_1 + \phi_0)$ $- \gamma x^2] + 4 A_0^2 h_0 [b_{10} e^{2\gamma t} + 2 a_1 \gamma (c_0 - t_0)]\} + \phi_0$, γ is an arbitrary real constant	(5.55)

2.
$$\Phi(x,t) = c_2 A_0 \sqrt{\frac{-2q}{a_2}} e^{\gamma t} \tanh[A_0 (\frac{c_4 e^{2\gamma t}}{2\gamma} + e^{\gamma t} x - x_0 + c_3)]$$
$$\times e^{i \phi(x,t)}$$

$$\phi(x,t) = \frac{-1}{8 b_{10} \gamma} [(c_4 e^{\gamma t} (c_4 e^{\gamma t} + 4\gamma x) + 2\gamma [4 b_{10} (\phi_0 - c_1)$$
$$+ \gamma x^2] + 8 A_0^2 b_{10} [b_{10} e^{2\gamma t} + 2 a_1 \gamma (c_0 - t_0)]] + \phi_0,$$ $\quad \gamma$ is an arbitrary real constant

Dark soliton (5.59)

3.
$$\Phi(x,t) = \sqrt{\frac{i}{a_2}} [\psi_1(x,t) + \psi_2(x,t)] e^{-i \frac{1}{8}(2 e^t + 4 e^{\frac{t}{2}} x + x^2)}$$

See text

Two-bright-soliton (5.62)

Constant Dispersion and Real Linear Potential

Transformation:
$$\Phi(x,t) = \sqrt{\frac{c_2^2 c_5}{a_2 c_7}} \psi[\frac{c_5 x}{c_7} + g_4(t), c_0 + \frac{b_{10} c_5^2 t}{a_1 c_7^2}] e^{i [c_1 - \frac{c_7^2}{4 b_{10} c_5^2} \int g_4'^2(t) dt] dt - \frac{c_7 g_4(0) x}{2 b_{10} c_5}}$$

Equation (1): $i\Phi_t + b_{10} \Phi_{xx} + \frac{a_2 b_{10} c_5}{a_1 c_7^2} |\Phi|^2 \Phi - \frac{c_7 g_4''(t)}{2 b_{10} c_5} x \Phi = 0$

1.
$$\Phi(x,t) = c_2 A_0 \sqrt{\frac{2 a_1 c_5}{a_2 c_7}} \operatorname{sech}[A_0 [\frac{c_5}{c_7} x - x_0 + g_4(t)]] e^{i \phi(x,t)}$$

$$\phi(x,t) = A_0^2 a_1 [c_0 + \frac{b_{10} c_5^2}{a_1 c_7} (t - t_0)] - \frac{c_7^2}{4 b_{10} c_5^2} \int g_4'^2(t) dt$$
$$- \frac{c_7 g_4'(t)}{2 b_{10} c_5} x + c_1 + \phi_0,$$

$a_1 a_2 c_5 c_7 > 0$, A_0, x_0, t_0, and ϕ_0 are arbitrary real constants

Bright soliton (5.67)

2.
$$\Phi(x,t) = c_2 A_0 \sqrt{\frac{-2 a_1 c_5}{a_2 c_7}} \tanh[A_0 [\frac{c_5}{c_7} x - x_0 + g_4(t)]] e^{-i \phi(x,t)}$$

$$\phi(x,t) = 2 A_0^2 a_1 [c_0 + \frac{b_{10} c_5^2}{a_1 c_7} (t - t_0)] + \frac{c_7^2}{4 b_{10} c_5^2} \int g_4'^2(t) dt$$
$$+ \frac{c_7 g_4'(t)}{2 b_{10} c_5} x - c_1 + \phi_0$$

Dark soliton (5.70)

3.
$$\Phi(x,t) = A_0 \sqrt{\frac{c_2^2 c_5}{a_2 c_7}} (1 - \frac{\sqrt{8} \lambda_r}{A_0} p(x,t))$$
$$\times e^{i [A_0^2 (c_0 + \frac{b_{10} c_5^2}{a_1 c_7^2} t_0)] - \frac{c_7^2}{4 b_{10} c_5^2} \int g_4'^2(t) dt \, (\gamma d \, t - \frac{c_7 g_4(0) x}{2 b_{10} c_5} + c_1 + \phi_0]}$$

See text

Generalized first-order breather (5.72)

Equation (2): $i\Phi_t + b_{10} \Phi_{xx} + \frac{a_2 b_{10} c_5}{a_1 c_7^2} |\Phi|^2 \Phi - \frac{c_7 \alpha}{b_{10} c_5} x \Phi = 0$

1.
$$\Phi(x,t) = c_2 A_0 \sqrt{\frac{2 a_1 c_5}{a_2 c_7}} \operatorname{sech}[A_0 [\frac{c_5}{c_7} x - x_0 + \alpha t^2]]$$
$$\times e^{i [A_0^2 a_1 [c_0 + \frac{b_{10} c_5^2}{a_1 c_7} (t - t_0)] - \frac{c_7^2 \alpha^2 t^3}{3 b_{10} c_5^2} - \frac{c_7 \alpha t}{b_{10} c_5} x + c_1 + \phi_0]}$$

Bright soliton (5.68)

(Continued)

2.	$\Phi(x,t) = c_2 A_0 \sqrt{\dfrac{-2 a_1 c_5}{a_2 c_7}}\, \tanh\{A_0 [\frac{c_5}{c_7} x - x_0 + \alpha\, t^2]\}$ $\times e^{-i\{2 A_6^2 a_1 [c_0 + \frac{b_{10} c_5^2}{a_1 c_7^2}(t-t_0)] + \frac{c_7^2 \alpha^2 t^3}{3 b_{10} c_5^2} + \frac{c_7 \alpha t}{b_{10} c_5} x - c_1 + \phi_0\}}$	Dark soliton	(5.71)
3.	$\Phi(x,t) = A_0 \sqrt{\dfrac{c_5^2 c_5}{a_2 c_7}}\, [1 - \frac{\sqrt{8}\,\lambda_r}{A_0} p(x,t)]$ $\times e^{i[A_0^2 (c_0 + \frac{b_{10} c_5^2 t}{a_1 c_7^2} - t_0) - \frac{c_7 \alpha c_5 s x + c_7 \alpha^2 t_0^2 t}{3 b_{10} c_5^2} + c_1 + \phi_0]}$	See text — Generalized first-order breather (case I)	(5.73)

Equation (3): $i\,\Phi_t + b_{10}\,\Phi_{xx} + \dfrac{a_2 b_{10} c_5}{c_5^2 c_7} |\Phi|^2\,\Phi - \dfrac{c_7 \alpha \beta^2 \cos(\beta t)}{2 b_{10} c_5}\, x\,\Phi = 0$

1.	$\Phi(x,t) = A_0 \sqrt{\dfrac{c_5^2 c_5}{a_2 c_7}}\, [1 - \frac{\sqrt{8}\,\lambda_r}{A_0} p(x,t)]\, e^{i\frac{A_0^2 (c_0 + \frac{b_{10} c_5^2 t}{a_1 c_7^2} - t_0)}{\cdot}}$ $\times e^{i[-\frac{c_7 \alpha^2 \beta[2\beta t - \sin(2\beta t)]}{16 b_{10} c_5^2} - \frac{c_7 \alpha \beta \sin(\beta t) x}{2 b_{10} c_5} + c_1 + \phi_0]}$	See text — Generalized first-order breather (case II)	(5.74)

Constant Nonlinearity and Complex Potential

Transformation: $\Phi(x,t) = A(x,t)\, e^{i\,B(x,t)}\, \psi[X(x,t), T(x,t)]$

Equation: $i\,\Phi_t + b_1(x,t)\,\Phi_{xx} + b_{20}|\Phi|^2\,\Phi + [b_{3r}(x,t) + i\, b_{3i}(x,t)]\,\Phi = 0$

Constant Nonlinearity and Real Quadratic Potential

Transformation:
$$\Phi(x,t) = c_2 \sqrt{g_5(t)}\; e^{i\,\{c_1 - \frac{a_2[2 a_4 g_5(t) x + g_5'(t) x^2 + \frac{c_4^2}{4}\int g_5(t)\,d t]}{4 a_1 b_{20} c_5^2}\}}$$
$$\times \psi\left[c_3 + g_5(t)\, x + c_4 \int g_5(t)\, d t,\; c_0 + \frac{b_{20} c_5^2}{a_2} \int g_5(t)\, d t\right]$$

Equation: $i\,\Phi_t + \dfrac{a_1 b_{20} c_5^2}{a_2 g_5(t)}\,\Phi_{xx} + b_{20}|\Phi|^2\,\Phi + \dfrac{a_2[g_5^2(t) - g_5(t) g_5''(t)]}{4 a_1 b_{20} c_5^2 g_5(t)}\, x^2\,\Phi = 0$

Constant Nonlinearity and Real Linear Potential

Transformation:
$$\Phi(x, t) = c_2 \sqrt{c_5}\, e^{c_6 t}\, e^{i\left\{c_1 - \frac{a_2[c_5^2 c_6 e^{c_6 t} x^2 - 2 c_5 g_4'(t) x + \int g_4^2(t) e^{-c_6 t}\, dt]}{4 a_1 b_{20} c_5^2 c_5}\right\}}$$
$$\times \psi\left[c_5 e^{c_6 t} x + g_4(t),\ c_0 + \frac{b_{20} c_5^2 c_5 (e^{c_6 t} - e^{c_6})}{a_2 c_6}\right]$$

Equation:
$$i\,\Phi_t + \frac{q\, b_{20}^2 c_5^2 e^{-c_6 t}}{a_2 c_5}\,\Phi_{xx} + b_{20}\,|\Phi|^2\,\Phi + \frac{a_2[c_6 g_4'(t) - g_4''(t)]}{2 a_1 b_{20} c_5^2}\, x\, \Phi = 0$$

Solution-Dependent Transformation

Special Case 1: Stationary Solution, Constant Dispersion and Nonlinearity Coefficients

Transformation:
$$\Phi(x, t) = \sqrt{\frac{a_2}{b_{20}}}\, e^{i\, B(x, t)}\, \psi\!\left[\sqrt{\frac{q}{q_{10}}}\,(x - x_0),\ (t - t_0)\right] = A_0 \sqrt{\frac{a_2}{b_{20}}} \int 1\!\left[\sqrt{\frac{q}{q_{10}}}\,(x - x_0)\right] e^{i\,[B(x, t) + A_0^2\,(t - t_0)]}$$

Equation:
$$i\,\Phi_t + b_{10}\,\Phi_{xx} + b_{20}\,|\Phi|^2\,\Phi + [B_t(x, t) + b_{10}\,B_x^2(x, t)]\,\Phi - i\, b_{10}\left(\frac{\frac{4 q}{q_{10}} f'\!\left[\sqrt{\frac{q}{q_{10}}}(x - x_0)\right] B_x(x, t)}{f\!\left[\sqrt{\frac{q}{q_{10}}}(x - x_0)\right]} + B_{xx}(x, t)\right)\right] = 0$$

$$\Phi = 0$$

Special Case 2: PT-Symmetric Potential

Transformation 1:
$$\Phi(x, t) = \sqrt{\frac{a_2}{b_{20}}}\, e^{i\, B(x, t)}\, \psi\!\left[\sqrt{\frac{q}{q_{10}}}\,(x - x_0),\ (t - t_0)\right] = A_0 \sqrt{\frac{a_2}{b_{20}}} \int 1\!\left[\sqrt{\frac{q}{q_{10}}}\,(x - x_0)\right] e^{i\,[B(x, t) + A_0^2\,(t - t_0)]}$$

Equation:
$$i\,\Phi_t + b_{10}\,\Phi_{xx} + b_{20}\,|\Phi|^2\,\Phi + [V_{even}(x) + i\, V_{odd}(x)]\,\Phi = 0$$

Transformation 2:
$$\Phi(x, t) = \sqrt{\frac{a_2}{b_{20}}}\, e^{i\, \sin(x - x_0)}\, \psi\!\left[\sqrt{\frac{q}{q_{10}}}\,(x - x_0),\ (t - t_0)\right]$$

Equation:
$$i\,\Phi_t + b_{10}\,\Phi_{xx} + b_{20}\,|\Phi|^2\,\Phi + b_{10}\,[\cos^2(x - x_0) + i\,\sin(x - x_0)]\,\Phi = 0$$

Special Case 3: Stationary Solution, Constant Dispersion and Nonlinearity Coefficients, and Real Potential

Transformation:
$$\Phi(x, t) = \sqrt{\frac{a_2}{b_{20}}}\, e^{i\, B(x, t)}\, \psi\!\left[\sqrt{\frac{q}{q_{10}}}\,(x - x_0),\ (t - t_0)\right] = A_0 \sqrt{\frac{a_2}{b_{20}}} \int 1\!\left[\sqrt{\frac{q}{q_{10}}}\,(x - x_0)\right] e^{i\,[B(x, t) + A_0^2\,(t - t_0)]}$$

Equation:
$$i\,\Phi_t + b_{10}\,\Phi_{xx} + b_{20}\,|\Phi|^2\,\Phi + \left\{\frac{b_{10}\, g_1^2(t)}{f^4\!\left[\sqrt{\frac{q}{q_{10}}}(x - x_0)\right]} + g_1'(t)\int \frac{dx}{f^2\!\left[\sqrt{\frac{q}{q_{10}}}(x - x_0)\right]} + g_2'(t)\right\}\Phi = 0$$

5.8 Other Equations

5.8.1 NLSE with Periodic Potentials

5.8.1.1 *Specific Case:* $sn^2(x, m)$ *Potential*
Equation:

$$i\,\psi_t + \frac{1}{2}\,\psi_{xx} - |\psi|^2\,\psi + V_0\,sn^2(x, m)\,\psi = 0, \qquad (5.112)$$

where $\psi = \psi(x, t)$ is the complex function profile, x and t are its two independent variables, V_0 is a real constant.

Solutions:

Solution 1.   *Solitary wave (SW)*

$$\psi(x, t) = \sqrt{V_0 + m}\; sn(x - x_0, m)\; e^{-i\,[\frac{(1+m)\,(t-t_0)}{2} + \phi_0]}, \qquad (5.113)$$

where $V_0 + m \geqslant 0$, x_0, t_0, and ϕ_0 are arbitrary real constants.
* *Reference:* [3], *we corrected the constant prefactor and the exponential term.*

Solution 2.   *SW*

$$\psi(x, t) = \sqrt{-(V_0 + m)}\; cn(x - x_0, m)\; e^{i\,[(V_0+m-\frac{1}{2})\,(t-t_0)+\phi_0]}, \qquad (5.114)$$

where $V_0 + m \leqslant 0$, x_0, t_0, and ϕ_0 are arbitrary real constants.
* *Reference:* [3], *we corrected the constant prefactor and the exponential term.*

Solution 3.   *SW*

$$\psi(x, t) = \sqrt{-\left(1 + \frac{V_0}{m}\right)}\; dn(x - x_0, m)\; e^{i\,[(1+\frac{V_0}{m}-\frac{m}{2})\,(t-t_0)+\phi_0]}, \qquad (5.115)$$

where $1 + \frac{V_0}{m} \leqslant 0$, x_0, t_0, and ϕ_0 are arbitrary real constants.
* *Reference:* [3], *we corrected the constant prefactor and the exponential term.*

5.8.1.2 *Specific Case:* $sin^2(x)$ *Potential*
Equation:

$$i\,\psi_t + \frac{1}{2}\,\psi_{xx} - |\psi|^2\,\psi + V_0\,sin^2(x)\,\psi = 0, \qquad (5.116)$$

where V_0 is a real constant.
Solutions:
Solution 1.

$$\psi(x, t) = \sqrt{V_0}\; sin(x - x_0)\; e^{-i\left(\frac{t-t_0}{2} + \phi_0\right)}, \qquad (5.117)$$

where $V_0 > 0$, x_0, t_0, and ϕ_0 are arbitrary real constants.
* *Reference:* [3].

Solution 2.

$$\psi(x,\ t) = \sqrt{-V_0}\ \cos(x - x_0)\ e^{-i\left[\left(V_0 - \frac{1}{2}\right)(t - t_0) + \phi_0\right]},\qquad(5.118)$$

where $V_0 < 0$, x_0, t_0, and ϕ_0 are arbitrary real constants.
- *Reference*: [3].

5.8.2 NLSE with Pöschl–Teller Potential

Equation:

$$i\ \psi_t + a_1\ \psi_{xx} + a_2\ |\psi|^2\ \psi + V_0\ \text{sech}^2(V_1\ x)\ \psi = 0,\qquad(5.119)$$

where $\psi = \psi(x,\ t)$ is the complex function profile, x and t are its two independent variables, a_1, a_2, V_0, and V_1 are real constants.

Solutions:

Solution 1.   *Bright soliton*

$$\psi(x,\ t) = \sqrt{\frac{V_0}{a_2}}\ \text{sech}(V_1\ x)\ e^{i\ [V_0\ (t - t_0) + \phi_0]},\qquad(5.120)$$

where $V_1 = \sqrt{2\ V_0}$, $a_1 = 1/2$, $V_0 > 0$, $a_2 > 0$, t_0 and ϕ_0 are arbitrary real constants.
- *Reference*: [4].

Solution 2.   *Dark soliton*

$$\psi(x,\ t) = \sqrt{\frac{2\ V_0}{a_2}}\ \tanh(V_1\ x)\ e^{i\ [2\ V_0\ (t - t_0) + \phi_0]},\qquad(5.121)$$

where $V_1 = \sqrt{-V_0}$, $a_1 = 1/2$, $V_0 < 0$, $a_2 < 0$, t_0 and ϕ_0 are arbitrary real constants.
- *Reference*: [4].

Solution 3.

$$\psi(x,\ t) = \sqrt{1 + \text{sech}(x)}\ e^{i\ [2\ (V_0 - a_1)\ (t - t_0) + \phi_0]},\qquad(5.122)$$

where $V_0 = (3\ a_1)/4$, $V_1 = 1$, $a_1 = 1/2$, $a_2 = -a_1/2$, t_0 and ϕ_0 are arbitrary real constants.
- *Obtained by the authors.*

Solution 4.

$$\psi(x,\ t) = A_0\sqrt{1 + A_1\ \text{sech}(V_1\ x) + A_2\ \text{sech}^2(V_1\ x)}\ e^{i\left[\frac{1}{2}\ a_2\ V_1^2\ A_3^2\ (t - t_0) + \phi_0\right]},\qquad(5.123)$$

where $V_0 = -3\ a_2\ V_1^2\ A_3^2$, $a_1 = -A_3^2\ a_2$, $A_0 = -(V_1\ A_3)/\sqrt{2}$, $A_1 = -2\sqrt{2}$, $A_2 = 2$, V_1, A_3, a_2, t_0, and ϕ_0 are arbitrary real constants.
- *Obtained by the authors.*

Solution 5.

$$\psi(x, t) = A_0 \sqrt{1 + \tanh(V_1 x)} \; e^{i[A_1 (t-t_0)+\phi_0]}, \tag{5.124}$$

where $A_1 = -2 A_0^2 a_2 + (2 V_0)/3 + 2 V_0$, $a_1 = (2 A_0^2 a_2)/V_1^2$, $a_2 = (2 V_0)/(3 A_0^2)$, V_0, V_1, A_0, t_0, and ϕ_0 are arbitrary real constants.
  • *Obtained by the authors.*

Solution 6.

$$\psi(x, t) = A_0 \sqrt{1 + A_1 \tanh(V_1 x) + A_2 \tanh^2(V_1 x)} \; e^{i[4 A_0^2 a_2 (t-t_0)+\phi_0]}, \tag{5.125}$$

where $V_0 = 3 A_0^2 a_2$, $a_1 = (A_0^2 a_2)/V_1^2$, $A_1 = 2$, $A_2 = 1$, V_1, A_0, a_2, t_0, and ϕ_0 are arbitrary real constants.
  • *Obtained by the authors.*

5.9 Summary of Section 5.8

Equation

$$i\,\psi_t + \tfrac{1}{2}\,\psi_{xx} - |\psi|^2\,\psi + V_0\,\mathrm{sn}^2(x,m)\,\psi = 0$$

#	Solution	Conditions	Name	Equation #
1.	$\psi(x,t) = \sqrt{V_0+m}\,\mathrm{sn}(x-x_0,m)\,e^{-i\left[\frac{(1+m)(t-t_0)}{2}+\phi_0\right]}$	$V_0+m \geq 0$, x_0, t_0, and ϕ_0 are arbitrary real constants	Solitary wave	(5.113)
2.	$\psi(x,t) = \sqrt{-(V_0+m)}\,\mathrm{cn}(x-x_0,m)\,e^{i\left[(V_0+m-\frac{1}{2})(t-t_0)+\phi_0\right]}$	$V_0+m \leq 0$, x_0, t_0, and ϕ_0 are arbitrary real constants	Solitary wave	(5.114)
3.	$\psi(x,t) = \sqrt{-(1+\frac{V_0}{m})}\,\mathrm{dn}(x-x_0,m)\,e^{i\left[(1+\frac{V_0}{m}-\frac{m}{2})(t-t_0)+\phi_0\right]}$	$1+\frac{V_0}{m} \leq 0$, x_0, t_0, and ϕ_0 are arbitrary real constants	Solitary wave	(5.115)

Equation

$$i\,\psi_t + \tfrac{1}{2}\,\psi_{xx} - |\psi|^2\,\psi + V_0\sin^2(x)\,\psi = 0$$

#	Solution	Conditions	Name	Equation #
1.	$\psi(x,t) = \sqrt{V_0}\,\sin(x-x_0)\,e^{-i\left[\frac{(t-t_0)}{2}+\phi_0\right]}$	$V_0 > 0$, x_0, t_0, and ϕ_0 are arbitrary real constants	–	(5.117)
2.	$\psi(x,t) = \sqrt{-V_0}\,\cos(x-x_0)\,e^{-i\left[(V_0-\frac{1}{2})(t-t_0)+\phi_0\right]}$	$V_0 < 0$, x_0, t_0, and ϕ_0 are arbitrary real constants	–	(5.118)

Equation

$$i\,\psi_t + a_1\,\psi_{xx} + a_2\,|\psi|^2\,\psi + V_0\,\mathrm{sech}^2(V_1\,x)\,\psi = 0$$

(Continued)

#	Solution	Name	Conditions	Equation #
1.	$\psi(x,t) = \sqrt{\dfrac{V_0}{a_2}}\,\text{sech}(V_1 x)\, e^{i[V_0(t-t_0)+\phi_0]}$	Bright soliton	$V_1 = \sqrt{2 V_0},\, a_1 = 1/2,\, V_0 > 0,\, a_2 > 0,\, t_0$ and ϕ_0 are arbitrary real constants	(5.120)
2.	$\psi(x,t) = \sqrt{\dfrac{2V_0}{a_2}}\,\tanh(V_1 x)\, e^{i[2 V_0(t-t_0)+\phi_0]}$	Dark soliton	$V_1 = \sqrt{-V_0},\, a_1 = 1/2,\, V_0 < 0,\, a_2 < 0,\, t_0$ and ϕ_0 are arbitrary real constants	(5.121)
3.	$\psi(x,t) = \sqrt{1 + \text{sech}(x)}\, e^{i[2(V_0-a_1)(t-t_0)+\phi_0]}$	–	$V_0 = (3\,a_1)/4,\, V_1 = 1,\, a_1 = 1/2,\, a_2 = -a_1/2,\, t_0$ and ϕ_0 are arbitrary real constants	(5.122)
4.	$\psi(x,t) = A_0 \sqrt{1 + A_1 \text{sech}(V_1 x) + A_2 \text{sech}^2(V_1 x)}$ $\times e^{i[\frac{1}{2} a_2 V_1^2 A_3^2 (t-t_0)+\phi_0]}$	–	$V_0 = -3\, a_2\, V_1^2\, A_3^2,\, a_1 = -A_3^2\, a_2,\, A_0 = -(V_1\, A_3)/\sqrt{2},\, A_1 = -2\sqrt{2},\, A_2 = 2,$ $V_1, A_3, a_2, t_0,$ and ϕ_0 are arbitrary real constants	(5.123)
5.	$\psi(x,t) = A_0 \sqrt{1 + \tanh(V_1 x)}\, e^{i[A_1(t-t_0)+\phi_0]}$	–	$A_1 = -2 A_0^2\, a_2 + (2 V_0)/3 + 2 V_0,\, a_1 = (2 A_0^2\, a_2)/V_1^2,$ $a_2 = (2 V_0)/(3 A_0^2),\, V_0, V_1, A_0, t_0,$ and ϕ_0 are arbitrary real constants	(5.124)
6.	$\psi(x,t) = A_0 \sqrt{1 + A_1 \tanh(V_1 x) + A_2 \tanh^2(V_1 x)}$ $\times e^{i[4 A_0^2\, a_2 (t-t_0)+\phi_0]}$	–	$V_0 = 3 A_0^2\, a_2,\, a_1 = (A_0^2\, a_2)/V_1^2,\, A_1 = 2,\, A_2 = 1,\, V_1, A_0, a_2, t_0,$ and ϕ_0 are arbitrary real constants	(5.125)

References

[1] Al Khawaja U 2010 A comparative analysis of Painlevé, Lax pair, and similarity transformation methods in obtaining the integrability conditions of nonlinear Schrödinger equations *J. Math. Phys.* **51** 053506

[2] Al Khawaja U and Bahlouli H 2019 Integrability conditions and solitonic solutions of the nonlinear Schrödinger equation with generalized dual-power nonlinearities, PT-symmetric potentials, and space- and time-dependent coefficients *Commun. Nonlinear Sci. Numer. Simul.* **69** 248–60

[3] Bronski J C, Carr L D, Deconinck B and Kutz J N 2001 Bose-Einstein condensates in standing waves: the cubic nonlinear Schrödinger equation with a periodic potential *Phys. Rev. Lett.* **86** 1402–5

[4] Al Sakkaf L, Uthayakumar T and Al Khawaja U 2022 Quantum reflection of dark solitons scattered by reflectionless potential barrier and position-dependent dispersion *Phys. Rev.* **E105** 064207

IOP Publishing

Handbook of Exact Solutions to the Nonlinear Schrödinger
Equations (Second Edition)

Usama Al Khawaja and Laila Al Sakkaf

Chapter 6

Nonlinear Schrödinger Equation in $(N + 1)$-Dimensions

A Glance at Chapter 6

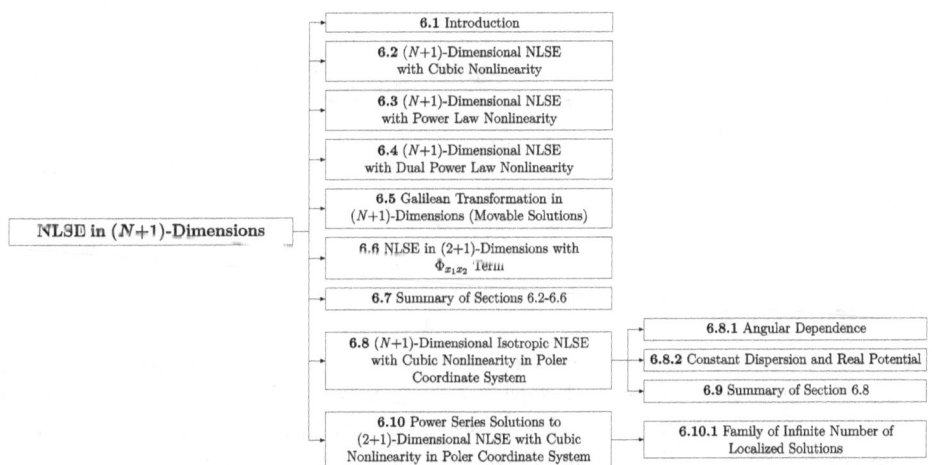

A Statistical View of Chapter 6

	Equation	Solutions
1	$i\,\Phi_t + \sum_{k=1}^{N} \alpha_k\, \Phi_{x_k x_k} + a_2\, \lvert \Phi \rvert^2\, \Phi = 0$	14
2	$i\,\Phi_t + \sum_{k=1}^{N} \alpha_k\, \Phi_{x_k x_k} + a_2\, \lvert \Phi \rvert^n\, \Phi = 0$	4
3	$i\,\Phi_t + \sum_{k=1}^{N} \alpha_k\, \Phi_{x_k x_k} + a_2\, \lvert \Phi \rvert^n\, \Phi + a_3\, \lvert \Phi \rvert^m\, \Phi = 0$	8
4	$i\,\Phi_t + b_1\, \Phi_{x_1 x_1} + b_2\, \Phi_{x_2 x_2} + b_3\, \Phi_{x_1 x_2} + b_4\, \lvert \Phi \rvert^2\, \Phi = 0$	2
5	$i\,\Phi_t + b_1(r,\,t)\,(\Phi_{rr} + \frac{N-1}{r}\,\Phi_r) + b_2(r,\,t)\,\lvert \Phi \rvert^2\,\Phi + [b_{3r}(r,\,t) + i\,b_{3i}(r,\,t)]\Phi = 0$	0
6	$i\,\Phi_t + a_1\,(\Phi_{rr} + \frac{N-1}{r}\,\Phi_r + \frac{1}{r^2}\,\Phi_{\theta\theta}) + a_2\, r^{N-1}\,\lvert \Phi \rvert^2\,\Phi = 0,\ \text{for } N=1:\ \Phi = \Phi(r)$	1
7	$i\,\Phi_t + a_1\,(\Phi_{rr} + \frac{1}{r}\,\Phi_r - \frac{1}{4r^2}\,\Phi) + a_2\, r\,\lvert \Phi \rvert^2\,\Phi = 0$	1
8	$i\,\Phi_t + a_1\,(\Phi_{rr} + \frac{2}{r}\,\Phi_r) + a_2\, r^2\,\lvert \Phi \rvert^2\,\Phi = 0$	1
9	$i\,\Phi_t + b_{10}\,\Phi_{rr} + \frac{b_{10}\,(N-1)}{r}\,\Phi_r + \frac{a_2\,\sqrt{g_1'(t)}}{c_0^2\, r^{1-N}}\,\lvert \Phi \rvert^2\,\Phi$ $+ \left\{ \frac{b_{10}\,(N-1)\,(N-3)}{4\,r^2} + \frac{g_2'^2(t)}{4\,a_1\,g_1'(t)} + g_4'(t) + \frac{[g_2'(t)\,g_1''(t) - g_1'(t)\,g_2''(t)]\,r}{2\,\sqrt{a_1\,b_{10}}\;g_1'^{3/2}(t)} + \frac{[3\,g_1''^2(t) - 2\,g_1'(t)\,g_1'''(t)]\,r^2}{16\,b_{10}\,g_1'^2(t)} \right\} \Phi = 0$	2
10	$-\lambda\, Z(r) + a_1\,[Z''(r) + \frac{1}{r}\,Z'(r) - \frac{\alpha^2}{r^2}\,Z(r)] + a_2\, Z^3(r) = 0$	1
Total 10		34

6.1 Introduction

The NLSE in most physical systems is essentially three- or two-dimensional. This is the case, for instance, for optical pulses in nonlinear media and for Bose–Einstein condensates in parabolic magnetic traps or optical lattices. Reductions to one-dimensional NLSE are common due to specific boundary conditions that, essentially, 'freeze' the degrees of freedom in two spatial dimensions such as optical fibers where the radial confinement is provided by the geometry of the fiber and the only degree of freedom for pulse propagation is along the fiber axis, as shown in Figure 2.2 of Chapter 2. A similar situation is for Bose–Einstein condensates confined by a tight magnetic trap in two spatial directions and much weaker confinement in the third, which results in an elongated cigar-shaped condensate, as shown in Figure 2.1 of Chapter 2. In these cases, the system is quasi one-dimensional such that it can be described accurately by the one-dimensional NLSE.

Localized excitations of the higher-dimensional NLSE are generally unstable [1]. This can be seen via the following energy scaling argument. Consider the NLSE

$$i\frac{\partial}{\partial t}\psi(\mathbf{r}, t) + a_1 \nabla^2 \psi(\mathbf{r}, t) + a_2 |\psi(\mathbf{r}, t)|\psi(\mathbf{r}, t) = 0. \tag{6.1}$$

The corresponding energy functional is

$$E = \int \left[a_1 |\nabla \psi(\mathbf{r}, t)|^2 - \frac{1}{2} a_2 |\psi(\mathbf{r}, t)|^4 \right] d\mathbf{r}. \tag{6.2}$$

The solution, $\psi(\mathbf{r}, t)$, is assumed for simplicity to be a spherically-symmetric localized excitation of size R, and it is normalized to a constant N,

$$\int_0^\infty |\psi(\mathbf{r}, t)|^2 r^{D-1} dr \sim N, \tag{6.3}$$

with $D = 2$ or 3 for two- or three-dimensions, respectively. This leads to

$$\psi(\mathbf{r}, t) \sim \frac{1}{R^{D/2}}. \tag{6.4}$$

The energy functional scales then as

$$E \sim a_1 R^{-2} - \frac{a_2}{2} R^{-D}. \tag{6.5}$$

Existence of stable localized excitations with finite size occurs for a minimum of E with respect to R. For such localized excitations to exist, the constants a_1 and a_2 should satisfy $a_1 a_2 > 0$. This condition is obtained from exact solutions with vanishing background, such as the bright soliton. For instance, $a_1 = \hbar^2/(2m) > 0$ and $a_2 = 4\pi a_s \hbar/m < 0$ for the case of Gross–Pitaevskii equation with a_s the magnitude of the s-wave scattering length. Here $\hbar = h/(2\pi)$, where h is Plank's constant. For the one-dimensional case, $D = 1$, it is always possible to stabilize the excitation since the energy scales as $E \sim a_1 R^{-2} - \frac{a_2}{2} R^{-1}$. For the two-dimensional case, $E \sim (a_1 - a_2/2) R^{-2}$, it is not possible to stabilize the excitation, except for the very specific case when the two terms of energy cancel each other, namely $a_1 = a_2/2$. This results in the so-called Townes soliton which is generally unstable, but was experimentally-realized for the multicomponent systems [2]. In three dimensions, $D = 3$, the energy, $E \sim a_1 R^{-2} - \frac{a_2}{2} R^{-3}$, does not have a local minimum for $a_1 > 0$ and $a_2 > 0$.

The above-described energetic instability was established for the fundamental two- and three-dimensional excitations with cubic nonlinearity and with no external potentials. In other situations, higher-dimensional excitations may be stabilized. For instance, the NLSE with dual nonlinearity supports stable localized excitations in higher dimensions. Confining potentials, such as the parabolic potential of magnetic trap confining Bose–Einstein condensates, stabilize two- and three-dimensional localized excitations [3]. Another approach to stabilize two- and three-dimensional excitations is by using an oscillating nonlinearity strength [4].

Higher dimensionality has a more profound effect by supporting excitations that would not exist in one-dimensional systems, such as vortex solitons and topological excitations [5–9]. However, many of these excitations cannot be obtained in analytical form since two- and three-dimensional NLSEs are in general not integrable, as is the case for the one-dimensional NLSE. Some higher-dimensionality

NLSEs can be reduced through functional transformations into the fundamental one-dimensional NLSE, which of course implies their integrability. As we stated in the preface of this book, we do not include solutions which are obtained only in numerical form, although, for the present case, there are many such numerical solutions with important and interesting features.

6.2 (N + 1)-Dimensional NLSE with Cubic Nonlinearity

If $\psi(x, t; a_1, a_2)$ is a solution to the fundamental NLSE (2.160),

$$i\,\psi_t + a_1\,\psi_{xx} + a_2\,|\psi|^2\,\psi = 0,$$

then

$$\Phi(x_1, x_2, \ldots, x_N, t; \alpha_1, \alpha_2, \ldots, \alpha_N)$$
$$= \psi\Big[c_1\,(x_1 - x_{01}) + c_2\,(x_2 - x_{02}) + \cdots + c_N\,(x_N - x_{0N}),\ t;\ c_1^2\,\alpha_1 + c_1^2\,\alpha_2 + \cdots + c_N^2\,\alpha_N,\ a_2\Big] \quad (6.6)$$

is a solution to

$$i\,\Phi_t + \sum_{k=1}^{N}\alpha_k\,\Phi_{x_k x_k} + a_2\,|\Phi|^2\,\Phi = 0, \qquad (6.7)$$

with the replacements $x \to \sum_{k=1}^{N} c_k\,(x_k - x_{0k})$, $t \to t$, $a_1 \to \sum_{k=1}^{N} c_k^2\,\alpha_k$, $a_2 \to a_2$, where α_k, a_2, c_k, and x_{0k} are arbitrary real constants.

***Example* 1:** *2D bright soliton*
(Figure 6.1)

Given $\psi(x, t) = A_0 \sqrt{\dfrac{2\,a_1}{a_2}}\ \text{sech}[A_0\,(x - x_0)]\ e^{i\,[a_1\,A_0^2\,(t - t_0) + \phi_0]}$ is a solution to (2.160), then

$$\Phi(x_1, x_2, t) = A_0 \sqrt{\dfrac{2\,(c_1^2\,\alpha_1 + c_2^2\,\alpha_2)}{a_2}}\ \text{sech}\{A_0\,[c_1\,(x_1 - x_{01}) + c_2\,(x_2 - x_{02})]\}$$
$$\times e^{i\left[A_0^2\,\left(c_1^2\,\alpha_1 + c_2^2\,\alpha_2\right)\,(t - t_0) + \phi_0\right]} \qquad (6.8)$$

is a solution to

$$i\,\Phi_t + \alpha_1\,\Phi_{x_1 x_1} + \alpha_2\,\Phi_{x_2 x_2} + a_2\,|\Phi|^2\,\Phi = 0, \qquad (6.9)$$

where $a_2\,(c_1^2\,\alpha_1 + c_2^2\,\alpha_2) > 0$, A_0, t_0, and ϕ_0 are arbitrary real constants.

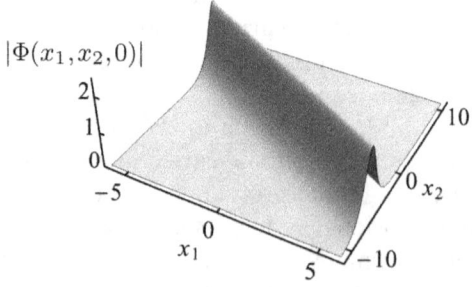

Figure 6.1. 2D bright soliton (6.8) at $t = 0$, with $\alpha_1 = \alpha_2 = A_0 = c_1 = c_2 = a_2 = 1$ and $x_{01} = x_{02} = t_0 = \phi_0 = 0$.

Example 2: *3D bright soliton*

Given $\psi(x, t) = A_0 \sqrt{\frac{2 a_1}{a_2}} \, \text{sech}[A_0 \, (x - x_0)] \, e^{i \, [a_1 \, A_0^2 \, (t-t_0)+\phi_0]}$ is a solution to (2.160), then

$$
\Phi(x_1, x_2, x_3, t) = A_0 \sqrt{\frac{2 \, (c_1^2 \, \alpha_1 + c_2^2 \, \alpha_2 + c_3^2 \, \alpha_3)}{a_2}} \, \text{sech}\{A_0 \, [c_1 \, (x_1 - x_{01}) + c_2 \, (x_2 - x_{02})
$$
$$
+ \, c_3 \, (x_3 - x_{03})]\} \, e^{i \left[A_0^2 \left(c_1^2 \, \alpha_1 + c_2^2 \, \alpha_2 + c_3^2 \, \alpha_3 \right) (t-t_0)+\phi_0 \right]} \tag{6.10}
$$

is a solution to

$$
i \, \Phi_t + \alpha_1 \, \Phi_{x_1 x_1} + \alpha_2 \, \Phi_{x_2 x_2} + \alpha_3 \, \Phi_{x_3 x_3} + a_2 \, |\Phi|^2 \, \Phi = 0, \tag{6.11}
$$

where $a_2 \, (c_1^2 \, \alpha_1 + c_2^2 \, \alpha_2 + c_3^2 \, \alpha_3) > 0$, A_0, t_0, and ϕ_0 are arbitrary real constants.

Example 3: *2D dark soliton*
(Figure 6.2)

Given $\psi(x, t) = A_0 \sqrt{\frac{-2 \, a_1}{a_2}} \, \tanh[A_0 \, (x - x_0)] \, e^{-i \, [2 \, a_1 \, A_0^2 \, (t-t_0)+\phi_0]}$ is a solution to (2.160), then

$$
\Phi(x_1, x_2, t) = A_0 \sqrt{\frac{-2 \, (c_1^2 \, \alpha_1 + c_2^2 \, \alpha_2)}{a_2}} \, \tanh\{A_0 \, [c_1 \, (x_1 - x_{01}) + c_2 \, (x_2 - x_{02})]\}
$$
$$
\times \, e^{-i \left[2 \, A_0^2 \left(c_1^2 \, \alpha_1 + c_2^2 \, \alpha_2 \right) (t-t_0)+\phi_0 \right]} \tag{6.12}
$$

is a solution to (6.9), where $a_2 \, (c_1^2 \, \alpha_1 + c_2^2 \, \alpha_2) < 0$, A_0, t_0, and ϕ_0 are arbitrary real constants.

Example 4: *3D dark soliton*

Given $\psi(x, t) = A_0 \sqrt{\frac{-2 \, a_1}{a_2}} \, \tanh[A_0 \, (x - x_0)] \, e^{-i \, [2 \, a_1 \, A_0^2 \, (t-t_0)+\phi_0]}$ is a solution to (2.160), then

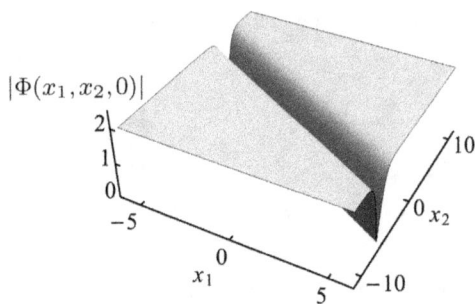

Figure 6.2. 2D dark soliton (6.12) at $t = 0$, with $\alpha_1 = \alpha_2 = A_0 = c_1 = c_2 = 1$, $a_2 = -1$ and $x_{01} = x_{02} = t_0 = \phi_0 = 0$.

$$\Phi(x_1, x_2, x_3, t) = A_0 \sqrt{\frac{-2 (c_1^2 \alpha_1 + c_2^2 \alpha_2 + c_3^2 \alpha_3)}{a_2}} \tanh\{A_0 [c_1 (x_1 - x_{01}) + c_2 (x_2 - x_{02})$$

$$+ c_3 (x_3 - x_{03})]\} e^{-i \left[2 A_0^2 \left(c_1^2 \alpha_1 + c_2^2 \alpha_2 + c_3^2 \alpha_3\right)(t - t_0) + \phi_0\right]} \tag{6.13}$$

is a solution to (6.11), where $a_2 (c_1^2 \alpha_1 + c_2^2 \alpha_2 + c_3^2 \alpha_3) < 0$, A_0, t_0, and ϕ_0 are arbitrary real constants.

***Example* 5: Periodicity in *t* and Localization in x_1 and x_2 2D Kuznetsov–Ma breather (Figure 6.3)**

Given $\psi(x, t) = \dfrac{1}{\sqrt{a_2}} \left\{ \dfrac{-p^2 \cos[\omega (t - t_0)] - 2 i p \nu \sin[\omega (t - t_0)]}{2 \cos[\omega (t - t_0)] - 2 \nu \cosh[\frac{p}{\sqrt{2 a_1}} (x - x_0)]} - 1 \right\} e^{i (t - t_0 + \phi_0)}$ is a solution to (2.160), then

$$\Phi(x_1, x_2, t) = \frac{-1}{\sqrt{a_2}} \left\{ \frac{p^2 \cos[\omega (t - t_0)] + 2 i p \nu \sin[\omega (t - t_0)]}{2 \cos[\omega (t - t_0)] - q(x_1, x_2)} + 1 \right\} e^{i (t - t_0 + \phi_0)} \tag{6.14}$$

is a solution to (6.9), where

$q(x_1, x_2) = 2 \nu \cosh\{\dfrac{p}{\sqrt{2 (c_1^2 \alpha_1 + c_2^2 \alpha_2)}} [c_1 (x_1 - x_{01}) + c_2 (x_2 - x_{02})]\}$, $p = 2 \sqrt{\nu^2 - 1}$,

$\omega = p \nu$, $\nu > 1$, $a_2 > 0$, $(c_1^2 \alpha_1 + c_2^2 \alpha_2) > 0$, t_0 and ϕ_0 are arbitrary real constants.

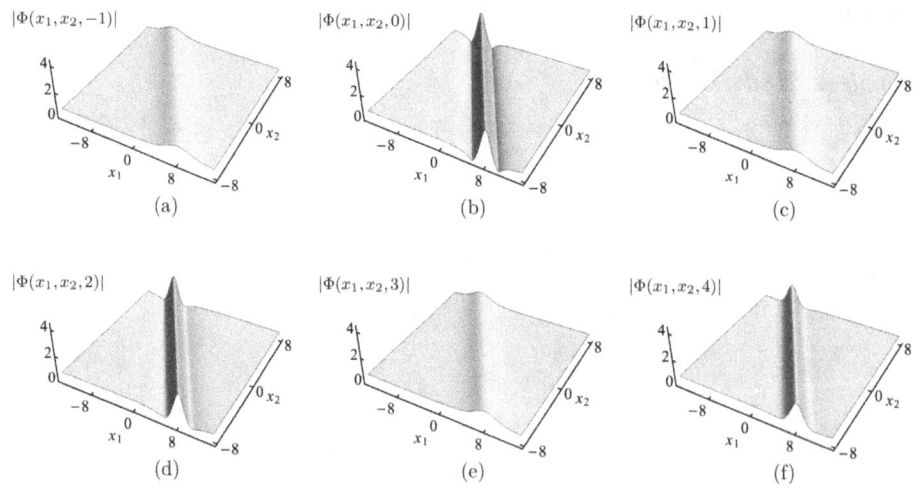

Figure 6.3. 2D Kuznetsov–Ma breather (6.14). (a) at $t = -1$, (b) at $t = 0$, (c) at $t = 1$, (d) at $t = 2$, (e) at $t = 3$, and (f) at $t = 4$. Used parameters: $\alpha_1 = \alpha_2 = c_1 = c_2 = 3$, $a_2 = 2$, $x_{01} = x_{02} = t_0 = \phi_0 = 0$, and $\nu = 1.5$. Animation available online at http://doi.org/10.1088/978-0-7503-5954-2.

Example 6: Periodicity in t and Localization in x_1, x_2, and x_3 *3D Kuznetsov–Ma breather*

Given $\psi(x, t) = \dfrac{1}{\sqrt{a_2}} \left\{ \dfrac{-p^2 \cos[\omega\,(t - t_0)] - 2\,i\,p\,\nu\,\sin[\omega\,(t - t_0)]}{2\cos[\omega\,(t - t_0)] - 2\,\nu\,\cosh[-\frac{p}{\sqrt{2}\,a_1}\,(x - x_0)]} - 1 \right\} e^{i\,(t - t_0 + \phi_0)}$ is a solution to (2.160), then

$$\Phi(x_1, x_2, x_3, t) = \dfrac{-1}{\sqrt{a_2}} \left\{ \dfrac{p^2 \cos[\omega\,(t - t_0)] + 2\,i\,p\,\nu\,\sin[\omega\,(t - t_0)]}{2\cos[\omega\,(t - t_0)] - q(x_1, x_2, x_3)} + 1 \right\} e^{i\,(t - t_0 + \phi_0)} \quad (6.15)$$

is a solution to (6.11), where

$$q(x_1, x_2, x_3) = 2\,\nu\,\cosh\left\{ \frac{p}{\sqrt{2\,(c_1^2\,a_1 + c_2^2\,a_2 + c_3^2\,a_3)}} \,[c_1\,(x_1 - x_{01}) + c_2\,(x_2 - x_{02}) + c_3\,(x_3 - x_{03})] \right\},$$

$p = 2\sqrt{\nu^2 - 1}$, $\omega = p\,\nu$, $\nu > 1$, $a_2 > 0$, $(c_1^2\,a_1 + c_2^2\,a_2 + c_3^2\,a_3) > 0$, t_0 and ϕ_0 are arbitrary real constants.

Example 7: Periodicity in x_1 and x_2 and Localization in t *2D Akhmediev breather* (Figure 6.4)

Given $\psi(x, t) = \dfrac{1}{\sqrt{a_2}} \left\{ \dfrac{\kappa^2 \cosh[\delta\,(t - t_0)] + 2\,i\,\kappa\,\nu\,\sinh[\delta\,(t - t_0)]}{2\cosh[\delta\,(t - t_0)] - 2\,\nu\,\cos[-\frac{\kappa}{\sqrt{2}\,a_1}\,(x - x_0)]} - 1 \right\} e^{i\,(t - t_0 + \phi_0)}$ is a solution to (2.160), then

$$\Phi(x_1, x_2, t) = \dfrac{1}{\sqrt{a_2}} \left\{ \dfrac{\kappa^2 \cosh[\delta\,(t - t_0)] + 2\,i\,\kappa\,\nu\,\sinh[\delta\,(t - t_0)]}{2\cosh[\delta\,(t - t_0)] - q(x_1, x_2)} - 1 \right\} e^{i\,(t - t_0 + \phi_0)} \quad (6.16)$$

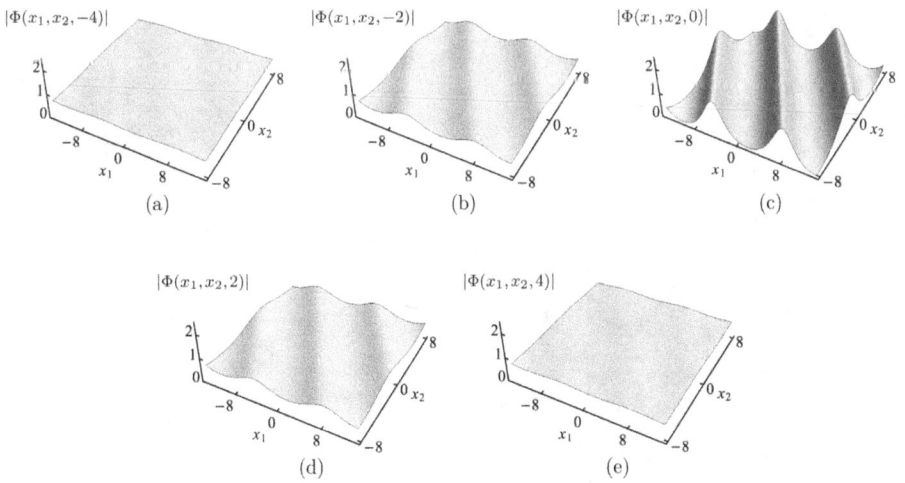

$|\Phi(x_1, x_2, -4)|$ $|\Phi(x_1, x_2, -2)|$ $|\Phi(x_1, x_2, 0)|$

(a)   (b)   (c)

$|\Phi(x_1, x_2, 2)|$ $|\Phi(x_1, x_2, 4)|$

(d)   (e)

Figure 6.4. 2D Akhmediev breather (6.16). (a) at $t = -4$, (b) at $t = -2$, (c) at $t = 0$, (d) at $t = 2$, and (e) at $t = 4$. Used parameters: $\alpha_1 = \alpha_2 = c_1 = c_2 = 3$, $a_2 = 2$, $x_{01} = x_{02} = t_0 = \phi_0 = 0$, and $\nu = 0.5$. Animation available online at http://doi.org/10.1088/978-0-7503-5954-2.

is a solution to (6.9), where

$$q(x_1, x_2) = 2\,\nu\,\cos\left\{\frac{\kappa}{\sqrt{2\,(c_1^2\,\alpha_1 + c_2^2\,\alpha_2)}}\,[c_1\,(x_1 - x_{01}) + c_2\,(x_2 - x_{02})]\right\},\ \kappa = 2\,\sqrt{1 - \nu^2},$$

$\delta = \kappa\,\nu,\ \nu < 1,\ a_2 > 0\ (c_1^2\,\alpha_1 + c_2^2\,\alpha_2) > 0,\ t_0$ and ϕ_0 are arbitrary real constants.

***Example* 8: Periodicity in x_1, x_2, and x_3 and Localization in t** *3D Akhmediev breather*

Given $\psi(x, t) = \dfrac{1}{\sqrt{a_2}}\left\{\dfrac{\kappa^2\cosh[\delta\,(t - t_0)] + 2\,i\,\kappa\,\nu\,\sinh[\delta\,(t - t_0)]}{2\cosh[\delta\,(t - t_0)] - 2\,\nu\,\cos[\frac{\kappa}{\sqrt{2}\,a_1}(x - x_0)]} - 1\right\}e^{i\,(t - t_0 + \phi_0)}$ is a solution to (2.160), then

$$\Phi(x_1, x_2, x_3, t) = \frac{1}{\sqrt{a_2}}\left\{\frac{\kappa^2\cosh[\delta\,(t - t_0)] + 2\,i\,\kappa\,\nu\,\sinh[\delta\,(t - t_0)]}{2\cosh[\delta\,(t - t_0)] - q(x_1, x_2, x_3)} - 1\right\}e^{i\,(t - t_0 + \phi_0)} \quad (6.17)$$

is a solution to (6.11), where
$q(x_1, x_2, x_3) =$

$$2\,\nu\,\cos\left\{\frac{\kappa}{\sqrt{2\,(c_1^2\,\alpha_1 + c_2^2\,\alpha_2 + c_3^2\,\alpha_3)}}\,[c_1\,(x_1 - x_{01}) + c_2\,(x_2 - x_{02}) + c_3\,(x_3 - x_{03})]\right\},$$

$\kappa = 2\,\sqrt{1 - \nu^2}$, $\delta = \kappa\,\nu,\ \nu < 1,\ a_2 > 0,\ (c_1^2\,\alpha_1 + c_2^2\,\alpha_2 + c_3^2\,\alpha_3) > 0,\ t_0$ and ϕ_0 are arbitrary real constants.

***Example* 9: Localization in t, x_1, and x_2** *2D Peregrine soliton*
(Figure 6.5)

Given $\psi(x, t) = \dfrac{1}{\sqrt{a_2}}\left[\dfrac{4 + i\,8\,(t - t_0)}{1 + 4\,(t - t_0)^2 + \frac{2}{a_1}(x - x_0)^2} - 1\right]e^{i\,[t - t_0 + \phi_0]}$ is a solution to (2.160), then

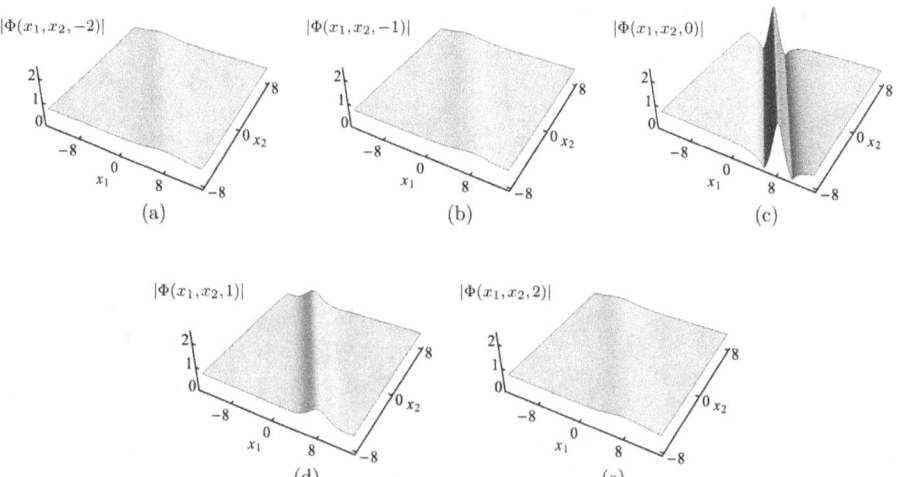

Figure 6.5. 2D Peregrine soliton (6.18). (a) at $t = -2$, (b) at $t = -1$, (c) at $t = 0$, (d) at $t = 1$, and (e) at $t = 2$. Used parameters: $\alpha_1 = \alpha_2 = c_1 = c_2 = a_2 = 2$, and $x_{01} = x_{02} = t_0 = \phi_0 = 0$. Animation available online at http://doi.org/10.1088/978-0-7503-5954-2.

$$\Phi(x_1, x_2, t) = \frac{1}{\sqrt{a_2}} \left\{ \frac{4 + i\,8\,(t - t_0)}{1 + 4\,(t - t_0)^2 + \dfrac{2}{c_1^2\,\alpha_1 + c_2^2\,\alpha_2}\,[c_1\,(x_1 - x_{01}) + c_2\,(x_2 - x_{02})]^2} - 1 \right\} \tag{6.18}$$
$$\times\, e^{i\,(t - t_0 + \phi_0)},$$

is a solution to (6.9), where $a_2 > 0$, t_0 and ϕ_0 are arbitrary real constants.

Example* 10: Localization in t, x_1, x_2, and x_3 $3D$ *Peregrine soliton

Given $\psi(x, t) = \dfrac{1}{\sqrt{a_2}} \left[\dfrac{4 + i\,8\,(t - t_0)}{1 + 4\,(t - t_0)^2 + \frac{2}{a_1}\,(x - x_0)^2} - 1 \right] e^{i\,(t - t_0 + \phi_0)}$ is a solution to (2.160),

then

$$\Phi(x_1, x_2, x_3, t) = \frac{1}{\sqrt{a_2}} \left[\frac{4 + i\,8\,(t - t_0)}{1 + 4\,(t - t_0)^2 + q(x_1, x_2, x_3)} - 1 \right] e^{i\,(t - t_0 + \phi_0)} \tag{6.19}$$

is a solution to (6.11), where
$$q(x_1, x_2, x_3) = \frac{2}{c_1^2\,\alpha_1 + c_2^2\,\alpha_2 + c_3^2\,\alpha_3} \quad [c_1\,(x_1 - x_{01}) + c_2\,(x_2 - x_{02}) + c_3\,(x_3 - x_{03})]^2,$$
$a_2 > 0$, t_0 and ϕ_0 are arbitrary real constants.

6.3 $(N + 1)$-Dimensional NLSE with Power Law Nonlinearity

If $\psi(x, t; a_1, a_2)$ is a solution to the NLSE with power law nonlinearity (3.9), $i\,\psi_t + a_1\,\psi_{xx} + a_2\,|\psi|^n\,\psi = 0$, then (6.6) is a solution to

$$i\,\Phi_t + \sum_{k=1}^{N} \alpha_k\,\Phi_{x_k x_k} + a_2\,|\Phi|^n\,\Phi = 0, \tag{6.20}$$

with the replacements $x \to \sum_{k=1}^{N} c_k\,(x_k - x_{0k})$, $t \to t$, $a_1 \to \sum_{k=1}^{N} c_k^2\,\alpha_k$, where α_k, a_2, and n are real constants.

***Example* 1:** *2D bright soliton*

Given $\psi(x, t) = \left\{ \dfrac{2\,A_0^2\,a_1\,(n + 2)}{a_2\,n^2}\,\text{sech}^2\,[A_0\,(x - x_0)] \right\}^{\frac{1}{n}} e^{i\,[\frac{4\,a_1\,A_0^2}{n^2}\,(t - t_0) + \phi_0]}$ is a solution to
(3.9), then

$$\Phi(x_1, x_2, t) = \left(\frac{2\,A_0^2\,(n + 2)\,(c_1^2\,\alpha_1 + c_2^2\,\alpha_2)}{a_2\,n^2}\,\text{sech}^2\{A_0\,[\,c_1\,(x_1 - x_{01}) + c_2\,(x_2 - x_{02})]\} \right)^{\frac{1}{n}}$$
$$\times\, e^{i\left[\frac{4\,A_0^2\,\left(c_1^2\,\alpha_1 + c_2^2\,\alpha_2\right)}{n^2}\,(t - t_0) + \phi_0 \right]} \tag{6.21}$$

is a solution to

$$i\,\Phi_t + \alpha_1\,\Phi_{x_1 x_1} + \alpha_2\,\Phi_{x_2 x_2} + a_2\,|\Phi|^n\,\Phi = 0, \tag{6.22}$$

where $a_2\,(n + 2)\,(c_1^2\,\alpha_1 + c_2^2\,\alpha_2) > 0$, A_0, t_0, and ϕ_0 are arbitrary real constants.

***Example* 2**: *3D bright soliton*

Given $\psi(x, t) = \left\{ \dfrac{2 A_0^2 a_1 (n+2)}{a_2 n^2} \operatorname{sech}^2[A_0 (x - x_0)] \right\}^{\frac{1}{n}} e^{i \left[\frac{4 a_1 A_0^2}{n^2} (t-t_0)+\phi_0 \right]}$ is a solution to (3.9), then

$$
\begin{aligned}
\Phi(x_1, x_2, x_3, t) = &\left(\frac{2 A_0^2 (n+2) (c_1^2 \alpha_1 + c_2^2 \alpha_2 + c_3^2 \alpha_3)}{a_2 n^2} \operatorname{sech}^2 \{ A_0 [c_1 (x_1 - x_{01}) + c_2 (x_2 - x_{02}) \right. \\
&\left. + c_3 (x_3 - x_{03})] \} \right)^{\frac{1}{n}} e^{i \left[\frac{4 A_0^2 \left(c_1^2 \alpha_1 + c_2^2 \alpha_2 + c_3^2 \alpha_3 \right)}{n^2} (t-t_0)+\phi_0 \right]}
\end{aligned}
\tag{6.23}
$$

is a solution to

$$
i \, \Phi_t + \alpha_1 \, \Phi_{x_1 x_1} + \alpha_2 \, \Phi_{x_2 x_2} + \alpha_3 \, \Phi_{x_3 x_3} + a_2 \, |\Phi|^n \, \Phi = 0,
\tag{6.24}
$$

where $a_2 (n + 2) (c_1^2 \alpha_1 + c_2^2 \alpha_2 + c_3^2 \alpha_3) > 0$, A_0, t_0, and ϕ_0 are arbitrary real constants.

6.4 (N + 1)-Dimensional NLSE with Dual Power Law Nonlinearity

If $\psi(x, t; a_1, a_2, a_3)$ is a solution to the NLSE with dual power law nonlinearity (3.27), $i \, \psi_t + a_1 \, \psi_{xx} + a_2 \, |\psi|^n \, \psi + a_3 \, |\psi|^m \, \psi = 0$, then (6.6) is a solution to

$$
i \, \Phi_t + \sum_{k=1}^{N} \alpha_k \, \Phi_{x_k x_k} + a_2 \, |\Phi|^n \, \Phi + a_3 \, |\Phi|^m \, \Phi = 0,
\tag{6.25}
$$

with the replacements $x \to \sum_{k=1}^{N} c_k (x_k - x_{0k})$, $t \to t$, $a_1 \to \sum_{k=1}^{N} c_k^2 \alpha_k$, $a_2 \to a_2$, $a_3 \to a_3$, where α_k, a_2, a_3, n, and m are real constants.

***Example* 1**: *2D flat-top soliton*
(Figure 6.6)

Given $\psi(x, t) = \left\{ \dfrac{A_0 (n+2)}{a_2 + a_2 \sin(\theta) \cosh[n \sqrt{\frac{A_0}{a_1}} (x - x_0)]} \right\}^{\frac{1}{n}} e^{i \, [A_0 (t-t_0)+\phi_0]}$, is a solution to (3.27),

then

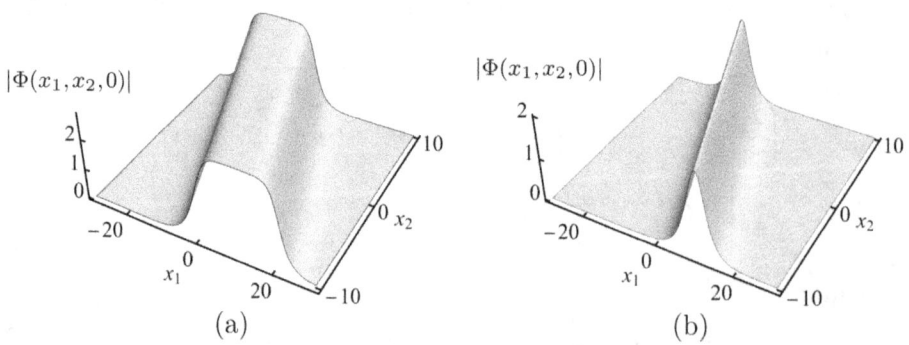

Figure 6.6. Plot of solution (6.26) at $t = 0$. (a) 2D flat-top soliton with $\theta = \pi/9998$, (b) 2D bright soliton with $\theta = \pi/2$. Values of the other parameters are: $A_0 = 2$, $n = 2$, $\alpha_1 = \alpha_2 = 5$, $c_1 = c_2 = a_2 = 1$ and $x_{01} = x_{02} = t_0 = \phi_0 = 0$.

$$\Phi(x_1, x_2, t) = \left(\frac{A_0 (n + 2)}{a_2 + a_2 \sin(\theta) \cosh\left\{ n \sqrt{\dfrac{A_0}{c_1^2 \alpha_1 + c_2^2 \alpha_2}} [c_1 (x_1 - x_{01}) + c_2 (x_2 - x_{02})] \right\}} \right)^{\frac{1}{n}} \tag{6.26}$$

$$\times \, e^{i \, [A_0 \, (t - t_0) + \phi_0]},$$

is a solution to

$$i \, \Phi_t + \alpha_1 \, \Phi_{x_1 x_1} + \alpha_2 \, \Phi_{x_2 x_2} + a_2 \, |\Phi|^n \, \Phi + a_3 \, |\Phi|^m \, \Phi = 0, \tag{6.27}$$

where $a_3 = \gamma \, a_{03}$, $\gamma = -\cos^2(\theta)$, $0 < \theta < \pi/2$, $a_{03} = a_2^2 (n + 1)/[A_0 (n + 2)^2]$, $A_0 (c_1^2 \alpha_1 + c_2^2 \alpha_2) > 0$, $A_0 (n + 1) > 0$, $m = 2n$, x_{01}, x_{02}, t_0, A_0, and ϕ_0 are arbitrary real constants.

Example 2: *3D flat-top soliton*

Given $\psi(x, t) = \left\{ \dfrac{A_0 (n + 2)}{a_2 + a_2 \sin(\theta) \cosh[n \sqrt{\frac{A_0}{a_1}} (x - x_0)]} \right\}^{\frac{1}{n}} e^{i \, [A_0 \, (t - t_0) + \phi_0]}$ is a solution to (3.27),

then

$$\Phi(x_1, x_2, x_3, t) = \left\{ \frac{A_0 (n + 2)}{a_2 + a_2 \sin(\theta) \cosh[q(x_1, x_2, x_3)]} \right\}^{\frac{1}{n}} e^{i \, [A_0 \, (t - t_0) + \phi_0]}, \tag{6.28}$$

is a solution to

$$i \, \Phi_t + \alpha_1 \, \Phi_{x_1 x_1} + \alpha_2 \, \Phi_{x_2 x_2} + \alpha_3 \, \Phi_{x_3 x_3} + a_2 \, |\Phi|^n \, \Phi + a_3 \, |\Phi|^m \, \Phi = 0, \tag{6.29}$$

where

$$q(x_1, x_2, x_3) = n \sqrt{\frac{A_0}{c_1^2 \alpha_1 + c_2^2 \alpha_2 + c_3^2 \alpha_3}} [c_1 (x_1 - x_{01}) + c_2 (x_2 - x_{02}) + c_3 (x_3 - x_{03})],$$

$a_3 = \gamma \, a_{03}$, $\gamma = -\cos^2(\theta)$, $0 < \theta < \pi/2$, $u_{03} = a_2^2 (n + 1)/[A_0 (n + 2)^2]$, $A_0 (c_1^2 \alpha_1 + c_2^2 \alpha_2 + c_3^2 \alpha_3) > 0$, $A_0 (n + 1) > 0$, $m = 2n$, x_{01}, x_{02}, x_{03}, t_0, A_0, and ϕ_0 are arbitrary real constants.

Example 3: *2D dark soliton*
(Figure 6.7)

Given $\psi(x, t) = \left(\dfrac{2 a_1 A_0^2 (n + 2)}{a_2 \, n^2} \{1 - \tanh[A_0 (x - x_0)]\} \right)^{\frac{1}{n}} e^{i [\frac{4 a_1 A_0^2}{n^2} (t - t_0) + \phi_0]}$ is a solu-

tion to (3.27), then

$$\Phi(x_1, x_2, t) = \left[\frac{2 A_0^2 (n + 2) (c_1^2 \alpha_1 + c_2^2 \alpha_2)}{a_2 \, n^2} (1 - \tanh\{A_0 [c_1 (x_1 - x_{01}) + c_2 (x_2 - x_{02})]\}) \right]^{\frac{1}{n}} \tag{6.30}$$

$$\times \, e^{i \left[\frac{4 A_0^2 (c_1^2 \alpha_1 + c_2^2 \alpha_2)}{n^2} (t - t_0) + \phi_0 \right]}$$

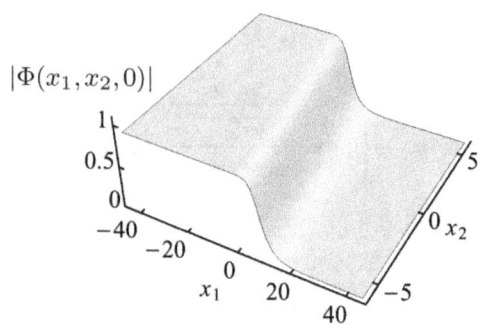

$|\Phi(x_1, x_2, 0)|$

Figure 6.7. 2D dark soliton (6.30) at $t = 0$, with $\alpha_1 = \alpha_2 = c_1 = c_2 = a_2 = 1$, $a3 = -1$, $x_{01} = x_{02} = t_0 = \phi_0 = 0$, and $n = 2.5$.

is a solution to (6.27), where $A_0 = \dfrac{a_2 n}{2(n+2)} \sqrt{\dfrac{-(n+1)}{a_3 (c_1^2 \alpha_1 + c_2^2 \alpha_2)}}$, $a_2 (n + 2)$

$(c_1^2 \alpha_1 + c_2^2 \alpha_2) > 0$, $a_3 (n + 1)(c_1^2 \alpha_1 + c_2^2 \alpha_2) < 0$, $m = 2n$, t_0 and ϕ_0 are arbitrary real constants.

***Example* 4**: *3D dark soliton*

Given $\psi(x, t) = \left(\dfrac{2 a_1 A_0^2 (n+2)}{a_2 n^2} \{1 - \tanh[A_0 (x - x_0)]\} \right)^{\frac{1}{n}} e^{i \left[\frac{4 a_1 A_0^2}{n^2} (t - t_0) + \phi_0 \right]}$ is a solution to (3.27), then

$$\Phi(x_1, x_2, x_3, t) = \left[\frac{2 A_0^2 (n+2)(c_1^2 \alpha_1 + c_2^2 \alpha_2 + c_3^2 \alpha_3)}{a_2 n^2} (1 - \tanh\{A_0 [c_1 (x_1 - x_{01}) + c_2 (x_2 - x_{02}) \right.$$

$$\left. + c_3 (x_3 - x_{03})]\})]^{\frac{1}{n}} e^{i \left[\frac{4 A_0^2 (c_1^2 \alpha_1 + c_2^2 \alpha_2 + c_3^2 \alpha_3)}{n^2} (t - t_0) + \phi_0 \right]} \right] \tag{6.31}$$

is a solution to (6.29), where $A_0 = \dfrac{a_2 n}{2(n+2)} \sqrt{\dfrac{-(n+1)}{a_3 (c_1^2 \alpha_1 + c_2^2 \alpha_2 + c_3^2 \alpha_3)}}$,

$a_2 (n + 2)(c_1^2 \alpha_1 + c_2^2 \alpha_2 + c_3^2 \alpha_3) > 0$, $a_3 (n + 1)(c_1^2 \alpha_1 + c_2^2 \alpha_2 + c_3^2 \alpha_3) < 0$, $m = 2n$, t_0 and ϕ_0 are arbitrary real constants.

6.5 Galilean Transformation in $(N + 1)$-Dimensions (Moving Solutions)

If $\psi(x, t; a_1)$ is a solution to one of the three equations, fundamental NLSE (2.160), NLSE with power law nonlinearity (3.9), and NLSE with dual power law nonlinearity (3.27), $i \psi_t + a_1 \psi_{xx} + a_2 |\psi|^2 \psi = 0$, $i \psi_t + a_1 \psi_{xx} + a_2 |\psi|^n \psi = 0$, $i \psi_t + a_1 \psi_{xx} + a_2 |\psi|^n \psi + a_3 |\psi|^m \psi = 0$, then $\Phi(x_1, x_2, \ldots, x_N, t; \alpha_1, \alpha_2, \ldots, \alpha_N) =$

$$\psi\{c_1 [x_1 - (x_{01} + v_1 t)] + c_2 [x_2 - (x_{02} + v_2 t)] + \cdots$$

$$+ c_N [x_N - (x_{0N} + v_N t)], t; c_1^2 \alpha_1 + c_1^2 \alpha_2 + \cdots + c_N^2 \alpha_N \}$$

$$\times e^{i \left[\frac{v_1}{2 \alpha_1} (x_1 - x_{01}) + \frac{v_2}{2 \alpha_2} (x_2 - x_{02}) + \cdots + \frac{v_N}{2 \alpha_N} (x_N - x_{0N}) - (\frac{v_1^2}{4 \alpha_1} + \frac{v_2^2}{4 \alpha_2} + \cdots + \frac{v_N^2}{4 \alpha_N}) (t - t_0) \right]} \tag{6.32}$$

is a moving solution to the $i\,\Phi_t + \sum_{k=1}^{N}\alpha_k\,\Phi_{x_k x_k} + a_2\,|\Phi|^2\,\Phi = 0$, $i\,\Phi_t + \sum_{k=1}^{N}\alpha_k\,\Phi_{x_k x_k} + a_2\,|\Phi|^n\,\Phi = 0$, $i\,\Phi_t + \sum_{k=1}^{N}\alpha_k\,\Phi_{x_k x_k} + a_2\,|\Phi|^n\,\Phi + a_3\,|\Phi|^m\,\Phi = 0$, respectively, with the replacements $x \rightarrow \sum_{k=1}^{N} c_k\,[x_k - (x_{0k} + v_k\,t)]$, $t \rightarrow t$, $a_1 \rightarrow \sum_{k=1}^{N} c_k^2\,\alpha_k$, $a_2 \rightarrow a_2$, $a_3 \rightarrow a_3$, $\psi \rightarrow \psi\,e^{i\,\sum_{k=1}^{N}\frac{v_k}{2\alpha_k}(x_k - x_{0k}) - \frac{v_k^2}{4\alpha_k}(t - t_0)}$ where α_k, c_k, x_{0k}, v_k, a_2, a_3, n, m, and t_0 are real constants.

***Example* 1:** *moving 2D bright soliton*
(Figure 6.8)

Given $\psi(x, t) = A_0\,\sqrt{\frac{2\,a_1}{a_2}}\,\text{sech}[A_0\,(x - x_0)]\,e^{i\,[a_1\,A_0^2\,(t - t_0) + \phi_0]}$ is a static solution to (2.160), then

$$\Phi(x_1, x_2, t) = A_0\,\sqrt{\frac{2\,(c_1^2\,\alpha_1 + c_2^2\,\alpha_2)}{a_2}}\,\text{sech}(A_0\,\{c_1\,[x_1 - (x_{01} + v_1\,t)] + c_2\,[x_2 - (x_{02} + v_2\,t)]\})$$

$$\times\,e^{i\,\left\{\frac{v_1\,(x_1 - x_{01})}{2\,\alpha_1} + \frac{v_2\,(x_2 - x_{02})}{2\,\alpha_2} + \left[A_0^2\,(c_1^2\,\alpha_1 + c_2^2\,\alpha_2) - \frac{v_1^2\,\alpha_2 + v_2^2\,\alpha_1}{4\,\alpha_1\,\alpha_2}\right](t - t_0) + \phi_0\right\}}$$
(6.33)

is a moving solution to (6.9), where $a_2\,(c_1^2\,\alpha_1 + c_2^2\,\alpha_2) > 0$, A_0, t_0, and ϕ_0 are arbitrary real constants.

***Example* 2:** *moving 3D bright soliton*
Given $\psi(x, t) = A_0\,\sqrt{\frac{2\,a_1}{a_2}}\,\text{sech}[A_0\,(x - x_0)]\,e^{i\,[a_1\,A_0^2\,(t - t_0) + \phi_0]}$ is a static solution to (2.160), then

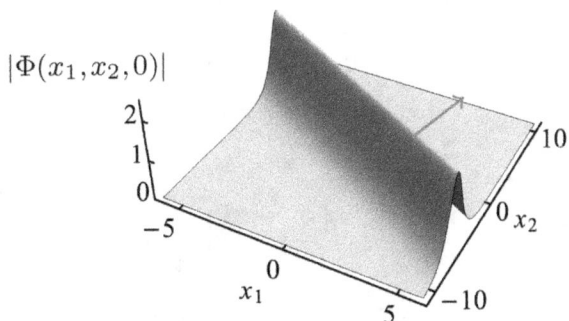

Figure 6.8. Moving 2D bright soliton (6.33) at $t = 0$, with $a_2 = \alpha_1 = \alpha_2 = A_0 = c_1 = c_2 = 1$, $v_1 = v_2 = 1/2$, and $x_{01} = x_{02} = t_0 = \phi_0 = 0$. The arrow shows direction of motion. Animation available online at http://doi.org/10.1088/978-0-7503-5954-2.

$$\Phi(x_1, x_2, x_3, t) = A_0 \sqrt{\frac{2\left(c_1^2\, \alpha_1 + c_2^2\, \alpha_2 + c_3^2\, \alpha_3\right)}{a_2}}\ \text{sech}(A_0\, \{c_1\,[x_1 - (x_{01} + v_1\, t)]$$

$$+ c_2\,[x_2 - (x_{02} + v_2\, t)] + c_3\,[x_3 - (x_{03} + v_3\, t)]\})\, e^{i\,\phi(x_1,\, x_2,\, x_3,\, t)} \tag{6.34}$$

is a moving solution to (6.11), where

$$\phi(x_1, x_2, x_3, t) = \frac{v_1\,(x_1 - x_{01})}{2\,\alpha_1} + \frac{v_2\,(x_2 - x_{02})}{2\,\alpha_2} + \frac{v_3\,(x_3 - x_{03})}{2\,\alpha_3}$$

$$+ \left[A_0^2\left(c_1^2\,\alpha_1 + c_2^2\,\alpha_2 + c_3^2\,\alpha_3\right) - \frac{v_1^2\,\alpha_2\,\alpha_3 + v_2^2\,\alpha_1\,\alpha_3 + v_3^2\,\alpha_1\,\alpha_2}{4\,\alpha_1\,\alpha_2\,\alpha_3}\right](t - t_0) + \phi_0, \tag{6.35}$$

$a_2\,(c_1^2\,\alpha_1 + c_2^2\,\alpha_2 + c_3^2\,\alpha_3) > 0$, A_0, t_0, and ϕ_0 are arbitrary real constants.

***Example* 3:** *moving 2D dark soliton*
(Figure 6.9)

Given $\psi(x, t) = A_0 \sqrt{\frac{-2\,a_1}{a_2}}\, \tanh[A_0\,(x - x_0)]\, e^{-i\,[2\,a_1\,A_0^2\,(t-t_0)+\phi_0]}$ is a static solution to (2.160), then

$$\Phi(x_1, x_2, t) = A_0 \sqrt{\frac{-2\left(c_1^2\,\alpha_1 + c_2^2\,\alpha_2\right)}{a_2}}\ \tanh(A_0\,\{c_1\,[x_1 - (x_{01} + v_1\, t)] + c_2\,[x_2 - (x_{02} + v_2\, t)]\})$$

$$\times\, e^{i\left\{\frac{v_1\,(x_1-x_{01})}{2\,\alpha_1} + \frac{v_2\,(x_2-x_{02})}{2\,\alpha_2} - \left[2\,A_0^2\,(c_1^2\,\alpha_1+c_2^2\,\alpha_2)+\frac{v_1^2\,\alpha_2+v_2^2\,\alpha_1}{4\,\alpha_1\,\alpha_2}\right](t-t_0)+\phi_0\right\}} \tag{6.36}$$

is a moving solution to (6.9), where $a_2\,(c_1^2\,\alpha_1 + c_2^2\,\alpha_2) < 0$, A_0, t_0, and ϕ_0 are arbitrary real constants.

***Example* 4:** *moving 3D dark soliton*
Given $\psi(x, t) = A_0 \sqrt{\frac{-2\,a_1}{a_2}}\, \tanh[A_0\,(x - x_0)]\, e^{-i\,[2\,a_1\,A_0^2\,(t-t_0)+\phi_0]}$ is a static solution to (2.160), then

$$\Phi(x_1, x_2, x_3, t) = A_0 \sqrt{\frac{-2\left(c_1^2\,\alpha_1 + c_2^2\,\alpha_2 + c_3^2\,\alpha_3\right)}{a_2}}\ \tanh(A_0\,\{c_1\,[x_1 - (x_{01} + v_1\, t)]$$

$$+ c_2\,[x_2 - (x_{02} + v_2\, t)] + c_3\,[x_3 - (x_{03} + v_3\, t)]\})\, e^{i\,\phi(x_1,\, x_2,\, x_3,\, t)} \tag{6.37}$$

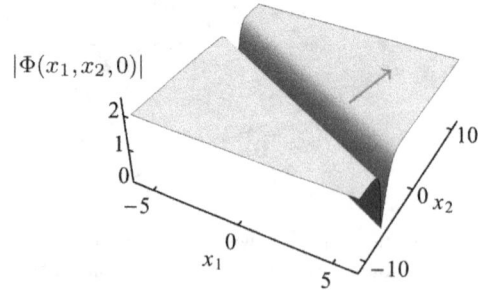

Figure 6.9. Moving 2D dark soliton (6.36) at $t = 0$, with $\alpha_1 = \alpha_2 = A_0 = c_1 = c_2 = 1$, $a_2 = -1$, $v_1 = v_2 = 1/2$, and $x_{01} = x_{02} = t_0 = \phi_0 = 0$. The arrow shows direction of motion. Animation available online at http://doi.org/10.1088/978-0-7503-5954-2.

is a moving solution to (6.11), where

$$\phi(x_1, x_2, x_3, t) = \phi_0 + \frac{v_1 (x_1 - x_{01})}{2 \, \alpha_1} + \frac{v_2 (x_2 - x_{02})}{2 \, \alpha_2} + \frac{v_3 (x_3 - x_{03})}{2 \, \alpha_3}$$

$$- \left[2 A_0^2 (c_1^2 \, \alpha_1 + c_2^2 \, \alpha_2 + c_3^2 \, \alpha_3) + \frac{v_1^2 \, \alpha_2 \, \alpha_3 + v_2^2 \, \alpha_1 \, \alpha_3 + v_3^2 \, \alpha_1 \, \alpha_2}{4 \, \alpha_1 \, \alpha_2 \, \alpha_3} \right] (t - t_0),$$

$a_2 (c_1^2 \, \alpha_1 + c_2^2 \, \alpha_2 + c_3^2 \, \alpha_3) < 0$, A_0, t_0, and ϕ_0 are arbitrary real constants.

Example 5: *moving 2D bright soliton*

Given $\psi(x, t) = \left\{ \dfrac{2 A_0^2 \, a_1 \, (n + 2)}{a_2 \, n^2} \, \text{sech}^2[A_0 \, (x - x_0)] \right\}^{\frac{1}{n}} e^{i \, [\frac{4 a_1 A_0^2}{n^2} \, (t - t_0) + \phi_0]}$ is a static solution to (3.9), then

$$\Phi(x_1, x_2, t) = \left(\frac{2 A_0^2 \, (n + 2) \, (c_1^2 \, \alpha_1 + c_2^2 \, \alpha_2)}{a_2 \, n^2} \right.$$

$$\times \, \text{sech}^2(A_0 \, \{c_1 \, [x_1 - (x_{01} + v_1 \, t)] + c_2 \, [x_2 - (x_{02} + v_2 \, t)]\}))^{\frac{1}{n}} \qquad (6.38)$$

$$\times \, e^{i \left\{ \frac{v_1 (x_1 - x_{01})}{2 \, \alpha_1} + \frac{v_2 (x_2 - x_{02})}{2 \, \alpha_2} + \left[\frac{4 A_0^2 (c_1^2 \, \alpha_1 + c_2^2 \, \alpha_2)}{n^2} - \frac{v_1^2 \, \alpha_2 + v_2^2 \, \alpha_1}{4 \, \alpha_1 \, \alpha_2} \right] (t - t_0) + \phi_0 \right\}}$$

is a moving solution to (6.22), where $a_2 (n + 2) (c_1^2 \, \alpha_1 + c_2^2 \, \alpha_2) > 0$, A_0, t_0, and ϕ_0 are arbitrary real constants.

Example 6: *moving 3D bright soliton*

Given $\psi(x, t) = \{ \dfrac{2 A_0^2 \, a_1 \, (n + 2)}{a_2 \, n^2} \, \text{sech}^2[A_0 \, (x - x_0)] \}^{\frac{1}{n}} \, e^{i \, [\frac{4 a_1 A_0^2}{n^2} \, (t - t_0) + \phi_0]}$ is a static solution to (3.9), then

$$\Phi(x_1, x_2, x_3, t) = \left(\frac{2 A_0^2 \, (n + 2) \, (c_1^2 \, \alpha_1 + c_2^2 \, \alpha_2 + c_3^2 \, \alpha_3)}{a_2 \, n^2} \, \text{sech}^2[q(x_1, x_2, x_3, t)] \right)^{\frac{1}{n}} \qquad (6.39)$$

$$\times \, e^{i \, \phi(x_1, x_2, x_3, t)},$$

is a moving solution to (6.24), where

$$\phi(x_1, x_2, x_3, t) = \phi_0 + \frac{v_1 (x_1 - x_{01})}{2 \, \alpha_1} + \frac{v_2 (x_2 - x_{02})}{2 \, \alpha_2} + \frac{v_3 (x_3 - x_{03})}{2 \, \alpha_3}$$

$$+ \left[\frac{4 A_0^2 (c_1^2 \, \alpha_1 + c_2^2 \, \alpha_2 + c_3^2 \, \alpha_3)}{n^2} - \frac{v_1^2 \, \alpha_2 \, \alpha_3 + v_2^2 \, \alpha_1 \, \alpha_3 + v_3^2 \, \alpha_1 \, \alpha_2}{4 \, \alpha_1 \, \alpha_2 \, \alpha_3} \right] (t - t_0),$$

$q(x_1, x_2, x_3, t) = A_0 \, \{c_1 \, [x_1 - (x_{01} + v_1 \, t)] + c_2 \, [x_2 - (x_{02} + v_2 \, t)] + c_3 \, [x_3 - (x_{03} + v_3 \, t)]\}$,
$a_2 (n + 2) (c_1^2 \, \alpha_1 + c_2^2 \, \alpha_2 + c_3^2 \, \alpha_3) > 0$, A_0, t_0, and ϕ_0 are arbitrary real constants.

Example 7: *moving 2D flat-top soliton*
(Figure 6.10)

Given $\psi(x, t) = \left\{ \dfrac{A_0 \, (n + 2)}{a_2 + a_2 \, \sin(\theta) \, \cosh[n \sqrt{\frac{A_0}{a_1}} \, (x - x_0)]} \right\}^{\frac{1}{n}} e^{i \, [A_0 \, (t - t_0) + \phi_0]}$ is a static solution tok
(3.27), then

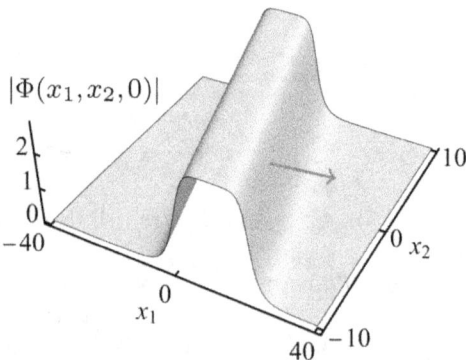

Figure 6.10. Moving 2D flat-top soliton (6.40) at $t = 0$, with $a_2 = 1$, $A_1 = 2$, $c_1 = c_2 = \alpha_1 = \alpha_2 = 1$, $a_3 = -0.069\,444\,444\,44$, $n = 4$, $v_1 = v_2 = 1/2$, and $x_{01} = x_{02} = t_0 = \phi_0 = 0$. The arrow shows direction of motion. Animation available online at http://doi.org/10.1088/978-0-7503-5954-2.

$$\Phi(x_1, x_2, t) = \left\{ \frac{A_0\,(n + 2)}{a_2 + a_2\,\sin(\theta)\,\cosh[q(x_1, x_2, t)]} \right\}^{\frac{1}{n}} e^{i\,[A_0\,(t - t_0) + \phi_0]}$$

$$\times\, e^{i\left[\frac{v_1\,(x_1 - x_{01})}{2\,\alpha_1} + \frac{v_2\,(x_2 - x_{02})}{2\,\alpha_2} - \frac{v_1^2\,\alpha_2 + v_2^2\,\alpha_1}{4\,\alpha_1\,\alpha_2}\,(t - t_0) \right]} \tag{6.40}$$

is a moving solution to (6.27), where $q(x_1, x_2, t) = n \sqrt{\dfrac{A_0}{c_1^2\,\alpha_1 + c_2^2\,\alpha_2}}$

$\{c_1\,[x_1 - (x_{01} + v_1\,t)] + c_2\,[x_2 - (x_{02} + v_2\,t)]\}$, $\quad a_3 = \gamma\,a_{03}$, $\quad \gamma = -\cos^2(\theta)$, $0 < \theta < \pi/2$, $\quad a_{03} = a_2^2\,(n + 1)/[A_0\,(n + 2)^2]$, $\quad A_0\,(c_1^2\,\alpha_1 + c_2^2\,\alpha_2) > 0$, $A_0\,(n + 1) > 0$, $m = 2\,n$, x_{01}, x_{02}, t_0, A_0, and ϕ_0 are arbitrary real constants.

Example 8: *moving 3D flat-top soliton*

Given $\psi(x, t) = \left\{ \dfrac{A_0\,(n + 2)}{a_2 + a_2\,\sin(\theta)\,\cosh[n\,\sqrt{\frac{A_0}{q}}\,(x - x_0)]} \right\}^{\frac{1}{n}} e^{i\,[A_0\,(t - t_0) + \phi_0]}$ is a static solution to

(3.27), then

$$\Phi(x_1, x_2, x_3, t) = \left\{ \frac{A_0\,(n + 2)}{a_2 + a_2\,\sin(\theta)\,\cosh[q(x_1, x_2, x_3, t)]} \right\}^{\frac{1}{n}} e^{i\,[A_0\,(t - t_0) + \phi_0]}$$

$$\times\, e^{i\left[\frac{v_1\,(x_1 - x_{01})}{2\,\alpha_1} + \frac{v_2\,(x_2 - x_{02})}{2\,\alpha_2} + \frac{v_3\,(x_3 - x_{03})}{2\,\alpha_3} - \frac{v_1^2\,\alpha_2\,\alpha_3 + v_2^2\,\alpha_1\,\alpha_3 + v_3^2\,\alpha_1\,\alpha_2}{4\,\alpha_1\,\alpha_2\,\alpha_3}\,(t - t_0) \right]} \tag{6.41}$$

is a moving solution to (6.29), where

$q(x_1, x_2, x_3, t) = n \sqrt{\dfrac{A_0}{c_1^2\,\alpha_1 + c_2^2\,\alpha_2 + c_3^2\,\alpha_3}}\,\{c_1\,[x_1 - (x_{01} + v_1\,t)]$

$+\, c_2\,[x_2 - (x_{02} + v_2\,t)] + c_3\,[x_3 - (x_{03} + v_3\,t)]\}$,

$a_3 = \gamma\,a_{03}$, $\quad \gamma = -\cos^2(\theta)$, $\quad 0 < \theta < \pi/2$, $\quad a_{03} = a_2^2\,(n + 1)/[A_0\,(n + 2)^2]$, $A_0\,(c_1^2\,\alpha_1 + c_2^2\,\alpha_2 + c_3^2\,\alpha_3) > 0$, $A_0\,(n + 1) > 0$, $m = 2\,n$, x_{01}, x_{02}, x_{03}, t_0, A_0, and ϕ_0 are arbitrary real constants.

6.6 NLSE in (2 + 1)-Dimensions with $\Phi_{x_1 x_2}$ Term

If $\psi(x, t)$ is a solution to the fundamental NLSE (2.160), $i\,\psi_t + a_1\,\psi_{xx} + a_2\,|\psi|^2\,\psi = 0$, then

$$\Phi(x_1, x_2, t) = \sqrt{\frac{a_2}{b_4}}\;\psi[X(x_1, x_2), t], \tag{6.42}$$

is a solution to

$$i\,\Phi_t + b_1\,\Phi_{x_1 x_1} + b_2\,\Phi_{x_2 x_2} + b_3\,\Phi_{x_1 x_2} + b_4\,|\Phi|^2\,\Phi = 0, \tag{6.43}$$

where

$$X(x_1, x_2) = c_0 - \frac{1}{2\,b_1}\left(b_3\,c_1 - \sqrt{4\,a_1\,b_1 - 4\,b_1\,b_2\,c_1^2 + b_3^2\,c_1^2}\right)x_1 + c_1\,x_2, \tag{6.44}$$

$a_2\,b_4 > 0$, $4\,a_1\,b_1\,(1 - b_2\,c_1^2) > b_3^2\,c_1^2$, b_1, b_2, b_3, b_4, c_0 and c_1 are arbitrary real constants.

***Example* 1**: *bright soliton*

Given $\psi(x, t) = A_0\,\sqrt{\frac{2\,a_1}{a_2}}\,\text{sech}[A_0\,(x - x_0)]\,e^{i\,[a_1\,A_0^2\,(t-t_0)+\phi_0]}$ is a solution to (2.160), then

$$\Phi(x_1, x_2, t) = A_0\,\sqrt{\frac{2\,a_1}{b_4}}\,\text{sech}\left\{A_0\left[c_0 - \frac{(b_3\,c_1 - \sqrt{4\,a_1\,b_1 - 4\,b_1\,b_2\,c_1^2 + b_3^2\,c_1^2})\,(x_1 - x_{01})}{2\,b_1}\right.\right.$$
$$\left.\left. + c_1\,(x_2 - x_{02})\right]\right\}\,e^{i\left[a_1\,A_0^2\,(t-t_0)+\phi_0\right]} \tag{6.45}$$

is a solution to (6.43), where $a_1\,b_4 > 0$, $(4\,a_1\,b_1 - 4\,b_1\,b_2\,c_1^2 + b_3^2\,c_1^2) > 0$, A_0, t_0, and ϕ_0 are arbitrary real constants.

***Example* 2**: *dark soliton*

Given $\psi(x, t) = A_0\,\sqrt{\frac{-2\,a_1}{a_2}}\,\tanh[A_0\,(x - x_0)]\,e^{-i\,[2\,a_1\,A_0^2\,(t-t_0)+\phi_0]}$ is a solution to (2.160), then

$$\Phi(x_1, x_2, t) = A_0\,\sqrt{\frac{-2\,a_1}{b_4}}\,\tanh\left\{A_0\left[c_0 + \frac{\left(\sqrt{4\,a_1\,b_1 - 4\,b_1\,b_2\,c_1^2 + b_3^2\,c_1^2} - b_3\,c_1\right)(x_1 - x_{01})}{2\,b_1}\right.\right.$$
$$\left.\left. + c_1\,(x_2 - x_{02})\right]\right\}\,e^{-i\,[2\,a_1\,A_0^2\,(t-t_0)+\phi_0]} \tag{6.46}$$

is a solution to (6.43), where $a_1\,b_4 < 0$, $(4\,a_1\,b_1 - 4\,b_1\,b_2\,c_1^2 + b_3^2\,c_1^2) > 0$, A_0, t_0, and ϕ_0 are arbitrary real constants.

6.7 Summary of Sections 6.2–6.6

(N+1)-Dimensional NLSE with Cubic Nonlinearity

Transformation: $\Phi(x_1, x_2, \ldots, x_N, t; \alpha_1, \alpha_2, \ldots, \alpha_N)$
$$= \Phi[c_1(x_1 - x_{01}) + c_2(x_2 - x_{02}) + \cdots + c_N(x_N - x_{0N}), t$$
$$; c_1^2\alpha_1 + c_1^2\alpha_2 + \cdots + c_N^2\alpha_N, a_2]$$

Equation: $i\,\Phi_t + \sum_{k=1}^{N} \alpha_k\, \Phi_{x_k x_k} + a_2\, |\Phi|^2\, \Phi = 0$

#	Example	Conditions	Name	Equation #
1.	$\Phi(x_1, x_2, t) = A_0 \sqrt{\dfrac{2(c_1^2\alpha_1 + c_2^2\alpha_2)}{a_2}}$ $\times \operatorname{sech}\{A_0[c_1(x_1 - x_{01}) + c_2(x_2 - x_{02})]\}$ $\times e^{i[A_0^2(c_1^2\alpha_1 + c_2^2\alpha_2)(t-t_0)+\phi_0]}$	$a_2(c_1^2\alpha_1 + c_2^2\alpha_2) > 0$, A_0, t_0, and ϕ_0 are arbitrary real constants	2D bright soliton	(6.8)
2.	$\Phi(x_1, x_2, x_3, t) = A_0 \sqrt{\dfrac{2(c_1^2\alpha_1 + c_2^2\alpha_2 + c_3^2\alpha_3)}{a_2}}$ $\times \operatorname{sech}\{A_0[c_1(x_1 - x_{01}) + c_2(x_2 - x_{02}) + c_3(x_3 - x_{03})]\}$ $\times e^{i[A_0^2(c_1^2\alpha_1 + c_2^2\alpha_2 + c_3^2\alpha_3)(t-t_0)+\phi_0]}$	$a_2(c_1^2\alpha_1 + c_2^2\alpha_2 + c_3^2\alpha_3) > 0$, A_0, t_0, and ϕ_0 are arbitrary real constants	3D bright soliton	(6.10)
3.	$\Phi(x_1, x_2, t) = A_0 \sqrt{\dfrac{-2(c_1^2\alpha_1 + c_2^2\alpha_2)}{a_2}}$ $\times \tanh\{A_0[c_1(x_1 - x_{01}) + c_2(x_2 - x_{02})]\}$ $\times e^{-i[2A_0^2(c_1^2\alpha_1 + c_2^2\alpha_2)(t-t_0)+\phi_0]}$	$a_2(c_1^2\alpha_1 + c_2^2\alpha_2) < 0$, A_0, t_0, and ϕ_0 are arbitrary real constants	2D dark soliton	(6.12)
4.	$\Phi(x_1, x_2, x_3, t) = A_0 \sqrt{\dfrac{-2(c_1^2\alpha_1 + c_2^2\alpha_2 + c_3^2\alpha_3)}{a_2}}$ $\times \tanh\{A_0[c_1(x_1 - x_{01}) + c_2(x_2 - x_{02}) + c_3(x_3 - x_{03})]\}$ $\times e^{-i[2A_0^2(c_1^2\alpha_1 + c_2^2\alpha_2 + c_3^2\alpha_3)(t-t_0)+\phi_0]}$	$a_2(c_1^2\alpha_1 + c_2^2\alpha_2 + c_3^2\alpha_3) < 0$, A_0, t_0, and ϕ_0 are arbitrary real constants	3D dark soliton	(6.13)

5.
$$\Phi(x_1, x_2, t) = \frac{-1}{\sqrt{a_2}} \left(\frac{p^2 \cos[\omega(t-t_0)] + 2 i p \nu \sin[\omega(t-t_0)]}{2\cos[\omega(t-t_0)] - q(x_1,x_2)} + 1 \right) e^{i[(t-t_0)+\phi_0]},$$
$$q(x_1, x_2) = 2\nu\cosh\left\{ \frac{p}{\sqrt{2(c_1^2 a_1 + c_2^2 a_2)}} [c_1(x_1 - x_{01}) + c_2(x_2 - x_{02})] \right\}$$

2D Kuznetsov–Ma breather (6.14)

$p = 2\sqrt{\nu^2 - 1}$, $\omega = p\nu$, $\nu > 1$, $a_2 > 0$, $(c_1^2 a_1 + c_2^2 a_2) > 0$, t_0 and ϕ_0 are arbitrary real constants

6.
$$\Phi(x_1, x_2, x_3, t) = \frac{-1}{\sqrt{a_2}} \left(\frac{p^2 \cos[\omega(t-t_0)] + 2 i p \nu \sin[\omega(t-t_0)]}{2\cos[\omega(t-t_0)] - q(x_1,x_2,x_3)} + 1 \right)$$
$$\times e^{i[(t-t_0)+\phi_0]},$$
$$q(x_1, x_2, x_3) = 2\nu\cosh\left\{ \frac{p}{\sqrt{2(c_1^2 a_1 + c_2^2 a_2 + c_3^2 a_3)}} [c_1(x_1 - x_{01}) + c_2(x_2 - x_{02}) + c_3(x_3 - x_{03})] \right\}$$

3D Kuznetsov–Ma breather (6.15)

$p = 2\sqrt{\nu^2 - 1}$, $\omega = p\nu$, $\nu > 1$, $a_2 > 0$, $(c_1^2 a_1 + c_2^2 a_2 + c_3^2 a_3) > 0$, t_0 and ϕ_0 are arbitrary real constants

7.
$$\Phi(x_1, x_2, t) = \frac{1}{\sqrt{a_2}} \left(\frac{\kappa^2 \cosh[\delta(t-t_0)] + 2 i \kappa\nu \sinh[\delta(t-t_0)]}{2\cosh[\delta(t-t_0)] - q(x_1,x_2)} - 1 \right)$$
$$\times e^{i[(t-t_0)+\phi_0]},$$
$$q(x_1, x_2) = 2\nu\cos\left\{ \frac{\kappa}{\sqrt{2(c_1^2 a_1 + c_2^2 a_2)}} [c_1(x_1 - x_{01}) + c_2(x_2 - x_{02})] \right\}$$

2D Akhmediev breather (6.16)

$\kappa = 2\sqrt{1 - \nu^2}$, $\delta = \kappa\nu$, $\nu < 1$, $a_2 > 0$, $(c_1^2 a_1 + c_2^2 a_2) > 0$, t_0 and ϕ_0 are arbitrary real constants

8.
$$\Phi(x_1, x_2, x_3, t) = \frac{1}{\sqrt{a_2}} \left(\frac{\kappa^2 \cosh[\delta(t-t_0)] + 2 i \kappa\nu \sinh[\delta(t-t_0)]}{2\cosh[\delta(t-t_0)] - q(x_1,x_2,x_3)} - 1 \right)$$
$$\times e^{i[(t-t_0)+\phi_0]},$$
$$q(x_1, x_2, x_3) = 2\nu\cos\left\{ \frac{\kappa}{\sqrt{2(c_1^2 a_1 + c_2^2 a_2 + c_3^2 a_3)}} [c_1(x_1 - x_{01}) + c_2(x_2 - x_{02}) + c_3(x_3 - x_{03})] \right\}$$

3D Akhmediev breather (6.17)

$\kappa = 2\sqrt{1 - \nu^2}$, $\delta = \kappa\nu$, $\nu < 1$, $a_2 > 0$, $(c_1^2 a_1 + c_2^2 a_2 + c_3^2 a_3) > 0$, t_0 and ϕ_0 are arbitrary real constants

9.
$$\Phi(x_1, x_2, t) = \frac{1}{\sqrt{a_2}} \left(\frac{4 + i 8(t-t_0)}{1 + 4(t-t_0)^2 + q(x_1,x_2)} - 1 \right) e^{i[(t-t_0)+\phi_0]},$$
$$q(x_1, x_2) = \frac{2}{c_1^2 a_1 + c_2^2 a_2} [c_1(x_1 - x_{01}) + c_2(x_2 - x_{02})]^2$$

2D Peregrine soliton (6.18)

$a_2 > 0$, t_0 and ϕ_0 are arbitrary real constants

10.
$$\Phi(x_1, x_2, x_3, t) = \frac{1}{\sqrt{a_2}} \left(\frac{4 + i 8(t-t_0)}{1 + 4(t-t_0)^2 + q(x_1,x_2,x_3)} - 1 \right) e^{i[(t-t_0)+\phi_0]},$$
$$q(x_1, x_2, x_3) = \frac{2}{c_1^2 a_1 + c_2^2 a_2 + c_3^2 a_3} [c_1(x_1 - x_{01}) + c_2(x_2 - x_{02}) + c_3(x_3 - x_{03})]^2$$

3D Peregrine soliton (6.19)

$a_2 > 0$, t_0 and ϕ_0 are arbitrary real constants

(Continued)

(N +1)-Dimensional NLSE with Power Law Nonlinearity

Transformation: $\Phi(x_1, x_2, \ldots, x_N, t;\ \alpha_1, \alpha_2, \ldots, \alpha_N)$
$$= \psi[c_1 (x_1 - x_{01}) + c_2 (x_2 - x_{02}) + \cdots + c_N (x_N - x_{0N}),\ t$$
$$;\ c_1^2 \alpha_1 + c_1^2 \alpha_2 + \cdots + c_N^2 \alpha_N,\ a_2]$$

Equation: $i\,\Phi_t + \sum_{k=1}^{N} \alpha_k\,\Phi_{x_k x_k} + a_2\,|\Phi|^n\,\Phi = 0$

#	Example	Conditions	Name	Equation #
1.	$\Phi(x_1, x_2, t) = \left(\dfrac{2 A_0^2 (n+2)(c_1^2 \alpha_1 + c_2^2 \alpha_2)}{a_2\, n^2}\ \mathrm{sech}^2\{A_0\,[\,c_1\,(x_1 - x_{01})\right.$ $\left. + c_2\,(x_2 - x_{02})]\}\right)^{\frac{1}{n}} e^{i\left[\frac{4 A_0^2 (c_1^2 \alpha_1 + c_2^2 \alpha_2)}{n^2}\,(t-t_0)+\phi_0\right]}$	$a_2\,(n+2)\,(c_1^2\,\alpha_1 + c_2^2\,\alpha_2) > 0.$ $N = 2$, A_0, t_0, and ϕ_0 are arbitrary real constants	2D bright soliton	(6.21)
2.	$\Phi(x_1, x_2, x_3, t) = \left(\dfrac{2 A_0^2 (n+2)(c_1^2 \alpha_1 + c_2^2 \alpha_2 + c_3^2 \alpha_3)}{a_2\, n^2}\right.$ $\times\,\mathrm{sech}^2\{A_0\,[c_1\,(x_1 - x_{01}) + c_2\,(x_2 - x_{02}) + c_3\,(x_3 - x_{03})]\}\big)^{\frac{1}{n}}$ $\times\, e^{i\left[\frac{4 A_0^2 (c_1^2 \alpha_1 + c_2^2 \alpha_2 + c_3^2 \alpha_3)}{n^2}\,(t-t_0)+\phi_0\right]}$	$N = 3.$ $a_2\,(n+2)\,(c_1^2\,\alpha_1 + c_2^2\,\alpha_2 + c_3^2\,\alpha_3) > 0.$ A_0, t_0, and ϕ_0 are arbitrary real constants	3D bright soliton	(6.23)

(N +1)-Dimensional NLSE with Dual Power Law Nonlinearity

Transformation: $\Phi(x_1, x_2, \ldots, x_N, t;\ \alpha_1, \alpha_2, \ldots, \alpha_N)$
$$= \psi[c_1 (x_1 - x_{01}) + c_2 (x_2 - x_{02}) + \cdots + c_N (x_N - x_{0N}),\ t$$
$$;\ c_1^2 \alpha_1 + c_1^2 \alpha_2 + \cdots + c_N^2 \alpha_N,\ a_2]$$

Equation: $i\,\Phi_t + \sum_{k=1}^{N} \alpha_k\,\Phi_{x_k x_k} + a_2\,|\Phi|^n\,\Phi + a_3\,|\Phi|^m\,\Phi = 0$

#	Example	Conditions	Name	Equation #
1.	$$\Phi(x_1, x_2, t) = \left\{ \frac{A_0(n+2)}{a_2 + a_2 \sin(\theta)\cosh\left[n\sqrt{\frac{A_0}{c_1^2 \alpha_1 + c_2^2 \alpha_2}}\,[c_1(x_1 - x_{01}) + c_2(x_2 - x_{02})]\right]} \right\}^{\frac{1}{n}}$$ $$\times e^{i[A_0(t-t_0)+\phi_0]}$$	$a_3 = \gamma\, a_{03}, \gamma = -\cos^2(\theta).$ $0 < \theta < \pi/2.$ $a_{03} = a_2^2(n+1)/[A_0(n+2)^2],$ $A_0(c_1^2 \alpha_1 + c_2^2 \alpha_2) > 0.$ $A_0(n+1) > 0.\ m = 2n.\ x_{01}, x_{02}, t_0,$ A_0 and ϕ_0 are arbitrary real constants	2D flat-top soliton	(6.26)
2.	$$\Phi(x_1, x_2, x_3, t) = \left\{ \frac{A_0(n+2)}{a_2 + a_2 \sin(\theta)\cosh[q(x_1, x_2, x_3)]} \right\}^{\frac{1}{n}} e^{i[A_0(t-t_0)+\phi_0]},$$ $$q(x_1, x_2, x_3) = n\sqrt{\frac{A_0}{c_1^2 \alpha_1 + c_2^2 \alpha_2 + c_3^2 \alpha_3}}\,[c_1(x_1 - x_{0.}) + c_2(x_2 - x_{02}) + c_3(x_3 - x_{03})]$$	$a_3 = \gamma\, a_{03}, \gamma = -\cos^2(\theta).$ $0 < \theta < \pi/2.$ $a_{03} = a_2^2(n+1)/[A_0(n+2)^2].$ $A_0(c_1^2 \alpha_1 + c_2^2 \alpha_2 + c_3^2 \alpha_3) > 0.$ $A_0(n+1) > 0.\ m = 2n.\ x_{01}, x_{02}, x_{03},$ $t_0, A_0,$ and ϕ_0 are arbitrary real constants.	3D flat-top soliton	(6.28)
3.	$$\Phi(x_1, x_2, t) = \left[\frac{2 A_0^2(n+2)(c_1^2 \alpha_1 + c_2^2 \alpha_2)}{a_2 n^2}\,(1 - \tanh\{A_0[c_1(x_1 - x_{01}) + c_2(x_2 - x_{02})]\}) \right]^{\frac{1}{n}}$$ $$\times e^{i\left[\frac{4 A_0^2(c_1^2 \alpha_1 + c_2^2 \alpha_2)}{n^2}(t-t_0)+\phi_0\right]}$$	$A_0 = \frac{a_2 n}{2(n+2)}\sqrt{\frac{-(n+1)}{a_3(c_1^2 \alpha_1 + c_2^2 \alpha_2)}},$ $a_2(n+2)(c_1^2 \alpha_1 + c_2^2 \alpha_2) > 0,$ $N = 2.$ $a_3(n+1)(c_1^2 \alpha_1 + c_2^2 \alpha_2) < 0.$ $m = 2n.\ t_0$ and ϕ_0 are arbitrary real constants	2D dark soliton	(6.30)
4.	$$\Phi(x_1, x_2, x_3, t) = \left[\frac{2 A_0^2(n+2)(c_1^2 \alpha_1 + c_2^2 \alpha_2 + c_3^2 \alpha_3)}{a_2 n^2}\,(1 - \tanh\{A_0[c_1(x_1 - x_{01}) + c_2(x_2 - x_{02})\right.$$ $$\left. + c_3(x_3 - x_{03})]\}) \right]^{\frac{1}{n}} e^{i\left[\frac{4 A_0^2(c_1^2 \alpha_1 + c_2^2 \alpha_2 + c_3^2 \alpha_3)}{n^2}(t-t_0)+\phi_0\right]}$$	$A_0 = \frac{a_2 n}{2(n+2)}\sqrt{\frac{-(n+1)}{a_3(c_1^2 \alpha_1 + c_2^2 \alpha_2 + c_3^2 \alpha_3)}},$ $a_2(n+2)(c_1^2 \alpha_1 + c_2^2 \alpha_2 + c_3^2 \alpha_3) > 0,$ $a_3(n+1)(c_1^2 \alpha_1 + c_2^2 \alpha_2 + c_3^2 \alpha_3) < 0.$ $m = 2n.\ N = 3.$ t_0 and ϕ_0 are arbitrary real constants	3D dark soliton	(6.31)

(Continued)

Galilean Transformation in $(N + 1)$-Dimensions (Movable Solutions)

Transformation:

$$\Phi(x_1, x_2, \ldots, x_N, t; \alpha_1, \alpha_2, \ldots, \alpha_N, a_2) = \Phi\{c_1 [x_1 - (x_{01} + v_1 t)] + c_2 [x_2 - (x_{02} + v_2 t)] + \cdots$$
$$+ c_N [x_N - (x_{0N} + v_N t)],$$
$$t; c_1^2 \alpha_1 + c_2^2 \alpha_2 + \cdots + c_N^2 \alpha_N, a_2\} \times$$
$$e^{i [\frac{v_1}{2\alpha_1} (x_1 - x_{01}) + \frac{v_2}{2\alpha_2} (x_2 - x_{02}) + \cdots + \frac{v_N}{2\alpha_N} (x_N - x_{0N}) - (\frac{v_1^2}{4\alpha_1} + \frac{v_2^2}{4\alpha_2} + \cdots + \frac{v_N^2}{4\alpha_N})(t - t_0)]}$$

Equation: $i \Phi_t + \sum_{k=1}^{N} \alpha_k \Phi_{x_k x_k} + a_2 |\Phi|^2 \Phi = 0$

#	Example	Conditions	Name	Equation #
1.	$\Phi(x_1, x_2, t) = A_0 \sqrt{\dfrac{2(c_1^2 \alpha_1 + c_2^2 \alpha_2)}{a_2}}$ $\times \text{sech}(A_0 \{c_1 [x_1 - (x_{01} + v_1 t)] + c_2 [x_2 - (x_{02} + v_2 t)]\})$ $\times e^{i [\frac{v_1 (x_1 - x_{01})}{2\alpha_1} + \frac{v_2 (x_2 - x_{02})}{2\alpha_2} + [A_0^2 (c_1^2 \alpha_1 + c_2^2 \alpha_2) - \frac{v_1^2 \alpha_2 + v_2^2 \alpha_1}{4\alpha_1 \alpha_2}](t - t_0) + \phi_0]}$	$a_2 (c_1^2 \alpha_1 + c_2^2 \alpha_2) > 0. N = 2. A_0, t_0,$ and ϕ_0 are arbitrary real constants	Moving 2D bright soliton	(6.33)
2.	$\Phi(x_1, x_2, x_3, t) = A_0 \sqrt{\dfrac{2(c_1^2 \alpha_1 + c_2^2 \alpha_2 + c_3^2 \alpha_3)}{a_2}}$ $\times \text{sech}(A_0 \{c_1 [x_1 - (x_{01} + v_1 t)]$ $+ c_2 [x_2 - (x_{02} + v_2 t)] + c_3 [x_3 - (x_{03} + v_3 t)]\}) e^{i \phi(x_1, x_2, x_3, t)},$ $\phi(x_1, x_2, x_3, t) = \phi_0 + \frac{v_1 (x_1 - x_{01})}{2\alpha_1} + \frac{v_2 (x_2 - x_{02})}{2\alpha_2} + \frac{v_3 (x_3 - x_{03})}{2\alpha_3}$ $+ [A_0^2 (c_1^2 \alpha_1 + c_2^2 \alpha_2 + c_3^2 \alpha_3) - \frac{v_1^2 \alpha_2 \alpha_3 + v_2^2 \alpha_1 \alpha_3 + v_3^2 \alpha_1 \alpha_2}{4\alpha_1 \alpha_2 \alpha_3}](t - t_0)$	$a_2 (c_1^2 \alpha_1 + c_2^2 \alpha_2 + c_3^2 \alpha_3) > 0.$ $N = 3. A_0, t_0,$ and ϕ_0 are arbitrary real constants	Moving 3D bright soliton	(6.34)
3.	$\Phi(x_1, x_2, t) = A_0 \sqrt{\dfrac{-2(c_1^2 \alpha_1 + c_2^2 \alpha_2)}{a_2}}$ $\times \tanh(A_0 \{c_1 [x_1 - (x_{01} + v_1 t)] + c_2 [x_2 - (x_{02} + v_2 t)]\})$ $\times e^{i \{\frac{v_1 (x_1 - x_{01})}{2\alpha_1} + \frac{v_2 (x_2 - x_{02})}{2\alpha_2} - [2 A_0^2 (c_1^2 \alpha_1 + c_2^2 \alpha_2) + \frac{v_1^2 \alpha_2 + v_2^2 \alpha_1}{4\alpha_1 \alpha_2}](t - t_0) + \phi_0\}}$	$a_2 (c_1^2 \alpha_1 + c_2^2 \alpha_2) < 0. N = 2. A_0, t_0,$ and ϕ_0 are arbitrary real constants	Moving 2D dark soliton	(6.36)

Handbook of Exact Solutions to the Nonlinear Schrödinger Equations (Second Edition)

4.
$$\Phi(x_1, x_2, x_3, t) = A_0 \sqrt{\frac{-2(c_1^2 \alpha_1 + c_2^2 \alpha_2 + c_3^2 \alpha_3)}{a_2}}$$
$$\times \tanh(A_0 \{c_1 [x_1 - (x_{01} + v_1 t)]$$
$$+ c_2 [x_2 - (x_{02} + v_2 t)] + c_3 [x_3 - (x_{03} + v_3 t)]\}) e^{i \phi(x_1, x_2, x_3, t)},$$

$$\phi(x_1, x_2, x_3, t) = \frac{v_1 (x_1 - x_{01})}{2\alpha_1} + \frac{v_2 (x_2 - x_{02})}{2\alpha_2} + \frac{v_3 (x_3 - x_{03})}{2\alpha_3}$$
$$- \left[2 A_0^2 (c_1^2 \alpha_1 + c_2^2 \alpha_2 + c_3^2 \alpha_3) + \frac{v_1^2 \alpha_2 \alpha_3 + v_2^2 \alpha_1 \alpha_3 + v_3^2 \alpha_1 \alpha_2}{4 \alpha_1 \alpha_2 \alpha_3}\right]$$
$$\times (t - t_0) + \phi_0,$$

Conditions: $a_2 (c_1^2 \alpha_1 + c_2^2 \alpha_2 + c_3^2 \alpha_3) < 0,$ $N = 3, A_0, t_0,$ and ϕ_0 are arbitrary real constants

Name: Moving 3D dark soliton

Equation #: (6.37)

Equation: $i \Phi_t + \sum_{k=1}^{N} \alpha_k \Phi_{x_k x_k} + a_2 |\Phi|^n \Phi = 0$

#	Example	Conditions	Name	Equation #

1.
$$\Phi(x_1, x_2, t) = \left(\frac{2 A_0^2 (n+2)(c_1^2 \alpha_1 + c_2^2 \alpha_2)}{a_2 n^2}\right)^{\frac{1}{n}}$$
$$\times \operatorname{sech}^2(A_0 \{c_1 [x_1 - (x_{01} + v_1 t)] + c_2 [x_2 - (x_{02} + v_2 t)]\})^{\frac{1}{n}}$$
$$\times e^{i \left[\frac{v_1 (x_1 - x_{01})}{2\alpha_1} + \frac{v_2 (x_2 - x_{02})}{2\alpha_2} - \left[\frac{4 A_0^2 (c_1^2 \alpha_1 + c_2^2 \alpha_2)}{n^2} - \frac{v_2^2 \alpha_1 + v_2^2 \alpha_2}{4 \alpha_1 \alpha_2}\right](t - t_0) + \phi_0\right]}$$

Conditions: $a_2 (c_1^2 \alpha_1 + c_2^2 \alpha_2) > 0,$ $N = 2, A_0, t_0,$ and ϕ_0 are arbitrary real constants

Name: Moving 2D bright soliton

Equation #: (6.38)

2.
$$\Phi(x_1, x_2, x_3, t) = \left(\frac{2 A_0^2 (n+2)(c_1^2 \alpha_1 + c_2^2 \alpha_2 + c_3^2 \alpha_3)}{a_2 n^2}\right)^{\frac{1}{n}} \operatorname{sech}^2[q(x_1, x_2, x_3, t)]$$
$$\times e^{i \phi(x_1, x_2, x_3, t)},$$

$$q(x_1, x_2, x_3, t) = A_0 \{c_1 [x_1 - (x_{01} + v_1 t)] + c_2 [x_2 - (x_{02} + v_2 t)] + c_3 [x_3 - (x_{03} + v_3 t)]\}$$

$$\phi(x_1, x_2, x_3, t) = \phi_0 + \frac{v_1 (x_1 - x_{01})}{2\alpha_1} + \frac{v_2 (x_2 - x_{02})}{2\alpha_2} + \frac{v_3 (x_3 - x_{03})}{2\alpha_3}$$
$$- \left[\frac{4 A_0^2 (c_1^2 \alpha_1 + c_2^2 \alpha_2 + c_3^2 \alpha_3)}{n^2} - \frac{v_1^2 \alpha_2 \alpha_3 + v_2^2 \alpha_1 \alpha_3 + v_3^2 \alpha_1 \alpha_2}{4 \alpha_1 \alpha_2 \alpha_3}\right](t - t_0)$$

Conditions: $a_2 (n+2)(c_1^2 \alpha_1 + c_2^2 \alpha_2 + c_3^2 \alpha_3) > 0,$ $N = 3, A_0, t_0,$ and ϕ_0 are arbitrary real constants

Name: Moving 3D bright soliton

Equation #: (6.39)

(Continued)

Equation: $i\,\Phi_t + \sum_{k=1}^{N} \alpha_k\,\Phi_{x_k x_k} + a_2\,|\Phi|^n\,\Phi + a_3\,|\Phi|^m\,\Phi = 0$

#	Example	Conditions	Name	Equation #
1.	$\Phi(x_1, x_2, t) = \left\{ \dfrac{A_0(n+2)}{a_2 + a_2 \sin(\theta)\cosh[q(x_1, x_2, t)]} \right\}^{\frac{1}{n}} e^{i[A_0(t-t_0)+\phi_0]}$ $\times e^{i\left[\frac{v_1(x_1-x_{01})}{2\alpha_1} + \frac{v_2(x_2-x_{02})}{2\alpha_2} - \frac{v_1^2\alpha_2+v_2^2\alpha_1}{4\alpha_1\alpha_2}(t-t_0)\right]},$ $q(x_1, x_2, t) = n\sqrt{\dfrac{A_0}{c_1^2 a_1+c_2^2 a_2}}\left\{c_1[x_1-(x_{01}+v_1 t)] + c_2[x_2-(x_{02}+v_2 t)]\right\}$	$a_3 = \gamma\,a_{03}; \gamma = -\cos^2(\theta),$ $0 < \theta < \pi/2,$ $a_{03} = a_2^2(n+1)/[A_0(n+2)^2].$ $A_0(c_1^2\,\alpha_1 + c_2^2\,\alpha_2) > 0.$ $A_0(n+1) > 0, m = 2n, x_{01}, x_{02}, t_0,$ $A_0,$ and ϕ_0 are arbitrary real constants	Moving 2D flat-top soliton	(6.40)
2.	$\Phi(x_1, x_2, x_3, t) = \left\{ \dfrac{A_0(n+2)}{a_2 + a_2 \sin(\theta)\cosh[q(x_1, x_2, x_3, t)]} \right\}^{\frac{1}{n}} e^{i[A_0(t-t_0)+\phi_0]}$ $\times e^{i\left[\frac{v_1(x_1-x_{01})}{2\alpha_1} + \frac{v_2(x_2-x_{02})}{2\alpha_2} + \frac{v_3(x_3-x_{03})}{2\alpha_3} - \frac{v_1^2\alpha_2\alpha_3+v_2^2\alpha_1\alpha_3+v_3^2\alpha_1\alpha_2}{4\alpha_1\alpha_2\alpha_3}(t-t_0)\right]}$ $q(x_1, x_2, x_3, t) = n\sqrt{\dfrac{A_0}{c_1^2 a_1+c_2^2 a_2+c_3^2 a_3}}\{c_1[x_1-(x_{01}+v_1 t)]$ $+ c_2[x_2-(x_{02}+v_2 t)] + c_3[x_3-(x_{03}+v_3 t)]\}$	$a_3 = \gamma\,a_{03}; \gamma = -\cos^2(\theta),$ $0 < \theta < \pi/2,$ $a_{03} = a_2^2(n+1)/[A_0(n+2)^2].$ $A_0(c_1^2\,\alpha_1 + c_2^2\,\alpha_2 + c_3^2\,\alpha_3) > 0.$ $A_0(n+1) > 0, m = 2n, x_{01}, x_{02}, x_{03},$ $t_0, A_0,$ and ϕ_0 are arbitrary real constants	Moving 3D flat-top soliton	(6.41)

NLSE in (2 + 1)-Dimensions with $\Phi_{x_1 x_2}$ Term

Transformation: $\Phi(x_1, x_2, t) = \sqrt{\dfrac{a_2}{b_4}}\,\Phi[X(x_1, x_2), t], \quad X(x_1, x_2)$

$$= c_0 - \frac{1}{2b}\left(b_3 c_1 - \sqrt{4a_1 b - 4b b_2 c_1^2 + b_3^2 c_1^2}\right)x_1 + c_1 x_2$$

Equation: $i\,\Phi_t + b\,\Phi_{x_1, x_1} + b_2\,\Phi_{x_2, x_2} + b_3\,\Phi_{x_1, x_2} + b_4\,|\Phi|^2\,\Phi = 0$

#	Example	Conditions	Name	Equation #
1.	$\Phi(x_1,\ x_2,\ t) = A_0 \sqrt{\dfrac{2 a_1}{b_4}}$ $\times \operatorname{sech}\left\{A_0 \left[c_0 - \dfrac{\left(b_3 c_1 - \sqrt{4 a_1 b_1 - 4 b_1 b_2 c_1^2 + b_3^2 c_1^2}\right)(x_1 - x_{01})}{2 b_1}\right.\right.$ $\left.\left. + c_1 (x_2 - x_{02})\right]\right\} e^{i\,[a_1 A_0^2\,(t - t_0) + \phi_0]}$	$a_1 b_4 > 0,$ $(4 a_1 b_1 - 4 b_1 b_2 c_1^2 + b_3^2 c_1^2) > 0,$ A_0, t_0 and ϕ_0 are arbitrary real constants	Bright soliton	(6.45)
2.	$\Phi(x_1,\ x_2,\ t) = A_0 \sqrt{\dfrac{-2 a_1}{b_4}}$ $\times \tanh\left\{A_0 \left[c_0 + \dfrac{\left(\sqrt{4 a_1 b_1 - 4 b_1 b_2 c_1^2 + b_3^2 c_1^2} - b_3 c_1\right)(x_1 - x_{01})}{2 b_1}\right.\right.$ $\left.\left. + c_1 (x_2 - x_{02})\right]\right\} e^{i\,[2 a_1 A_0^2\,(t - t_0) + \phi_0]}$	$a_1 b_4 < 0,$ $(4 a_1 b_1 - 4 b_1 b_2 c_1^2 + b_3^2 c_1^2) > 0,$ A_0, t_0 and ϕ_0 are arbitrary real constants	Dark soliton	(6.46)

6.8 (N + 1)-Dimensional Isotropic NLSE with Cubic Nonlinearity in Polar Coordinate System

If $\psi(x, t)$ is a solution to the fundamental NLSE in 1D (2.160),

$$i \, \psi_t + a_1 \, \psi_{xx} + a_2 \, |\psi|^2 \, \psi = 0,$$

then

$$\Phi(r, t) = A(r, t) \, e^{i \, B(r, t)} \, \psi[R(r, t), T(r, t)] \qquad (6.47)$$

is a solution to

$$i \, \Phi_t + b_1(r, t) \left(\Phi_{rr} + \frac{N - 1}{r} \, \Phi_r \right) + b_2(r, t) \, |\Phi|^2 \, \Phi + [b_{3r}(r, t) + i \, b_{3i}(r, t)]\Phi = 0, \qquad (6.48)$$

where

$$T(r, t) = g_1(t), \qquad (6.49)$$

$$A(r, t) = \frac{g_2(t) \, r^{\frac{1-N}{2}}}{\sqrt{R_r(r, t)}}, \qquad (6.50)$$

$$B(r, t) = g_3(t) - \int \frac{R_t(r, t) \, R_r(r, t)}{2 \, a_1 \, g_1'(t)}, \qquad (6.51)$$

$$b_1(r, t) = \frac{a_1 \, g_1'(t)}{R_r^2(r, t)}, \qquad (6.52)$$

$$b_2(r, t) = \frac{a_2 \, g_1'(t)}{A^2(r, t)}, \qquad (6.53)$$

$$b_{3i}(r, t) = \frac{R_{rt}(r, t)}{R_r(r, t)} - \frac{g_2'(t)}{g_2(t)}, \qquad (6.54)$$

$$
\begin{aligned}
b_{3r}(r, t) = \frac{1}{4 \, a_1 \, g_1'^2(t)} \Big\{ & 2 \, g_1''(t) \int R_t(r, t) \, R_r(r, t) \, dr \\
& - 2 \, g_1'(t) \int [R_{tt}(r, t) \, R_r(r, t) + R_t(r, t) \, R_{rt}(r, t)] \, dr + g_1'(t) \, R_t^2(r, t) \\
& + \frac{a_1^2 \, g_1'^3(t)}{r^2 \, R_r^4(r, t)}[(N - 1) \, (N - 3) \, R_r^2(r, t) \\
& - 3 \, r^2 \, R_{rr}^2(r, t) + 2 \, r^2 \, R_r(r, t) \, R_{rrr}(r, t)] \Big\},
\end{aligned}
\qquad (6.55)
$$

$g_1(t)$, $g_2(t)$, and $R(r, t)$ are arbitrary real functions.

6.8.1 Angular Dependence

If $\psi(x, t)$ is a solution to the fundamental NLSE (2.160), then

$$\Phi(r, \theta, t) = r^{(1-N)/2} \, e^{i \frac{1}{2} \sqrt{-(N-1)(N-3)} \, \theta} \psi(r, t) \tag{6.56}$$

is a solution to the NLSE

$$i \, \Phi_t + a_1 \left(\Phi_{rr} + \frac{N-1}{r} \, \Phi_r + \frac{1}{r^2} \, \Phi_{\theta\theta} \right) + a_2 \, r^{N-1} \, |\Phi|^2 \, \Phi = 0. \tag{6.57}$$

This is obtained from the previous section with the special choices: $g_1 = t$, $g_2 = 1$, $g_3 = 0$, $R = r$.

Remarks:
1. The angular term $\Phi_{\theta\theta}/r^2$ vanishes in 1D and 3D.
2. The prefactor $r^{(1-N)/2}$ in Φ diverges at $r = 0$ as $r^{-1/2}$ and r^{-1} for 2D and 3D, respectively. This divergence may be removed with certain solutions of the fundamental NLSE, ψ, such as the tanh(r) solution:

***Example* 1. General Case**
Given $\psi(x, t) = A_0 \sqrt{\frac{-2a_1}{a_2}} \, \tanh[A_0 \, (x - x_0)] \, e^{-i \, [2 \, a_1 \, A_0^2 \, (t - t_0) + \phi_0]}$ is a solution to (2.160), then

$$\Phi(r, \theta, t) = r^{(1-N)/2} \, e^{i \frac{1}{2} \sqrt{-(N-1)(N-3)} \, \theta} \left[A_0 \sqrt{\frac{-2a_1}{a_2}} \, e^{-i \left(2a_1 \, A_0^2 \, t + \phi_0\right)} \tanh(A_0 \, r) \right] \tag{6.58}$$

is a solution to (6.57), where $a_1 \, a_2 < 0$, A_0 and ϕ_0 are arbitrary real constants.

Example* 2. 2D *vortex soliton
(Figure 6.11)

$$\Phi(r, \theta, t) = r^{-1/2} \, e^{i \sqrt{a_1} \, \theta} \left[A_0 \sqrt{\frac{-2a_1}{a_2}} \, e^{-i \left(2a_1 \, A_0^2 \, t + \phi_0\right)} \tanh(A_0 \, r) \right] \tag{6.59}$$

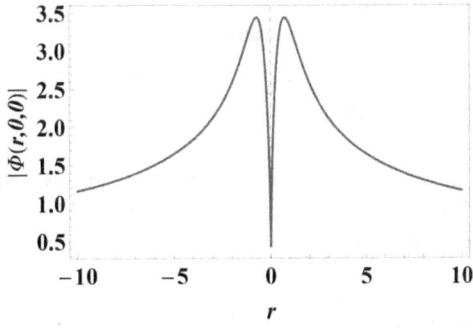

Figure 6.11. Vortex soliton (6.59) at $\theta = t = 0$, with $a_1 = A_0 = 3/2$, $a_2 = -1/2$, and $\phi_0 = 0$.

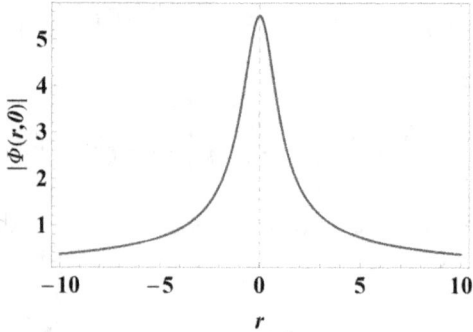

Figure 6.12. Plot of solution (6.61) at $t = 0$ with $a_1 = A_0 = 3/2$, $a_2 = -1/2$, and $\phi_0 = 0$.

is a solution to

$$i\,\Phi_t + a_1\left(\Phi_{rr} + \frac{1}{r}\,\Phi_r - \frac{1}{4r^2}\right) + a_2\,r\,|\Phi|^2\,\Phi = 0, \tag{6.60}$$

where $a_1\,a_2 < 0$, A_0 and ϕ_0 are arbitrary real constants.

***Example* 3. 3D**
(Figure 6.12)

$$\Phi(r,\,t) = r^{-1}\left[A_0\,\sqrt{\frac{-2a_1}{a_2}}\;e^{-i\left(2a_1\,A_0^2\,t+\phi_0\right)}\tanh(A_0\,r)\right] \tag{6.61}$$

is a solution to

$$i\,\Phi_t + a_1\left(\Phi_{rr} + \frac{2}{r}\,\Phi_r\right) + a_2\,r^2\,|\Phi|^2\,\Phi = 0, \tag{6.62}$$

where $a_1\,a_2 < 0$, A_0 and ϕ_0 are arbitrary real constants.

6.8.2 Constant Dispersion and Real Potential

If $\psi(x,\,t)$ is a solution to the fundamental NLSE (2.160), $i\,\psi_t + a_1\,\psi_{xx} + a_2\,|\psi|^2\,\psi = 0$, then

$$\Phi(r,\,t) = c_0\,g_1^{1/4}(t)\,r^{\frac{1-N}{2}}\,e^{i\left\{g_4(t) - \frac{\left[4\sqrt{b_{10}}\,g_1'(t)\,g_2'(t) + \sqrt{a_1}\,g_1''(t)\,r\right]r}{8\,b_{10}\,\sqrt{a_1}\,g_1'(t)}\right\}}$$

$$\times\,\psi\left[g_2(t) + \sqrt{\frac{a_1\,g_1'(t)}{b_{10}}}\,r,\,g_1(t)\right] \tag{6.63}$$

is a solution to

$$
i \, \Phi_t + b_{10} \, \Phi_{rr} + \frac{b_{10} \, (N-1)}{r} \, \Phi_r + \frac{a_2 \, \sqrt{g_1'(t)}}{c_0^2 \, r^{1-N}} \, |\Phi|^2 \, \Phi
$$

$$
+ \left\{ \frac{b_{10} \, (N-1) \, (N-3)}{4 \, r^2} + \frac{g_2'^2(t)}{4 \, a_1 \, g_1'(t)} + g_4'(t) + \frac{\left[g_2'(t) \, g_1''(t) - g_1'(t) \, g_2''(t) \right] r}{2 \, \sqrt{a_1 \, b_{10}} \, g_1'^{3/2}(t)} \right. \tag{6.64}
$$

$$
\left. + \frac{\left[3 \, g_1''^2(t) - 2 \, g_1'(t) \, g_1'''(t) \right] r^2}{16 \, b_{10} \, g_1'^2(t)} \right\} \Phi = 0,
$$

where

$$
T(r, \, t) = g_1(t), \tag{6.65}
$$

$$
R(r, \, t) = g_2(t) + \sqrt{\frac{a_1 \, g_1'(t)}{b_{10}}} \, r, \tag{6.66}
$$

$$
A(r, \, t) = g_3(t) \, r^{\frac{1-N}{2}}, \tag{6.67}
$$

$$
B(r, \, t) = g_4(t) - \frac{\left[4 \, \sqrt{b_{10} \, g_1'(t)} \, g_2'(t) + \sqrt{a_1} \, g_1''(t) \, r \right] r}{8 \, b_{10} \, \sqrt{a_1} \, g_1'(t)}, \tag{6.68}
$$

$$
b_1(r, \, t) = b_{10}, \tag{6.69}
$$

$$
b_2(r, \, t) = \frac{a_2 \, g_1'(t)}{A^2(r, \, t)}, \tag{6.70}
$$

$$
b_3(r, \, t) = \frac{b_{10} \, (N-1) \, (N-3)}{4 \, r^2} + \frac{g_2'^2(t)}{4 \, a_1 \, g_1'(t)} + g_4'(t) + \frac{\left[g_2'(t) \, g_1''(t) - g_1'(t) \, g_2''(t) \right] r}{2 \, \sqrt{a_1 \, b_{10}} \, g_1'^{3/2}(t)}
$$

$$
+ \frac{\left[3 \, g_1''^2(t) - 2 \, g_1'(t) \, g_1'''(t) \right] r^2}{16 \, b_{10} \, g_1'^2(t)}, \tag{6.71}
$$

$$
g_3(t) = c_0 \, g_1'^{1/4}(t), \tag{6.72}
$$

$g_1(t)$, $g_2(t)$, and $g_4(t)$ are arbitrary real functions, c_0 and b_{10} are arbitrary real constants.

Example 1: *bright soliton*

Given $\psi(x,\ t) = A_0 \sqrt{\frac{2\,a_1}{a_2}}\ \mathrm{sech}[A_0\ (x - x_0)]\ e^{i\ [a_1\ A_0^2\ (t-t_0)+\phi_0]}$ is a solution to (2.160), then

$$\Phi(r,\ t) = c_0\ A_0\ \sqrt{\frac{2\,a_1}{a_2}}\ g_1'^{1/4}(t)\ r^{\frac{1-N}{2}}\ \mathrm{sech}\left\{A_0\left[g_2(t) + \sqrt{\frac{a_1\ g_1'(t)}{b_{10}}}\ r\right]\right\}$$

$$\times\ e^{i\left\{A_0^2\ a_1\ [g_1(t)-t_0]+g_4(t)-\frac{g_2'(t)}{2\ \sqrt{a_1\ b_{10}\ g_1'(t)}}\ r-\frac{g_1''(t)}{8\ b_{10}\ g_1'(t)}\ r^2+\phi_0\right\}} \tag{6.73}$$

is a solution to (6.64), where $a_1\ a_2 > 0$, $a_1\ b_{10} > 0$, A_0, t_0, ϕ_0 are arbitrary real constants.

Example 2: *dark soliton*

Given $\psi(x,\ t) = A_0 \sqrt{\frac{-2\,a_1}{a_2}}\ \tanh[A_0\ (x - x_0)]\ e^{-i\ [2\,a_1\ A_0^2\ (t-t_0)+\phi_0]}$ is a solution to (2.160), then

$$\Phi(r,\ t) = c_0\ A_0\ \sqrt{\frac{-2\,a_1}{a_2}}\ g_1'^{1/4}(t)\ r^{\frac{1-N}{2}}\ \tanh\left\{A_0\left[g_2(t) + \sqrt{\frac{a_1\ g_1'(t)}{b_{10}}}\ r\right]\right\}$$

$$\times\ e^{-i\left\{2\,A_0^2\ a_1\ [g_1(t)-t_0]-g_4(t)+\frac{g_2'(t)}{2\ \sqrt{a_1\ b_{10}\ g_1'(t)}}\ r+\frac{g_1''(t)}{8\ b_{10}\ g_1'(t)}\ r^2+\phi_0\right\}} \tag{6.74}$$

is a solution to (6.64), where $a_1\ a_2 < 0$, $a_1\ b_{10} > 0$, A_0, t_0, and ϕ_0 are arbitrary real constants.

6.9 Summary of Section 6.8

$(N+1)$-Dimensional NLSE with Cubic Nonlinearity in Polar Coordinate System

Transformation: $\Phi(r, t) = A(r, t) \, e^{i\, B(r, t)} \, \psi[R(r, \hat{t}_1, \hat{t}_2), T(r, t)]$

Equation: $i\, \Phi_t + b_1(r, t)\left(\Phi_{rr} + \frac{N-1}{r}\, \Phi_r\right) + b_2(r, t)\, |\Phi|^2\, \Phi + [b_{3r}(r, t) + i\, b_{3i}(r, t)]\Phi = 0$

Angular dependence

Transformation: $\Phi(r, \theta, t) = r^{(1-N)/2}\, e^{i\,\frac{1}{2}\sqrt{-(N-1)(N-3)}\;\theta}\, \psi(r, t)$

Equation (1): $i\, \Phi_t + a_1\left(\Phi_{rr} + \frac{N-1}{r}\, \Phi_r + \frac{1}{r^2}\, \Phi_{\theta\theta}\right) + a_2\, r^{N-1}\, |\Phi|^2\, \Phi = 0$

#		Name	Equation #
Example: general case	**Conditions**	**Name**	**Equation #**
1. $\Phi(r, \theta, t) = r^{(1-N)/2}\, e^{i\,\frac{1}{2}\sqrt{-(N-1)(N-3)}\;\theta}$ $\times \left[A_0\sqrt{\frac{-2a_1}{a_2}}\; e^{-i\,(2a_1\, A_0^2\, t+\phi_0)}\tanh(A_0\, r)\right]$	$a_1\, a_2 < 0, A_0$ and ϕ_0 are arbitrary real constants	Dark soliton	(6.58)

Equation (2): $i\, \Phi_t + a_1\left(\Phi_{rr} + \frac{1}{r}\, \Phi_r - \frac{1}{4r^2}\right) + a_2\, r\, |\Phi|^2\, \Phi = 0$

#		Name	Equation #
Example: 2D	**Conditions**	**Name**	**Equation #**
1. $\Phi(r, \theta, t) = r^{-1/2}\, e^{i\,\sqrt{a_1}\,\theta}\left[A_0\sqrt{\frac{-2a_1}{a_2}}\; e^{-i\,(2a_1\, A_0^2\, t+\phi_0)}\tanh(A_0\, r)\right]$	$a_1\, a_2 < 0, A_0$ and ϕ_0 are arbitrary real constants	Dark (vortex) soliton	(6.59)

(Continued)

Equation (3): $i\,\Phi_t + a_1\,(\Phi_{rr} + \frac{2}{r}\,\Phi_r) + a_2\,r^2\,|\Phi|^2\,\Phi = 0$

#	Example: 3D	Conditions	Name	Equation #
1.	$\Phi(r,t) = r^{-1}\,[A_0\,\sqrt{\dfrac{2a_1}{a_2}}\,e^{-i\,(2a_1\,A_0^2\,t + \phi_0)}\tanh(A_0\,r)]$	$a_1\,a_2 < 0, A_0$ and ϕ_0 are arbitrary real constants	Dark soliton	(6.61)

Constant dispersion and real potential

Transformation: $\Phi(r,t) = c_0\,g_1^{1/4}(t)\,r^{\frac{1-N}{2}}\,e^{\,i\,\{g_4(t) - \frac{[4\sqrt{b_{10}\,g_1(t)}\,g_2(t) + \sqrt{a_1}\,g_1''(t)]\,r^2}{8\,b_{10}\,\sqrt{a_1}\,g_1(t)}\}}\;\psi\,[g_2(t) + \sqrt{\dfrac{a_1\,g_1'(t)}{b_{10}}}\,r,\;g_1(t)]$

$$i\,\Phi_t + b_{10}\,\Phi_{rr} + \frac{b_{10}\,(N-1)}{r}\,\Phi_r + \frac{a_2\sqrt{g_1'(t)}}{c_0^2\,r^{1-N}}\,|\Phi|^2$$

Equation: $\Phi + \{\dfrac{b_{10}\,(N-1)\,(N-3)}{4\,r^2} + \dfrac{g_2'^{\,2}(t)}{4\,a_1\,g_1'(t)} + g_4'(t) + \dfrac{[g_2'(t)\,g_1''(t) - g_1'(t)\,g_2''(t)]\,r}{2\,\sqrt{a_1\,b_{10}\,g_1'^{\,3/2}(t)}}$

$+ \dfrac{[3\,g_1''^{\,2}(t) - 2\,g_1'(t)\,g_1'''(t)]\,r^2}{16\,b_{10}\,g_1'^{\,2}(t)}\}\,\Phi = 0$

#	Example	Conditions	Name	Equation #
1.	$\Phi(r,t) = c_0\,A_0\,\sqrt{\dfrac{2a_1}{a_2}}\,g_1^{1/4}(t)\,r^{\frac{1-N}{2}}\,\text{sech}\{A_0\,[g_2(t) + \sqrt{\dfrac{a_1\,g_1'(t)}{b_{10}}}\,r]\}$ $\times\,e^{\,i\,\{A_0^2\,a_1\,[g_2(t) - t_0] + g_4(t) - \frac{g_2'(t)}{2\sqrt{a_1\,b_{10}\,g_1'(t)}}\,r - \frac{g_1''(t)}{8\,b_{10}\,g_1'(t)}\,r^2 + \phi_0\}}$	$a_1\,a_2 > 0, a_1\,b_{10} > 0, A_0, t_0, \phi_0$ are arbitrary real constants	Bright soliton	(6.73)
2.	$\Phi(r,t) = c_0\,A_0\,\sqrt{\dfrac{-2a_1}{a_2}}\,g_1^{1/4}(t)\,r^{\frac{1-N}{2}}\,\tanh\{A_0\,[g_2(t) + \sqrt{\dfrac{a_1\,g_1'(t)}{b_{10}}}\,r]\}$ $\times\,e^{\,-i\,\{2\,A_0^2\,a_1\,[g_2(t) - t_0] - g_4(t) + \frac{g_2'(t)}{2\sqrt{a_1\,b_{10}\,g_1'(t)}}\,r + \frac{g_1''(t)}{8\,b_{10}\,g_1'(t)}\,r^2 + \phi_0\}}$	$a_1\,a_2 < 0, a_1\,b_{10} > 0, A_0, t_0,$ and ϕ_0 are arbitrary real constants	Dark soliton	(6.74)

6.10 Power Series Solutions to (2 + 1)-Dimensional NLSE with Cubic Nonlinearity in Polar Coordinate System

The function

$$\Phi(r, \theta, t) = Z(r)\, e^{i\,(\lambda\, t + \alpha\, \theta)} \tag{6.75}$$

with

$$
\begin{aligned}
Z(r) = a_0 &+ \left(\frac{5\, b_0}{4}\right) r + \left(\frac{25\, [-a_2\, (a_0)^3 - a_1\, b_0 + a_1\, a_0\, \alpha^2 + a_0\, \lambda]}{32\, a_1}\right) r^2 \\
&+ \left(\frac{15\, [a_2\, (a_0)^2\, (a_0 - 3\, b_0) - 3\, a_1\, a_0\, \alpha^2 + a_1\, b_0\, (2 + \alpha^2) + (b_0 - a_0)\, \lambda]}{64\, a_1}\right) r^3 \\
&+ \frac{5}{512\, a_1^2} \Big(a_1^2\, [a_0\, \alpha^2\, (68 + 7\, \alpha^2) \\
&\quad - 3\, b_0\, (12 + 13\, \alpha^2)] + a_1\, \{-a_2\, a_0\, [-33\, a_0\, b_0 + 42\, (b_0)^2 + 2\, (a_0)^2\, (9 + 14\, \alpha^2)] \\
&\quad + [-11\, b_0 + 2\, a_0\, (9 + 7\, \alpha^2)]\, \lambda\} + 7\, a_0\, [3\, a_2^2\, (a_0)^4 - 4\, a_2\, (a_0)^2\, \lambda + \lambda^2]\Big) r^4 + O^5(r)
\end{aligned}
\tag{6.76}
$$

is a stationary solution to

$$i\, \Phi_t + a_1 \left(\Phi_{rr} + \frac{1}{r}\, \Phi_r + \frac{1}{r^2}\, \Phi_{\theta\theta}\right) + a_2\, |\Phi|^2\, \Phi = 0, \tag{6.77}$$

where λ, α, a_0, and b_0 are arbitrary real constants.

The solution is obtained using an *iterative power series* (IPS) method [10, 11], which is briefly described as follows: The function $Z(r)$ obeys the ordinary differential equation

$$-\lambda\, Z(r) + a_1 \left[Z''(r) + \frac{1}{r}\, Z'(r) - \frac{\alpha^2}{r^2}\, Z(r)\right] + a_2\, Z^3(r) = 0. \tag{6.78}$$

A convergent power series solution is obtained by the following algorithm·

1. Set initial values a_0 and b_0.
2. Expand $Z(r)$ in power series around the arbitrary real r_0: $Z(r) = a_0 + b_0\, (r - r_0) + \sum_{j=2}^{n_{max}} c_j (r - r_0)^j$.
3. Substitute in (6.76) to obtain the recursion relation for c_n in terms of a_0 and b_0.
4. Calculate $Z(\Delta)$ and $Z'(\Delta)$, where $\Delta = (r - r_0)/I$, and I is an integer larger than 1.
5. Assign: $a_0 = Z(\Delta)$ and $b_0 = Z'(\Delta)$.
6. Obtain c_n in terms of a_0 and b_0.
7. Repeat steps 2-6 I times.
8. At the Ith step, a_0 will correspond to the power series of Z.

The solution above is obtained with $I = 4$, $n_{max} = 2$, and for arbitrary $a_1, a_2, \lambda, \alpha, a_0$, and b_0.

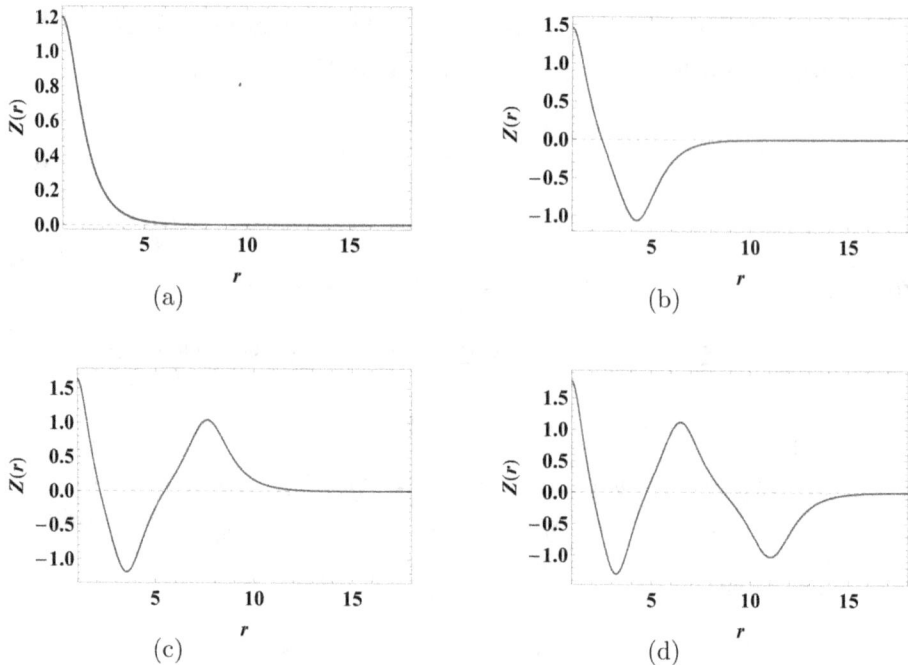

Figure 6.13. Stationary power series solutions of (6.78) with different number of nodes. (a) nodeless solution with $a_0 = 1.198\,664\,4033$, (b) single-node solution with $a_0 = 1.460\,514\,6251$, root at $r = 2.46$, (c) double-node solution with $a_0 = 1.632\,667\,0699$, roots at $r = 2.17, 5.46$, and (d) triple-node solution with $a_0 = 1.764\,200\,3085$, roots at $r = 2.02, 4.75, 8.64$. IPS parameters used are: $n_{\max} = 2$, $I = 1700$, and $\Delta = 0.01$.

6.10.1 Family of Infinite Number of Localized Solutions

Using the IPS method, described above, we tune the parameter a_0 to obtain a family of infinite number of solutions differing by the number of nodes. In Figure 6.13, we present some plots showing the nodeless, single-node, double-node, and triple-node solutions of (6.78). Other parameters are fixed to $b_0 = 0$, $a_1 = 1$, $a_2 = 2$, $\lambda = 1$, and $\alpha = 0$.

References

[1] Sulem C and Sulem P L 1999 *Nonlinear Schrödinger equations: Self-focusing Instability and Wave Collapse* (New York: Springer)

[2] Bakkali-Hassani B, Maury C, Zou Y-Q, Cerf É, Saint-Jalm R, Castilho P C M, Nascimbene S, Dalibard J and Beugnon J 2021 Realization of a townes soliton in a two-component planar Bose gas *Phys. Rev. Lett.* **127** 023603

[3] Pethick C J and Smith H 2008 *Bose–Einstein Condensation in Dilute Gases* (Cambridge: Cambridge University Press)

[4] Montesinos G D, Pérez-García V M and Torres P J 2004 Stabilization of solitons of the multidimensional nonlinear Schrödinger equation: matter-wave breathers *Physica* **D191** 193–210

[5] Brand J and Reinhardt W P 2001 Generating ring currents, solitons and svortices by stirring a Bose–Einstein condensate in a toroidal trap *J. Phys. B: At. Mol. Opt. Phys.* **34** L113

[6] Brand J and Reinhardt W P 2002 Solitonic vortices and the fundamental modes of the 'snake instability': possibility of observation in the gaseous Bose–Einstein condensate *Phys. Rev. A* **65** 043612

[7] Anderson B P, Haljan P C, Regal C A, Feder D L, Collins L A, Clark C W and Cornell E A 2001 Watching dark solitons decay into vortex rings in a Bose-Einstein condensate *Phys. Rev. Lett.* **86** 2926

[8] Al Khawaja U and Stoof H 2001 Skyrmions in a ferromagnetic Bose-Einstein condensate *Nature* **411** 918–20

[9] Malomed B A, Mihalache D, Wise F and Torner L 2005 Spatiotemporal optical solitons *J. Opt. B: Quantum Semiclass. Opt.* **7** R53

[10] Al Khawaja U and Al-Mdallal Q M 2018 Convergent power series of and solutions to nonlinear differential equations *Int. J. Differ. Equ.* **2018** 1–10

[11] Al Sakkaf L Y, Al-Mdallal Q M and Al Khawaja U 2018 A numerical algorithm for solving higher-order nonlinear BVPs with an application on fluid flow over a shrinking permeable infinite long cylinder *Complexity* **2018** 1–11

IOP Publishing

Handbook of Exact Solutions to the Nonlinear Schrödinger Equations (Second Edition)

Usama Al Khawaja and Laila Al Sakkaf

Chapter 7

Coupled Nonlinear Schrödinger Equations

A Glance at Chapter 7

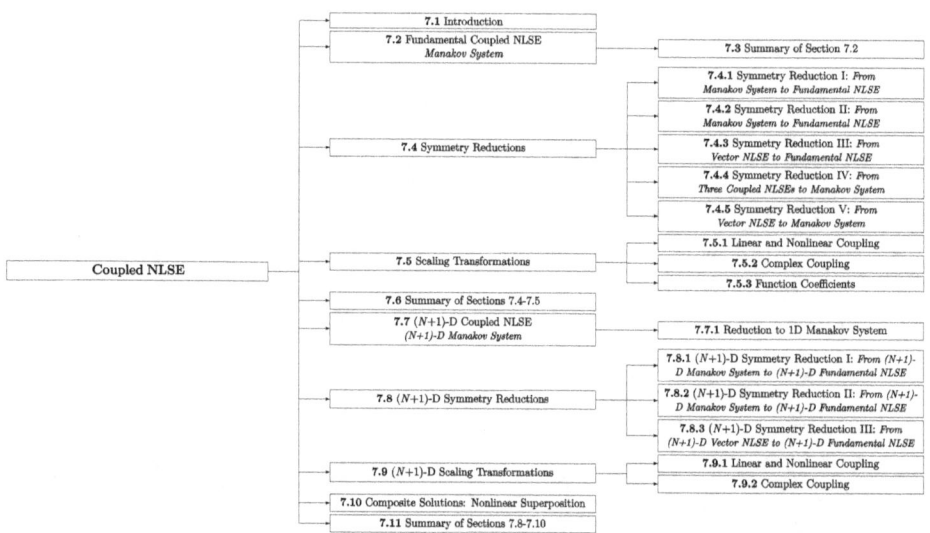

A Statistical View of Chapter 7

	Equation	Solutions
1	$i\psi_{1t} + b_0\psi_{1xx} + (b_1|\psi_1|^2 + b_2|\psi_2|^2)\psi_1 = 0,$ $i\psi_{2t} + c_0\psi_{2xx} + (c_1|\psi_1|^2 + c_2|\psi_2|^2)\psi_2 = 0$ *Manakov system*	28
2	$i\psi_{1t} + b_0\psi_{1xx} + (c_1 + c_2|\sigma|^2)|\psi_1|^2\psi_1 = 0$ *reduced from Manakov system*	0
3	$i\psi_{1t} + b_0\psi_{1xx} - (c_1 + c_2)|\psi_1|^2\psi_1 = 0$ *reduced from Manakov system*	0
4	$i\psi_{1t} + b_{0j}\psi_{1xx} + \sum_{k=1}^{N} b_{1j}|\sigma_j|^2|\psi_1|^2\psi_1 = 0, j = 1, 2, \ldots, N$	0
5	$i\Phi_{1t} + \Phi_{1xx} + (g_1|\Phi_1|^2 - g_2|\Phi_2|^2)\Phi_1 + g_0(g_1 + g_2)\Phi_1 - 2g_0 g_2\Phi_2 = 0,$ $i\Phi_{2t} + \Phi_{2xx} + (g_1|\Phi_1|^2 - g_2|\Phi_2|^2)\Phi_2 - g_0(g_1 + g_2)\Phi_2 + 2g_0 g_1\Phi_1 = 0$	0
6	$i\psi_{1t} + a_1\psi_{1xx} + (b_1|\psi_1|^2 + b_3|\psi_3|^2)\psi_1 = 0,$ $i\psi_{3t} + a_3\psi_{3xx} + (d_1|\psi_1|^2 + d_3|\psi_3|^2)\psi_3 = 0$ *three coupled NLSEs → Manakov system*	2
7	$i\psi_{1t} + a_1\psi_{1xx} + (b_1|\psi_1|^2 + b_2|\psi_2|^2)\psi_1 = 0,$ $i\psi_{2t} + a_2\psi_{2xx} + (c_1|\psi_1|^2 + c_2|\psi_2|^2)\psi_2 = 0$ *vector NLSE → Manakov system*	0
8	$i\Phi_{1t} + \Phi_{1xx} + (g_1|\Phi_1|^2 - g_2|\Phi_2|^2)\Phi_1 = 0,$ $i\Phi_{2t} + \Phi_{2xx} + (g_1|\Phi_1|^2 - g_2|\Phi_2|^2)\Phi_2 = 0$	1
9	$i\Phi_{1t} + \Phi_{1xx} + 2(a_{11}|\Phi_1|^2 + a_{12}|\Phi_2|^2)\Phi_1 + 2(b_{11}\Phi_1\Phi_2^* + b_{12}\Phi_2\Phi_1^*)\Phi_1 = 0,$ $i\Phi_{2t} + \Phi_{2xx} + 2(a_{21}|\Phi_1|^2 + a_{22}|\Phi_2|^2)\Phi_2 + 2(b_{21}\Phi_1\Phi_2^* + b_{22}\Phi_2\Phi_1^*)\Phi_2 = 0$	0
10	$i\Phi_{1t} + \Phi_{1xx} - 2(a + b)(|\Phi_1|^2 + |\Phi_2|^2)\Phi_1 + 2((a + i\,b)\Phi_1\Phi_2^* + (a - i\,b)\Phi_2\Phi_1^*)\Phi_1$ $= 0,$ $i\Phi_{2t} + \Phi_{2xx} - 2(a + b)(|\Phi_1|^2 + |\Phi_2|^2)\Phi_2 + 2((a + i\,b)\Phi_1\Phi_2^* + (a - i\,b)\Phi_2\Phi_1^*)\Phi_2$ $= 0$	0

11	$i\psi_{1t} + \sum_{k=1}^N b_{0k}\,\psi_{1x_kx_k} + (b_1\,|\psi_1|^2 + b_2\,|\psi_2|^2)\,\psi_1 = 0,$ $i\psi_{2t} + \sum_{k=1}^N c_{0k}\,\psi_{2x_kx_k} + (c_1\,|\psi_1|^2 + c_2\,|\psi_2|^2)\,\psi_2 = 0$	1
12	$i\psi_{1t} + \sum_{k=1}^N c_{0k}\,\psi_{1x_kx_k} + (c_1 + c_2)\,|\psi_1|^2\,\psi_1 = 0$ *three coupled NLSEs → Mana*	0
13	$i\psi_{1t} - \sum_{k=1}^N c_{0k}\,\psi_{1x_kx_k} - 2\,(c_1 + c_2)\,|\psi_1|^2\,\psi_1 = 0$	0
14	$i\psi_{1t} + \sum_{k=1}^N b_{10k}\,\psi_{1x_kx_k} + \sum_{k=1}^N b_{1j}\,|\sigma_{j+1}|^2\,|\psi_1|^2\,\psi_1 = 0$	0
15	$i\Phi_{1t} + \sum_{k=1}^N \Phi_{1x_kx_k} + (g_1\,|\Phi_1|^2 - g_2\,|\Phi_2|^2)\,\Phi_1 + g_0\,(g_1 + g_2)\,\Phi_1 - 2\,g_0\,g_2\,\Phi_2 = 0,$ $i\Phi_{2t} + \sum_{k=1}^N \Phi_{2x_kx_k} + (g_1\,|\Phi_1|^2 - g_2\,|\Phi_2|^2)\,\Phi_2 - g_0\,(g_1 + g_2)\,\Phi_2 + 2\,g_0\,g_1\,\Phi_1 = 0$	0
16	$i\Phi_{1t} + \sum_{k=1}^N \Phi_{1x_kx_k} + (g_1\,|\Phi_1|^2 - g_2\,|\Phi_2|^2)\,\Phi_1 = 0,$ $i\Phi_{2t} + \sum_{k=1}^N \Phi_{2x_kx_k} + (g_1\,|\Phi_1|^2 - g_2\,|\Phi_2|^2)\,\Phi_2 = 0$	0
17	$i\Phi_{1t} + \sum_{k=1}^N \Phi_{1x_kx_k} + 2\,(a_{11}\,|\Phi_1|^2 + a_{12}\,|\Phi_2|^2)\,\Phi_1 + 2\,(b_{11}\,\Phi_1\,\Phi_2^* + b_{12}\,\Phi_2\,\Phi_1^*)\,\Phi_1 = 0,$ $i\Phi_{2t} + \sum_{k=1}^N \Phi_{2x_kx_k} + 2\,(a_{21}\,|\Phi_1|^2 + a_{22}\,|\Phi_2|^2)\,\Phi_2 + 2\,(b_{21}\,\Phi_1\,\Phi_2^* + b_{22}\,\Phi_2\,\Phi_1^*)\,\Phi_2 = 0$	0
18	$i\Phi_{1t} + \sum_{k=1}^N \Phi_{1x_kx_k} - 2(a + b)\,(|\Phi_1|^2 + |\Phi_2|^2)$ $\Phi_1 + 2\,((a + i\,b)\,\Phi_1\,\Phi_2^* + (a - i\,b)\,\Phi_2\,\Phi_1^*)\,\Phi_1 = 0,$ $i\Phi_{2t} + \sum_{k=1}^N \Phi_{2x_kx_k} - 2(a + b)\,(|\Phi_1|^2 + |\Phi_2|^2)$ $\Phi_2 + 2\,((a + i\,b)\,\Phi_1\,\Phi_2^* + (a - i\,b)\,\Phi_2\,\Phi_1^*)\,\Phi_2 = 0$	0
19	$i\frac{\partial}{\partial t}\phi_1(x,\,t) + \frac{\partial^2}{\partial x^2}\phi_1(x,\,t) + (b_{11}\,|\phi_1(x,\,t)|^2 + b_{12}\,|\phi_2(x,\,t)|^2)\,\phi_1(x,\,t) = 0,$ $i\frac{\partial}{\partial t}\phi_2(x,\,t) + \frac{\partial^2}{\partial x^2}\phi_2(x,\,t) + (b_{11}\,|\phi_1(x,\,t)|^2 + b_{12}\,|\phi_2(x,\,t)|^2)\,\phi_2(x,\,t) = 0$	3
Total	19	35

7.1 Introduction

Coupled NLSEs describe a number of physical systems including two orthogonally-polarized states of single mode fibers, multi-mode pulses in fibers, and mixtures of different Bose–Einstein condensate species or different hyperfine states of the same species. In the following, we present these coupled NLSEs for each case. The stress will be on the nonlinear coupling in its simplest form leading to the so-called Manakov system [1].

7.1.1 Single-Mode Optical Fibers

In isotropic optical fibers, the refractive index is the same for any axis of polarization perpendicular to the axis of the fiber. In such ideal fibers the two orthogonally-polarized single modes are degenerate. The degeneracy is lifted for birefringent fibers, where the refractive index is not the same for the two perpendicular directions of polarization. For light pulses propagating along the z-axis and electric fields polarized along the x- and y-axis, the corresponding refractive indices n_x and n_y, are different. The pulse is in this case described by two nondegenerate and coupled modes. Birefringence may occur as a result of three main sources: (1) random imperfections in the geometry and material of the fiber, (2) intentionally-made birefringence, and (3) self-induced birefringence by nonlinearity. The latter, in particular, leads to cross-phase modulation (XPM), where the intensity of one pulse affects the phase of another. This is the relevant case to the present chapter.

   The electric field of two single-mode pulses linearly polarized in the x- and y-directions is written as

$$\mathbf{E}(\mathbf{r},\, t) = \operatorname{Re}\left[E_x(\mathbf{r},\, t)\hat{x} + E_y(\mathbf{r},\, t)\hat{y}\right]e^{-i\omega_0 t}, \quad E_z = 0, \tag{7.1}$$

with ω_0 being the single mode frequency. The induced polarization is similarly written as

$$\mathbf{P}(\mathbf{r},\, t) = \operatorname{Re}\left[P_x(\mathbf{r},\, t)\hat{x} + P_y(\mathbf{r},\, t)\hat{y}\right]e^{-i\omega_0 t}, \quad P_z = 0. \tag{7.2}$$

As shown in Chapter 2, the NLSE may be derived within the wide-pulse special case with

$$E_x = F(x,\, y)A_x(z)e^{i\beta_{0x}z}, \tag{7.3}$$

$$E_y = F(x,\, y)A_y(z)e^{i\beta_{0y}z}, \tag{7.4}$$

where $A_x(z)$ and $A_y(z)$ are the envelope profiles and $F(x,\, y)$ is the spacial distribution in the radial direction, and β_{0x} and β_{0y} are the propagation constants. It should be recalled that due to the difference in refractive indices, the two polarized pulses will have different propagation constants. Following the derivation of the scalar NLSE in Chapter 2, the pulse profiles, Equations (7.3) and (7.4) are substituted in the wave equation derived from Maxwell's equations, Equation (2.69), which results in the coupled NLSE

$$i\frac{\partial}{\partial z}A_x + a_1\frac{\partial^2}{\partial z^2}A_x + a_2(|A_x|^2 + c|A_y|^2)A_x + a_3A_y^2A_x^* = 0, \qquad (7.5)$$

$$i\frac{\partial}{\partial z}A_y + a_1\frac{\partial^2}{\partial z^2}A_y + a_2(|A_x|^2 + c|A_y|^2)A_y + a_3A_x^2A_y^* = 0, \qquad (7.6)$$

where a_1 and a_2 are constants given in terms of the fiber's parameters, similar to the scalar NLSE in Chapter 2. The constant c ranges from 2/3 for linearly polarized pulses to 2 for circular polarization [2]. It should be noted that the last coupling terms, corresponding to four-wave mixing, can be neglected for high birefringence fibers since they will oscillate much faster than the propagation rate of the pulse and thus their effect will average to zero. The system (7.5) and (7.6) is known as the coupled NLSE, or sometimes the vector NLSE.

7.1.2 Coupled Modes Optical Fibers

Another possibility for coupled NLSE to describe pulses in optical fibers is through the coupling of different fiber modes. For instance, the electric field of two pulses both polarized along the x-axis is written as

$$\mathbf{E}(\mathbf{r}, t) = \mathrm{Re}\,[E_1\,e^{-i\omega_1 t} + E_2\,e^{-i\omega_2 t}]\hat{x}, \qquad (7.7)$$

where ω_1 and ω_2 are the frequencies of the two modes. Note that, for simplicity, both modes are assumed to be polarized along the x-axis. In general, one may also consider different polarizations, which will just complicate further the nonlinear coupling. As in the previous section, we employ the wide-pulse approximation

$$E_1 = F_1(x, y)A_1(z)\,e^{i\beta_0 z}, \qquad (7.8)$$

$$E_2 = F_2(x, y)A_2(z)\,e^{i\beta_0 z}, \qquad (7.9)$$

where $A_{1,2}$ are the envelopes profiles, $F_{1,2}$ are their spacial radial distributions, and $\omega_{1,2}$ are the modes frequencies. Substituting the profiles (7.8) and (7.9) in the wave equation derived from Maxwell's equations, Equation (2.69), we get the coupled NLSE

$$i\frac{\partial}{\partial z}A_1 + a_{01}\frac{\partial}{\partial z}A_1 + a_{11}\frac{\partial^2}{\partial z^2}A_1 + a_{21}(|A_1|^2 + 2|A_2|^2)A_1 = 0, \qquad (7.10)$$

$$i\frac{\partial}{\partial z}A_2 + a_{02}\frac{\partial}{\partial z}A_1 + a_{12}\frac{\partial^2}{\partial z^2}A_2 + a_{22}(|A_1|^2 + 2|A_2|^2)A_2 = 0, \qquad (7.11)$$

where the dispersion coefficients, a_{11} and a_{12}, the nonlinearity strengths a_{21} and a_{22}, and the inverse group velocities a_{01} and a_{02} are now different because of the difference in mode frequencies.

Obviously, the two-coupled NLSE system (7.10) and (7.11), can be generalized to an N-coupled NLSE corresponding to N modes

$$i\frac{\partial}{\partial z}A_j + a_{0j}\frac{\partial}{\partial z}A_j + a_{1j}\frac{\partial^2}{\partial z^2}A_j + \sum_{i=1}^{N}a_{2i}|A_i|^2 A_j = 0, \quad j = 1, 2, \ldots, N. \quad (7.12)$$

7.1.3 Mixtures and Spinor Bose–Einstein Condensates

In Bose–Einstein condensates, it is possible to macroscopically condense two species including different atoms or different isotopes of the same atom. It is also possible to condense different hyperfine states of the same isotope. In all of these cases, the system will be described by a coupled NLSE, known as coupled Gross–Pitaevskii equation in the condensed matter community [3].

7.1.3.1 Mixtures of Bose–Einstein Condensates
The N-body quantum state, Equation (2.39), is now generalized to the two-component Bose–Einstein condensate as [3]

$$\Psi(\mathbf{r}_1, \mathbf{r}_2, \ldots, \mathbf{r}_{N_1}, \mathbf{r}'_1, \mathbf{r}'_2, \ldots, \mathbf{r}'_{N_2}, t) = \prod_{i=1}^{N_1} \phi_1(\mathbf{r}_i, t) \times \prod_{i=1}^{N_2} \phi_2(\mathbf{r}'_i, t), \quad (7.13)$$

where $\phi_1(\mathbf{r}_i, t)$ and $\phi_2(\mathbf{r}'_i, t)$ are the single-particle wavefunctions of the two components, and N_1 and N_2 are their corresponding number of atoms. It should be noted here that these numbers are conserved since there are no transitions between the different components. This is not the case for spinor condensates where transitions between hyperfine states are possible. Along the same lines as in Chapter 2, the generalized action can be constructed and the corresponding Euler–Lagrange equations result in the coupled NLSE (or coupled Gross–Pitaevskii equations)

$$i\hbar\frac{\partial}{\partial t}\phi_1(\mathbf{r}, t) + \frac{\hbar^2}{2m_1}\nabla^2\phi_1(\mathbf{r}, t) \quad V_{\text{ext}}(r)\phi_1(\mathbf{r}, t) - (g_{11}|\phi_1(\mathbf{r}, t)|^2 + g_{12}|\phi_2(\mathbf{r}, t)|^2)\phi_1(\mathbf{r}, t) = 0, \quad (7.14)$$

$$i\hbar\frac{\partial}{\partial t}\phi_2(\mathbf{r}, t) + \frac{\hbar^2}{2m_2}\nabla^2\phi_2(\mathbf{r}, t) - V_{\text{ext}}(r)\phi_2(\mathbf{r}, t) - (g_{21}|\phi_1(\mathbf{r}, t)|^2 + g_{22}|\phi_2(\mathbf{r}, t)|^2)\phi_2(\mathbf{r}, t) = 0, \quad (7.15)$$

where m_1 and m_2 are the atom masses of the two components. The constants g_{11}, g_{12}, g_{21}, and g_{22} characterize the strengths of inter- and intra-nonlinear interactions and they are directly proportional to the corresponding scattering lengths. There are no restrictions on the values of these constants, as in the previous cases for optical fibers, except that $g_{12} = g_{21}$.

7.1.3.2 Spinor Condensates
In the case of condensing hyperfine states of a total spin S, the coupled NLSE takes the form [3]

$$i\hbar\frac{\partial}{\partial t}\Psi_i + \frac{\hbar^2}{2m}\nabla^2\Psi_i - V(\mathbf{r})\Psi_i + g_0\left(\sum_j \Psi_j^\dagger \Psi_j\right)\Psi_i + g_1\sum_{j,k,l}\Psi_j^\dagger \mathbf{S}_{jk} \cdot \mathbf{S}_{il}\Psi_l\Psi_k = 0, \quad (7.16)$$

where $\mathbf{S}$ is the spin-S matrix, Ψ is the S-components spinor, and g_0 and g_1 are the nonlinear interaction strengths. Clearly, the spin–spin interaction introduces more complicated nonlinear interaction coupling.

We conclude with a final remark about the integrability of the coupled NLSE presented above. The equations are in general not integrable except for specific values of the coupling constants [4], where a Lax pair exists and many exact solutions are thus derived, as we show in the rest of this chapter.

7.2 Fundamental Coupled NLSE *Manakov System*

Equations:

$$i\,\psi_{1t} + b_0\,\psi_{1xx} + (b_1\,|\psi_1|^2 + b_2\,|\psi_2|^2)\,\psi_1 = 0,$$
$$i\,\psi_{2t} + c_0\,\psi_{2xx} + (c_1\,|\psi_1|^2 + c_2\,|\psi_2|^2)\,\psi_2 = 0,$$
(7.17)

where $\psi_j = \psi_j(x, t)$ is the complex function profile, $j = 1, 2$. x and t are its two independent variables, b_0, c_0, b_1, c_1, b_2, and c_2 are real constants.

Solutions:

Solution **1. Constant Amplitude I** *continuous wave (CW), t- and x-independent phase*

$$\psi_1(x, t) = A_0\,e^{i\,\phi_1},$$
$$\psi_2(x, t) = B_0\,e^{i\,\phi_2},$$
(7.18)

where $b_1 = -\dfrac{b_2\,B_0^2}{A_0^2}$, $c_2 = -\dfrac{c_1\,A_0^2}{B_0^2}$, A_0, B_0, ϕ_1, and ϕ_2 are arbitrary real constants.

Solution **2. Constant Amplitude II** *CW, t-dependent phase*

$$\psi_1(x, t) = A_0\,e^{i\,[A_1\,(t-t_0)+\phi_1]},$$
$$\psi_2(x, t) = B_0\,e^{i\,[B_1\,(t-t_0)+\phi_2]},$$
(7.19)

where $A_1 = A_0^2\,b_1 + B_0^2\,b_2$, $B_1 = A_0^2\,c_1 + B_0^2\,c_2$, A_0, B_0, t_0, ϕ_1, and ϕ_2 are arbitrary real constants.

Solution **3. Constant Amplitude III** *CW, x-dependent phase*

$$\psi_1(x, t) = A_0\,e^{i\,[A_1\,(x-x_0)+\phi_1]},$$
$$\psi_2(x, t) = B_0\,e^{i\,[B_1\,(x-x_0)+\phi_2]},$$
(7.20)

where $\quad A_1 = \pm\sqrt{\dfrac{A_0^2\,b_1 + B_0^2\,b_2}{b_0}}, \quad B_1 = \pm\sqrt{\dfrac{A_0^2\,c_1 + B_0^2\,c_2}{c_0}}, \quad b_0\,(A_0^2\,b_1 + B_0^2\,b_2) > 0,$ $c_0\,(A_0^2\,c_1 + B_0^2\,c_2) > 0$, A_0, B_0, x_0, ϕ_1, and ϕ_2 are arbitrary real constants.

Solution 4. Constant Amplitude IV *CW, t- and x-dependent phase*

$$\psi_1(x, t) = A_0 \, e^{i \left[A_1 \, (t - t_0) + A_2 \, (x - x_0) + \phi_1 \right]},$$

$$\psi_2(x, t) = B_0 \, e^{i \left[B_1 \, (t - t_0) + B_2 \, (x - x_0) + \phi_2 \right]},$$ (7.21)

where $A_1 = -A_2^2 \, b_0 + A_0^2 \, b_1 + B_0^2 \, b_2$, $B_1 = -B_2^2 \, c_0 + A_0^2 \, c_1 + B_0^2 \, c_2$, A_0, B_0, A_2, B_2, x_0, t_0, ϕ_1, and ϕ_2 are arbitrary real constants.

Solution 5. Rational Solution *decaying wave (DW)*

$$\psi_1(x, t) = \frac{A_0}{\sqrt{2 \, [A_1 + t - t_0]}} \, e^{i \, \phi_1(x, \, t)},$$

$$\psi_2(x, t) = \frac{B_0}{\sqrt{2 \, [B_1 + t - t_0]}} \, e^{i \, \phi_2(x, \, t)},$$ (7.22)

where

$$\phi_1(x, t) = \frac{(b_0 \, A_2 + x - x_0)^2}{4 \, b_0 \, (A_1 + t - t_0)} + \frac{b_1 \, A_0^2}{2} \ln[2(A_1 + t - t_0)] + \frac{b_2 \, B_0^2}{2} \ln[2 \, (B_1 + t - t_0)] + \phi_{01},$$

$$\phi_2(x, t) = \frac{(c_0 \, B_2 + x - x_0)^2}{4 \, c_0 \, (B_1 + t - t_0)} + \frac{c_1 \, A_0^2}{2} \ln[2(A_1 + t - t_0)] + \frac{c_2 \, B_0^2}{2} \ln[2 \, (B_1 + t - t_0)] + \phi_{02},$$

A_0, A_1, A_2, B_0, B_1, B_2, x_0, t_0, ϕ_{01} and ϕ_{02} are arbitrary real constants.

Solution 6.

$$\psi_1(x, t) = \frac{A_0 \, [\beta + (x - x_0)^2]}{\alpha + (x - x_0)^2} \, e^{i \left[A_1 \, (t - t_0) + \phi_1 \right]},$$

$$\psi_2(x, t) = \frac{B_0 \, (x - x_0)}{\alpha + (x - x_0)^2} \, e^{i \left[B_1 \, (t - t_0) + \phi_2 \right]},$$ (7.23)

where $A_1 = 3 \, b_0 / \beta$, $B_1 = A_0^2 \, c_1$, $\alpha = -3 \, \beta$, $c_0 = 4 \, A_0^2 \, c_1 \, \beta / 9$, $b_1 = 3 \, b_0 / (A_0^2 \, \beta)$, $b_2 = -24 \, b_0 / B_0^2$, $c_2 = -80 \, A_0^2 \, c_1 \, \beta / (9 \, B_0^2)$, A_0, B_0, β, b_0, x_0, t_0, ϕ_{01} and ϕ_{02} are arbitrary real constants.

  • *Reference*: [5], *taken from the nonlocal case, corrected.* For the nonlocal case, replace $b_2 \to -b_2$ and $c_2 \to -c_2$.

Solution 7.

$$\psi_1(x, t) = \frac{A_0 \, [\beta + (x - x_0)^2]}{\alpha + (x - x_0)^2} \, e^{i \left[A_1 \, (t - t_0) + \phi_1 \right]},$$

$$\psi_2(x, t) = \frac{B_0 \, (x - x_0)}{\sqrt{\alpha + (x - x_0)^2}} \, e^{i \left[B_1 \, (t - t_0) + \phi_2 \right]},$$ (7.24)

where $A_1 = 5 \, b_0 / (2 \, \beta)$, $B_1 = -5 \, A_0^2 \, c_1 / 3$, $\alpha = -3 \, \beta$, $c_0 = -16 \, A_0^2 \, c_1 \, \beta / 9$, $b_1 = -3 \, b_0 / (2 \, A_0^2 \, \beta)$, $b_2 = 4 \, b_0 / (B_0^2 \, \beta)$, $c_2 = -8 \, A_0^2 \, c_1 / (3 \, B_0^2)$, A_0, B_0, β, b_0, x_0, t_0, ϕ_{01} and ϕ_{02} are arbitrary real constants.

- *Reference*: [5], *taken from the nonlocal case, corrected.* For the nonlocal case, replace $b_2 \to -b_2$ and $c_2 \to -c_2$.

Solution 8.

$$\psi_1(x,\,t) = \frac{A_0\,(x - x_0)}{\alpha + (x - x_0)^2}\, e^{i\,[A_1\,(t-t_0)+\phi_1]},$$

$$\psi_2(x,\,t) = \frac{B_0\,(x - x_0)}{\sqrt{\alpha + (x - x_0)^2}}\, e^{i\,[B_1\,(t-t_0)+\phi_2]},$$

(7.25)

where $A_1 = -6\,b_0/\alpha$, $B_1 = -3\,c_0/\alpha$, $b_1 = -8\,b_0/A_0^2$, $b_2 = -6\,b_0/(B_0^2\,\alpha)$, $c_1 = -3\,c_0/A_0^2$, $c_2 = -3\,c_0/(B_0^2\,\alpha)$, A_0, B_0, α, b_0, c_0, x_0, t_0, ϕ_{01} and ϕ_{02} are arbitrary real constants.

- *Reference*: [5], *taken from the nonlocal case, corrected.* For the nonlocal case, replace $b_1 \to -b_1$, $c_1 \to -c_1$, $b_2 \to -b_2$, and $c_2 \to -c_2$.

Solution 9.

$$\psi_1(x,\,t) = \frac{A_0}{\alpha + \cos\left[\dfrac{\beta}{\sqrt{b_0}}(x - x_0)\right]}\, e^{i\,[A_1\,(t-t_0)+\phi_1]},$$

$$\psi_2(x,\,t) = \frac{B_0\,\sin\left[\dfrac{\beta}{\sqrt{c_0}}(x - x_0)\right]}{\alpha + \cos\left[\dfrac{\beta}{\sqrt{b_0}}(x - x_0)\right]}\, e^{i\,[B_1\,(t-t_0)+\phi_2]},$$

(7.26)

where $A_1 = B_1 = \beta^2/2$, $b_0 = c_0$, $b_1 = c_1/3$, $b_2 = -3\,\beta^2/(2\,B_0^2)$, $c_1 = 3\,(\alpha^2 - 1)\,\beta^2/(2\,A_0^2)$, $c_2 = b_2/3$, A_0, B_0, α, β, c_0, x_0, t_0, ϕ_{01} and ϕ_{02} are arbitrary real constants.

- *Reference*: [5], *taken from the nonlocal case.* For the nonlocal case, replace $b_2 \to -b_2$ and $c_2 \to -c_2$.

Solution 10.

$$\psi_1(x,\,t) = \frac{A_0\,\cos\left[\dfrac{\beta}{\sqrt{c_0}}(x - x_0)\right]}{1 + \alpha\,\cos^2\left[\dfrac{\beta}{\sqrt{b_0}}(x - x_0)\right]}\, e^{i\,[A_1\,(t-t_0)+\phi_1]},$$

$$\psi_2(x,\,t) = \frac{B_0\,\sin\left[\dfrac{\beta}{\sqrt{c_0}}(x - x_0)\right]}{1 + \alpha\,\cos^2\left[\dfrac{\beta}{\sqrt{b_0}}(x - x_0)\right]}\, e^{i\,[B_1\,(t-t_0)+\phi_2]},$$

(7.27)

where $A_1 = B_1 = -\beta^2$, $b_0 = c_0$, $b_1 = c_1/3$, $b_2 = 6\,\alpha\,\beta^2/B_0^2$, $c_1 = -6\,\alpha\,\beta^2(1 + \alpha)/A_0^2$, $c_2 = b_2/3$, A_0, B_0, α, β, c_0, x_0, t_0, ϕ_{01} and ϕ_{02} are arbitrary real constants.

- *Reference*: [5], *taken from the nonlocal case*. For the nonlocal case, replace $b_2 \to -b_2$ and $c_2 \to -c_2$.

Solution 11.

$$
\psi_1(x, t) = \frac{A_0 \cos\left[\dfrac{\beta}{\sqrt{c_0}}(x - x_0)\right]}{1 + \alpha \cos^2\left[\dfrac{\beta}{\sqrt{b_0}}(x - x_0)\right]}\, e^{i\left[A_1\,(t - t_0) + \phi_1\right]},
$$

$$
\psi_2(x, t) = \frac{B_0}{1 + \alpha \cos^2\left[\dfrac{\beta}{\sqrt{b_0}}(x - x_0)\right]}\, e^{i\left[B_1\,(t - t_0) + \phi_2\right]},
$$

(7.28)

where $A_1 = -\beta^2$, $B_1 = -4\,\beta^2$, $b_0 = c_0$, $b_1 = -2\,\alpha\,(\alpha + 4)\,\beta^2/A_0^2$, $b_2 = 6\,\alpha\,\beta^2/B_0^2$, $c_1 = -6\,\alpha\,\beta^2(2 + \alpha)/A_0^2$, $c_2 = -2\,(2 - \alpha)\,\beta^2/B_0^2$, A_0, B_0, α, β, c_0, x_0, t_0, ϕ_{01} and ϕ_{02} are arbitrary real constants.

- *Reference*: [5], *taken from the nonlocal case*.

Solution 12.

$$
\psi_1(x, t) = \frac{A_0}{1 + \alpha \cos^2\left[\dfrac{\beta}{\sqrt{b_0}}(x - x_0)\right]}\, e^{i\left[A_1\,(t - t_0) + \phi_1\right]},
$$

$$
\psi_2(x, t) = \frac{B_0 \sin\left[\dfrac{\beta}{\sqrt{c_0}}(x - x_0)\right]}{1 + \alpha \cos^2\left[\dfrac{\beta}{\sqrt{b_0}}(x - x_0)\right]}\, e^{i\left[B_1\,(t - t_0) + \phi_2\right]},
$$

(7.29)

where $A_1 = -4\,\beta^2$, $B_1 = -\beta^2$, $b_0 = c_0$, $b_1 = -2\,(\alpha + 1)\,(3\,\alpha + 2)\,\beta^2/A_0^2$, $b_2 = 6\,\alpha\,(\alpha + 2)\,\beta^2/B_0^2$, $c_1 = -6\,\alpha\,\beta^2\,(1 + \alpha)/A_0^2$, $c_2 = 2\,\alpha\,(3\,\alpha + 4)\,\beta^2/B_0^2$, A_0, B_0, α, β, c_0, x_0, t_0, ϕ_{01} and ϕ_{02} are arbitrary real constants.

- *Reference*: [5], *taken from the nonlocal case*. For the nonlocal case, replace $b_2 \to -b_2$ and $c_2 \to -c_2$.

Solution 13.

$$\psi_1(x,\,t) = \frac{A_0}{1 + \alpha \cos^2\left[\dfrac{\beta}{\sqrt{b_0}}(x - x_0)\right]}\, e^{i\left[A_1\,(t - t_0) + \phi_1\right]},$$

$$\psi_2(x,\,t) = \frac{B_0 \sin\left[\dfrac{\beta}{\sqrt{c_0}}(x - x_0)\right] \cos\left[\dfrac{\beta}{\sqrt{c_0}}(x - x_0)\right]}{1 + \alpha \cos^2\left[\dfrac{\beta}{\sqrt{b_0}}(x - x_0)\right]}\, e^{i\left[B_1\,(t - t_0) + \phi_2\right]}, \tag{7.30}$$

where $A_1 = B_1 = 2\,\beta^2$, $b_0 = c_0$, $b_1 = 2\,(\alpha + 1)\beta^2/A_0^2$, $b_2 = -6\,\alpha^2\,\beta^2/B_0^2$, $c_1 = 6\,\beta^2\,(1 + \alpha)/A_0^2$, $c_2 = -2\,\alpha^2\,\beta^2/B_0^2$, A_0, B_0, α, β, c_0, x_0, t_0, ϕ_{01} and ϕ_{02} are arbitrary real constants.

• *Reference*: [5], *taken from the nonlocal case*. For the nonlocal case, replace $b_2 \to -b_2$ and $c_2 \to -c_2$.

Solution 14.

$$\psi_1(x,\,t) = \frac{A_0 \sqrt{m}\, \mathrm{sn}\left[\dfrac{\beta}{\sqrt{b_0}}(x - x_0),\, m\right]}{\alpha + \mathrm{dn}\left[\dfrac{\beta}{\sqrt{b_0}}(x - x_0),\, m\right]}\, e^{i\left[A_1\,(t - t_0) + \phi_1\right]},$$

$$\psi_2(x,\,t) = \frac{B_0 \sqrt{m}\, \mathrm{cn}\left[\dfrac{\beta}{\sqrt{b_0}}(x - x_0),\, m\right]}{\alpha + \mathrm{dn}\left[\dfrac{\beta}{\sqrt{b_0}}(x - x_0),\, m\right]}\, e^{i\left[B_1\,(t - t_0) + \phi_2\right]}, \tag{7.31}$$

where $A_1 = B_1 = -(2 - m)\,\beta^2/2$, $b_0 = c_0$, $b_1 = c_1/3$, $b_2 = 3\,\beta^2\,(\alpha^2 - 1)/(2\,B_0^2)$, $c_1 = -3\,\beta^2\,(\alpha^2 + m - 1)/(2\,A_0^2)$, $c_2 = b_2/3$, A_0, B_0, m, α, β, c_0, x_0, t_0, ϕ_{01} and ϕ_{02} are arbitrary real constants.

• *Reference*: [5], *taken from the nonlocal case*.

Solution 15. (Solution 14. for $m = 1$)

$$\psi_1(x,\,t) = \frac{A_0 \tanh\left[\dfrac{\beta}{\sqrt{b_0}}(x - x_0)\right]}{\alpha + \mathrm{sech}\left[\dfrac{\beta}{\sqrt{b_0}}(x - x_0)\right]}\, e^{i\left[A_1\,(t - t_0) + \phi_1\right]},$$

$$\psi_2(x,\,t) = \frac{B_0 \,\mathrm{sech}\left[\dfrac{\beta}{\sqrt{b_0}}(x - x_0)\right]}{\alpha + \mathrm{sech}\left[\dfrac{\beta}{\sqrt{b_0}}(x - x_0)\right]}\, e^{i\left[B_1\,(t - t_0) + \phi_2\right]}, \tag{7.32}$$

where $A_1 = B_1 = -\beta^2/2$, $b_0 = c_0$, $b_1 = c_1/3$, $b_2 = 3\,\beta^2\,(\alpha^2 - 1)/(2\,B_0^2)$, $c_1 = -3\,\beta^2\,\alpha^2/(2\,A_0^2)$, $c_2 = b_2/3$, A_0, B_0, α, β, c_0, x_0, t_0, ϕ_{01} and ϕ_{02} are arbitrary real constants.

- *Reference*: [5], *taken from the nonlocal case.*

Solution 16.

$$\psi_1(x,\,t) = \frac{A_0\,\mathrm{sn}\left[\dfrac{\beta}{\sqrt{b_0}}\,(x - x_0),\,m\right]}{\alpha + \mathrm{cn}\left[\dfrac{\beta}{\sqrt{b_0}}\,(x - x_0),\,m\right]}\,e^{i\left[A_1\,(t - t_0) + \phi_1\right]},$$

$$\psi_2(x,\,t) = \frac{B_0\,\mathrm{dn}\left[\dfrac{\beta}{\sqrt{b_0}}\,(x - x_0),\,m\right]}{\alpha + \mathrm{cn}\left[\dfrac{\beta}{\sqrt{b_0}}\,(x - x_0),\,m\right]}\,e^{i\left[B_1\,(t - t_0) + \phi_2\right]},$$

(7.33)

where $A_1 = B_1 = -(2\,m - 1)\,\beta^2/2$, $b_0 = c_0$, $b_1 = c_1/3$, $b_2 = 3\,\beta^2\,(\alpha^2 - 1)/(2\,B_0^2)$, $c_1 = -3\,\beta^2\,(1 - m + \alpha^2\,m)/(2\,A_0^2)$, $c_2 = b_2/3$, A_0, B_0, α, β, m, c_0, x_0, t_0, ϕ_{01} and ϕ_{02} are arbitrary real constants.

- *Reference*: [5], *taken from the nonlocal case.*

Solution 17.

$$\psi_1(x,\,t) = \frac{A_0\,\mathrm{dn}\left[-\dfrac{\beta}{\sqrt{b_0}}\,(x - x_0),\,m\right]}{\alpha + \mathrm{sn}\left[\dfrac{\beta}{\sqrt{b_0}}\,(x - x_0),\,m\right]}\,e^{i\left[A_1\,(t - t_0) + \phi_1\right]},$$

$$\psi_2(x,\,t) = \frac{B_0\,\mathrm{cn}\left[\dfrac{\beta}{\sqrt{b_0}}\,(x - x_0),\,m\right]}{\alpha + \mathrm{sn}\left[\dfrac{\beta}{\sqrt{b_0}}\,(x - x_0),\,m\right]}\,e^{i\left[B_1\,(t - t_0) + \phi_2\right]},$$

(7.34)

where $A_1 = B_1 = (m + 1)\,\beta^2/2$, $b_0 = c_0$, $b_1 = (\alpha^2 - 1)\,\beta^2/(2\,A_0^2)$, $b_2 = 3\,\beta^2\,(m\,\alpha^2 - 1)/(2\,B_0^2)$, $c_1 = 3\,\beta^2\,(\alpha^2 - 1)/(2\,A_0^2)$, $c_2 = (m\,\alpha^2 - 1)\,\beta^2/(2\,B_0^2)$, A_0, B_0, α, β, m, c_0, x_0, t_0, ϕ_{01} and ϕ_{02} are arbitrary real constants.

- *Reference*: [5].

Solution 18. *Single bright soliton*

$$\psi_1(x,\, t) = \alpha_{11}\, \frac{e^{\eta(x,\, t)}}{1 + e^{\eta(x,\, t) + \eta*(x,\, t) + r}},$$

$$\psi_2(x,\, t) = \alpha_{12}\, \frac{e^{\eta(x,\, t)}}{1 + e^{\eta(x,\, t) + \eta*(x,\, t) + r}},$$

$$(7.35)$$

where $\eta(x,\, t) = k\, (x + i\, k\, t)$, $r = \ln[\frac{\mu\, (\mid \alpha_{11}\mid^2 + \mid \alpha_{12}\mid^2)}{(k + k^*)^2}]$, $b_0 = c_0 = 1$, $b_1 = c_1 = b_2 = c_2 = 2\,\mu$, μ is an arbitrary real constants, α_{11}, α_{12}, are k, are arbitrary complex constants.

• *Reference*: [6].

Solution 19. *Two bright solitons*

$$\psi_1(x,\, t) = \frac{\Omega_1(x,\, t)}{q(x,\, t)},$$

$$\psi_2(x,\, t) = \frac{\Omega_2(x,\, t)}{q(x,\, t)},$$

$$(7.36)$$

where

$\Omega_1(x,\, t) = \alpha_{11}\, e^{\eta_1(x,\, t)} + \alpha_{21}\, e^{\eta_2(x,\, t)} + e^{\eta_1(x,\, t) + \eta_1^*(x,\, t) + \eta_2(x,\, t) + \delta_1} + e^{\eta_1(x,\, t) + \eta_2^*(x,\, t) + \eta_2(x,\, t) + \delta_2}$,

$\Omega_2(x,\, t) = \alpha_{12}\, e^{\eta_1(x,\, t)} + \alpha_{22}\, e^{\eta_2(x,\, t)} + e^{\eta_1(x,\, t) + \eta_1^*(x,\, t) + \eta_2(x,\, t) + \delta_{1p}} + e^{\eta_1(x,\, t) + \eta_2^*(x,\, t) + \eta_2(x,\, t) + \delta_{2p}}$,

$q(x,\, t) = 1 + e^{\eta_1(x,\, t) + \eta_1^*(x,\, t) + r_1} + e^{\eta_1(x,\, t) + \eta_2^*(x,\, t) + \delta_0} + e^{\eta_2(x,\, t) + \eta_1^*(x,\, t) + \delta_0^*} + e^{\eta_2(x,\, t) + \eta_2^*(x,\, t) + r_2}$
$\quad + e^{\eta_1(x,\, t) + \eta_1^*(x,\, t) + \eta_2(x,\, t) + \eta_2^*(x,\, t) + r_3}$,

$\eta_1(x,\, t) = k_1\, (x + i\, k_1\, t)$, $\eta_2(x,\, t) = k_2\, (x + i\, k_2\, t)$,

$\kappa_{11} = \frac{\mu\, (\mid \alpha_{11}\mid^2 + \mid \alpha_{12}\mid^2)}{k_1 + k_1^*}$, $\kappa_{12} = \frac{\mu\, (\alpha_{11}\, \alpha_{21}^* + \alpha_{12}\, \alpha_{22}^*)}{k_1 + k_2^*}$, $\kappa_{21} = \frac{\mu\, (\alpha_{21}\, \alpha_{11}^* + \alpha_{22}\, \alpha_{12}^*)}{k_2 + k_1^*}$,

$\kappa_{22} = \frac{\mu\, (\mid \alpha_{22}\mid^2 + \mid \alpha_{21}\mid^2)}{k_2 + k_2^*}$, $\delta_0 = \ln\left(\frac{\kappa_{12}}{k_1 + k_2^*}\right)$,

$\delta_1 = \ln\left[\frac{(k_1 - k_2)\, (\alpha_{11}\, \kappa_{21} - \alpha_{21}\, \kappa_{11})}{(k_1 + k_1^*)\, (k_2 + k_1^*)}\right]$, $\delta_{1p} = \ln\left[\frac{(k_1 - k_2)\, (\alpha_{12}\, \kappa_{21} - \alpha_{22}\, \kappa_{11})}{(k_1 + k_1^*)\, (k_2 + k_1^*)}\right]$,

$\delta_2 = \ln\left[\frac{(k_2 - k_1)\, (\alpha_{21}\, \kappa_{12} - \alpha_{11}\, \kappa_{22})}{(k_2 + k_2^*)\, (k_1 + k_2^*)}\right]$, $\delta_{2p} = \ln\left[\frac{(k_2 - k_1)\, (\alpha_{22}\, \kappa_{12} - \alpha_{12}\, \kappa_{22})}{(k_2 + k_2^*)\, (k_1 + k_2^*)}\right]$, $r_1 = \ln\left(\frac{\kappa_{11}}{k_1 + k_1^*}\right)$,

$r_2 = \ln\left(\frac{\kappa_{22}}{k_2 + k_2^*}\right)$, $r_3 = \ln\left[\frac{(k_1 - k_2)\, (k_1^* - k_2^*)}{(k_1 + k_1^*)\, (k_2 + k_2^*)\, (k_1 + k_2^*)\, (k_2 + k_1^*)}\right]$, $b_0 = c_0 = 1$, $b_1 = c_1 = b_2 = c_2 = 2\,\mu$, μ is an arbitrary real constants, α_{11}, α_{12}, α_{21}, α_{22}, k_1, and k_2 are arbitrary complex constants.

• *Reference*: [6].

Solution 20. *Three bright solitons*

$$\psi_1(x,\, t) = \frac{\Omega_{11}(x,\, t) + \Omega_{12}(x,\, t) + \Omega_{13}(x,\, t) + \Omega_{14}(x,\, t)}{q(x,\, t)},$$

$$\psi_2(x,\, t) = \frac{\Omega_{21}(x,\, t) + \Omega_{22}(x,\, t) + \Omega_{23}(x,\, t) + \Omega_{24}(x,\, t)}{q(x,\, t)},$$

$$(7.37)$$

where $\quad \Omega_{11}(x,\,t) = \alpha_{11}\,e^{\eta_1(x,\,t)} + \alpha_{21}\,e^{\eta_2(x,\,t)} + \alpha_{31}\,e^{\eta_3(x,\,t)} + e^{\eta_1(x,\,t)+\eta_1^*(x,\,t)+\eta_2(x,\,t)+\delta_{11}}$

$\quad + e^{\eta_1(x,\,t)+\eta_1^*(x,\,t)+\eta_3(x,\,t)+\delta_{21}},$

$\Omega_{12}(x,\,t) = e^{\eta_2(x,\,t)+\eta_2^*(x,\,t)+\eta_1(x,\,t)+\delta_{31}} + e^{\eta_2(x,\,t)+\eta_2^*(x,\,t)+\eta_3(x,\,t)+\delta_{41}}$

$\quad + e^{\eta_3(x,\,t)+\eta_3^*(x,\,t)+\eta_1(x,\,t)+\delta_{51}} + e^{\eta_3(x,\,t)+\eta_3^*(x,\,t)+\eta_2(x,\,t)+\delta_{61}},$

$\Omega_{13}(x,\,t) = e^{\eta_1^*(x,\,t)+\eta_2(x,\,t)+\eta_3(x,\,t)+\delta_{71}} + e^{\eta_1(x,\,t)+\eta_2^*(x,\,t)+\eta_3(x,\,t)+\delta_{81}}$

$\quad + e^{\eta_1(x,\,t)+\eta_2(x,\,t)+\eta_3^*(x,\,t)+\delta_{91}} + e^{\eta_1(x,\,t)+\eta_1^*(x,\,t)+\eta_2(x,\,t)+\eta_2^*(x,\,t)+\eta_3(x,\,t)+\tau_{11}},$

$\Omega_{14}(x,\,t) = e^{\eta_1(x,\,t)+\eta_1^*(x,\,t)+\eta_3(x,\,t)+\eta_3^*(x,\,t)+\eta_2(x,\,t)+\tau_{21}} + e^{\eta_2(x,\,t)+\eta_2^*(x,\,t)+\eta_3(x,\,t)+\eta_3^*(x,\,t)+\eta_1(x,\,t)+\tau_{31}},$

$\Omega_{21}(x,\,t) = \alpha_{12}\,e^{\eta_1(x,\,t)} + \alpha_{22}\,e^{\eta_2(x,\,t)} + \alpha_{32}\,e^{\eta_3(x,\,t)} + e^{\eta_1(x,\,t)+\eta_1^*(x,\,t)+\eta_2(x,\,t)+\delta_{12}}$

$\quad + e^{\eta_1(x,\,t)+\eta_1^*(x,\,t)+\eta_3(x,\,t)+\delta_{22}},$

$\Omega_{22}(x,\,t) = e^{\eta_2(x,\,t)+\eta_2^*(x,\,t)+\eta_1(x,\,t)+\delta_{32}} + e^{\eta_2(x,\,t)+\eta_2^*(x,\,t)+\eta_3(x,\,t)+\delta_{42}}$

$\quad + e^{\eta_3(x,\,t)+\eta_3^*(x,\,t)+\eta_1(x,\,t)+\delta_{52}} + e^{\eta_3(x,\,t)+\eta_3^*(x,\,t)+\eta_2(x,\,t)+\delta_{62}},$

$\Omega_{23}(x,\,t) = e^{\eta_1^*(x,\,t)+\eta_2(x,\,t)+\eta_3(x,\,t)+\delta_{72}} + e^{\eta_1(x,\,t)+\eta_2^*(x,\,t)+\eta_3(x,\,t)+\delta_{82}}$

$\quad + e^{\eta_1(x,\,t)+\eta_2(x,\,t)+\eta_3^*(x,\,t)+\delta_{92}} + e^{\eta_1(x,\,t)+\eta_1^*(x,\,t)+\eta_2(x,\,t)+\eta_2^*(x,\,t)+\eta_3(x,\,t)+\tau_{12}},$

$\Omega_{24}(x,\,t) = e^{\eta_1(x,\,t)+\eta_1^*(x,\,t)+\eta_3(x,\,t)+\eta_3^*(x,\,t)+\eta_2(x,\,t)+\tau_{22}} + e^{\eta_2(x,\,t)+\eta_2^*(x,\,t)+\eta_3(x,\,t)+\eta_3^*(x,\,t)+\eta_1(x,\,t)+\tau_{32}},$

$q(x,\,t) = 1 + e^{\eta_1(x,\,t)+\eta_1^*(x,\,t)+r_1} + e^{\eta_2(x,\,t)+\eta_2^*(x,\,t)+r_2} + e^{\eta_3(x,\,t)+\eta_3^*(x,\,t)+r_3} + e^{\eta_1(x,\,t)+\eta_2^*(x,\,t)+\delta_{10}}$

$\quad + e^{\eta_1^*(x,\,t)+\eta_2(x,\,t)+\delta_{10}^*} + e^{\eta_1(x,\,t)+\eta_3^*(x,\,t)+\delta_{20}} + e^{\eta_1^*(x,\,t)+\eta_3(x,\,t)+\delta_{20}^*} + e^{\eta_2(x,\,t)+\eta_3^*(x,\,t)+\delta_{30}}$

$\quad + e^{\eta_2^*(x,\,t)+\eta_3(x,\,t)+\delta_{30}^*} + e^{\eta_1(x,\,t)+\eta_1^*(x,\,t)+\eta_2(x,\,t)+\eta_2^*(x,\,t)+r_4}$

$\quad + e^{\eta_1(x,\,t)+\eta_1^*(x,\,t)+\eta_3(x,\,t)+\eta_3^*(x,\,t)+r_5} + e^{\eta_2(x,\,t)+\eta_2^*(x,\,t)+\eta_3(x,\,t)+\eta_3^*(x,\,t)+r_6}$

$\quad + e^{\eta_1(x,\,t)+\eta_1^*(x,\,t)+\eta_2(x,\,t)+\eta_3^*(x,\,t)+\tau_{10}} + e^{\eta_1(x,\,t)+\eta_1^*(x,\,t)+\eta_3(x,\,t)+\eta_2^*(x,\,t)+\tau_{10}^*}$

$\quad + e^{\eta_2(x,\,t)+\eta_2^*(x,\,t)+\eta_1(x,\,t)+\eta_3^*(x,\,t)+\tau_{20}} + e^{\eta_2(x,\,t)+\eta_2^*(x,\,t)+\eta_1^*(x,\,t)+\eta_3(x,\,t)+\tau_{20}^*}$

$\quad + e^{\eta_3(x,\,t)+\eta_3^*(x,\,t)+\eta_1(x,\,t)+\eta_2^*(x,\,t)+\tau_{30}} + e^{\eta_3(x,\,t)+\eta_3^*(x,\,t)+\eta_1^*(x,\,t)+\eta_2(x,\,t)+\tau_{30}^*}$

$\quad + e^{\eta_1(x,\,t)+\eta_1^*(x,\,t)+\eta_2(x,\,t)+\eta_2^*(x,\,t)+\eta_3(x,\,t)+\eta_3^*(x,\,t)+r_7},$

$\eta_1(x,\,t) = k_1\,(x + i\,k_1\,t),\ \eta_2(x,\,t) = k_2\,(x + i\,k_2\,t),\ \eta_3(x,\,t) = k_3\,(x + i\,k_3\,t),$

$\kappa_{11} = \dfrac{\mu\left(|\alpha_{11}|^2 + |\alpha_{12}|^2\right)}{k_1 + k_1^*},\ \kappa_{12} = \dfrac{\mu\left(\alpha_{11}\,\alpha_{21}^* + \alpha_{12}\,\alpha_{22}^*\right)}{k_1 + k_2^*},\ \kappa_{13} = \dfrac{\mu\left(\alpha_{11}\,\alpha_{31}^* + \alpha_{12}\,\alpha_{32}^*\right)}{k_1 + k_3^*},$

$\kappa_{21} = \dfrac{\mu\left(\alpha_{21}\,\alpha_{11}^* + \alpha_{22}\,\alpha_{12}^*\right)}{k_2 + k_1^*},\ \kappa_{22} = \dfrac{\mu\left(|\alpha_{22}|^2 + |\alpha_{21}|^2\right)}{k_2 + k_2^*},\ \kappa_{23} = \dfrac{\mu\left(\alpha_{21}\,\alpha_{31}^* + \alpha_{22}\,\alpha_{32}^*\right)}{k_2 + k_3^*},$

$\kappa_{31} = \dfrac{\mu\left(\alpha_{31}\,\alpha_{11}^* + \alpha_{32}\,\alpha_{12}^*\right)}{k_3 + k_1^*},\ \kappa_{32} = \dfrac{\mu\left(\alpha_{31}\,\alpha_{21}^* + \alpha_{32}\,\alpha_{22}^{\ast\ast}\right)}{k_3 + k_2^*},\ \kappa_{33} = \dfrac{\mu\left(|\alpha_{31}|^2 + |\alpha_{32}|^2\right)}{k_3 + k_3^*},$

$\delta_{10} = \ln\left(\dfrac{\kappa_{12}}{k_1 + k_2^*}\right),\ \delta_{20} = \ln\left(\dfrac{\kappa_{13}}{k_1 + k_3^*}\right),\ \delta_{30} = \ln\left(\dfrac{\kappa_{23}}{k_2 + k_3^*}\right),\ \delta_{11} = \ln\left[\dfrac{(k_1 - k_2)\,(\alpha_{11}\,\kappa_{21} - \alpha_{21}\,\kappa_{11})}{\left(k_1 + k_1^*\right)\left(k_2 + k_1^*\right)}\right],$

$\delta_{12} = \ln\left[\dfrac{(k_1 - k_2)\,(\alpha_{12}\,\kappa_{21} - \alpha_{22}\,\kappa_{11})}{\left(k_1 + k_1^*\right)\left(k_2 + k_1^*\right)}\right],\ \delta_{21} = \ln\left[\dfrac{(k_1 - k_3)\,(\alpha_{11}\,\kappa_{31} - \alpha_{31}\,\kappa_{11})}{\left(k_1 + k_1^*\right)\left(k_3 + k_1^*\right)}\right],$

$\delta_{22} = \ln\left[\dfrac{(k_1 - k_3)\,(\alpha_{12}\,\kappa_{31} - \alpha_{32}\,\kappa_{11})}{\left(k_1 + k_1^*\right)\left(k_3 + k_1^*\right)}\right],\ \delta_{31} = \ln\left[\dfrac{(k_1 - k_2)\,(\alpha_{11}\,\kappa_{22} - \alpha_{21}\,\kappa_{12})}{\left(k_1 + k_2^*\right)\left(k_2 + k_2^*\right)}\right],$

$\delta_{32} = \ln\left[\dfrac{(k_1 - k_2)\,(\alpha_{12}\,\kappa_{22} - \alpha_{22}\,\kappa_{12})}{\left(k_1 + k_2^*\right)\left(k_2 + k_2^*\right)}\right],\ \delta_{41} = \ln\left[\dfrac{(k_2 - k_3)\,(\alpha_{21}\,\kappa_{32} - \alpha_{31}\,\kappa_{22})}{\left(k_2 + k_2^*\right)\left(k_3 + k_2^*\right)}\right],$

$\delta_{42} = \ln\left[\dfrac{(k_2 - k_3)\,(\alpha_{22}\,\kappa_{32} - \alpha_{32}\,\kappa_{22})}{\left(k_2 + k_2^*\right)\left(k_3 + k_2^*\right)}\right],\ \delta_{51} = \ln\left[\dfrac{(k_1 - k_3)\,(\alpha_{11}\,\kappa_{33} - \alpha_{31}\,\kappa_{13})}{\left(k_3 + k_3^*\right)\left(k_1 + k_3^*\right)}\right],$

$$\delta_{52} = \ln\left[\frac{(k_1 - k_3)\,(\alpha_{12}\,\kappa_{33} - \alpha_{32}\,\kappa_{13})}{\left(k_3 + k_3^*\right)\left(k_1 + k_3^*\right)}\right], \quad \delta_{61} = \ln\left[\frac{(k_2 - k_3)\,(\alpha_{21}\,\kappa_{33} - \alpha_{31}\,\kappa_{23})}{\left(k_2 + k_3^*\right)\left(k_3 + k_3^*\right)}\right],$$

$$\delta_{62} = \ln\left[\frac{(k_2 - k_3)\,(\alpha_{22}\,\kappa_{33} - \alpha_{32}\,\kappa_{23})}{\left(k_2 + k_3^*\right)\left(k_3 + k_3^*\right)}\right], \quad \delta_{71} = \ln\left[\frac{(k_2 - k_3)\,(\alpha_{21}\,\kappa_{31} - \alpha_{31}\,\kappa_{21})}{\left(k_2 + k_1^*\right)\left(k_3 + k_1^*\right)}\right],$$

$$\delta_{72} = \ln\left[\frac{(k_2 - k_3)\,(\alpha_{22}\,\kappa_{31} - \alpha_{32}\,\kappa_{21})}{\left(k_2 + k_1^*\right)\left(k_3 + k_1^*\right)}\right], \quad \delta_{81} = \ln\left[\frac{(k_1 - k_3)\,(\alpha_{11}\,\kappa_{32} - \alpha_{31}\,\kappa_{12})}{\left(k_1 + k_2^*\right)\left(k_3 + k_2^*\right)}\right],$$

$$\delta_{82} = \ln\left[\frac{(k_1 - k_3)\,(\alpha_{12}\,\kappa_{32} - \alpha_{32}\,\kappa_{12})}{\left(k_1 + k_2^*\right)\left(k_3 + k_2^*\right)}\right], \quad \delta_{91} = \ln\left[\frac{(k_1 - k_2)\,(\alpha_{11}\,\kappa_{23} - \alpha_{21}\,\kappa_{13})}{\left(k_1 + k_3^*\right)\left(k_2 + k_3^*\right)}\right],$$

$$\delta_{92} = \ln\left[\frac{(k_1 - k_2)\,(\alpha_{12}\,\kappa_{23} - \alpha_{22}\,\kappa_{13})}{\left(k_1 + k_3^*\right)\left(k_2 + k_3^*\right)}\right],$$

$$\tau_{10} = \ln\left[\frac{\left(k_2 - k_1\right)\left(k_3^* - k_1^*\right)}{\left(k_1 + k_1^*\right)\left(k_2 + k_1^*\right)\left(k_1 + k_3^*\right)\left(k_2 + k_3^*\right)}\,(\kappa_{11}\,\kappa_{23} - \kappa_{21}\,\kappa_{13})\right],$$

$$\tau_{20} = \ln\left[\frac{\left(k_1 - k_2\right)\left(k_3^* - k_2^*\right)}{\left(k_1 + k_2^*\right)\left(k_2 + k_2^*\right)\left(k_1 + k_3^*\right)\left(k_2 + k_3^*\right)}\,(\kappa_{22}\,\kappa_{13} - \kappa_{12}\,\kappa_{23})\right],$$

$$\tau_{30} = \ln\left[\frac{\left(k_3 - k_1\right)\left(k_3^* - k_2^*\right)}{\left(k_1 + k_2^*\right)\left(k_3 + k_2^*\right)\left(k_1 + k_3^*\right)\left(k_3 + k_3^*\right)}\,(\kappa_{33}\,\kappa_{12} - \kappa_{13}\,\kappa_{32})\right],$$

$$\tau_{11} = \ln\Bigg\{\frac{\left(k_2 - k_1\right)\left(k_3 - k_1\right)\left(k_3 - k_2\right)\left(k_2^* - k_1^*\right)}{\left(k_1 + k_1^*\right)\left(k_2 + k_1^*\right)\left(k_3 + k_1^*\right)\left(k_1 + k_2^*\right)\left(k_2 + k_2^*\right)\left(k_3 + k_2^*\right)}$$
$$\times\left[\alpha_{11}\,(\kappa_{21}\,\kappa_{32} - \kappa_{22}\,\kappa_{31}) + \alpha_{21}\,(\kappa_{12}\,\kappa_{31} - \kappa_{32}\,\kappa_{11}) + \alpha_{31}\,(\kappa_{11}\,\kappa_{22} - \kappa_{12}\,\kappa_{21})\right]\Bigg\},$$

$$\tau_{12} = \ln\Bigg\{\frac{\left(k_2 - k_1\right)\left(k_3 - k_1\right)\left(k_3 - k_2\right)\left(k_2^* - k_1^*\right)}{\left(k_1 + k_1^*\right)\left(k_2 + k_1^*\right)\left(k_3 + k_1^*\right)\left(k_1 + k_2^*\right)\left(k_2 + k_2^*\right)\left(k_3 + k_2^*\right)}$$
$$\times\left[\alpha_{12}\,(\kappa_{21}\,\kappa_{32} - \kappa_{22}\,\kappa_{31}) + \alpha_{22}\,(\kappa_{12}\,\kappa_{31} - \kappa_{32}\,\kappa_{11}) + \alpha_{32}\,(\kappa_{11}\,\kappa_{22} - \kappa_{12}\,\kappa_{21})\right]\Bigg\},$$

$$\tau_{21} = \ln\Bigg\{\frac{\left(k_2 - k_1\right)\left(k_3 - k_1\right)\left(k_3 - k_2\right)\left(k_3^* - k_1^*\right)}{\left(k_1 + k_1^*\right)\left(k_2 + k_1^*\right)\left(k_3 + k_1^*\right)\left(k_1 + k_3^*\right)\left(k_2 + k_3^*\right)\left(k_3 + k_3^*\right)}$$
$$\times\left[\alpha_{11}\,(\kappa_{33}\,\kappa_{21} - \kappa_{31}\,\kappa_{23}) + \alpha_{21}\,(\kappa_{31}\,\kappa_{13} - \kappa_{11}\,\kappa_{33}) + \alpha_{31}\,(\kappa_{23}\,\kappa_{11} - \kappa_{13}\,\kappa_{21})\right]\Bigg\},$$

$$\tau_{22} = \ln\Bigg\{\frac{\left(k_2 - k_1\right)\left(k_3 - k_1\right)\left(k_3 - k_2\right)\left(k_3^* - k_1^*\right)}{\left(k_1 + k_1^*\right)\left(k_2 + k_1^*\right)\left(k_3 + k_1^*\right)\left(k_1 + k_3^*\right)\left(k_2 + k_3^*\right)\left(k_3 + k_3^*\right)}$$
$$\times\left[\alpha_{12}\,(\kappa_{33}\,\kappa_{21} - \kappa_{31}\,\kappa_{23}) + \alpha_{22}\,(\kappa_{31}\,\kappa_{13} - \kappa_{11}\,\kappa_{33}) + \alpha_{32}\,(\kappa_{23}\,\kappa_{11} - \kappa_{13}\,\kappa_{21})\right]\Bigg\},$$

$$\tau_{31} = \ln\Bigg\{\frac{\left(k_2 - k_1\right)\left(k_3 - k_1\right)\left(k_3 - k_2\right)\left(k_3^* - k_2^*\right)}{\left(k_1 + k_2^*\right)\left(k_2 + k_2^*\right)\left(k_3 + k_2^*\right)\left(k_1 + k_3^*\right)\left(k_2 + k_3^*\right)\left(k_3 + k_3^*\right)}$$
$$\times\left[\alpha_{11}\,(\kappa_{22}\,\kappa_{33} - \kappa_{23}\,\kappa_{32}) + \alpha_{21}\,(\kappa_{13}\,\kappa_{32} - \kappa_{33}\,\kappa_{12}) + \alpha_{31}\,(\kappa_{12}\,\kappa_{23} - \kappa_{22}\,\kappa_{13})\right]\Bigg\},$$

$$\tau_{32} = \ln\Bigg\{\frac{\left(k_2 - k_1\right)\left(k_3 - k_1\right)\left(k_3 - k_2\right)\left(k_3^* - k_2^*\right)}{\left(k_1 + k_2^*\right)\left(k_2 + k_2^*\right)\left(k_3 + k_2^*\right)\left(k_1 + k_3^*\right)\left(k_2 + k_3^*\right)\left(k_3 + k_3^*\right)}$$
$$\times\left[\alpha_{12}\,(\kappa_{22}\,\kappa_{33} - \kappa_{23}\,\kappa_{32}) + \alpha_{22}\,(\kappa_{13}\,\kappa_{32} - \kappa_{33}\,\kappa_{12}) + \alpha_{32}\,(\kappa_{12}\,\kappa_{23} - \kappa_{22}\,\kappa_{13})\right]\Bigg\},$$

$$r_1 = \ln\left(\frac{\kappa_{11}}{k_1 + k_1^*}\right), \quad r_2 = \ln\left(\frac{\kappa_{22}}{k_2 + k_2^*}\right), \quad r_3 = \ln\left(\frac{\kappa_{33}}{k_3 + k_3^*}\right),$$

$$r_4 = \ln\left[\frac{\left(k_2 - k_1\right)\left(k_2^* - k_1^*\right)}{\left(k_1 + k_1^*\right)\left(k_2 + k_1^*\right)\left(k_1 + k_2^*\right)\left(k_2 + k_2^*\right)}\,(\kappa_{11}\,\kappa_{22} - \kappa_{12}\,\kappa_{21})\right],$$

$$r_5 = \ln\left[\frac{\left(k_3 - k_1\right)\left(k_3^* - k_1^*\right)}{\left(k_1 + k_1^*\right)\left(k_3 + k_1^*\right)\left(k_1 + k_3^*\right)\left(k_3 + k_3^*\right)}\left(\kappa_{33}\,\kappa_{11} - \kappa_{13}\,\kappa_{31}\right)\right],$$

$$r_6 = \ln\left[\frac{\left(k_3 - k_2\right)\left(k_3^* - k_2^*\right)}{\left(k_2 + k_2^*\right)\left(k_3 + k_2^*\right)\left(k_2 + k_3^*\right)\left(k_3 + k_3^*\right)}\left(\kappa_{22}\,\kappa_{33} - \kappa_{23}\,\kappa_{32}\right)\right],$$

$$r_7 = \ln\left\{\frac{(k_1 - k_2)\left(k_1^* - k_2^*\right)\left(k_2 - k_3\right)\left(k_2^* - k_3^*\right)\left(k_3 - k_1\right)\left(k_3^* - k_1^*\right)}{\left(k_1 + k_1^*\right)\left(k_2 + k_2^*\right)\left(k_3 + k_3^*\right)\left(k_1 + k_2^*\right)\left(k_2^* + k_1^*\right)\left(k_2 + k_3^*\right)\left(k_3 + k_2^*\right)\left(k_3 + k_1^*\right)\left(k_1 + k_3^*\right)}\right.$$

$$\times \left[\left(\kappa_{11}\,\kappa_{22}\,\kappa_{33} - \kappa_{11}\,\kappa_{23}\,\kappa_{32}\right) + \left(\kappa_{12}\,\kappa_{23}\,\kappa_{31} - \kappa_{12}\,\kappa_{21}\,\kappa_{33}\right)\right.$$

$$\left.\left. + \left(\kappa_{21}\,\kappa_{13}\,\kappa_{32} - \kappa_{22}\,\kappa_{13}\,\kappa_{31}\right)\right]\right\}, \quad b_0 = c_0 = 1, \quad b_1 = c_1 = b_2 = c_2 = 2\,\mu, \quad \mu \text{ is an}$$

arbitrary real constant, α_{11}, α_{12}, α_{13}, α_{21}, α_{22}, α_{23}, α_{31}, α_{32}, α_{33}, k_1, k_2, and k_3 are arbitrary complex constants.
- *Reference*: [6].

***Solution* 21.** *Dark–bright soliton*
  (Figure 7.1)

$$\psi_1(x,\,t) = \left(A_0 \tanh\{A_1\,[x - x_0 - v\,(t - t_0)]\} + i\,A_2\right)e^{i\,[b_1\,(t - t_0) + \phi_1]},$$

$$\psi_2(x,\,t) = B_0 \operatorname{sech}\{A_1\,[x - x_0 - v\,(t - t_0)]\}\,e^{i\,[(A_1^2 + b_2)\,(t - t_0) + \theta(x,\,t) + \phi_2]}, \tag{7.38}$$

where $\quad A_0 = \cos\left[\tan^{-1}\left(\dfrac{-v}{2\,A_1}\right)\right], \quad A_1 = \sqrt{\dfrac{b_1\,(1 - b_2^2)}{2\,(1 + b_1\,b_2)} - \dfrac{v^2}{4}}, \quad A_2 = \sin\left[\tan^{-1}\left(\dfrac{-v}{2\,A_1}\right)\right],$

$B_0 = \sqrt{\dfrac{2\,A_1^2\,(b_1 + b_2)}{1 - b_2^2}}, \quad \theta(x,\,t) = \dfrac{v}{2}\,[x - x_0 - \dfrac{v}{2}\,(t - t_0)], \quad b_0 = -1, \quad c_0 = 1, \quad c_1 = 1/b_1,$

$c_2 = b_2, \; b_1\,b_2 = 1, \; \dfrac{b_1\,(1 - b_2^2)}{2\,(1 + b_1\,b_2)} > \dfrac{v^2}{4}, \; 2\,A_1^2\,(b_1 + b_2)\,(1 - b_2^2) > 0, \; x_0, \; t_0, \; \phi_1, \; \phi_2, \text{ and}$

v are arbitrary real constants.
- *Reference*: [7].

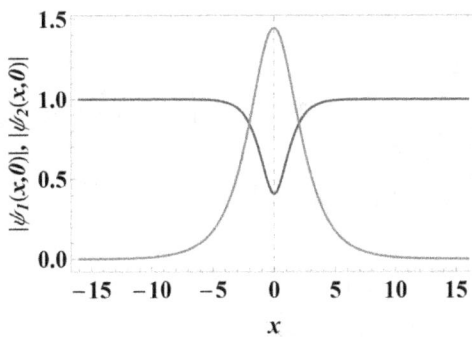

Figure 7.1. Dark-bright soliton (7.38) at $t = 0$. Blue is ψ_1 and red is ψ_2 with $b_2 = 1/2$, $v = 1/2$, and $x_0 = t_0 = \phi_1 = \phi_2 = 0$. Animation available online at http://doi.org/10.1088/978-0-7503-5954-2.

Solution 22.

(Figure 7.2)

$$\psi_1(x, t) = A_0 \left(1 - 4 \left[\alpha_1^2(x, t) + \alpha_2^2(x, t) - \alpha_1(x, t) + i \alpha_2(x, t)\right] e^{-\delta(x, t)}\right.$$

$$\times \left\{\left[2 \alpha_1^2(x, t) + 2 \alpha_2^2(x, t) - 2 \alpha_1(x, t) + 1\right] e^{-\delta(x, t)} + (\beta_3^2 + \beta_4^2) e^{2 \delta(x, t)}\right\}^{-1}\right)$$

$$\times e^{i [A_1 x + (2 A_0^2 - A_1^2) t]},$$

$$\psi_2(x, t) = - 4 A_0 \{\beta_3 [\alpha_1(x, t) - 1] - \beta_4 \alpha_2(x, t) + i [\beta_3 \alpha_2(x, t) + \beta_4 \alpha_1(x, t) - \beta_4]\} \qquad (7.39)$$

$$\times \left(\left\{2 \left[\alpha_1^2(x, t) + \alpha_2^2(x, t)\right] - 2 \alpha_1(x, t) + 1\right\} e^{-\delta(x, t)} + (\beta_3^2 + \beta_4^2) e^{2 \delta(x, t)}\right)^{-1}$$

$$\times e^{i \left[A_1 x + (3 A_0^2 - A_1^2)\right] + \frac{\delta(x, t)}{2}},$$

where $\qquad \delta(x, t) = \frac{2 A_0 (x - 2 A_1 t)}{3}, \qquad \alpha_1(x, t) = \beta_1 A_0 + A_0 (x - 2 A_1 t),$
$\alpha_2(x, t) = \beta_2 A_0 - 2 A_0^2 t$, $b_0 = c_0 = 1$, $b_1 = b_2 = c_1 = c_2 = 2$, A_0, A_1, β_1, β_2, β_3,
and β_4 are arbitrary real constants.

- *Reference*: [8].

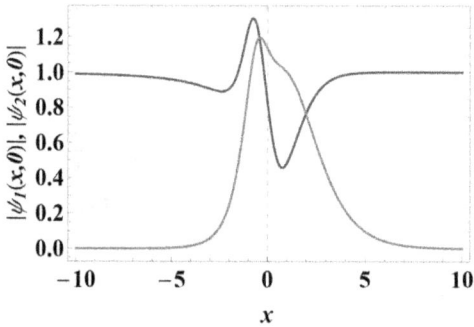

Figure 7.2. Plot of solution (7.39) at $t = 0$. Blue is ψ_1 and red is ψ_2 with $A_0 = A_1 = \beta_1 = \beta_2 = \beta_3 = \beta_4 = 1$.
Animation available online at http://doi.org/10.1088/978-0-7503-5954-2.

Solution 23.

(Figure 7.3)

$$\psi_1(x, t) = A_0 \left(\frac{-6 \sqrt{3} A_0 \delta(x, t) - 36 \sqrt{3} A_0^2 t + i \left[36 A_0^2 t + 6 A_0 \delta(x, t) + 5 \sqrt{3}\right] - 3}{12 A_0^2 \delta^2(x, t) + 8 \sqrt{3} A_0 \delta(x, t) + 144 A_0^4 t^2 + 5}\right.$$

$$\left. - (1 + i \sqrt{3})\right) \times e^{i [A_1 x + (16 A_0^2 - A_1^2) t]},$$

$$\qquad (7.40)$$

$$\psi_2(x, t) = A_0 \left(\frac{-6 \sqrt{3} A_0 \delta(x, t) + 36 \sqrt{3} A_0^2 t + i \left[36 A_0^2 t - 6 A_0 \delta(x, t) - 5 \sqrt{3}\right] - 3}{12 A_0^2 \delta^2(x, t) + 8 \sqrt{3} A_0 \delta(x, t) + 144 A_0^4 t^2 + 5}\right.$$

$$\left. - (1 - i \sqrt{3})\right) \times e^{i [A_2 x + (16 A_0^2 - A_2^2) t]},$$

where $\delta(x, t) = x + 6 A_3 t$, $b_0 = c_0 = 1$, $b_1 = b_2 = c_1 = c_2 = 2$, $A_0 = A_2 + 3 A_3$, $A_1 = A_2 - 2 A_0$, A_2 and A_3 are arbitrary real constants.

- *Reference*: [8].

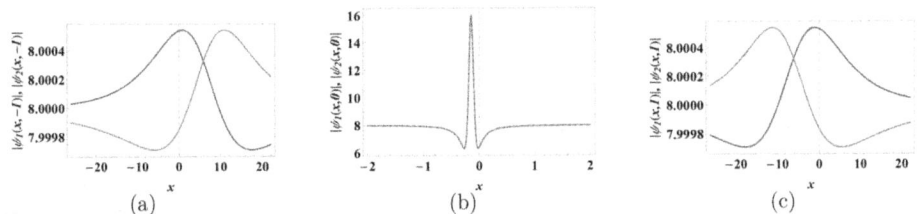

Figure 7.3. Plot of solution (7.40). Blue is ψ_1 and red is ψ_2 with $A_2 = A_3 = 1$. (a) at $t = -1$, (b) at $t = 0$, and (c) at $t = 1$.

Solution 24.
(Figure 7.4)

$$\psi_1(x, t) = A_0 \left[\frac{\alpha_1(x, t) + i\,\beta_1(x, t)}{\gamma(x, t)} - (1 + i\,\sqrt{3}) \right] e^{i\,[A_1\,x + (16\,A_0^2 - A_1^2)\,t]},$$

$$\psi_2(x, t) = A_0 \left[\frac{\alpha_2(x, t) + i\,\beta_2(x, t)}{\gamma(x, t)} - (1 - i\,\sqrt{3}) \right] e^{i\,[A_2\,x + (16\,A_0^2 - A_2^2)\,t]},$$

(7.41)

where

$$\alpha_1(x, t) = -864\sqrt{3}\,A_0^6\,t^3 - 144\sqrt{3}\,A_0^5\,\delta(x, t)\,t^2 - 72\sqrt{3}\,A_0^4\,\delta^2(x, t)\,t - 216\,A_0^4\,t^2$$
$$- 12\sqrt{3}\,A_0^3\,\delta^3(x, t) - 144\,A_0^3\,\delta(x, t)\,t - 18\,A_0^2\,\delta^2(x, t) - 12\sqrt{3}\,A_0^2\,t + 3,$$

$$\alpha_2(x, t) = 864\,A_0^6\,\sqrt{3}\,t^3 - 144\sqrt{3}\,A_0^5\,\delta(x, t)\,t^2 + 72\sqrt{3}\,A_0^4\,\delta^2(x, t)\,t - 216\,A_0^4\,t^2$$
$$- 12\sqrt{3}\,A_0^3\,\delta^3(x, t) + 144\,A_0^3\,\delta(x, t)\,t - 18\,A_0^2\,\delta^2(x, t) + 12\sqrt{3}\,A_0^2\,t + 3,$$

$$\beta_1(x, t) = 864\,A_0^6\,t^3 + 144\,A_0^5\,\delta(x, t)\,t^2 + 72\,A_0^4\,\delta^2(x, t)\,t + 312\sqrt{3}\,A_0^4\,t^2$$
$$+ 12\,A_0^3\,\delta^3(x, t) + 96\sqrt{3}\,A_0^3\,\delta(x, t)\,t + 18\sqrt{3}\,A_0^2\,\delta^2(x, t) + 108\,A_0^2\,t$$
$$+ 12\,A_0\,\delta(x, t) + \sqrt{3},$$

$$\beta_2(x, t) = 864\,A_0^6\,t^3 - 144\,A_0^5\,\delta(x, t)\,t^2 + 72\,A_0^4\,\delta^2(x, t)\,t - 312\sqrt{3}\,A_0^4\,t^2$$
$$- 12\,A_0^3\,\delta^3(x, t) + 96\sqrt{3}\,A_0^3\,\delta(x, t)\,t - 18\sqrt{3}\,A_0^2\,\delta^2(x, t) + 108\,A_0^2\,t$$
$$- 12\,A_0\,\delta(x, t) - \sqrt{3},$$

$$\gamma(x, t) = 1728\,A_0^8\,t^4 + 288\,A_0^6\,\delta^2(x, t)\,t^2 + 384\sqrt{3}\,A_0^5\,\delta(x, t)\,t^2 + 12\,A_0^4\,\delta^4(x, t)$$
$$+ 432\,A_0^4\,t^2 + 16\sqrt{3}\,A_0^3\,\delta^3(x, t) + 24\,A_0^2\,\delta^2(x, t) + 4\sqrt{3}\,A_0\,\delta(x, t) + 1,$$

$A_0 = A_2 + 3 A_3$, $A_1 = A_2 - 2 A_0$, $A_2 = -6A_3$, $\delta(x, t) = x + 6 A_3 t$, $b_0 = c_0 = 1$, $b_1 = b_2 = c_1 = c_2 = 2$, A_3 is an arbitrary real constant.

• *Reference*: [8].

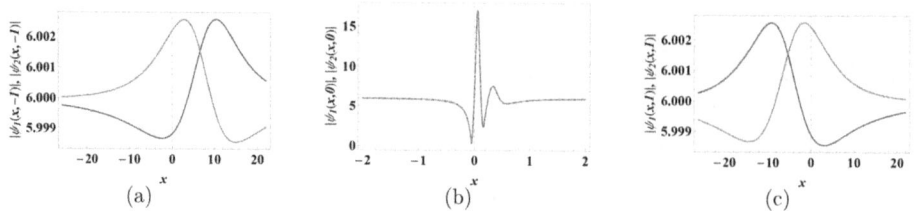

(a) (b) (c)

Figure 7.4. Plot of solution (7.41). Blue is ψ_1 and red is ψ_2 with $A_2 = A_3 = 1$. (a) at $t = -1$, (b) at $t = 0$, and (c) at $t = 1$.

Solution 25.

(Figure 7.5)

$$\psi_1(x, t) = \{A_0 \operatorname{sech}^2[A_1 (x - x_0)] + A_3\} e^{-i \left[\omega_1 (t-t_0)+\phi_1\right]},$$

$$\psi_2(x, t) = B_0 \operatorname{sech}^2[A_1 (x - x_0)] e^{-i \left[\omega_2 (t-t_0)+\phi_2\right]}, \tag{7.42}$$

where $A_3 = \dfrac{-2 A_0}{3}$, $\omega_1 = 2 A_1^2$, $\omega_2 = -2 A_1^2$, $b_1 = \dfrac{-9 A_1^2}{2 A_0^2}$, $b_2 = \dfrac{9 A_1^2}{2 B_0^2}$, $b_0 = c_0 = 1$, $b_1 = c_1$, $b_2 = c_2$, $A_0 \neq 0$, $B_0 \neq 0$, A_1, B_1, x_0, t_0, ϕ_1, and ϕ_2 are arbitrary real constants.

• *Reference*: [9], *taken from the nonlocal case.*

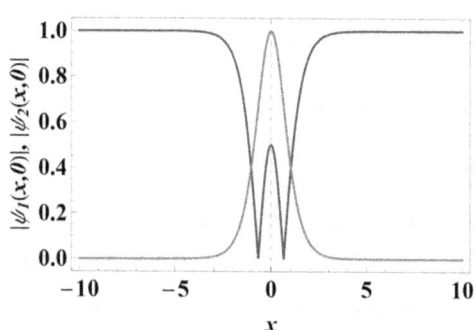

Figure 7.5. Plot of solution (7.42) at $t = 0$. Blue is ψ_1 and red is ψ_2 with $A_0 = 3/2$, $B_0 = B_1 = A_1 = 1$, $b_1 = b_2 = 2$, and $x_0 = t_0 = \phi_1 = \phi_2 = 0$.

Solution 26. *Solitary wave (SW)*

(Figure 7.6)

$$\psi_1(x, t) = A_0 \sqrt{m} \operatorname{cd}[A_1 (x - x_0), m] e^{-i \left[\omega_1 (t-t_0)+\phi_1\right]},$$

$$\psi_2(x, t) = B_0 \sqrt{1 - m} \operatorname{nd}[A_1 (x - x_0), m] e^{-i \left[\omega_2 (t-t_0)+\phi_2\right]}, \tag{7.43}$$

where $\omega_1 = -(1-m)\,A_1^2 - b_1\,A_0^2$, $\omega_2 = -(2-m)\,A_1^2 - c_1\,A_0^2$, $b_0 = c_0 = 1$,
$b_1 = \dfrac{-2\,A_1^2 + b_2\,B_0^2}{A_0^2}, c_1 = \dfrac{-2\,A_1^2 + c_2\,B_0^2}{A_0^2}, 0 \leqslant m \leqslant 1$, A_0, B_0, x_0, t_0, ϕ_1, and ϕ_2 are arbitrary
real constants.

- *Reference*: [9], *taken from the nonlocal case.*

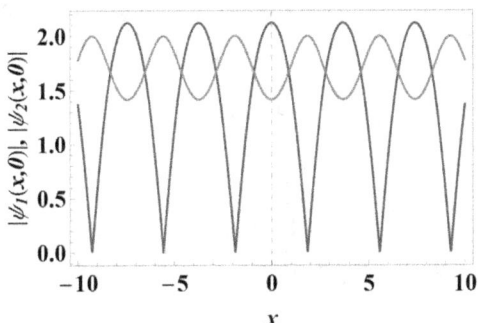

Figure 7.6. Solitary wave (7.43) at $t = 0$. Blue is ψ_1 and red is ψ_2 with $A_0 = 3$, $B_0 = 2$, $A_1 = B_1 = 1$,
$b_2 = c_2 = 2$, $m = 1/2$, and $x_0 = t_0 = \phi_1 = \phi_2 = 0$.

Solution **27. Weierstrass Elliptic Function I:** $\wp(z, g_2, g_3)$ *SW*

$$\psi_1(x,\,t) = \Phi_1(z)\,e^{i\,\phi_1(z,\,t)},$$
$$\psi_2(x,\,t) = \Phi_2(z)\,e^{i\,\phi_2(z,\,t)},$$
(7.44)

where

$$\Phi_1(z) = \sqrt{F\,\wp^2(z,\,g_2,\,g_3) + A_0\,\wp(z,\,g_2,\,g_3) + A_1}\,,$$
$$\Phi_2(z) = \sqrt{G\,\wp^2(z,\,g_2,\,g_3) + B_0\,\wp(z,\,g_2,\,g_3) + B_1}\,,$$
(7.45)

which satisfy

$$-A_2^2\,b_0 + [b_1\,\Phi_1^2(z) + b_2\,\Phi_2^2(z) - \gamma]\,\Phi_1^4(z) + b_0\,\Phi_1^3(z)\,\Phi_1^{''}(z) = 0,$$
$$-B_2^2\,c_0 + [b_2\,\Phi_1^2(z) + b_1^{-1}\,\Phi_2^2(z) - \beta]\,\Phi_2^4(z) + c_0\,\Phi_2^3(z)\,\Phi_2^{''}(z) = 0,$$
(7.46)

with

$\phi_1(z,\,t) = \dfrac{b_0\,\alpha}{2}\,z - b_0\,A_2\left[\int \dfrac{d z}{\Phi_1^2(z)}\right] + \phi_{01}(t),$

$\phi_2(z,\,t) = \dfrac{c_0\,\alpha}{2}\,z - c_0\,B_2\left[\int \dfrac{d z}{\Phi_2^2(z)}\right] + \phi_{02}(t), c_1 = b_2, c_2 = b_1^{-1}, z = x - c\,t,$

$\phi_{01}(t) = \left(\gamma + \dfrac{b_0\,\alpha^2}{4}\right)t + \phi_{011}, \phi_{02}(t) = \left(\beta + \dfrac{c_0\,\alpha^2}{4}\right)t + \phi_{022},$

$\gamma = b_1\,A_1 + b_2\,B_1 - 3\,b_0\,\dfrac{A_0}{F}, \beta = b_2\,A_1 + b_1^{-1}\,B_1 - 3\,c_0\,\dfrac{B_0}{G}, A_1 = \dfrac{4\,A_0^2 - F^2\,g_2}{4\,F},$

$A_2 = -\sqrt{\dfrac{3\,F^2}{4}\left(\dfrac{A_0^2}{F^2} - \dfrac{g_2}{3}\right)\left(4\,\dfrac{A_0^3}{F^3} - \dfrac{g_2\,A_0}{F} - g_3\right)}, B_0 = \dfrac{-6\,b_0 - b_1\,A_0}{b_2},$

$$B_2 = -\sqrt{\frac{3\,G^2}{4}\left(\frac{B_0^2}{G^2} - \frac{g_2}{3}\right)\left(4\,\frac{B_0^3}{G^3} - \frac{g_2\,B_0}{G} - g_3\right)},\ g_2 = \frac{4\,(B_0^2 - B_1\,G)}{G^2},\ G = \frac{-b_1\,F}{b_2},\ F \neq 0,$$

$b_1 = \frac{b_0\,b_2}{c_0}$, $b_2 = 1$, b_0, c_0, A_0, B_1, g_3, α, c, ϕ_{011}, and ϕ_{022} are arbitrary real constants.

- *Reference*: [10].

Solution 28. Weierstrass Elliptic Function II: $\wp(z,g_2,g_3)$ *SW*

$$\begin{aligned}
\psi_1(x,\ t) &= \Phi_1(z)\ e^{i\,\phi_1(z,\ t)}, \\
\psi_2(x,\ t) &= \Phi_2(z)\ e^{i\,\phi_2(z,\ t)},
\end{aligned} \tag{7.47}$$

where

$$\begin{aligned}
\Phi_1(z) &= \sqrt{A_0\ \wp(z,\ g_2,\ g_3) + A_1}, \\
\Phi_2(z) &= \sqrt{B_0\ \wp(z,\ g_2,\ g_3) + B_1},
\end{aligned} \tag{7.48}$$

which satisfy (7.46), with $\phi_1(z,\ t) = \frac{b_0\,\alpha}{2}\,z - b_0\,A_2\left[\int \frac{d\,z}{\Phi_1^2(z)}\right] + \phi_{01}(t)$, $\phi_2(z,\ t) =$

$\frac{c_0\,\alpha}{2}\,z - c_0\,B_2\left[\int \frac{d\,z}{\Phi_2^2(z)}\right] + \phi_{02}(t)$, $c_1 = b_2$, $c_2 = b_1^{-1}$, $z = x - c\,t$, $\phi_{01}(t) =$

$\left(\gamma + \frac{b_0\,\alpha^2}{4}\right)t + \phi_{011}$, $\phi_{02}(t) = \left(\beta + \frac{c_0\,\alpha^2}{4}\right)t + \phi_{022}$, $\gamma = \frac{b_2\,(B_1\,A_0 - A_1\,B_0) - 3\,b_0\,A_1}{A_0}$,

$\beta = \frac{b_2\,(A_1\,B_0 - B_1\,A_0) - 3\,c_0\,B_1}{B_0}$, $A_0 = \frac{2\,(b_0\,c_0\,b_1 - b_0)}{b_1\,(1 - b_2^2)}$, $A_2 = -\sqrt{\frac{A_0^2}{4}\left(\frac{-A_1\,g_2}{A_0} + g_3 + \frac{4\,A_1^3}{A_0^3}\right)}$,

$B_0 = \frac{2\,(b_0\,b_2 - c_0\,b_1)}{1 - b_2^2}$, $B_2 = -\sqrt{\frac{B_0^2}{4}\left(\frac{-B_1\,g_2}{B_0} + g_3 + \frac{4\,B_1^3}{B_0^3}\right)}$, b_0, b_1, b_2, c_0, A_1, B_1, g_2, g_3, α,

c, ϕ_{011}, and ϕ_{022} are arbitrary real constants.

- *Reference*: [10].

7.3 Summary of Section 7.2

Note: For lengthy conditions, the reader is referred to the solutions in Section 7.2.

Equation (1): $i\psi_{1t} + b_0\psi_{1xx} + (b_1|\psi_1|^2 + b_2|\psi_2|^2)\psi_1 = 0, \; i\psi_{2t} + c_0\psi_{2xx} + (c_1|\psi_1|^2 + c_2|\psi_2|^2)\psi_2 = 0$

#	Solution	Conditions	Name	Equation #
1.	$\psi_1(x,t) = A_0 e^{i\phi_1}, \; \psi_2(x,t) = B_0 e^{i\phi_2}$	$b_1 = -\frac{b_2 B_0^2}{A_0^2}, \; c_2 = -\frac{c_1 A_0^2}{B_0^2}, A_0, B_0, \phi_1,$ and ϕ_2 are arbitrary real constants	Continuous wave, t- and x-independent phase	(7.18)
2.	$\psi_1(x,t) = A_0 e^{i[A_1(t-t_0)+\phi_1]}, \; \psi_2(x,t) = B_0 e^{i[B_1(t-t_0)+\phi_2]}$	$A_1 = A_0^2 b_1 + B_0^2 b_2, \; B_1 = A_0^2 c_1 + B_0^2 c_2, A_0, B_0, t_0, \phi_1,$ and ϕ_2 are arbitrary real constants	Continuous wave, t-dependent phase	(7.19)
3.	$\psi_1(x,t) = A_0 e^{i[A_1(x-x_0)+\phi_1]},$ $\psi_2(x,t) = B_0 e^{i[B_1(x-x_0)+\phi_2]}$	$A_1 = \pm\sqrt{\frac{A_0^2 b_1 + B_0^2 b_2}{b_0}}, \; B_1 = \pm\sqrt{\frac{A_0^2 c_1 + B_0^2 c_2}{c_0}}$, $b_0(A_0^2 b_1 + B_0^2 b_2) > 0, c_0(A_0^2 c_1 + B_0^2 c_2) > 0, A_0, B_0, x_0, \phi_1,$ and ϕ_2 are arbitrary real constants	Continuous wave, x-dependent phase	(7.20)
4.	$\psi_1(x,t) = A_0 e^{i[A_1(t-t_0)+A_2(x-x_0)+\phi_1]},$ $\psi_2(x,t) = B_0 e^{i[B_1(t-t_0)+B_2(x-x_0)+\phi_2]}$	$A_1 = -A_2^2 b_0 + A_0^2 b_1 + B_0^2 b_2, B_1 = -B_2^2 c_0 + A_0^2 c_1 + B_0^2 c_2, A_0, B_0, A_2, B_2, x_0, t_0, \phi_1,$ and ϕ_2 are arbitrary real constants	Continuous wave, t- and x-dependent phase	(7.21)
5.	$\psi_1(x,t) = \frac{A_0}{\sqrt{2}|A_1+t-t_0|} e^{i\phi_1(x,t)},$ $\psi_2(x,t) = \frac{B_0}{\sqrt{2}|B_1+t-t_0|} e^{i\phi_2(x,t)}$	$\phi_1(x,t) = \frac{|b_0 A_2 + (x-x_0)|^2}{4 b_0[A_1+t-t_0]} + \frac{b_1 A_0^2}{2}\ln[2(A_1$ $+ t - t_0)] + \frac{b_2 B_0^2}{2}\ln[2(B_1 + t - t_0)]$ $+ \phi_{01},$ $\phi_2(x,t) = \frac{|c_0 B_2 + (x-x_0)|^2}{4 c_0[B_1+t-t_0]} + \frac{c_1 A_0^2}{2}\ln[2(A_1$ $+ t - t_0)] + \frac{c_2 B_0^2}{2}\ln[2(B_1 + t - t_0)]$ $+ \phi_{02},$ $A_0, A_1, A_2, B_0, B_1, B_2, x_0, t_0, \phi_{01}, \phi_{02}$ are arbitrary real constants	Decaying wave	(7.22)
6.	$\psi_1(x,t) = \frac{A_0[3+(x-x_0)^2]}{\alpha+(x-x_0)^2} e^{i[A_1(t-t_0)+\phi_1]},$ $\psi_2(x,t) = \frac{B_0(x-x_0)}{\alpha+(x-x_0)^2} e^{i[B_1(t-t_0)+\phi_2]}$	$A_1 = 3 b_0/\beta, \; B_1 = A_0^2 c_1, \; \alpha = -3\beta, \; c_0 = 4 A_0^2 c_1 \beta/9,$ $b_1 = 3 b_0/(A_0^2 \beta), \; b_2 = -24 b_0/B_0^2,$ $c_2 = -80 A_0^2 c_1 \beta/(9 B_0^2), A_0, B_0, \beta, b_0, b_0, x_0, t_0, \phi_{01}$ and ϕ_{02} are arbitrary real constants	—	(7.23)

7.

$$\psi_1(x,t) = \frac{A_0[\beta + (x-x_0)^2]}{a + (x-x_0)^2}\, e^{i[A_1(t-t_0)+\phi_1]},$$
$$\psi_2(x,t) = \frac{B_0(x-x_0)}{\sqrt{a + (x-x_0)^2}}\, e^{i[B_1(t-t_0)+\phi_2]}$$

$A_1 = 5b_0/(2\beta)$, $B_1 = -5A_0^2 c_1/3$, $\alpha = -3\beta$, $c_0 = -16A_0^2 c_1\beta/9$, $b_1 = -3b_0/(2A_0^2\beta)$, $b_2 = 4b_0/(B_0^2\beta)$, $c_2 = -8A_0^2 c_1/(3B_0^2)$, $A_0, B_0, \beta, b_0, x_0, t_0, \phi_{01}$ and ϕ_{02} are arbitrary real constants — (7.24)

8.

$$\psi_1(x,t) = \frac{A_0(x-x_0)}{a + (x-x_0)^2}\, e^{i[A_1(t-t_0)+\phi_1]},$$
$$\psi_2(x,t) = \frac{B_0(x-x_0)}{\sqrt{a + (x-x_0)^2}}\, e^{i[B_1(t-t_0)+\phi_2]}$$

$A_1 = -6b_0/\alpha$, $B_1 = -3c_0/\alpha$, $b_1 = -8b_0/A_0^2$, $b_2 = -6b_0/(B_0^2\alpha)$, $c_1 = -3c_0/A_0^2$, $c_2 = -3c_0/(B_0^2\alpha)$, $A_0, B_0, \alpha, b_0, c_0, x_0, t_0, \phi_{01}$ and ϕ_{02} are arbitrary real constants — (7.25)

9.

$$\psi_1(x,t) = \frac{A_0}{a + \cos[\frac{\beta}{\sqrt{b_0}}(x-x_0)]}\, e^{i[A_1(t-t_0)+\phi_1]},$$
$$\psi_2(x,t) = \frac{B_0\sin[\frac{\beta}{\sqrt{b_0}}(x-x_0)]}{a + \cos[\frac{\beta}{\sqrt{b_0}}(x-x_0)]}\, e^{i[B_1(t-t_0)+\phi_2]}$$

$A_1 = B_1 = \beta^2/2$, $b_0 = c_0$, $b_1 = c_1/3$, $b_2 = -3\beta^2/(2B_0^2)$, $c_1 = 3(\alpha^2-1)\beta^2/(2A_0^2)$, $c_2 = b_2/3$, $A_0, B_0, \alpha, \beta, c_0, x_0, t_0, \phi_{01}$ and ϕ_{02} are arbitrary real constants — (7.26)

10.

$$\psi_1(x,t) = \frac{A_0\cos[\frac{\beta}{\sqrt{b_0}}(x-x_0)]}{1 + a\cos^2[\frac{\beta}{\sqrt{b_0}}(x-x_0)]}\, e^{i[A_1(t-t_0)+\phi_1]},$$
$$\psi_2(x,t) = \frac{B_0\sin[\frac{\beta}{\sqrt{b_0}}(x-x_0)]}{1 + a\cos^2[\frac{\beta}{\sqrt{b_0}}(x-x_0)]}\, e^{i[B_1(t-t_0)+\phi_2]}$$

$A_1 = B_1 = -\beta^2$, $b_0 = c_0$, $b_1 = c_1/3$, $b_2 = 6\alpha\beta^2/B_0^2$, $c_1 = -6\alpha\beta^2(1+\alpha)/A_0^2$, $c_2 = b_2/3$, $A_0, B_0, \alpha, \beta, c_0, x_0, t_0, \phi_{01}$ and ϕ_{02} are arbitrary real constants — (7.27)

11.

$$\psi_1(x,t) = \frac{A_0\cos[\frac{\beta}{\sqrt{b_0}}(x-x_0)]}{1 + a\cos^2[\frac{\beta}{\sqrt{b_0}}(x-x_0)]}\, e^{i[A_1(t-t_0)+\phi_1]},$$
$$\psi_2(x,t) = \frac{B_0}{1 + a\cos^2[\frac{\beta}{\sqrt{b_0}}(x-x_0)]}\, e^{i[B_1(t-t_0)+\phi_2]}$$

$A_1 = -\beta^2$, $B_1 = -4\beta^2$, $b_0 = c_0$, $b_1 = -2\alpha(\alpha+4)\beta^2/A_0^2$, $b_2 = 6\alpha\beta^2/B_0^2$, $c_1 = -6\alpha\beta^2(2+\alpha)/A_0^2$, $c_2 = -2(2-\alpha)\beta^2/B_0^2$, $A_0, B_0, \alpha, \beta, c_0, x_0, t_0, \phi_{01}$ and ϕ_{02} are arbitrary real constants — (7.28)

12.

$$\psi_1(x,t) = \frac{A_0}{1 + a\cos^2[\frac{\beta}{\sqrt{b_0}}(x-x_0)]}\, e^{i[A_1(t-t_0)+\phi_1]},$$
$$\psi_2(x,t) = \frac{B_0\sin[\frac{\beta}{\sqrt{b_0}}(x-x_0)]}{1 + a\cos^2[\frac{\beta}{\sqrt{b_0}}(x-x_0)]}\, e^{i[B_1(t-t_0)+\phi_2]}$$

$A_1 = -4\beta^2$, $B_1 = -\beta^2$, $b_0 = c_0$, $b_1 = -2(\alpha+1)(3\alpha+2)\beta^2/A_0^2$, $b_2 = 6\alpha(\alpha+2)\beta^2/B_0^2$, $c_1 = -6\alpha\beta^2(1+\alpha)/A_0^2$, $c_2 = 2\alpha(3\alpha+4)\beta^2/B_0^2$, $A_0, B_0, \alpha, \beta, c \leq 0, x_0, t_0, \phi_{01}$ and ϕ_{02} are arbitrary real constants — (7.29)

13.

$$\psi_1(x,t) = \frac{A_0}{1 + a\cos^2[\frac{\beta}{\sqrt{b_0}}(x-x_0)]}\, e^{i[A_1(t-t_0)+\phi_1]},$$
$$\psi_2(x,t) = \frac{B_0\sin[\frac{\beta}{\sqrt{b_0}}(x-x_0)]\cos[\frac{\beta}{\sqrt{b_0}}(x-x_0)]}{1 + a\cos^2[\frac{\beta}{\sqrt{b_0}}(x-x_0)]}\, e^{i[B_1(t-t_0)+\phi_2]}$$

$A_1 = B_1 = 2\beta^2$, $b_0 = c_0$, $b_1 = 2(\alpha+1)\beta^2/A_0^2$, $b_2 = -6\alpha^2\beta^2/B_0^2$, $c_1 = 6\beta^2(1+\alpha)/A_0^2$, $c_2 = -2\alpha^2\beta^2/B_0^2$, $A_0, B_0, \alpha, \beta, c_0, x_0, t_0, \phi_{01}$ and ϕ_{02} are arbitrary real constants — (7.30)

(Continued)

#	Solution	Conditions	Type	Eq.
14.	$\psi_1(x,t) = \dfrac{A_0\sqrt{m}\,\mathrm{sn}[\frac{\beta}{\sqrt{b_0}}(x-x_0),m]}{\alpha + \mathrm{dn}[\frac{\beta}{\sqrt{b_0}}(x-x_0),m]}\, e^{i[A_1(t-t_0)+\phi_1]},$ $\psi_2(x,t) = \dfrac{B_0\sqrt{m}\,\mathrm{cn}[\frac{\beta}{\sqrt{b_0}}(x-x_0),m]}{\alpha + \mathrm{dn}[\frac{\beta}{\sqrt{b_0}}(x-x_0),m]}\, e^{i[B_1(t-t_0)+\phi_2]}$	$A_1 = B_1 = -(2-m)\,\beta^2/2$, $b_0 = c_0$, $b_1 = c_1/3$, $b_2 = 3\,\beta^2(\alpha^2-1)/(2\,B_0^2)$, $c_1 = -3\,\beta^2(\alpha^2+m-1)/(2\,A_0^2)$, $c_2 = b_2/3$, $A_0, B_0, m, \alpha, \beta, c_0, x_0, t_0, \phi_{01}$ and ϕ_{02} are arbitrary real constants	—	(7.31)
15.	$\psi_1(x,t) = \dfrac{A_0\tanh[\frac{\beta}{\sqrt{b_0}}(x-x_0)]}{\alpha + \mathrm{sech}[\frac{\beta}{\sqrt{b_0}}(x-x_0)]}\, e^{i[A_1(t-t_0)+\phi_1]},$ $\psi_2(x,t) = \dfrac{B_0\,\mathrm{sech}[\frac{\beta}{\sqrt{b_0}}(x-x_0)]}{\alpha + \mathrm{sech}[\frac{\beta}{\sqrt{b_0}}(x-x_0)]}\, e^{i[B_1(t-t_0)+\phi_2]}$	$A_1 = B_1 = -\beta^2/2$, $b_0 = c_0$, $b_1 = c_1/3$, $b_2 = 3\,\beta^2(\alpha^2-1)/(2\,B_0^2)$, $c_1 = -3\,\beta^2\alpha^2/(2\,A_0^2)$, $c_2 = b_2/3$, $A_0, B_0, \alpha, \beta, m, c_0, x_0, t_0, \phi_{01}$ and ϕ_{02} are arbitrary real constants	—	(7.32)
16.	$\psi_1(x,t) = \dfrac{A_0\,\mathrm{sn}[\frac{\beta}{\sqrt{b_0}}(x-x_0),m]}{\alpha + \mathrm{cn}[\frac{\beta}{\sqrt{b_0}}(x-x_0),m]}\, e^{i[A_1(t-t_0)+\phi_1]},$ $\psi_2(x,t) = \dfrac{B_0\,\mathrm{dn}[\frac{\beta}{\sqrt{b_0}}(x-x_0),m]}{\alpha + \mathrm{cn}[\frac{\beta}{\sqrt{b_0}}(x-x_0),m]}\, e^{i[B_1(t-t_0)+\phi_2]}$	$A_1 = B_1 = -(2m-1)\,\beta^2/2$, $b_0 = c_0$, $b_1 = c_1/3$, $b_2 = 3\,\beta^2(\alpha^2-1)/(2\,B_0^2)$, $c_1 = -3\,\beta^2(1-m+\alpha^2 m)/(2\,A_0^2)$, $c_2 = b_2/3$, $A_0, B_0, \alpha, \beta, m, c_0, x_0, t_0, \phi_{01}$ and ϕ_{02} are arbitrary real constants	—	(7.33)
17.	$\psi_1(x,t) = \dfrac{A_0\,\mathrm{dn}[\frac{\beta}{\sqrt{b_0}}(x-x_0),m]}{\alpha + \mathrm{sn}[\frac{\beta}{\sqrt{b_0}}(x-x_0),m]}\, e^{i[A_1(t-t_0)+\phi_1]},$ $\psi_2(x,t) = \dfrac{B_0\,\mathrm{dn}[\frac{\beta}{\sqrt{b_0}}(x-x_0),m]}{\alpha + \mathrm{sn}[\frac{\beta}{\sqrt{b_0}}(x-x_0),m]}\, e^{i[B_1(t-t_0)+\phi_2]}$	$A_1 = B_1 = (m+1)\,\beta^2/2$, $b_0 = c_0$, $b_1 = (\alpha^2-1)\,\beta^2/(2\,A_0^2)$, $b_2 = 3\,\beta^2(m\,\alpha^2-1)/(2\,B_0^2)$, $c_1 = 3\,\beta^2(\alpha^2-1)/(2\,A_0^2)$, $c_2 = (m\,\alpha^2-1)\,\beta^2/(2\,B_0^2)$, $A_0, B_0, \alpha, \beta, m, c_0, x_0, t_0, \phi_{01}$ and ϕ_{02} are arbitrary real constants	—	(7.34)
18.	$\psi_1(x,t) = \alpha_{11}\dfrac{e^{\eta(x,t)}}{1+e^{\eta(x,t)+\eta*(x,t)+r}},$ $\psi_2(x,t) = \alpha_{12}\dfrac{e^{\eta(x,t)}}{1+e^{\eta(x,t)+\eta*(x,t)+r}}$	$\eta(x,t) = k\,(x+i\,k\,t)$, $r = \ln[\frac{\mu(|\alpha_1|^2+|\alpha_2|^2)}{(k+k*)^2}]$, $b_0 = c_0 = 1$, $b_1 = c_1 = b_2 = c_2 = 2\,\mu$, μ is an arbitrary real constant, α_{11}, α_{12}, are k, are arbitrary complex constants	Single bright soliton	(7.35)
19.	$\psi_1(x,t) = \dfrac{\Omega_1(x,t)}{q(x,t)},\quad \psi_2(x,t) = \dfrac{\Omega_2(x,t)}{q(x,t)}$	See text	Two bright solitons	(7.36)
20.	$\psi_1(x,t) = \dfrac{\Omega_1(x,t)+\Omega_{12}(x,t)+\Omega_{13}(x,t)+\Omega_{14}(x,t)}{q(x,t)},$ $\psi_2(x,t) = \dfrac{\Omega_{21}(x,t)+\Omega_{22}(x,t)+\Omega_{23}(x,t)+\Omega_{24}(x,t)}{q(x,t)}$	See text	Three bright solitons	(7.37)

21. Dark-bright soliton (7.38)

$$\psi_1(x,t) = (A_0 \tanh[A_1((x − x_0) − v(t − t_0))] + A_2)\, e^{i[b_1(t−t_0)+\phi_1]},$$
$$\psi_2(x,t) = B_0 \operatorname{sech}[A_1((x − x_0) − v(t − t_0))]$$
$$\times e^{i[(A_1^2+b_2)(t−t_0)+\theta(x,t)+\phi_2]}$$

$$A_0 = \cos[\tan^{-1}(\tfrac{-v}{2A_1})],\quad A_1 = \sqrt{\frac{b_1(1−b_2^2)}{2(1+b_1 b_2)}},\quad A_2 = \sin[\tan^{-1}(\tfrac{-v}{2A_1})],$$
$$B_0 = \sqrt{\frac{2A_1^2(b_1+b_2)}{1−b_2^2}},\quad \theta(x,t) = \tfrac{v}{2}[x − x_0 − \tfrac{v}{2}(t − t_0)],\quad b_0 = −1,$$
$$\frac{b_1(1−b_2^2)}{2(1+b_1 b_2)} > \frac{v^2}{4},\quad \frac{2A_1^2(b_1+b_2)}{1−b_2^2} > 0,\quad c_0 = 1,\ c_1 = 1/b_1,$$
$$c_2 = b_2,\ b_1 b_2 = 1,\ x_0, t_0, \phi_1, \phi_2, \text{ and } v \text{ are arbitrary real constants}$$

22. — (7.39)

$$\psi_1(x,t) = (1 − 4[\alpha_1^2(x,t) + \alpha_2^2(x,t) − \alpha(x,t) + i\alpha_2(x,t)])$$
$$\times e^{−\delta(x,t)}[(2\alpha_1^2(x,t) + 2\alpha_2^2(x,t) − 2\alpha_1(x,t) + 1)e^{−\delta(x,t)}$$
$$+ (\beta_3^2+\beta_4^2)e^{2\delta(x,t)} − b_1]e^{i[A_1 x+(2A_0^2−A_2^2)t]},$$
$$\psi_2(x,t) = −4A_0[\beta_3 [\alpha(x,t) − 1] − \beta_4\,\alpha_2(x,t) + i\beta_3\,\alpha_2(x,t)$$
$$+ \beta_4\,\alpha(x,t) − \beta_4]([2[\alpha_1^2(x,t) + \alpha_2^2(x,t) − 2\alpha(x,t) + 1]$$
$$\times e^{−\delta(x,t)} + (\beta_3^2 + \beta_4^2)e^{2\delta(x,t)}])^{-1} e^{i[A_1 x+(3A_0^2−A_2^2)t]+\frac{\theta(x,t)}{2}}$$

$$\delta(x,t) = \frac{2A_0(x−2A_1 t)}{3},\quad \alpha_1(x,t) = \beta_1 A_0 + A_0(x − 2A_1 t),$$
$$\alpha_2(x,t) = \beta_2 A_0 − 2A_0^2 t,\quad b_0 = c_0 = 1,\ b_1 = b_2 = c_1 = c_2 = 2,$$
$$A_0, A_1, \beta_1, \beta_2, \beta_3, \beta_4 \text{ are arbitrary real constants}$$

23. — (7.40)

$$\psi_1(x,t) = A_0\left(\frac{−6\sqrt{3}\,A_0^8(x,t) − 36\sqrt{3}\,A_0^4+(36\,A_0^2+4\,A_0\delta(x,t)+ 5\sqrt{3})−3}{12\,A_0^2\delta^2(x,t)+8\sqrt{3}\,A_0\delta(x,t)+144\,A_0^4−2·5}\right.$$
$$− (1 + i\sqrt{3})\Big)e^{i[A_1 x+(16\,A_0^2−A_2^2)t]},$$
$$\psi_2(x,t) = A_0\left(\frac{−6\sqrt{3}\,A_0\delta(x,t)+36\sqrt{3}\,A_0^4+(36\,A_0^2+i\,A_0\delta(x,t)−6\,A_0\delta(x,t)−5\sqrt{3})−3}{12\,A_0^2\delta^2(x,t)+8\sqrt{3}\,A_0\delta(x,t)+144\,A_0^4−2·5}\right.$$
$$− (1 − i\sqrt{3})\Big)e^{i[A_1 x+(16\,A_0^2−A_2^2)t]}$$

$$\delta(x,t) = x + 6A_3 t,\quad b_0 = c_0 = 1,\ b_1 = b_2 = c_1 = c_2 = 2,$$
$$A_0 = A_2 + 3A_3,\ A_1 = A_2 − 2A_0 A_2 \text{ and } A_3 \text{ are arbitrary real constants}$$

24. — (7.41)

$$\psi_1(x,t) = A_0\left[\frac{−9[q(x,t) + i\beta q(x,t)]}{q(x,t)} − (1 + i\sqrt{3})\right]e^{i[A_1 x+(16\,A_0^2−A_2^2)t]},$$
$$\psi_2(x,t) = A_0\left[\frac{(9p(x,t) + i\beta p(x,t))}{q(x,t)} − (1 − i\sqrt{3})\right]e^{i[A_1 x+(16\,A_0^2−A_2^2)t]}$$

See text

25. — (7.42)

$$\psi_1(x,t) = \{A_0 \operatorname{sech}^2[A_1(x − x_0)] + A_3\}\, e^{−i[\omega_1(t−t_0)+\phi_1]},$$
$$\psi_2(x,t) = B_0 \operatorname{sech}^2[A_1(x − x_0)]\, e^{−i[\omega_2(t−t_0)+\phi_2]}$$

$$A_3 = \frac{−2A_0}{3},\ \omega_1 = 2A_1^2,\ \omega_2 = −2A_1^2,\ b_1 = \frac{−9A_1^2}{2A_0^2},\ b_2 = \frac{9A_1^2}{2B_0^2},$$
$$b_0 = c_0 = 1,\ b_1 = c_1 = c_2,\ A_0 \neq 0,\ B_0 \neq 0,\ A_1, B_1, x_0, t_0, \phi_1,$$
$$\text{and } \phi_2 \text{ are arbitrary real constants}$$

26. Solitary wave (7.43)

$$\psi_1(x,t) = A_0\sqrt{m}\,\operatorname{cd}[A_1(x − x_0), m]\, e^{−i[\omega_1(t−t_0)+\phi_1]},$$
$$\psi_2(x,t) = B_0\sqrt{1 − m}\,\operatorname{nd}[A_1(x − x_0), m]\, e^{−i[\omega_2(t−t_0)+\phi_2]}$$

$$\omega_1 = −(1 − m)\,A_1^2 − b_1\,A_0^2,\quad \omega_2 = −(2 − m)\,A_1^2 − c_1\,A_0^2,$$
$$b_0 = c_0 = 1,\ b_1 = \frac{−2A_1^2+b_2 B_0^2}{A_0^2},\ c_1 = \frac{−2A_1^2+c_2 B_0^2}{A_0^2},\ 0 < m \leq 1,\ A_0,$$
$$B_0, x_0, t_0, \phi_1, \text{ and } \phi_2 \text{ are arbitrary real constants}$$

(Continued)

equation (2): $-A_2^2 b_0 + [h_1 \Phi_1^2(z) + b_2 \Phi_2^2(z) - \gamma_1 \Phi_1^4(z) + b_0 \Phi_1^3(z) \Phi_1''(z)] = 0, \ -B_2^2 c_0 + [b_2 \Phi_1^2(z) + b_1^{-1} \Phi_2^2(z) - \beta_1 \Phi_2^4(z) + c_0 \Phi_2^3(z) \Phi_2''(z) = 0$

27.	$\Phi_1(z) = \sqrt{F \wp^2(z, g_2, g_3) + A_0 \wp(z, g_2, g_3) + A_1}$, $\Phi_2(z) = \sqrt{G \wp^2(z, g_2, g_3) + B_0 \wp(z, g_2, g_3) + B_1}$	Weierstrass elliptic function I	(7.45)

$\phi_{01}(t) = (\gamma + \frac{b_0 \alpha^2}{4}) t + \phi_{011}, \ \phi_{02}(t) = (\beta + \frac{c_0 \alpha^2}{4}) t + \phi_{022},$

$\gamma = b_1 A_1 + b_2 B_1 - 3 b_0 \frac{A_0}{F}, \ \beta = b_2 A_1 + b_1^{-1} B_1 - 3 c_0 \frac{B_0}{G},$

$A_1 = \frac{4 A_0^2 - F^2 g_2}{4 F}, \ A_2 = -\sqrt{\frac{3 F^2}{4} (\frac{A_0^2}{F^2} - \frac{g_2}{3}) (4 \frac{A_0^3}{F^3} - \frac{g_2 A_0}{F} - g_3)},$

$B_0 = \frac{-6 b_0 - b_1 A_0}{b_2}, \ B_2 = -\sqrt{\frac{3 G^2}{4} (\frac{B_0^2}{G^2} - \frac{g_2}{3}) (4 \frac{B_0^3}{G^3} - \frac{g_2 B_0}{G} - g_3)},$

$g_2 = \frac{4 (B_0^2 - B_1 G)}{G^2}, \ G = \frac{-b_1 F}{b_2}, \ F \neq 0, \ b_1 = \frac{b_0 b_2}{c_0}, \ b_2 = 1, \ b_0, \ c_0, \ c, \ A_0,$

$B_1, \ g_3, \ \alpha, \ \phi_{011}, \ \text{and} \ \phi_{022} \ \text{are arbitrary real constants}$

28.	$\Phi_1(z) = \sqrt{A_0 \wp(z, g_2, g_3) + A_1}$, $\Phi_2(z) = \sqrt{B_0 \wp(z, g_2, g_3) + B_1}$	Weierstrass elliptic function II	(7.48)

$\phi_{01}(t) = (\gamma + \frac{b_0 \alpha^2}{4}) t + \phi_{011}, \ \phi_{02}(t) = (\beta + \frac{c_0 \alpha^2}{4}) t + \phi_{022},$

$\gamma = \frac{b_2 (B_1 A_0 - A_1 B_0) - 3 b_0 A_1}{A_0}, \ \beta = \frac{b_2 (A_1 B_0 - B_1 A_0) - 3 c_0 B_1}{B_0},$

$A_0 = \frac{2 (b_0 - c_0 b_1 - b_0)}{b_1 (1 - b_2^2)}, \ A_2 = -\sqrt{\frac{A_0^2}{4} (\frac{-A_1 g_2}{A_0} + g_3) + \frac{4 A_1^3}{A_0^3}},$

$B_0 = \frac{2 (b_0 b_2 - c_0 b_1)}{1 - b_2^2}, \ B_2 = -\sqrt{\frac{B_0^2}{4} (\frac{-B_1 g_2}{B_0} + g_3) + \frac{4 B_1^3}{B_0^3}}, \ b_0, b_1, b_2, c_0,$

$c, A_1, B_1, g_2, g_3, \alpha, \phi_{011}, \ \text{and} \ \phi_{022} \ \text{are arbitrary real constants}$

7.4 Symmetry Reductions

7.4.1 Symmetry Reduction I

From Manakov System to Fundamental NLSE

The CNLSE (7.17),

$$i\,\psi_{1t} + b_0\,\psi_{1xx} + (b_1\,|\psi_1|^2 + b_2\,|\psi_2|^2)\,\psi_1 = 0,$$
$$i\,\psi_{2t} + b_0\,\psi_{2xx} + (c_1\,|\psi_1|^2 + c_2\,|\psi_2|^2)\,\psi_2 = 0,$$

transforms to the scalar NLSE

$$i\,\psi_{1t} + b_0\,\psi_{1xx} + (c_1 + c_2|\sigma|^2)\,|\psi_1|^2\,\psi_1 = 0, \tag{7.49}$$

with the replacements
1. $\psi_2(x,\,t) = \sigma\,\psi_1(x,\,t)$,
2. $b_1 = c_1 + (c_2 - b_2)\,|\sigma|^2$,

where σ is an arbitrary complex constant.

Conclusion:

If $\psi_1(x,\,t)$ is a solution to the fundamental NLSE

$$i\,\psi_{1t} + a_1\,\psi_{1xx} + a_2\,|\psi_1|^2\,\psi_1 = 0,$$

then

$$(\psi_1,\,\psi_2) = (\psi_1,\,\sigma\,\psi_1) \tag{7.50}$$

is a solution to the CNLSE

$$i\,\psi_{1t} + b_0\,\psi_{1xx} + (b_1\,|\psi_1|^2 + b_2\,|\psi_2|^2)\,\psi_1 = 0,$$
$$i\,\psi_{2t} + b_0\,\psi_{2xx} + (c_1\,|\psi_1|^2 + c_2\,|\psi_2|^2)\,\psi_2 = 0,$$

with $a_1 = b_0$, $a_2 = c_1 + c_2|\sigma|^2$, $b_1 = c_1 + (c_2 - b_2)\,|\sigma|^2$.

7.4.2 Symmetry Reduction II

From Manakov System to Fundamental NLSE

The CNLSE (7.17),

$$i\,\psi_{1t} + b_0\,\psi_{1xx} + (b_1\,|\psi_1|^2 + b_2\,|\psi_2|^2)\,\psi_1 = 0,$$
$$i\,\psi_{2t} - b_0\,\psi_{2xx} + (c_1\,|\psi_1|^2 + c_2\,|\psi_2|^2)\,\psi_2 = 0,$$

transforms to the scalar NLSE

$$i\,\psi_{1t} + b_0\,\psi_{1xx} - (c_1 + c_2)\,|\psi_1|^2\,\psi_1 = 0, \tag{7.51}$$

with the replacements
1. $\psi_2(x,\,t) = e^{i\,\phi}\,\psi_1^*(x,\,t)$,
2. $b_1 = -(c_1 + c_2 + b_2)$,

where φ is an arbitrary real constant.

Conclusion:

If $\psi_1(x, t)$ is a solution to the fundamental NLSE

$$i\,\psi_{1_t} + b_0\,\psi_{1xx} + a_2\,|\psi_1|^2\,\psi_1 = 0,$$

then

$$(\psi_1,\,\psi_2) = (\psi_1,\,e^{i\,\phi}\,\psi_1^*) \tag{7.52}$$

is a solution to the CNLSE

$$i\,\psi_{1_t} + b_0\,\psi_{1xx} + (b_1\,|\psi_1|^2 + b_2\,|\psi_2|^2)\,\psi_1 = 0,$$
$$i\,\psi_{2_t} - b_0\,\psi_{2xx} + (c_1\,|\psi_1|^2 + c_2\,|\psi_2|^2)\,\psi_2 = 0,$$

with $a_2 = -(c_1 + c_2)$.

7.4.3 Symmetry Reduction III

From Vector NLSE to Fundamental NLSE

The vector CNLSE

$$i\,\psi_{j_t} + b_{0j}\,\psi_{j_{xx}} + \left(\sum_{k=1}^{N} b_{1k}\,|\psi_k|^2\right)\psi_j = 0, \quad j = 1, 2, \ldots, N, \tag{7.53}$$

transforms to the scalar NLSE

$$i\,\psi_{1_t} + b_{0j}\,\psi_{1xx} + \sum_{k=1}^{N} b_{1j}|\sigma_j|^2\,|\psi_1|^2\,\psi_1 = 0,$$

with the replacement: $\psi_j(x, t) - \sigma_j\,\psi_1(x, t)$, where σ_j are arbitrary complex constants, $\sigma_1 = 1$.

Conclusion:

If $\psi_1(x, t)$ is a solution to the fundamental NLSE

$$i\,\psi_{1_t} + a_1\,\psi_{1xx} + a_2\,|\psi_1|^2\,\psi_1 = 0,$$

then

$$(\psi_1,\,\psi_2,\,\psi_3,\,\ldots) = (\psi_1,\,\sigma_2\,\psi_1,\,\sigma_3\,\psi_1,\,\ldots) \tag{7.54}$$

is a solution to the vector CNLSE

$$i\,\psi_{j_t} + b_{0j}\,\psi_{j_{xx}} + \left(\sum_{k=1}^{N} b_{1k}\,|\psi_k|^2\right)\psi_j = 0, \quad j = 1, 2, \ldots, N,$$

with $b_{0j} = a_1$, $a_2 = \sum_{j=1}^{N} b_{1j}|\sigma_j|^2$.

7.4.4 Symmetry Reduction IV

From Three Coupled NLSEs to Manakov System

The three CNLSEs,

$$i\,\psi_{1t} + a_1\,\psi_{1xx} + (b_1\,|\psi_1|^2 + b_2\,|\psi_2|^2 + b_3\,|\psi_3|^2)\,\psi_1 = 0,$$
$$i\,\psi_{2t} + a_2\,\psi_{2xx} + (c_1\,|\psi_1|^2 + c_2\,|\psi_2|^2 + c_3\,|\psi_3|^2)\,\psi_2 = 0, \qquad (7.55)$$
$$i\,\psi_{3t} + a_3\,\psi_{3xx} + (d_1\,|\psi_1|^2 + d_2\,|\psi_2|^2 + d_3\,|\psi_3|^2)\,\psi_3 = 0,$$

transforms to the CNLSE

$$i\,\psi_{1t} + a_1\,\psi_{1xx} + (b_1\,|\psi_1|^2 + b_3\,|\psi_3|^2)\,\psi_1 = 0,$$
$$i\,\psi_{3t} + a_3\,\psi_{3xx} + (d_1\,|\psi_1|^2 + d_3\,|\psi_3|^2)\,\psi_3 = 0, \qquad (7.56)$$

with the replacements
  1. $\psi_2(x,\,t) = \sigma\,\psi_1(x,\,t)$,
  2. $c_1 = b_1 + (b_2 - c_2)\,|\sigma|^2$,
  3. $a_1 = a_2$,
  4. $c_3 = b_3$,

where $a_{1,2,3}$, $b_{1,2,3}$, $c_{1,2,3}$, and $d_{1,2,3}$ are real constants, σ is an arbitrary complex constant.

***Example* 1.** *dark–dark–bright soliton*
  (Figure 7.7)

$$\psi_1(x,\,t) = (A_0 \tanh\{A_1\,[x - x_0 - v\,(t - t_0)]\} + i\,A_2)\,e^{i\,[b_1\,(t-t_0)+\phi_1]},$$
$$\psi_2(x,\,t) = \sigma_2\,\psi_1(x,\,t), \qquad (7.57)$$
$$\psi_3(x,\,t) = B_0 \operatorname{sech}\{A_1\,[x - x_0 - v\,(t - t_0)]\}\,e^{i\,[(A_1^2+b_2)\,(t-t_0)+\theta(x,\,t)+\phi_2]},$$

are a solution to

$$i\,\psi_{1t} + a_1\,\psi_{1xx} + (b_{11}\,|\psi_1|^2 + b_{12}\,|\psi_2|^2 + b_{13}\,|\psi_3|^2)\,\psi_1 = 0,$$
$$i\,\psi_{2t} + a_2\,\psi_{2xx} + (c_{11}\,|\psi_1|^2 + c_{12}\,|\psi_2|^2 + c_{13}\,|\psi_3|^2)\,\psi_2 = 0, \qquad (7.58)$$
$$i\,\psi_{3t} + a_3\,\psi_{3xx} + (d_{11}\,|\psi_1|^2 + d_{12}\,|\psi_2|^2 + d_{13}\,|\psi_3|^2)\,\psi_3 = 0,$$

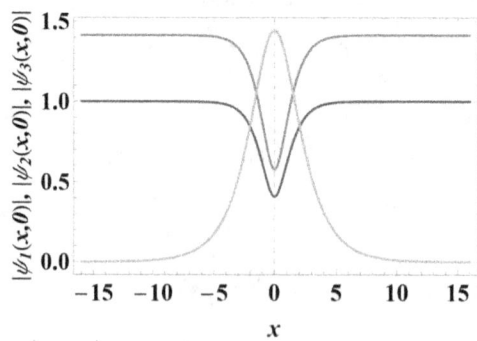

Figure 7.7. Dark-dark-bright soliton (7.57) at $t = 0$. Blue is ψ_1, red is ψ_2, and green is ψ_3 with $b_2 = 1/2$, $v = 1/2$, $\sigma_{2r} = \sigma_{2i} = 1$, and $x_0 = t_0 = \phi_1 = \phi_2 = 0$. Animation available online at http://doi.org/10.1088/978-0-7503-5954-2.

where $A_0 = \cos\left[\tan^{-1}\left(\frac{-v}{2A_1}\right)\right]$, $A_1 = \sqrt{\frac{b_1(1-b_2^2)}{2(1+b_1 b_2)} - \frac{v^2}{4}}$, $A_2 = \sin\left[\tan^{-1}\left(\frac{-v}{2A_1}\right)\right]$,

$B_0 = \sqrt{\frac{2A_1^2(b_1+b_2)}{1-b_2^2}}$, $\theta(x,t) = \frac{v}{2}[x - x_0 - \frac{v}{2}(t-t_0)]$, $b_0 = -1$, $c_0 = 1$, $c_1 = 1/b_1$,

$c_2 = b_2$, $b_1 b_2 = 1$, $a_1 = a_2 = b_0$, $a_3 = c_0$, $b_{11} = b_1 - b_{12}|\sigma_2|^2$, $b_{13} = b_2$,
$d_{11} = c_1 - d_{12}|\sigma_2|^2$, $d_{13} = c_2$, $c_{13} = b_{13}$, $c_{11} = b_{11} + (b_{12} - c_{12})|\sigma_2|^2$,

$\sigma_2 = \sigma_{2r} + i\,\sigma_{2i}$, $\frac{b_1(1-b_2^2)}{2(1+b_1 b_2)} > \frac{v^2}{4}$, $2A_1^2(b_1+b_2)(1-b_2^2) > 0$, x_0, t_0, ϕ_1, ϕ_2, σ_{2r}, σ_{2i},
and v are arbitrary real constants.

Example 2.   *dark bright soliton*
  (Figure 7.8)

$$\psi_1(x,t) = (A_0 \tanh\{A_1[x - x_0 - v(t-t_0)]\} + i A_2)\,e^{i\,[b_1(t-t_0)+\phi_1]},$$
$$\psi_2(x,t) = \sigma_2\,\psi_3(x,t), \qquad\qquad\qquad (7.59)$$
$$\psi_3(x,t) = B_0 \operatorname{sech}\{A_1[x - x_0 - v(t-t_0)]\}\,e^{i\,[(A_1^2+b_2)(t-t_0)+\theta(x,t)+\phi_2]},$$

are a solution to (7.58), where $A_0 = \cos\left[\tan^{-1}\left(\frac{-v}{2A_1}\right)\right]$, $A_1 = \sqrt{\frac{b_1(1-b_2^2)}{2(1+b_1 b_2)} - \frac{v^2}{4}}$,

$A_2 = \sin\left[\tan^{-1}\left(\frac{-v}{2A_1}\right)\right]$, $B_0 = \sqrt{\frac{2A_1^2(b_1+b_2)}{1-b_2^2}}$, $\theta(x,t) = \frac{v}{2}[(x-x_0) - \frac{v}{2}(t-t_0)]$,

$b_0 = -1$, $c_0 = 1$, $c_1 = 1/b_1$, $c_2 = b_2$, $b_1 b_2 = 1$, $a_1 = a_2 = b_0$, $a_3 = c_0$,
$b_{11} = b_1 - b_{12}|\sigma_2|^2$, $b_{13} = b_2$, $d_{11} = c_1 - d_{12}|\sigma_2|^2$, $d_{13} = c_2$, $c_{13} = b_{13}$,
$c_{11} = b_{11} + (b_{12} - c_{12})|\sigma_2|^2$, $\sigma = \sigma_{2r} + i\,\sigma_{2i}$, $\frac{b_1(1-b_2^2)}{2(1+b_1 b_2)} > \frac{v^2}{4}$, $2A_1^2(b_1+b_2)$
$(1-b_2^2) > 0$, x_0, t_0, ϕ_1, ϕ_2, σ_{2r}, σ_{2i}, and v are arbitrary real constants.

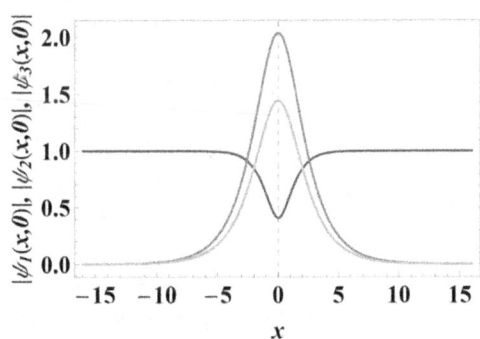

Figure 7.8. Dark-bright-bright soliton (7.59) at $t = 0$. Blue is ψ_1, red is ψ_2, and green is ψ_3 with $b_2 = 1/2$, $v = 1/2$, $\sigma_{2r} = \sigma_{2i} = 1$, and $x_0 = t_0 = \phi_1 = \phi_2 = 0$. Animation available online at http://doi.org/10.1088/978-0-7503-5954-2.

7.4.5 Symmetry Reduction V

From Vector NLSE to Manakov System

The vector CNLSE

$$i\,\psi_{j_t} + a_{1j}\,\psi_{j_{xx}} + \left(\sum_{k=1}^{N} b_{jk}\,|\psi_k|^2\right)\psi_j = 0, \quad j = 1, 2, \dots, N, \qquad (7.60)$$

transforms to the CNLSE

$$i\,\psi_{1t} + a_1\,\psi_{1xx} + (b_1\,|\psi_1|^2 + b_2\,|\psi_2|^2)\,\psi_1 = 0,$$
$$i\,\psi_{2t} + a_2\,\psi_{2xx} + (c_1\,|\psi_1|^2 + c_2\,|\psi_2|^2)\,\psi_2 = 0,$$

(7.61)

with the replacements:

1. $\psi_k(x,\,t) = \begin{cases} \sigma_k\,\psi_1(x,\,t), & k = 1,\,2,\,\dots,\,m,\ \sigma_1 = 1, \\ \sigma_k\,\psi_{m+1}(x,\,t), & k = m+1,\,\dots,\,N,\ \sigma_{m+1} = 1, \end{cases}$

2. $b_1 = \sum_{k=1}^{m} b_{j_k}\,|\sigma_k|^2, \qquad j = 1,\,2,\,\dots,\,m,$

3. $b_2 = \sum_{k=m+1}^{N} b_{j_k}\,|\sigma_k|^2, \qquad j = 1,\,2,\,\dots,\,m,$

4. $c_1 = \sum_{k=1}^{m} b_{j_k}\,|\sigma_k|^2, \qquad j = m,\,m+1,\,\dots,\,N,$

5. $c_2 = \sum_{k=m+1}^{N} b_{j_k}\,|\sigma_k|^2, \qquad j = m,\,m+1,\,\dots,\,N,$

6. $a_{1j} = \begin{cases} a_1, & j = 1,\,2,\,\dots,\,m, \\ a_2, & j = m+1,\,m+2,\,\dots,\,N, \end{cases}$

where a_1 and b_{1k} are real constants, σ_k is an arbitrary complex constant.

7.5 Scaling Transformations

7.5.1 Linear and Nonlinear Coupling

7.5.1.1 General Case
If $(\psi_1,\,\psi_2)$ is a solution to

$$i\,\psi_{1t} + \psi_{1xx} + (b_1\,|\psi_1|^2 + b_2\,|\psi_2|^2)\,\psi_1 = 0,$$
$$i\,\psi_{2t} + \psi_{2xx} + (b_1\,|\psi_1|^2 + b_2\,|\psi_2|^2)\,\psi_2 = 0,$$

(7.62)

then

$$\Phi_1(x,\,t) = \sqrt{\frac{b_1}{g_1 - g_2}}\,\psi_1(x,\,t)\,e^{i\,g_0\,(g_1 - g_2)\,t} + \sqrt{\frac{g_2\,b_2}{g_1\,(g_2 - g_1)}}\,\psi_2(x,\,t)\,e^{-i\,g_0\,(g_1 - g_2)\,t},$$

$$\Phi_2(x,\,t) = \sqrt{\frac{b_1}{g_1 - g_2}}\,\psi_1(x,\,t)\,e^{i\,g_0\,(g_1 - g_2)\,t} + \sqrt{\frac{g_1\,b_2}{g_2\,(g_2 - g_1)}}\,\psi_2(x,\,t)\,e^{-i\,g_0\,(g_1 - g_2)\,t}$$

(7.63)

is a solution to

$$i\,\Phi_{1t} + \Phi_{1xx} + (g_1\,|\Phi_1|^2 - g_2\,|\Phi_2|^2)\,\Phi_1 + g_0\,(g_1 + g_2)\,\Phi_1 - 2\,g_0\,g_2\,\Phi_2 = 0,$$
$$i\,\Phi_{2t} + \Phi_{2xx} + (g_1\,|\Phi_1|^2 - g_2\,|\Phi_2|^2)\,\Phi_2 - g_0\,(g_1 + g_2)\,\Phi_2 + 2\,g_0\,g_1\,\Phi_1 = 0,$$

(7.64)

where $b_1\,(g_1 - g_2) > 0,\,b_2\,g_1\,g_2\,(g_2 - g_1) > 0,\,b_1,\,b_2,\,g_0,\,g_1,$ and g_2 are real constants.

7.5.1.2 Specific Case I: Manakov System to Another Manakov System
If $(\psi_1,\,\psi_2)$ is a solution to

$$i\,\psi_{1t} + \psi_{1xx} + (b_1\,|\psi_1|^2 + b_2\,|\psi_2|^2)\,\psi_1 = 0,$$
$$i\,\psi_{2t} + \psi_{2xx} + (b_1\,|\psi_1|^2 + b_2\,|\psi_2|^2)\,\psi_2 = 0,$$

(7.65)

then

$$\Phi_1(x, t) = \sqrt{\frac{b_1}{g_1 - g_2}} \, \psi_1(x, t) + \sqrt{\frac{g_2 \, b_2}{g_1 \, (g_2 - g_1)}} \, \psi_2(x, t),$$

$$\Phi_2(x, t) = \sqrt{\frac{b_1}{g_1 - g_2}} \, \psi_1(x, t) + \sqrt{\frac{g_1 \, b_2}{g_2 \, (g_2 - g_1)}} \, \psi_2(x, t)$$

(7.66)

is a solution to

$$i \, \Phi_{1t} + \Phi_{1xx} + (g_1 \, |\Phi_1|^2 - g_2 \, |\Phi_2|^2) \, \Phi_1 = 0,$$

$$i \, \Phi_{2t} + \Phi_{2xx} + (g_1 \, |\Phi_1|^2 - g_2 \, |\Phi_2|^2) \, \Phi_2 = 0,$$

(7.67)

where $b_1 \, (g_1 - g_2) > 0$, $b_2 \, g_1 \, g_2 \, (g_2 - g_1) > 0$, b_1, b_2, g_1 and g_2 are real constants.

7.5.1.3 Specific Case II: Manakov System to the same Manakov System.
Superposition principle for a nonlinear system

If (ψ_1, ψ_2) is a solution to

$$i \, \psi_{1t} + \psi_{1xx} + (b_1 \, |\psi_1|^2 + b_2 \, |\psi_2|^2) \, \psi_1 = 0,$$

$$i \, \psi_{2t} + \psi_{2xx} + (b_1 \, |\psi_1|^2 + b_2 \, |\psi_2|^2) \, \psi_2 = 0,$$

(7.68)

then

$$\Phi_1(x, t) = \sqrt{\frac{b_1}{b_1 + b_2}} \left[\psi_1(x, t) - \frac{b_2}{b_1} \, \psi_2(x, t) \right],$$

$$\Phi_2(x, t) = \sqrt{\frac{b_1}{b_1 + b_2}} \, [\psi_1(x, t) + \psi_2(x, t)]$$

(7.69)

is also a solution to (7.68), where $b_1 + b_2 \neq 0$, $b_1 \, (b_1 + b_2) > 0$.

Example 1.
(Figure 7.9)

Given

$\psi_1(x, t) = \{A_0 \, \text{sech}^2[A_1 \, (x - x_0)] + A_3\} \, e^{-i \, [\omega_1 \, (t - t_0) + \phi_1]}$,

$\psi_2(x, t) = B_0 \, \text{sech}^2[A_1 \, (x - x_0)] \, e^{-i \, [\omega_2 \, (t - t_0) + \phi_2]}$ is a solution to (7.17), then

$$\psi_1(x, t) = \sqrt{\frac{b_1}{b_1 + b_2}} \left(\{A_0 \, \text{sech}^2[A_1 \, (x - x_0)] + A_3\} \, e^{-i \, [\omega_1 \, (t - t_0) + \phi_1]} \right.$$

$$\left. - \frac{b_2}{b_1} \, B_0 \, \text{sech}^2[A_1 \, (x - x_0)] \, e^{-i \, [\omega_2 \, (t - t_0) + \phi_2]} \right),$$

(7.70)

$$\psi_2(x, t) = \sqrt{\frac{b_1}{b_1 + b_2}} \left(\{A_0 \, \text{sech}^2[A_1 \, (x - x_0)] + A_3\} \, e^{-i \, [\omega_1 \, (t - t_0) + \phi_1]} \right.$$

$$\left. + B_0 \, \text{sech}^2[A_1 \, (x - x_0)] \, e^{-i \, [\omega_2 \, (t - t_0) + \phi_2]} \right)$$

is a solution to (7.68), where $A_3 = \frac{-2\,A_0}{3}$, $\omega_1 = 2\,A_1^2$, $\omega_2 = -2\,A_1^2$, $b_1 = \frac{-9\,A_1^2}{2\,A_0^2}$,

$b_2 = \frac{9\,A_1^2}{2\,B_0^2}$, $b_0 = c_0 = 1$, $b_1 = c_1$, $b_2 = c_2$, $b_1 + b_2 \neq 0$, $b_1\,(b_1 + b_2) > 0$, A_0, A_1, B_0, B_1, x_0, t_0, ϕ_1, and ϕ_2 are arbitrary real constants.

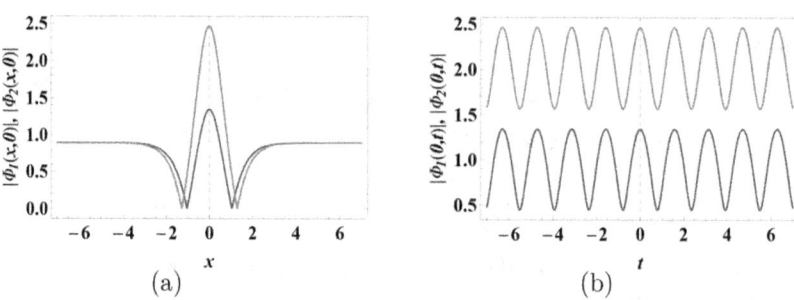

Figure 7.9. Plot of solution (7.70). (a) $t = 0$, (b) $x = 0$, with $A_0 = A_1 = B_1 = 1$, $B_0 = 3/2$, and $x_0 = t_0 = \phi_1 = \phi_2 = 0$. Blue is ϕ_1 and red is ϕ_2.

7.5.2 Complex Coupling

7.5.2.1 General Case

If (ψ_1, ψ_2) is a solution to

$$i\,\psi_{1t} + \psi_{1xx} + q_1\,(q_2\,|\psi_1|^2 + q_3\,|\psi_2|^2)\,\psi_1 = 0,$$
$$i\,\psi_{2t} + \psi_{2xx} + q_1\,(q_2\,|\psi_1|^2 + q_3\,|\psi_2|^2)\,\psi_2 = 0, \qquad (7.71)$$

then

$$\Phi_1(x,\,t) = c_1\,\psi_1(x,\,t) + c_2\,\psi_2(x,\,t),$$
$$\Phi_2(x,\,t) = c_3\,\psi_1(x,\,t) + c_4\,\psi_2(x,\,t) \qquad (7.72)$$

is a solution to

$$i\,\Phi_{1t} + \Phi_{1xx} + 2\,(a_{11}\,|\Phi_1|^2 + a_{12}\,|\Phi_2|^2)\,\Phi_1 + 2\,(b_{11}\,\Phi_1\,\Phi_2^* + b_{12}\,\Phi_2\,\Phi_1^*)\,\Phi_1 = 0,$$
$$i\,\Phi_{2t} + \Phi_{2xx} + 2\,(a_{21}\,|\Phi_1|^2 + a_{22}\,|\Phi_2|^2)\,\Phi_2 + 2\,(b_{21}\,\Phi_1\,\Phi_2^* + b_{22}\,\Phi_2\,\Phi_1^*)\,\Phi_2 = 0, \qquad (7.73)$$

where $\quad q_1 = \dfrac{2\,(c_1\,c_4 - c_2\,c_3)\,(c_2^*\,c_3^* - c_1^*\,c_4^*)}{c_1^*\,c_2\,c_3\,c_4^* - c_1\,c_2^*\,c_3^*\,c_4}, \qquad q_2 = (a - i\,b)\,c_1^*\,c_3 - (a + i\,b)\,c_1\,c_3^*,$

$q_3 = a\,(c_2\,c_4^* - c_2^*\,c_4) + i\,b\,(c_2^*\,c_4 + c_2\,c_4^*),$

$a_{12} = \dfrac{b_{12}\,c_1^*\,c_2^*\,(c_2\,c_3 - c_1\,c_4) + b_{11}\,c_1\,c_2\,(c_1^*\,c_4^* - c_2^*\,c_3^*)}{c_1\,c_2^*\,c_3^*\,c_4 - c_1^*\,c_2\,c_3\,c_4^*},$

$a_{11} = \dfrac{b_{11}\,c_3^*\,c_4^*\,(c_2\,c_3 - c_1\,c_4) + b_{12}\,c_3\,c_4\,(c_1^*\,c_4^* - c_2^*\,c_3^*)}{c_1\,c_2^*\,c_3^*\,c_4 - c_1^*\,c_2\,c_3\,c_4^*},\ a_{22} = a_{12},\ b_{12} = a - i\,b,\ b_{21} = a + i\,b,$

$b_{11} = b_{21}$, $b_{22} = b_{12}$, $c_1 c_2^* c_3^* c_4$ should not be pure real or pure imaginary, $c_2 c_3 - c_1 c_4 \neq 0$, c_j, $j = 1, 2, 3, 4$ are complex constants, a and b are real constants.

7.5.2.2 Specific Case
If (ψ_1, ψ_2) is a solution to

$$
\begin{aligned}
i\,\psi_{1t} + \psi_{1xx} - 4\,(b\,|\psi_1|^2 + a\,|\psi_2|^2)\,\psi_1 = 0, \\
i\,\psi_{2t} + \psi_{2xx} - 4\,(b\,|\psi_1|^2 + a\,|\psi_2|^2)\,\psi_2 = 0,
\end{aligned}
\tag{7.74}
$$

then

$$
\begin{aligned}
\Phi_1(x,\,t) = \psi_1(x,\,t) + \psi_2(x,\,t), \\
\Phi_2(x,\,t) = \psi_1(x,\,t) + i\,\psi_2(x,\,t)
\end{aligned}
\tag{7.75}
$$

is a solution to

$$
\begin{aligned}
i\,\Phi_{1t} + \Phi_{1xx} - 2(a+b)(|\Phi_1|^2 + |\Phi_2|^2)\,\Phi_1 + 2\,((a+ib)\,\Phi_1\,\Phi_2^* + (a-ib)\,\Phi_2\,\Phi_1^*)\,\Phi_1 = 0, \\
i\,\Phi_{2t} + \Phi_{2xx} - 2(a+b)(|\Phi_1|^2 + |\Phi_2|^2)\,\Phi_2 + 2\,((a+ib)\,\Phi_1\,\Phi_2^* + (a-ib)\,\Phi_2\,\Phi_1^*)\,\Phi_2 = 0,
\end{aligned}
\tag{7.76}
$$

where a and b are real constants.

7.5.3 Function Coefficients

7.5.3.1 General Case
If (ψ_1, ψ_2) is a solution to

$$
\begin{aligned}
i\,\psi_{1t} + a_{11}\,\psi_{1xx} + (a_{12}\,|\psi_1|^2 + a_{13}\,|\psi_2|^2)\,\psi_1 = 0, \\
i\,\psi_{2t} + a_{21}\,\psi_{2xx} + (a_{22}\,|\psi_1|^2 + a_{23}\,|\psi_2|^2)\,\psi_2 = 0,
\end{aligned}
\tag{7.77}
$$

then

$$
\begin{aligned}
\Phi_1(x,\,t) = A(x,\,t)\,e^{i\,B_1(x,\,t)}\,\psi_1[X(x,\,t),\,T(x,\,t)], \\
\Phi_2(x,\,t) = A(x,\,t)\,e^{i\,B_2(x,\,t)}\,\psi_2[X(x,\,t),\,T(x,\,t)]
\end{aligned}
\tag{7.78}
$$

is a solution to

$$
\begin{aligned}
i\Phi_{1t} + b_{11}(x,\,t)\Phi_{1xx} + [b_{12}(x,\,t)|\Phi_1|^2 + b_{13}(x,\,t)|\Phi_2|^2]\Phi_1 \\
+ [b_{14r}(x,\,t) + ib_{14i}(x,\,t)]\Phi_1 = 0, \\
i\Phi_{2t} + b_{21}(x,\,t)\Phi_{2xx} + [b_{22}(x,\,t)|\Phi_1|^2 + b_{23}(x,\,t)|\Phi_2|^2]\Phi_2 \\
+ [b_{24r}(x,\,t) + ib_{24i}(x,\,t)_i]\Phi_2 = 0,
\end{aligned}
\tag{7.79}
$$

where

$$
T(x,\,t) = g_1(t),
\tag{7.80}
$$

$$
A(x,\,t) = \frac{g_3(t)}{\sqrt{X_x(x,\,t)}},
\tag{7.81}
$$

$$B_1(x, t) = g_2(t) - \int \frac{X_t(x, t)}{2 b_{11}(x, t) X_x(x, t)} \, dx, \tag{7.82}$$

$$b_{11}(x, t) = \frac{a_{11} g_1'(t)}{X_x^2(x, t)}, \tag{7.83}$$

$$b_{12}(x, t) = \frac{a_{12} g_1'(t)}{A^2(x, t)}, \tag{7.84}$$

$$b_{13}(x, t) = \frac{a_{13} g_1'(t) X_x(x, t)}{g_3^2(t)}, \tag{7.85}$$

$$B_2(x, t) = g_2(t) - \int \frac{X_t(x, t)}{2 b_{21}(x, t) X_x(x, t)} \, dx, \tag{7.86}$$

$$b_{21}(x, t) = \frac{a_{21} g_1'(t)}{X_x^2(x, t)}, \tag{7.87}$$

$$b_{22}(x, t) = \frac{a_{22} g_1'(t)}{A^2(x, t)}, \tag{7.88}$$

$$b_{23}(x, t) = \frac{a_{23} g_1'(t) X_x(x, t)}{g_3^2(t)}, \tag{7.89}$$

$$b_{14r}(x, t) = g_2'(t) + \frac{1}{4 a_{11} g_1'^2(t)} \left(2 g_1''(t) \int X_t(x, t) X_x(x, t) \, dx \right.$$
$$+ g_1'(t) \left\{ -2 \int \left[X_{tt}(x, t) X_x(x, t) + X_t(x, t) X_{xt}(x, t) \right] dx \right. \tag{7.90}$$
$$\left. \left. + X_t^2(x, t) + \frac{a_{11}^2 g_1'^2(t)}{X_x^4(x, t)} \left[-3 X_{xx}^2(x, t) + 2 X_x(x, t) X_{xxx}(x, t) \right] \right\} \right),$$

$$b_{14i}(x, t) = \frac{X_{xt}(x, t)}{X_x(x, t)} - \frac{g_3'(t)}{g_3(t)}, \tag{7.91}$$

$$b_{24r}(x, t) = g_2'(t) + \frac{1}{4 a_{21} g_1'^2(t)} \left(2 g_1''(t) \int X_t(x, t) X_x(x, t) \, dx \right.$$

$$+ g_1'(t) \left\{ -2 \int \left[X_{tt}(x, t) X_x(x, t) + X_t(x, t) X_{xt}(x, t) \right] dx \right. \tag{7.92}$$

$$\left. \left. + X_t^2(x, t) + \frac{a_{21}^2 g_1'^2(t)}{X_x^4(x, t)} \left[-3 X_{xx}^2(x, t) + 2 X_x(x, t) X_{xxx}(x, t) \right] \right\} \right),$$

$$b_{24i}(x, t) = \frac{X_{xt}(x, t)}{X_x(x, t)} - \frac{g_3'(t)}{g_3(t)}, \tag{7.93}$$

a_{11}, a_{12}, a_{13}, a_{21}, a_{22}, and a_{23} are arbitrary real constants.

7.5.3.2 Specific Case: Constant Dispersion and Real Quadratic Potential

If (ψ_1, ψ_2) is a solution to

$$i \, \psi_{1t} + a_{11} \, \psi_{1xx} + (a_{12} \, |\psi_1|^2 + a_{13} \, |\psi_2|^2) \, \psi_1 = 0,$$
$$i \, \psi_{2t} + a_{21} \, \psi_{2xx} + (a_{22} \, |\psi_1|^2 + a_{23} \, |\psi_2|^2) \, \psi_2 = 0, \tag{7.94}$$

then

$$\Phi_1(x, t) = e^{\frac{\gamma(t)}{2}} e^{-\frac{i x^2 \gamma'(t)}{a_{11}}} \psi_1 \left[e^{\gamma(t)} x, \frac{1}{4} \int e^{2 \gamma(t)} \, dt \right],$$

$$\Phi_2(x, t) = e^{\frac{\gamma(t)}{2}} e^{-\frac{i x^2 \gamma'(t)}{a_{21}}} \psi_2 \left[e^{\gamma(t)} x, \frac{1}{4} \int e^{2 \gamma(t)} \, dt \right] \tag{7.95}$$

is a solution to

$$i \, \Phi_{1t} + a_{11} \, \Phi_{1xx} + e^{\gamma(t)} \left[a_{12} \, |\Phi_1|^2 + a_{13} \, |\Phi_2|^2 \right] \Phi_1 + \frac{4}{a_{11}} \left[\gamma'^2(t) - \gamma''(t) \right] x^2 \, \Phi_1 = 0,$$

$$i \, \Phi_{2t} + a_{21} \, \Phi_{2xx} + e^{\gamma(t)} \left[a_{22} \, |\Phi_1|^2 + a_{23} \, |\Phi_2|^2 \right] \Phi_2 + \frac{4}{a_{21}} \left[\gamma'^2(t) - \gamma''(t) \right] x^2 \, \Phi_2 = 0, \tag{7.96}$$

where $\gamma(t)$ is an arbitrary real function.

Remark: This case is obtained from the general case with the specifications: $X(x, t) = c(t) x$, $g_1(t) = \int c^2(t) \, dt$, $g_2(t) = 0$, $g_3(t) = c(t)$, $c(t) = e^{\gamma(t)}$.

7.6 Summary of Sections 7.4–7.5

Symmetry reductions

Symmetry Reduction I: *From Manakov system to fundamental NLSE*

Transformation: $\psi_2(x,t) = \sigma\,\psi_1(x,t)$, $b_1 = c_1 + (c_2 - b_2)|\sigma|^2$, $\quad (\psi_1, \psi_2)$ **is a solution to the fundamental CNLSE**

Equation: $i\,\psi_{1t} + b_0\,\psi_{1xx} + (c_1 + c_2|\sigma|^2)\,|\psi_1|^2\,\psi_1 = 0$

Symmetry Reduction II: *From Manakov system to fundamental NLSE*

Transformation: $\psi_2(x,t) = e^{i\phi}\,\psi_1^*(x,t)$, $b_1 = -(c_1 + c_2 + b_2)$

Equation: $i\,\psi_{1t} + b_0\,\psi_{1xx} - (c_1 + c_2)\,|\psi_1|^2\,\psi_1 = 0$

Symmetry Reduction III: *From vector NLSE to fundamental NLSE*

Transformation: $\psi_j(x,t) = \sigma_j\,\psi_1(x,t)$

Equation: $i\,\psi_{1t} + b_{0j}\,\psi_{1xx} + \sum_{k=1}^{N} b_{1j}|\sigma_j|^2\,|\psi_1|^2\,\psi_1 = 0$

Symmetry Reduction IV: *From three coupled NLSEs to Manakov system*

Transformation: $\psi_2(x,t) = \sigma\,\psi_1(x,t)$, $c_1 = b_1 + (b_2 - c_2)|\sigma|^2, a_1 = a_2, c_3 = b_3$

Equation: $i\,\psi_{1t} + a_1\,\psi_{1xx} + (b_1|\psi_1|^2 + b_3|\psi_3|^2)\,\psi_1 = 0, \; i\,\psi_{3t} + a_3\,\psi_{3xx} + (d_1|\psi_1|^2 + d_3|\psi_3|^2)\,\psi_3 = 0$

(Continued)

Symmetry Reduction V: From vector NLSE to Manakov system

Transformation:

$$\psi_k(x,t) = \begin{cases} \sigma_k\,\psi_1(x,t), & k = 1, 2, \ldots, m, \ \sigma_1 = 1, \\ \sigma_k\,\psi_{m+1}(x,t), & k = m+1, \ldots, N, \ \sigma_{m+1} = 1, \end{cases}$$

$$b_1 = \sum_{k=1}^{m} b_{jk}\,|\sigma_k|^2, j = 1, 2, \ldots, m,$$

$$b_2 = \sum_{k=m+1}^{N} b_{jk}\,|\sigma_k|^2, j = 1, 2, \ldots, m,$$

$$c_1 = \sum_{k=1}^{m} b_{jk}\,|\sigma_k|^2, j = m, m+1, \ldots, N,$$

$$c_2 = \sum_{k=m+1}^{N} b_{jk}\,|\sigma_k|^2, j = m, m+1, \ldots, N,$$

$$a_{1j} = \begin{cases} a_1, & j = 1, 2, \ldots, m, \\ a_2, & j = m+1, m+2, \ldots, N \end{cases}$$

Equation: $i\,\psi_{1t} + a_1\,\psi_{1xx} + (b_1\,|\psi_1|^2 + b_2\,|\psi_2|^2)\,\psi_1 = 0,\ i\,\psi_{2t} + c_2\,\psi_{2xx} + (c_1\,|\psi_1|^2 + c_2\,|\psi_2|^2)\,\psi_2 = 0$

Scaling Transformations

Linear and Nonlinear Coupling

General case

Transformation: $\Phi_1(x,t) = \sqrt{\dfrac{b_1}{g_1 - g_2}}\,\psi_1(x,t)\,e^{i\,g_0\,(g_1 - g_2)\,t} + \sqrt{\dfrac{g_2\,b_2}{g_1\,(g_2 - g_1)}}\,\psi_2(x,t)\,e^{-i\,g_0\,(g_1 - g_2)\,t},$

$\Phi_2(x,t) = \sqrt{\dfrac{b_1}{g_1 - g_2}}\,\psi_1(x,t)\,e^{i\,g_0\,(g_1 - g_2)\,t} + \sqrt{\dfrac{g_1\,b_2}{g_2\,(g_2 - g_1)}}\,\psi_2(x,t)\,e^{-i\,g_0\,(g_1 - g_2)\,t}$

Equation: $i\,\Phi_{1t} + \Phi_{1xx} + (g_1\,|\Phi_1|^2 - g_2\,|\Phi_2|^2)\,\Phi_1 + g_0\,(g_1 + g_2)\,\Phi_1 - 2\,g_0\,g_2\,\Phi_2 = 0,\ i\,\Phi_{2t} + \Phi_{2xx} + (g_1\,|\Phi_1|^2 - g_2\,|\Phi_2|^2)\,\Phi_2 - g_0\,(g_1 + g_2)\,\Phi_2 + 2\,g_0\,g_1\,\Phi_1 = 0$

Specific Case I: Manakov system to another Manakov system

Transformation: $\Phi_1(x,t) = \sqrt{\dfrac{b_1}{g_1 - g_2}}\,\psi_1(x,t) + \sqrt{\dfrac{g_2\,b_2}{g_1\,(g_2 - g_1)}}\,\psi_2(x,t),\ \Phi_2(x,t) = \sqrt{\dfrac{b_1}{g_1 - g_2}}\,\psi_1(x,t) + \sqrt{\dfrac{g_1\,b_2}{g_2\,(g_2 - g_1)}}\,\psi_2(x,t)$

equation: $i\Phi_{1t} + \Phi_{1xx} + (g_1|\Phi_1|^2 - g_2|\Phi_2|^2)\Phi_1 = 0,\ i\Phi_{2t} + \Phi_{2xx} + (g_1|\Phi_1|^2 - g_2|\Phi_2|^2)\Phi_2 = 0$

Specific Case II: Manakov system to the same Manakov system   *Superposition principle for a nonlinear system*

Transformation: $\Phi_1(x,t) = \sqrt{\frac{b_1}{b_1+b_2}}\,[\psi_1(x,t) - \frac{b_2}{b_1}\psi_2(x,t)],\ \Phi_2(x,t) = \sqrt{\frac{b_1}{b_1+b_2}}\,[\psi_1(x,t) + \psi_2(x,t)]$

Complex Coupling

General Case

Transformation: $\Phi_1(x,t) = c_1\,\psi_1(x,t) + c_2\,\psi_2(x,t),\ \Phi_2(x,t) = c_3\,\psi_1(x,t) + c_4\,\psi_2(x,t)$

Equation: $i\Phi_{1t} + \Phi_{1xx} + 2(a_{11}|\Phi_1|^2 + a_{12}|\Phi_2|^2)\Phi_1 + 2(b_{11}\Phi_1\Phi_2^* + b_{12}\Phi_2\Phi_1^*)\Phi_1 = 0,\ i\Phi_{2t} + \Phi_{2xx} + 2(a_{21}|\Phi_1|^2 + a_{22}|\Phi_2|^2)\Phi_2 + 2(b_{21}\Phi_1\Phi_2^* + b_{22}\Phi_2\Phi_1^*)\Phi_2 = 0$

Specific Case

Transformation: $\Phi_1(x,t) = \psi_1(x,t) + \psi_2(x,t),\ \Phi_2(x,t) = \psi_1(x,t) + i\,\psi_2(x,t)$

Equation: $i\Phi_{1t} + \Phi_{1xx} - 2(a+b)(|\Phi_1|^2 + |\Phi_2|^2)\Phi_1 + 2[(a+ib)\Phi_1\Phi_2^* + (a-ib)\Phi_2\Phi_1^*]\Phi_1 = 0,$

$i\Phi_{2t} + \Phi_{2xx} - 2(a+b)(|\Phi_1|^2 + |\Phi_2|^2)\Phi_2 + 2[(a+ib)\Phi_1\Phi_2^* + (a-ib)\Phi_2\Phi_1^*]\Phi_2 = 0$

Function Coefficients

General Case

Transformation: $\Phi_1(x,t) = A(x,t)\,e^{iB_1(x,t)}\,\psi_1[X(x,t),\,T(x,t)],\ \Phi_2(x,t) = A(x,t)\,e^{iB_2(x,t)}\,\psi_2[X(x,t),\,T(x,t)]$

equation: $i\Phi_{1t} + b_{11}(x,t)\,\Phi_{1xx} + [b_{12}(x,t)|\Phi_1|^2 + b_{13}(x,t)|\Phi_2|^2]\Phi_1 + [b_{14r}(x,t) + i\,b_{14i}(x,t)]\Phi_1 = 0,$

$i\Phi_{2t} + b_{21}(x,t)\,\Phi_{2xx} + [b_{22}(x,t)|\Phi_1|^2 + b_{23}(x,t)|\Phi_2|^2]\Phi_2 + [b_{24r}(x,t) + i\,b_{24i}(x,t)]\Phi_2 = 0,$

Specific Case: constant dispersion and real quadratic potential

Transformation: $\Phi_1(x,t) = e^{\frac{\gamma(t)}{2}}\,e^{-\frac{ix^2\gamma'(t)}{a_{11}}}\,\psi_1\!\left[e^{\gamma(t)}x,\,\frac{1}{4}\int e^{2\gamma(t)}\,dt\right],\ \Phi_2(x,t) = e^{\frac{\gamma(t)}{2}}\,e^{-\frac{ix^2\gamma'(t)}{a_{21}}}\,\psi_2\!\left[e^{\gamma(t)}x,\,\frac{1}{4}\int e^{2\gamma(t)}\,dt\right]$

Equation: $i\Phi_{1t} + a_{11}\Phi_{1xx} + e^{\gamma(t)}[a_{12}|\Phi_1|^2 + a_{13}|\Phi_2|^2]\Phi_1 + \frac{4}{a_{11}}[\gamma'^2(t) - \gamma''(t)]\,x^2\Phi_1 = 0,$

$i\Phi_{2t} + a_{21}\Phi_{2xx} + e^{\gamma(t)}[a_{22}|\Phi_1|^2 + a_{23}|\Phi_2|^2]\Phi_2 + \frac{4}{a_{21}}[\gamma'^2(t) - \gamma''(t)]\,x^2\Phi_2 = 0$

7.7 $(N + 1)$-Dimensional Coupled NLSE $(N + 1)$-*Dimensional Manakov System*

7.7.1 Reduction to 1D Manakov System

If (ψ_1, ψ_2) is a solution to the Manakov system

$$i\,\psi_{1t} + a_1\,\psi_{1xx} + (b_1\,|\psi_1|^2 + b_2\,|\psi_2|^2)\,\psi_1 = 0, \quad \psi_1 = \psi_1(x, t; a_1, b_1, b_2),$$
$$i\,\psi_{2t} + a_2\,\psi_{2xx} + (c_1\,|\psi_1|^2 + c_2\,|\psi_2|^2)\,\psi_2 = 0, \quad \psi_2 = \psi_2(x, t; a_2, c_1, c_2), \tag{7.97}$$

then

$$\Phi_1(x_1, x_2, \dots, x_N) = \psi_1\left[\sum_{k=1}^{N} d_k\,(x_k - x_{k0}), \, t; \, \sum_{k=1}^{N} d_k^2\,b_{0k}, \, b_1, \, b_2\right],$$
$$\Phi_2(x_1, x_2, \dots, x_N) = \psi_2\left[\sum_{k=1}^{N} d_k\,(x_k - x_{k0}), \, t; \, \sum_{k=1}^{N} d_k^2\,c_{0k}, \, c_1, \, c_2\right] \tag{7.98}$$

is a solution to

$$i\,\Phi_{1t} + \sum_{k=1}^{N} b_{0k}\,\Phi_{1x_k x_k} + (b_1\,|\Phi_1|^2 + b_2\,|\Phi_2|^2)\,\Phi_1 = 0,$$
$$i\,\Phi_{2t} + \sum_{k=1}^{N} c_{0k}\,\Phi_{2x_k x_k} + (c_1\,|\Phi_1|^2 + c_2\,|\Phi_2|^2)\,\Phi_2 = 0, \tag{7.99}$$

where $\Phi_j = \Phi_j(x_1, x_2, \dots, x_N, t)$ is the complex function profile, $j = 1, 2, b_{0k}, c_{0k}, d_k,$ $b_1, c_1, b_2,$ and c_2 are real constants.

Example 1. *2D Dark-bright soliton*
(Figure 7.10)

Given $\psi_1(x, t) = (A_0\,\tanh\{A_1\,[x - x_0 - v\,(t - t_0)]\} + i\,A_2)\,e^{i\,[b_1\,(t-t_0)+\phi_1]}$,
$\psi_2(x, t) = B_0\,\mathrm{sech}\{A_1\,[x - x_0 - v\,(t - t_0)]\}\,e^{i\,[(A_1^2+b_2)\,(t-t_0)+\theta(x,\,t)+\phi_2]}$ is a solution to
(7.17), then

$$\Phi_1(x_1, x_2, t) = (A_0\,\tanh\left\{A_1\left[d_1\,(x_1 - x_{01}) + d_2\,(x_2 - x_{02}) - v\,(t - t_0)\right]\right\} + i\,A_2)\,e^{i\,[b_1\,(t-t_0)+\Phi_1]},$$
$$\Phi_2(x_1, x_2, t) = B_0\,\mathrm{sech}\{A_1\,[d_1\,(x_1 - x_{01}) + d_2\,(x_2 - x_{02}) - v\,(t - t_0)]\}$$
$$\times\,e^{i\left[(A_1^2+b_2)\,(t-t_0)+\theta(x_1,\,x_2,\,t)+\phi_2\right]} \tag{7.100}$$

is a solution to

$$i\,\Phi_{1t} + b_{01}\,\Phi_{1x_1x_1} + b_{02}\,\Phi_{1x_2x_2} + (b_1\,|\Phi_1|^2 + b_2\,|\Phi_2|^2)\,\Phi_1 = 0,$$
$$i\,\Phi_{2t} + c_{01}\,\Phi_{2x_1x_1} + c_{02}\,\Phi_{2x_2x_2} + (c_1\,|\Phi_1|^2 + c_2\,|\Phi_2|^2)\,\Phi_2 = 0, \tag{7.101}$$

where $\quad A_0 = \cos\left[\tan^{-1}\left(\dfrac{-v}{2\,A_1}\right)\right], \quad A_1 = \sqrt{\dfrac{b_1\,(1 - b_2^2)}{2\,(1 + b_1\,b_2)} - \dfrac{v^2}{4}}, \quad A_2 = \sin\left[\tan^{-1}\left(\dfrac{-v}{2\,A_1}\right)\right],$

$B_0 = \sqrt{\dfrac{2\,A_1^2\,(b_1 + b_2)}{1 - b_2^2}}, \qquad \theta(x_1, x_2, t) = \dfrac{v}{2}[d_1\,(x_1 - x_{01}) + d_2\,(x_2 - x_{02}) - \dfrac{v}{2}\,(t - t_0)],$

$$b_{01} = \frac{-1 - d_2^2\, b_{02}}{d_1^2}, \quad c_{01} = \frac{1 - d_2^2\, c_{02}}{d_1^2}, \quad c_1 = 1/b_1, \quad c_2 = b_2, \quad b_1\, b_2 = 1, \quad \frac{b_1\,(1 - b_2^2)}{2\,(1 + b_1\, b_2)} > \frac{v^2}{4},$$

$2\, A_1^2\, (b_1 + b_2)\, (1 - b_2^2) > 0$, $d_1 \neq 0$, x_{01}, x_{02}, t_0, ϕ_1, ϕ_2, d_2, and v are arbitrary real constants.

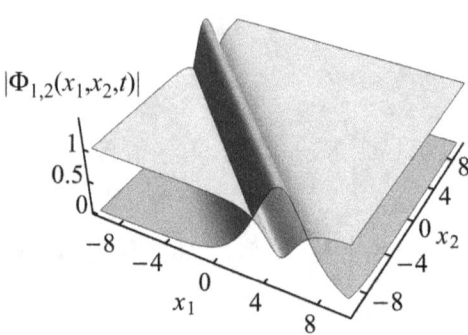

Figure 7.10. 2D dark-bright soliton (7.100) at $t = -10$ with $v = b_2 = 1/2$, $d_1 = d_2 = 1$, $b_{02} = c_{02} = 1/4$, and $x_{01} = x_{02} = t_0 = \phi_1 = \phi_2 = 0$. Yellow is Φ_1 and green is Φ_2. Animation available online at http://doi.org/10.1088/978-0-7503-5954-2.

7.8 Symmetry Reductions of $(N + 1)$-Dimensional CNLSE to Scalar NLSE

7.8.1 Symmetry Reduction I *From $(N + 1)$-Dimensional Manakov System to $(N + 1)$-Dimensional Fundamental NLSE*

The $(N + 1)$-dimensional CNLSE (7.99), $i\,\psi_{1t} + \sum_{k=1}^{N} b_{0k}\, \psi_{1x_k x_k} + (b_1\, |\psi_1|^2 + b_2\, |\psi_2|^2)\, \psi_1 = 0$, $i\,\psi_{2t} + \sum_{k=1}^{N} c_{0k}\, \psi_{2x_k x_k} + (c_1\, |\psi_1|^2 + c_2\, |\psi_2|^2)\, \psi_2 = 0$, transforms to the scalar $(N + 1)$-dimensional NLSE

$$i\,\psi_{1t} + \sum_{k=1}^{N} c_{0k}\, \psi_{1x_k x_k} + (c_1 + c_2|\sigma|^2)\, |\psi_1|^2\, \psi_1 = 0, \tag{7.102}$$

with the replacements:
1. $\psi_2(x_1, x_2, \ldots, x_N, t) = \sigma\, \psi_1(x_1, x_2, \ldots, x_N, t)$,
2. $b_1 = c_1 + (c_2 - b_2)\, |\sigma|^2$,
3. $b_{0k} = c_{0k}$,

where σ is an arbitrary complex constant.

Conclusion:

If $\psi_1(x_1, x_2, \ldots, x_N, t)$ is a solution to the $(N + 1)$-dimensional NLSE

$$i\,\psi_{1t} + \sum_{k=1}^{N}\alpha_k\,\psi_{1x_k x_k} + a_2\,|\psi_1|^2\,\psi_1 = 0, \tag{7.103}$$

then

$$(\psi_1, \psi_2) = (\psi_1, \sigma\,\psi_1) \tag{7.104}$$

is a solution to the $(N + 1)$-dimensional CNLSE $i\,\psi_{1t} + \sum_{k=1}^{N} b_{0k}\,\psi_{1x_k x_k} + (b_1\,|\psi_1|^2 + b_2\,|\psi_2|^2)\,\psi_1 = 0,$ $\qquad i\,\psi_{2t} + \sum_{k=1}^{N} c_{0k}\,\psi_{2x_k x_k} + (c_1\,|\psi_1|^2 + c_2\,|\psi_2|^2)\,\psi_2 = 0,$ with $\alpha_k = b_{0k} = c_{0k},\, a_2 = c_1 + c_2|\sigma|^2,\, b_1 = c_1 + (c_2 - b_2)\,|\sigma|^2$.

7.8.2 Symmetry Reduction II *From (N + 1)-Dimensional Manakov System to (N + 1)-Dimensional Fundamental NLSE*

The $(N + 1)$-dimensional CNLSE (7.99), $i\,\psi_{1t} + \sum_{k=1}^{N} b_{0k}\,\psi_{1x_k x_k} + (b_1\,|\psi_1|^2 + b_2\,|\psi_2|^2)\,\psi_1 = 0,\, i\,\psi_{2t} + \sum_{k=1}^{N} c_{0k}\,\psi_{2x_k x_k} + (c_1\,|\psi_1|^2 + c_2\,|\psi_2|^2)\,\psi_2 = 0$, transforms to the scalar $(N + 1)$-dimensional NLSE

$$i\,\psi_{1t} - \sum_{k=1}^{N} c_{0k}\,\psi_{1x_k x_k} - 2\,(c_1 + c_2)\,|\psi_1|^2\,\psi_1 = 0, \tag{7.105}$$

with the replacements:
1. $\psi_2(x_1, x_2, \ldots, x_N, t) = e^{i\,\phi}\,\psi_1^*(x_1, x_2, \ldots, x_N, t),$
2. $b_1 = -(c_1 + c_2 + b_2),$
3. $b_{0k} = -c_{0k},$

where ϕ is an arbitrary real constant.

Conclusion:

If $\psi_1(x_1, x_2, \ldots, x_N, t)$ is a solution to the $(N + 1)$-dimensional NLSE

$$i\,\psi_{1t} + \sum_{k=1}^{N}\alpha_k\,\psi_{1x_k x_k} + a_2\,|\psi_1|^2\,\psi_1 = 0, \tag{7.106}$$

then

$$(\psi_1, \psi_2) = (\psi_1, e^{i\,\phi}\,\psi_1^*) \tag{7.107}$$

is a solution to the $(N + 1)$-dimensional CNLSE $i\,\psi_{1t} + \sum_{k=1}^{N} b_{0k}\,\psi_{1x_k x_k} + (b_1\,|\psi_1|^2 + b_2\,|\psi_2|^2)\,\psi_1 = 0,\, i\,\psi_{2t} + \sum_{k=1}^{N} c_{0k}\,\psi_{2x_k x_k} + (c_1\,|\psi_1|^2 + c_2\,|\psi_2|^2)\,\psi_2 = 0,$ with $\alpha_k = b_{0k} = -c_{0k},\, a_2 = -2\,(c_1 + c_2)$.

7.8.3 Symmetry Reduction III *From (N+1)-Dimensional Vector NLSE to (N + 1)-Dimensional Fundamental NLSE*

The generalized $(N + 1)$-dimensional CNLSE

$$i\,\psi_{j_t} + \sum_{k=1}^{N} b_{j0k}\,\psi_{j_{x_k x_k}} + \left(\sum_{k=1}^{M} b_{1k}\,|\psi_k|^2\right)\psi_j = 0,\, j = 1, 2, \dots, M, \qquad (7.108)$$

transforms to the scalar $(N+1)$-dimensional NLSE

$$i\,\psi_{1_t} + \sum_{k=1}^{N} b_{10k}\,\psi_{1_{x_k x_k}} + \sum_{k=1}^{N} b_{1j}|\sigma_{j+1}|^2\,|\psi_1|^2\,\psi_1 = 0, \qquad (7.109)$$

with the replacement: $\psi_j(x_1, x_2, \dots, x_N, t) = \sigma_j\,\psi_1(x_1, x_2, \dots, x_N, t)$, where σ_j are arbitrary complex constants, $\sigma_1 = 1$.

Conclusion:

If $\psi_1(x_1, x_2, \dots, x_N, t)$ is a solution to the $(N+1)$-dimensional NLSE $i\,\psi_{1t} + \sum_{k=1}^{N}\alpha_k\,\psi_{1_{x_k x_k}} + a_2\,|\psi_1|^2\,\psi_1 = 0$, then

$$(\psi_1, \psi_2, \psi_3, \dots) = (\psi_1, \sigma_2\,\psi_1, \sigma_3\,\psi_1, \dots) \qquad (7.110)$$

is a solution to the generalized $(N+1)$-dimensional CNLSE $i\,\psi_{j_t} + \sum_{k=1}^{N} b_{0kj}\,\psi_{j_{x_k x_k}} + (\sum_{k=1}^{M} b_{1k}\,|\psi_k|^2)\,\psi_j = 0,\, j = 1, 2, \dots, M$, with $\alpha_k = b_{10k}$, $a_2 = \sum_{j=1}^{N} b_{1j}|\sigma_{j+1}|^2$.

Notes:

1. The $(N+1)$-dimensional three coupled NLSEs reduce to $(N+1)$-dimensional Manakov system in a similar manner to the above described symmetry reductions (7.8.1)–(7.8.3).
2. The $(N+1)$-dimensional vector NLSE reduces to $(N+1)$-dimensional Manakov system in a similar manner to the above described symmetry reductions (7.8.1)–(7.8.3).

7.9 $(N+1)$-Dimensional Scaling Transformations

7.9.1 Linear and Nonlinear Coupling

7.9.1.1 General Case

If (ψ_1, ψ_2) is a solution to

$$i\,\psi_{1t} + \sum_{k=1}^{N}\psi_{1_{x_k x_k}} + (b_1\,|\psi_1|^2 + b_2\,|\psi_2|^2)\,\psi_1 = 0,$$

$$i\,\psi_{2t} + \sum_{k=1}^{N}\psi_{2_{x_k x_k}} + (b_1\,|\psi_1|^2 + b_2\,|\psi_2|^2)\,\psi_2 = 0, \qquad (7.111)$$

then

$$\Phi_1(x_1, x_2, \ldots, x_N, t) = \sqrt{\frac{b_1}{g_1 - g_2}}\ \psi_1(x_1, x_2, \ldots, x_N, t)\ e^{i\,g_0\,(g_1 - g_2)\,t}$$

$$+ \sqrt{\frac{g_2\,b_2}{g_1\,(g_2 - g_1)}}\ \psi_2(x_1, x_2, \ldots, x_N, t)\ e^{-i\,g_0\,(g_1 - g_2)\,t},$$

$$\Phi_2(x_1, x_2, \ldots, x_N, t) = \sqrt{\frac{b_1}{g_1 - g_2}}\ \psi_1(x_1, x_2, \ldots, x_N, t)\ e^{i\,g_0\,(g_1 - g_2)\,t}$$

$$+ \sqrt{\frac{g_1\,b_2}{g_2\,(g_2 - g_1)}}\ \psi_2(x_1, x_2, \ldots, x_N, t)\ e^{-i\,g_0\,(g_1 - g_2)\,t}$$

$$(7.112)$$

is a solution to

$$i\,\Phi_{1t} + \sum_{k=1}^{N}\Phi_{1x_kx_k} + (g_1\,|\Phi_1|^2 - g_2\,|\Phi_2|^2)\,\Phi_1 + g_0\,(g_1 + g_2)\,\Phi_1 - 2\,g_0\,g_2\,\Phi_2 = 0,$$

$$i\,\Phi_{2t} + \sum_{k=1}^{N}\Phi_{2x_kx_k} + (g_1\,|\Phi_1|^2 - g_2\,|\Phi_2|^2)\,\Phi_2 - g_0\,(g_1 + g_2)\,\Phi_2 + 2\,g_0\,g_1\,\Phi_1 = 0,$$

$$(7.113)$$

where $b_1\,(g_1 - g_2) > 0$, $b_2\,g_1\,g_2\,(g_2 - g_1) > 0$, b_1, b_2, g_0, g_1, and g_2 are real constants.

7.9.1.2 Specific Case I: $(N + 1)$-Dimensional Manakov System to Another $(N + 1)$-Dimensional Manakov System

If (ψ_1, ψ_2) is a solution to $i\,\psi_{1t} + \sum_{k=1}^{N}\psi_{1x_kx_k} + (b_1\,|\psi_1|^2 + b_2\,|\psi_2|^2)\,\psi_1 = 0$, $i\,\psi_{2t} + \sum_{k=1}^{N}\psi_{2x_kx_k} + (b_1\,|\psi_1|^2 + b_2\,|\psi_2|^2)\,\psi_2 = 0$, then

$$\Phi_1(x_1, x_2, \ldots, x_N, t) = \sqrt{\frac{b_1}{g_1 - g_2}}\ \psi_1(x_1, x_2, \ldots, x_N, t)$$

$$+ \sqrt{\frac{g_2\,b_2}{g_1\,(g_2 - g_1)}}\ \psi_2(x_1, x_2, \ldots, x_N, t),$$

$$\Phi_2(x_1, x_2, \ldots, x_N, t) = \sqrt{\frac{b_1}{g_1 - g_2}}\ \psi_1(x_1, x_2, \ldots, x_N, t)$$

$$+ \sqrt{\frac{g_1\,b_2}{g_2\,(g_2 - g_1)}}\ \psi_2(x_1, x_2, \ldots, x_N, t)$$

$$(7.114)$$

is a solution to

$$i\,\Phi_{1t} + \sum_{k=1}^{N}\Phi_{1x_kx_k} + (g_1\,|\Phi_1|^2 - g_2\,|\Phi_2|^2)\,\Phi_1 = 0,$$

$$i\,\Phi_{2t} + \sum_{k=1}^{N}\Phi_{2x_kx_k} + (g_1\,|\Phi_1|^2 - g_2\,|\Phi_2|^2)\,\Phi_2 = 0,$$

$$(7.115)$$

where $b_1\,(g_1 - g_2) > 0$, $b_2\,g_1\,g_2\,(g_2 - g_1) > 0$, b_1, b_2, g_1 and g_2 are real constants.

7.9.1.3 Specific Case II: $(N + 1)$-Dimensional Manakov System to the same $(N + 1)$-Dimensional Manakov System

If (ψ_1, ψ_2) is a solution to $i\,\psi_{1t} + \sum_{k=1}^{N} \psi_{1x_k x_k} + (b_1\,|\psi_1|^2 + b_2\,|\psi_2|^2)\,\psi_1 = 0$, $i\,\psi_{2t} + \sum_{k=1}^{N} \psi_{2x_k x_k} + (b_1\,|\psi_1|^2 + b_2\,|\psi_2|^2)\,\psi_2 = 0$, then

$$\Phi_1(x_1, x_2, \ldots, x_N, t) = \sqrt{\frac{b_1}{b_1 + b_2}}\left[\psi_1(x_1, x_2, \ldots, x_N, t) - \frac{b_2}{b_1}\,\psi_2(x_1, x_2, \ldots, x_N, t)\right],$$

$$\Phi_2(x_1, x_2, \ldots, x_N, t) = \sqrt{\frac{b_1}{b_1 + b_2}}\,[\psi_1(x_1, x_2, \ldots, x_N, t) + \psi_2(x_1, x_2, \ldots, x_N, t)]$$

(7.116)

is also a solution to (7.65), where $b_1 + b_2 \neq 0$.

7.9.2 Complex Coupling

7.9.2.1 General Case

If (ψ_1, ψ_2) is a solution to

$$i\,\psi_{1t} + \sum_{k=1}^{N} \psi_{1x_k x_k} + q_1\,\{q_2\,|\psi_1|^2 + q_3\,|\psi_2|^2\}\,\psi_1 = 0,$$

$$i\,\psi_{2t} + \sum_{k=1}^{N} \psi_{2x_k x_k} + q_1\,\{q_2\,|\psi_1|^2 + q_3\,|\psi_2|^2\}\,\psi_2 = 0,$$

(7.117)

then

$$\Phi_1(x_1, x_2, \ldots, x_N, t) = c_1\,\psi_1(x_1, x_2, \ldots, x_N, t) + c_2\,\psi_2(x_1, x_2, \ldots, x_N, t),$$

$$\Phi_2(x_1, x_2, \ldots, x_N, t) = c_3\,\psi_1(x_1, x_2, \ldots, x_N, t) + c_4\,\psi_2(x_1, x_2, \ldots, x_N, t)$$

(7.118)

is a solution to

$$i\,\Phi_{1t} + \sum_{k=1}^{N} \Phi_{1x_k x_k} + 2\,(a_{11}\,|\Phi_1|^2 + a_{12}\,|\Phi_2|^2)\,\Phi_1 + 2\,(b_{11}\,\Phi_1\,\Phi_2^* + b_{12}\,\Phi_2\,\Phi_1^*)\,\Phi_1 = 0,$$

$$i\,\Phi_{2t} + \sum_{k=1}^{N} \Phi_{2x_k x_k} + 2\,(a_{21}\,|\Phi_1|^2 + a_{22}\,|\Phi_2|^2)\,\Phi_2 + 2\,(b_{21}\,\Phi_1\,\Phi_2^* + b_{22}\,\Phi_2\,\Phi_1^*)\,\Phi_2 = 0,$$

(7.119)

where $\quad q_1 = \dfrac{2\,(c_1\,c_4 - c_2\,c_3)\,(c_2^*\,c_3^* - c_1^*\,c_4^*)}{c_1^*\,c_2\,c_3\,c_4^* - c_1\,c_2^*\,c_3^*\,c_4}, \qquad q_2 = (a - i\,b)\,c_1^*\,c_3 - (a + i\,b)\,c_1\,c_3^*,$

$q_3 = a\,(c_2\,c_4^* - c_2^*\,c_4) + i\,b\,(c_2^*\,c_4 + c_2\,c_4^*),$

$a_{12} = \dfrac{b_{12}\,c_1^*\,c_2^*\,(c_2\,c_3 - c_1\,c_4) + b_{11}\,c_1\,c_2\,(c_1^*\,c_4^* - c_2^*\,c_3^*)}{c_1\,c_2^*\,c_3^*\,c_4 - c_1^*\,c_2\,c_3\,c_4^*},$

$a_{11} = \dfrac{b_{11}\,c_3^*\,c_4^*\,(c_2\,c_3 - c_1\,c_4) + b_{12}\,c_3\,c_4\,(c_1^*\,c_4^* - c_2^*\,c_3^*)}{c_1\,c_2^*\,c_3^*\,c_4 - c_1^*\,c_2\,c_3\,c_4^*},\; a_{22} = a_{12},\; b_{12} = a - i\,b,\; b_{21} = a + i\,b,$

$b_{11} = b_{21}$, $b_{22} = b_{12}$, $c_1\,c_2^*\,c_3^*\,c_4$ should not be pure real or pure imaginary, $c_2\,c_3 - c_1\,c_4 \neq 0$, $c_j, j = 1, 2, 3, 4$ are complex constants, a and b are real constants.

7.9.2.2 Specific Case

If (ψ_1, ψ_2) is a solution to

$$i\,\psi_{1t} + \sum_{k=1}^{N}\psi_{1x_k x_k} - 4\,(b\,|\psi_1|^2 + a\,|\psi_2|^2)\,\psi_1 = 0,$$

$$i\,\psi_{2t} + \sum_{k=1}^{N}\psi_{2x_k x_k} - 4\,(b\,|\psi_1|^2 + a\,|\psi_2|^2)\,\psi_2 = 0,$$

(7.120)

then

$$\Phi_1(x_1, x_2, \ldots, x_N, t) = \psi_1(x_1, x_2, \ldots, x_N, t) + \psi_2(x_1, x_2, \ldots, x_N, t),$$

$$\Phi_2(x_1, x_2, \ldots, x_N, t) = \psi_1(x_1, x_2, \ldots, x_N, t) + i\,\psi_2(x_1, x_2, \ldots, x_N, t)$$

(7.121)

is a solution to

$$i\,\Phi_{1t} + \sum_{k=1}^{N}\Phi_{1x_k x_k} - 2(a + b)\,(|\Phi_1|^2 + |\Phi_2|^2)\,\Phi_1$$

$$+ 2\,[(a + i\,b)\,\Phi_1\,\Phi_2^* + (a - i\,b)\,\Phi_2\,\Phi_1^*]\,\Phi_1 = 0,$$

$$i\,\Phi_{2t} + \sum_{k=1}^{N}\Phi_{2x_k x_k} - 2(a + b)\,(|\Phi_1|^2 + |\Phi_2|^2)\,\Phi_2$$

$$+ 2\,[(a + i\,b)\,\Phi_1\,\Phi_2^* + (a - i\,b)\,\Phi_2\,\Phi_1^*]\,\Phi_2 = 0,$$

(7.122)

where a and b are real constants.

7.10 Composite Solutions: Nonlinear Superposition

If (ψ_1, ψ_2) is a solution to the CNLSE, which is denoted as the *seed solution*,

$$i\,\frac{\partial}{\partial t}\psi_1(x, t) + \frac{\partial^2}{\partial x^2}\psi_1(x, t) + [b_{11}\,|\psi_1(x, t)|^2 + b_{12}\,|\psi_2(x, t)|^2]\,\psi_1(x, t) = 0,$$

$$i\,\frac{\partial}{\partial t}\psi_2(x, t) + \frac{\partial^2}{\partial x^2}\psi_2(x, t) + [b_{21}\,|\psi_1(x, t)|^2 + b_{22}\,|\psi_2(x, t)|^2]\,\psi_2(x, t) = 0,$$

(7.123)

then, the superposition, which is denoted as the *composite solution*,

$$\begin{pmatrix} \phi_1 \\ \phi_2 \end{pmatrix} = \begin{pmatrix} \cos\theta & \dfrac{\sin\theta}{r} \\ -r\sin\theta & \cos\theta \end{pmatrix}\begin{pmatrix} \psi_1 \\ \psi_2 \end{pmatrix}$$

(7.124)

is a solution to the same system, with $b_{21} = b_{11}$, $b_{22} = b_{12}$, namely

$$i\,\frac{\partial}{\partial t}\phi_1(x, t) + \frac{\partial^2}{\partial x^2}\phi_1(x, t) + [b_{11}\,|\phi_1(x, t)|^2 + b_{12}\,|\phi_2(x, t)|^2]\,\phi_1(x, t) = 0,$$

$$i\,\frac{\partial}{\partial t}\phi_2(x, t) + \frac{\partial^2}{\partial x^2}\phi_2(x, t) + [b_{11}\,|\phi_1(x, t)|^2 + b_{12}\,|\phi_2(x, t)|^2]\,\phi_2(x, t) = 0,$$

(7.125)

where $r = \sqrt{b_{11}/b_{12}}$, and θ is an arbitrary real constant [11].

***Example* 1. Constant Amplitude** *CW, x-dependent phase*
(Figure 7.11)

Given

$$\psi_1(x,\,t) = A_0\,e^{i\,(A_1t+A_2x+\varphi_1)},$$
$$\psi_2(x,\,t) = B_0\,e^{i(B_1t+B_2x+\varphi_2)},$$

(7.126)

is a seed solution to the CNLSE, Equation (7.123), then

$$\phi_1(x,\,t) = A_0\,e^{i\,(A_1t+A_2x+\varphi_1)}\cos\theta\, + B_0\sqrt{\frac{b_{12}}{b_{11}}}\,e^{i(B_1t+B_2x+\varphi_2)}\sin\theta,$$

$$\phi_2(x,\,t) = -\,A_0\sqrt{\frac{b_{11}}{b_{12}}}\,e^{i\,(A_1t+A_2x+\varphi_1)}\sin\theta\, + B_0\,e^{i(B_1t+B_2x+\varphi_2)}\cos\theta,$$

(7.127)

is a composite superposition solution to the CNLSE, Equation (7.125), where
$A_1 = -A_2^2 + A_0^2\,b_{11} + B_0^2b_{12}$, $B_1 = -B_2^2 + A_0^2\,b_{11} + B_0^2b_{12}$, A_0, A_2, B_0, B_2, φ_1, φ_2
are arbitrary real constants.

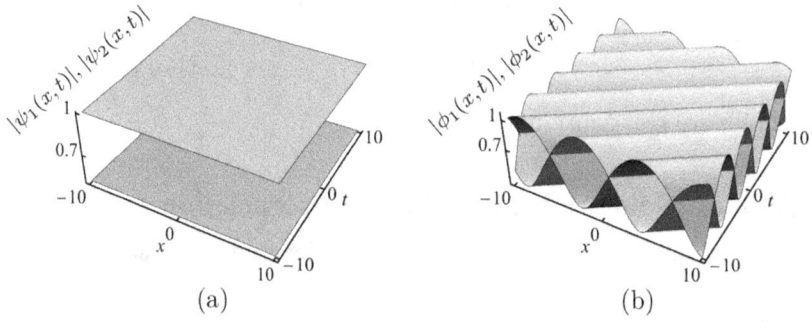

(a) (b)

Figure 7.11. The norm of the CW seed solution (left) and composite solution (right), given by Eqs. (7.128) and (7.129). Orange corresponds to ψ_1 and ϕ_1 and blue corresponds to ψ_2 and ϕ_2. Parameters used: $A_0 = A_2 = b_{11} = b_{12} = 1$, $B_0 = B_2 = 1/2$, $\varphi_1 = \varphi_2 = 0$, and $\theta = \pi/4$ for the composite solution.

***Example* 2.** *Bright–dark soliton*
(Figure 7.12)

Given

$$\psi_1(x,\,t) = \sqrt{\frac{2}{b_{11}}}\,A_0\,e^{2iA_0^2t}\,\tanh(A_0x),$$

$$\psi_2(x,\,t) = -\,2i\sqrt{\frac{2}{b_{12}}}\,A_0\,e^{3iA_0^2t}\,\mathrm{sech}(A_0x),$$

(7.128)

is a seed solution to CNLSE, Equation (7.123), then

$$\phi_1(x,\ t) = A_0\ \sqrt{\frac{2}{b_{11}}}\ e^{2iA_0^2 t}\Big[\tanh(A_0 x)\cos\theta - i\sqrt{2}\ e^{iA_0^2 t}\ \mathrm{sech}(A_0 x)\sin\theta\Big],$$

$$\phi_2(x,\ t) = -A_0\ \sqrt{\frac{2}{b_{11}}}\ e^{2iA_0^2 t}\Big[\tanh(A_0 x)\sin\theta + i\sqrt{2}\ e^{iA_0^2 t}\ \mathrm{sech}(A_0 x)\cos\theta\Big],$$

$$(7.129)$$

is a composite superposition solution to the CNLSE, Equation (7.125), where A_0 is an arbitrary real constant.

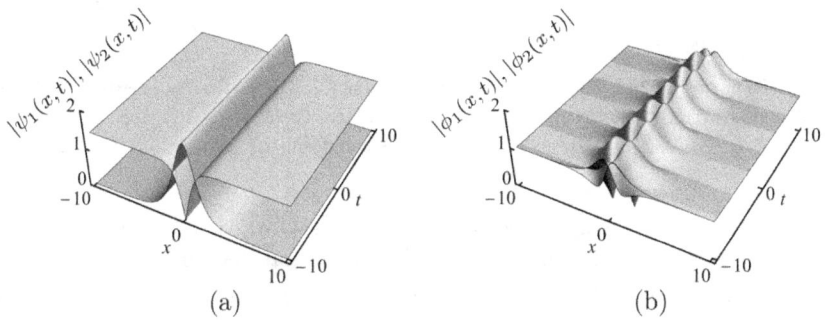

Figure 7.12. The norm of seed solution (left) and its composite solution (right), given by Eqs. (7.128) and (7.129). Orange corresponds to ψ_1 and ϕ_1 and blue corresponds to ψ_2 and ϕ_2. Parameters used: $A_0 = 1$ and $\theta = \pi/4$ for the composite solution.

Example 3. *Solitary wave (SW)*
(Figure 7.13)

Given

$$\psi_1(x,\ t) = A_0\ e^{-i(\omega_1 t + \varphi_1)}\sqrt{m}\ \mathrm{cd}(A_1 x|m),$$
$$\psi_2(x,\ t) = B_0\ e^{-i(\omega_2 t + \varphi_2)}\sqrt{1-m}\ \mathrm{nd}(A_1 x|m),$$

$$(7.130)$$

is a seed solution to the CNLSE, Equation (7.123), then

$$\phi_1(x,\ t) = A_0\sqrt{m}\ e^{-i\left\{\left[A_1^2(1+m) - B_0^2 b_{12}\right]t + \varphi_1\right\}}\mathrm{cd}(A_1 x|m)\cos\theta$$

$$+ B_0\sqrt{1-m}\ \sqrt{\frac{A_0^2 b_{12}}{B_0^2 b_{12} - 2A_1^2}}\ e^{-i\left[(A_1^2 m - B_0^2 b_{12})t + \varphi_2\right]}$$

$$\times\ \mathrm{nd}(A_1 x|m)\ \sin\theta,$$

$$\phi_2(x,\ t) = -A_0\sqrt{m}\ \sqrt{\frac{B_0^2 b_{12} - 2A_1^2}{A_0^2 b_{12}}}\ e^{-i\left\{\left[A_1^2(1+m) - B_0^2 b_{12}\right]t + \varphi_1\right\}}$$

$$\times\ \mathrm{cd}(A_1 x|m)\sin\theta$$

$$+ B_0\sqrt{1-m}\ e^{-i\left[(A_1^2 m - B_0^2 b_{12})t + \varphi_2\right]}$$

$$\times\ \mathrm{nd}(A_1 x|m)\ \cos\theta,$$

$$(7.131)$$

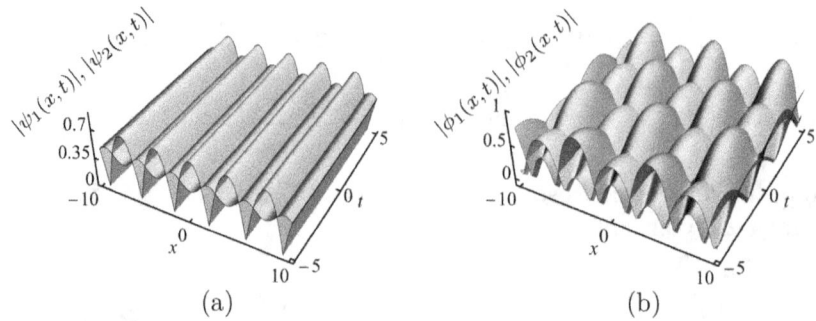

Figure 7.13. The norm of seed (left) and its composite solution, (used in the front cover page), (right) given by Eqs. (7.130) and (7.131). Orange corresponds to ψ_1 and ϕ_1 and blue corresponds to ψ_2 and ϕ_2. Parameters used: $A_1 = A_0 = 1$, $B_0 = m = 1/2$, $b_{12} = 1$, and $\theta = \pi/4$ for the composite solution.

is a composite superposition solution to the CNLSE, Equation (7.125), where $cd(x|m)$ and $nd(x|m)$ are the Jacobi elliptic functions with modulus $k^2 = m$, $\omega_1 = -(1-m)A_1^2 - b_{11}A_0^2$, $\omega_2 = -(2-m)A_1^2 - b_{11}A_0^2$, $b_{11} = (-2A_1^2 + b_{12}B_0^2)/A_0^2$.

7.11 Summary of Sections 7.8–7.10

(N + 1)-dimensional symmetry reductions

(N + 1)-dimensional symmetry reduction I: *From (N + 1)-dimensional Manakov system to (N + 1)-dimensional fundamental NLSE*

Transformation: $\psi_2(x_1, x_2, \ldots, x_N, t) = \sigma \, \psi_1(x_1, x_2, \ldots, x_N, t)$, $b_1 = c_1 + (c_2 - b_2) |\sigma|^2$, $b_{0k} = c_{0k}$, (ψ_1, ψ_2) **is a solution to the (N + 1)-dimensional CNLSE**

Equation: $i \, \psi_{1t} + \sum_{k=1}^{N} c_{0k} \, \psi_{1x_k x_k} + (c_1 + c_2) |\sigma|^2 |\psi_1|^2 \, \psi_1 = 0$

(N + 1)-dimensional symmetry reduction II: *From (N + 1)-dimensional Manakov system to (N + 1)-dimensional fundamental NLSE*

Transformation: $\psi_2(x_1, x_2, \ldots, x_N, t) = e^{j \, \phi} \psi_1^*(x_1, x_2, \ldots, x_N, t)$, $b_1 = -(c_1 + c_2 + b_2)$, $b_{0k} = -c_{0k}$

Equation: $i \, \psi_{1t} - \sum_{k=1}^{N} c_{0k} \, \psi_{1x_k x_k} - 2 \, (c_1 + c_2) |\psi_1|^2 \, \psi_1 = 0$

(N + 1)-dimensional symmetry reduction III: *From (N + 1)-dimensional vector NLSE to (N + 1)-dimensional fundamental NLSE*

Transformation: $\psi_j(x_1, x_2, \ldots, x_N, t) = \sigma_j \, \psi_1(x_1, x_2, \ldots, x_N, t)$

Equation: $i \, \psi_{1t} + \sum_{k=1}^{N} b_{10k} \, \psi_{1x_k x_k} + \sum_{k=1}^{N} b_{1j} |\sigma_{j+1}|^2 |\psi_1|^2 \, \psi_1 = 0$

(N + 1)-dimensional scaling transformations

Linear and nonlinear coupling

General case

Transformation:

$$\Phi_1(x_1, x_2, \ldots, x_N, t) = \sqrt{\frac{g_1}{g_1 - g_2}} \, \psi_1(x_1, x_2, \ldots, x_N, t) \, e^{j \, g_0 \, (g_1 - g_2) \, t}$$

$$\qquad + \sqrt{\frac{g_2 \, b_2}{g_1 (g_2 - g_1)}} \, \psi_2(x_1, x_2, \ldots, x_N, t) \, e^{-i \, g_0 \, (g_1 - g_2) \, t},$$

$$\Phi_2(x_1, x_2, \ldots, x_N, t) = \sqrt{\frac{g_1}{g_1 - g_2}} \, \psi_1(x_1, x_2, \ldots, x_N, t) \, e^{j \, g_0 \, (g_1 - g_2) \, t}$$

$$\qquad + \sqrt{\frac{g_1 \, b_2}{g_2 (g_2 - g_1)}} \, \psi_2(x_1, x_2, \ldots, x_N, t) \, e^{-i \, g_0 \, (g_1 - g_2) \, t}$$

Equation: $i \, \Phi_{1t} + \sum_{k=1}^{N} \Phi_{1x_k x_k} + (g_1 |\Phi_1|^2 - g_2 |\Phi_2|^2) \, \Phi_1 - 2 \, g_0 \, g_2 \, \Phi_2 = 0$, $i \, \Phi_{2t} + \sum_{k=1}^{N} \Phi_{2x_k x_k} + (g_1 |\Phi_1|^2 - g_2 |\Phi_2|^2) \, \Phi_2 - g_0 \, (g_1 + g_2) \, \Phi_2 + 2 \, g_0 \, g_1 \, \Phi_1 = 0$

Specific Case I: (N + 1)-dimensional Manakov system to another (N + 1)-dimensional Manakov system

Transformation:

$$\Phi_1(x_1, x_2, \ldots, x_N, t) = \sqrt{\frac{g_1}{g_1 - g_2}} \, \psi_1(x_1, x_2, \ldots, x_N, t)$$

$$\qquad + \sqrt{\frac{g_2 \, b_2}{g_1 (g_2 - g_1)}} \, \psi_2(x_1, x_2, \ldots, x_N, t),$$

$$\Phi_2(x_1, x_2, \ldots, x_N, t) = \sqrt{\frac{g_1}{g_1 - g_2}} \, \psi_1(x_1, x_2, \ldots, x_N, t)$$

$$\qquad + \sqrt{\frac{g_1 \, b_2}{g_2 (g_2 - g_1)}} \, \psi_2(x_1, x_2, \ldots, x_N, t)$$

Equation: $i\,\Phi_{1t} + \sum_{k=1}^{N}\Phi_{1x_kx_k} + (g_1|\Phi_1|^2 - g_2|\Phi_2|^2)\Phi_1 = 0,\; i\,\Phi_{2t} + \sum_{k=1}^{N}\Phi_{2x_kx_k} + (g_1|\Phi_1|^2 - g_2|\Phi_2|^2)\Phi_2 = 0$

Specific Case II: (N + 1)-dimensional Manakov system to the same (N + 1)-dimensional Manakov system

Superposition principle to a nonlinear system

Transformation:
$$\Phi_1(x_1, x_2, \ldots, x_N, t) = \sqrt{\frac{t_1}{t_1+b_2}}\,[\psi_1(x_1, x_2, \ldots, x_N, t)\; \Phi_2(x_1, x_2, \ldots, x_N, t) = \sqrt{\frac{t_1}{t_1+b_2}}\,[\psi_1(x_1, x_2, \ldots, x_N, t)$$
$$+\tfrac{b_2}{t_1}\psi_2(x_1, x_2, \ldots, x_N, t)],\qquad\qquad + \psi_2(x_1, x_2, \ldots, x_N, t)]$$

Complex coupling

General case

Transformation: $\Phi_1(x_1, x_2, \ldots, x_N, t) = c_1\,\psi_1(x_1, x_2, \ldots, x_N, t)\qquad \Phi_2(x_1, x_2, \ldots, x_N, t) = c_3\,\psi_1(x_1, x_2, \ldots, x_N, t)$
$+ c_2\,\psi_2(x_1, x_2, \ldots, x_N, t),\qquad\qquad\qquad\quad + c_4\,\psi_2(x_1, x_2, \ldots, x_N, t)$

Equation: $i\,\Phi_{1t} + \sum_{k=1}^{N}\Phi_{1x_kx_k} + 2\,(a_{11}|\Phi_1|^2 + a_{12}|\Phi_2|^2)\,\Phi_1 + 2\,(b_{11}\Phi_1\Phi_2^* + b_{12}\Phi_2\Phi_1^*)\,\Phi_1 = 0,$

$i\,\Phi_{2t} + \sum_{k=1}^{N}\Phi_{2x_kx_k} + 2\,(a_{21}|\Phi_1|^2 + a_{22}|\Phi_2|^2)\,\Phi_2 + 2\,(b_{21}\Phi_1\Phi_2^* + b_{22}\Phi_2\Phi_1^*)\,\Phi_2 = 0$

Specific case

Transformation: $\Phi_1(x_1, x_2, \ldots, x_N, t) = \psi_1(x_1, x_2, \ldots, x_N, t) + \psi_2(x_1, x_2, \ldots, x_N, t),\; \Phi_2(x_1, x_2, \ldots, x_N, t) = \psi_1(x_1, x_2, \ldots, x_N, t) + i\,\psi_2(x_1, x_2, \ldots, x_N, t)$

Equation: $i\,\Phi_{1t} + \sum_{k=1}^{N}\Phi_{1x_kx_k} - 2(a+b)\,(|\Phi_1|^2 + |\Phi_2|^2)\,\Phi_1 + 2\,((a+i\,b)\,\Phi_1\Phi_2^* + (a-i\,b)\,\Phi_2\Phi_1^*)\,\Phi_1 = 0,$

$i\,\Phi_{2t} + \sum_{k=1}^{N}\Phi_{2x_kx_k} - 2(a+b)\,(|\Phi_1|^2 + |\Phi_2|^2)\,\Phi_2 + 2\,((a+i\,b)\,\Phi_1\Phi_2^* + (a-i\,b)\,\Phi_2\Phi_1^*)\,\Phi_2 = 0$

Composite solutions: nonlinear superposition

Transformation: $\begin{pmatrix}\phi_1\\[2pt]\phi_2\end{pmatrix} = \begin{pmatrix}\cos\theta & \dfrac{\sin\theta}{r}\\[6pt] -r\sin\theta & \cos\theta\end{pmatrix}\begin{pmatrix}\psi_1\\[2pt]\psi_2\end{pmatrix},\qquad (\psi_1,\psi_2)\;$ **is a seed solution to the Manakov system**

Equation: $i\,\dfrac{\partial}{\partial t}\phi(x, t) + \dfrac{\partial^2}{\partial x^2}\phi(x, t) + (b_{11}|\phi(x, t)|^2 + b_{12}|\phi_2(x, t)|^2)\,\phi_1(x, t) = 0,$

$i\,\dfrac{\partial}{\partial t}\phi_2(x, t) + \dfrac{\partial^2}{\partial x^2}\phi_2(x, t) + (b_{11}|\phi(x, t)|^2 + b_{12}|\phi_2(x, t)|^2)\,\phi_2(x, t) = 0$

References

[1] Manakov S V 1974 On the theory of two-dimensional stationary self focussing of electromagnetic waves *Sov. Phys. JETP* **38** 248–53

[2] Agrawal G 2001 *Nonlinear fiber optics* 3rd ed (San Diego, CA: Academic)

[3] Pethick C J and Smith H 2008 *Bose–Einstein condensation in dilute gases* (Cambridge: Cambridge University Press)

[4] Akhmediev N and Ankiewicz A 1997 *Solitons: Nonlinear Pulses and Beams* (London: Chapman and Hall)

[5] Khare A and Saxena A 2023 New Solutions of coupled nonlocal NLS and coupled nonlocal mKdV equations *Ann. Phys.* **463** 169627

[6] Porsezian K and Kuriakose V C 2003 *Optical Solitons: Theoretical and Experimental Challenges* (Springer)

[7] Buryak A V, Kivshar Y S and Parker D F 1996 Coupling between dark and bright solitons *Phys. Lett.* **A215** 57–62

[8] Guo B L and Ling L M 2011 Rogue wave, breathers and bright-dark-rogue solutions for the coupled Schrödinger equations *Chin. Phys. Lett.* **28** 110202

[9] Khare A and Saxena A 2015 Periodic and hyperbolic soliton solutions of a number of nonlocal nonlinear equations *J. Math. Phys.* **56** 032104

[10] Porubov A V and Parker D F 1999 Some general periodic solutions to coupled nonlinear Schrödinger equations *Wave Motion* **29** 97–109

[11] Al Sakkaf L and Al Khawaja U 2020 Superposition principle and composite solutions to coupled nonlinear Schrödinger equations *Math. Methods Appl. Sci.* **43** 10168–89

IOP Publishing

Handbook of Exact Solutions to the Nonlinear Schrödinger Equations (Second Edition)

Usama Al Khawaja and Laila Al Sakkaf

Chapter 8

Discrete Nonlinear Schrödinger Equation

A Glance at Chapter 8

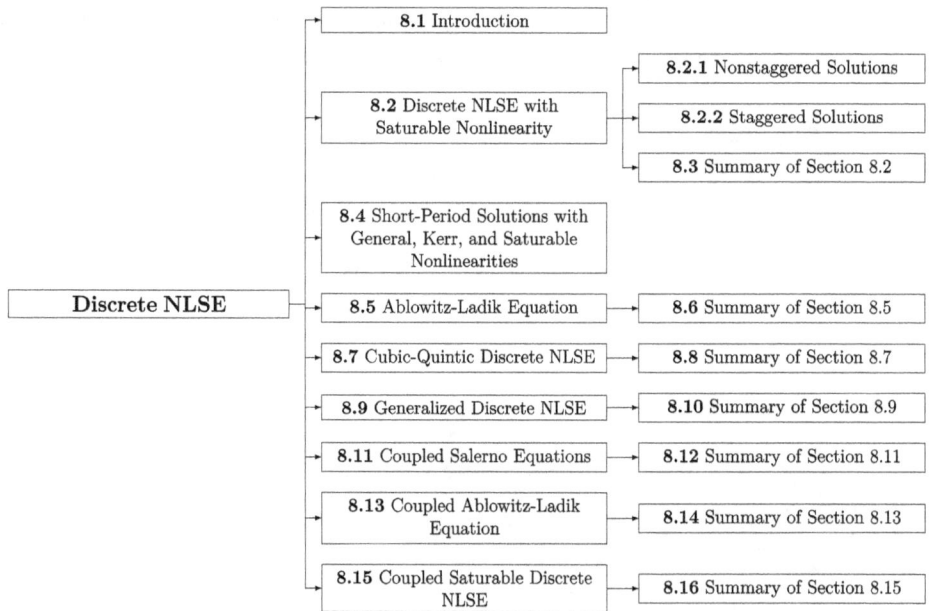

A Statistical View of Chapter 8

	Equation	Solutions
1	$i\,\psi_{nt} + \psi_{n+1} + \psi_{n-1} - 2\,\psi_n + \dfrac{a_2\,\|\psi_n\|^2\,\psi_n}{1 + \mu\,\|\psi_n\|^2} = 0$	35
2	$i\,\psi_{nt} + \psi_{n+1} + \psi_{n-1} - 2\,\psi_n + a_2\,F[\|\psi_n\|^2]\,\psi_n = 0$	6
3	$i\,\psi_{nt} + \psi_{n+1} + \psi_{n-1} - 2\,\psi_n + a_2\,\|\psi_n\|^2\,\psi_n = 0$	6
4	$i\,\psi_{nt} + \psi_{n+1} + \psi_{n-1} - 2\,\psi_n + a_2\,(\psi_{n+1} + \psi_{n-1})\,\|\psi_n\|^2 = 0$	12
5	$i\,\psi_{nt} + a_1\,(\psi_{n+1} + \psi_{n-1} - 2\,\psi_n) + a_2\,\|\psi_n\|^2\,\psi_n + (a_3\,\|\psi_n\|^2 + a_4\,\|\psi_n\|^4)(\psi_{n+1} + \psi_{n-1}) = 0$	5
6	$i\,\psi_{nt} + a_1\,(\psi_{n+1} + \psi_{n-1} - 2\,\psi_n) + f[\psi_{n-1}, \psi_n, \psi_{n+1}] = 0$	7
7	$i\,\psi_{1nt} + \psi_{1n+1} + \psi_{1n-1} - 2\,\psi_{1n} + (\mu_1\,\|\psi_{1n}\|^2 + \mu_2\,\|\psi_{2n}\|^2)\,(\psi_{1n+1} + \psi_{1n-1} + \dfrac{\nu_1 - 2\,\mu_1}{\mu_1}\,\psi_{1n})$ $= 0,$ $i\,\psi_{2nt} + [\psi_{2n+1} + \psi_{2n-1} - (2 + \dfrac{\nu_1\,\mu_2}{\mu_1^2} - \dfrac{\nu_2}{\mu_2})\,\psi_{2n}] + (\mu_1\,\|\psi_{1n}\|^2 + \mu_2\,\|\psi_{2n}\|^2)$ $[\psi_{2n+1} + \psi_{2n-1} + (\dfrac{\nu_2 - 2\,\mu_2}{\mu_2})\,\psi_{2n}] = 0$	14
8	$i\,\psi_{1nt} + \psi_{1n+1} + \psi_{1n-1} - 2\,\psi_{1n} + (\mu_1\,\|\psi_{1n}\|^2 + \mu_2\,\|\psi_{2n}\|^2)\,(\psi_{1n+1} + \psi_{1n-1}) = 0,$ $i\,\psi_{2nt} + \psi_{2n+1} + \psi_{2n-1} - \dfrac{2\,\mu_2}{\mu_1}\,\psi_{2n} + (\mu_1\,\|\psi_{1n}\|^2 + \mu_2\,\|\psi_{2n}\|^2)\,(\psi_{2n+1} + \psi_{2n-1}) = 0$	18
9	$i\,\psi_{1nt} + \psi_{1n+1} + \psi_{1n-1} - 2\,\psi_{1n} + \dfrac{\nu_1\,(\mu_1\,\|\psi_{1n}\|^2 + \nu_2\,\|\psi_{2n}\|^2)\,\psi_{1n}}{\mu_1\,(1 + \mu_1\,\|\psi_{1n}\|^2 + \mu_2\,\|\psi_{2n}\|^2)} = 0,$ $i\,\psi_{2nt} + \psi_{2n+1} + \psi_{2n-1} - 2\,\psi_{2n} + \dfrac{[\nu_2 - \dfrac{\nu_1\,\mu_2}{\mu_1^2} + \nu_2\,(\mu_1\,\|\psi_{1n}\|^2 + \nu_2\,\|\psi_{2n}\|^2)]\,\psi_{2n}}{\mu_2\,(1 + \mu_1\,\|\psi_{1n}\|^2 + \nu_2\,\|\psi_{2n}\|^2)} = 0$	3
Total	9	106

8.1 Introduction

Optical pulses in waveguide arrays [1, 2] and matter waves of Bose–Einstein condensates in optical lattices are among the most prominent physical systems described by the discrete NLSE [3]. In general, almost any physical system of periodic anharmonic oscillators will be described by a certain version of the discrete NLSE. In waveguide arrays, as shown in Figure 8.1, hundreds of grooves per centimeter forming an array are carved using a highly focused laser beam on a transparent substrate [2]. A broad beam is injected at one end of the waveguide. After a transient distance, a single mode will be localized at each waveguide. The evanescent overlap between neighboring modes leads to a weak interaction between the neighboring modes which results in the nonlinearity in the discrete NLSE. In Bose–Einstein condensates, the atoms are trapped by an optical lattice, as shown in Figure 8.1. Optical lattices are created by two counter-propagating laser beams forming a standing wave with local minima working as potential wells. A number of atoms is trapped at each local potential well of the periodic structure and the interaction is provided by the overlap between the corresponding wavefunctions.

There are many versions of the NLSE which differ mainly by how the nonlinear term is discretized [4]. The most common approach to derive the discrete NLSE is the so-called *tight-binding* model—a term borrowed from solid state physics.

Optical pulse in waveguide array

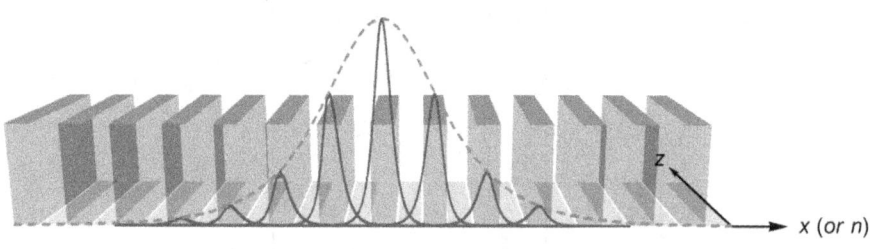

BEC in optical lattice

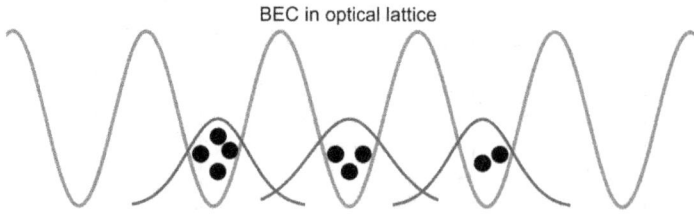

Figure 8.1. Upper subfigure: Light pulse propagating in a waveguide array. Blue curves are the single-mode field intensity in each waveguide. Lower subfigure: BEC in an optical lattice. Black dots correspond to BEC atoms. Red curve is the optical lattice field formed by standing waves of two counter propagating laser beams. Blue curves are the matter-wavefunctions.

Here, the collective excitation is written as a superposition of the single-mode excitations in each waveguide or local potential well.

In the following, we outline the derivation of the discrete NLSE starting from the NLSE for the continuum, Equation (2.1). Consider the linear eigenvalue problem

$$a_1 \frac{d}{dx}\phi_n(x) + V(x)\phi_n(x) = \lambda_n\phi_n(x), \qquad (8.1)$$

with normalized orthogonal eigenfunctions

$$\int \phi_n(x)\phi_m(x)dx = \delta_{nm}, \qquad (8.2)$$

where n and m are integers, $V(x)$ is an arbitrary function, a_1 is an arbitrary constant and λ_n is an eigenvalue. Integration here is over the whole space domain. Expanding the solution of the NLSE, Equation (2.1), in the set of orthogonal functions, $\phi_n(x)$, as

$$u(x, t) = \sum_n c_n(t)\phi_n(x), \qquad (8.3)$$

and then substituting in Equation (2.1), we get

$$i\sum_n \dot{c}_n(t)\phi_n(x) + a_1\sum_n c_n(t)\phi_n''(x) + a_2\sum_{n_1}\sum_{n_2}\sum_{n_3} c_{n_1}c_{n_2}c_{n_3}^*\phi_{n_1}(x)\phi_{n_2}(x)\phi_{n_3}^*(x) = 0. \qquad (8.4)$$

Multiplying by $\phi_m^*(x)$, integrating over x, and then employing the orthogonality condition (8.2), we obtain

$$i\dot{c}_m(t) + a_1\sum_n c_n(t)A_{nm} + a_2\sum_{n_1}\sum_{n_2}\sum_{n_3} c_{n_1}c_{n_2}c_{n_3}^*B_{m\,n_1\,n_2\,n_3} = 0, \qquad (8.5)$$

where $A_{nm} = \int \phi_m^*(x)\phi_n''(x)dx$ and $B_{m\,n_1\,n_2\,n_3} = \int \phi_m^*(x)\phi_{n_1}(x)\phi_{n_2}(x)\phi_{n_3}^*(x)dx$. In view of the fact that the overlap between the neighboring modes is small and decays as the separation between them increases, the nearest neighbor approximation will be justified for the linear terms in the last equation, and keeping only the self-interaction for the nonlinear term will be sufficient. The last equation then reduces to

$$i\dot{c}_m(t) + a_1A_{m-1\,m}[c_{m-1}(t) + c_{m+1}(t)] + a_1A_{m\,m}c_m(t) + a_2B_{m\,m\,m\,m}|c_m|^2c_m = 0, \quad (8.6)$$

where we have used $A_{m-1\,m} = A_{m\,m-1}$. The central term can be removed using the scaling $c_m(t) = \psi_m(t)e^{ia_1A_{m\,m}t}$, which finally gives the discrete version of the NLSE

$$i\dot{\psi}_m(t) + \alpha_m[\psi_{m-1}(t) + \psi_{m+1}(t)] + \gamma_m|\psi_m|^2\psi_m = 0, \qquad (8.7)$$

where $\alpha_m = a_1A_{m-1\,m}$ and $\gamma_m = a_2B_{m\,m\,m\,m}$ are coupling constants that depend on the geometry and material of the waveguides.

While the discrete NLSE (8.7), is not integrable, another similar discrete equation, namely the Ablowitz–Ladik equation is often considered

$$i\dot{\psi}_m(t) + \alpha_m[\psi_{m-1}(t) + \psi_{m+1}(t)] + \gamma_m|\psi_m|^2(\psi_{m-1} + \psi_{m+1}) = 0, \qquad (8.8)$$

since it is integrable. However it corresponds to a non-Hermitian Hamiltonian, hence less interesting from a physical point of view. Many other versions of the discrete NLSE have been considered including higher dimensionality and various forms of nonlinearity [4]. The solutions presented in this chapter are an attempt to cover all such various cases.

8.2 Discrete NLSE with Saturable Nonlinearity

Equation:

$$i\,\psi_{nt} + \psi_{n+1} + \psi_{n-1} - 2\,\psi_n + \frac{a_2\,|\psi_n|^2\,\psi_n}{1 + \mu\,|\psi_n|^2} = 0, \tag{8.9}$$

where $\psi_n = \psi(n, t)$ is the complex function profile, the integer site index, n, and t are its two independent variables, a_2 and μ are real constants.

Solutions:

8.2.1 Nonstaggered Solutions

Solution 1. Constant Amplitude *discrete continuous wave (CW), t- and n-dependent phase*

$$\psi(n, t) = A_0\, e^{i\left[A_1\,(n-n_0) - A_2\,(t-t_0) + \phi_0\right]}, \tag{8.10}$$

where $A_2 = 4\sin^2(A_1/2) - \dfrac{a_2\,A_0^2}{1 + \mu\,A_0^2}$, A_0, A_1, t_0, n_0, and ϕ_0 are arbitrary real constants.

- *Reference*: [6].

Solution 2.

$$\psi(n, t) = A_0\, \sec[A_1\,(n - n_0)]\, e^{-i\left[A_2\,(t-t_0) + \phi_0\right]}, \tag{8.11}$$

where $A_0 = \dfrac{\sin(A_1)}{\sqrt{-\mu}}$, $A_2 = \dfrac{2\,\mu - a_2}{\mu}$, $a_2 = 2\,\mu\cos(A_1)$, $\mu < 0$, A_1, t_0, n_0, and ϕ_0 are arbitrary real constants.

- *Reference*: [7].

Solution 3.

$$\psi(n, t) = A_0\, \tan[A_1\,(n - n_0)]\, e^{-i\left[A_2\,(t-t_0) + \phi_0\right]}, \tag{8.12}$$

where $A_0 = \dfrac{\tan(A_1)}{\sqrt{-\mu}}$, $A_2 = 2 - 2\sec^2(A_1)$, $a_2 = 2\,\mu\sec^2(A_1)$, $\mu < 0$, A_1, t_0, n_0, and ϕ_0 are arbitrary real constants.

- *Reference*: [7], we corrected the expression of A_0.

Solution 4. *Discrete bright soliton*
  (Figure 8.2)

$$\psi(n, t) = A_0\, \mathrm{sech}[A_1\,(n - n_0)]\, e^{-i\left[A_2\,(t-t_0) + \phi_0\right]}, \tag{8.13}$$

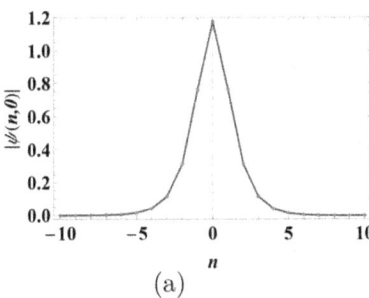 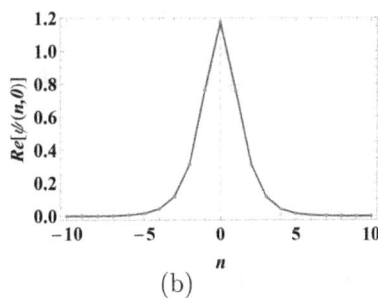

(a)   (b)

Figure 8.2. Discrete bright soliton (8.13) at $t = 0$. (a) Absolute value, (b) Real part, with $a_2 = A_1 = 1$ and $n_0 = t_0 = \phi_0 = 0$. The lines are guides for the eye.

where $A_0 = \frac{\sinh(A_1)}{\sqrt{\mu}}$, $A_2 = \frac{2\mu - a_2}{\mu}$, $\mu = \frac{a_2 \operatorname{sech}(A_1)}{2} > 0$, A_1, t_0, n_0, and ϕ_0 are arbitrary real constants.

- *Reference*: [6].

Solution 5.

$$\psi(n, t) = A_0 \operatorname{csch}[A_1 (n - n_0)] \, e^{-i \left[A_2 (t - t_0) + \phi_0 \right]}, \tag{8.14}$$

where $A_0 = \frac{\sinh(A_1)}{\sqrt{-\mu}}$, $A_2 = \frac{2\mu - a_2}{\mu}$, $a_2 = 2\mu \cosh(A_1)$, $\mu < 0$, A_1, t_0, n_0, and ϕ_0 are arbitrary real constants.

- *Reference*: [7].

Solution 6. *Discrete dark soliton*
(Figure 8.3)

$$\psi(n, t) = A_0 \tanh[A_1 (n - n_0)] \, e^{-i \left[A_2 (t - t_0) + \phi_0 \right]}, \tag{8.15}$$

where $A_0 = \frac{\tanh(A_1)}{\sqrt{-\mu}}$, $A_2 = \frac{2\mu \quad a_2}{\mu}$, $a_2 = 2\mu \operatorname{sech}^2(A_1)$, $\mu < 0$, A_1, t_0, n_0, and ϕ_0 are arbitrary real constants.

- *Reference*: [7].

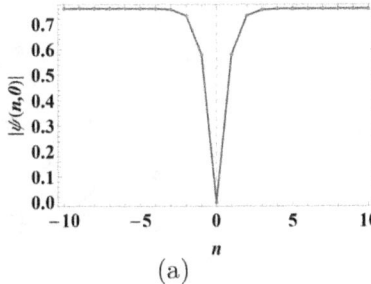 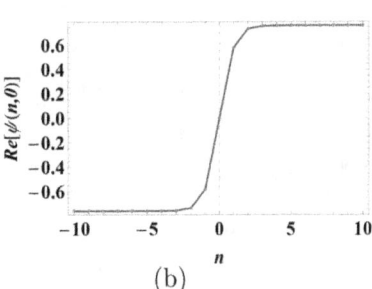

(a)   (b)

Figure 8.3. Discrete dark soliton (8.15) at $t = 0$. (a) Absolute value, (b) Real part, with $A_1 = 1$, $\mu = -1$, and $n_0 = t_0 = \phi_0 = 0$. The lines are guides for the eye.

Solution 7.

$$\psi(n,\,t) = A_0 \coth[A_1 \,(n - n_0)]\, e^{-i\left[A_2\,(t-t_0)+\phi_0\right]}, \tag{8.16}$$

where $A_0 = \frac{\tanh(A_1)}{\sqrt{-\mu}}$, $A_2 = \frac{2\mu - a_2}{\mu}$, $a_2 = 2\,\mu\,\text{sech}^2(A_1)$, $\mu < 0$, A_1, t_0, n_0, and ϕ_0 are arbitrary real constants.
  • *Reference*: [7].

Solution 8. *Discrete solitary wave (SW)*
   (Figure 8.4)

$$\psi(n,\,t) = A_0 \,\text{sn}[A_1 \,(n - n_0),\, m]\, e^{-i\left[A_2\,(t-t_0)+\phi_0\right]}, \tag{8.17}$$

where $A_0 = \frac{\sqrt{m}\,\text{sn}(A_1,m)}{\sqrt{-\mu}}$, $A_2 = \frac{2\mu - a_2}{\mu}$, $a_2 = 2\,\mu\,\text{cn}(A_1,\,m)\,\text{dn}(A_1,\,m)$, $\mu < 0$, $0 \leqslant m \leqslant 1$, A_1, t_0, n_0, and ϕ_0 are arbitrary real constants.
  • *Reference*: [7], *we corrected the expression A_0.*

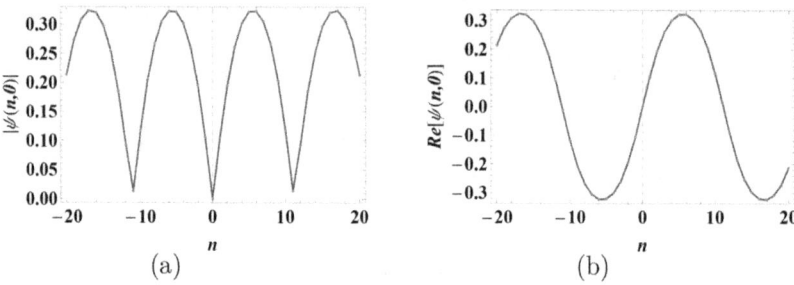

(a) (b)

Figure 8.4. Discrete solitary wave (8.17) at $t = 0$. (a) Absolute value, (b) Real part, with $A_1 = 1/3$, $\mu = -1/2$, $m = 1/2$, and $n_0 = t_0 = \phi_0 = 0$. The lines are guides for the eye.

Solution 9. *Discrete SW*
   (Figure 8.5)

$$\psi(n,\,t) = A_0 \,\text{cn}[A_1 \,(n - n_0),\, m]\, e^{-i\left[A_2\,(t-t_0)+\phi_0\right]}, \tag{8.18}$$

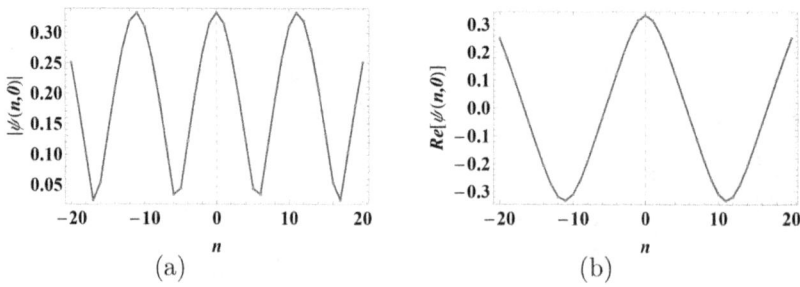

(a) (b)

Figure 8.5. Discrete solitary wave (8.18) at $t = 0$. (a) Absolute value, (b) Real part, with $a_2 = 1$, $A_1 = 1/3$, $m = 1/2$, and $n_0 = t_0 = \phi_0 = 0$. The lines are guides for the eye.

where $A_0 = \frac{\sqrt{m}\,\mathrm{sn}(A_1, m)}{\sqrt{\mu}\,\mathrm{dn}(A_1, m)}$, $A_2 = \frac{2\mu - a_2}{\mu}$, $\mu = \frac{a_2\,\mathrm{dn}^2(A_1, m)}{2\,\mathrm{cn}(A_1, m)} > 0$, $0 \leqslant m \leqslant 1$, A_1, t_0, n_0,
and ϕ_0 are arbitrary real constants.
- *Reference*: [6].

Solution 10. *Discrete SW*
(Figure 8.6)

$$\psi(n, t) = A_0\,\mathrm{dn}[A_1\,(n - n_0), m]\,e^{-i\,[A_2\,(t-t_0)+\phi_0]}, \qquad (8.19)$$

where $A_0 = \frac{\mathrm{sn}(A_1, m)}{\sqrt{\mu}\,\mathrm{cn}(A_1, m)}$, $A_2 = \frac{2\mu - a_2}{\mu}$, $\mu = \frac{a_2\,\mathrm{cn}^2(A_1, m)}{2\,\mathrm{dn}(A_1, m)} > 0$, $0 \leqslant m \leqslant 1$, A_1, t_0, n_0,
and ϕ_0 are arbitrary real constants.
- *Reference*: [6].

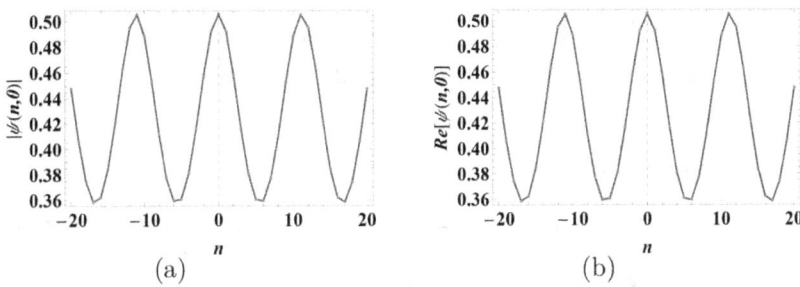

Figure 8.6. Discrete solitary wave (8.19) at $t = 0$. (a) Absolute value, (b) Real part, with $a_2 = 1$, $A_1 = 1/3$, $m = 1/2$, and $n_0 = t_0 = \phi_0 = 0$. The lines are guides for the eye.

Solution 11. *Discrete SW*

$$\psi(n, t) = A_0\,\mathrm{ns}[A_1\,(n - n_0), m]\,e^{-i\,[A_2\,(t-t_0)+\phi_0]}, \qquad (8.20)$$

where $A_0 = \frac{\mathrm{sn}(A_1, m)}{\sqrt{-\mu}}$, $A_2 = \frac{2\mu - a_2}{\mu}$, $a_2 = 2\mu\,\mathrm{dn}(A_1, m)\,\mathrm{cn}(A_1, m)$, $\mu < 0$,
$0 \leqslant m \leqslant 1$, A_1, t_0, n_0, and ϕ_0 are arbitrary real constants.
- *Reference*: [7].

Solution 12. *Discrete SW*

$$\psi(n, t) = A_0\,\mathrm{cs}[A_1\,(n - n_0), m]\,e^{-i\,[A_2\,(t-t_0)+\phi_0]}, \qquad (8.21)$$

where $A_0 = \frac{\mathrm{sn}(A_1, m)}{\sqrt{-\mu}\,\mathrm{cn}(A_1, m)}$, $A_2 = \frac{2\mu - a_2}{\mu}$, $a_2 = \frac{2\mu\,\mathrm{dn}(A_1, m)}{\mathrm{cn}^2(A_1, m)}$, $\mu < 0$, $0 \leqslant m \leqslant 1$, A_1, t_0,
n_0, and ϕ_0 are arbitrary real constants.
- *Reference*: [7].

Solution 13. *Discrete SW*

$$\psi(n, t) = A_0\,\mathrm{ds}[A_1\,(n - n_0), m]\,e^{-i\,[A_2\,(t-t_0)+\phi_0]}, \qquad (8.22)$$

where $A_0 = \frac{\text{sn}(A_1, m)}{\sqrt{-\mu}\ \text{dn}(A_1, m)}$, $A_2 = \frac{2\mu - a_2}{\mu}$, $a_2 = \frac{2\mu\ \text{cn}(A_1, m)}{\text{dn}^2(A_1, m)}$, $\mu < 0$, $0 \leqslant m \leqslant 1$, A_1, t_0, n_0, and ϕ_0 are arbitrary real constants.

- Reference: [7].

Solution 14. *Discrete SW*

$$\psi(n, t) = A_0\ \text{cd}[A_1\ (n - n_0), m]\ e^{-i\left[A_2\ (t - t_0) + \phi_0\right]}, \tag{8.23}$$

where $A_0 = \frac{\sqrt{m}\ \text{sn}(A_1, m)}{\sqrt{-\mu}}$, $A_2 = \frac{2\mu - a_2}{\mu}$, $a_2 = 2\mu\ \text{cn}(A_1, m)\ \text{dn}(A_1, m)$, $\mu < 0$, $0 \leqslant m \leqslant 1$, A_1, t_0, n_0, and ϕ_0 are arbitrary real constants.

- Reference: [7], *we corrected the expression of A_0.*

Solution 15. *Discrete SW*

$$\psi(n, t) = A_0\ \text{dc}[A_1\ (n - n_0), m]\ e^{-i\left[A_2\ (t - t_0) + \phi_0\right]}, \tag{8.24}$$

where $A_0 = \frac{\text{sn}(A_1, m)}{\sqrt{-\mu}}$, $A_2 = \frac{2\mu - a_2}{\mu}$, $a_2 = 2\mu\ \text{cn}(A_1, m)\ \text{dn}(A_1, m)$, $\mu < 0$, $0 \leqslant m \leqslant 1$, A_1, t_0, n_0, and ϕ_0 are arbitrary real constants.

- Reference: [7].

Solution 16. *Discrete SW*

$$\psi(n, t) = \left\{ \frac{A_0}{2}\ \text{dn}[A_1\ (n - n_0), m] + \frac{B_0}{2}\ \sqrt{m}\ \text{cn}[A_1\ (n - n_0), m] \right\} e^{-i\left[A_2\ (t - t_0) + \phi_0\right]}, \tag{8.25}$$

where $A_0 = \frac{2}{\text{cs}(A_1, m) + \text{ds}(A_1, m)}$, $B_0 = \pm A_0$, $A_2 = 2 - a_2$, $A_2 = \frac{4}{\text{cn}(A_1, m) + \text{dn}(A_1, m)}$, $\mu = 1$, $0 \leqslant m \leqslant 1$, A_1, t_0, and ϕ_0 are arbitrary real constants, n_0 is an arbitrary real integer.

- Reference: [5], *taken from the nonlocal case.*

8.2.2 Staggered Solutions

If $\psi(n, t; a_2)$ is a nonstaggered solution to (8.9), then

$$\psi_s(n, t, a_2) = (-1)^n\ \psi^*(n, t, -a_2)\ e^{-4\,i\,(t - t_0)} \tag{8.26}$$

is a staggered solution to the same equation, where ψ^* is the complex conjugate of the nonstaggered solution.

Solution 1. Constant Amplitude *staggered discrete CW, t- and n-dependent phase*

$$\psi_s(n, t) = (-1)^n\ A_0\ e^{-i\left[A_1\ (n - n_0) - (A_2 - 4)\ (t - t_0) + \phi_0\right]}, \tag{8.27}$$

where $A_2 = 4\sin^2(A_1/2) + \frac{a_2\ A_0^2}{1 + \mu\ A_0^2}$, A_0, A_1, t_0, n_0, and ϕ_0 are arbitrary real constants.

Solution 2.

$$\psi_s(n, t) = A_0\ \cos[\pi\ (n - n_0)]\ e^{-i\left[A_2\ (t - t_0) + \phi_0\right]}, \tag{8.28}$$

where $A_2 = 4 - \frac{a_2 A_0^2}{1 + \mu A_0^2}$, A_0, t_0, n_0, and ϕ_0 are arbitrary real constants.

- *Reference*: [8].

Solution 3.

$$\psi_s(n, t) = A_0 \cos\left[\frac{\pi}{2}(n - n_0)\right] e^{\pm i\left[A_2 (t - t_0) + \phi_0\right]}, \qquad (8.29)$$

where $A_2 = -2 + \frac{A_0^2 a_2}{A_0^2 \mu + \sec^2\left[\pi (n - n_0)/2\right]}$, A_0, t_0, n_0, and ϕ_0 are arbitrary real constants.

- *Reference*: [8].

Solution 4.

$$\psi_s(n, t) = A_0 \left\{\cos\left[\frac{\pi}{2}(n - n_0)\right] - \sin\left[\frac{\pi}{2}(n - n_0)\right]\right\} e^{-i\left[A_2 (t - t_0) + \phi_0\right]}, \qquad (8.30)$$

where $A_2 = 2 - \frac{a_2 A_0^2}{1 + \mu A_0^2}$, A_0, t_0, and ϕ_0 are arbitrary real constants, n_0 is an arbitrary real integer.

- *Reference*: [8].

Solution 5.

$$\psi_s(n, t) = (-1)^n A_0 \sec[A_1 (n - n_0)] e^{i\left[A_2 (t - t_0) - 4 (t - t_0) + \phi_0\right]}, \qquad (8.31)$$

where $A_0 = \frac{\sin(A_1)}{\sqrt{-\mu}}$, $A_2 = \frac{2\mu + a_2}{\mu}$, $a_2 = -2\mu \cos(A_1)$, $\mu < 0$, A_1, t_0, n_0, and ϕ_0 are arbitrary real constants.

Solution 6.

$$\psi_s(n, t) = (-1)^n A_0 \tan[A_1 (n - n_0)] e^{i\left[A_2 (t - t_0) - 4 (t - t_0) + \phi_0\right]}, \qquad (8.32)$$

where $A_0 = \frac{\tan(A_1)}{\sqrt{-\mu}}$, $A_2 = 2 - 2\sec^2(A_1)$, $a_2 = -2\mu \sec^2(A_1)$, $\mu < 0$, A_1, t_0, n_0, and ϕ_0 are arbitrary real constants.

Solution 7. *Staggered discrete bright soliton*
(Figure 8.7)

$$\psi_s(n, t) = (-1)^n A_0 \operatorname{sech}[A_1 (n - n_0)] e^{i\left[A_2 (t - t_0) - 4 (t - t_0) + \phi_0\right]}, \qquad (8.33)$$

where $A_0 = \frac{\sinh(A_1)}{\sqrt{\mu}}$, $A_2 = \frac{2\mu + a_2}{\mu}$, $\mu = -\frac{a_2 \operatorname{sech}(A_1)}{2} > 0$, A_1, t_0, n_0, and ϕ_0 are arbitrary real constants.

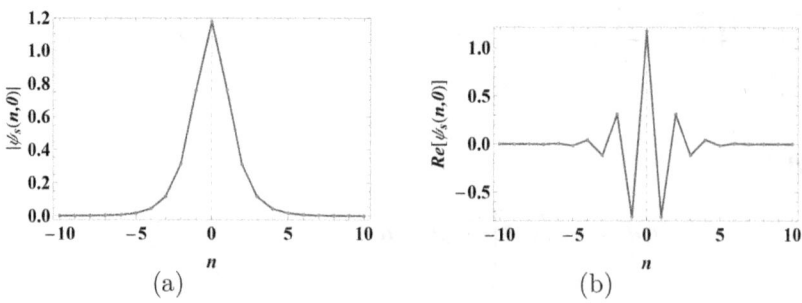

Figure 8.7. Staggered discrete bright soliton (8.33) at $t = 0$. (a) Absolute value, (b) Real part, with $a_2 = -1$, $A_1 = 1$, and $n_0 = t_0 = \phi_0 = 0$. The lines are guides for the eye.

Solution 8.

$$\psi_s(n, t) = (-1)^n A_0 \operatorname{csch}[A_1 (n - n_0)] \, e^{i \left[A_2 (t-t_0) - 4 (t-t_0) + \phi_0 \right]}, \qquad (8.34)$$

where $A_0 = \dfrac{\sinh(A_1)}{\sqrt{-\mu}}$, $A_2 = \dfrac{2\mu + a_2}{\mu}$, $a_2 = -2\mu \cosh(A_1)$, $\mu < 0$, A_1, t_0, n_0, and ϕ_0 are arbitrary real constants.

Solution 9. *Staggered discrete dark soliton*
  (Figure 8.8)

$$\psi_s(n, t) = (-1)^n A_0 \tanh[A_1 (n - n_0)] \, e^{i \left[A_2 (t-t_0) - 4 (t-t_0) + \phi_0 \right]}, \qquad (8.35)$$

where $A_0 = \dfrac{\tanh(A_1)}{\sqrt{-\mu}}$, $A_2 = \dfrac{2\mu + a_2}{\mu}$, $a_2 = -2\mu \operatorname{sech}^2(A_1)$, $\mu < 0$, A_1, t_0, n_0, and ϕ_0 are arbitrary real constants.

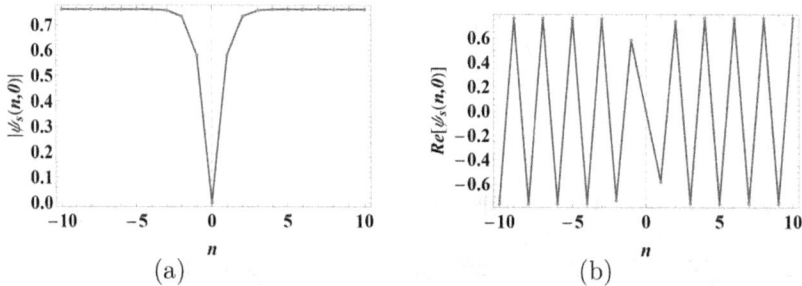

Figure 8.8. Staggered discrete dark soliton (8.35) at $t = 0$. (a) Absolute value, (b) Real part, with $A_1 = 1$, $\mu = -1$, and $n_0 = t_0 = \phi_0 = 0$. The lines are guides for the eye.

Solution 10.

$$\psi_s(n, t) = (-1)^n A_0 \coth[A_1 (n - n_0)] \, e^{i \left[A_2 (t-t_0) - 4 (t-t_0) + \phi_0 \right]}, \qquad (8.36)$$

where $A_0 = \dfrac{\tanh(A_1)}{\sqrt{-\mu}}$, $A_2 = \dfrac{2\mu + a_2}{\mu}$, $a_2 = -2\mu \operatorname{sech}^2(A_1)$, $\mu < 0$, A_1, t_0, n_0, and ϕ_0 are arbitrary real constants.

Solution 11. *Staggered discrete SW*
(Figure 8.9)

$$\psi_s(n, t) = (-1)^n A_0 \operatorname{sn}[A_1 (n - n_0), m]\, e^{i\left[A_2 (t-t_0)-4 (t-t_0)+\phi_0\right]}, \qquad (8.37)$$

where $A_0 = \dfrac{\sqrt{m}\,\operatorname{sn}(A_1, m)}{\sqrt{-\mu}}$, $A_2 = \dfrac{2\,\mu+a_2}{\mu}$, $a_2 = -2\,\mu\,\operatorname{cn}(A_1, m)\,\operatorname{dn}(A_1, m)$, $\mu < 0$,
$0 \leqslant m \leqslant 1$, A_1, t_0, n_0, and ϕ_0 are arbitrary real constants.

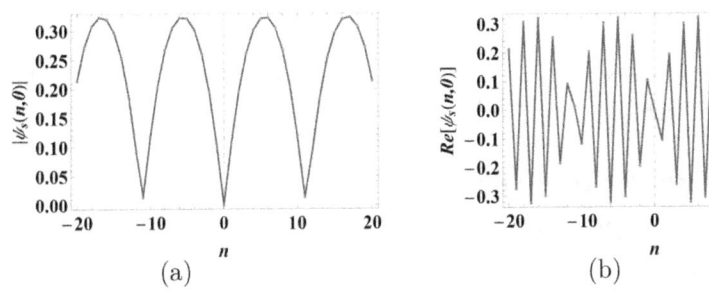

(a) (b)

Figure 8.9. Staggered discrete solitary wave (8.37) at $t = 0$. (a) Absolute value, (b) Real part, with $A_1 = 1/3$, $\mu = -1/2$, $m = 1/2$, and $n_0 = t_0 = \phi_0 = 0$. The lines are guides for the eye.

Solution 12. *Staggered discrete SW*
(Figure 8.10)

$$\psi_s(n, t) = (-1)^n A_0 \operatorname{cn}[A_1 (n - n_0), m]\, e^{i\left[A_2 (t-t_0)-4 (t-t_0)+\phi_0\right]}, \qquad (8.38)$$

where $A_0 = \dfrac{\sqrt{m}\,\operatorname{sn}(A_1, m)}{\sqrt{\mu}\,\operatorname{dn}(A_1, m)}$, $A_2 = \dfrac{2\,\mu+a_2}{\mu}$, $\mu = -\dfrac{a_2\,\operatorname{dn}^2(A_1, m)}{2\,\operatorname{cn}(A_1, m)} > 0$, $0 \leqslant m \leqslant 1$, A_1, t_0, n_0,
and ϕ_0 are arbitrary real constants.

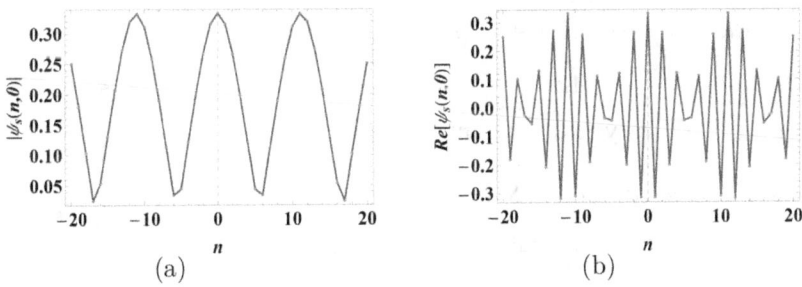

(a) (b)

Figure 8.10. Staggered discrete solitary wave (8.38) at $t = 0$. (a) Absolute value, (b) Real part, with $a_2 = -1$, $A_1 = 1/3$, $m = 1/2$, and $n_0 = t_0 = \phi_0 = 0$. The lines are guides for the eye.

Solution 13. *Staggered discrete SW*
(Figure 8.11)

$$\psi_s(n, t) = (-1)^n A_0 \operatorname{dn}[A_1 (n - n_0), m]\, e^{i\left[A_2 (t-t_0)-4 (t-t_0)+\phi_0\right]}, \qquad (8.39)$$

where $A_0 = \dfrac{\operatorname{sn}(A_1, m)}{\sqrt{\mu}\,\operatorname{cn}(A_1, m)}$, $A_2 = \dfrac{2\,\mu+a_2}{\mu}$, $\mu = -\dfrac{a_2\,\operatorname{cn}^2(A_1, m)}{2\,\operatorname{dn}(A_1, m)} > 0$, $0 < m \leqslant 1$, A_1, t_0, n_0,
and ϕ_0 are arbitrary real constants.

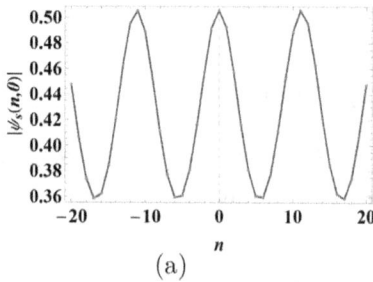 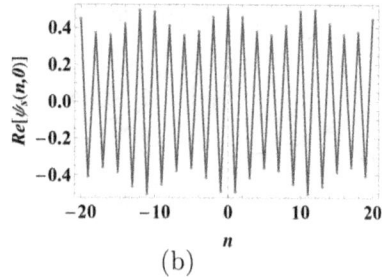

(a)　　　　　　　　　　　　　(b)

Figure 8.11. Staggered discrete solitary wave (8.39) at $t = 0$. (a) Absolute value, (b) Real part, with $a_2 = -1$, $A_1 = 1/3$, $m = 1/2$, and $n_0 = t_0 = \phi_0 = 0$. The lines are guides for the eye.

Solution 14. *Staggered discrete SW*

$$\psi_s(n,\, t) = (-1)^n\, A_0\, \text{ns}[A_1\, (n - n_0),\, m]\, e^{i\,[A_2\,(t-t_0)-4\,(t-t_0)+\phi_0]}, \qquad (8.40)$$

where $\quad A_0 = \dfrac{\text{sn}(A_1, m)}{\sqrt{-\mu}}, \qquad A_2 = \dfrac{2\,\mu + a_2}{\mu}, \qquad a_2 = -2\,\mu\,\text{dn}(A_1,\, m)\,\text{cn}(A_1,\, m), \qquad \mu < 0,$

$0 \leqslant m \leqslant 1$, A_1, t_0, n_0, and ϕ_0 are arbitrary real constants.

Solution 15. *Staggered discrete SW*

$$\psi_s(n,\, t) = (-1)^n\, A_0\, \text{cs}[A_1\, (n - n_0),\, m]\, e^{i\,[A_2\,(t-t_0)-4\,(t-t_0)+\phi_0]}, \qquad (8.41)$$

where $A_0 = \dfrac{\text{sn}(A_1, m)}{\sqrt{-\mu}\,\text{cn}(A_1, m)}$, $A_2 = \dfrac{2\,\mu + a_2}{\mu}$, $a_2 = -\dfrac{2\,\mu\,\text{dn}(A_1, m)}{\text{cn}^2(A_1, m)}$, $\mu < 0$, $0 \leqslant m \leqslant 1$, A_1, t_0,

n_0, and ϕ_0 are arbitrary real constants.

Solution 16. *Staggered discrete SW*

$$\psi_s(n,\, t) = (-1)^n\, A_0\, \text{ds}[A_1\, (n - n_0),\, m]\, e^{i\,[A_2\,(t-t_0)-4\,(t-t_0)+\phi_0]}, \qquad (8.42)$$

where $A_0 = \dfrac{\text{sn}(A_1, m)}{\sqrt{-\mu}\,\text{dn}(A_1, m)}$, $A_2 = \dfrac{2\,\mu + a_2}{\mu}$, $a_2 = -\dfrac{2\,\mu\,\text{cn}(A_1, m)}{\text{dn}^2(A_1, m)}$, $\mu < 0$, $0 \leqslant m \leqslant 1$, A_1, t_0,

n_0, and ϕ_0 are arbitrary real constants.

Solution 17. *Staggered discrete SW*

$$\psi_s(n,\, t) = (-1)^n\, A_0\, \text{cd}[A_1\, (n - n_0),\, m]\, e^{i\,[A_2\,(t-t_0)-4\,(t-t_0)+\phi_0]}, \qquad (8.43)$$

where $\quad A_0 = \dfrac{\sqrt{m}\,\text{sn}(A_1, m)}{\sqrt{-\mu}}, \qquad A_2 = \dfrac{2\,\mu + a_2}{\mu}, \qquad a_2 = -2\,\mu\,\text{cn}(A_1,\, m)\,\text{dn}(A_1,\, m), \qquad \mu < 0,$

$0 \leqslant m \leqslant 1$, A_1, t_0, n_0, and ϕ_0 are arbitrary real constants.

Solution 18. *Staggered discrete SW*

$$\psi_s(n,\, t) = (-1)^n\, A_0\, \text{dc}[A_1\, (n - n_0),\, m]\, e^{i\,[A_2\,(t-t_0)-4\,(t-t_0)+\phi_0]}, \qquad (8.44)$$

where $\quad A_0 = \dfrac{\text{sn}(A_1, m)}{\sqrt{-\mu}}, \qquad A_2 = \dfrac{2\,\mu + a_2}{\mu}, \qquad a_2 = -2\,\mu\,\text{cn}(A_1,\, m)\,\text{dn}(A_1,\, m), \qquad \mu < 0,$

$0 \leqslant m \leqslant 1$, A_1, t_0, n_0, and ϕ_0 are arbitrary real constants.

8.3 Summary of Section 8.2

Equation: $i\,\psi_{nt} + \psi_{n+1} + \psi_{n-1} - 2\,\psi_n + \dfrac{a_2\,|\psi_n|^2\,\psi_n}{1+\mu\,|\psi_n|^2} = 0$

Nonstaggered Solutions

#	Solution	Conditions	Name	Equation #
1.	$\psi(n,t) = A_0\, e^{[iA_1(n-n_0)-A_2(t-t_0)+\phi_0]}$	$A_2 = 4\sin^2(A_1/2) - \dfrac{a_2 A_0^2}{1+\mu A_0^2}$, A_0, A_1, t_0, and ϕ_0 are arbitrary real constants, n_0 is an arbitrary real integer	Discrete constant wave, t- and n-dependent phase	(8.10)
2.	$\psi(n,t) = A_0\,\sec[A_1(n-n_0)]\,e^{-i\,[A_2(t-t_0)+\phi_0]}$	$A_0 = \dfrac{\sin(A_1)}{\sqrt{\mu}}$, $A_2 = \dfrac{2\mu-a_2}{\mu}$, $a_2 = 2\,\mu\cos(A_1)$, $\mu < 0$, A_1, t_0, and ϕ_0 are arbitrary real constants, n_0 is an arbitrary real integer	—	(8.11)
3.	$\psi(n,t) = A_0\,\tan[A_1(n-n_0)]\,e^{-i\,[A_2(t-t_0)+\phi_0]}$	$A_0 = \dfrac{\tan(A_1)}{\sqrt{-\mu}}$, $A_2 = 2 - 2\sec^2(A_1)$, $\mu < 0$, A_1, t_0, ϕ_0 are arbitrary real constants, n_0 is an arbitrary real integer	—	(8.12)
4.	$\psi(n,t) = A_0\,\text{sech}[A_1(n-n_0)]\,e^{-i\,[A_2(t-t_0)+\phi_0]}$	$A_0 = \dfrac{\sinh(A_1)}{\sqrt{\mu}}$, $A_2 = \dfrac{2\mu-a_2}{\mu}$, $\mu = \dfrac{a_2\,\text{sech}(A_1)}{2} > 0$, A_1, t_0, and ϕ_0 are arbitrary real constants, n_0 is an arbitrary real integer	Discrete bright soliton	(8.13)
5.	$\psi(n,t) = A_0\,\text{csch}[A_1(n-n_0)]\,e^{-i\,[A_2(t-t_0)+\phi_0]}$	$A_0 = \dfrac{\sinh(A_1)}{\sqrt{-\mu}}$, $A_2 = \dfrac{2\mu-a_2}{\mu}$, $a_2 = 2\,\mu\cosh(A_1)$, $\mu < 0$, A_1, t_0, and ϕ_0 are arbitrary real constants, n_0 is an arbitrary real integer	—	(8.14)

#	Solution	Parameters	Description	Eq.
6.	$\psi(n, t) = A_0 \tanh[A_1 (n - n_0)]\, e^{-i\,[A_2 (t - t_0) + \phi_0]}$	$A_0 = \dfrac{\tanh(A_1)}{\sqrt{-\mu}}$, $A_2 = \dfrac{2\,\mu - a_2}{\mu}$, $a_2 = 2\,\mu\,\text{sech}^2(A_1)$, $\mu < 0$, A_1, t_0, and ϕ_0 are arbitrary real constants, n_0 is an arbitrary real integer	Discrete dark soliton	(8.15)
7.	$\psi(n, t) = A_0 \coth[A_1 (n - n_0)]\, e^{-i\,[A_2 (t - t_0) + \phi_0]}$	$A_0 = \dfrac{\tanh(A_1)}{\sqrt{-\mu}}$, $A_2 = \dfrac{2\,\mu - a_2}{\mu}$, $a_2 = 2\,\mu\,\text{sech}^2(A_1)$, $\mu < 0$, A_1, t_0, and ϕ_0 are arbitrary real constants, n_0 is an arbitrary real integer	—	(8.16)
8.	$\psi(n, t) = A_0 \,\text{sn}[A_1 (n - n_0), m]\, e^{-i\,[A_2 (t - t_0) + \phi_0]}$	$A_0 = \dfrac{\sqrt{m}\,\text{sn}(A_1, m)}{\sqrt{-\mu}}$, $A_2 = \dfrac{2\,\mu - a_2}{\mu}$, $a_2 = 2\,\mu\,\text{cn}(A_1, m)\,\text{dn}(A_1, m)$, $\mu < 0$, $0 \leqslant m \leqslant 1$, A_1, t_0, and ϕ_0 are arbitrary real constants, n_0 is an arbitrary real integer	discrete solitary wave	(8.17)
9.	$\psi(n, t) = A_0 \,\text{cn}[A_1 (n - n_0), m]\, e^{-i\,[A_2 (t - t_0) + \phi_0]}$	$A_0 = \dfrac{\sqrt{m}\,\text{sn}(A_1, m)}{\sqrt{\mu}\,\text{dn}(A_1, m)}$, $A_2 = \dfrac{2\,\mu - a_2}{\mu}$, $\mu = \dfrac{a_2\,\text{dn}^2(A_1, m)}{2\,\text{cn}(A_1, m)} > 0$, $0 \leqslant m \leqslant 1$, A_1, t_0, and ϕ_0 are arbitrary real constants, n_0 is an arbitrary real integer	Discrete solitary wave	(8.18)
10.	$\psi(n, t) = A_0 \,\text{dn}[A_1 (n - n_0), m]\, e^{-i\,[A_2 (t - t_0) + \phi_0]}$	$A_0 = \dfrac{\text{sn}(A_1, m)}{\sqrt{\mu}\,\text{cn}(A_1, m)}$, $A_2 = \dfrac{2\,\mu - a_2}{\mu}$, $\mu = \dfrac{a_2\,\text{cn}^2(A_1, m)}{2\,\text{dn}(A_1, m)} > 0$, $0 \leqslant m \leqslant 1$, A_1, t_0, and ϕ_0 are arbitrary real constants, n_0 is an arbitrary real integer	Discrete solitary wave	(8.19)

(*Continued*)

11. $\psi(n, t) = A_0 \, \text{ns}[A_1 (n - n_0), m] \, e^{-i [A_2 (t-t_0)+\phi_0]}$

$A_0 = \frac{\text{sn}(A_1, m)}{\sqrt{-\mu}}$, $A_2 = \frac{2\mu - a_2}{\mu}$, $a_2 = 2\mu \, \text{dn}(A_1, m) \, \text{cn}(A_1, m)$, $\mu < 0$, $0 \le m \le 1$, $A_1, t_0,$ and ϕ_0 are arbitrary real constants, n_0 is an arbitrary real integer

Discrete solitary wave (8.20)

12. $\psi(n, t) = A_0 \, \text{cs}[A_1 (n - n_0), m] \, e^{-i [A_2 (t-t_0)+\phi_0]}$

$A_0 = \frac{\sqrt{-\mu}\,\text{cn}(A_1, m)}{2\,\mu\,\text{dn}(A_1,m)}$, $A_2 = \frac{2\mu - a_2}{\mu}$, $a_2 = \frac{2\,\mu\,\text{dn}(A_1, m)}{\text{cn}^2(A_1, m)}$, $\mu < 0$, $0 \le m \le 1$, $A_1, t_0,$ and ϕ_0 are arbitrary real constants, n_0 is an arbitrary real integer

Discrete solitary wave (8.21)

13. $\psi(n, t) = A_0 \, \text{ds}[A_1 (n - n_0), m] \, e^{-i [A_2 (t-t_0)+\phi_0]}$

$A_0 = \frac{\sqrt{-\mu}\,\text{dn}(A_1, m)}{2\,\mu\,\text{cn}(A_1,m)}$, $A_2 = \frac{2\mu - a_2}{\mu}$, $a_2 = \frac{2\,\mu\,\text{cn}(A_1, m)}{\text{dn}^2(A_1, m)}$, $\mu < 0$, $0 \le m \le 1$, $A_1, t_0,$ and ϕ_0 are arbitrary real constants, n_0 is an arbitrary real integer

Discrete solitary wave (8.22)

14. $\psi(n, t) = A_0 \, \text{cd}[A_1 (n - n_0), m] \, e^{-i [A_2 (t-t_0)+\phi_0]}$

$A_0 = \frac{\sqrt{m}\,\text{sn}(A_1, m)}{\sqrt{-\mu}}$, $A_2 = \frac{2\mu - a_2}{\mu}$, $a_2 = 2\,\mu\,\text{cn}(A_1, m)\,\text{dn}(A_1, m)$, $\mu < 0$, $0 \le m \le 1$, $A_1, t_0,$ and ϕ_0 are arbitrary real constants, n_0 is an arbitrary real integer

Discrete solitary wave (8.23)

15. $\psi(n, t) = A_0 \, \text{dc}[A_1 (n - n_0), m] \, e^{-i [A_2 (t-t_0)+\phi_0]}$

$A_0 = \frac{\text{sn}(A_1, m)}{\sqrt{-\mu}}$, $A_2 = \frac{2\mu - a_2}{\mu}$, $a_2 = 2\,\mu\,\text{cn}(A_1, m)\,\text{dn}(A_1, m)$, $\mu < 0$, $0 \le m \le 1$, $A_1, t_0,$ and ϕ_0 are arbitrary real constants, n_0 is an arbitrary real integer

Discrete solitary wave (8.24)

#	Solution	Conditions	Name	Equation #
16.	$\psi(n,t) = \{\frac{A_0}{2}\,\text{dn}[A_1(n-n_0),m] + \frac{B_0}{2}\sqrt{m}\,\text{cn}[A_1(n-n_0),m]\}$ $\times e^{-i[A_2(t-t_0)+\phi_0]}$	$A_0 = \frac{2}{\text{cs}(A_1,m)+\text{ds}(A_1,m)}$, $B_0 = \pm A_0$, $A_2 = 2 - a_2$, $A_2 = \frac{4}{\text{cn}(A_1,m)+\text{dn}(A_1,m)}$, $\mu = 1$, $0 \leqslant m \leqslant 1$, A_1, t_0, and ϕ_0 are arbitrary real constants, n_0 is an arbitrary real integer	Discrete solitary wave	(8.25)

Staggered Solutions: $\psi_s(n,t,a_2) = (-1)^n \psi^*(n,t,-a_2)\, e^{-4i(t-t_0)}$

#	Solution	Conditions	Name	Equation #
1.	$\psi_s(n,t) = (-1)^n A_0\, e^{-i[A_1(n-n_0)-(A_2-4)(t-t_0)+\phi_0]}$	$A_2 = 4\sin^2(A_1/2) + \frac{a_2 A_0^2}{1+\mu A_0^2}$, A_0, A_1, t_0, ϕ_0 are arbitrary real constants, n_0 is an arbitrary real integer	Staggered discrete constant wave, t- and n-dependent phase	(8.27)
2.	$\psi_s(n,t) = A_0 \cos[\pi(n-n_0)]\, e^{-i[A_2(t-t_0)+\phi_0]}$	$A_2 = 4 - \frac{a_2 A_0^2}{1+\mu A_0^2}$, A_0, t_0, and ϕ_0 are arbitrary real constants, n_0 is an arbitrary real integer	—	(8.28)
3.	$\psi_s(n,t) = A_0 \cos[\frac{\pi}{2}(n-n_0)]\, e^{\pm i[A_2(t-t_0)+\phi_0]}$	$A_2 = -2 + \frac{A_0^2 a_2}{A_0^2\mu + \sec^2[\pi(n-n_0)/2]}$, A_0, t_0, and ϕ_0 are arbitrary real constants, n_0 is an arbitrary real integer	—	(8.29)
4.	$\psi_s(n,t) = A_0\{\cos[\frac{\pi}{2}(n-n_0)] - \sin[\frac{\pi}{2}(n-n_0)]\}\, e^{-i[A_2(t-t_0)+\phi_0]}$	$A_2 = 2 - \frac{a_2 A_0^2}{1+\mu A_0^2}$, A_0, t_0, and ϕ_0 are arbitrary real constants, n_0 is an arbitrary real integer	—	(8.30)

(*Continued*)

5.	$\psi_s(n,t) = (-1)^n A_0 \sec[A_1(n-n_0)]\, e^{i[A_2(t-t_0)-4(t-t_0)+\phi_0]}$	$A_0 = \dfrac{\sin(A_1)}{\sqrt{-\mu}}$, $A_2 = \dfrac{2\mu+a_2}{\mu}$, $a_2 = -2\mu\cos(A_1)$, $\mu<0$, A_1, t_0, and ϕ_0 are arbitrary real constants, n_0 is an arbitrary real integer	—	(8.31)
6.	$\psi_s(n,t) = (-1)^n A_0 \tan[A_1(n-n_0)]\, e^{i[A_2(t-t_0)-4(t-t_0)+\phi_0]}$	$A_0 = \dfrac{\tan(A_1)}{\sqrt{-\mu}}$, $A_2 = 2-2\sec^2(A_1)$, $a_2 = -2\mu\sec^2(A_1)$, $\mu<0$, A_1, t_0, and ϕ_0 are arbitrary real constants, n_0 is an arbitrary real integer	—	(8.32)
7.	$\psi_s(n,t) = (-1)^n A_0 \operatorname{sech}[A_1(n-n_0)]\, e^{i[A_2(t-t_0)-4(t-t_0)+\phi_0]}$	$A_0 = \dfrac{\sinh(A_1)}{\sqrt{\mu}}$, $A_2 = \dfrac{2\mu+a_2}{\mu}$, $\mu = -\dfrac{a_2\operatorname{sech}(A_1)}{2}>0$, A_1, t_0, and ϕ_0 are arbitrary real constants, n_0 is an arbitrary real integer	Staggered discrete bright soliton	(8.33)
8.	$\psi_s(n,t) = (-1)^n A_0 \operatorname{csch}[A_1(n-n_0)]\, e^{i[A_2(t-t_0)-4(t-t_0)+\phi_0]}$	$A_0 = \dfrac{\sinh(A_1)}{\sqrt{-\mu}}$, $A_2 = \dfrac{2\mu+a_2}{\mu}$, $a_2 = -2\mu\cosh(A_1)$, $\mu<0$, A_1, t_0, and ϕ_0 are arbitrary real constants, n_0 is an arbitrary real integer	—	(8.34)
9.	$\psi_s(n,t) = (-1)^n A_0 \tanh[A_1(n-n_0)]\, e^{i[A_2(t-t_0)-4(t-t_0)+\phi_0]}$	$A_0 = \dfrac{\tanh(A_1)}{\sqrt{-\mu}}$, $A_2 = \dfrac{2\mu+a_2}{\mu}$, $a_2 = -2\mu\operatorname{sech}^2(A_1)$, $\mu<0$, A_1, t_0, and ϕ_0 are arbitrary real constants, n_0 is an arbitrary real integer	Staggered discrete dark soliton	(8.35)
10.	$\psi_s(n,t) = (-1)^n A_0 \coth[A_1(n-n_0)]\, e^{i[A_2(t-t_0)-4(t-t_0)+\phi_0]}$	$A_0 = \dfrac{\tanh(A_1)}{\sqrt{-\mu}}$, $A_2 = \dfrac{2\mu+a_2}{\mu}$, $a_2 = -2\mu\operatorname{sech}^2(A_1)$, $\mu<0$, A_1, t_0, and ϕ_0 are arbitrary real constants, n_0 is an arbitrary real integer	—	(8.36)

11.	$\psi_s(n,t) = (-1)^n A_0\, \text{sn}[A_1(n-n_0),m]\, e^{i[A_2(t-t_0)-4(t-t_0)+\phi_0]}$	$A_0 = \dfrac{\sqrt{m}\,\text{sn}(A_1,m)}{\sqrt{-\mu}}$, $A_2 = \dfrac{2\mu+a_2}{\mu}$, $a_2 = -2\,\mu\,\text{cn}(A_1,m)\,\text{dn}(A_1,m)$, $\mu<0$, $0\le m\le 1$, A_1, t_0, and ϕ_0 are arbitrary real constants, n_0 is an arbitrary real integer	Staggered discrete solitary wave (8.37)
12.	$\psi_s(n,t) = (-1)^n A_0\, \text{cn}[A_1(n-n_0),m]\, e^{i[A_2(t-t_0)-4(t-t_0)+\phi_0]}$	$A_0 = \dfrac{\sqrt{m}\,\text{sn}(A_1,m)}{\sqrt{\mu}\,\text{dn}(A_1,m)}$, $A_2 = \dfrac{2\mu+a_2}{\mu}$, $\mu = -\dfrac{a_2\,\text{dn}^2(A_1,m)}{2\,\text{cn}(A_1,m)} > 0$, $0\le m\le 1$, A_1, t_0, and ϕ_0 are arbitrary real constants, n_0 is an arbitrary real integer	Staggered discrete solitary wave (8.38)
13.	$\psi_s(n,t) = (-1)^n A_0\, \text{dn}[A_1(n-n_0),m]\, e^{i[A_2(t-t_0)-4(t-t_0)+\phi_0]}$	$A_0 = \dfrac{\text{sn}(A_1,m)}{\sqrt{\mu}\,\text{cn}(A_1,m)}$, $A_2 = \dfrac{2\mu+a_2}{\mu}$, $\mu = -\dfrac{a_2\,\text{cn}^2(A_1,m)}{2\,\text{dn}(A_1,m)} > 0$, $0< m\le 1$, A_1, t_0, and ϕ_0 are arbitrary real constants, n_0 is an arbitrary real integer	Staggered discrete solitary wave (8.39)
14.	$\psi_s(n,t) = (-1)^n A_0\, \text{ns}[A_1(n-n_0),m]\, e^{i[A_2(t-t_0)-4(t-t_0)+\phi_0]}$	$A_0 = \dfrac{\text{sn}(A_1,m)}{\sqrt{-\mu}}$, $A_2 = \dfrac{2\mu+a_2}{\mu}$, $a_2 = -2\,\mu\,\text{dn}(A_1,m)\,\text{cn}(A_1,m)$, $\mu<0$, $0\le m\le 1$, A_1, t_0, and ϕ_0 are arbitrary real constants, n_0 is an arbitrary real integer	Staggered discrete solitary wave (8.40)
15.	$\psi_s(n,t) = (-1)^n A_0\, \text{cs}[A_1(n-n_0),m]\, e^{i[A_2(t-t_0)-4(t-t_0)+\phi_0]}$	$A_0 = \dfrac{\text{sn}(A_1,m)}{\sqrt{-\mu}\,\text{cn}(A_1,m)}$, $A_2 = \dfrac{2\mu+a_2}{\mu}$, $a_2 = -\dfrac{2\,\mu\,\text{dn}(A_1,m)}{\text{cn}^2(A_1,m)}$, $\mu<0$, $0\le m\le 1$, A_1, t_0, and ϕ_0 are arbitrary real constants, n_0 is an arbitrary real integer	Staggered discrete solitary wave (8.41)

(Continued)

16.	$\psi_s(n, t) = (-1)^n A_0 \, ds[A_1 (n - n_0), m] \, e^{i [A_2 (t-t_c) - 4 (t-t_0) + \phi_0]}$	$A_0 = \frac{sn(A_1, m)}{\sqrt{-\mu} \, dn(A_1, m)}, \quad A_2 = \frac{2\mu + a_2}{\mu},$ $a_2 = -\frac{2\mu \, cn(A_1, m)}{dn^2(A_1, m)}, \quad \mu < 0, \ 0 \leq m \leq 1,$ $A_1, t_0, \text{ and } \phi_0 \text{ are arbitrary real constants, } n_0 \text{ is an arbitrary real integer}$	Staggered discrete solitary wave (8.42)
17.	$\psi_s(n, t) = (-1)^n A_0 \, cd[A_1 (n - n_0), m] \, e^{i [A_2 (t-t_0) - 4 (t-t_0) + \phi_0]}$	$A_0 = \frac{\sqrt{m} \, sn(A_1, m)}{\sqrt{-\mu}}, \quad A_2 = \frac{2\mu + a_2}{\mu},$ $a_2 = -2\mu \, cn(A_1, m) \, dn(A_1, m),$ $\mu < 0, \ 0 \leq m \leq 1, \ A_1, t_0, \text{ and } \phi_0$ are arbitrary real constants, n_0 is an arbitrary real integer	Staggered discrete solitary wave (8.43)
18.	$\psi_s(n, t) = (-1)^n A_0 \, dc[A_1 (n - n_0), m] \, e^{i [A_2 (t-t_0) - 4 (t-t_0) + \phi_0]}$	$A_0 = \frac{sn(A_1, m)}{\sqrt{-\mu}}, \quad A_2 = \frac{2\mu + a_2}{\mu},$ $a_2 = -2\mu \, cn(A_1, m) \, dn(A_1, m),$ $\mu < 0, \ 0 \leq m \leq 1, \ A_1, t_0, \text{ and } \phi_0$ are arbitrary real constants, n_0 is an arbitrary real integer	Staggered discrete solitary wave (8.44)

8.4 Short-period Solutions with General, Kerr, and Saturable Nonlinearities

Equations:

Case I: DNLS with general nonlinearity (GN)

$$i\,\psi_{nt} + \psi_{n+1} + \psi_{n-1} - 2\,\psi_n + a_2\,F\big[\,|\psi_n|^2\,\big]\,\psi_n = 0, \tag{8.45}$$

Case II: DNLS with Kerr nonlinearity (KN)

$$i\,\psi_{nt} + \psi_{n+1} + \psi_{n-1} - 2\,\psi_n + a_2\,|\psi_n|^2\,\psi_n = 0, \tag{8.46}$$

Case III: DNLS with saturable nonlinearity (SN)

$$i\,\psi_{nt} + \psi_{n+1} + \psi_{n-1} - 2\,\psi_n + a_2\,\frac{|\psi_n|^2\,\psi_n}{1 + \mu\,|\psi_n|^2} = 0, \tag{8.47}$$

where a_2 and μ are real constants, $\psi_n = \psi(n,\,t)$ is the complex function profile, the integer site index, n, and t are its two independent variables, F is a general real function.

For other solutions of Case III, see Section 8.2.

General Solutions:

$$\psi(n,\,t) = A_0\,(\ldots,\,c_0,\,c_1,\,c_2,\,c_3,\,\ldots)\,e^{i\,[A_2\,(t-t_0)+\phi_0]}, \tag{8.48}$$

where A_0, ϕ_0, t_0, c_j, $j = 0, 1, 2, 3, \ldots$ are arbitrary real constants. For specific values of A_2, short-period solutions are obtained as summarized in Table 8.1.

Table 8.1. Short-period solutions to general, Kerr, and saturable nonlinearities of the discrete NLSE (8.45), (8.46) and (8.47), where F is a general real function and, A_0, t_0, and ϕ_0 are arbitrary real constants.

Period	Nonlinearity*	Condition on A_2	Solution
	GN KN SN	$A_2 = 0 - a_2\,F[A_0^2]$	$\psi(n,\,t) = A_0\,(\ldots,\,1,\,1,\,1,\,1,\,\ldots)\,e^{i\,[A_2\,(t-t_0)+\phi_0]}$
		$A_2 = 0 - a_2 A_0^2$	
		$A_2 = 0 - \dfrac{a_2\,A_0^2}{1 + \mu\,A_0^2}$	
2	GN KN SN	$A_2 = 4 - a_2\,F[A_0^2]$	$\psi(n,\,t) = A_0\,(\ldots,\,1,\,-1,\,\ldots)\,e^{i\,[A_2\,(t-t_0)+\phi_0]}$
		$A_2 = 4 - a_2 A_0^2$	
		$A_2 = 4 - \dfrac{a_2\,A_0^2}{1 + \mu\,A_0^2}$	
3	GN KN SN	$A_2 = 3 - a_2\,F[A_0^2]$	$\psi(n,\,t) = A_0\,(\ldots,\,1,\,0,\,-1,\,\ldots)\,e^{i\,[A_2\,(t-t_0)+\phi_0]}$
		$A_2 = 3 - a_2 A_0^2$	
		$A_2 = 3 - \dfrac{a_2\,A_0^2}{1 + \mu\,A_0^2}$	

Table 8.1. (*Continued*)

4	GN KN SN	$A_2 = 2 - a_2\,F[A_0^2]$	$\psi(n,\,t) = A_0\,(\ldots,\,1,\,1,\,-1,\,-1,\,\ldots)\,e^{i\,[A_2(t-t_0)+\phi_0]}$
		$A_2 = 2 - a_2 A_0^2$	$\psi(n,\,t) = A_0\,(\ldots,\,1,\,0,\,-1,\,0,\,\ldots)\,e^{i\,[A_2\,(t-t_0)+\phi_0]}$
		$A_2 = 2 - \dfrac{a_2\,A_0^2}{1+\mu\,A_0^2}$	
6	GN KN SN	$A_2 = 1 - a_2\,F[A_0^2]$	$\psi(n,\,t) = A_0\,(\ldots,\,1,\,1,\,0,\,-1,\,-1,\,0,\,\ldots)$
		$A_2 = 1 - a_2 A_0^2$	$e^{i\,[A_2\,(t-t_0)+\phi_0]}$
		$A_2 = 1 - \dfrac{a_2\,A_0^2}{1+\mu\,A_0^2}$	

* GN: general nonlinearity, $a_2\,F[|\psi|^2]$, KN: Kerr nonlinearity, $a_2\,|\psi|^2$, SN: saturable nonlinearity, $\dfrac{a_2\,|\psi|^2}{1+\mu\,|\psi|^2}$.

8.5 Ablowitz–Ladik Equation

Equation:

$$i\,\psi_{nt} + \psi_{n+1} + \psi_{n-1} - 2\,\psi_n + a_2\,(\psi_{n+1} + \psi_{n-1})\,|\psi_n|^2 = 0, \tag{8.49}$$

where a_2 is a real constant.

Solutions:

***Solution 1.* Constant Amplitude** *discrete CW, t- and n-dependent phase*

$$\psi(n,\,t) = A_0\,e^{i[A_1\,(n-n_0)+(A_2-2)\,(t-t_0)+\phi_0]}, \tag{8.50}$$

where $A_2 = 2\cos(A_1)\,(1 + a_2\,A_0^2)$, A_0, A_1, t_0, n_0, and ϕ_0 are arbitrary real constants.

Solution 2. *Discrete bright soliton*
(Figure 8.12)

$$\psi(n,\,t) = A_0\,\mathrm{sech}[A_1\,(n - n_0)]\,e^{\,i[(A_2+2)\,(t-t_0)+\phi_0]}, \tag{8.51}$$

where $A_2 = -2\cosh(A_1)$, $a_2 = \dfrac{\sinh^2(A_1)}{A_0^2}$, $A_0 \neq 0$, A_1, t_0, n_0, and ϕ_0 are arbitrary real constants.

• *Reference:* [5], *taken from the nonlocal case.*

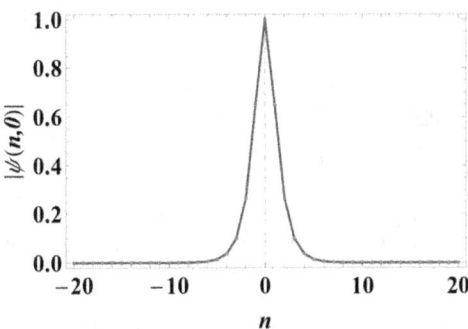

Figure 8.12. Discrete bright soliton (8.51) at $t = 0$ with $A_0 = A_1 = 1$, and $n_0 = t_0 = \phi_0 = 0$. The lines are guides for the eye.

Solution 3. *Discrete dark soliton*
(Figure 8.13)

$$\psi(n,\, t) = A_0 \tanh[A_1\, (n - n_0)]\, e^{-i\left[(A_2 + 2)\, (t - t_0) + \phi_0\right]}, \tag{8.52}$$

where $A_2 = -2\, \text{sech}^2(A_1)$, $a_2 = \dfrac{-\tanh^2(A_1)}{A_0^2}$, $A_0 \neq 0$, A_1, t_0, n_0, and ϕ_0 are arbitrary real constants.

- *Reference*: [5], taken from the nonlocal case.

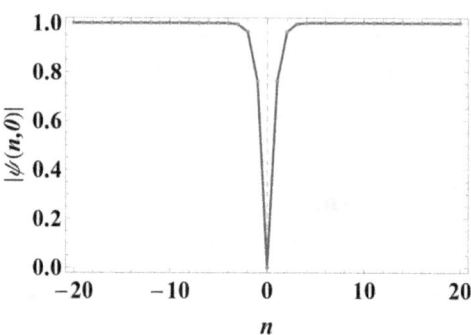

Figure 8.13. Discrete dark soliton (8.52) at $t = 0$ with $A_0 = A_1 = 1$, and $n_0 = t_0 = \phi_0 = 0$. The lines are guides for the eye.

Solution 4. *Discrete SW*
(Figure 8.14)

$$\psi(n,\, t) = A_0\, \sqrt{m}\, \text{sn}[A_1\, (n - n_0),\, m]\, e^{-i\left[(A_2 + 2)\, (t - t_0) + \phi_0\right]}, \tag{8.53}$$

where $A_2 = -2\, \text{cn}(A_1,\, m)\, \text{dn}(A_1,\, m)$, $a_2 = \dfrac{-1}{A_0^2\, \text{ns}^2(A_1,\, m)}$, $0 \leqslant m \leqslant 1$, $A_0 \neq 0$, A_1, t_0, n_0, and ϕ_0 are arbitrary real constants.

- *Reference*: [5], taken from the nonlocal case.

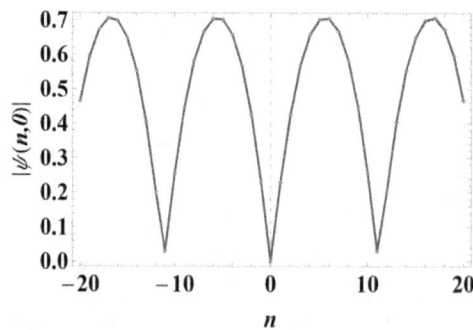

Figure 8.14. Discrete solitary wave (8.53) at $t = 0$ with $A_0 = 1$, $A_1 = 1/3$, $m = 1/2$, and $n_0 = t_0 = \phi_0 = 0$. The lines are guides for the eye.

Solution 5. *Discrete SW*
(Figure 8.15)

$$\psi(n,\, t) = A_0 \sqrt{m}\ \text{cn}[A_1\, (n - n_0),\, m]\, e^{-i\left[(A_2 + 2)\, (t - t_0) + \phi_0\right]}, \tag{8.54}$$

where $A_2 = \dfrac{-2\,\text{cn}(A_1, m)}{\text{dn}^2(A_1, m)}$, $a_2 = \dfrac{1}{A_0^2\, \text{ds}^2(A_1, m)}$, $0 \leqslant m \leqslant 1$, $A_0 \neq 0$, A_1, t_0, n_0, and ϕ_0 are arbitrary real constants.

• *Reference*: [5], taken from the nonlocal case.

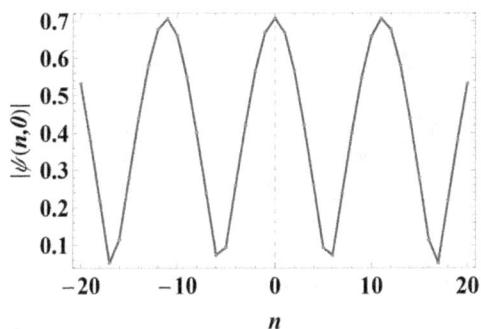

Figure 8.15. Discrete solitary wave (8.54) at $t = 0$ with $A_0 = 1$, $A_1 = 1/3$, $m = 1/2$, and $n_0 = t_0 = \phi_0 = 0$. The lines are guides for the eye.

Solution 6. *Discrete SW*
(Figure 8.16)

$$\psi(n,\, t) = A_0\, \text{dn}[A_1\, (n - n_0),\, m]\, e^{-i\left[(A_2 + 2)\, (t - t_0) + \phi_0\right]}, \tag{8.55}$$

where $A_2 = \dfrac{2\,\text{dn}(A_1, m)}{\text{cn}^2(A_1, m)}$, $a_2 = \dfrac{1}{A_0^2\, \text{cs}^2(A_1, m)}$, $0 \leqslant m \leqslant 1$, $A_0 \neq 0$, A_1, t_0, n_0, and ϕ_0 are arbitrary real constants.

• *Reference*: [5], taken from the nonlocal case.

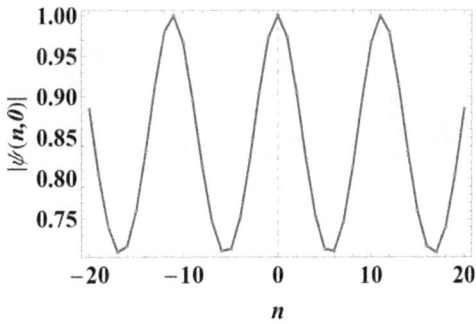

Figure 8.16. Discrete solitary wave (8.55) at $t = 0$ with $A_0 = 1$, $A_1 = 1/3$, $m = 1/2$, and $n_0 = t_0 = \phi_0 = 0$. The lines are guides for the eye.

Solution 7. *Discrete SW*

$$\psi(n, t) = \left\{ \frac{A_0}{2} \, \mathrm{dn}[A_1 \, (n - n_0), m] + \frac{B_0 \, \sqrt{m}}{2} \, \mathrm{cn}[A_1 \, (n - n_0), m] \right\} e^{-i \left[(A_2 + 2) \, (t - t_0) + \phi_0 \right]}, \quad (8.56)$$

where $A_2 = \dfrac{-4}{\mathrm{cn}(A_1, m) + \mathrm{dn}(A_1, m)}$, $a_2 = \dfrac{4}{A_0^2 \, [\mathrm{ds}(A_1, m) + \mathrm{cs}(A_1, m)]^2}$, $B_0 = \pm A_0$, $0 \leqslant m \leqslant 1$, $A_0 \neq 0$, A_1, t_0, n_0, and ϕ_0 are arbitrary real constants.

- *Reference*: [5], taken from the nonlocal case.

Solution 8. *Discrete SW*

$$\psi(n, t) = A_0 \, \sqrt{m} \, \mathrm{cd}[A_1 \, (n - n_0), m] \, e^{-i \left[(A_2 + 2) \, (t - t_0) + \phi_0 \right]}, \quad (8.57)$$

where $A_2 = \dfrac{2 \, \mathrm{ns}(A_1, m) \, [\mathrm{cs}^2(A_1, m) - \mathrm{ds}^2(A_1, m)] \, [\mathrm{cs}(2 \, A_1, m) + \mathrm{ds}(2 \, A_1, m)]}{\mathrm{ds}(A_1, m) \, \mathrm{cs}(A_1, m) \, [\mathrm{cs}(A_1, m) \mathrm{ds}(A_1, m) - 2 \, \mathrm{cs}(A_1, m) \, \mathrm{ns}(A_1, m)]}$,

$a_2 = \dfrac{2 \, \mathrm{cs}(2 \, A_1, m) \, \mathrm{ds}(A_1, m) \, \mathrm{ns}(A_1, m) - \mathrm{cs}^3(A_1, m)}{A_0^2 \, [\mathrm{cs}(A_1, m) \, \mathrm{ds}^2(A_1, m) - 2 \, \mathrm{cs}(2 \, A_1, m) \, \mathrm{ds}(A_1, m) \, \mathrm{ns}(A_1, m)]}$, $0 \leqslant m \leqslant 1$, $A_0 \neq 0$, A_1, t_0, n_0, and ϕ_0 are arbitrary real constants.

- *Reference*: [5], taken from the nonlocal case.

Solution 9. Periodicity in *n* and Localization in *t* *discrete Akhmediev breather* (Figure 8.17)

$$\psi(n, t) = \kappa \left\{ \frac{\cos\left[\alpha \, (n - n_0) \right] + i \, \sqrt{\dfrac{2 + \kappa^2}{1 + \kappa^2}} \, \sinh\left[2 \, \kappa^2 \, (t - t_0) \right]}{\sqrt{\dfrac{2 + \kappa^2}{1 + \kappa^2}} \, \cosh\left[2 \, \kappa^2 \, (t - t_0) \right] - \cos\left[\alpha \, (n - n_0) \right]} \right\} e^{i \left[2 \, \kappa^2 \, (t - t_0) + \phi_0 \right]}, \quad (8.58)$$

where $\alpha = \cos^{-1}\left(\dfrac{1}{1 + \kappa^2} \right)$, t_0, κ, n_0, and ϕ_0 are arbitrary real constants.

- *Reference*: [9].

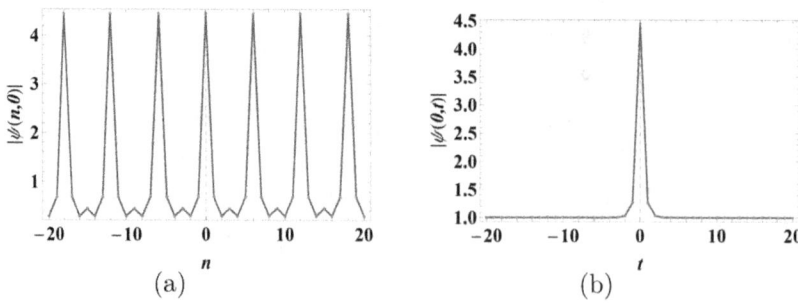

Figure 8.17. Discrete Akhmediev breather (8.58). (a) $t = 0$, (b) $n = 0$, with $\kappa = 1$ and $n_0 = t_0 = \phi_0 = 0$. The lines are guides for the eye. Animation available online at http://doi.org/10.1088/978-0-7503-5954-2

Solution 10. Localization in n and t discrete Peregrine soliton
(Figure 8.18)

$$\psi(n, t) = \left\{ \frac{8 \left[1 + 4 i \left(t - t_0\right)\right]}{1 + 4 \left(n - n_0\right)^2 + 32 \left(t - t_0\right)^2} - 1 \right\} e^{i \left[2 \left(t-t_0\right)+\phi_0\right]}, \qquad (8.59)$$

where t_0, n_0, and ϕ_0 are arbitrary real constants.

- *Reference*: [9].

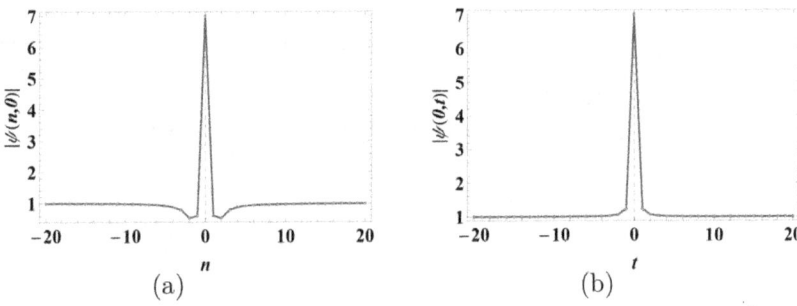

Figure 8.18. Discrete Peregrine soliton (8.59). (a) $t = 0$, (b) $n = 0$, with $n_0 = t_0 = \phi_0 = 0$. The lines are guides for the eye. Animation available online at http://doi.org/10.1088/978-0-7503-5954-2.

Solution 11. Periodicity in n and t
(Figure 8.19)

$$\psi(n, t) = \kappa \left\{ \frac{\sqrt{m}\, \mathrm{dn}\left[2\,\kappa^2\,(t - t_0),\, \sin^2(\theta)\right] \mathrm{cn}\left[A_0\,(n - n_0),\, m^2\right] + i\, A_1\, \sqrt{\sin(\theta)}\, \mathrm{sn}\left[2\,\kappa^2\,(t - t_0),\, \sin^2(\theta)\right]}{A_1 - \sqrt{m}\, \sin(\theta)\, \mathrm{cn}\left[2\,\kappa^2\,(t - t_0),\, \sin^2(\theta)\right] \mathrm{cn}\left[A_0\,(n - n_0),\, m^2\right]} \right\}$$
$$\times e^{i \left[2\,\kappa^2\, \sin(\theta)\,(t-t_0)+\phi_0\right]}, \qquad (8.60)$$

where $\qquad A_1 = \left\{(1 - m^2)\left[1 - \sin^2(\theta)\right]\right\}^{1/4}, \qquad \kappa = \sqrt{\dfrac{m\,\sqrt{1 - m^2}\,\mathrm{sn}^2(A_0, m^2)}{\sqrt{1 - \sin^2(\theta)}\,\mathrm{cn}(A_0, m^2)}},$

$\theta = \tan^{-1}\left\{\dfrac{1}{m\,\sqrt{1 - m^2}}\left[\dfrac{1}{1 + \mathrm{cn}^2(A_0, m^2)} - m^2\right]\right\}, \quad 0 < m < 1,\ A_0,\ t_0,\ n_0,\ \text{and} \quad \phi_0 \text{ are}$
arbitrary real constants.

- *Reference*: [9].

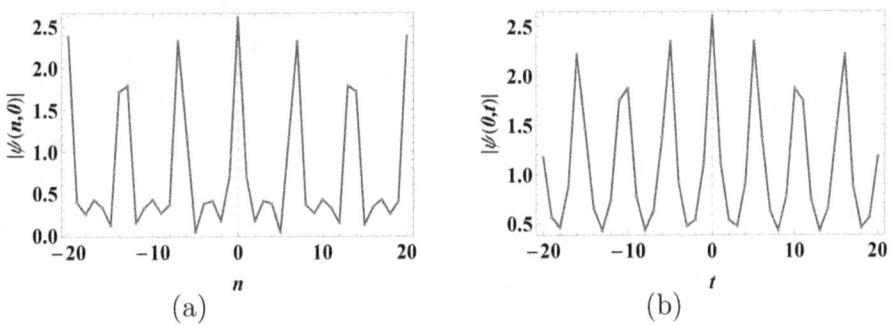

Figure 8.19. Plot of solution (8.60). (a) $t = 0$, (b) $n = 0$, with $A_0 = 1$, $m = 1/2$, and $n_0 = t_0 = \phi_0 = 0$. The lines are guides for the eye.

Solution 12. Periodicity in t and Localization in n *discrete Kuznetsov–Ma breather* (Figure 8.20)

$$\psi(n, t) = \left\{ \frac{(1 + \kappa\,\alpha_1)\cos[\alpha_4\,(t - t_0)] + \kappa\,\alpha_1\,\alpha_2\cosh[\alpha_5\,(n - n_0)] + i\,\alpha_3\sin[\alpha_4\,(t - t_0)]}{-\alpha_1\cos[\alpha_4\,(t - t_0)] - \alpha_1\,\alpha_2\cosh[\alpha_5\,(n - n_0)]} \right\}$$
$$\times\, e^{i\left[2\,\kappa^2\,(t - t_0) + \phi_0\right]}, \tag{8.61}$$

where $\quad \alpha_1 = \dfrac{\kappa}{(1 + \kappa^2)\,[\cosh(\alpha_5) - 1]}, \qquad \alpha_2 = \sqrt{1 + \dfrac{\tanh^2(\frac{\alpha_5}{2})}{\kappa^2}}, \qquad \alpha_3 = \sqrt{1 + 2\,\kappa\,\alpha_1},$

$\alpha_4 = 4\,(1 + \kappa^2)\sinh\left(\frac{\alpha_5}{2}\right)\sqrt{\dfrac{\kappa^2}{1 + \kappa^2} + \sinh^2\left(\frac{\alpha_5}{2}\right)}$, $\alpha_5 > 0$, κ, t_0, n_0, and ϕ_0 are arbitrary real constants.

- *Reference*: [9].

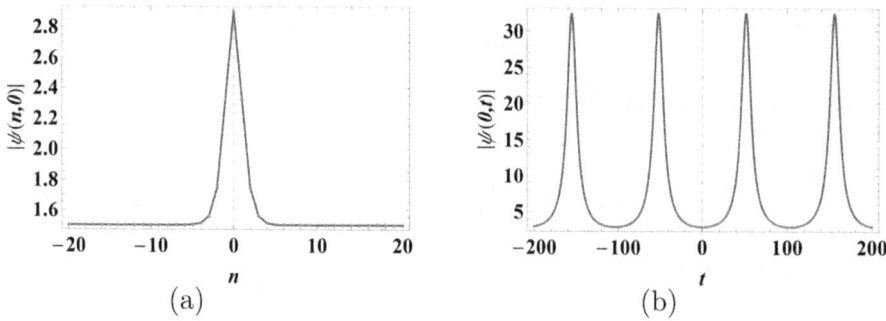

(a)   (b)

Figure 8.20. Discrete Kuznetsov–Ma breather (8.61). (a) $t = 0$, (b) $n = 0$, with $\kappa = \alpha_5 = 3/2$, $n_0 = t_0 = \phi_0 = 0$. The lines are guides for the eye. Animation available online at http://doi.org/10.1088/978-0-7503-5954-2.

8.6 Summary of Section 8.5

Note: For lengthy conditions, the reader is referred to the solutions in Section 8.5.

Equation

$$i\,\psi_{nt} + \psi_{n+1} + \psi_{n-1} - 2\,\psi_n + a_2\,(\psi_{n+1} + \psi_{n-1})\,|\psi_n|^2 = 0$$

#	Solution	Name	Conditions	Equation #
1.	$\psi(n,\,t) = A_0\,e^{i[A_1\,(n-n_0)+(A_2-2)\,(t-t_0)+\phi_0]}$	Discrete continuous wave, t- and n-dependent phase	$A_2 = 2\cos(A_1)\,(1 + a_2\,A_0^2)$, A_0, A_1, t_0, and ϕ_0 are arbitrary real constants, n_0 is an arbitrary real integer	(8.50)
2.	$\psi(n,\,t) = A_0\,\mathrm{sech}[A_1\,(n-n_0)]\,e^{-i[(A_2+2)\,(t-t_0)+\phi_0]}$	Discrete bright soliton	$A_2 = -2\cosh(A_1)$, $a_2 = \dfrac{\sinh^2(A_1)}{A_0^2}$, $A_0 \neq 0$, A_1, t_0, and ϕ_0 are arbitrary real constants, n_0 is an arbitrary real integer	(8.51)
3.	$\psi(n,\,t) = A_0\,\tanh[A_1\,(n-n_0)]\,e^{-i[(A_2+2)\,(t-t_0)+\phi_0]}$	Discrete dark soliton	$A_2 = -2\,\mathrm{sech}^2(A_1)$, $a_2 = \dfrac{-\tanh^2(A_1)}{A_0^2}$, $A_0 \neq 0$, A_1, t_0, and ϕ_0 are arbitrary real constants, n_0 is an arbitrary real integer	(8.52)
4.	$\psi(n,\,t) = A_0\,\sqrt{m}\,\mathrm{sn}[A_1\,(n - n_0),\,m]\,e^{-i[(A_2+2)\,(t-t_0)+\phi_0]}$	Discrete solitary wave	$A_2 = -2\,\mathrm{cn}(A_1,\,m)\,\mathrm{dn}(A_1,\,m)$, $a_2 = \dfrac{-1}{A_0^2\,\mathrm{ns}^2(A_1,\,m)}$, $0 \leqslant m \leqslant 1$, $A_0 \neq 0$, A_1, t_0, and ϕ_0 are arbitrary real constants, n_0 is an arbitrary real integer	(8.53)

5.	$\psi(n, t) = A_0 \sqrt{m}\, \mathrm{cn}[A_1 (n - n_0), m]\, e^{-i[(A_2+2)(t-t_0)+\phi_0]}$	$A_2 = \dfrac{-2\,\mathrm{cn}(A_1,m)}{\mathrm{dn}^2(A_1,m)}$, $\quad a_2 = \dfrac{1}{A_0^2\, \mathrm{ds}^2(A_1,m)}$, $0 \leqslant m \leqslant 1$, $A_0 \neq 0$, A_1, t_0, and ϕ_0 are arbitrary real constants, n_0 is an arbitrary real integer	Discrete solitary wave	(8.54)
6.	$\psi(n, t) = A_0\, \mathrm{dn}[A_1 (n - n_0), m]\, e^{-i[(A_2+2)(t-t_0)+\phi_0]}$	$A_2 = \dfrac{-2\,\mathrm{dn}(A_1,m)}{\mathrm{cn}^2(A_1,m)}$, $\quad a_2 = \dfrac{1}{A_0^2\, \mathrm{cs}^2(A_1,m)}$, $0 \leqslant m \leqslant 1$, $A_0 \neq 0$, A_1, t_0, and ϕ_0 are arbitrary real constants, n_0 is an arbitrary real integer	Discrete solitary wave	(8.55)
7.	$\psi(n, t) = \{\tfrac{A_0}{2}\, \mathrm{dn}[A_1 (n - n_0), m] + \tfrac{B_0 \sqrt{m}}{2}\, \mathrm{cn}[A_1 (n - n_0), m]\}$ $\times e^{-i[(A_2+2)(t-t_0)+\phi_0]}$	$A_2 = \dfrac{-4}{\mathrm{cn}(A_1,m) + \mathrm{dn}(A_1,m)}$, $a_2 = \dfrac{4}{A_0^2\,[\mathrm{ds}(A_1,m) + \mathrm{cs}(A_1,m)]^2}$, $B_0 = \pm A_0$, $\; 0 \leqslant m \leqslant 1$, $A_0 \neq 0$, A_1, t_0, and ϕ_0 are arbitrary real constants, n_0 is an arbitrary real integer	Discrete solitary wave	(8.56)
8.	$\psi(n, t) = A_0 \sqrt{m}\, \mathrm{cd}[A_1 (n - n_0), m]\, e^{-i[(A_2+2)(t-t_0)+\phi_0]}$	See text.	Discrete solitary wave	(8.57)
9.	$\psi(n, t) = \kappa \left\{ \dfrac{\cos[\alpha(n-n_0)] + i\sqrt{\frac{2+\kappa^2}{1+\kappa^2}}\, \sinh[2\kappa^2(t-t_0)]}{\sqrt{\frac{2+\kappa^2}{1+\kappa^2}}\, \cosh[2\kappa^2(t-t_0)] - \cos[\alpha(n-n_0)]} \right\} e^{i[2\kappa^2(t-t_0)+\phi_0]}$	$\alpha = \cos^{-1}\!\left(\dfrac{1}{1+\kappa^2}\right)$, t_0, κ, and ϕ_0 are arbitrary real constants, n_0 is an arbitrary real integer	Discrete Akhmediev breather	(8.58)
10.	$\psi(n, t) = \left\{ \dfrac{8[1+4i(t-t_0)]}{1+4(n-n_0)^2+32(t-t_0)^2} - 1 \right\} e^{i[2(t-t_0)+\phi_0]}$	t_0 and ϕ_0 are arbitrary real constants, n_0 is an arbitrary real integer	Discrete Peregrine soliton	(8.59)

(Continued)

11.
$$\psi(n, t) = \kappa \left\{ \frac{\sqrt{m}\, \mathrm{dn}[2\,\kappa^2\,(t-t_0),\sin^2(\theta)]\, \mathrm{cn}[A_0\,(n-n_0),m^2] + i\, A_1 \sqrt{\sin(\theta)}\, \mathrm{sn}[2\,\kappa^2\,(t-t_0),\sin^2(\theta)]}{A_1 - \sqrt{m}\,\sin(\theta)\, \mathrm{cn}[2\,\kappa^2\,(t-t_0),\sin^2(\theta)]\, \mathrm{cn}[A_0\,(n-n_0),m^2]} \right\}$$
$$\times e^{i\,[2\,\kappa^2\sin(\theta)\,(t-t_0)+\phi_0]}$$

$$A_1 = \{(1-m^2)\,[1-\sin^2(\theta)]\}^{1/4},$$

$$\kappa = \sqrt{\frac{m\sqrt{1-m^2}\,\mathrm{sn}^2(A_0,m^2)}{\sqrt{1-\sin^2(\theta)}\,\mathrm{cn}(A_0,m^2)}},$$

$$0 < m < 1,$$

$$\theta = \tan^{-1}\!\left[\frac{1}{m\sqrt{1-m^2}}\left[\frac{1}{1+\mathrm{cn}^2(A_0,m^2)} - m^2\right]\right],$$

A_0, t_0, and ϕ_0 are arbitrary real constants, n_0 is an arbitrary real integer

— (8.60)

12.
$$\psi(n, t) = \left\{ \frac{(1+\alpha_1)\cos[\alpha_4\,(t-t_0)] + \kappa\,\alpha_2\cosh[\alpha_5\,(n-n_0)] + i\,\alpha_3\sin[\alpha_4\,(t-t_0)]}{-\alpha_1\cos[\alpha_4\,(t-t_0)] - \alpha_1\,\alpha_2\cosh[\alpha_5\,(n-n_0)]} \right\}$$
$$\times e^{i\,[2\,\kappa^2\,(t-t_0)+\phi_0]}$$

$$\alpha_1 = \frac{\kappa}{(1+\kappa^2)\,[\cosh(\alpha_5)-1]},$$

$$\alpha_2 = \sqrt{1 + \frac{\tanh^2\!\left(\frac{\alpha_5}{2}\right)}{\kappa^2}}, \qquad \alpha_3 = \sqrt{1 + 2\,\kappa\,\alpha_1},$$

$$\alpha_4 = 4\,(1+\kappa^2)\sinh\!\left(\frac{\alpha_5}{2}\right)$$
$$\times \sqrt{\frac{\kappa^2}{1+\kappa^2} + \sinh^2\!\left(\frac{\alpha_5}{2}\right)}, \qquad \alpha_5 > 0,\ \kappa,\ t_0,$$

and ϕ_0 are arbitrary real constants, n_0 is an arbitrary real integer

Discrete Kuznetsov–Ma breather

(8.61)

8.7 Cubic-Quintic Discrete NLSE

Equation:

$$i\,\psi_{nt} + a_1\,(\psi_{n+1} + \psi_{n-1} - 2\,\psi_n) + a_2\,|\psi_n|^2\,\psi_n + (a_3\,|\psi_n|^2 + a_4\,|\psi_n|^4)(\psi_{n+1} + \psi_{n-1}) = 0,\quad (8.62)$$

where $\psi_n = \psi(n,\,t)$ is the complex function profile, the integer site index, n, and t are its two independent variables, a_1, a_2, a_3, and a_4 are real constants.

Solutions:

Solution **1. Constant Amplitude** *discrete CW, t- and n-dependent phase*

$$\psi(n,\,t) = A_0\,e^{i\left[A_1\,(n-n_0)+A_2\,(t-t_0)+\phi_0\right]},\quad (8.63)$$

where $A_2 = -2\,a_1 + A_0^2\,a_2 + 2\cos(A_1)\,(a_1 + A_0^2\,a_3 + A_0^4\,a_4)$, A_0, A_1, t_0, n_0, and ϕ_0 are arbitrary real constants.

Solution **2.** *Discrete bright soliton*
(Figure 8.21)

$$\psi(n,\,t) = A_0\,\mathrm{sech}[A_1\,(n - n_0)]\,e^{i\left[A_2\,(t-t_0)+\phi_0\right]},\quad (8.64)$$

where $A_0 = \pm\dfrac{\sqrt{a_3\,(a_2^2 - 4\,a_1\,a_4) + \sqrt{a_3^2 - 4\,a_1\,a_4}\,(a_2^2 + 4\,a_1\,a_4)}}{2\,a_4\,\sqrt{2\,a_1}}$, $A_1 = \mathrm{sech}^{-1}\!\left(\dfrac{-a_3 + \sqrt{a_3^2 - 4\,a_1\,a_4}}{a_2}\right)$,

$A_2 = -2\,a_1 - \dfrac{a_2\,(a_3 + \sqrt{a_3^2 - 4\,a_1\,a_4})}{2\,a_4}$, $\qquad a_3^2 - 4\,a_4\,a_1 \geqslant 0$, $\qquad a_1 > 0$,

$a_3\,(a_2^2 - 4\,a_1\,a_4) + \sqrt{a_3^2 - 4\,a_1\,a_4}\,(a_2^2 + 4\,a_1\,a_4) > 0$, t_0, n_0, and ϕ_0 are arbitrary real constants.
- *Reference*: [10].

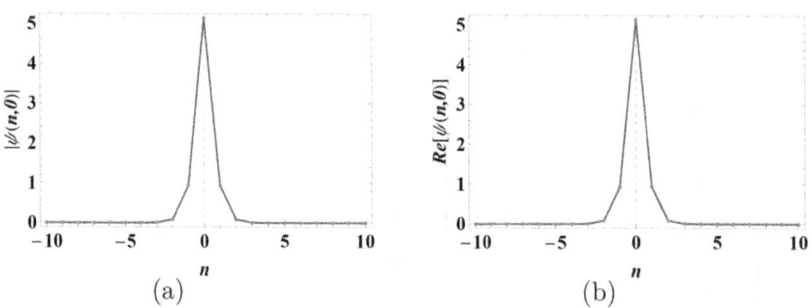

Figure 8.21. Discrete bright soliton (8.64) $t = 0$. (a) Absolute value, (b) Real part, with $n_0 = \phi_0 = 0$, $a_1 = a_2 = a_3 = 1$, and $a_4 = -1/10$. The lines are guides for the eye.

Solution **3.** *Staggered discrete bright soliton*
(Figure 8.22)

$$\psi(n,\,t) = (-1)^n\,A_0\,\mathrm{sech}[A_1\,(n - n_0)]\,e^{i\left[A_2\,(t-t_0)+\phi_0\right]},\quad (8.65)$$

where $\quad A_0 = \pm \dfrac{\sqrt{a_3\left(a_2^2 - 4\,a_1\,a_4\right) + \sqrt{a_3^2 - 4\,a_1\,a_4}\,\left(a_2^2 + 4\,a_1\,a_4\right)}}{2\,a_4\,\sqrt{2\,a_1}}, \quad A_1 = \operatorname{sech}^{-1}\left(\dfrac{a_3 - \sqrt{a_3^2 - 4\,a_1\,a_4}}{a_2}\right),$

$A_2 = -2\,a_1 - \dfrac{a_2\left(a_3 + \sqrt{a_3^2 - 4\,a_1\,a_4}\right)}{2\,a_4}, \quad a_3^2 - 4\,a_4\,a_1 \geqslant 0, \quad a_1 > 0, \quad a_3\left(a_2^2 - 4\,a_1\,a_4\right) +$

$\sqrt{a_3^2 - 4\,a_1\,a_4}\,\left(a_2^2 + 4\,a_1\,a_4\right) > 0,\ t_0,\ n_0,$ and ϕ_0 are arbitrary real constants.

• *Reference*: [10].

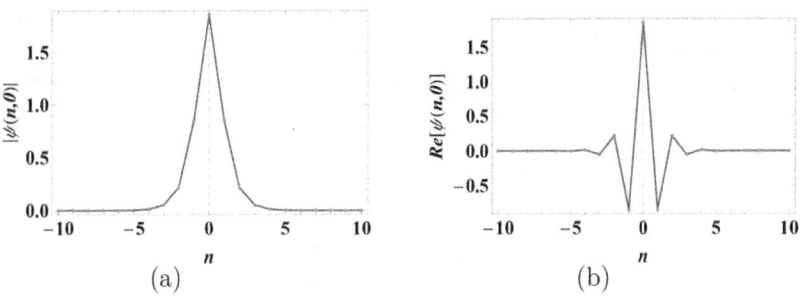

(a)   (b)

Figure 8.22. Staggered discrete bright soliton (8.65) $t = 0$. (a) Absolute value, (b) Real part, with $n_0 = \phi_0 = 0$, $a_1 = a_3 = 1$, $a_2 = -4/10$, and $a_4 = -1/10$. The lines are guides for the eye.

***Solution* 4.** *Discrete dark soliton*
(Figure 8.23)

$$\psi(n,\,t) = A_0 \tanh[A_1\,(n - n_0)]\,e^{i\left[A_2\,(t - t_0) + \phi_0\right]}, \tag{8.66}$$

where $\quad A_0 = \pm\sqrt{\dfrac{a_2 + a_3 + \sqrt{a_3^2 - 4\,a_1\,a_4}}{-2\,a_4}}, \quad A_1 = \cosh^{-1}\left(\sqrt{\dfrac{a_3 + \sqrt{a_3^2 - 4\,a_1\,a_4}}{-a_2}}\right), \quad A_2 = -2\,a_1 -$

$\dfrac{a_2\left(a_3 - \sqrt{a_3^2 - 4\,a_1\,a_4}\right)}{2\,a_4}, \qquad a_3^2 - 4\,a_4\,a_1 \geqslant 0, \qquad a_4\left(a_2 + a_3 + \sqrt{a_3^2 - 4\,a_1\,a_4}\right) < 0,$

$a_2\left(a_3 + \sqrt{a_3^2 - 4\,a_1\,a_4}\right) < 0,\ t_0,\ n_0,$ and ϕ_0 are arbitrary real constants.

• *Reference*: [10].

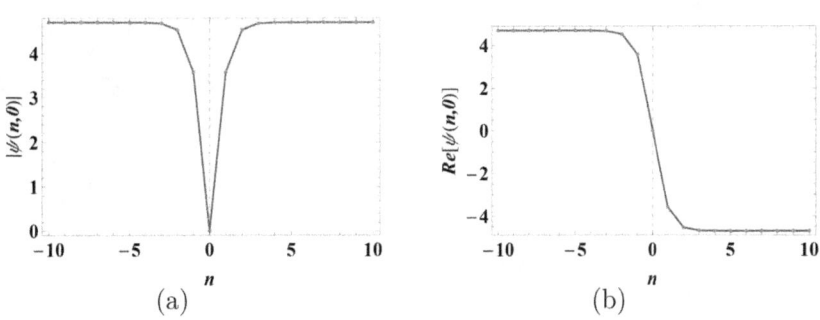

(a)   (b)

Figure 8.23. Discrete dark soliton (8.66) $t = 0$. (a) Absolute value, (b) Real part, with $n_0 = \phi_0 = 0$, $a_1 = -18/10$, $a_2 = -8/10$, $a_3 = 1$, and $a_4 = -18/10$. The lines are guides for the eye.

Solution 5. *Staggered discrete dark soliton*
(Figure 8.24)

$$\psi(n, t) = (-1)^n A_0 \tanh[A_1 (n - n_0)] \, e^{i \left[A_2 (t - t_0) + \phi_0 \right]}, \qquad (8.67)$$

where
$$A_0 = \pm \sqrt{\frac{a_2 - a_3 - \sqrt{a_3^2 - 4 a_1 a_4}}{2 a_4}}, \qquad A_1 = \cosh^{-1}\left(\sqrt{\frac{a_3 + \sqrt{a_3^2 - 4 a_1 a_4}}{a_2}} \right),$$

$$A_2 = -2 a_1 - \frac{a_2 \left(a_3 - \sqrt{a_3^2 - 4 a_1 a_4} \right)}{2 a_4}, \qquad\qquad a_3^2 - 4 a_4 a_1 \geqslant 0,$$

$(a_2 - a_3 - \sqrt{a_3^2 - 4 a_1 a_4}) > 0$, $\quad a_2 \left(a_3 + \sqrt{a_3^2 - 4 a_1 a_4} \right) > 0$, $\quad t_0$, n_0, and ϕ_0 are arbitrary real constants.

- *Reference*: [10].

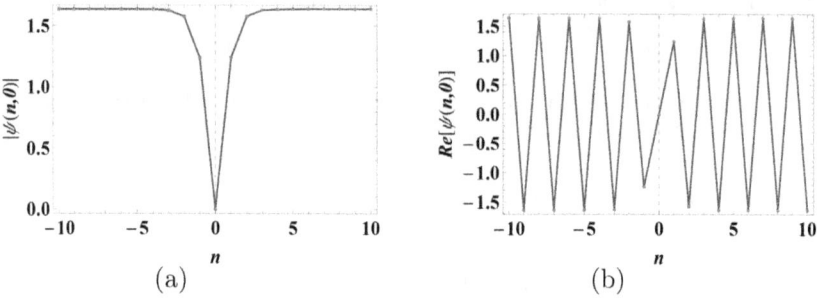

(a)   (b)

Figure 8.24. Staggered discrete dark soliton (8.67) at $t = 0$. (a) Absolute value, (b) Real part, with $n_0 = \phi_0 = 0$, $a_1 = 8/10$, $a_2 = a_3 = 1$, and $a_4 = -25/100$. The lines are guides for the eye.

8.8 Summary of Section 8.7

Equation

$$i\psi_{nt} + a_1(\psi_{n+1} + \psi_{n-1} - 2\psi_n) + a_2|\psi_n|^2\psi_n + (a_3|\psi_n|^2 + a_4|\psi_n|^4)(\psi_{n+1} + \psi_{n-1}) = 0$$

#	Solution	Conditions	Name	Equation #
1.	$\psi(n,t) = A_0\, e^{i[A_1(n-n_0)+A_2(t-t_0)+\phi_0]}$	$A_2 = -2a_1 + A_0^2 a_2 + 2\cos(A_1)(a_1 + A_0^2 a_3 + A_0^4 a_4)$, $A_0, A_1, t_0,$ and ϕ_0 are arbitrary real constants, n_0 is an arbitrary real integer	Discrete continuous wave, t- and n-dependent phase	(8.63)
2.	$\psi(n,t) = A_0\, \mathrm{sech}[A_1(n-n_0)]\, e^{i[A_2(t-t_0)+\phi_0]}$	$A_0 = \pm\sqrt{\dfrac{a_3(a_2^2-4a_1a_4)+\sqrt{a_3^2-4a_1a_4}\,(a_2^2+4a_1a_4)}{2a_4\sqrt{2a_1}}}$, $A_1 = \mathrm{sech}^{-1}\left(\dfrac{-a_3+\sqrt{a_3^2-4a_1a_4}}{a_2}\right)$, $A_2 = -2a_1 - \dfrac{a_2(a_3+\sqrt{a_3^2-4a_1a_4})}{2a_4}$, $a_3^2 - 4a_4a_1 \geqslant 0$, $a_1 > 0$, $a_3(a_2^2-4a_1a_4)+\sqrt{a_3^2-4a_1a_4}\,(a_2^2+4a_1a_4) > 0$, t_0 and ϕ_0 are arbitrary real constants, n_0 is an arbitrary real integer	Discrete bright soliton	(8.64)
3.	$\psi(n,t) = (-1)^n A_0\, \mathrm{sech}[A_1(n-n_0)]\, e^{i[A_2(t-t_0)+\phi_0]}$	$A_0 = \pm\sqrt{\dfrac{a_3(a_2^2-4a_1a_4)+\sqrt{a_3^2-4a_1a_4}\,(a_2^2+4a_1a_4)}{2a_4\sqrt{2a_1}}}$, $A_1 = \mathrm{sech}^{-1}\left(\dfrac{a_3-\sqrt{a_3^2-4a_1a_4}}{a_2}\right)$, $A_2 = -2a_1 - \dfrac{a_2(a_3+\sqrt{a_3^2-4a_1a_4})}{2a_4}$, $a_3^2 - 4a_4a_1 \geqslant 0$,	Staggered discrete bright soliton	(8.65)

$a_1 > 0$,

$a_3(a_2^2 - 4a_1a_4) + \sqrt{a_3^2 - 4a_1a_4}\,(a_2^2 + 4a_1a_4) > 0$, t_0 and ϕ_0 are arbitrary real constants, n_0 is an arbitrary real integer

Discrete dark soliton (8.66)

4. $\psi(n,t) = A_0 \tanh[A_1(n-n_0)]\, e^{i[A_2(t-t_0)+\phi_0]}$

$A_0 = \pm\sqrt{\dfrac{a_2+a_3+\sqrt{a_3^2-4a_1a_4}}{-2a_4}}$,

$A_1 = \cosh^{-1}\!\left(\sqrt{\dfrac{a_3+\sqrt{a_3^2-4a_1a_4}}{-a_2}}\right)$,

$A_2 = -2a_1 - \dfrac{a_2(a_3-\sqrt{a_3^2-4a_1a_4})}{2a_4}$, $\quad a_3^2 - 4a_4a_1 \geqslant 0$,

$a_4(a_2+a_3+\sqrt{a_3^2-4a_1a_4}) < 0$,

$a_2(a_3+\sqrt{a_3^2-4a_1a_4}) < 0$, t_0 and ϕ_0 are arbitrary real constants, n_0 is an arbitrary real integer

Staggered discrete dark soliton (8.67)

5. $\psi(n,t) = (-1)^n A_0 \tanh[A_1(n-n_0)]\, e^{i[A_2(t-t_0)+\phi_0]}$

$A_0 = \pm\sqrt{\dfrac{a_2-a_3-\sqrt{a_3^2-4a_1a_4}}{2a_4}}$,

$A_1 = \cosh^{-1}\!\left(\sqrt{\dfrac{a_3+\sqrt{a_3^2-4a_1a_4}}{a_2}}\right)$,

$A_2 = -2a_1 - \dfrac{a_2(a_3-\sqrt{a_3^2-4a_1a_4})}{2a_4}$, $\quad a_3^2 - 4a_4a_1 \geqslant 0$,

$(a_2-a_3-\sqrt{a_3^2-4a_1a_4}) > 0$,

$a_2(a_3+\sqrt{a_3^2-4a_1a_4}) > 0$, t_0 and ϕ_0 are arbitrary real constants, n_0 is an arbitrary real integer

8.9 Generalized Discrete NLSE

Equation:

$$i\,\psi_{nt} + a_1\,(\psi_{n+1} + \psi_{n-1} - 2\,\psi_n) + f[\psi_{n-1},\,\psi_n,\,\psi_{n+1}] = 0, \qquad (8.68)$$

where

$$
\begin{aligned}
f[\psi_{n-1},\,\psi_n,\,\psi_{n+1}] =\;& \alpha_1\,|\psi_n|^2\,\psi_n + \alpha_2\,|\psi_n|^2\,(\psi_{n+1} + \psi_{n-1}) + \alpha_3\,\psi_n^2\,(\psi_{n+1}^* + \psi_{n-1}^*) \\
& + \alpha_4\,\psi_n\,(|\psi_{n+1}|^2 + |\psi_{n-1}|^2) + \alpha_5\,\psi_n\,(\psi_{n+1}^*\,\psi_{n-1} + \psi_{n-1}^*\,\psi_{n+1}) \\
& + \alpha_6\,\psi_n^*\,(\psi_{n+1}^2 + \psi_{n-1}^2) + \alpha_7\,\psi_n^*\,\psi_{n+1}\,\psi_{n-1} \\
& + \alpha_8\,(|\psi_{n+1}|^2\,\psi_{n+1} + |\psi_{n-1}|^2\,\psi_{n-1}) + \alpha_9\,(\psi_{n-1}^*\,\psi_{n+1}^2 + \psi_{n+1}^*\,\psi_{n-1}^2) \\
& + \alpha_{10}\,(|\psi_{n+1}|^2\,\psi_{n-1} + |\psi_{n-1}|^2\,\psi_{n+1}) + \alpha_{11}\,(|\psi_{n-1}\,\psi_n| + |\psi_n\,\psi_{n+1}|)\,\psi_n \\
& + \alpha_{12}\,(\psi_{n+1}\,|\psi_{n+1}\,\psi_n| + \psi_{n-1}\,|\psi_n\,\psi_{n-1}|) \\
& + \alpha_{13}\,(\psi_{n+1}\,|\psi_{n-1}\,\psi_n| + \psi_{n-1}\,|\psi_n\,\psi_{n+1}|) \\
& + \alpha_{14}\,(\psi_{n+1}\,|\psi_{n-1}\,\psi_{n+1}| + \psi_{n-1}\,|\psi_{n-1}\,\psi_{n+1}|),
\end{aligned}
$$

$\psi_n = \psi(n,\,t)$ is the complex function profile, the integer site index, n, and t are its two independent variables, a_1 and $\alpha_1,\,\ldots,\,\alpha_{14}$ are real constants.

Solutions:

Solution **1.** *Discrete bright soliton*
  (Figure 8.25)

$$\psi(n,\,t) = A_0\,\mathrm{sech}[\beta\,(n+\gamma)]\,e^{-i\,(A_2\,t+\phi_0)}, \qquad (8.69)$$

where $\alpha_1 = \alpha_8 = 0$, $\quad A_2 = 2\,a_1\,[1 - \cosh(\beta)]$, $\quad \alpha_4 = -\alpha_{12} + \alpha_{13} + \alpha_5 - \alpha_6 +$
$\frac{\alpha_7}{2} + \cosh(\beta)\left[\alpha_2 + \alpha_3 + \alpha_{11} - \dfrac{a_1 \sinh^2(\beta)}{A_0^2}\right]$, $\quad \alpha_{10} = -\alpha_9 - \alpha_{14} - (2\,\alpha_{13} + 2\,\alpha_5 + \alpha_7)$
$\cosh(\beta) - 2\,(\alpha_2 + \alpha_3 + \alpha_{11})\cosh^2(\beta) + \dfrac{2\,a_1 \cosh^2(\beta)\,\sinh^2(\beta)}{A_0^2}$, $\quad \alpha_7 = -2\,\alpha_{13} - 2\,\alpha_5$,
$\alpha_{11} = -\alpha_2 - \alpha_3$, a_1, A_0, A_1, α_2, α_3, α_5, α_6, α_9, α_{12}, α_{13}, α_{14}, γ, and ϕ_0 are arbitrary real constants.
  • *Reference*: [11].

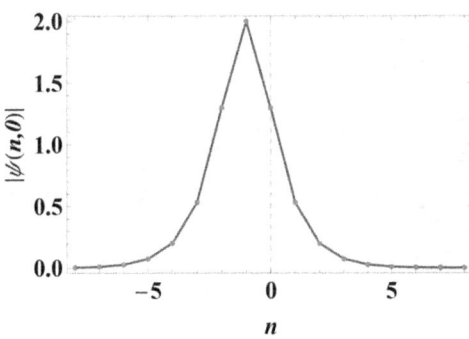

Figure 8.25. Discrete bright soliton (8.69) at $t = 0$ with $a_1 = A_0 = A_1 = 2$, $\alpha_2 = \alpha_3 = \alpha_5 = \alpha_7 = \alpha_9 = \alpha_{11} = \alpha_{12} = \alpha_{13} = \alpha_{14} = \beta = \gamma = 1$, and $\phi_0 = 0$. The lines are guides for the eye.

Solution 2. *Moving discrete bright soliton*
(Figure 8.26)

$$\psi(n,\, t) = A_0 \, \mathrm{sech}[\beta\, (n - v\, t + \gamma)] \, e^{i\, (A_1 n - A_2 t + \phi_0)}, \tag{8.70}$$

where $a_1 = 1$, $\alpha_1 = \alpha_8 = 0$, $v = \dfrac{2\, a_1 \sin(A_1) \sinh(\beta)}{\beta}$, $A_2 = 2\, a_1\, [1 - \cos(A_1)\cosh(\beta)]$,

$\alpha_2 = \alpha_3 + \dfrac{\sinh^2(\beta)}{A_0^2}$, $\qquad \alpha_4 = \alpha_6 - (\alpha_{10} - \alpha_9)\cos(A_1)\,\mathrm{sech}(\beta)$, $\qquad \alpha_6 = -\dfrac{1}{4}\sec(A_1)$

$[2\,\alpha_{12} - \alpha_{10}\,\mathrm{sech}(\beta) + \alpha_{14}\,\mathrm{sech}(\beta) + \alpha_9 \csc(A_1)\,\mathrm{sech}(\beta)\sin(3\,A_1)]$, $\qquad \alpha_7 = -2$

$[\alpha_{13}\cos(A_1) + \alpha_5\cos(2\,A_1)] - [(\alpha_{10} + \alpha_{14})\cos(A_1) + \alpha_9\cos(3\,A_1)]\,\mathrm{sech}(\beta) + 2\cosh(\beta)$

$\left[-\alpha_{11} - 2\,\alpha_3\cos(A_1) + \dfrac{(a_1 - 1)\cos(A_1)\sinh^2(\beta)}{A_0^2} \right]$, A_0, A_1, α_3, α_5, α_9, α_{10}, α_{11}, α_{12}, α_{13},

α_{14}, γ, and ϕ_0 are arbitrary real constants.
- *Reference*: [11].

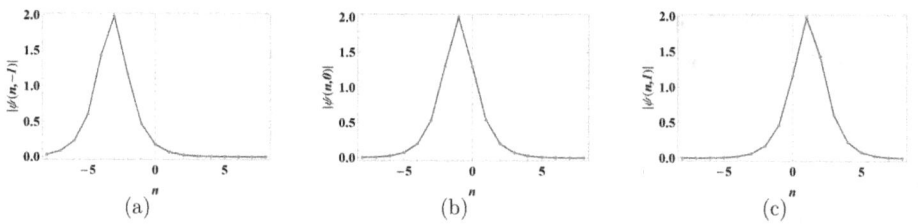

Figure 8.26. Moving discrete bright soliton (8.70) with $A_0 = A_1 = 2$, $a_1 = \alpha_3 = \alpha_5 = \alpha_9 = \alpha_{10} = \alpha_{11} = \alpha_{12} = \alpha_{13} = \alpha_{14} = \beta = \gamma = 1$, and $\phi_0 = 0$. (a) at $t = -1$, (b) at $t = 0$, and (c) at $t = 1$. The lines are guides for the eye. Animation available online at http://doi.org/10.1088/978-0-7503-5954-2.

Solution 3. *Discrete dark soliton*
(Figure 8.27)

$$\psi(n,\, t) = A_0 \tanh[\beta\, (n + \gamma)] \, e^{-i\, (A_2 t + \phi_0)}, \tag{8.71}$$

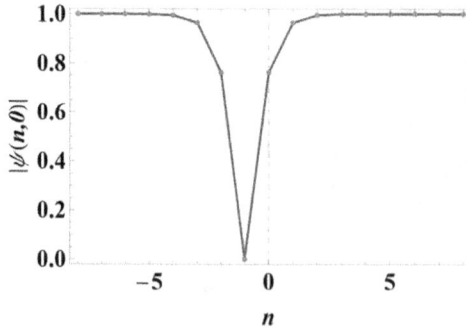

Figure 8.27. Discrete dark soliton (8.71) at $t = 0$ with $a_1 = 2$, $A_0 = \alpha_2 = \alpha_3 = \alpha_6 = \alpha_7 = \alpha_9 = \alpha_{10} = \alpha_{11} = \beta = \gamma = 1$, $\alpha_{12} = 7$, $\alpha_{13} = \alpha_{14} = 2$, and $\phi_0 = 0$. The lines are guides for the eye.

where $\alpha_1 = \alpha_8 = 0$, $A_2 = \dfrac{2}{1 + \coth^2(\beta)} [A_0^2 (\alpha_{10} - \alpha_{11} + \alpha_{14} - \alpha_2 - \alpha_3 + \alpha_9) + a_1]$,

$\alpha_4 = \dfrac{-1}{1 + \cosh(2\,\beta)} [\alpha_6 + (\alpha_{10} + \alpha_{14} + \alpha_6 + \alpha_9) \cosh(2\,\beta) + 2\,\alpha_{12} \cosh^2(\beta)]$,

$\alpha_5 = \dfrac{-1}{2\,A_0^2\,[1 + \tanh^2(\beta)]} \{A_0^2 (\alpha_{10} + 2\,\alpha_{11} + 2\,\alpha_{13} + \alpha_{14} + 2\,\alpha_2 + 2\,\alpha_3 + \alpha_7 + \alpha_9)$

$\qquad + [A_0^2 (2\,\alpha_{10} + 2\,\alpha_{14} + 2\,\alpha_{13} + \alpha_7 + 2\,\alpha_9) + 2\,a_1] \tanh^2(\beta)$ $\qquad a_1$,

$\qquad - A_0^2 (\alpha_{10} + \alpha_{14} + \alpha_9) \tanh^4(\beta)\}$,

A_0, A_1, γ, α_2, α_3, α_6, α_7, α_9, α_{10}, α_{11}, α_{12}, α_{13}, α_{14}, and ϕ_0 are arbitrary real constants.

- *Reference*: [11].

***Solution* 4.** *Moving discrete dark soliton*
(Figure 8.28)
$$\psi(n,\,t) = A_0 \tanh[\beta\,(n - v\,t + \gamma)]\, e^{i\,(A_1\,n - A_2\,t + \phi_0)}, \qquad (8.72)$$
where $\alpha_1 = \alpha_8 = \alpha_{14} = \gamma = 0$, $v = \dfrac{2\,A_0^2\,(\alpha_2 - \alpha_3)\coth(\beta)\sin(A_1)}{\beta}$,

$A_2 = \dfrac{2}{1 + \coth^2(\beta)} (a_1 + 2\,A_0^2\,\alpha_{10}\cos(A_1) - 2\,A_0^2\,\alpha_9\cos(A_1) - A_0^2\,[\alpha_{11}$

$\qquad + 2\,\alpha_2\cos(A_1)]\coth^4(\beta) + A_0^2\,[\alpha_{11} + (\alpha_2 + \alpha_3)\cos(A_1)]\operatorname{csch}^2(\beta)$

$\qquad + \coth^2(\beta)\,\{a_1 + A_0^2\,[\alpha_{11} + (\alpha_2 + \alpha_3)\cos(A_1)]\operatorname{csch}^2(\beta)\})$,

$\alpha_3 = \dfrac{1}{A_0^2}\,[-A_0^2\,\alpha_{10} + A_0^2\,\alpha_{14} - a_1\coth^2(\beta) + A_0^2\,\alpha_2\coth^4(\beta) + A_0^2\,\alpha_9\csc(A_1)$,

$\quad \sin(3\,A_1)]\,\tanh^4(\beta)$

$\alpha_4 = \tfrac{1}{2}\,\{\alpha_6\operatorname{sech}^2(\beta) + [\alpha_6 - 2\,(\alpha_{10} - \alpha_9)\cos(A_1)]\,[1 + \tanh^2(\beta)]\}$,

$\alpha_6 = \dfrac{1}{4\,A_0^2}\,[a_1 + 2\,A_0^2\,\alpha_{12} + a_1\coth^2(\beta) - A_0^2\,\alpha_2\coth^2(\beta) + A_0^2\,\alpha_3\coth^2(\beta)$

$\qquad - A_0^2\,\alpha_2\coth^4(\beta) + A_0^2\,\alpha_3\coth^4(\beta)]\sec(A_1)$,

$\alpha_7 = \dfrac{1}{A_0^2\,(1 + \coth^2(\beta))}\,(-4\,A_0^2\,\alpha_{10}\cos(A_1) - 2\,A_0^2\,\alpha_{13}\cos(A_1) + 4\,A_0^2\,\alpha_9\cos(A_1)$

$\quad -a_1\cos(A_1) - 2\,A_0^2\,\alpha_5\cos(2\,A_1) - \{[A_0^2\,(2\,\alpha 10 + 2\,\alpha_{13} + 3\,\alpha_2 + \alpha_3 - 2\,\alpha_9) + 2\,a_1]$

$\quad \times\cos(A_1) + 2\,A_0^2\,[\alpha_{11} + \alpha_5\cos(2\,A_1)]\}\coth^2(\beta) + [2\,A_0^2\,(\alpha_2 - \alpha_3) - a_1]$

$\quad \times\cos(A_1)\coth^4(\beta) + A_0^2\,(\alpha_2 - \alpha_3)\cos(A_1)\coth^6(\beta) + 2\,A_0^2\,(\alpha_{10} - \alpha_9)$

$\quad \times\cos(A_1)\tanh^2(\beta))$,

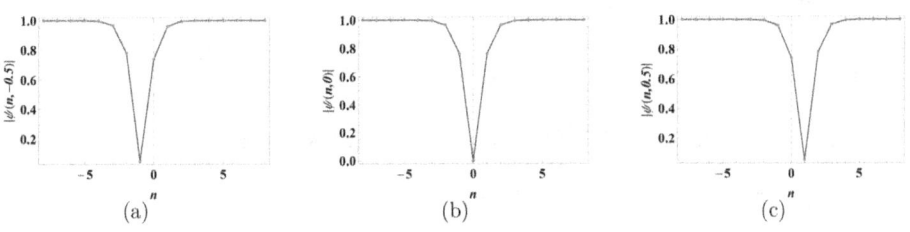

Figure 8.28. Moving discrete dark soliton (8.72) with $a_1 = A_0 = A_1 = \alpha_2 = \alpha_5 = \alpha_9 = \alpha_{10} = \alpha_{11} = \alpha_{12} = \alpha_{13} = \beta = 1$, and $\gamma = \phi_0 = 0$. (a) at $t = -0.5$, (b) at $t = 0$, and (c) at $t = 0.5$. The lines are guides for the eye. Animation available online at http://doi.org/10.1088/978-0-7503-5954-2.

a_1, A_0, A_1, α_2, α_5, α_9, α_{10}, α_{11}, α_{12}, α_{13}, γ, and ϕ_0 are arbitrary real constants.
  • *Reference*: [11].

Solution 5.

$$\psi(n, t) = A_0 \sin[\beta\,(n - v\,t + \gamma)]\, e^{i\,(A_1\,n - A_2\,t + \phi_0)}, \qquad (8.73)$$

where $\alpha_1 = \alpha_8 = 0$,

$$v = \tfrac{2\sin(\beta)}{\beta}\{A_0^2\,\alpha_6 \sin(2\,A_1)\sin(\beta)\sin(2\,\beta)$$
$$+ \sin(A_1)\,[a_1 + A_0^2\,(\alpha_2 - \alpha_3 + 4\,\alpha_8)\sin^2(\beta)$$
$$- 4\,A_0^2\,\alpha_8 \sin^4(\beta) + A_0^2\,\alpha_{12}\sin(\beta)\sin(2\,\beta)]\},$$

$A_2 = 2\,a_1 - A_0^2\cos(A_1)$

$$\cos(\beta)[\alpha_2 + \alpha_3 + \alpha_8 + \tfrac{2\,a_1}{A_0^2} + \alpha_{12}\cos(\beta) - (\alpha_2 + \alpha_3)\cos(2\,\beta)$$
$$- \alpha_{12}\cos(3\,\beta) - \alpha_8\cos(4\,\beta)] - A_0^2\sin^2(\beta)\{\alpha_1 + 2\,\alpha_4 + 2\,\alpha_6\cos(2\,A_1)$$
$$+ \alpha_6\cos[2\,(A_1 - \beta)] + 2\,\alpha_{11}\cos(\beta) + 2\,\alpha_4\cos(2\,\beta) + \alpha_6\cos[2\,(A_1 + \beta)]\},$$

$\alpha_7 = -\alpha_1 + \alpha_5 + \alpha_6 - 2\,(\alpha_5 - \alpha_6)\cos^2(A_1) + (\alpha_6 - \alpha_5)\cos(2\,A_1) - 2\,\alpha_{11}\cos(\beta)$
$$+ 2\cos(A_1)\,[\alpha_{12} - \alpha_{13} - 2\cos(\beta)\,(\alpha_{10} + \alpha_3 - \alpha_8 - \alpha_9)]$$
$$- 2\alpha_4\cos(2\beta) + 2\alpha_6\cos(2\beta),$$

$\alpha_{14} = \alpha_{10} - \alpha_2 + \alpha_3 - 3\,\alpha_8 - 2\cos(\beta)\,[\alpha_{12} + 2\,\alpha_6\cos(A_1)] - \alpha_9\csc(A_1)\sin(3\,A_1)$
$$+ 4\,\alpha_8\sin^2(\beta),$$

$\gamma = \beta = 1$, a_1, A_0, A_1, α_2, α_3, α_4, α_5, α_6, α_9, α_{10}, α_{11}, α_{12}, α_{13}, and ϕ_0 are arbitrary real constants.
  • *Reference*: [11].

Solution 6. *Moving discrete SW*
  (Figure 8.29)

$$\psi(n, t) = A_0\,\mathrm{dn}[\beta\,(n - v\,t + \gamma), m]\, e^{i\,(A_1\,n - A_2\,t + \phi_0)}, \qquad (8.74)$$

where $a_1 = 1$, $\alpha_1 = \alpha_8 = \alpha_{11} = \alpha_{12} = \alpha_{13} = \alpha_{14} = 0$,

$v = \dfrac{2\,A_0^2}{\beta}\,(\alpha_2 - \alpha_3)\sin(A_1)\,\mathrm{cs}(\beta, m)$,

$A_0 = \sqrt{\dfrac{\sin(A_1)}{q_1 - q_2 - q_3\,q_4}}$, $q_1 = (\alpha_2 - \alpha_3)\sin(A_1)\,\mathrm{cs}^2(\beta, m)$,

$q_2 = \alpha_6\sin(2\,A_1)\,\mathrm{ds}(\beta, m)\,\mathrm{ns}(\beta, m)$, $q_3 = \alpha_9\sin(3\,A_1) - \alpha_{10}\sin(A_1)$,

$$q_4 = \mathrm{cs}^2(2\,\beta, m) + \mathrm{ds}(2\,\beta, m)\,\mathrm{ns}(2\,\beta, m),$$

$A_2 = A_0^2\,\{-2\,(\alpha_2 + \alpha_3)\cos(A_1)\,\mathrm{ds}(\beta, m)\,\mathrm{ns}(\beta, m)$

$$+ 2\,[\alpha_4 + \alpha_6\cos(2\,A_1)]\,\mathrm{cs}^2(\beta, m) - [2\,\alpha_5\cos(2\,A_1) + \alpha_7]\,\mathrm{cs}^2(\beta, m) + \tfrac{2}{A_0^2}\Big\},$$

$\alpha_4 = -\alpha_6\cos(2\,A_1) - \dfrac{[\alpha_9\cos(3\,A_1) + \alpha_{10}\cos(A_1)]\,\mathrm{cs}(2\,\beta, m)}{\mathrm{cs}(\beta, m)}$,

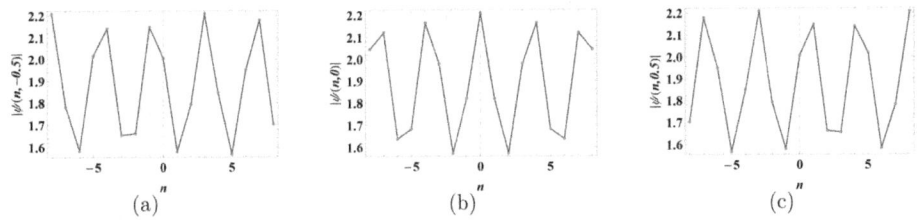

Figure 8.29. Moving discrete solitary wave (8.74) with $a_1 = A_1 = \alpha_2 = \alpha_9 = \alpha_{10} = \beta = 1$, $\alpha_3 = 2$, $m = 1/2$, and $\gamma = \phi_0 = 0$. (a) at $t = -0.5$, (b) at $t = 0$, and (c) at $t = 0.5$. The lines are guides for the eye.

$$\alpha_5 = -\frac{\alpha_7}{\cos(2\,A_1)}$$

$$+ \frac{1}{2\,\mathrm{cs}(\beta, m)\,\mathrm{cs}(2\,\beta, m)\,\cos(2\,A_1)} \left\{ \frac{\cos(A_1)}{A_0^2} - (\alpha_2 + \alpha_3)\cos(A_1)\,\mathrm{cs}^2(\beta, m) \right.$$

$$+ [\alpha_4 + \alpha_6 \cos(2\,A_1)]\,\mathrm{ds}(\beta, m)\,\mathrm{ns}(\beta, m)$$

$$\left. + [\alpha_9 \cos(3\,A_1) + \alpha_{10} \cos(A_1)]\,[\mathrm{ds}(2\,\beta, m)\,\mathrm{ns}(2\,\beta, m) - \mathrm{cs}^2(2\,\beta, m)] \right\},$$

$$\alpha_6 = -\frac{[\alpha_9 \sin(3\,A_1) - \alpha_{10} \sin(A_1)]\,\mathrm{cs}(2\beta, m)}{\sin(2\,A_1)\,\mathrm{cs}(\beta, m)},$$

$$\alpha_7 = \frac{1}{\mathrm{cs}^2(\beta, m)\,\mathrm{cs}(2\,\beta, m)\,[\sec(2\,A_1) - 1]}\,(2\,(\alpha_2 + \alpha_3 + \alpha_9 + \alpha_{10} - 1)\,\mathrm{cs}^2(\beta, m)\,\mathrm{cs}(2\,\beta, m)$$

$$- 2\,\alpha_3 \cos(A_1)\,\mathrm{cs}^3(\beta, m)\,\sec(2\,A_1)$$

$$+ 2\,(\alpha_9 - \alpha_{10})\cos(A_1)\,\mathrm{cs}(2\,\beta, m)\,\mathrm{ds}(\beta, m)\,\mathrm{ns}(\beta, m)\,\sec(2\,A_1) \qquad ,$$

$$+ \mathrm{cs}(\beta, m)\,\{-2\,[\alpha_9 + (\alpha_{10} + 2\,\alpha_9)\cos(2\,A_1)]\,\mathrm{cs}^2(2\,\beta, m)\,\sec(A_1)$$

$$+ 2(\alpha_{10} - \alpha_9)\cos(A_1)\,\mathrm{ds}(2\,\beta, m)\,\mathrm{ns}(2\,\beta, m)\,\sec(2\,A_1)\}),$$

$0 \leqslant m \leqslant 1$, A_1, β, γ, α_2, α_3, α_9, α_{10}, and ϕ_0 are arbitrary real constants.

- *Reference*: [12].

Solution 7. *Moving discrete SW*
(Figure 8.30)

$$\psi(n, t) = A_0 \sqrt{m}\,\mathrm{cn}[\beta\,(n - v\,t + \gamma), m]\,e^{i\,(A_1\,n - A_2\,t + \phi_0)}, \qquad (8.75)$$

where $a_1 = 1$, $\alpha_1 = \alpha_8 = \alpha_{11} = \alpha_{12} = \alpha_{13} = \alpha_{14} = 0$,

$$v = \frac{2\,A_0^2}{\beta}\,(\alpha_2 - \alpha_3)\sin(A_1)\,\mathrm{ds}(\beta, m), \quad A_0 = \sqrt{\frac{\sin(A_1)}{p_1 - p_2 - p_3\,p_4}},$$

$$p_1 = (\alpha_2 - \alpha_3)\sin(A_1)\,\mathrm{ds}^2(\beta, m), \quad p_2 = \alpha_6 \sin(2\,A_1)\,\mathrm{cs}(\beta, m)\,\mathrm{ns}(\beta, m),$$

$$p_3 = \alpha_9 \sin(3\,A_1) - \alpha_{10} \sin(A_1), \quad p_4 = \mathrm{ds}^2(2\,\beta, m) + \mathrm{cs}(2\,\beta, m)\,\mathrm{ns}(2\,\beta, m),$$

$$A_2 = A_0^2\,\{-2\,(\alpha_2 + \alpha_3)\cos(A_1)\,\mathrm{cs}(\beta, m)\,\mathrm{ns}(\beta, m) + 2\,[\alpha_4 + \alpha_6 \cos(2\,A_1)]\,\mathrm{ds}^2(\beta, m)$$

$$- [2\,\alpha_5 \cos(2\,A_1) + \alpha_7]\,\mathrm{ds}^2(\beta, m) + \frac{2}{A_0^2}\},$$

$$\alpha_4 = -\alpha_6 \cos(2\,A_1) - \frac{[\alpha_9 \cos(3\,A_1) + \alpha_{10} \cos(A_1)]\,\mathrm{ds}(2\,\beta, m)}{\mathrm{ds}(\beta, m)},$$

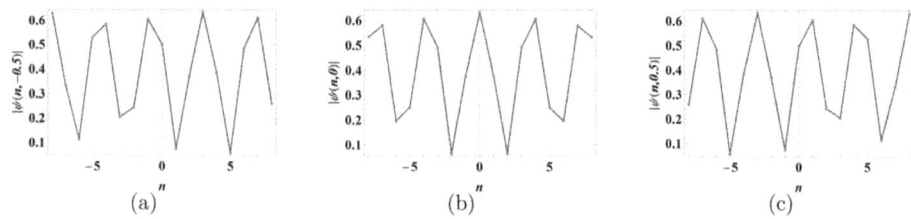

Figure 8.30. Moving discrete solitary wave (8.75) with $a_1 = A_1 = \alpha_2 = \alpha_9 = \alpha_{10} = \beta = 1$, $\alpha_3 = 2$, $m = 1/2$, and $\gamma = \phi_0 = 0$. (a) at $t = -0.5$, (b) at $t = 0$, and (c) at $t = 0.5$. The lines are guides for the eye.

$$\alpha_5 = -\frac{\alpha_7}{\cos(2\,A_1)}$$

$$+ \frac{1}{2\,\mathrm{ds}(\beta,m)\,\mathrm{ds}(2\,\beta,m)\cos(2\,A_1)} \left\{ \frac{\cos(A_1)}{A_0^2} - (\alpha_2 + \alpha_3)\cos(A_1)\,\mathrm{ds}^2(\beta,m) \right.$$

$$+ [\alpha_4 + \alpha_6\cos(2\,A_1)]\,\mathrm{cs}(\beta,m)\,\mathrm{ns}(\beta,m)$$

$$+ [\alpha_9\cos(3\,A_1) + \alpha_{10}\cos(A_1)]\,[\mathrm{cs}(2\,\beta,m)\,\mathrm{ns}(2\,\beta,m)$$

$$\left. -\,\mathrm{ds}^2(2\,\beta,m)] \right\},$$

$$\alpha_6 = -\frac{[\alpha_9\sin(3\,A_1) - \alpha_{10}\sin(A_1)]\,\mathrm{ds}(2\beta,m)}{\sin(2\,A_1)\,\mathrm{ds}(\beta,m)},$$

$$\alpha_7 = \frac{1}{\mathrm{ds}^2(\beta,m)\,\mathrm{ds}(2\,\beta,m)\,[\sec(2\,A_1) - 1]}\,(2\,(\alpha_2 + \alpha_3 + \alpha_9 + \alpha_{10} - 1)\,\mathrm{ds}^2(\beta,m)\,\mathrm{ds}(2\,\beta,m)$$

$$-\,2\,\alpha_3\cos(A_1)\,\mathrm{ds}^3(\beta,m)\sec(2\,A_1)$$

$$+\,2\,(\alpha_9 - \alpha_{10})\cos(A_1)\,\mathrm{ds}(2\,\beta,m)\,\mathrm{cs}(\beta,m)\,\mathrm{ns}(\beta,m)\sec(2\,A_1)$$

$$+\,\mathrm{ds}(\beta,m)\,\{-2\,[\alpha_9 + (\alpha_{10} + 2\,\alpha_9)\cos(2\,A_1)]\,\mathrm{ds}^2(2\,\beta,m)\sec(A_1)$$

$$+\,2(\alpha_{10} - \alpha_9)\cos(A_1)\,\mathrm{cs}(2\,\beta,m)\,\mathrm{ns}(2\,\beta,m)\sec(2\,A_1)\}),$$

$0 \leqslant m \leqslant 1$, A_1, β, γ, α_2, α_3, α_9, α_{10}, and ϕ_0 are arbitrary real constants.
- *Reference*: [12].

8.10 Summary of Section 8.9

Note: For lengthy conditions, the reader is referred to the solutions in section 8.9.

Equation		
$i\,\psi_{nt} + a_1\,(\psi_{n+1} + \psi_{n-1} - 2\,\psi_n) + f[\psi_{n-1},\,\psi_n,\,\psi_{n+1}] = 0,$		

$$f[\psi_{n-1},\,\psi_n,\,\psi_{n+1}] = \alpha_1\,|\psi_n|^2\,\psi_n + \alpha_2\,|\psi_n|^2\,(\psi_{n+1} + \psi_{n-1}) + \alpha_3\,\psi_n^2\,(\psi_{n+1}^* + \psi_{n-1}^*)$$
$$+\,\alpha_4\,\psi_n\,(|\psi_{n+1}|^2 + |\psi_{n-1}|^2) + \alpha_5\,\psi_n\,(\psi_{n+1}^*\,\psi_{n-1} + \psi_{n-1}^*\,\psi_{n+1})$$
$$+\,\alpha_6\,\psi_n^*\,(\psi_{n+1}^2 + \psi_{n-1}^2) + \alpha_7\,\psi_n^*\,\psi_{n+1}\,\psi_{n-1}$$
$$+\,\alpha_8\,(|\psi_{n+1}|^2\,\psi_{n+1} + |\psi_{n-1}|^2\,\psi_{n-1}) + \alpha_9\,(\psi_{n-1}^*\,\psi_{n+1}^2 + \psi_{n+1}^*\,\psi_{n-1}^2)$$
$$+\,\alpha_{10}\,(|\psi_{n+1}|^2\,\psi_{n-1} + |\psi_{n-1}|^2\,\psi_{n+1}) + \alpha_{11}\,(|\psi_{n-1}\,\psi_n| + |\psi_n\,\psi_{n+1}|)\,\psi_n$$
$$+\,\alpha_{12}\,(\psi_{n+1}\,|\psi_{n+1}\,\psi_n| + \psi_{n-1}\,|\psi_n\,\psi_{n-1}|)$$
$$+\,\alpha_{13}\,(\psi_{n+1}\,|\psi_{n-1}\,\psi_n| + \psi_{n-1}\,|\psi_n\,\psi_{n+1}|)$$
$$+\,\alpha_{14}\,(\psi_{n+1}\,|\psi_{n-1}\,\psi_{n+1}| + \psi_{n-1}\,|\psi_{n-1}\,\psi_{n+1}|)$$

#	Solution	Conditions	Name	Equation #
1.	$\psi(n, t) = A_0\,\mathrm{sech}[\beta\,(n + \gamma)]\,e^{-i\,(A_2\,t+\phi_0)}$	See text	Discrete bright soliton	(8.69)
2.	$\psi(n, t) = A_0\,\mathrm{sech}[\beta\,(n - v\,t + \gamma)]\,e^{i\,(A_1\,n - A_2\,t+\phi_0)}$	See text.	Moving discrete bright soliton	(8.70)
3.	$\psi(n, t) = A_0\,\tanh[\beta\,(n + \gamma)]\,e^{-i\,(A_2\,t+\phi_0)}$	See text	Discrete dark soliton	(8.71)
4.	$\psi(n, t) = A_0\,\tanh[\beta\,(n - v\,t + \gamma)]\,e^{i\,(A_1\,n - A_2\,t+\phi_0)}$	See text	Moving discrete dark soliton	(8.72)
5.	$\psi(n, t) = A_0\,\sin[\beta\,(n - v\,t + \gamma)]\,e^{i\,(A_1\,n - A_2\,t+\phi_0)}$	See text	–	(8.73)
6.	$\psi(n, t) = A_0\,\mathrm{dn}[\beta\,(n - v\,t + \gamma),\,m]\,e^{i\,(A_1\,n - A_2\,t+\phi_0)}$	See text	Moving discrete solitary wave	(8.74)
7.	$\psi(n, t) = A_0\,\sqrt{m}\,\mathrm{cn}[\beta\,(n - v\,t + \gamma),\,m]\,e^{i\,(A_1\,n - A_2\,t+\phi_0)}$	See text	Moving discrete solitary wave	(8.75)

8.11 Coupled Salerno Equations

Equation:

$$i\,\psi_{1nt} + \psi_{1n+1} + \psi_{1n-1} - 2\,\psi_{1n}$$

$$+ (\mu_1\,|\psi_{1n}|^2 + \mu_2\,|\psi_{2n}|^2)\left(\psi_{1n+1} + \psi_{1n-1} + \frac{\nu_1 - 2\,\mu_1}{\mu_1}\,\psi_{1n}\right) = 0,$$

$$i\,\psi_{2nt} + \left[\psi_{2n+1} + \psi_{2n-1} - \left(2 + \frac{\nu_1\,\mu_2}{\mu_1^2} - \frac{\nu_2}{\mu_2}\right)\psi_{2n}\right] \tag{8.76}$$

$$+ (\mu_1\,|\psi_{1n}|^2 + \mu_2\,|\psi_{2n}|^2)\left[\psi_{2n+1} + \psi_{2n-1} + \left(\frac{\nu_2 - 2\,\mu_2}{\mu_2}\right)\psi_{2n}\right] = 0,$$

where $\psi_j = \psi_j(n,\,t)$ is the complex function profile, $j = 1, 2$, the integer site index, n, and t are its two independent variables, μ_1, μ_2, ν_1, and ν_2 are real constants.

Solutions:

Solution 1. *Discrete SW*
   (Figure 8.31)

$$\psi_1(n,\,t) = A_0\,\mathrm{dn}[A_1\,(n + n_0),\,m]\,e^{-i\,(\omega_1 t + \phi_1)},$$
$$\psi_2(n,\,t) = B_0\,\sqrt{m}\,\mathrm{sn}[A_1\,(n + n_0),\,m]\,e^{-i\,(\omega_2 t + \phi_2)}, \tag{8.77}$$

where $\omega_1 = \frac{\nu_1}{\mu_1}$, $\omega_2 = \frac{\nu_1\,\mu_2}{\mu_1^2}$, $\mu_1 = \frac{-1}{A_0^2}$, $\mu_2 = \frac{\mu_1\,A_0^2}{B_0^2}$, $0 \leqslant m \leqslant 1$, $A_0, A_1, B_0, n_0, \phi_1$, and ϕ_2 are arbitrary real constants.
   • *Reference*: [13].

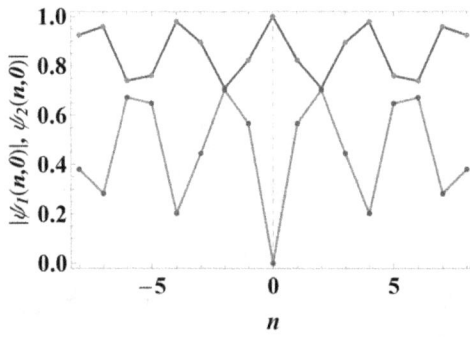

Figure 8.31. Discrete solitary wave (8.77) at $t = 0$. Blue is ψ_1 and red is ψ_2 with $A_0 = B_0 = A_1 = 1$, $\nu_1 = \nu_2 = 2$, $m = 1/2$, and $n_0 = t_0 = \phi_1 = \phi_2 = 0$. The lines are guides for the eye.

Solution 2. *Discrete SW*
   (Figure 8.32)

$$\psi_1(n,\,t) = A_0\,\sqrt{m}\,\mathrm{cn}[A_1\,(n + n_0),\,m]\,e^{-i\,(\omega_1 t + \phi_1)},$$
$$\psi_2(n,\,t) = B_0\,\sqrt{m}\,\mathrm{sn}[A_1\,(n + n_0),\,m]\,e^{-i\,(\omega_2 t + \phi_2)}, \tag{8.78}$$

where $\omega_1 = \frac{\nu_1}{\mu_1}$, $\omega_2 = \frac{\nu_1 \mu_2}{\mu_1^2}$, $\mu_1 = \frac{-1}{m A_0^2}$, $\mu_2 = \frac{\mu_1 A_0^2}{B_0^2}$, $0 \leqslant m \leqslant 1$, A_0, A_1, B_0, n_0, ϕ_1, and ϕ_2 are arbitrary real constants.

- *Reference*: [13].

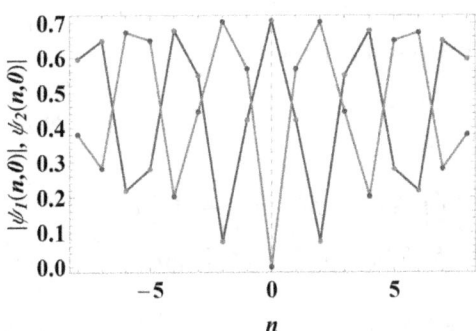

Figure 8.32. Discrete solitary wave (8.78) at $t = 0$. Blue is ψ_1 and red is ψ_2 with $A_0 = B_0 = A_1 = 1$, $\nu_1 = \nu_2 = 2$, $m = 1/2$, and $n_0 = t_0 = \phi_1 = \phi_2 = 0$. The lines are guides for the eye.

Solution 3. *Discrete bright–dark soliton*
  (Figure 8.33)

$$\psi_1(n, t) = A_0 \, \text{sech}[A_1 \, (n + n_0)] \, e^{-i \, (\omega_1 t + \phi_1)},$$
$$\psi_2(n, t) = B_0 \, \tanh[A_1 \, (n + n_0)] \, e^{-i \, (\omega_2 t + \phi_2)}, \tag{8.79}$$

where $\omega_1 = \frac{\nu_1}{\mu_1}$, $\omega_2 = \frac{\nu_1 \mu_2}{\mu_1^2}$, $\mu_1 = \frac{-1}{A_0^2}$, $\mu_2 = \frac{\mu_1 A_0^2}{B_0^2}$, A_0, A_1, B_0, n_0, ϕ_1, and ϕ_2 are arbitrary real constants.

- *Reference*: [13].

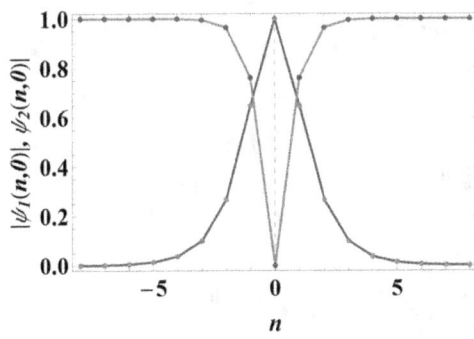

Figure 8.33. Discrete bright–dark soliton (8.79) at $t = 0$. Blue is ψ_1 and red is ψ_2 with $A_0 = B_0 = A_1 = 1$, $\nu_1 = \nu_2 = 2$, and $n_0 = t_0 = \phi_1 = \phi_2 = 0$. The lines are guides for the eye.

Solution 4. *Discrete SW*

$$\psi_1(n,\ t) = \{A_0\ \mathrm{dn}^2[A_1\ (n\ +\ n_0),\ m]\ +\ A_2\}\ e^{-i\ (\omega_1\ t+\phi_1)},$$
$$\psi_2(n,\ t) = B_0\ \sqrt{m}\ \mathrm{sn}[A_1\ (n\ +\ n_0),\ m]\ \mathrm{dn}[A_1\ (n\ +\ n_0),\ m]\ e^{-i\ (\omega_2\ t+\phi_2)},$$

(8.80)

where $A_0 = -2\ A_2$, $\omega_1 = \frac{\nu_1}{\mu_1}$, $\omega_2 = \frac{\nu_1\ \mu_2}{\mu_1^2}$, $\mu_1 = \frac{-4}{A_0^2}$, $\mu_2 = \frac{\mu_1\ A_0^2}{B_0^2}$, $0 \leqslant m \leqslant 1$, A_1, A_2, B_0, n_0, ϕ_1, and ϕ_2 are arbitrary real constants.
  • *Reference*: [13].

Solution 5. *Discrete SW*

$$\psi_1(n,\ t) = \left\{ A_0\ \mathrm{dn}^2\Big[A_1\ (n\ +\ n_0),\ m \Big]\ +\ A_2 \right\}\ e^{-i\ (\omega_1\ t+\phi_1)},$$
$$\psi_2(n,\ t) = B_0\ m\ \mathrm{sn}[A_1\ (n\ +\ n_0),\ m]\ \mathrm{cn}[A_1\ (n\ +\ n_0),\ m]\ e^{-i\ (\omega_2\ t+\phi_2)},$$

(8.81)

where $A_0 = \frac{-2\ A_2}{2-m}$, $\omega_1 = \frac{\nu_1}{\mu_1}$, $\omega_2 = \frac{\nu_1\ \mu_2}{\mu_1^2}$, $\mu_1 = \frac{-4}{m^2\ A_0^2}$, $\mu_2 = \frac{\mu_1\ A_0^2}{B_0^2}$, $0 \leqslant m \leqslant 1$, A_1, A_2, B_0, n_0, ϕ_1, and ϕ_2 are arbitrary real constants.
  • *Reference*: [13].

Solution 6.
  (Figure 8.34)

$$\psi_1(n,\ t) = \left\{ A_0\ \mathrm{sech}^2\Big[A_1\ (n\ +\ n_0) \Big]\ +\ A_2 \right\}\ e^{-i\ (\omega_1\ t+\phi_1)},$$
$$\psi_2(n,\ t) = B_0\ \mathrm{sech}[A_1\ (n\ +\ n_0)]\ \tanh[A_1\ (n\ +\ n_0)]\ e^{-i\ (\omega_2\ t+\phi_2)},$$

(8.82)

where $A_0 = -2\ A_2$, $\omega_1 = \frac{\nu_1}{\mu_1}$, $\omega_2 = \frac{\nu_1\ \mu_2}{\mu_1^2}$, $\mu_1 = \frac{-4}{A_0^2}$, $\mu_2 = \frac{\mu_1\ A_0^2}{B_0^2}$, A_1, A_2, B_0, n_0, ϕ_1, and ϕ_2 are arbitrary real constants.
  • *Reference*: [13].

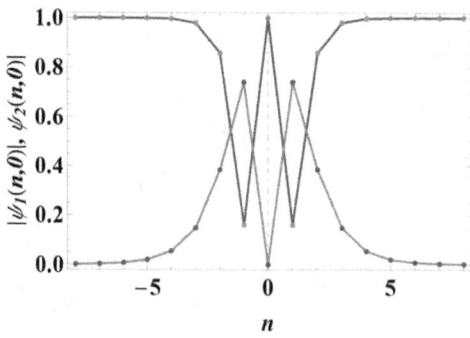

Figure 8.34. Plot of solution (8.82) at $t = 0$. Blue is ψ_1 and red is ψ_2 with $A_1 = A_2 = 1$, $B_0 = 3/2$, $\nu_1 = \nu_2 = 2$, $m = 1/2$, and $n_0 = t_0 = \phi_1 = \phi_2 = 0$. The lines are guides for the eye.

Solution 7. **Rational Solution I**

$$\psi_1(n, t) = \frac{A_0}{\sqrt{1 + n^2}}\, e^{-i\,(\omega_1 t + \phi_1)},$$

$$\psi_2(n, t) = \frac{B_0\, n}{\sqrt{1 + n^2}}\, e^{-i\,(\omega_2 t + \phi_2)},$$

(8.83)

where $\omega_1 = \frac{\nu_1}{\mu_1}$, $\omega_2 = \frac{\nu_1\,\mu_2}{\mu_1^2}$, $\mu_1 = \frac{-1}{A_0^2}$, $\mu_2 = \frac{\mu_1\,A_0^2}{B_0^2}$, A_0, B_0, ϕ_1, and ϕ_2 are arbitrary real constants.

• *Reference*: [13].

Solution 8. **Rational Solution II**

$$\psi_1(n, t) = A_0\,\sqrt{\frac{1 + n^2}{1 + n^2 + n^4}}\, e^{-i\,(\omega_1 t + \phi_1)},$$

$$\psi_2(n, t) = \frac{B_0\, n^2}{\sqrt{1 + n^2 + n^4}}\, e^{-i\,(\omega_2 t + \phi_2)},$$

(8.84)

where $\omega_1 = \frac{\nu_1}{\mu_1}$, $\omega_2 = \frac{\nu_1\,\mu_2}{\mu_1^2}$, $\mu_1 = \frac{-1}{A_0^2}$, $\mu_2 = \frac{\mu_1\,A_0^2}{B_0^2}$, A_0, B_0, ϕ_1, and ϕ_2, are arbitrary real constants.

• *Reference*: [13].

Solution 9. **Rational Solution III**

$$\psi_1(n, t) = A_0\,\sqrt{\frac{2 + n^2}{1 + n^2}}\, e^{-i\,(\omega_1 t + \phi_1)},$$

$$\psi_2(n, t) - \frac{B_0}{\sqrt{1 + n^2}}\, e^{-i\,(\omega_2 t + \phi_2)},$$

(8.85)

where $\omega_1 = \frac{\nu_1}{\mu_1}$, $\omega_2 = \frac{\nu_1\,\mu_2}{\mu_1^2}$, $\mu_1 = \frac{-1}{A_0^2}$, $\mu_2 = 1$, A_0, B_0, ϕ_1, and ϕ_2 are arbitrary real constants.

• *Reference*: [13].

Solution 10.

$$\psi_1(n, t) = A_0\,\cos[A_1\,(n + n_0)]\, e^{-i\,(\omega_1 t + \phi_1)},$$
$$\psi_2(n, t) = B_0\,\sin[A_1\,(n + n_0)]\, e^{-i\,(\omega_2 t + \phi_2)},$$

(8.86)

where $\omega_1 = \frac{\nu_1}{\mu_1}$, $\omega_2 = \frac{\nu_1\,\mu_2}{\mu_1^2}$, $\mu_1 = \frac{-1}{A_0^2}$, $\mu_2 = \frac{\mu_1\,A_0^2}{B_0^2}$, A_0, B_0, n_0, ϕ_1, and ϕ_2 are arbitrary real constants.

• *Reference*: [13].

Solution 11. *Discrete SW*

$$\psi_1(n, t) = A_0 \, \text{nd}[A_1 (n + n_0), m] \, e^{-i (\omega_1 t + \phi_1)},$$
$$\psi_2(n, t) = B_0 \, \sqrt{m} \, \text{sd}[A_1 (n + n_0), m] \, e^{-i (\omega_2 t + \phi_2)},$$

(8.87)

where $A_0 = \dfrac{\mu_2 B_0^2}{|\mu_1|}$, $\omega_1 = \dfrac{\nu_1}{\mu_1}$, $\omega_2 = \dfrac{\nu_1 \mu_2}{\mu_1^2}$, $\mu_1 = -1$, $\mu_2 = 1$, $0 \leqslant m \leqslant 1$, A_1, A_2, B_0, n_0, ϕ_1, and ϕ_2 are arbitrary real constants.
  • *Reference*: [13].

Solution 12.

$$\psi_1(n, t) = A_0 \cosh[A_1 (n + n_0)] \, e^{-i (\omega_1 t + \phi_1)},$$
$$\psi_2(n, t) = B_0 \sinh[A_1 (n + n_0)] \, e^{-i (\omega_2 t + \phi_2)},$$

(8.88)

where $A_0 = \dfrac{\mu_2 B_0^2}{|\mu_1|}$, $\omega_1 = \dfrac{\nu_1}{\mu_1}$, $\omega_2 = \dfrac{\nu_1 \mu_2}{\mu_1^2}$, $\mu_1 = -1$, $\mu_2 = 1$, A_1, B_0, n_0, ϕ_1, and ϕ_2 are arbitrary real constants.
  • *Reference*: [13].

Solution 13. *Discrete SW*

$$\psi_1(n, t) = \left\{ A_0 \, \text{nd}^2\!\left[A_1 (n + n_0), m\right] + A_2 \right\} e^{-i (\omega_1 t + \phi_1)},$$
$$\psi_2(n, t) = \frac{B_0 \, \sqrt{m} \, \text{sn}[A_1 (n + n_0), m]}{\text{dn}^2[A_1 (n + n_0), m]} \, e^{-i (\omega_2 t + \phi_2)},$$

(8.89)

where $A_0 = -2 A_2$, $B_0 = \sqrt{\dfrac{|\mu_1| A_0^2}{\mu_2}}$, $\omega_1 = \dfrac{\nu_1}{\mu_1}$, $\omega_2 = \dfrac{\nu_1 \mu_2}{\mu_1^2}$, $\mu_1 = \dfrac{-4}{A_0^2}$, $\mu_2 > 0$, $0 \leqslant m \leqslant 1$, A_1, A_2, n_0, ϕ_1, and ϕ_2 are arbitrary real constants.
  • *Reference*: [13].

Solution 14.

$$\psi_1(n, t) = \left\{ A_0 \cosh^2\!\left[A_1 (n + n_0)\right] + A_2 \right\} e^{-i (\omega_1 t + \phi_1)},$$
$$\psi_2(n, t) = B_0 \sinh[A_1 (n + n_0)] \cosh[A_1 (n + n_0)] \, e^{-i (\omega_2 t + \phi_2)},$$

(8.90)

where $A_0 = -2 A_2$, $B_0 = \sqrt{\dfrac{|\mu_1| A_0^2}{\mu_2}}$, $\omega_1 = \dfrac{\nu_1}{\mu_1}$, $\omega_2 = \dfrac{\nu_1 \mu_2}{\mu_1^2}$, $\mu_1 = \dfrac{-4}{A_0^2}$, $\mu_2 > 0$, A_1, A_2, n_0, ϕ_1, and ϕ_2 are arbitrary real constants.
  • *Reference*: [13].

8.12 Summary of Section 8.11

Equation

$$i\psi_{1nt} + (\psi_{1n+1} + \psi_{1n-1} - 2\psi_{1n}) + (\mu_1|\psi_{1n}|^2 + \mu_2|\psi_{2n}|^2)\left(\psi_{1n+1} + \psi_{1n-1} + \frac{\nu_1 - 2\mu_1}{\mu_1}\psi_{1n}\right) = 0,$$

$$i\psi_{2nt} + \left[\psi_{2n+1} + \psi_{2n-1} - \left(2 + \frac{\nu_1\mu_2}{\mu_1^2} - \frac{\nu_2}{\mu_2}\right)\psi_{2n}\right] + [\mu_1|\psi_{1n}|^2 + \mu_2|\psi_{2n}|^2]\left(\psi_{2n+1} + \psi_{2n-1} + \left(\frac{\nu_2 - 2\mu_2}{\mu_2}\right)\psi_{2n}\right) = 0$$

#	Solution	Conditions	Name	Equation #
1.	$\psi_1(n,t) = A_0\,\text{dn}[A_1(n+n_0), m]\,e^{-i(\omega_1 t + \phi_1)}$, $\psi_2(n,t) = B_0\sqrt{m}\,\text{sn}[A_1(n+n_0), m]\,e^{-i(\omega_2 t + \phi_2)}$	$\omega_1 = \frac{\nu_1}{\mu_1}$, $\omega_2 = \frac{\nu_1\mu_2}{\mu_1^2}$, $\mu_1 = \frac{-1}{A_0^2}$, $\mu_2 = \frac{\mu_1 A_0^2}{B_0^2}$, $0 \le m \le 1$, $A_0, A_1, B_0, \phi_1,$ and ϕ_2 are arbitrary real constants, n_0 is an arbitrary real integer	Discrete solitary wave	(8.77)
2.	$\psi_1(n,t) = A_0\sqrt{m}\,\text{cn}[A_1(n+n_0), m]\,e^{-i(\omega_1 t + \phi_1)}$, $\psi_2(n,t) = B_0\sqrt{m}\,\text{sn}[A_1(n+n_0), m]\,e^{-i(\omega_2 t + \phi_2)}$	$\omega_1 = \frac{\nu_1}{\mu_1}$, $\omega_2 = \frac{\nu_1\mu_2}{\mu_1^2}$, $\mu_1 = \frac{-1}{m A_0^2}$, $\mu_2 = \frac{\mu_1 A_0^2}{B_0^2}$, $0 \le m \le 1$, $A_0, A_1, B_0, \phi_1,$ and ϕ_2 are arbitrary real constants, n_0 is an arbitrary real integer	Discrete solitary wave	(8.78)
3.	$\psi_1(n,t) = A_0\,\text{sech}[A_1(n+n_0)]\,e^{-i(\omega_1 t + \phi_1)}$, $\psi_2(n,t) = B_0\tanh[A_1(n+n_0)]\,e^{-i(\omega_2 t + \phi_2)}$	$\omega_1 = \frac{\nu_1}{\mu_1}$, $\omega_2 = \frac{\nu_1\mu_2}{\mu_1^2}$, $\mu_1 = \frac{-1}{A_0^2}$, $\mu_2 = \frac{\mu_1 A_0^2}{B_0^2}$, A_0, $A_1, B_0, \phi_1,$ and ϕ_2 are arbitrary real constants, n_0 is an arbitrary real integer	Discrete bright-dark soliton	(8.79)
4.	$\psi_1(n,t) = \{A_0\,\text{dn}^2[A_1(n+n_0), m] + A_2\}\,e^{-i(\omega_1 t + \phi_1)}$, $\psi_2(n,t) = B_0\sqrt{m}\,\text{sn}[A_1(n+n_0), m]\,\text{dn}[A_1(n+n_0), m]\,e^{-i(\omega_2 t + \phi_2)}$	$A_0 = -2 A_2$, $\omega_1 = \frac{\nu_1}{\mu_1}$, $\omega_2 = \frac{\nu_1\mu_2}{\mu_1^2}$, $\mu_1 = \frac{-4}{A_0^2}$, $\mu_2 = \frac{\mu_1 A_0^2}{B_0^2}$, $0 \le m \le 1$, $A_1, A_2, B_0, \phi_1,$ and ϕ_2 are arbitrary real constants, n_0 is an arbitrary real integer	Discrete solitary wave	(8.80)

	Solution	Parameters		Eq.
5.	$\psi_1(n,t) = \{A_0 \, dn^2[A_1(n+n_0), m] + A_2\} e^{-i(\omega_1 t+\phi_1)}$, $\psi_2(n,t) = B_0 \, m \, sn[A_1(n+n_0), m] \, cn[A_1(n+n_0), m] e^{-i(\omega_2 t+\phi_2)}$	$A_0 = \frac{-2A_2}{2-m}$, $\omega_1 = \frac{\eta}{\mu_1}$, $\omega_2 = \frac{\eta \mu_2}{\mu_1^2}$, $\mu_1 = \frac{-4}{m^2 A_0^2}$, $\mu_2 = \frac{\mu_1 A_0^2}{B_0^2}$, $0 \le m \le 1$, A_1, A_2, B_0, ϕ_1, and ϕ_2 are arbitrary real constants, n_0 is an arbitrary real integer	Discrete solitary wave	(8.81)
6.	$\psi_1(n,t) = \{A_0 \, sech^2[A_1(n+n_0)] + A_2\} e^{-i(\omega_1 t+\phi_1)}$, $\psi_2(n,t) = B_0 \, sech[A_1(n+n_0)] \, tanh[A_1(n+n_0)] e^{-i(\omega_2 t+\phi_2)}$	$A_0 = -2A_2$, $\omega_1 = \frac{\eta}{\mu_1}$, $\omega_2 = \frac{\eta \mu_2}{\mu_1^2}$, $\mu_1 = \frac{-4}{A_0^2}$, $\mu_2 = \frac{\mu_1 A_0^2}{B_0^2}$, A_1, A_2, B_0, ϕ_1, and ϕ_2 are arbitrary real constants, n_0 is an arbitrary real integer	—	(8.82)
7.	$\psi_1(n,t) = \frac{A_0}{\sqrt{1+n^2}} e^{-i(\omega_1 t+\phi_1)}$, $\psi_2(n,t) = \frac{B_0 n}{\sqrt{1+n^2}} e^{-i(\omega_2 t+\phi_2)}$	$\omega_1 = \frac{\eta}{\mu_1}$, $\omega_2 = \frac{\eta \mu_2}{\mu_1^2}$, $\mu_1 = \frac{-1}{A_0^2}$, $\mu_2 = \frac{\mu_1 A_0^2}{B_0^2}$, A_0, B_0, ϕ_1, and ϕ_2 are arbitrary real constants	—	(8.83)
8.	$\psi_1(n,t) = A_0 \sqrt{\frac{1+n^2}{1+n^2+n^4}} \, e^{-i(\omega_1 t+\phi_1)}$, $\psi_2(n,t) = \frac{B_0 n^2}{\sqrt{1+n^2+n^4}} e^{-i(\omega_2 t+\phi_2)}$	$\omega_1 = \frac{\eta}{\mu_1}$, $\omega_2 = \frac{\eta \mu_2}{\mu_1^2}$, $\mu_1 = \frac{-1}{A_0^2}$, $\mu_2 = \frac{\mu_1 A_0^2}{B_0^2}$, A_0, B_0, ϕ_1, and ϕ_2 are arbitrary real constants	—	(8.84)
9.	$\psi_1(n,t) = A_0 \sqrt{\frac{2+n^2}{1+n^2}} \, e^{-i(\omega_1 t+\phi_1)}$, $\psi_2(n,t) = \frac{B_0}{\sqrt{1+n^2}} e^{-i(\omega_2 t+\phi_2)}$	$\omega_1 = \frac{\eta}{\mu_1}$, $\omega_2 = \frac{\eta \mu_2}{\mu_1^2}$, $\mu_1 = \frac{-1}{A_0^2}$, $\mu_2 = 1$, A_0, B_0, ϕ_1, and ϕ_2 are arbitrary real constants	—	(8.85)
10.	$\psi_1(n,t) = A_0 \cos[A_1(n+n_0)] \, e^{-i(\omega_1 t+\phi_1)}$, $\psi_2(n,t) = B_0 \sin[A_1(n+n_0)] \, e^{-i(\omega_2 t+\phi_2)}$	$\omega_1 = \frac{\eta}{\mu_1}$, $\omega_2 = \frac{\eta \mu_2}{\mu_1^2}$, $\mu_1 = \frac{-1}{A_0^2}$, $\mu_2 = \frac{\mu_1 A_0^2}{B_0^2}$, A_0, B_0, ϕ_1, and ϕ_2 are arbitrary real constants, n_0 is an arbitrary real integer	—	(8.86)

(Continued)

#	Solution	Parameters	Type	Eq.
11.	$\psi_1(n, t) = A_0 \, \mathrm{nd}[A_1 (n + n_0), m] \, e^{-i(\omega_1 t + \phi_1)}$, $\psi_2(n, t) = B_0 \sqrt{m} \, \mathrm{sd}[A_1 (n + n_0), m] \, e^{-i(\omega_2 t - \phi_2)}$	$A_0 = \frac{\mu_2 B_0^2}{|\mu_1|}$, $\omega_1 = \frac{\eta}{\mu_1}$, $\omega_2 = \frac{\eta \mu_2}{\mu_1^2}$, $\mu_1 = -1$, $\mu_2 = 1$, $0 \leqslant m \leqslant 1$, $A_1, A_2, B_0,$ ϕ_1, and ϕ_2 are arbitrary real constants, n_0 is an arbitrary real integer	Discrete solitary wave	(8.87)
12.	$\psi_1(n, t) = A_0 \cosh[A_1 (n + n_0)] \, e^{-i(\omega_1 t + \phi_1)}$, $\psi_2(n, t) = B_0 \sinh[A_1 (n + n_0)] \, e^{-i(\omega_2 t + \phi_2)}$	$A_0 = \frac{\mu_2 B_0^2}{|\mu_1|}$, $\omega_1 = \frac{\eta}{\mu_1}$, $\omega_2 = \frac{\eta \mu_2}{\mu_1^2}$, $\mu_1 = -1$, $\mu_2 = 1$, $A_1, B_0,$ $\phi_1,$ and ϕ_2 are arbitrary real constants, n_0 is an arbitrary real integer	—	(8.88)
13.	$\psi_1(n, t) = \{A_0 \, \mathrm{nd}^2[A_1 (n + n_0), m] + A_2\} \, e^{-i(\omega_1 t + \phi_1)}$, $\psi_2(n, t) = \frac{B_0 \sqrt{m} \, \mathrm{sn}[A_1 (n+n_0), m]}{\mathrm{dn}^2[A_1 (n+n_0), m]} \, e^{-i(\omega_2 t + \phi_2)}$	$A_0 = -2 A_2$, $B_0 = \sqrt{\frac{|\mu_1| A_0^2}{\mu_2}}$, $\omega_1 = \frac{\eta}{\mu_1}$, $\omega_2 = \frac{\eta \mu_2}{\mu_1^2}$, $\mu_1 = \frac{-4}{A_0^2}$, $\mu_2 > 0$, $0 \leqslant m \leqslant 1$, $A_1, A_2,$ ϕ_1, and ϕ_2 are arbitrary real constants, n_0 is an arbitrary real integer	Discrete solitary wave	(8.89)
14.	$\psi_1(n, t) = \{A_0 \cosh^2[A_1 (n + n_0)] + A_2\} \, e^{-i(\omega_1 t + \phi_1)}$, $\psi_2(n, t) = B_0 \sinh[A_1 (n + n_0)] \cosh[A_1 (n + n_0)] \, e^{-i(\omega_2 t + \phi_2)}$	$A_0 = -2 A_2$, $B_0 = \sqrt{\frac{|\mu_1| A_0^2}{\mu_2}}$, $\omega_1 = \frac{\eta}{\mu_1}$, $\omega_2 = \frac{\eta \mu_2}{\mu_1^2}$, $\mu_1 = \frac{-4}{A_0^2}$, $\mu_2 > 0$, $A_1, A_2,$ ϕ_1, and ϕ_2 are arbitrary real constants, n_0 is an arbitrary real integer	—	(8.90)

8.13 Coupled Ablowitz–Ladik Equation

Equation

$$i\,\psi_{1nt} + \psi_{1n+1} + \psi_{1n-1} - 2\,\psi_{1n} + (\mu_1\,|\psi_{1n}|^2 + \mu_2\,|\psi_{2n}|^2)\,(\psi_{1n+1} + \psi_{1n-1}) = 0,$$

$$i\,\psi_{2nt} + \psi_{2n+1} + \psi_{2n-1} - \frac{2\,\mu_2}{\mu_1}\,\psi_{2n} + (\mu_1\,|\psi_{1n}|^2 + \mu_2\,|\psi_{2n}|^2)\,(\psi_{2n+1} + \psi_{2n-1}) = 0,$$

(8.91)

where μ_1 and μ_2 are real constants, $\psi_j = \psi_j(n, t)$ is the complex function profile, $j = 1, 2$, the integer site index, n, and t are its two independent variables.

Solutions:

***Solution* 1.** *Moving discrete SW*
(Figure 8.35)

$$\psi_1(n, t) = A_0\,\mathrm{dn}[A_1\,(n - v\,t + n_0), m]\,e^{-i\,(\omega_1 t - \kappa_1 n + \phi_1)},$$

$$\psi_2(n, t) = B_0\,\sqrt{m}\,\mathrm{sn}[A_1\,(n - v\,t + n_0), m]\,e^{-i\,(\omega_2 t - \kappa_2 n + \phi_2)},$$

(8.92)

where $A_0 = \sqrt{\dfrac{1 + \mu_2\,B_0^2\,\mathrm{ns}^2(A_1, m)}{\mu_1\,\mathrm{cs}^2(A_1, m)}}$, $\kappa_2 = \sin^{-1}[\mathrm{cn}(A_1, m)\sin(\kappa_1)]$, $v = \dfrac{2\sin(\kappa_1)\,(1 + \mu_2\,B_0^2)}{A_1\,\mathrm{cs}(A_1, m)}$,

$\omega_1 = 2 - \dfrac{2\,(1 + \mu_2\,B_0^2)\cos(\kappa_1)\,\mathrm{dn}(A_1, m)}{\mathrm{cn}^2(A_1, m)}$, $\omega_2 = \dfrac{2\,\mu_2}{\mu_1} - \dfrac{2\,(1 + \mu_2\,B_0^2)\cos(\kappa_2)\,\mathrm{dn}(A_1, m)}{\mathrm{cn}(A_1, m)}$,

- *Reference*: [13].

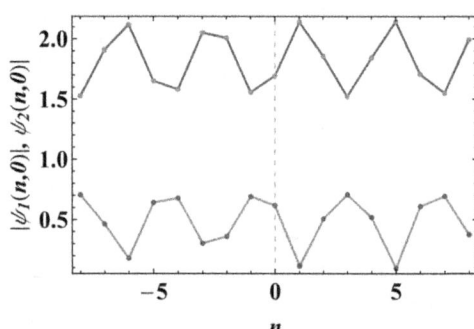

Figure 8.35. Moving discrete solitary wave (8.84) at $t = 0$. Blue is ψ_1 and red is ψ_2 with $B_0 = A_1 = 1$, $\mu_1 = -1$, $\mu_2 = \kappa_1 = 1$, $m = 1/2$, $n_0 = -1/3$, and $\phi_1 = \phi_2 = 0$. Lines are guide to the eye. Animation available online at https://doi.org/10.1088/978-0-7503-5954-2.

***Solution* 2.** *Moving discrete SW*

$$\psi_1(n, t) = A_0\,\sqrt{m}\,\mathrm{cn}[A_1\,(n - v\,t + n_0), m]\,e^{-i\,(\omega_1 t - \kappa_1 n + \phi_1)},$$

$$\psi_2(n, t) = B_0\,\sqrt{m}\,\mathrm{sn}[A_1\,(n - v\,t + n_0), m]\,e^{-i\,(\omega_2 t - \kappa_2 n + \phi_2)},$$

(8.93)

where $\qquad A_0 = \sqrt{\dfrac{1 + \mu_2\, B_0^2\, \text{ns}^2(A_1, m)}{\mu_1\, \text{ds}^2(A_1, m)}}$, $\qquad \kappa_2 = \sin^{-1}[\text{dn}(A_1, m)\sin(\kappa_1)]$,

$$v = \frac{2\sin(\kappa_1)\,(1 + m\,\mu_2\, B_0^2)}{A_1\, \text{ds}(A_1, m)},$$

$$\omega_1 = 2 - \frac{2\,(1 + m\,\mu_2\, B_0^2)\cos(\kappa_1)\,\text{cn}(A_1, m)}{\text{dn}^2(A_1, m)}, \quad \omega_2 = \frac{2\,\mu_2}{\mu_1} - \frac{2\,(1 + m\,\mu_2\, B_0^2)\cos(\kappa_2)\,\text{cn}(A_1, m)}{\text{dn}(A_1, m)},$$

$\dfrac{1 + \mu_2\, B_0^2\, \text{ns}^2(A_1, m)}{\mu_1\, \text{ds}^2(A_1, m)} > 0$, $0 \leqslant m \leqslant 1$, A_1, B_0, κ_1, ϕ_1, ϕ_2, μ_1, μ_2, and n_0 are arbitrary real constants.

- *Reference*: [13].

Solution 3. *Moving discrete bright-dark soliton*
(Figure 8.36)

$$\psi_1(n, t) = A_0 \,\text{sech}[A_1\,(n - v\,t + n_0)]\, e^{-i\,(\omega_1 t - \kappa_1 n + \phi_1)},$$
$$\psi_2(n, t) = B_0 \,\text{tanh}[A_1\,(n - v\,t + n_0)]\, e^{-i\,(\omega_2 t - \kappa_2 n + \phi_2)},$$
$$(8.94)$$

where $\quad A_0 = \sqrt{\dfrac{\sinh^2(A_1) + \mu_2\, B_0^2\, \cosh^2(A_1)}{\mu_1}}$, $\quad \kappa_2 = \sin^{-1}[\text{sech}(A_1)\sin(\kappa_1)]$,

$$v = \frac{2\sin(\kappa_1)\sinh(A_1)\,(1 + \mu_2\, B_0^2)}{A_1}, \quad \omega_1 = 2 - 2\,(1 + \mu_2\, B_0^2)\cos(\kappa_1)\cosh(A_1),$$

$\omega_2 = \dfrac{2\,\mu_2}{\mu_1} - 2\,(1 + \mu_2\, B_0^2)\cos(\kappa_2)$, $\dfrac{\sinh^2(A_1) + \mu_2\, B_0^2\, \cosh^2(A_1)}{\mu_1} > 0$, A_1, B_0, κ_1, ϕ_1, ϕ_2, μ_1, μ_2, and n_0 are arbitrary real constants.

- *Reference*: [13].

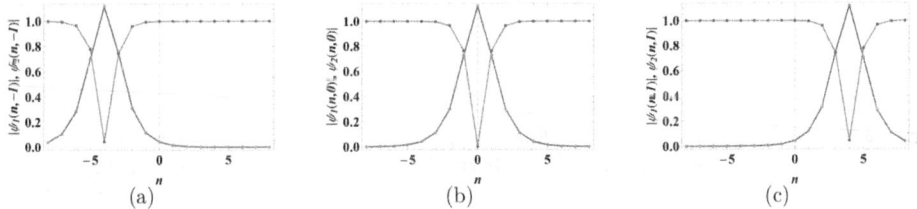

Figure 8.36. Moving discrete bright-dark soliton (8.94). Blue is ψ_1 and red is ψ_2 with $B_0 = A_1 = \mu_2 = \kappa_1 = 1$, $\mu_1 = 3$, and $n_0 = \phi_1 = \phi_2 = 0$. (a) at $t = -1$, (b) at $t = 0$, and (c) at $t = 1$. The lines are guides for the eye. Animation available online at http://doi.org/10.1088/978-0-7503-5954-2.

Solution 4. *Moving discrete SW*

$$\psi_1(n, t) = A_0 \,\text{dn}[A_1\,(n - v\,t + n_0), m]\, e^{-i\,(\omega_1 t - \kappa_1 n + \phi_1)},$$
$$\psi_2(n, t) = B_0 \,\sqrt{m}\, \text{cn}[A_1\,(n - v\,t + n_0), m]\, e^{-i\,(\omega_2 t - \kappa_2 n + \phi_2)},$$
$$(8.95)$$

where $A_0 = \sqrt{\dfrac{1 - \mu_2\, B_0^2\, \mathrm{ds}^2(A_1, m)}{\mu_1\, \mathrm{cs}^2(A_1, m)}}$, $\kappa_2 = \sin^{-1}[\dfrac{\sin(\kappa_1)\, \mathrm{cn}(A_1, m)}{\mathrm{dn}(A_1, m)}]$, $v = \dfrac{2\sin(\kappa_1)\,[1 - (1 - m)\,\mu_2\, B_0^2]}{A_1\, \mathrm{cs}(A_1, m)}$,

$\omega_1 = 2 - 2\,[1 - (1 - m)\,\mu_2\, B_0^2][\dfrac{\cos(\kappa_1)\, \mathrm{dn}(A_1, m)}{\mathrm{cn}^2(A_1, m)}]$,

$\omega_2 = \dfrac{2\mu_2}{\mu_1} - 2\,[1 - (1 - m)\,\mu_2\, B_0^2][\dfrac{\cos(\kappa_2)}{\mathrm{cn}(A_1, m)}]$, $\dfrac{1 - \mu_2\, B_0^2\, \mathrm{ds}^2(A_1, m)}{\mu_1\, \mathrm{cs}^2(A_1, m)} > 0$, $0 < m \leqslant 1$, A_1,

B_0, κ_1, ϕ_1, ϕ_2, μ_1, μ_2, and n_0 are arbitrary real constants.

• *Reference*: [13].

Note: Solutions (5–18) below can be obtained from solutions (1–14) in section 8.11 with the replacements: $\omega_1 = 2$ and $\omega_2 = \dfrac{2\mu_2}{\mu_1}$.

Solution 5. *Discrete SW*
  (Figure 8.37)

$$\psi_1(n, t) = A_0\, \mathrm{dn}[A_1\,(n + n_0), m]\, e^{-i\,(\omega_1 t + \phi_1)},$$
$$\psi_2(n, t) = B_0\, \sqrt{m}\, \mathrm{sn}[A_1\,(n + n_0), m]\, e^{-i\,(\omega_2 t + \phi_2)}, \tag{8.96}$$

where $\omega_1 = 2$, $\omega_2 = \dfrac{2\mu_2}{\mu_1}$, $\mu_1 = \dfrac{-1}{A_0^2}$, $\mu_2 = \dfrac{\mu_1\, A_0^2}{B_0^2}$, $0 \leqslant m \leqslant 1$, A_0, A_1, B_0, ϕ_1, and ϕ_2 are arbitrary real constants, n_0 is an arbitrary real integer.

• *Reference*: [13].

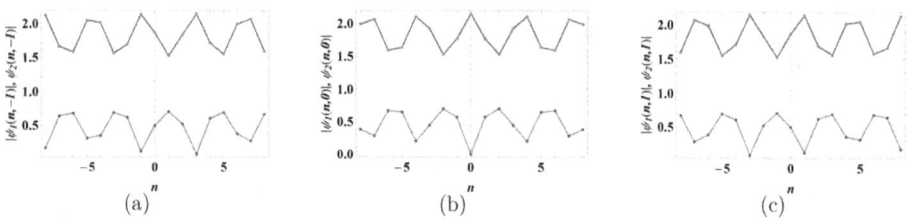

Figure 8.37. Discrete solitary wave (8.96). Blue is ψ_1 and red is ψ_2 with $B_0 = A_1 = \mu_1 = \mu_2 = \kappa_1 = 1$, $m = 1/2$, and $n_0 = \phi_1 = \phi_2 = 0$. (a) at $t = -1$, (b) at $t = 0$, and (c) at $t = 1$. The lines are guides for the eye.

Solution 6. *Discrete SW*
  (Figure 8.38)

$$\psi_1(n, t) = A_0\, \sqrt{m}\, \mathrm{cn}[A_1\,(n + n_0), m]\, e^{-i\,(\omega_1 t + \phi_1)},$$
$$\psi_2(n, t) = B_0\, \sqrt{m}\, \mathrm{sn}[A_1\,(n + n_0), m]\, e^{-i\,(\omega_2 t + \phi_2)}, \tag{8.97}$$

where $\omega_1 = 2$, $\omega_2 = \dfrac{2\mu_2}{\mu_1}$, $\mu_1 = \dfrac{-1}{m\, A_0^2}$, $\mu_2 = \dfrac{\mu_1\, A_0^2}{B_0^2}$, $0 \leqslant m \leqslant 1$, A_0, A_1, B_0, n_0, ϕ_1, and ϕ_2 are arbitrary real constants.

• *Reference*: [13].

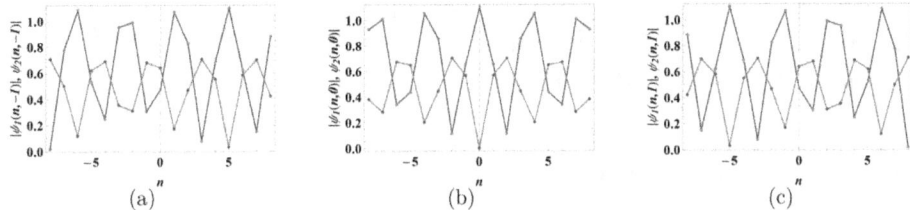

Figure 8.38. Discrete solitary wave (8.97). Blue is ψ_1 and red is ψ_2 with $B_0 = A_1 = \mu_1 = \mu_2 = \kappa_1 = 1$, $m = 1/2$, and $n_0 = \phi_1 = \phi_2 = 0$. (a) at $t = -1$, (b) at $t = 0$, and (c) at $t = 1$. The lines are guides for the eye.

Solution 7. *Discrete bright–dark soliton*

$$\psi_1(n,\,t) = A_0 \operatorname{sech}[A_1\,(n + n_0)]\,e^{-i\,(\omega_1 t + \phi_1)},$$
$$\psi_2(n,\,t) = B_0 \tanh[A_1\,(n + n_0)]\,e^{-i\,(\omega_2 t + \phi_2)},$$

(8.98)

where $\omega_1 = 2$, $\omega_2 = \dfrac{2\,\mu_2}{\mu_1}$, $\mu_1 = \dfrac{-1}{A_0^2}$, $\mu_2 = \dfrac{\mu_1\,A_0^2}{B_0^2}$, A_0, A_1, B_0, n_0, ϕ_1, and ϕ_2 are arbitrary real constants.

• *Reference*: [13].

Solution 8. *Discrete SW*

$$\psi_1(n,\,t) = \left\{ A_0 \operatorname{dn}^2\!\Big[A_1\,(n + n_0),\, m\Big] + A_2 \right\} e^{-i\,(\omega_1 t + \phi_1)},$$
$$\psi_2(n,\,t) = B_0\,\sqrt{m}\,\operatorname{sn}[A_1\,(n + n_0),\, m]\,\operatorname{dn}[A_1\,(n + n_0),\, m]\,e^{-i\,(\omega_2 t + \phi_2)},$$

(8.99)

where $A_0 = -2\,A_2$, $\omega_1 = 2$, $\omega_2 = \dfrac{2\,\mu_2}{\mu_1}$, $\mu_1 = \dfrac{-4}{A_0^2}$, $\mu_2 = \dfrac{\mu_1\,A_0^2}{B_0^2}$, $0 \leqslant m \leqslant 1$, A_1, A_2, B_0, n_0, ϕ_1, and ϕ_2 are arbitrary real constants.

• *Reference*: [13].

Solution 9. *Discrete SW*

$$\psi_1(n,\,t) = \left\{ A_0 \operatorname{dn}^2\!\Big[A_1\,(n + n_0),\, m\Big] + A_2 \right\} e^{-i\,(\omega_1 t + \phi_1)},$$
$$\psi_2(n,\,t) = B_0\,m\,\operatorname{sn}[A_1\,(n + n_0),\, m]\,\operatorname{cn}[A_1\,(n + n_0),\, m]\,e^{-i\,(\omega_2 t + \phi_2)},$$

(8.100)

where $A_0 = \dfrac{-2\,A_2}{2 - m}$, $\omega_1 = 2$, $\omega_2 = \dfrac{2\,\mu_2}{\mu_1}$, $\mu_1 = \dfrac{-4}{m^2\,A_0^2}$, $\mu_2 = \dfrac{\mu_1\,A_0^2}{B_0^2}$, $0 \leqslant m \leqslant 1$, A_1, A_2, B_0, n_0, ϕ_1, and ϕ_2 are arbitrary real constants.

• *Reference*: [13].

Solution 10.

$$\psi_1(n,\,t) = \left\{ A_0 \operatorname{sech}^2\!\Big[A_1\,(n + n_0)\Big] + A_2 \right\} e^{-i\,(\omega_1 t + \phi_1)},$$
$$\psi_2(n,\,t) = B_0 \operatorname{sech}[A_1\,(n + n_0)]\,\tanh[A_1\,(n + n_0)]\,e^{-i\,(\omega_2 t + \phi_2)},$$

(8.101)

where $A_0 = -2\,A_2$, $\omega_1 = 2$, $\omega_2 = \frac{2\,\mu_2}{\mu_1}$, $\mu_1 = \frac{-4}{A_0^2}$, $\mu_2 = \frac{\mu_1\,A_0^2}{B_0^2}$, A_1, A_2, B_0, n_0, ϕ_1, and ϕ_2 are arbitrary real constants.

- *Reference*: [13].

Solution 11. Rational Solution I

$$\psi_1(n,\,t) = \frac{A_0}{\sqrt{1 + n^2}}\, e^{-i\,(\omega_1 t + \phi_1)},$$

$$\psi_2(n,\,t) = \frac{B_0\,n}{\sqrt{1 + n^2}}\, e^{-i\,(\omega_2 t + \phi_2)},$$

(8.102)

where $\omega_1 = 2$, $\omega_2 = \frac{2\,\mu_2}{\mu_1}$, $\mu_1 = \frac{-1}{A_0^2}$, $\mu_2 = \frac{\mu_1\,A_0^2}{B_0^2}$, A_0, B_0, ϕ_1, and ϕ_2 are arbitrary real constants.

- *Reference*: [13].

Solution 12. Rational Solution II

$$\psi_1(n,\,t) = A_0\,\sqrt{\frac{1 + n^2}{1 + n^2 + n^4}}\, e^{-i\,(\omega_1 t + \phi_1)},$$

$$\psi_2(n,\,t) = \frac{B_0\,n^2}{\sqrt{1 + n^2 + n^4}}\, e^{-i\,(\omega_2 t + \phi_2)},$$

(8.103)

where $\omega_1 = 2$, $\omega_2 = \frac{2\,\mu_2}{\mu_1}$, $\mu_1 = \frac{-1}{A_0^2}$, $\mu_2 = \frac{\mu_1\,A_0^2}{B_0^2}$, A_0, B_0, ϕ_1, and ϕ_2 are arbitrary real constants.

- *Reference*: [13].

Solution 13. Rational Solution III

$$\psi_1(n,\,t) = A_0\,\sqrt{\frac{2 + n^2}{1 + n^2}}\, e^{-i\,(\omega_1 t + \phi_1)},$$

$$\psi_2(n,\,t) = \frac{B_0}{\sqrt{1 + n^2}}\, e^{-i\,(\omega_2 t + \phi_2)},$$

(8.104)

where $\omega_1 = 2$, $\omega_2 = \frac{2\,\mu_2}{\mu_1}$, $\mu_1 = \frac{-1}{A_0^2}$, $\mu_2 = 1$, A_0, B_0, ϕ_1, and ϕ_2 are arbitrary real constants.

- *Reference*: [13].

Solution 14.

$$\psi_1(n,\,t) = A_0\,\cos[A_1\,(n + n_0)]\, e^{-i\,(\omega_1 t + \phi_1)},$$

$$\psi_2(n,\,t) = B_0\,\sin[A_1\,(n + n_0)]\, e^{-i\,(\omega_2 t + \phi_2)},$$

(8.105)

where $\omega_1 = 2$, $\omega_2 = \frac{2\mu_2}{\mu_1}$, $\mu_1 = \frac{-1}{A_0^2}$, $\mu_2 = \frac{\mu_1 A_0^2}{B_0^2}$, A_0, B_0, n_0, ϕ_1, and ϕ_2 are arbitrary real constants.

- *Reference*: [13].

Solution 15. *Discrete SW*

$$\psi_1(n,\ t) = A_0 \ \mathrm{nd}[A_1\ (n + n_0),\ m]\ e^{-i\ (\omega_1 t + \phi_1)},$$
$$\psi_2(n,\ t) = B_0\ \sqrt{m}\ \mathrm{sd}[A_1\ (n + n_0),\ m]\ e^{-i\ (\omega_2 t + \phi_2)},$$

(8.106)

where $A_0 = \frac{\mu_2\ B_0^2}{|\mu_1|}$, $\omega_1 = 2$, $\omega_2 = \frac{2\mu_2}{\mu_1}$, $\mu_1 = -1$, $\mu_2 = 1$, $0 \leqslant m \leqslant 1$, A_1, A_2, B_0, n_0, ϕ_1, and ϕ_2 are arbitrary real constants.

- *Reference*: [13].

Solution 16.

$$\psi_1(n,\ t) = A_0 \cosh[A_1\ (n + n_0)]\ e^{-i\ (\omega_1 t + \phi_1)},$$
$$\psi_2(n,\ t) = B_0 \sinh[A_1\ (n + n_0)]\ e^{-i\ (\omega_2 t + \phi_2)},$$

(8.107)

where $A_0 = \frac{\mu_2\ B_0^2}{|\mu_1|}$, $\omega_1 = 2$, $\omega_2 = \frac{2\mu_2}{\mu_1}$, $\mu_1 = -1$, $\mu_2 = 1$, A_1, B_0, n_0, ϕ_1, and ϕ_2 are arbitrary real constants.

- *Reference*: [13].

Solution 17. *Discrete SW*

$$\psi_1(n,\ t) = \left\{ A_0\ \mathrm{nd}^2\Big[A_1\ (n + n_0),\ m \Big] + A_2 \right\} e^{-i\ (\omega_1 t + \phi_1)},$$
$$\psi_2(n,\ t) = \frac{B_0\ \sqrt{m}\ \mathrm{sn}[A_1\ (n + n_0),\ m]}{\mathrm{dn}^2[A_1\ (n + n_0),\ m]}\ e^{-i\ (\omega_2 t + \phi_2)},$$

(8.108)

where $A_0 = -2\ A_2$, $B_0 = \sqrt{\frac{|\mu_1|\ A_0^2}{\mu_2}}$, $\omega_1 = 2$, $\omega_2 = \frac{2\mu_2}{\mu_1}$, $\mu_1 = \frac{-4}{A_0^2}$, $\mu_2 > 0$, $0 \leqslant m \leqslant 1$, A_1, A_2, n_0, ϕ_1, and ϕ_2 are arbitrary real constants.

- *Reference*: [13].

Solution 18.

$$\psi_1(n,\ t) = \left\{ A_0 \cosh^2\Big[A_1\ (n + n_0) \Big] + A_2 \right\} e^{-i\ (\omega_1 t + \phi_1)},$$
$$\psi_2(n,\ t) = B_0 \sinh[A_1\ (n + n_0)] \cosh[A_1\ (n + n_0)]\ e^{-i\ (\omega_2 t + \phi_2)},$$

(8.109)

where $A_0 = -2\ A_2$, $B_0 = \sqrt{\frac{|\mu_1|\ A_0^2}{\mu_2}}$, $\omega_1 = 2$, $\omega_2 = \frac{2\mu_2}{\mu_1}$, $\mu_1 = \frac{-4}{A_0^2}$, $\mu_2 > 0$, A_1, A_2, n_0, ϕ_1, and ϕ_2 are arbitrary real constants.

- *Reference*: [13].

8.14 Summary of Section 8.13

Equation

$$i\psi_{1nt} + \psi_{1n+1} + \psi_{1n-1} - 2\psi_{1n} + (\mu_1|\psi_{1n}|^2 + \mu_2|\psi_{2n}|^2)(\psi_{1n+1} + \psi_{1n-1}) = 0,$$

$$i\psi_{2nt} + \psi_{2n+1} + \psi_{2n-1} - \frac{2\mu_2}{\mu_1}\psi_{2n} + (\mu_1|\psi_{1n}|^2 + \mu_2|\psi_{2n}|^2)(\psi_{2n+1} + \psi_{2n-1}) = 0$$

#	Solution	Conditions	Name	Equation #
1.	$\psi_1(n, t) = A_0\, dn[A_1(n - v t + n_0), m]\, e^{-i(\omega_1 t - \kappa_1 n + \phi_1)}$, $\psi_2(n, t) = B_0\sqrt{m}\, sn[A_1(n - v t + n_0), m]\, e^{-i(\omega_2 t - \kappa_2 n + \phi_2)}$	$A_0 = \sqrt{\dfrac{1 + \mu_2 B_0^2\, ns^2(A_1, m)}{\mu_1\, cs^2(A_1, m)}}$, $v = \dfrac{2\sin(\kappa_1)(1 + \mu_2 B_0^2)}{A_1\, cs(A_1, m)}$, $\kappa_2 = \sin^{-1}[cn(A_1, m)\sin(\kappa_1)]$, $\dfrac{1 + \mu_2 B_0^2\, ns^2(A_1, m)}{\mu_1\, cs^2(A_1, m)} > 0$, $\omega_1 = 2 - \dfrac{2(1 + \mu_2 B_0^2)\cos(\kappa_1)\, dn(A_1, m)}{cn^2(A_1, m)}$, $\omega_2 = \dfrac{2\mu_2}{\mu_1} - \dfrac{2(1 + \mu_2 B_0^2)\cos(\kappa_2)\, dn(A_1, m)}{cn(A_1, m)}$, $0 \leq m \leq 1$, A_1, $B_0, \kappa_1, \phi_1, \phi_2, \mu_1, \mu_2$, and n_0 are arbitrary real constants	Moving discrete solitary wave	(8.92)
2.	$\psi_1(n, t) = A_0\sqrt{m}\, cn[A_1(n - v t + n_0), m]\, e^{-i(\omega_1 t - \kappa_1 n + \phi_1)}$, $\psi_2(n, t) = B_0\sqrt{m}\, sn[A_1(n - v t + n_0), m]\, e^{-i(\omega_2 t - \kappa_2 n + \phi_2)}$	$A_0 = \sqrt{\dfrac{1 + \mu_2 B_0^2\, ns^2(A_1, m)}{\mu_1\, ds^2(A_1, m)}}$, $v = \dfrac{2\sin(\kappa_1)(1 + m\,\mu_2 B_0^2)}{A_1\, ds(A_1, m)}$, $\kappa_2 = \sin^{-1}[dn(A_1, m)\sin(\kappa_1)]$, $\dfrac{1 + \mu_2 B_0^2\, ns^2(A_1, m)}{\mu_1\, ds^2(A_1, m)} > 0$, $\omega_1 = 2 - \dfrac{2(1 + m\,\mu_2 B_0^2)\cos(\kappa_1)\, cn(A_1, m)}{dn^2(A_1, m)}$, $\omega_2 = \dfrac{2\mu_2}{\mu_1} - \dfrac{2(1 + m\,\mu_2 B_0^2)\cos(\kappa_2)\, cn(A_1, m)}{dn(A_1, m)}$, $0 \leq m \leq 1$, A_1, $B_0, \kappa_1, \phi_1, \phi_2, \mu_1, \mu_2$, and n_0 are arbitrary real constants	Moving discrete solitary wave	(8.93)

(Continued)

3. $\psi_1(n, t) = A_0 \operatorname{sech}[A_1(n - v\, t + n_0)]\, e^{-i(\omega_1 t - \kappa_1 n + \phi_1)}$,

$\psi_2(n, t) = B_0 \tanh[A_1(n - v\, t + n_0)]\, e^{-i(\omega_2 t - \kappa_2 n + \phi_2)}$

$A_0 = \sqrt{\dfrac{\sinh^2(A_1) + \mu_2 B_0^2 \cosh^2(A_1)}{\mu_1}}$,

$\kappa_2 = \sin^{-1}[\operatorname{sech}(A_1)\sin(\kappa_1)]$,

$v = \dfrac{2\sin(\kappa_1)\sinh(A_1)(1 + \mu_2 B_0^2)}{A_1}$,

$\omega_1 = 2 - 2(1 + \mu_2 B_0^2)\cos(\kappa_1)\cosh(A_1)$.

$\omega_2 = \dfrac{2\mu_2}{\mu_1} - 2(1 + \mu_2 B_0^2)\cos(\kappa_2)$,

$\dfrac{\sinh^2(A_1) + \mu_2 B_0^2 \cosh^2(A_1)}{\mu_1} > 0$, $A_1, B_0, \kappa_1, \phi_1, \phi_2, \mu_1, \mu_2$, and n_0 are arbitrary real constants

Moving discrete bright-dark soliton (8.94)

4. $\psi_1(n, t) = A_0 \operatorname{dn}[A_1(n - v\, t + n_0), m]\, e^{-i(\omega_1 t - \kappa_1 n + \phi_1)}$,

$\psi_2(n, t) = B_0\sqrt{m}\,\operatorname{cn}[A_1(n - v\, t + n_0), m]\, e^{-i(\omega_2 t - \kappa_2 n + \phi_2)}$

$A_0 = \sqrt{\dfrac{1 - \mu_2 B_0^2 \operatorname{ds}^2(A_1, m)}{\mu_1 \operatorname{cs}^2(A_1, m)}}$, $\kappa_2 = \sin^{-1}[\dfrac{\sin(\kappa_1)\operatorname{cn}(A_1, m)}{\operatorname{dn}(A_1, m)}]$,

$\omega_1 = 2 - 2[1 - (1 - m)\mu_2 B_0^2][\dfrac{\cos(\kappa_1)\operatorname{dn}(A_1, m)}{\operatorname{cn}^2(A_1, m)}]$,

$\omega_2 = \dfrac{2\mu_2}{\mu_1} - 2[1 - (1 - m)\mu_2 B_0^2][\dfrac{\cos(\kappa_2)}{\operatorname{cn}(A_1, m)}]$,

$0 < m \le 1$, $v = \dfrac{2\sin(\kappa_1)[1 - (1 - m)\mu_2 B_0^2]}{A_1 \operatorname{cs}(A_1, m)}$,

$\dfrac{1 - \mu_2 B_0^2 \operatorname{ds}^2(A_1, m)}{\mu_1 \operatorname{cs}^2(A_1, m)} > 0$, $A_1, B_0, \kappa_1, \phi_1, \phi_2, \mu_1, \mu_2$, and n_0 are arbitrary real constants

Moving discrete solitary wave (8.95)

5. $\psi_1(n, t) = A_0 \operatorname{dn}[A_1(n + n_0), m]\, e^{-i(\omega_1 t + \phi_1)}$,

$\psi_2(n, t) = B_0\sqrt{m}\,\operatorname{sn}[A_1(n + n_0), m]\, e^{-i(\omega_2 t + \phi_2)}$

$\omega_1 = 2$, $\omega_2 = \dfrac{2\mu_2}{\mu_1}$, $\mu_1 = \dfrac{-1}{A_0^2}$, $\mu_2 = \dfrac{\mu_1 A_0^2}{B_0^2}$, $0 \le m \le 1$, A_0, A_1, B_0, ϕ_1, and ϕ_2 are arbitrary real constants, n_0 is an arbitrary real integer

Discrete solitary wave (8.96)

6. $\psi_1(n, t) = A_0\sqrt{m}\,\operatorname{cn}[A_1(n + n_0), m]\, e^{-i(\omega_1 t + \phi_1)}$,

$\psi_2(n, t) = B_0\sqrt{m}\,\operatorname{sn}[A_1(n + n_0), m]\, e^{-i(\omega_2 t + \phi_2)}$

$\omega_1 = 2$, $\omega_2 = \dfrac{2\mu_2}{\mu_1}$, $\mu_1 = \dfrac{-1}{m A_0^2}$, $\mu_2 = \dfrac{\mu_1 A_0^2}{B_0^2}$, $0 \le m \le 1$, A_0, A_1, B_0, ϕ_1, and ϕ_2 are arbitrary real constants, n_0 is an arbitrary real integer

Discrete solitary wave (8.97)

7.	$\psi_1(n, t) = A_0 \operatorname{sech}[A_1 (n + n_0)] e^{-i (\omega_1 t + \phi_1)}$, $\psi_2(n, t) = B_0 \tanh[A_1 (n + n_0)] e^{-i (\omega_2 t + \phi_2)}$	$\omega_1 = 2$, $\omega_2 = \frac{2 \mu_2}{\mu_1}$, $\mu_1 = \frac{-1}{A_0^2}$, $\mu_2 = \frac{\mu_1 A_0^2}{B_0^2}$, A_0, A_1, B_0, ϕ_1, and ϕ_2 are arbitrary real constants, n_0 is an arbitrary real integer	Discrete bright-dark soliton	(8.98)
8.	$\psi_1(n, t) = \{A_0 \operatorname{dn}^2[A_1 (n + n_0), m] + A_2\} e^{-i (\omega_1 t + \phi_1)}$, $\psi_2(n, t) = B_0 \sqrt{m} \operatorname{sn}[A_1 (n + n_0), m] \operatorname{dn}[A_1 (n + n_0), m]$ $e^{-i (\omega_2 t + \phi_2)}$	$A_0 = -2 A_2$, $\omega_1 = 2$, $\omega_2 = \frac{2 \mu_2}{\mu_1}$, $\mu_1 = \frac{-4}{A_0^2}$, $\mu_2 = \frac{\mu_1 A_0^2}{B_0^2}$, $0 \leq m \leq 1$, A_1, A_2, B_0, ϕ_1, and ϕ_2 are arbitrary real constants, n_0 is an arbitrary real integer	Discrete solitary wave	(8.99)
9.	$\psi_1(n, t) = \{A_0 \operatorname{dn}^2[A_1 (n + n_0), m] + A_2\} e^{-i (\omega_1 t + \phi_1)}$, $\psi_2(n, t) = B_0 m \operatorname{sn}[A_1 (n + n_0), m] \operatorname{cn}[A_1 (n + n_0), m]$ $e^{-i (\omega_2 t + \phi_2)}$	$A_0 = \frac{-2 A_2}{2 - m}$, $\omega_1 = 2$, $\omega_2 = \frac{2 \mu_2}{\mu_1}$, $\mu_1 = \frac{-4}{m^2 A_0^2}$, $\mu_2 = \frac{\mu_1 A_0^2}{B_0^2}$, $0 \leq m \leq 1$, A_1, A_2, B_0, ϕ_1, and ϕ_2 are arbitrary real constants, n_0 is an arbitrary real integer	Discrete solitary wave	(8.100)
10.	$\psi_1(n, t) = \{A_0 \operatorname{sech}^2[A_1 (n + n_0)] + A_2\} e^{-i (\omega_1 t + \phi_1)}$, $\psi_2(n, t) = B_0 \operatorname{sech}[A_1 (n + n_0)] \tanh[A_1 (n + n_0)] e^{-i (\omega_2 t + \phi_2)}$	$A_0 = -2 A_2$, $\omega_1 = 2$, $\omega_2 = \frac{2 \mu_2}{\mu_1}$, $\mu_1 = \frac{-4}{A_0^2}$, $\mu_2 = \frac{\mu_1 A_0^2}{B_0^2}$, A_1, A_2, B_0, ϕ_1, and ϕ_2 are arbitrary real constants, n_0 is an arbitrary real integer	—	(8.101)
11.	$\psi_1(n, t) = \frac{A_0}{\sqrt{1 + n^2}} e^{-i (\omega_1 t + \phi_1)}$, $\psi_2(n, t) = \frac{B_0 n}{\sqrt{1 + n^2}} e^{-i (\omega_2 t + \phi_2)}$	$\omega_1 = 2$, $\omega_2 = \frac{2 \mu_2}{\mu_1}$, $\mu_1 = \frac{-1}{A_0^2}$, $\mu_2 = \frac{\mu_1 A_0^2}{B_0^2}$, $A_0, B_0,$ ϕ_1, and ϕ_2 are arbitrary real constants	—	(8.102)
12.	$\psi_1(n, t) = A_0 \sqrt{\frac{1 + n^2}{1 + n^2 + n^4}} \, e^{-i (\omega_1 t + \phi_1)}$, $\psi_2(n, t) = \frac{B_0 n^2}{\sqrt{1 + n^2 + n^4}} e^{-i (\omega_2 t + \phi_2)}$	$\omega_1 = 2$, $\omega_2 = \frac{2 \mu_2}{\mu_1}$, $\mu_1 = \frac{-1}{A_0^2}$, $\mu_2 = \frac{\mu_1 A_0^2}{B_0^2}$, $A_0, B_0,$ ϕ_1, and ϕ_2 are arbitrary real constants	—	(8.103)
13.	$\psi_1(n, t) = A_0 \sqrt{\frac{2 + n^2}{1 + n^2}} \, e^{-i (\omega_1 t + \phi_1)}$, $\psi_2(n, t) = \frac{B_0}{\sqrt{1 + n^2}} e^{-i (\omega_2 t + \phi_2)}$	$\omega_1 = 2$, $\omega_2 = \frac{2 \mu_2}{\mu_1}$, $\mu_1 = \frac{-1}{A_0^2}$, $\mu_2 = 1$, A_0, B_0, ϕ_1, and ϕ_2 are arbitrary real constants	—	(8.104)

(Continued)

14. $\psi_1(n, t) = A_0 \cos[A_1 (n + n_0)] \, e^{-i \, (\omega_1 t + \phi_1)}$,

$\psi_2(n, t) = B_0 \sin[A_1 (n + n_0)] \, e^{-i \, (\omega_2 t + \phi_2)}$

$\omega_1 = 2, \ \omega_2 = \frac{2 \, \mu_2}{\mu_1}, \ \mu_1 = \frac{-1}{A_0^2}, \ \mu_2 = \frac{\mu_1 A_0^2}{B_0^2}, \ A_0, B_0,$

ϕ_1, and ϕ_2 are arbitrary real constants, n_0 is an arbitrary real integer

(8.105) —

15. $\psi_1(n, t) = A_0 \, \text{nd}[A_1 (n + n_0), m] \, e^{-i \, (\omega_1 t + \phi_1)}$,

$\psi_2(n, t) = B_0 \sqrt{m} \, \text{sd}[A_1 (n + n_0), m] \, e^{-i \, (\omega_2 t + \phi_2)}$

$A_0 = \frac{\mu_2 \, B_0^2}{|\mu_1|}, \ \omega_1 = 2, \ \omega_2 = \frac{2 \, \mu_2}{\mu_1}, \ \mu_1 = -1,$

$\mu_2 = 1, \ 0 \leqslant m \leqslant 1, \ A_1, A_2, B_0, \ \phi_1$, and

ϕ_2 are arbitrary real constants, n_0 is an arbitrary real integer

(8.106) Discrete solitary wave

16. $\psi_1(n, t) = A_0 \cosh[A_1 (n + n_0)] \, e^{-i \, (\omega_1 t + \phi_1)}$,

$\psi_2(n, t) = B_0 \sinh[A_1 (n + n_0)] \, e^{-i \, (\omega_2 t + \phi_2)}$

$A_0 = \frac{\mu_2 \, B_0^2}{|\mu_1|}, \ \omega_1 = 2, \ \omega_2 = \frac{2 \, \mu_2}{\mu_1}, \ \mu_1 = -1,$

$\mu_2 = 1, \ A_1, B_0, \ \phi_1$, and ϕ_2 are arbitrary real constants, n_0 is an arbitrary real integer

(8.107) —

17. $\psi_1(n, t) = \{A_0 \, \text{nd}^2[A_1 (n + n_0), m] + A_2\} \, e^{-i \, (\omega_1 t + \phi_1)}$,

$\psi_2(n, t) = \frac{B_0 \sqrt{m} \, \text{sn}[A_1 (n + n_0), m]}{\text{dn}^2[A_1 (n + n_0), m]} \, e^{-i \, (\omega_2 t + \phi_2)}$

$A_0 = -2 \, A_2, \ B_0 = \sqrt{\frac{|\mu_1| A_0^2}{\mu_2}}, \ \omega_1 = 2,$

$\omega_2 = \frac{2 \, \mu_2}{\mu_1}, \ \mu_1 = \frac{-4}{A_0^2}, \ \mu_2 > 0, \ 0 \leqslant m \leqslant 1, \ A_1, A_2,$

ϕ_1, and ϕ_2 are arbitrary real constants, n_0 is an arbitrary real integer

(8.108) Discrete solitary wave

18. $\psi_1(n, t) = \{A_0 \cosh^2[A_1 (n + n_0)] + A_2\} \, e^{-i \, (\omega_1 t + \phi_1)}$,

$\psi_2(n, t) = B_0 \sinh[A_1 (n + n_0)] \cosh[A_1 (n + n_0)] \, e^{-i \, (\omega_2 t + \phi_2)}$

$A_0 = -2 \, A_2, \ B_0 = \sqrt{\frac{|\mu_1| A_0^2}{\mu_2}}, \ \omega_1 = 2,$

$\omega_2 = \frac{2 \, \mu_2}{\mu_1}, \ \mu_1 = \frac{-4}{A_0^2}, \ \mu_2 > 0, \ A_1, A_2,$

ϕ_1, and ϕ_2 are arbitrary real constants, n_0 is an arbitrary real integer

(8.109) —

8.15 Coupled Saturable Discrete NLSE

Equation:

$$i\,\psi_{1nt} + \psi_{1n+1} + \psi_{1n-1} - 2\,\psi_{1n} + \frac{\nu_1\,(\mu_1\,|\psi_{1n}|^2 + \mu_2\,|\psi_{2n}|^2)\,\psi_{1n}}{\mu_1\,(1 + \mu_1\,|\psi_{1n}|^2 + \mu_2\,|\psi_{2n}|^2)} = 0,$$

$$i\,\psi_{2nt} + \psi_{2n+1} + \psi_{2n-1} - 2\,\psi_{2n} + \frac{\left[\nu_2 - \dfrac{\nu_1\,\mu_2^2}{\mu_1^2} + \nu_2\,(\mu_1\,|\psi_{1n}|^2 + \mu_2\,|\psi_{2n}|^2)\right]\psi_{2n}}{\mu_2\,(1 + \mu_1\,|\psi_{1n}|^2 + \mu_2\,|\psi_{2n}|^2)} = 0, \qquad (8.110)$$

where $\psi_j = \psi_j(n,\,t)$ is the complex function profile, $j = 1,\,2$, the integer site index, n, and t are its two independent variables, μ_1, μ_2, ν_1, and ν_2 are real constants.

Solutions:

Solution 1. *Discrete SW*

$$\psi_1(n,\,t) = A_0\,\mathrm{dn}[A_1\,(n + n_0),\,m]\,e^{-i\,(\omega_1 t + \phi_1)},$$
$$\psi_2(n,\,t) = B_0\,\sqrt{m}\,\mathrm{sn}[A_1\,(n + n_0),\,m]\,e^{-i\,(\omega_2 t + \phi_2)}, \qquad (8.111)$$

where $A_0 = \sqrt{\dfrac{\nu_1}{2\,\mu_1^2\,\mathrm{dn}(A_1,m)} - \dfrac{1}{\mu_1}}$, $B_0 = \sqrt{\dfrac{\nu_1\,\mathrm{cn}^2(A_1,m)}{2\,\mu_1\,\mu_2\,\mathrm{dn}(A_1,m)} - \dfrac{1}{\mu_2}}$, $A_1 = \mu_1$, $\omega_1 = 2 - \dfrac{\nu_1}{\mu_1}$,

$\omega_2 = 2 - \dfrac{\nu_2}{\mu_2}$, $\mu_2 = \mu_1\,\mathrm{cn}(A_1,\,m)$, $\dfrac{\nu_1}{2\,\mu_1^2\,\mathrm{dn}(A_1,m)} > \dfrac{1}{\mu_1}$, $\dfrac{\nu_1\,\mathrm{cn}^2(A_1,m)}{2\,\mu_1\,\mu_2\,\mathrm{dn}(A_1,m)} > \dfrac{1}{\mu_2}$, $0 \leqslant m \leqslant 1$,

ν_1, ν_2, μ_1, n_0, ϕ_1, and ϕ_2 are arbitrary real constants.
  • *Reference*: [13].

Solution 2. *Discrete SW*

$$\psi_1(n,\,t) = A_0\,\sqrt{m}\,\mathrm{cn}[A_1\,(n + n_0),\,m]\,e^{-i\,(\omega_1 t + \phi_1)},$$
$$\psi_2(n,\,t) = B_0\,\sqrt{m}\,\mathrm{sn}[A_1\,(n + n_0),\,m]\,e^{-i\,(\omega_2 t + \phi_2)}, \qquad (8.112)$$

where $A_0 = \sqrt{\dfrac{\nu_1}{2\,m\,\mu_1^2\,\mathrm{cn}(A_1,m)} - \dfrac{1}{m\,\mu_1}}$, $B_0 = \sqrt{\dfrac{\nu_1\,\mathrm{dn}^2(A_1,m)}{2\,m\,\mu_1\,\mu_2\,\mathrm{cn}(A_1,m)} - \dfrac{1}{m\,\mu_2}}$, $A_1 = \mu_1$,

$\omega_1 = 2 - \dfrac{\nu_1}{\mu_1}$, $\omega_2 = 2 - \dfrac{\nu_2}{\mu_2}$, $\mu_2 = \mu_1\,\mathrm{dn}(A_1,\,m)$, $\dfrac{\nu_1}{2\,m\,\mu_1^2\,\mathrm{cn}(A_1,m)} > \dfrac{1}{m\,\mu_1}$,

$\dfrac{\nu_1\,\mathrm{dn}^2(A_1,m)}{2\,m\,\mu_1\,\mu_2\,\mathrm{cn}(A_1,m)} > \dfrac{1}{m\,\mu_2}$, $0 \leqslant m \leqslant 1$, ν_1, ν_2, μ_1, n_0, ϕ_1, and ϕ_2 are arbitrary real constants.
  • *Reference*: [13].

Solution 3. *Discrete bright–dark soliton*

$$\psi_1(n,\,t) = A_0\,\mathrm{sech}[A_1\,(n + n_0)]\,e^{-i\,(\omega_1 t + \phi_1)},$$
$$\psi_2(n,\,t) = B_0\,\tanh[A_1\,(n + n_0)]\,e^{-i\,(\omega_2 t + \phi_2)}, \qquad (8.113)$$

where $A_0 = \sqrt{\dfrac{\nu_1 \cosh(A_1)}{2 \mu_1^2} - \dfrac{1}{\mu_1}}$, $B_0 = \sqrt{\dfrac{\nu_1}{2 \mu_1 \mu_2 \cosh(A_1)} - \dfrac{1}{\mu_2}}$, $A_1 = \mu_1$, $\omega_1 = 2 - \dfrac{\nu_1}{\mu_1}$,

$\omega_2 = 2 - \dfrac{\nu_2}{\mu_2}$, $\mu_2 = \mu_1 \operatorname{sech}(A_1)$, $\dfrac{\nu_1 \cosh(A_1)}{2 \mu_1^2} > \dfrac{1}{\mu_1}$, $\dfrac{\nu_1}{2 \mu_1 \mu_2 \cosh(A_1)} > \dfrac{1}{\mu_2}$, ν_1, ν_2, μ_1, n_0, ϕ_1,

and ϕ_2 are arbitrary real constants.
- *Reference*: [13].

8.16 Summary of Section 8.15

Equation

$$i\,\psi_{1nt} + \psi_{1n+1} + \psi_{1n-1} - 2\,\psi_{1n} + \frac{\eta\,(\mu_1\,|\psi_{1n}|^2 + \mu_2\,|\psi_{2n}|^2)\,\psi_{1n}}{\mu_1\,(1 + \mu_1\,|\psi_{1n}|^2 + \mu_2\,|\psi_{2n}|^2)} = 0,$$

$$i\,\psi_{2nt} + \psi_{2n+1} + \psi_{2n-1} - 2\,\psi_{2n} + \frac{\left[\nu_2 - \frac{\eta\,\mu_2^2}{\mu_1^2} + \nu_2\,(\mu_1\,|\psi_{1n}|^2 + \mu_2\,|\psi_{2n}|^2)\right]\psi_{2n}}{\mu_2\,(1 + \mu_1\,|\psi_{1n}|^2 + \mu_2\,|\psi_{2n}|^2)} = 0$$

#	Solution	Conditions	Name	Equation #
1.	$\psi_1(n,t) = A_0\,\mathrm{dn}[A_1\,(n+n_0),m]\,e^{-i\,(\omega_1 t+\phi_1)},$ $\psi_2(n,t) = B_0\,\sqrt{m}\,\mathrm{sn}[A_1\,(n+n_0),m]\,e^{-i\,(\omega_2 t+\phi_2)}$	$A_0 = \sqrt{\frac{\eta}{2\,\mu_1^2\,\mathrm{dn}(A_1,m)} - \frac{1}{\mu_1}}$, $B_0 = \sqrt{\frac{\eta\,\mathrm{cn}^2(A_1,m)}{2\,\mu_1\,\mu_2\,\mathrm{dn}(A_1,m)} - \frac{1}{\mu_2}}$, $A_1 = \mu_1$, $\omega_1 = 2 - \frac{\eta}{\mu_1}$, $\omega_2 = 2 - \frac{\nu_2}{\mu_2}$, $\mu_2 = \mu_1\,\mathrm{cn}(A_1,m)$, $\frac{\eta}{2\,\mu_1^2\,\mathrm{dn}(A_1,m)} > \frac{1}{\mu_1}$, $\frac{\eta\,\mathrm{cn}^2(A_1,m)}{2\,\mu_1\,\mu_2\,\mathrm{dn}(A_1,m)} > \frac{1}{\mu_2}$, $0 \le m \le 1$, ν_1, ν_2, μ_1, ϕ_1, and ϕ_2 are arbitrary real constants, n_0 is an arbitrary real integer	Discrete solitary wave	(8.111)
2.	$\psi_1(n,t) = A_0\,\sqrt{m}\,\mathrm{cn}[A_1\,(n+n_0),m]\,e^{-i\,(\omega_1 t+\phi_1)},$ $\psi_2(n,t) = B_0\,\sqrt{m}\,\mathrm{sn}[A_1\,(n+n_0),m]\,e^{-i\,(\omega_2 t+\phi_2)}$	$A_0 = \sqrt{\frac{\eta}{2\,m\,\mu_1^2\,\mathrm{cn}(A_1,m)} - \frac{1}{m\,\mu_1}}$, $B_0 = \sqrt{\frac{\eta\,\mathrm{dn}^2(A_1,m)}{2\,m\,\mu_1\,\mu_2\,\mathrm{cn}(A_1,m)} - \frac{1}{m\,\mu_2}}$, $A_1 = \mu_1$, $\omega_1 = 2 - \frac{\eta}{\mu_1}$, $\omega_2 = 2 - \frac{\nu_2}{\mu_2}$, $\mu_2 = \mu_1\,\mathrm{dn}(A_1,m)$, $\frac{\eta}{2\,m\,\mu_1^2\,\mathrm{cn}(A_1,m)} > \frac{1}{m\,\mu_1}$, $\frac{\eta\,\mathrm{dn}^2(A_1,m)}{2\,m\,\mu_1\,\mu_2\,\mathrm{cn}(A_1,m)} > \frac{1}{m\,\mu_2}$, $0 \le m \le 1$, ν_1, ν_2, μ_1, ϕ_1, and ϕ_2 are arbitrary real constants, n_0 is an arbitrary real integer	Discrete solitary wave	(8.112)
3.	$\psi_1(n,t) = A_0\,\mathrm{sech}[A_1\,(n+n_0)]\,e^{-i\,(\omega_1 t+\phi_1)},$ $\psi_2(n,t) = B_0\,\tanh[A_1\,(n+n_0)]\,e^{-i\,(\omega_2 t+\phi_2)}$	$A_0 = \sqrt{\frac{\eta\,\cosh(A_1)}{2\,\mu_1^2} - \frac{1}{\mu_1}}$, $B_0 = \sqrt{\frac{\eta}{2\,\mu_1\,\mu_2\,\cosh(A_1)} - \frac{1}{\mu_2}}$, $A_1 = \mu_1$, $\omega_1 = 2 - \frac{\eta}{\mu_1}$, $\omega_2 = 2 - \frac{\nu_2}{\mu_2}$, $\mu_2 = \mu_1\,\mathrm{sech}(A_1)$, $\frac{\eta\,\cosh(A_1)}{2\,\mu_1^2} > \frac{1}{\mu_1}$, $\frac{\eta}{2\,\mu_1\,\mu_2\,\cosh(A_1)} > \frac{1}{\mu_2}$, ν_1, ν_2, μ_1, ϕ_1, and ϕ_2 are arbitrary real constants, n_0 is an arbitrary real integer	Discrete bright-dark soliton	(8.113)

References

[1] Lederer F, Stegeman G I, Demetri Christodoulides N, Assanto G, Segev M and Silberberg Y 2008 Discrete solitons in optics *Phys. Rep.* **463** 1–126

[2] Garanovich I L, Longhi S, Sukhorukov A A and Kivshar Y S 2012 Light propagation and localization in modulated photonic lattices and waveguides *Phys. Rep.* **518** 1–78

[3] Pethick C J and Smith H 2008 *Bose–Einstein Condensation in Dilute Gases* (Cambridge: Cambridge University Press)

[4] Kevrekidis P G 2009 The Discrete Nonlinear Schrödinger equation: Mathematical Analysis, Numerical Computations and Physical Perspectives *The Discrete Nonlinear Schrödinger equation: Mathematical Analysis, Numerical Computations and Physical Perspectives* (Springer) (Springer Tracts in Modern Physics, vol 232)

[5] Khare A and Saxena A 2015 Periodic and hyperbolic soliton solutions of a number of nonlocal nonlinear equations *J. Math. Phys.* **56** 032104–27

[6] Khare A, Rasmussen K Ø, Samuelsen M R and Saxena A 2005 Exact solutions of the saturable discrete nonlinear Schrödinger equation *J. Phys. A: Math. Gen.* **38** 807–14

[7] Yan Z 2009 Envelope solution profiles of the discrete nonlinear Schrödinger equation with a saturable nonlinearity *Appl. Math. Lett.* **22** 448–52

[8] Khare A, Rasmussen K Ø, Samuelsen M R and Saxena A 2009 Staggered and short-period solutions of the saturable discrete nonlinear Schrödinger equation *J. Phys. A: Math. Theor.* **42** 085002–6

[9] Ankiewicz A, Akhmediev N and Lederer F 2011 Approach to first-order exact solutions of the Ablowitz–Ladik equation *Phys. Rev.* **E83** 056602–6

[10] Hua-Mei L and Feng-Min W 2005 Exact discrete soliton solutions of quintic discrete nonlinear Schrödinger equation *Chin. Phys.* **14** 1069–7

[11] Kevrekidis P G 2009 The Discrete Nonlinear Schrödinger equation: Mathematical Analysis *Numerical Computations and Physical Perspectives* (Springer) (Springer Tracts in Modern Physics vol 232)

[12] Khare A, Dmitriev S V and Saxena A 2007 Exact moving and stationary solutions of a generalized discrete nonlinear Schrödinger equation *J. Phys. A: Math. Theor.* **40** 11301–17

[13] Khare A and Saxena A 2012 Solutions of several coupled discrete models in terms of Lamé polynomials of order one and two *Pramana* **78** 187–213

IOP Publishing

Handbook of Exact Solutions to the Nonlinear Schrödinger Equations (Second Edition)

Usama Al Khawaja and Laila Al Sakkaf

Chapter 9

Nonlocal Nonlinear Schrödinger Equation

A Glance at Chapter 9

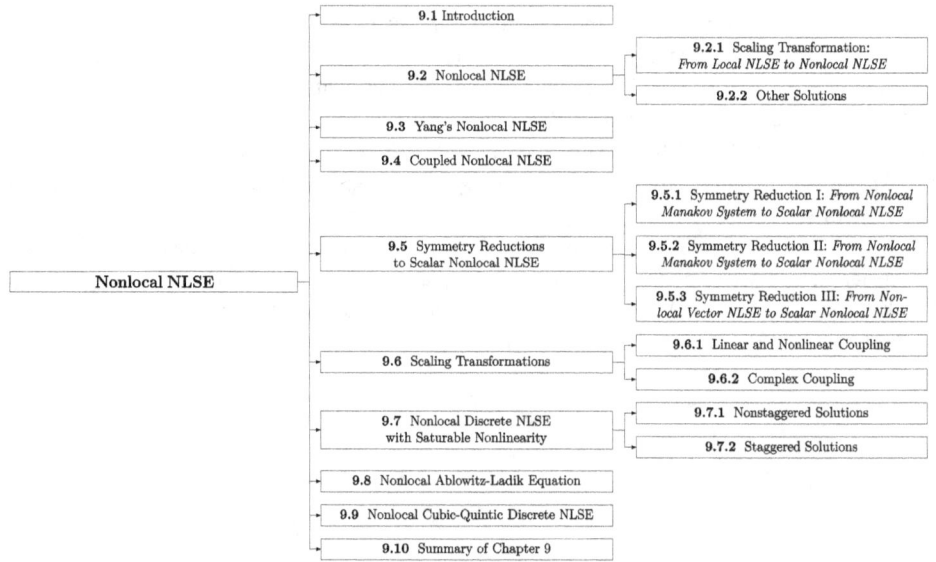

A Statistical View of Chapter 9

	Equation	Solutions
1	$i\,\Phi_t + a_1\,\Phi_{xx} + a_2\,\Phi^2\,\bar\Phi = 0$	11
2	$i\,\psi_t + a_1\,\psi_{xx} + a_2\,(|\psi|^2 + |\bar\psi|^2)\,\psi = 0$	4
3	$i\,\Phi_{1t} + \Phi_{1xx} + (a_1\,\Phi_1\,\bar\Phi_1 + a_2\,\Phi_2\,\bar\Phi_2)\,\Phi_1 = 0,$ $i\,\Phi_{2t} + \Phi_{2xx} + (b_1\,\Phi_1\,\bar\Phi_1 + b_2\,\Phi_2\,\bar\Phi_2)\,\Phi_2 = 0$	4
4	$i\,\Phi_{1t} + b_0\,\Phi_{1xx} + (c_1 + c_2|\sigma|^2)\,\Phi_1^2\,\bar\Phi_1 = 0$	0
5	$i\,\Phi_{1t} + b_0\,\Phi_{1xx} - (c_1 + c_2)\,\Phi_1^2\,\bar\Phi_1 = 0$	0
6	$i\,\Phi_{1t} + a_1\,\Phi_{1xx} + \sum_{k=1}^{N} b_{1k}|\sigma_k|^2\,\Phi_1^2\,\bar\Phi_1 = 0$	0
7	$i\,\Phi_{1t} + \Phi_{1xx} + (g_1\,\Phi_1\,\bar\Phi_1 - g_2\,\Phi_2\,\bar\Phi_2)\,\Phi_1 + g_0\,(g_1 + g_2)\,\Phi_1 - 2\,g_0\,g_2\,\Phi_2 = 0,$ $i\,\Phi_{2t} + \Phi_{2xx} + (g_1\,\Phi_1\,\bar\Phi_1 - g_2\,\Phi_2\,\bar\Phi_2)\,\Phi_2 - g_0\,(g_1 + g_2)\,\Phi_2 + 2\,g_0\,g_1\,\Phi_1 = 0$	0
8	$i\,\Phi_{1t} + \Phi_{1xx} + (g_1\,\Phi_1\,\bar\Phi_1 - g_2\,\Phi_2\,\bar\Phi_2)\,\Phi_1 = 0,$ $i\,\Phi_{2t} + \Phi_{2xx} + (g_1\,\Phi_1\,\bar\Phi_1 - g_2\,\Phi_2\,\bar\Phi_2)\,\Phi_2 = 0$	0
9	$i\,\Phi_{1t} + \Phi_{1xx} + 2\,(a_{11}\,\Phi_1\,\bar\Phi_1 + a_{12}\,\Phi_2\,\bar\Phi_2)\,\Phi_1 + 2\,(b_{11}\,\Phi_1\,\bar\Phi_2 + b_{12}\,\Phi_2\,\bar\Phi_1)\,\Phi_1 = 0,$ $i\,\Phi_{2t} + \Phi_{2xx} + 2\,(a_{21}\,\Phi_1\,\bar\Phi_1 + a_{22}\,\Phi_2\,\bar\Phi_2)\,\Phi_2 + 2\,(b_{21}\,\Phi_1\,\bar\Phi_2 + b_{22}\,\Phi_2\,\bar\Phi_1)\,\Phi_2 = 0,$	0
10	$i\,\Phi_{1t} + \Phi_{1xx} - 2(a + b)\,(\Phi_1\,\bar\Phi_1 + \Phi_2\,\bar\Phi_2)\,\Phi_1 + 2\,((a + i\,b)\,\Phi_1\,\bar\Phi_2 + (a - i\,b)\,\Phi_2\,\bar\Phi_1)\,\Phi_1 = 0,$ $i\,\Phi_{2t} + \Phi_{2xx} - 2(a + b)\,(\Phi_1\,\bar\Phi_1 + \Phi_2\,\bar\Phi_2)\,\Phi_2 + 2\,((a + i\,b)\,\Phi_1\,\bar\Phi_2 + (a - i\,b)\,\Phi_2\bar\Phi_1)\,\Phi_2 = 0$	0
11	$i\,\Phi_{nt} + \Phi_{n+1} + \Phi_{n-1} - 2\,\Phi_n + \dfrac{a_2\,\Phi_n^2\,\bar\Phi_n}{1 + \mu\,\Phi_n\,\bar\Phi_n} = 0$	3
12	$i\,\Phi_{nt} + \Phi_{n+1} + \Phi_{n-1} - 2\,\Phi_n + a_2\,(\Phi_{n+1} + \Phi_{n-1})\,\Phi_n\,\bar\Phi_n = 0$	2
13	$i\,\Phi_{nt} + a_1\,(\Phi_{n+1} + \Phi_{n-1} - 2\,\Phi_n) + a_2\,\Phi_n^2\,\bar\Phi_n + (a_3\,\Phi_n\,\bar\Phi_n + a_4\,\Phi_n^2\,\bar\Phi_n^2)(\Phi_{n+1} + \Phi_{n-1}) = 0$	4
Total	**13**	**28**

9.1 Introduction

Interest in the nonlocal NLSE essentially started and then expanded significantly with the discovery of the integrable class of nonlocal NLSE by Ablowitz and Musslimani [1]. The following form of nonlocal NLSE

$$i\frac{\partial}{\partial t}u(x,\,t) + a_1\frac{\partial^2}{\partial x^2}u(x,\,t) + a_2 u^2(x,\,t)u^*(-x,\,t) = 0 \tag{9.1}$$

was found to admit the Lax pair

$$\frac{\partial}{\partial x}\mathbf{\Phi}(x,\,t) = \mathbf{U}_0\mathbf{\Phi}(x,\,t) + \mathbf{U}_1\mathbf{\Phi}(x,\,t)\mathbf{\Lambda}, \tag{9.2}$$

$$\frac{\partial}{\partial t}\Phi(x,\ t) = \mathbf{V}_0\Phi(x,\ t) + \mathbf{V}_1\Phi(x,\ t)\mathbf{\Lambda} + \mathbf{V}_2\Phi(x,\ t)\mathbf{\Lambda}^2, \tag{9.3}$$

where

$$\Phi(x,\ t) = \begin{pmatrix} \psi_1(x,\ t) & \psi_2(x,\ t) \\ \phi_1(x,\ t) & \phi_2(x,\ t) \end{pmatrix} \tag{9.4}$$

is an auxiliary field of four unknown components, and

$$\mathbf{U}_0(x,\ t) = \sqrt{\frac{a_2}{2a_1}}\begin{pmatrix} 0 & iu(x,\ t) \\ iu^*(-x,\ t) & 0 \end{pmatrix}, \tag{9.5}$$

$$\mathbf{U}_1(x,\ t) = \begin{pmatrix} -1 & 0 \\ 0 & 1 \end{pmatrix}, \tag{9.6}$$

$$\mathbf{V}_0(x,\ t) = \frac{a_2}{2}\begin{pmatrix} -iu(x,\ t)u(-x,\ t) & \sqrt{\frac{2a_1}{a_2}}\frac{\partial}{\partial x}u(x,\ t) \\ -\sqrt{\frac{2a_1}{a_2}}\frac{\partial}{\partial x}u(-x,\ t) & iu(x,\ t)u(-x,\ t) \end{pmatrix}, \tag{9.7}$$

$$\mathbf{V}_1(x,\ t) = -a_2\sqrt{\frac{2a_1}{2a_2}}\begin{pmatrix} 0 & u(x,\ t) \\ u^*(-x,\ t) & 0 \end{pmatrix}, \tag{9.8}$$

$$\mathbf{V}_2(x,\ t) = -2a_1\begin{pmatrix} i & 0 \\ 0 & -i \end{pmatrix}, \tag{9.9}$$

$$\mathbf{\Lambda} = \begin{pmatrix} \lambda_1 & 0 \\ 0 & \lambda_2 \end{pmatrix}, \tag{9.10}$$

and $\lambda_{1,\ 2}$ are two arbitrary complex spectral parameters. The compatibility conditions

$$\frac{\partial}{\partial t}\mathbf{U}_0 - \frac{\partial}{\partial x}\mathbf{V}_0 + [\mathbf{U}_0,\ \mathbf{V}_0] = 0, \tag{9.11}$$

$$\frac{\partial}{\partial t}\mathbf{U}_1 - \frac{\partial}{\partial x}\mathbf{V}_1 + [\mathbf{U}_0,\ \mathbf{V}_1] + [\mathbf{U}_1,\ \mathbf{V}_0] = 0, \tag{9.12}$$

$$\frac{\partial}{\partial x}\mathbf{V}_2 + [\mathbf{V}_2,\ \mathbf{U}_0] - [\mathbf{V}_1,\ \mathbf{U}_1] = 0, \tag{9.13}$$

$$[\mathbf{U}_1,\ \mathbf{V}_2] = 0, \tag{9.14}$$

are all satisfied with above Lax pair, such that the first compatibility condition (9.11), is satisfied only if $u(x,\ t)$ is a solution of the nonlocal NLSE (9.1).

Admitting a Lax pair allows for the application of the inverse scattering transform and the Darboux transformation, where families of exact solutions can be found. Many other versions of the nonlocal NLSE have been considered [2] including discrete version of nonlocal NLSE [3]. It was also shown that the nonlocal NLSE can be obtained by a gauge transformation of a coupled NLSE [4, 5], which may describe a magnetic system [4].

In a more general setting, the nonlocal NLSE is obtained for optical materials with nonlocal nonlinear response,

$$i\frac{\partial}{\partial t}u(x, t) + a_1\frac{\partial^2}{\partial x^2}u(x, t) + a_2 n(x, t)u(x, t) = 0, \qquad (9.15)$$

where the response function, $G(x, t)$, defines the refractive index as $n(x, t) = \int_{-\infty}^{\infty} G(x - y)|u(y, t)|^2 dy$.

In another setting, dipolar Bose–Einstein condensates are described by a nonlocal Gross–Pitaevskii equation [6]

$$i\hbar\frac{\partial}{\partial t}\psi(\mathbf{r}, t) = -\frac{\hbar^2}{2m}\nabla^2\psi(\mathbf{r}, t) + V(\mathbf{r})\psi(\mathbf{r}, t) + g|\psi(\mathbf{r}, t)|^2 + d^2\left(\int\frac{1 - 3\cos^2\theta}{|\mathbf{r} - \mathbf{r}'|^2}|\psi(\mathbf{r}', t)|^2 d\mathbf{r}'\right)\psi(\mathbf{r}, t) = 0, \quad (9.16)$$

where $\psi(\mathbf{r}, t)$ is the wavefunction normalized to the number of atoms. The last term corresponds to the dipole–dipole interaction with strength, d, and angle between the dipole vectors, θ.

Since only the specific type of nonlocal NLSE given by Equation (9.1) is integrable, while Equations (9.15) and (9.16) are not, this chapter is devoted to exact solutions of this version of the nonlocal NLSE, including other variations such as coupled and discrete nonlocal NLSE.

9.2 Nonlocal NLSE

9.2.1 Scaling Transformation *From Local NLSE to Nonlocal NLSE*

If $\psi(x, t)$ is a solution to the fundamental NLSE (2.160), $i\psi_t + a_1\psi_{xx} + a_2|\psi|^2\psi = 0$, then

$$\Phi(\zeta, t; a_1, a_2) = \begin{cases} \psi(x, t, a_1, a_2), & \psi \text{ is an even function in } x, \\ \psi(x, t, a_1, -a_2), & \psi \text{ is an odd function in } x \end{cases}$$

is a solution to

$$i\Phi_t + a_1\Phi_{xx} + a_2\Phi^2\bar{\Phi} = 0, \qquad (9.17)$$

where $\psi = \psi(x, t; a_1, a_2)$, $\Phi = \Phi(x, t; a_1, a_2)$, $\bar{\Phi} = \Phi^*(-x, t; a_1, a_2)$, $\zeta = x - x_0$, a_1 and a_2 are arbitrary real constants.

Example 1. **Even function: sech(x)** *bright soliton*

Given $\psi(\zeta, t) = A_0\sqrt{\frac{2a_1}{a_2}}\,\text{sech}[A_0\,\zeta]\,e^{i\,[a_1\,A_0^2\,(t-t_0)+\phi_0]}$ is a solution to (2.160), then

$$\Phi(\zeta, t) = A_0\sqrt{\frac{2a_1}{a_2}}\,\text{sech}[A_0\,\zeta]\,e^{i\,[a_1\,A_0^2\,(t-t_0)+\phi_0]} \qquad (9.18)$$

is a solution to (9.17), where $\zeta = x - x_0$, $a_1 a_2 > 0$, A_0, x_0, t_0, and ϕ_0 are arbitrary real constants.

Example 2. Odd function: tanh(x) *dark soliton*

Given $\psi(\zeta, t) = A_0 \tanh[\frac{A_1}{\sqrt{a_1}} \zeta] \, e^{-i \, [A_2 \, (t - t_0) + \phi_0]}$ is a solution to (2.160), where $A_2 = 2 \, A_1^2$, $a_2 = -\frac{2 \, A_1^2}{A_0^2}$, $a_1 > 0$, $\zeta = x - x_0$, A_0, A_1, x_0, t_0, and ϕ_0 are arbitrary real constants, then

$$\Phi(\zeta, t) = A_0 \tanh\left[\frac{A_1}{\sqrt{a_1}} \zeta\right] e^{-i \, [A_2 \, (t - t_0) + \phi_0]} \tag{9.19}$$

is a solution to (9.17), where $A_2 = 2 \, A_1^2$, $a_2 = \frac{2 \, A_1^2}{A_0^2}$, *(notice the change of sign compared to the local case)*, $a_1 > 0$, $\zeta = x - x_0$, A_0, A_1, x_0, t_0, and ϕ_0 are arbitrary real constants.

9.2.2 Other Solutions

Solution 1. *Singular soliton molecule*
   (Figure 9.1)

$$\psi(x, t) = \frac{-\lambda_1^{*2} e^{q_1(x, \, t)} + \lambda_1^2 e^{q_2(x, \, t)} - 4 \, i \, \lambda_{1i} \, e^{2 \, q_3(x, \, t)}(-\lambda_{1r} + |\lambda_1|^2 \, x)}{2 \, i \, e^{q_3(x, \, t)}\{\lambda_{1i} \cos[q_4(x, \, t)] + \lambda_{1i} \cosh[2 \, x \, \lambda_{1r} - 2 \, i \, t(\lambda_{1i}^2 - \lambda_{1r}^2)] + |\lambda_1|^2 \, x \sin[q_4(x, \, t)]\}}, \tag{9.20}$$

where $\qquad q_1(x, t) = 2 \, \lambda_1^*(x + i \, t \, \lambda_1^*)$, $\qquad q_2(x, t) = 2 \, \lambda_1(x + i \, t \, \lambda_1)$, $q_3(x, t) = 2 \, x \, \lambda_{1r} - 2 \, i \, t \, (\lambda_{1i}^2 - \lambda_{1r}^2)$, $\qquad q_4(x, t) = 2 \, \lambda_{1i} \, (x + 2 \, i \, t \, \lambda_{1r})$, $\lambda_1 = \lambda_{1r} + i \, \lambda_{1i}$, $a_1 = 1/2$, $a_2 = 1$, λ_{1r} and λ_{1i} are arbitrary real constants.
   • *Reference:* [7].

Solution 2.

$$\psi(\zeta, t) = \frac{A_0 + i \, A_3 \sin\left[\dfrac{A_1}{\sqrt{a_1}} \zeta\right]}{A_4 + \cos\left[\dfrac{A_1}{\sqrt{a_1}} \zeta\right]} \, e^{-i \, [A_2 \, (t - t_0) + \phi_0]}, \tag{9.21}$$

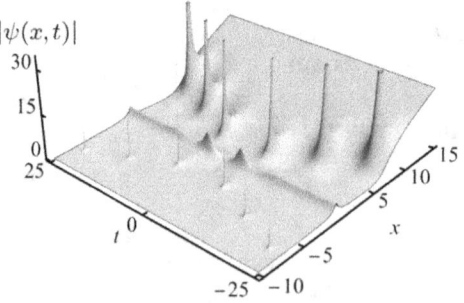

Figure 9.1. Singular soliton molecule (9.20) with $\lambda_{1r} = 0.6$ and $\lambda_{1i} = 0.1$.

where $A_2 = -A_1^2/2$, $A_3 = A_1/\sqrt{2\,a_2}$, $A_4 > 1$, $a_1 > 0$, $a_2 = (A_4^2 - 1)\,A_1^2/(2\,A_0^2)$, $\zeta = x - x_0$, A_0, A_1, x_0, t_0, and ϕ_0 are arbitrary real constants.

• *Reference*: [8].

Solution 3.

$$\psi(\zeta,\, t) = \frac{A_0 \sin\left[\dfrac{A_1}{\sqrt{a_1}}\,\zeta\right] + i\,A_3}{A_4 + \cos\left[\dfrac{A_1}{\sqrt{a_1}}\,\zeta\right]}\, e^{-i\,[A_2\,(t-t_0)+\phi_0]}, \tag{9.22}$$

where $A_2 = -A_1^2/2$, $A_3 = A_1\sqrt{A_4^2 - 1}\,/\sqrt{2\,a_2}$, $A_4 > 1$, $a_1 > 0$, $a_2 = A_1^2/(2\,A_0^2)$, $\zeta = x - x_0$, A_0, A_1, x_0, t_0, and ϕ_0 are arbitrary real constants.

• *Reference*: [8].

Solution 4.

$$\psi(\zeta,\, t) = \sqrt{m}\left\{\frac{A_0\,\mathrm{cn}\left[\dfrac{A_1}{\sqrt{a_1}}\,\zeta,\, m\right] + i\,A_3\,\mathrm{sn}\left[\dfrac{A_1}{\sqrt{a_1}}\,\zeta,\, m\right]}{A_4 + \mathrm{dn}\left[\dfrac{A_1}{\sqrt{a_1}}\,\zeta,\, m\right]}\right\}\, e^{-i\,[A_2\,(t-t_0)+\phi_0]}, \tag{9.23}$$

where $A_2 = (2 - m)\,A_1^2/2$, $A_3 = A_1\sqrt{A_4^2 - 1 + m}\,/\sqrt{2\,a_2}$, $A_4 > 1$, $a_1 > 0$, $a_2 = A_1^2\,(A_4^2 - 1)/(2\,A_0^2)$, $m > 0$, $\zeta = x - x_0$, A_0, A_1, x_0, t_0, and ϕ_0 are arbitrary real constants.

• *Reference*: [8].

Solution 5.

$$\psi(\zeta,\, t) = \left\{\frac{A_0\,\mathrm{sech}\left[\dfrac{A_1}{\sqrt{a_1}}\,\zeta\right] + i\,A_3\,\tanh\left[\dfrac{A_1}{\sqrt{a_1}}\,\zeta\right]}{A_4 + \mathrm{sech}\left[\dfrac{A_1}{\sqrt{a_1}}\,\zeta\right]}\right\}\, e^{-i\,[A_2\,(t-t_0)+\phi_0]}, \tag{9.24}$$

where $A_2 = A_1^2/2$, $A_3 = A_1\,A_4/\sqrt{2\,a_2}$, $A_4 > 1$, $a_1 > 0$, $a_2 = A_1^2\,(A_4^2 - 1)/(2\,A_0^2)$, $\zeta = x - x_0$, A_0, A_1, x_0, t_0, and ϕ_0 are arbitrary real constants.

• *Reference*: [8].

Solution 6.

$$\psi(\zeta,\, t) = \sqrt{m}\left\{\frac{A_0\,\mathrm{sn}\left[\dfrac{A_1}{\sqrt{a_1}}\,\zeta,\, m\right] + i\,A_3\,\mathrm{cn}\left[\dfrac{A_1}{\sqrt{a_1}}\,\zeta,\, m\right]}{A_4 + \mathrm{dn}\left[\dfrac{A_1}{\sqrt{a_1}}\,\zeta,\, m\right]}\right\}\, e^{-i\,[A_2\,(t-t_0)+\phi_0]}, \tag{9.25}$$

where $A_2 = (2 - m) A_1^2/2$, $A_3 = A_1 \sqrt{A_4^2 - 1}/\sqrt{2 a_2}$, $A_4 > 1$, $a_1 > 0$, $a_2 = A_1^2 (A_4^2 - 1 + m)/(2 A_0^2)$, $m > 0$, $\zeta = x - x_0$, A_0, A_1, x_0, t_0, and ϕ_0 are arbitrary real constants.
  • *Reference*: [8].

Solution 7.

$$\psi(\zeta, t) = \frac{A_0 \tanh\left[\dfrac{A_1}{\sqrt{a_1}} \zeta\right] + i A_3 \operatorname{sech}\left[\dfrac{A_1}{\sqrt{a_1}} \zeta\right]}{A_4 + \operatorname{sech}\left[\dfrac{A_1}{\sqrt{a_1}} \zeta\right]} e^{-i [A_2 (t - t_0) + \phi_0]}, \qquad (9.26)$$

where $A_2 = A_1^2/2$, $A_3 = A_1 \sqrt{A_4^2 - 1}/\sqrt{2 a_2}$, $A_4 > 1$, $a_1 > 0$, $a_2 = A_1^2 A_4^2/(2 A_0^2)$, $\zeta = x - x_0$, A_0, A_1, x_0, t_0, and ϕ_0 are arbitrary real constants.
  • *Reference*: [8].

Solution 8.

$$\psi(\zeta, t) = \frac{A_0 \operatorname{dn}\left[\dfrac{A_1}{\sqrt{a_1}} \zeta, m\right] + i A_3 \sqrt{m} \operatorname{sn}\left[\dfrac{A_1}{\sqrt{a_1}} \zeta, m\right]}{A_4 + \operatorname{cn}\left[\dfrac{A_1}{\sqrt{a_1}} \zeta, m\right]} e^{-i [A_2 (t - t_0) + \phi_0]}, \qquad (9.27)$$

where $A_2 = (2 m - 1) A_1^2/2$, $A_3 = A_1 \sqrt{m A_4^2 - m + 1}/\sqrt{2 m a_2}$, $A_4 > 1$, $a_1 > 0$, $a_2 = A_1^2 (A_4^2 - 1)/(2 A_0^2)$, $m > 0$, $\zeta = x - x_0$, A_0, A_1, x_0, t_0, and ϕ_0 are arbitrary real constants.
  • *Reference*: [8].

Solution 9.

$$\psi(\zeta, t) = \frac{A_0 \sqrt{m} \operatorname{sn}\left[\dfrac{A_1}{\sqrt{a_1}} \zeta, m\right] + i A_3 \operatorname{dn}\left[\dfrac{A_1}{\sqrt{a_1}} \zeta, m\right]}{A_4 + \operatorname{cn}\left[\dfrac{A_1}{\sqrt{a_1}} \zeta, m\right]} e^{-i [A_2 (t - t_0) + \phi_0]}, \qquad (9.28)$$

where $A_2 = (2 m - 1) A_1^2/2$, $A_3 = A_1 \sqrt{A_4^2 - 1}/\sqrt{2 a_2}$, $A_4 > 1$, $a_1 > 0$, $a_2 = A_1^2 (1 - m + m A_4^2)/(2 m A_0^2)$, $m > 0$, $\zeta = x - x_0$, A_0, A_1, x_0, t_0, and ϕ_0 are arbitrary real constants.
  • *Reference*: [8].

9.3 Yang's Nonlocal NLSE

Equation:

$$i\,\psi_t + a_1\,\psi_{xx} + a_2\,(|\psi|^2 + |\bar{\psi}|^2)\,\psi = 0, \tag{9.29}$$

where $\psi = \psi(\zeta, t)$ is the complex function profile, $\bar{\psi} = \psi(-\zeta, t)$, $\zeta = x - x_0$, x and t are its two independent variables, a_1, a_2, and x_0 are arbitrary constants.

Solutions:

Solution **1. Constant amplitude** *continuous wave (CW)*

$$\psi(\zeta, t) = A_0\,e^{i\,[A_1\,(t-t_0)+A_2\,(x-x_0)+\phi_0]}, \tag{9.30}$$

where $A_1 = 2\,A_0^2\,a_2 - A_2^2\,a_1$, A_0, A_2, x_0, t_0, and ϕ_0 are arbitrary real constants.

Solution **2.**

$$\psi(\zeta, t) = \frac{A_0\,\sqrt{m}\,\operatorname{sn}\left[\dfrac{A_1}{\sqrt{a_1}}\,\zeta, m\right]}{1 + \operatorname{dn}\left[\dfrac{A_1}{\sqrt{a_1}}\,\zeta, m\right]}\,e^{-i\,[A_2\,(t-t_0)+\phi_0]}, \tag{9.31}$$

where $A_2 = (2 - m)\,A_1^2/2$, $a_1 > 0$, $a_2 = -m\,A_1^2/(4\,A_0^2)$, $m > 0$, $\zeta = x - x_0$, A_0, A_1, x_0, t_0, and ϕ_0 are arbitrary real constants.
  • *Reference*: [8], corrected.

Solution **3.**

$$\psi(\zeta, t) = \frac{A_0\,\sqrt{m}\,\operatorname{cn}\left[\dfrac{A_1}{\sqrt{a_1}}\,\zeta, m\right]}{\sqrt{1 - m} + \operatorname{dn}\left[\dfrac{A_1}{\sqrt{a_1}}\,\zeta, m\right]}\,e^{-i\,[A_2\,(t-t_0)+\phi_0]}, \tag{9.32}$$

where $A_2 = (2 - m)\,A_1^2/2$, $a_1 > 0$, $a_2 = -m\,A_1^2/(4\,A_0^2)$, $m > 0$, $\zeta = x - x_0$, A_0, A_1, x_0, t_0, and ϕ_0 are arbitrary real constants.
  • *Reference*: [8].

Solution **4.**

$$\psi(\zeta, t) = A_0\left\{\frac{1}{2} - \frac{\operatorname{dn}\left[\dfrac{A_1}{\sqrt{a_1}}\,\zeta, m\right]}{(1 - m)^{1/4} + \operatorname{dn}\left[\dfrac{A_1}{\sqrt{a_1}}\,\zeta, m\right]}\right\}e^{-i\,[A_2\,(t-t_0)+\phi_0]}, \tag{9.33}$$

where $A_2 = [(2 - m) + 6\sqrt{1 - m}]\, A_1^2/2$, $a_1 > 0$, $a_2 = -(1 - \sqrt{1 - m})^2\, A_1^2/(4\,A_0^2)$, $m > 0$, $\zeta = x - x_0$, A_0, A_1, x_0, t_0, and ϕ_0 are arbitrary real constants.
- *Reference*: [8].

9.4 Nonlocal Coupled NLSE

If (ψ_1, ψ_2) is a solution to

$$i\,\psi_{1t} + \psi_{1xx} + (a_1\,\psi_1\,\psi_1^* + a_2\,\psi_2\,\psi_2^*)\,\psi_1 = 0,$$
$$i\,\psi_{2t} + \psi_{2xx} + (b_1\,\psi_1\,\psi_1^* + b_2\,\psi_2\,\psi_2^*)\,\psi_2 = 0,$$

(9.34)

then

$$\Phi_1(x, t; a_1, a_2, b_1, b_2) = \begin{cases} \psi_1(x, t, a_1, a_2, b_1, b_2), & \psi_1 \text{ is an even function in } x, \\ \psi_1(x, t, -a_1, a_2, -b_1, b_2), & \psi_1 \text{ is an odd function in } x, \end{cases}$$

$$\Phi_2(x, t; a_1, a_2, b_1, b_2) = \begin{cases} \psi_2(x, t, a_1, a_2, b_1, b_2), & \psi_2 \text{ is an even function in } x, \\ \psi_2(x, t, a_1, -a_2, b_1, -b_2), & \psi_2 \text{ is an odd function in } x \end{cases}$$

is a solution to

$$i\,\Phi_{1t} + \Phi_{1xx} + (a_1\,\Phi_1\,\bar{\Phi}_1 + a_2\,\Phi_2\,\bar{\Phi}_2)\,\Phi_1 = 0,$$
$$i\,\Phi_{2t} + \Phi_{2xx} + (b_1\,\Phi_1\,\bar{\Phi}_1 + b_2\,\Phi_2\,\bar{\Phi}_2)\,\Phi_2 = 0,$$

(9.35)

where $\psi_{1,2} = \psi_{1,2}(x, t; a_1, a_2, b_1, b_2)$, $\Phi_{1,2} = \Phi_{1,2}(x, t; a_1, a_2, b_1, b_2)$, $\bar{\Phi}_{1,2} = \Phi_{1,2}^*(-x, t; a_1, a_2, b_1, b_2)$, a_1, a_2, b_1, and b_2 are arbitrary real constants.

***Example* 1. Even–odd function: sech(x)-tanh(x)** *bright-dark soliton*
Given
$\psi_1(\zeta, t) = A_0\,\text{sech}[A_1\,\zeta]\,e^{-i\,[\omega_1\,(t-t_0)+\phi_1]}$, $\psi_2(\zeta, t) = B_0\,\tanh[A_1\,\zeta]\,e^{-i\,[\omega_2\,(t-t_0)+\phi_2]}$ is a solution to (9.34), where $\omega_1 = A_1^2 - a_1\,A_0^2$, $\omega_2 = 2\,A_1^2 - b_1\,A_0^2$, $a_1 = \dfrac{2\,A_1^2 + a_2\,B_0^2}{A_0^2}$, $b_1 = \dfrac{2\,A_1^2 + b_2\,B_0^2}{A_0^2}$, $\zeta = x - x_0$, $A_0 \neq 0$, A_0, A_1, B_0, x_0, t_0, ϕ_1, and ϕ_2 are arbitrary real constants, then

$$\Phi_1(\zeta, t) = A_0\,\text{sech}[A_1\,\zeta]\,e^{-i\,[\omega_1\,(t-t_0)+\phi_1]},$$
$$\Phi_2(\zeta, t) = B_0\,\tanh[A_1\,\zeta]\,e^{-i\,[\omega_2\,(t-t_0)+\phi_2]}$$

(9.36)

is a solution to (9.35), where $\omega_1 = A_1^2 - a_1\,A_0^2$, $\omega_2 = 2\,A_1^2 - b_1\,A_0^2$, $a_1 = \dfrac{2\,A_1^2 - a_2\,B_0^2}{A_0^2}$, *(changed the sign of a_2)*, $b_1 = \dfrac{2\,A_1^2 - b_2\,B_0^2}{A_0^2}$, *(changed the sign of b_2)*, $\zeta = x - x_0$, $A_0 \neq 0$, A_0, A_1, B_0, x_0, t_0, ϕ_1, and ϕ_2 are arbitrary real constants.

Example 2. Odd–odd function: cn(*x,m*)sn(*x,m*)-dn(*x,m*)sn(*x,m*) *solitary wave (SW)*
Given $\psi_1(\zeta, t) = A_0\, m\, \text{cn}[A_1\, \zeta, m]\, \text{sn}[A_1\, \zeta, m]\, e^{-i\,[\omega_1\,(t-t_0)+\phi_1]}$,
$\psi_2(\zeta, t) = B_0\, \sqrt{m}\, \text{dn}[A_1\, \zeta, m]\, \text{sn}[A_1\, \zeta, m]\, e^{-i\,[\omega_2\,(t-t_0)+\phi_2]}$ is a solution to (9.34),
where $\omega_1 = (4+m)\,A_1^2$, $\omega_2 = (1+4m)\,A_1^2$, $a_1 = \dfrac{6\,A_1^2}{A_0^2\,(1-m)}$, $a_2 = \dfrac{-6\,A_1^2}{B_0^2\,(1-m)}$, $b_1 = a_1$,
$b_2 = a_2, 0 \leqslant m < 1, \zeta = x - x_0, A_0 \neq 0, B_0 \neq 0, A_1, x_0, t_0, \phi_1,$ and ϕ_2 are arbitrary
real constants, then

$$\Phi_1(\zeta, t) = A_0\, m\, \text{cn}[A_1\, \zeta, m]\, \text{sn}[A_1\, \zeta, m]\, e^{-i\,[\omega_1\,(t-t_0)+\phi_1]},$$
$$\Phi_2(\zeta, t) = B_0\, \sqrt{m}\, \text{dn}[A_1\, \zeta, m]\, \text{sn}[A_1\, \zeta, m]\, e^{-i\,[\omega_2\,(t-t_0)+\phi_2]} \qquad (9.37)$$

is a solution to (9.35), where $\omega_1 = (4+m)\,A_1^2$, $\omega_2 = (1+4m)\,A_1^2$, $a_1 = \dfrac{-6\,A_1^2}{A_0^2\,(1-m)}$,
(*changed the sign of a_1*), $a_2 = \dfrac{6\,A_1^2}{B_0^2\,(1-m)}$, (*changed the sign of a_2*), $b_1 = a_1, b_2 = a_2,$
$0 \leqslant m < 1, \zeta = x - x_0, A_0 \neq 0, B_0 \neq 0, A_1, x_0, t_0, \phi_1,$ and ϕ_2 are arbitrary real constants.

Example 3. Odd–even function: sd(*x,m*)-nd(*x,m*) *SW*
Given
$\psi_1(\zeta, t) = A_0\, \sqrt{m\,(1-m)}\, \text{sd}[A_1\, \zeta, m]\, e^{-i\,[\omega_1\,(t-t_0)+\phi_1]}$,
$\psi_2(\zeta, t) = B_0\, \sqrt{1-m}\, \text{nd}[A_1\, \zeta, m]\, e^{-i\,[\omega_2\,(t-t_0)+\phi_2]}$ is a solution to (9.34), where
$\omega_1 = -(m-1)\,a_1\,A_0^2 - A_1^2$, $\qquad \omega_2 = (m-2)\,A_1^2 - (m-1)\,b_1\,A_0^2$,
$a_1 = -\dfrac{-2\,A_1^2 + a_2\,B_0^2}{A_0^2}$, $b_1 = -\dfrac{-2\,A_1^2 + b_2\,B_0^2}{A_0^2}, 0 \leqslant m \leqslant 1, \zeta = x - x_0, A_0 \neq 0, A_1, B_0, x_0,$
$t_0, \phi_1,$ and ϕ_2 are arbitrary real constants, then

$$\Phi_1(\zeta, t) = A_0\, \sqrt{m\,(1-m)}\, \text{sd}[A_1\, \zeta, m]\, e^{-i\,[\omega_1\,(t-t_0)+\phi_1]},$$
$$\Phi_2(\zeta, t) = B_0\, \sqrt{1-m}\, \text{nd}[A_1\, \zeta, m]\, e^{-i\,[\omega_2\,(t-t_0)+\phi_2]} \qquad (9.38)$$

is a solution to (9.35), where $\omega_1 = (m-1)\,a_1\,A_0^2 \quad A_1^2$, (*changed the sign of a_1*),
$\omega_2 = (m-2)\,A_1^2 + (m-1)\,b_1\,A_0^2$, (*changed the sign of b_1*), $a_1 = \dfrac{-2\,A_1^2 + a_2\,B_0^2}{A_0^2}$,
(*changed the sign of a_1*), $b_1 = \dfrac{-2\,A_1^2 + b_2\,B_0^2}{A_0^2}$, (*changed the sign of b_1*), $0 \leqslant m \leqslant 1,$
$\zeta = x - x_0, A_0 \neq 0, A_1, B_0, x_0, t_0, \phi_1,$ and ϕ_2 are arbitrary real constants.

Example 4. Even–even function: sech²(*x*)-sech²(*x*)
Given $\qquad\qquad\qquad\qquad \psi_1(\zeta, t) = [A_0\, \text{sech}^2(A_1\, \zeta) + A_3]\, e^{-i\,[\omega_1\,(t-t_0)+\phi_1]}$,
$\psi_2(\zeta, t) = B_0\, \text{sech}^2(A_1\, \zeta)\, e^{-i\,[\omega_2\,(t-t_0)+\phi_2]}$ is a solution to (9.34), where $A_3 = \dfrac{-2\,A_0}{3}$,
$\omega_1 = 2\,A_1^2, \omega_2 = -2\,A_1^2, a_1 = \dfrac{-9\,A_1^2}{2\,A_0^2}, a_2 = \dfrac{9\,A_1^2}{2\,B_0^2}, b_1 = a_1, b_2 = a_2, \zeta = x - x_0, A_0 \neq 0,$
$B_0 \neq 0, A_1, x_0, t_0, \phi_1,$ and ϕ_2 are arbitrary real constants, then

$$\Phi_1(\zeta, t) = \{A_0\, \text{sech}^2[A_1\, \zeta] + A_3\}\, e^{-i\,[\omega_1\,(t-t_0)+\phi_1]},$$
$$\Phi_2(\zeta, t) = B_0\, \text{sech}^2[A_1\, \zeta]\, e^{-i\,[\omega_2\,(t-t_0)+\phi_2]} \qquad (9.39)$$

is a solution to (9.35), where $A_3 = \frac{-2 A_0}{3}$, $\omega_1 = 2 A_1^2$, $\omega_2 = -2 A_1^2$, $a_1 = \frac{-9 A_1^2}{2 A_0^2}$,

$a_2 = \frac{9 A_1^2}{2 B_0^2}$, $b_1 = a_1$, $b_2 = a_2$, $\zeta = x - x_0$, $A_0 \neq 0$, $B_0 \neq 0$, A_1, x_0, t_0, ϕ_1, and ϕ_2 are arbitrary real constants.

9.5 Symmetry Reductions to Scalar Nonlocal NLSE

9.5.1 Symmetry Reduction I *From Nonlocal Manakov System to Scalar Nonlocal NLSE*

The nonlocal CNLSE (9.35), $i \Phi_{1t} + b_0 \Phi_{1xx} + (b_1 \Phi_1 \bar{\Phi}_1 + b_2 \Phi_2 \bar{\Phi}_2) \Phi_1 = 0$, $i \Phi_{2t} + b_0 \Phi_{2xx} + (c_1 \Phi_1 \bar{\Phi}_1 + c_2 \Phi_2 \bar{\Phi}_2) \Phi_2 = 0$, transforms to the scalar nonlocal NLSE

$$i \Phi_{1t} + b_0 \Phi_{1xx} + (c_1 + c_2 |\sigma|^2) \Phi_1^2 \bar{\Phi}_1 = 0, \tag{9.40}$$

with the replacements:
1. $\Phi_2(x, t) = \sigma \Phi_1(x, t)$,
2. $b_1 = c_1 + (c_2 - b_2) |\sigma|^2$,

where σ is an arbitrary complex constant.

Conclusion:

If $\Phi_1(x, t)$ is a solution to the nonlocal NLSE $i \Phi_{1t} + a_1 \Phi_{1xx} + a_2 \Phi_1^2 \bar{\Phi}_1 = 0$, then

$$(\Phi_1, \Phi_2) = (\Phi_1, \sigma \Phi_1) \tag{9.41}$$

is a solution to the nonlocal CNLSE $i \Phi_{1t} + b_0 \Phi_{1xx} + (b_1 \Phi_1 \bar{\Phi}_1 + b_2 \Phi_2 \bar{\Phi}_2) \Phi_1 = 0$, $i \Phi_{2t} + b_0 \Phi_{2xx} + (c_1 \Phi_1 \bar{\Phi}_1 + c_2 \Phi_2 \bar{\Phi}_2) \Phi_2 = 0$, with $a_1 = b_0$, $a_2 = c_1 + c_2 |\sigma|^2$, $b_1 = c_1 + (c_2 - b_2) |\sigma|^2$.

9.5.2 Symmetry Reduction II
From Nonlocal Manakov System to Scalar Nonlocal NLSE

The nonlocal CNLSE (9.35), $i \Phi_{1t} + b_0 \Phi_{1xx} + (b_1 \Phi_1 \bar{\Phi}_1 + b_2 \Phi_2 \bar{\Phi}_2) \Phi_1 = 0$, $i \Phi_{2t} - b_0 \Phi_{2xx} + (c_1 \Phi_1 \bar{\Phi}_1 + c_2 \Phi_2 \bar{\Phi}_2) \Phi_2 = 0$, transforms to the scalar nonlocal NLSE

$$i \Phi_{1t} + b_0 \Phi_{1xx} - (c_1 + c_2) \Phi_1^2 \bar{\Phi}_1 = 0, \tag{9.42}$$

with the replacements:
1. $\Phi_2(x, t; b_1, b_2, c_1, c_2) = e^{i \phi} \begin{cases} \bar{\Phi}_1(x, t, b_1, b_2, c_1, c_2) & \text{if } \Phi_1 \text{ is an even function in } x, \\ \bar{\Phi}_1(x, t, b_1, b_2, -c_1, -c_2) & \text{if } \Phi_1 \text{ is an odd function in } x, \end{cases}$
2. $b_1 = -(c_1 + c_2 + b_2)$,

where φ is an arbitrary real constant.

Conclusion:

I. If $\Phi_1(x, t)$ is an even solution to the nonlocal NLSE $i \Phi_{1t} + b_0 \Phi_{1xx} + a_2 \Phi_1^2 \bar{\Phi}_1 = 0$, then

$$(\Phi_1, \Phi_2) = \left[\Phi_1(x, t, b_1, b_2, c_1, c_2), e^{i\phi}\Phi_1^*(x, t, b_1, b_2, c_1, c_2)\right] \qquad (9.43)$$

is a solution to the nonlocal CNLSE $i\,\Phi_{1t} + b_0\,\Phi_{1xx} + (b_1\,\Phi_1\,\bar{\Phi}_1 + b_2\,\Phi_2\,\bar{\Phi}_2)\,\Phi_1 = 0$, $i\,\Phi_{2t} - b_0\,\Phi_{2xx} + (c_1\,\Phi_1\,\bar{\Phi}_1 + c_2\,\Phi_2\,\bar{\Phi}_2)\,\Phi_2 = 0$, with $a_2 = -(c_1 + c_2)$.

II. If $\Phi_1(x, t)$ is an odd solution to the nonlocal NLSE $i\,\Phi_{1t} + b_0\,\Phi_{1xx} + a_2\,\Phi_1^2\,\bar{\Phi}_1 = 0$, then

$$(\Phi_1, \Phi_2) = \left[\Phi_1(x, t, b_1, b_2, c_1, c_2), e^{i\phi}\Phi_1^*(-x, t, b_1, b_2, -c_1, -c_2)\right] \qquad (9.44)$$

is a solution to the nonlocal CNLSE $i\,\Phi_{1t} + b_0\,\Phi_{1xx} + (b_1\,\Phi_1\,\bar{\Phi}_1 + b_2\,\Phi_2\,\bar{\Phi}_2)\,\Phi_1 = 0$, $i\,\Phi_{2t} - b_0\,\Phi_{2xx} + (c_1\,\Phi_1\,\bar{\Phi}_1 + c_2\,\Phi_2\,\bar{\Phi}_2)\,\Phi_2 = 0$, with $a_2 = -(c_1 + c_2)$.

9.5.3 Symmetry Reduction III *From Nonlocal Vector NLSE to Scalar Nonlocal NLSE*

The generalized nonlocal CNLSE

$$i\,\Phi_{j_t} + a_1\,\Phi_{j_{xx}} + \left(\sum_{k=1}^{N} b_{1k}\,\Phi_k\,\bar{\Phi}_k\right)\Phi_j = 0, \, j = 1, 2, \ldots, N, \qquad (9.45)$$

transforms to the scalar nonlocal NLSE $i\,\Phi_{1t} + a_1\,\Phi_{1xx} + \sum_{k=1}^{N} b_{1k}|\sigma_k|^2\,\Phi_1^2\,\bar{\Phi}_1 = 0$, with the replacement: $\Phi_j(x, t) = \sigma_j\,\Phi_1(x, t)$, where σ_j are arbitrary complex constants and $\sigma_1 = 1$.

Conclusion:
If $\Phi_1(x, t)$ is a solution to the nonlocal NLSE $i\,\Phi_{1t} + a_1\,\Phi_{1xx} + a_2\,\Phi_1^2\,\Phi_1 = 0$, then

$$(\Phi_1, \Phi_2, \Phi_3, \ldots, \Phi_N) = (\Phi_1, \sigma_2\,\Phi_1, \sigma_3\,\Phi_1, \ldots, \sigma_N\,\Phi_N) \qquad (9.46)$$

is a solution to the generalized nonlocal CNLSE $i\,\Phi_{j_t} + a_1\,\Phi_{j_{xx}} + (\sum_{k=1}^{N} b_{1k}\,\Phi_k\,\bar{\Phi}_k)\,\Phi_j = 0, \, j = 1, 2, \ldots, N$, with $a_2 = \sum_{k=1}^{N} b_{1k}|\sigma_k|^2$.

9.6 Scaling Transformations

9.6.1 Linear and Nonlinear Coupling

9.6.1.1 General Case
If (ψ_1, ψ_2) is a solution to

$$i\,\psi_{1t} + \psi_{1xx} + (b_1\,\psi_1\,\bar{\psi}_1 + b_2\,\psi_2\,\bar{\psi}_2)\,\psi_1 = 0,$$
$$i\,\psi_{2t} + \psi_{2xx} + (b_1\,\psi_1\,\bar{\psi}_1 + b_2\,\psi_2\,\bar{\psi}_2)\,\psi_2 = 0, \qquad (9.47)$$

then

$$\Phi_1(x, t) = \sqrt{\frac{b_1}{g_1 - g_2}}\,\psi_1(x, t)\,e^{i\,g_0\,(g_1 - g_2)\,t} + \sqrt{\frac{g_2\,b_2}{g_1\,(g_2 - g_1)}}\,\psi_2(x, t)\,e^{-i\,g_0\,(g_1 - g_2)\,t},$$

$$\Phi_2(x, t) = \sqrt{\frac{b_1}{g_1 - g_2}}\,\psi_1(x, t)\,e^{i\,g_0\,(g_1 - g_2)\,t} + \sqrt{\frac{g_1\,b_2}{g_2\,(g_2 - g_1)}}\,\psi_2(x, t)\,e^{-i\,g_0\,(g_1 - g_2)\,t} \qquad (9.48)$$

is a solution to

$$i \, \Phi_{1t} + \Phi_{1xx} + (g_1 \, \Phi_1 \, \bar{\Phi}_1 - g_2 \, \Phi_2 \, \bar{\Phi}_2) \, \Phi_1 + g_0 \, (g_1 + g_2) \, \Phi_1 - 2 \, g_0 \, g_2 \, \Phi_2 = 0,$$
$$i \, \Phi_{2t} + \Phi_{2xx} + (g_1 \, \Phi_1 \, \bar{\Phi}_1 - g_2 \, \Phi_2 \, \bar{\Phi}_2) \, \Phi_2 - g_0 \, (g_1 + g_2) \, \Phi_2 + 2 \, g_0 \, g_1 \, \Phi_1 = 0,$$

(9.49)

where $b_1 \, (g_1 - g_2) > 0$, $b_2 \, g_1 \, g_2 \, (g_2 - g_1) > 0$, b_1, b_2, g_0, g_1, and g_2 are real constants.

9.6.1.2 Specific Case I: Nonlocal Manakov System to Another Nonlocal Manakov System

If (ψ_1, ψ_2) is a solution to $i \, \psi_{1t} + \psi_{1xx} + (b_1 \, \psi_1 \, \bar{\psi}_1 + b_2 \, \psi_2 \, \bar{\psi}_2) \, \psi_1 = 0$, $i \, \psi_{2t} + \psi_{2xx} + (b_1 \, \psi_1 \, \bar{\psi}_1 + b_2 \, \psi_2 \, \bar{\psi}_2) \, \psi_2 = 0$, then

$$\Phi_1(x, t) = \sqrt{\frac{b_1}{g_1 - g_2}} \, \psi_1(x, t) + \sqrt{\frac{g_2 \, b_2}{g_1 \, (g_2 - g_1)}} \, \psi_2(x, t),$$
$$\Phi_2(x, t) = \sqrt{\frac{b_1}{g_1 - g_2}} \, \psi_1(x, t) + \sqrt{\frac{g_1 \, b_2}{g_2 \, (g_2 - g_1)}} \, \psi_2(x, t)$$

(9.50)

is a solution to

$$i \, \Phi_{1t} + \Phi_{1xx} + (g_1 \, \Phi_1 \, \bar{\Phi}_1 - g_2 \, \Phi_2 \, \bar{\Phi}_2) \, \Phi_1 = 0,$$
$$i \, \Phi_{2t} + \Phi_{2xx} + (g_1 \, \Phi_1 \, \bar{\Phi}_1 - g_2 \, \Phi_2 \, \bar{\Phi}_2) \, \Phi_2 = 0,$$

(9.51)

where $b_1 \, (g_1 - g_2) > 0$, $b_2 \, g_1 \, g_2 \, (g_2 - g_1) > 0$, b_1, b_2, g_1 and g_2 are real constants.

9.6.1.3 Specific Case II: Nonlocal Manakov System to the Same Nonlocal Manakov System: Superposition Principle for a Nonlocal Nonlinear System

If (ψ_1, ψ_2) is a solution to (9.47) then

$$\Phi_1(x, t) = \sqrt{\frac{b_1}{b_1 + b_2}} \left[\psi_1(x, t) - \frac{b_2}{b_1} \, \psi_2(x, t) \right],$$
$$\Phi_2(x, t) = \sqrt{\frac{b_1}{b_1 + b_2}} \, [\psi_1(x, t) + \psi_2(x, t)]$$

(9.52)

is also a solution to (9.47), where $b_1 + b_2 \neq 0$.

9.6.2 Complex Coupling

9.6.2.1 General Case
If (ψ_1, ψ_2) is a solution to

$$i \, \psi_{1t} + \psi_{1xx} + q_1 \, (q_2 \, \psi_1 \, \bar{\psi}_1 + q_3 \, \psi_2 \, \bar{\psi}_2) \, \psi_1 = 0,$$
$$i \, \psi_{2t} + \psi_{2xx} + q_1 \, (q_2 \, \psi_1 \, \bar{\psi}_1 + q_3 \, \psi_2 \, \bar{\psi}_2) \, \psi_2 = 0,$$

(9.53)

then

$$\Phi_1(x, t) = c_1 \, \psi_1(x, t) + c_2 \, \psi_2(x, t),$$
$$\Phi_2(x, t) = c_3 \, \psi_1(x, t) + c_4 \, \psi_2(x, t) \tag{9.54}$$

is a solution to

$$i \, \Phi_{1t} + \Phi_{1xx} + 2 \, (a_{11} \, \Phi_1 \, \bar{\Phi}_1 + a_{12} \, \Phi_2 \, \bar{\Phi}_2) \, \Phi_1 + 2 \, (b_{11} \, \Phi_1 \, \bar{\Phi}_2 + b_{12} \, \Phi_2 \, \bar{\Phi}_1) \, \Phi_1 = 0,$$
$$i \, \Phi_{2t} + \Phi_{2xx} + 2 \, (a_{21} \, \Phi_1 \, \bar{\Phi}_1 + a_{22} \, \Phi_2 \, \bar{\Phi}_2) \, \Phi_2 + 2 \, (b_{21} \, \Phi_1 \, \bar{\Phi}_2 + b_{22} \, \Phi_2 \, \bar{\Phi}_1) \, \Phi_2 = 0, \tag{9.55}$$

where
$$q_1 = \frac{2 \, (c_1 \, c_4 - c_2 \, c_3) \, (c_2^* \, c_3^* - c_1^* \, c_4^*)}{c_1^* \, c_2 \, c_3 \, c_4^* - c_1 \, c_2^* \, c_3^* \, c_4}, \qquad q_2 = (a - i \, b) \, c_1^* \, c_3 - (a + i \, b) \, c_1 \, c_3^*,$$

$$q_3 = a \, (c_2 \, c_4^* - c_2^* \, c_4) + i \, b \, (c_2^* \, c_4 + c_2 \, c_4^*),$$

$$a_{12} = \frac{b_{12} \, c_1^* \, c_2^* \, (c_2 \, c_3 - c_1 \, c_4) + b_{11} \, c_1 \, c_2 \, (c_1^* \, c_4^* - c_2^* \, c_3^*)}{c_1 \, c_2^* \, c_3^* \, c_4 - c_1^* \, c_2 \, c_3 \, c_4^*},$$

$$a_{11} = \frac{b_{11} \, c_3^* \, c_4^* \, (c_2 \, c_3 - c_1 \, c_4) + b_{12} \, c_3 \, c_4 \, (c_1^* \, c_4^* - c_2^* \, c_3^*)}{c_1 \, c_2^* \, c_3^* \, c_4 - c_1^* \, c_2 \, c_3 \, c_4^*}, \quad a_{22} = a_{12}, \ b_{12} = a - i \, b, \ b_{21} = a + i \, b,$$

$b_{11} = b_{21}, \ b_{22} = b_{12}, \ c_1 \, c_2^* \, c_3^* \, c_4$ should not be pure real or pure imaginary, $c_2 \, c_3 - c_1 \, c_4 \neq 0, c_j, j = 1, 2, 3, 4$ are complex constants, a and b are real constants.

9.6.2.2 Specific Case

If (ψ_1, ψ_2) is a solution to

$$i \, \psi_{1t} + \psi_{1xx} - 4 \, (b \, \psi_1 \, \bar{\psi}_1 + a \, \psi_2 \, \bar{\psi}_2) \, \psi_1 = 0,$$
$$i \, \psi_{2t} + \psi_{2xx} - 4 \, (b \, \psi_1 \, \bar{\psi}_1 + a \, \psi_2 \, \bar{\psi}_2) \, \psi_2 = 0, \tag{9.56}$$

then

$$\Phi_1(x, t) = \psi_1(x, t) + \psi_2(x, t),$$
$$\Phi_2(x, t) = \psi_1(x, t) + i \, \psi_2(x, t) \tag{9.57}$$

is a solution to

$$i \, \Phi_{1t} + \Phi_{1xx} - 2(a + b) \, (\Phi_1 \, \bar{\Phi}_1 + \Phi_2 \, \bar{\Phi}_2) \, \Phi_1 + 2 \, [(a + i \, b) \, \Phi_1 \, \bar{\Phi}_2 + (a - i \, b) \, \Phi_2 \, \bar{\Phi}_1] \, \Phi_1 = 0,$$
$$i \, \Phi_{2t} + \Phi_{2xx} - 2(a + b) \, (\Phi_1 \, \bar{\Phi}_1 + \Phi_2 \, \bar{\Phi}_2) \, \Phi_2 + 2 \, [(a + i \, b) \, \Phi_1 \, \bar{\Phi}_2 + (a - i \, b) \, \Phi_2 \bar{\Phi}_1] \, \Phi_2 = 0, \tag{9.58}$$

where a and b are real constants.

9.7 Nonlocal Discrete NLSE with Saturable Nonlinearity

If $\psi(x, t)$ is a solution to the NLSE with saturable nonlinearity (8.9),

$$i \, \psi_{nt} + \psi_{n+1} + \psi_{n-1} - 2 \, \psi_n + \frac{a_2 \, |\psi_n|^2 \, \psi_n}{1 + \mu \, |\psi_n|^2} = 0, \text{ then}$$

$$\Phi_n(n, t; a_2, \mu) = \begin{array}{ll} \psi_n(n, t, a_2, \mu) & \text{if } \psi_n \text{ is an even function in } n, \\ \psi_n(n, t, -a_2, -\mu) & \text{if } \psi_n \text{ is an odd function in } n \end{array}$$

is a solution to

$$i \, \Phi_{nt} + \Phi_{n+1} + \Phi_{n-1} - 2 \, \Phi_n + \frac{a_2 \, \Phi_n^2 \, \bar{\Phi}_n}{1 + \mu \, \Phi_n \bar{\Phi}_n} = 0, \qquad (9.59)$$

where $\psi_n = \psi_n(n, t; a_2, \mu)$, $\Phi_n = \Phi_n(n, t; a_2, \mu)$, $\bar{\Phi}_n = \Phi_n^*(-n, t; a_2, \mu)$, a_2 and μ are real constants.

9.7.1 Nonstaggered Solutions

***Example* 1. Even function: sech(n)**   *discrete bright soliton*
Given $\psi(\zeta, t) = A_0 \, \text{sech}[A_1 \, \zeta] \, e^{-i \, [A_2 \, (t-t_0) + \phi_0]}$ is a solution to (8.9), where $A_0 = \frac{\sinh(A_1)}{\sqrt{\mu}}$, $A_2 = \frac{2 \, \mu - a_2}{\mu}$, $\mu = \frac{a_2 \, \text{sech}(A_1)}{2} > 0$, $\zeta = n - n_0$, A_1, t_0, n_0, and ϕ_0 are arbitrary real constants, then

$$\Phi(\zeta, t) = A_0 \, \text{sech}[A_1 \, \zeta] \, e^{-i \, [A_2 \, (t-t_0) + \phi_0]} \qquad (9.60)$$

is a solution to (9.59), where $A_0 = \frac{\sinh(A_1)}{\sqrt{\mu}}$, $A_2 = \frac{2 \, \mu - a_2}{\mu}$, $\mu = \frac{a_2 \, \text{sech}(A_1)}{2} > 0$, $\zeta = n - n_0$, A_1, t_0, n_0, and ϕ_0 are arbitrary real constants.

***Example* 2. Odd function: tanh(n)** *discrete dark soliton*
Given $\psi(\zeta, t) = A_0 \, \tanh[A_1 \, \zeta] \, e^{-i \, [A_2 \, (t-t_0) + \phi_0]}$ is a solution to (8.9), where $A_0 = \frac{\tanh(A_1)}{\sqrt{-\mu}}$, $A_2 = \frac{-2 \, \mu + a_2}{-\mu}$, $a_2 = 2 \, \mu \, \text{sech}^2(A_1)$, $\zeta = n - n_0$, $\mu < 0$, A_1, n_0, and ϕ_0 are arbitrary real constants, then

$$\Phi(\zeta, t) = A_0 \, \tanh[A_1 \, \zeta] \, e^{-i \, [A_2 \, (t-t_0) + \phi_0]} \qquad (9.61)$$

is a solution to (9.59), where $A_0 = \frac{\tanh(A_1)}{\sqrt{\mu}}$, (*changed the sign of μ*), $A_2 = \frac{2 \, \mu - a_2}{\mu}$, (*changed the sign of μ and a_2*), $a_2 = 2 \, \mu \, \text{sech}^2(A_1)$, $\zeta = n - n_0$, $\mu > 0$, A_1, n_0, and ϕ_0 are arbitrary real constants.

9.7.2 Staggered Solutions

If $\Phi_n(n, t; a_2)$ is a nonstaggered solution to (9.59), then

$$\Phi_{ns}(n, t, a_2) = (-1)^n \, \Phi_n^*(n, t, -a_2) \, e^{-4 \, i \, (t-t_0)}, \qquad (9.62)$$

is a staggered solution to the same equation.

***Example* 1.** *staggered discrete bright soliton*
   Given $\Phi(\zeta, t) = A_0 \, \text{sech}[A_1 \, \zeta] \, e^{i \, [A_2 \, (t-t_0) + \phi_0]}$ is a nonstaggered solution to (9.59), where $A_0 = \frac{\sinh(A_1)}{\sqrt{\mu}}$, $A_2 = \frac{2 \, \mu - a_2}{\mu}$, $\mu = \frac{a_2 \, \text{sech}(A_1)}{2} > 0$, $\zeta = n - n_0$, A_1, n_0, t_0, and ϕ_0 are arbitrary real constants, then

$$\Phi_s(\zeta, t) = (-1)^{\zeta + n_0} \, A_0 \, \text{sech}(A_1 \, \zeta) \, e^{i \, [A_2 \, (t-t_0) - 4 \, (t-t_0) + \phi_0]} \qquad (9.63)$$

is a staggered solution to (9.59), where $A_0 = \frac{\sinh(A_1)}{\sqrt{\mu}}$, $A_2 = \frac{2 \, \mu + a_2}{\mu}$, $\mu = -\frac{a_2 \, \text{sech}(A_1)}{2} > 0$, $\zeta = n - n_0$, A_1, n_0, t_0, and ϕ_0 are arbitrary real constants.

9.8 Nonlocal Ablowitz–Ladik Equation

If $\psi(x, t)$ is a solution to the Ablowitz–Ladik Equation (8.49), $i\,\psi_{nt} + \psi_{n+1} + \psi_{n-1} - 2\,\psi_n + a_2\,(\psi_{n+1} + \psi_{n-1})\,|\psi_n|^2 = 0$, then

$$\Phi_n(n, t; a_2) = \begin{cases} \psi_n(n, t; a_2) & \text{if } \psi_n \text{ is an even function in } n, \\ \psi_n(n, t; -a_2) & \text{if } \psi_n \text{ is an odd function in } n \end{cases}$$

is a solution to

$$i\,\Phi_{nt} + \Phi_{n+1} + \Phi_{n-1} - 2\,\Phi_n + a_2\,(\Phi_{n+1} + \Phi_{n-1})\,\Phi_n\,\bar{\Phi}_n = 0, \tag{9.64}$$

where $\psi_n = \psi_n(n, t; a_2)$, $\Phi_n = \Phi_n(n, t; a_2)$, $\bar{\Phi}_n = \Phi_n^*(-n, t; a_2)$, a_2 is an arbitrary real constant.

***Example* 1. Even function: sech(n)** *discrete bright soliton*
Given $\psi(\zeta, t) = A_0\,\text{sech}[A_1\,\zeta]\,e^{-i\,[(A_2+2)\,(t-t_0)+\phi_0]}$ is a solution to (8.49), where $A_2 = -2\cosh(A_1)$, $a_2 = \frac{\sinh^2(A_1)}{A_0^2}$, $\zeta = n - n_0$, $A_0 \neq 0$, A_1, n_0, t_0, and ϕ_0 are arbitrary real constants, then

$$\Phi(\zeta, t) = A_0\,\text{sech}[A_1\,\zeta]\,e^{-i\,[(A_2+2)\,(t-t_0)+\phi_0]} \tag{9.65}$$

is a solution to (9.64), where $A_2 = -2\cosh(A_1)$, $a_2 = \frac{\sinh^2(A_1)}{A_0^2}$, $\zeta = n - n_0$, $A_0 \neq 0$, A_1, n_0, t_0, and ϕ_0 are arbitrary real constants.

***Example* 2. Odd function: tanh(n)** *discrete dark soliton*
Given $\psi(\zeta, t) = A_0\,\tanh[A_1\,\zeta]\,e^{-i\,[(A_2+2)\,(t-t_0)+\phi_0]}$ is a solution to (8.49), where $A_2 = -2\,\text{sech}(A_1)$, $a_2 = -\frac{\tanh^2(A_1)}{A_0^2}$, $\zeta = n - n_0$, $A_0 \neq 0$, A_0, A_1, n_0, t_0, and ϕ_0 are arbitrary real constants, then

$$\Phi(\zeta, t) = A_0\,\tanh[A_1\,\zeta]\,e^{-i\,[(A_2+2)\,(t-t_0)+\phi_0]} \tag{9.66}$$

is a solution to (9.64), where $A_2 = -2\,\text{sech}(A_1)$, $a_2 = \frac{\tanh^2(A_1)}{A_0^2}$, (*changed the sign of* a_2), $\zeta = n - n_0$, $A_0 \neq 0$, A_0, A_1, n_0, t_0, and ϕ_0 are arbitrary real constants.

9.9 Nonlocal Cubic-Quintic Discrete NLSE

If $\psi(x, t)$ is a solution to the qubic-quintic discrete NLSE (8.62), $i\,\psi_{nt} + a_1\,(\psi_{n+1} + \psi_{n-1} - 2\,\psi_n) + a_2\,|\psi_n|^2\,\psi_n + (a_3\,|\psi_n|^2 + a_4\,|\psi_n|^4)(\psi_{n+1} + \psi_{n-1})$
$= 0$,
then

$$\Phi_n(n, t; a_1, a_2, a_3, a_4) = \begin{cases} \psi_n(n, t; a_1, a_2, a_3, a_4) & \text{if } \psi_n \text{ is an even function in } n, \\ \psi_n(n, t; a_1, -a_2, -a_3, a_4) & \text{if } \psi_n \text{ is an odd function in } n \end{cases}$$

is a solution to

$$i\,\Phi_{nt} + a_1\,(\Phi_{n+1} + \Phi_{n-1} - 2\,\Phi_n) + a_2\,\Phi_n{}^2\,\bar\Phi_n + (a_3\,\Phi_n\,\bar\Phi_n + a_4\,\Phi_n{}^2\,\bar\Phi_n{}^2)(\Phi_{n+1} + \Phi_{n-1}) = 0, \quad (9.67)$$

where $\quad \psi_n = \psi_n(n,\,t;\,a_1,\,a_2,\,a_3,\,a_4), \quad\qquad \Phi_n = \Phi_n(n,\,t;\,a_1,\,a_2,\,a_3,\,a_4),$
$\bar\Phi_n = \Phi_n^*(-n,\,t;\,a_1,\,a_2,\,a_3,\,a_4),\,a_1,\,a_2,\,a_3,$ and a_4 are arbitrary real constants.

***Example* 1. Even function: sech(*n*) I** *discrete bright soliton*
Given $\quad \psi(\zeta,\,t) = A_0\,\mathrm{sech}[A_1\,\zeta]\,e^{i\,[A_2\,(t-t_0)+\phi_0]}\quad$ is a solution to (8.62), where

$$A_0 = \pm\frac{\sqrt{a_3\,(a_2^2 - 4\,a_1\,a_4) + \sqrt{a_3^2 - 4\,a_1\,a_4}\,(a_2^2 + 4\,a_1\,a_4)}}{2\,a_4\,\sqrt{2\,a_1}}, \qquad A_1 = \mathrm{sech}^{-1}(\frac{-a_3 + \sqrt{a_3^2 - 4\,a_1\,a_4}}{a_2}),$$

$$A_2 = -2\,a_1 - \frac{a_2\,(a_3 + \sqrt{a_3^2 - 4\,a_1\,a_4})}{2\,a_4}, \qquad a_3^2 - 4\,a_4\,a_1 \geqslant 0, \qquad a_1 > 0, \qquad \zeta = n - n_0,$$

$a_3\,(a_2^2 - 4\,a_1\,a_4) + \sqrt{a_3^2 - 4\,a_1\,a_4}\,(a_2^2 + 4\,a_1\,a_4) > 0,\,t_0,\,n_0,$ and ϕ_0 are arbitrary real constants, then

$$\Phi(\zeta,\,t) = A_0\,\mathrm{sech}[A_1\,\zeta]\,e^{i\,[A_2\,(t-t_0)+\phi_0]} \qquad\qquad (9.68)$$

is a solution to (9.64), where $\quad A_0 = \pm\dfrac{\sqrt{a_3\,(a_2^2 - 4\,a_1\,a_4) + \sqrt{a_3^2 - 4\,a_1\,a_4}\,(a_2^2 + 4\,a_1\,a_4)}}{2\,a_4\,\sqrt{2\,a_1}},$

$A_1 = \mathrm{sech}^{-1}(\dfrac{-a_3 + \sqrt{a_3^2 - 4\,a_1\,a_4}}{a_2}),\quad A_2 = -2\,a_1 - \dfrac{a_2\,(a_3 + \sqrt{a_3^2 - 4\,a_1\,a_4})}{2\,a_4},\quad a_3^2 - 4\,a_4\,a_1 \geqslant 0,$

$a_1 > 0,\,a_3\,(a_2^2 - 4\,a_1\,a_4) + \sqrt{a_3^2 - 4\,a_1\,a_4}\,(a_2^2 + 4\,a_1\,a_4) > 0,\,\zeta = n - n_0,\,t_0,\,n_0,$
and ϕ_0 are arbitrary real constants.

***Example* 2. Even function: sech(*n*) II** *staggered discrete bright soliton*
Given $\psi(\zeta,\,t) = (-1)^{\zeta+n_0}\,A_0\,\mathrm{sech}[A_1\,\zeta]\,e^{i\,[A_2\,(t-t_0)+\phi_0]}$ is a solution to (8.62), where

$$A_0 = \pm\frac{\sqrt{a_3\,(a_2^2 - 4\,a_1\,a_4) + \sqrt{a_3^2 - 4\,a_1\,a_4}\,(a_2^2 + 4\,a_1\,a_4)}}{2\,a_4\,\sqrt{2\,a_1}}, \qquad A_1 = \mathrm{sech}^{-1}(\frac{a_3 - \sqrt{a_3^2 - 4\,a_1\,a_4}}{a_2}),$$

$$A_2 = -2\,a_1 - \frac{a_2\,(a_3 + \sqrt{a_3^2 - 4\,a_1\,a_4})}{2\,a_4}, \qquad\qquad a_3^2 - 4\,a_4\,a_1 \geqslant 0, \qquad\qquad a_1 > 0,$$

$a_3\,(a_2^2 - 4\,a_1\,a_4) + \sqrt{a_3^2 - 4\,a_1\,a_4}\,(a_2^2 + 4\,a_1\,a_4) > 0,\,\zeta = n - n_0,\,t_0,\,n_0,$ and ϕ_0
are arbitrary real constants then

$$\Phi(\zeta,\,t) = (-1)^{\zeta+n_0}\,A_0\,\mathrm{sech}[A_1\,\zeta]\,e^{i\,[A_2\,(t-t_0)+\phi_0]} \qquad\qquad (9.69)$$

is a solution to (9.64), where $\quad A_0 = \pm\dfrac{\sqrt{a_3\,(a_2^2 - 4\,a_1\,a_4) + \sqrt{a_3^2 - 4\,a_1\,a_4}\,(a_2^2 + 4\,a_1\,a_4)}}{2\,a_4\,\sqrt{2\,a_1}},$

$A_1 = \mathrm{sech}^{-1}(\dfrac{a_3 - \sqrt{a_3^2 - 4\,a_1\,a_4}}{a_2}),\quad A_2 = -2\,a_1 - \dfrac{a_2\,(a_3 + \sqrt{a_3^2 - 4\,a_1\,a_4})}{2\,a_4},\quad a_3^2 - 4\,a_4\,a_1 \geqslant 0,$

$a_1 > 0,\,a_3\,(a_2^2 - 4\,a_1\,a_4) + \sqrt{a_3^2 - 4\,a_1\,a_4}\,(a_2^2 + 4\,a_1\,a_4) > 0,\,\zeta = n - n_0,\,t_0,\,n_0,$
and ϕ_0 are arbitrary real constants.

***Example* 3. Odd function: tanh(*n*) I** *discrete dark soliton*

Given $\psi(\zeta, t) = A_0 \tanh[A_1 \zeta] e^{i [A_2 (t-t_0)+\phi_0]}$ is a solution to (8.62), where

$$A_0 = \pm\sqrt{\frac{a_2 + a_3 + \sqrt{a_3^2 - 4 a_1 a_4}}{-2 a_4}}, \qquad A_1 = \cosh^{-1}(\sqrt{\frac{a_3 + \sqrt{a_3^2 - 4 a_1 a_4}}{-a_2}}), \qquad A_2 = -2 a_1 - \frac{a_2 (a_3 - \sqrt{a_3^2 - 4 a_1 a_4})}{2 a_4},$$

$a_3^2 - 4 a_4 a_1 \geqslant 0$, $a_4 (a_2 + a_3 + \sqrt{a_3^2 - 4 a_1 a_4}) < 0$, $a_2 (a_3 + \sqrt{a_3^2 - 4 a_1 a_4}) < 0$, $\zeta = n - n_0$, t_0, n_0, and ϕ_0 are arbitrary real constants, then

$$\Phi(\zeta, t) = A_0 \tanh[A_1 \zeta] e^{i [A_2 (t-t_0)+\phi_0]} \tag{9.70}$$

is a solution to (9.64), where $A_0 = \pm\sqrt{\dfrac{-a_2 - a_3 + \sqrt{a_3^2 - 4 a_1 a_4}}{-2 a_4}}$, (*changed the sign of* a_2

and a_3), $A_1 = \cosh^{-1}(\sqrt{\dfrac{-a_3 + \sqrt{a_3^2 - 4 a_1 a_4}}{a_2}})$, (*changed the sign of* a_2 *and* a_3),

$A_2 = -2 a_1 - \dfrac{-a_2 (-a_3 - \sqrt{a_3^2 - 4 a_1 a_4})}{2 a_4}$, (*changed the sign of* a_2 *and* a_3), $a_3^2 - 4 a_4 a_1 \geqslant 0$,

$a_4 (-a_2 - a_3 + \sqrt{a_3^2 - 4 a_1 a_4}) < 0$, $a_2 (-a_3 + \sqrt{a_3^2 - 4 a_1 a_4}) > 0$, $\zeta = n - n_0$, t_0,
n_0, and ϕ_0 are arbitrary real constants.

***Example* 4. Odd function: tanh(*n*) II** *staggered discrete dark soliton*

Given $\psi(\zeta, t) = (-1)^{\zeta+n_0} A_0 \tanh[A_1 \zeta] e^{i [A_2 (t-t_0)+\phi_0]}$ is a solution to (8.62), where

$$A_0 = \pm\sqrt{\frac{a_2 - a_3 - \sqrt{a_3^2 - 4 a_1 a_4}}{2 a_4}}, \qquad A_1 = \cosh^{-1}(\sqrt{\frac{a_3 + \sqrt{a_3^2 - 4 a_1 a_4}}{a_2}}), \qquad A_2 = -2 a_1 - \frac{a_2 (a_3 - \sqrt{a_3^2 - 4 a_1 a_4})}{2 a_4},$$

$a_3^2 - 4 a_4 a_1 \geqslant 0$, $a_2 - a_3 - \sqrt{a_3^2 - 4 a_1 a_4} > 0$, $a_2 (a_3 + \sqrt{a_3^2 - 4 a_1 a_4}) > 0$,
$\zeta = n - n_0$, t_0, n_0, and ϕ_0 are arbitrary real constants, then

$$\Phi(\zeta, t) = (-1)^{\zeta+n_0} A_0 \tanh[A_1 \zeta] e^{i [A_2 (t-t_0)+\phi_0]} \tag{9.71}$$

is a solution to (9.64), where $A_0 = \pm\sqrt{\dfrac{-a_2 + a_3 - \sqrt{a_3^2 - 4 a_1 a_4}}{2 a_4}}$, (*changed the sign of* a_2

and a_3), $A_1 = \cosh^{-1}(\sqrt{\dfrac{-a_3 + \sqrt{a_3^2 - 4 a_1 a_4}}{-a_2}})$, (*changed the sign of* a_2 *and* a_3),

$A_2 = -2 a_1 - \dfrac{a_2 (a_3 + \sqrt{a_3^2 - 4 a_1 a_4})}{2 a_4}$, (*changed the sign of* a_2 *and* a_3), $a_3^2 - 4 a_4 a_1 \geqslant 0$,

$a_4 (-a_2 + a_3 - \sqrt{a_3^2 - 4 a_1 a_4}) > 0$, $a_2 (-a_3 + \sqrt{a_3^2 - 4 a_1 a_4}) < 0$, $\zeta = n - n_0$, t_0,
n_0, and ϕ_0 are arbitrary real constants.

9.10 Summary of Chapter 9

Nonlocal NLSE

Scaling Transformation: *From Local NLSE to Nonlocal NLSE*

$$\text{Transformation: } \Phi(x, t; a_1, a_2) = \begin{cases} \psi(x, t, a_1, a_2) & \text{if } \psi \text{ is an even function in } x, \\ \psi(x, t, a_1, -a_2) & \text{if } \psi \text{ is an odd function in } x \end{cases}$$

ψ is a solution to the fundamental NLSE (2.160).

Equation: $i\,\Phi_t + a_1\,\Phi_{xx} + a_2\,|\Phi|^2\,\Phi = 0$

#	Example	Conditions	Name	Equation #
1.	$\Phi(\zeta, t) = A_0 \sqrt{\dfrac{2a_1}{a_2}}\ \mathrm{sech}[A_0\,\zeta]\,e^{i\,[a_1\,A_0^2\,(t-t_0)+\phi_0]}$	$\zeta = x - x_0,\ a_1\,a_2 > 0,\ A_0,\,x_0,\,t_0,\ \text{and } \phi_0$ are arbitrary real constants	Bright soliton	(9.18)
2.	$\Phi(\zeta, t) = A_0 \tanh[\dfrac{A_1}{\sqrt{a_1}}\,\zeta]\,e^{-i\,[A_2\,(t-t_0)+\phi_0]}$	$A_2 = 2\,A_1^2,\ a_2 = \dfrac{2\,A_1^2}{A_0^2},\ a_1 > 0,\ \zeta = x - x_0,\, A_0,\, A_1,\, x_0,\, t_0,\ \text{and } \phi_0$ are arbitrary	Dark soliton	(9.19)

Other Solutions

Equation: $i\,\psi_t + a_1\,\psi_{xx} + a_2\,|\psi|^2\,\bar{\psi} = 0$

#	Solution	Conditions	Name	Equation #
1.	$\psi(x, t) = [-\lambda_1^{*2}\,e^{q_1(x,\,t)} + \lambda_1^2\,e^{q_2(x,\,t)}$ $- 4\,i\,\lambda_{li}\,e^2\,q_3(x,\,t)(-\lambda_{lr} + |\lambda_1|^2\,x)]$ $/\,[[2\,i\,e^{q_3(x,\,t)}\{\lambda_{li}\cos[q_4(x,\,t)] + \lambda_{li}\cosh[2\,x\,\lambda_{lr}$ $- 2\,i\,t(\lambda_{li}^2 - \lambda_{lr}^2)] + |\lambda_1|^2\,x\sin[q_4(x,\,t)]\}]]$	$q_1(x, t) = 2\,\lambda_1^*(x + i\,t\,\lambda_1^*),\ q_2(x, t) = 2\,\lambda_1(x + i\,t\,\lambda_1),$ $q_3(x, t) = 2\,x\,\lambda_{lr} - 2\,i\,t\,(\lambda_{li}^2 - \lambda_{lr}^2),$ $q_4(x, t) = 2\,\lambda_{li}\,(x + 2\,i\,t\,\lambda_{lr}),\ \lambda_1 = \lambda_{lr} + i\,\lambda_{li};\ a_1 = 1,$ $a_2 = 1/2.\ \lambda_{lr} \text{ and } \lambda_{li}$ are arbitrary real constants	Singular soliton molecule	(9.20)
2.	$\psi(\zeta, t) = \dfrac{A_0 + i\,A_3\sin[\frac{A_1}{\sqrt{a_1}}\,\zeta]}{A_4 + \cos[\frac{A_1}{\sqrt{a_1}}\,\zeta]}\,e^{-i\,[A_2\,(t-t_0)+\phi_0]}$	$A_2 = -A_1^2/2,\ A_3 = A_1/\sqrt{2\,a_2},\ A_4 > 1,\ a_1 > 0,$ $a_2 = (A_4^2 - 1)\,A_1^2/(2\,A_0^2),\ \zeta = x - x_0,\, A_0,\, A_1,\, x_0,\, t_0,\ \text{and } \phi_0$ are arbitrary real constants	—	(9.21)

(Continued)

3.
$$\psi(\zeta, t) = \frac{A_0 \sin[\frac{A_1}{\sqrt{a_1}}\zeta] + i A_3}{A_4 + \cos[\frac{A_1}{\sqrt{a_1}}\zeta]} e^{-i[A_2(t-t_0)+\phi_0]}$$

$A_2 = -A_1^2/2, A_3 = A_1\sqrt{A_4^2 - 1}/\sqrt{2 a_2}, A_4 > 1, a_1 > 0,$
$a_2 = A_1^2/(2 A_0^2), \zeta = x - x_0, A_0, A_1, x_0, t_0, \text{ and } \phi_0 \text{ are}$
arbitrary real constants

(9.22)

4.
$$\psi(\zeta, t) = \sqrt{m} \left\{ \frac{A_0 \, cn[\frac{A_1}{\sqrt{a_1}}\zeta, m] + i A_3 \, sn[\frac{A_1}{\sqrt{a_1}}\zeta, m]}{A_4 + dn[\frac{A_1}{\sqrt{a_1}}\zeta, m]} \right\}$$
$$\times e^{-i[A_2(t-t_0)+\phi_0]}$$

$A_2 = (2-m)A_1^2/2, A_3 = A_1\sqrt{A_4^2 - 1} + m/\sqrt{2 a_2},$
$A_4 > 1, a_1 > 0, a_2 = A_1^2(A_4^2 - 1)/(2 A_0^2), m > 0,$
$\zeta = x - x_0, A_0, A_1, x_0, t_0, \text{ and } \phi_0 \text{ are arbitrary real constants}$

(9.23)

5.
$$\psi(\zeta, t) = \left\{ \frac{A_0 \, sech[\frac{A_1}{\sqrt{a_1}}\zeta] + i A_3 \, tanh[\frac{A_1}{\sqrt{a_1}}\zeta]}{A_4 + sech[\frac{A_1}{\sqrt{a_1}}\zeta]} \right\}$$
$$\times e^{-i[A_2(t-t_0)+\phi_0]}$$

$A_2 = A_1^2/2, A_3 = A_1\sqrt{A_4^2}/\sqrt{2 a_2}, A_4 > 1, a_1 > 0,$
$a_2 = A_1^2(A_4^2 - 1)/(2 A_0^2), \zeta = x - x_0, A_0, A_1, x_0, t_0, \text{ and } \phi_0$
are arbitrary real constants

(9.24)

6.
$$\psi(\zeta, t) = \sqrt{m} \left\{ \frac{A_0 \, sn[\frac{A_1}{\sqrt{a_1}}\zeta, m] + i A_3 \, cn[\frac{A_1}{\sqrt{a_1}}\zeta, m]}{A_4 + dn[\frac{A_1}{\sqrt{a_1}}\zeta, m]} \right\}$$
$$\times e^{-i[A_2(t-t_0)+\phi_0]}$$

$A_2 = (2-m)A_1^2/2, A_3 = A_1\sqrt{A_4^2 - 1}/\sqrt{2 a_2}, A_4 > 1,$
$a_1 > 0, a_2 = A_1^2(A_4^2 - 1 + m)/(2 A_0^2), m > 0,$
$\zeta = x - x_0, A_0, A_1, x_0, t_0, \text{ and } \phi_0 \text{ are arbitrary real constants}$

(9.25)

7.
$$\psi(\zeta, t) = \frac{A_0 \, tanh[\frac{A_1}{\sqrt{a_1}}\zeta] + i A_3 \, sech[\frac{A_1}{\sqrt{a_1}}\zeta]}{A_4 + sech[\frac{A_1}{\sqrt{a_1}}\zeta]}$$
$$\times e^{-i[A_2(t-t_0)+\phi_0]}$$

$A_2 = A_1^2/2, A_3 = A_1\sqrt{A_4^2 - 1}/\sqrt{2 a_2}, A_4 > 1, a_1 > 0,$
$a_2 = A_1^2 A_4^2/(2 A_0^2), \zeta = x - x_0, A_0, A_1, x_0, t_0, \text{ and } \phi_0 \text{ are}$
arbitrary real constants

(9.26)

8.
$$\psi(\zeta, t) = \frac{A_0 \, dn[\frac{A_1}{\sqrt{a_1}}\zeta, m] + i A_3 \sqrt{m} \, sn[\frac{A_1}{\sqrt{a_1}}\zeta, m]}{A_4 + cn[\frac{A_1}{\sqrt{a_1}}\zeta, m]}$$
$$\times e^{-i[A_2(t-t_0)+\phi_0]}$$

$A_2 = (2m-1)A_1^2/2.$
$A_3 = A_1\sqrt{m A_4^2 - m + 1}/\sqrt{2 m a_2}, A_4 > 1, a_1 > 0,$
$a_2 = A_1^2(A_4^2 - 1)/(2 A_0^2), m > 0, \zeta = x - x_0, A_0, A_1, x_0,$
$t_0, \text{ and } \phi_0 \text{ are arbitrary real constants}$

(9.27)

9.
$$\psi(\zeta, t) = \frac{A_0 \sqrt{m} \, sn[\frac{A_1}{\sqrt{a_1}}\zeta, m] + i A_3 \, dn[\frac{A_1}{\sqrt{a_1}}\zeta, m]}{A_4 + cn[\frac{A_1}{\sqrt{a_1}}\zeta, m]}$$
$$\times e^{-i[A_2(t-t_0)+\phi_0]}$$

$A_2 = (2m-1)A_1^2/2, A_3 = A_1\sqrt{A_4^2 - 1}/\sqrt{2 a_2}, A_4 > 1,$
$a_1 > 0, a_2 = A_1^2(1 - m + m A_4^2)/(2 m A_0^2), m > 0,$
$\zeta = x - x_0, A_0, A_1, x_0, t_0, \text{ and } \phi_0 \text{ are arbitrary real constants}$

(9.28)

Yang's nonlocal NLSE

Equation: $i\,\psi_t + a_1\,\psi_{xx} + a_2\,(|\psi|^2 + |\bar\psi|^2)\,\psi = 0$

#	Solution	Conditions	Name	Equation #
1.	$\psi(\zeta, t) = A_0\, e^{i[A_1(t-t_0)+A_2(x-x_0)\phi_0]}$	$A_1 = 2A_0^2\,a_2 - A_2^2\,a_1$, $A_0, A_2, x_0, t_0,$ and ϕ_0 are arbitrary real constants	Continuous wave	(9.30)
2.	$\psi(\zeta, t) = \dfrac{A_0\sqrt{m}\,\mathrm{sn}[\frac{A_1}{\sqrt{a_1}}\zeta, m]}{1+\mathrm{dn}[\frac{A_1}{\sqrt{a_1}}\zeta, m]}\, e^{-i[A_2(t-t_0)+\phi_0]}$	$A_2 = (2-m)A_1^2/2$, $a_1 > 0$, $a_2 = -m\,A_1^2/(4A_0^2)$, $m > 0$. $\zeta = x - x_0$, $A_0, A_1, x_0, t_0,$ and ϕ_0 are arbitrary real constants	—	(9.31)
3.	$\psi(\zeta, t) = \dfrac{A_0\sqrt{m}\,\mathrm{cn}[\frac{A_1}{\sqrt{a_1}}\zeta, m]}{\sqrt{1-m}+\mathrm{dn}[\frac{A_1}{\sqrt{a_1}}\zeta, m]}\, e^{-i[A_2(t-t_0)+\phi_0]}$	$A_2 = (2-m)A_1^2/2$, $a_1 > 0$, $a_2 = -m\,A_1^2/(4A_0^2)$. $0 < m < 1$, $\zeta = x - x_0$, $A_0, A_1, x_0, t_0,$ and ϕ_0 are arbitrary real constants	—	(9.32)
4.	$\psi(\zeta, t) = A_0\left\{\dfrac{1}{2} - \dfrac{\mathrm{dn}[\frac{A_1}{\sqrt{a_1}}\zeta, m]}{(1-m)^{1/4}+\mathrm{dn}[\frac{A_1}{\sqrt{a_1}}\zeta, m]}\right\} e^{-i[A_2(t-t_0)+\phi_0]}$	$A_2 = [(2-m) + 6\sqrt{1-m}]\,A_1^2/2$, $a_1 > 0$, $a_2 = -(1-\sqrt{1-m})^2\,A_1^2/(4A_0^2)$, $0 < m < 1$, $\zeta = x - x_0$, $A_0, A_1, x_0, t_0,$ and ϕ_0 are arbitrary real constants	—	(9.33)

Nonlocal coupled NLSE

Transformation:

$$\Phi_1(x, t;\, a_1, a_2, b_1, b_2) = \begin{cases} \psi_1(x, t, a_1, a_2, b_1, b_2) & \text{if } \psi_1 \text{ is an even function in } x, \\ \psi_1(x, t, -a_1, a_2, -b_1, b_2) & \text{if } \psi_1 \text{ is an odd function in } x, \end{cases}$$

$$\Phi_2(x, t;\, a_1, a_2, b_1, b_2) = \begin{cases} \psi_2(x, t, a_1, a_2, b_1, b_2) & \text{if } \psi_2 \text{ is an even function in } x, \\ \psi_2(x, t, a_1, -a_2, b_1, -b_2) & \text{if } \psi_2 \text{ is an odd function in } x \end{cases}$$

Equation: $i\,\Phi_{1t} + \Phi_{1xx} + (a_1\,\Phi_1\bar\Phi_1 + a_2\,\Phi_2\bar\Phi_2)\,\Phi_1 = 0,\ i\,\Phi_{2t} + \Phi_{2xx} + (b_1\,\Phi_1\bar\Phi_1 + b_2\,\Phi_2\bar\Phi_2)\,\Phi_2 = 0$

(Continued)

#	Example	Conditions	Name	Equation #
1.	$\Phi_1(\zeta, t) = A_0 \operatorname{sech}[A_1 \zeta] e^{-i[\omega_1(t-t_0)+\phi_1]}$, $\Phi_2(\zeta, t) = B_0 \tanh[A_1 \zeta] e^{-i[\omega_2(t-t_0)+\phi_2]}$	$\omega_1 = A_1^2 - a_1 A_0^2, \omega_2 = 2 A_1^2 - b_1 A_0^2, a_1 = \dfrac{2 A_1^2 - a_2 B_0^2}{A_0^2}$, $b_1 = \dfrac{2 A_1^2 - b_2 B_0^2}{A_0^2}, \zeta = x - x_0, A_0 \neq 0, A_1, B_0, x_0, t_0, \phi_1,$ and ϕ_2 are arbitrary real constants	Bright-dark soliton	(9.36)
2.	$\Phi_1(\zeta, t) = A_0 m \operatorname{cn}[A_1 \zeta, m] \operatorname{sn}[A_1 \zeta, m] e^{-i[\omega_1(t-t_0)+\phi_1]}$, $\Phi_2(\zeta, t) = B_0 \sqrt{m} \operatorname{dn}[A_1 \zeta, m] \operatorname{sn}[A_1 \zeta, m] e^{-i[\omega_2(t-t_0)+\phi_2]}$	$\omega_1 = (4 + m) A_1^2, \omega_2 = (1 + 4 m) A_1^2$, $a_1 = \dfrac{-6 A_1^2}{A_0^2 (1 - m)}, a_2 = \dfrac{6 A_1^2}{B_0^2 (1 - m)}, b_1 = a_1, b_2 = a_2$, $0 \leq m < 1, \zeta = x - x_0$, $A_0 \neq 0$, $B_0 \neq 0, A_1, x_0, t_0, \phi_1,$ and ϕ_2 are arbitrary real constants	Solitary wave	(9.37)
3.	$\Phi_1(\zeta, t) = A_0 \sqrt{m(1-m)} \operatorname{sd}[A_1 \zeta, m] e^{-i[\omega_1(t-t_0)+\phi_1]}$, $\Phi_2(\zeta, t) = B_0 \sqrt{1 - m} \operatorname{nd}[A_1 \zeta, m] e^{-i[\omega_2(t-t_0)+\phi_2]}$	$\omega_1 = (m - 1) a_1 A_0^2 - A_1^2$, $\omega_2 = (m - 2) A_1^2 + (m - 1) b_1 A_0^2, a_1 = \dfrac{-2 A_1^2 + a_2 B_0^2}{A_0^2}$, $b_1 = \dfrac{-2 A_1^2 + b_2 B_0^2}{A_0^2}, 0 \leq m \leq 1, \zeta = x - x_0, A_0 \neq 0, A_1, B_0$, $x_0, t_0, \phi_1,$ and ϕ_2 are arbitrary real constants	Solitary wave	(9.38)
4.	$\Phi_1(\zeta, t) = \{A_0 \operatorname{sech}^2[A_1 \zeta] + A_3\} e^{-i[\omega_1(t-t_0)+\phi_1]}$, $\Phi_2(\zeta, t) = B_0 \operatorname{sech}^2[A_1 \zeta] e^{-i[\omega_2(t-t_0)+\phi_2]}$	$A_3 = \dfrac{-2 A_0}{3}, \omega_1 = 2 A_1^2, \omega_2 = -2 A_1^2, a_1 = \dfrac{-9 A_1^2}{2 A_0^2}, a_2 = \dfrac{9 A_1^2}{2 B_0^2}$, $b_1 = a_1, b_2 = a_2, \zeta = x - x_0, A_0 \neq 0, B_0 \neq 0, A_1, x_0, t_0, \phi_1,$ and ϕ_2 are arbitrary real constants	—	(9.39)

Symmetry Reductions to Scalar Nonlocal NLSE

Symmetry Reductions I: From Nonlocal Manakov System to Scalar Nonlocal NLSE

Transformation: $\Phi_2(x,\,t) = \sigma\,\Phi_1(x,\,t), b_1 = c_1 + (c_2 - b_2)\,|\sigma|^2$

Equation: $i\,\Phi_{1t} + b_0\,\Phi_{1xx} + (c_1 + c_2|\sigma|^2)\,\Phi_1^2\,\bar{\Phi}_1 = 0$

Symmetry Reductions II: From Nonlocal Manakov System to Scalar Nonlocal NLSE

Transformation: $\Phi_2(x;\,t;\,b_1,\,b_2,\,c_1,\,c_2) = e^{i\,\phi}\begin{cases}\Phi_1^*(-x,\,t,\,b_1,\,b_2,\,c_1,\,c_2) & \text{if } \Phi_1 \text{ is an even function in } x,\\ \Phi_1^*(-x,\,t,\,b_1,\,b_2,\,-c_1,\,-c_2) & \text{if } \Phi_1 \text{ is an odd function in } x,\end{cases}$

$b_1 = -(c_1 + c_2 + b_2)$

Equation: $i\,\Phi_{1t} + b_0\,\Phi_{1xx} - (c_1 + c_2)\,\Phi_1^2\,\bar{\Phi}_1 = 0$

Symmetry Reductions III: From Nonlocal Vector NLSE to Scalar Nonlocal NLSE

Transformation: $\Phi_j(x,\,t) = \sigma_j\,\Phi_1(x,\,t)$

Equation: $i\,\Phi_{1t} + a_1\,\Phi_{1xx} + \sum_{k=1}^{N} b_{1k}|\sigma_k|^2\,\Phi_1^2\,\bar{\Phi}_1 = 0$

Scaling Transformations

Linear and Nonlinear Coupling

General Case

Transformation: $\Phi_1(x,\,t) = \sqrt{\dfrac{b_1}{g_1 - g_2}}\,\psi_1(x,\,t)\,e^{i\,g_0\,(g_1 - g_2)\,t} + \sqrt{\dfrac{g_2\,b_2}{g_1\,(g_2 - g_1)}}\,\psi_2(x,\,t)\,e^{-i\,g_0\,(g_1 - g_2)\,t},$

$\Phi_2(x,\,t) = \sqrt{\dfrac{b_1}{g_1 - g_2}}\,\psi_1(x,\,t)\,e^{i\,g_0\,(g_1 - g_2)\,t} + \sqrt{\dfrac{g_1\,b_2}{g_2\,(g_2 - g_1)}}\,\psi_2(x,\,t)\,e^{-i\,g_0\,(g_1 - g_2)\,t}$

Equation: $i\,\Phi_{1t} + \Phi_{1xx} + (g_1\,\Phi_1\,\bar{\Phi}_1 - g_2\,\Phi_2\,\bar{\Phi}_2)\,\Phi_1 + g_0\,(g_1 + g_2)\,\Phi_1 - 2\,g_0\,g_2\,\Phi_2 = 0,$

$i\,\Phi_{2t} + \Phi_{2xx} + (g_1\,\Phi_1\,\bar{\Phi}_1 - g_2\,\Phi_2\,\bar{\Phi}_2)\,\Phi_2 - g_0\,(g_1 + g_2)\,\Phi_2 + 2\,g_0\,g_1\,\Phi_1 = 0$

(*Continued*)

Specific Case I: Nonlocal Manakov System to Another Nonlocal Manakov System

Transformation: $\Phi_1(x, t) = \sqrt{\dfrac{b_1}{g_1 - g_2}}\, \psi_1(x, t) + \sqrt{\dfrac{g_2\, b_2}{g_1\, (g_2 - g_1)}}\, \psi_2(x, t)$, $\Phi_2(x, t) = \sqrt{\dfrac{b_1}{g_1 - g_2}}\, \psi_1(x, t) + \sqrt{\dfrac{g_1\, b_2}{g_2\, (g_2 - g_1)}}\, \psi_2(x, t)$

Equation: $i\,\Phi_{1t} + \Phi_{1xx} + (g_1\, \Phi_1\, \bar{\Phi}_1 - g_2\, \Phi_2\, \bar{\Phi}_2)\, \Phi_1 = 0$, $i\,\Phi_{2t} + \Phi_{2xx} + (g_1\, \Phi_1\, \bar{\Phi}_1 - g_2\, \Phi_2\, \bar{\Phi}_2)\, \Phi_2 = 0$

Specific Case II: Nonlocal Manakov System to the Same System

Superposition Principle for a Nonlocal Nonlinear System

Transformation: $\Phi_1(x, t) = \sqrt{\dfrac{b_1}{b_1 + b_2}}\, [\psi_1(x, t) - \dfrac{b_2}{b_1}\, \psi_2(x, t)]$, $\Phi_2(x, t) = \sqrt{\dfrac{b_1}{b_1 + b_2}}\, [\psi_1(x, t) + \psi_2(x, t)]$

Complex coupling

General Case

Transformation: $\Phi_1(x, t) = c_1\, \psi_1(x, t) + c_2\, \psi_2(x, t)$, $\Phi_2(x, t) = c_3\, \psi_1(x, t) + c_4\, \psi_2(x, t)$

Equation: $i\,\Phi_{1t} + \Phi_{1xx} + 2\, (a_{11}\, \Phi_1\, \bar{\Phi}_1 + a_{12}\, \Phi_2\, \bar{\Phi}_2)\, \Phi_1 + 2\, (b_{11}\, \Phi_1\, \bar{\Phi}_2 + b_{12}\, \Phi_2\, \bar{\Phi}_1)\, \Phi_1 = 0$,

$i\,\Phi_{2t} + \Phi_{2xx} + 2\, (a_{21}\, \Phi_1\, \bar{\Phi}_1 + a_{22}\, \Phi_2\, \bar{\Phi}_2)\, \Phi_2 + 2\, (b_{21}\, \Phi_1\, \bar{\Phi}_2 + b_{22}\, \Phi_2\, \bar{\Phi}_1)\, \Phi_2 = 0$

Specific Case

Transformation: $\Phi_1(x, t) = \psi_1(x, t) + \psi_2(x, t)$, $\Phi_2(x, t) = \psi_1(x, t) + i\, \psi_2(x, t)$

Equation: $i\,\Phi_{1t} + \Phi_{1xx} - 2(a + b)\, (\Phi_1\, \bar{\Phi}_1 + \Phi_2\, \bar{\Phi}_2)\, \Phi_1 + 2\, [(a + i\, b)\, \Phi_1\, \bar{\Phi}_2 + (a - i\, b)\, \Phi_2\, \bar{\Phi}_1]\, \Phi_1 = 0$,

$i\,\Phi_{2t} + \Phi_{2xx} - 2(a + b)\, (\Phi_1\, \bar{\Phi}_1 + \Phi_2\, \bar{\Phi}_2)\, \Phi_2 + 2\, [(a + i\, b)\, \Phi_1\, \bar{\Phi}_2 + (a - i\, b)\, \Phi_2\, \bar{\Phi}_1]\, \Phi_2 = 0$

Nonlocal Discrete NLSE with Saturable Nonlinearity

Transformation: $\Phi_n(n, t; a_2, \mu) = \begin{cases} \psi_n(n, t, a_2, \mu) & \text{if } \psi_n \text{ is an even function in } n, \\ \psi_n(n, t, -a_2, -\mu) & \text{if } \psi_n \text{ is an odd function in } n \end{cases}$

Equation: $i\,\Phi_{nt} + \Phi_{n+1} + \Phi_{n-1} - 2\, \Phi_n + \dfrac{a_2\, \Phi_n^2\, \bar{\Phi}_n}{1 + \mu\, \Phi_n\, \bar{\Phi}_n} = 0$

#	Example	Conditions	Name	Equation #
1.	$\Phi(\zeta, t) = A_0 \, \text{sech}[A_1 \, \zeta] \, e^{-i \, [A_2 \, (t-t_0) + \phi_0]}$	$A_0 = \dfrac{\sinh(A_1)}{\sqrt{\mu}}, \ A_2 = \dfrac{2 \, \mu - a_2}{\mu}, \ \mu = \dfrac{a_2 \, \text{sech}(A_1)}{2} > 0, \ \zeta = n - n_0,$ $A_1, n_0, t_0, \text{ and } \phi_0 \text{ are arbitrary real constants}$	Bright soliton	(9.60)
2.	$\Phi(\zeta, t) = A_0 \, \tanh[A_1 \, \zeta] \, e^{-i \, [A_2 \, (t-t_0) + \phi_0]}$	$A_0 = \dfrac{\tanh(A_1)}{\sqrt{\mu}}, \ A_2 = \dfrac{2 \, \mu - a_2}{\mu}, \ a_2 = 2 \, \mu \, \text{sech}^2(A_1), \ \mu > 0,$ $\zeta = n - n_0, \ A_1, n_0, \text{ and } \phi_0 \text{ are arbitrary real constants}$	Dark soliton	(9.61)
3.	$\Phi_s(\zeta, t) = (-1)^{\zeta + n_0} A_0 \, \text{sech}[A_1 \, \zeta] \, e^{i \, [A_2 \, (t-t_0) - 4 \, (t-t_0) + \phi_0]}$	$A_0 = \dfrac{\sinh(A_1)}{\sqrt{\mu}}, \ A_2 = \dfrac{2 \, \mu + a_2}{\mu}, \ \mu = -\dfrac{a_2 \, \text{sech}(A_1)}{2} > 0, \ \zeta = n - n_0,$ $A_1, n_0, t_0, \text{ and } \phi_0 \text{ are arbitrary real constants}$	Staggered bright soliton	(9.63)

Nonlocal Ablowitz–Ladik Equation

Transformation: $\Phi_n(n, t; a_2) = \begin{cases} \psi_n(n, t; a_2) & \text{if } \psi_n \text{ is an even function in } n, \\ \psi_n(n, t; -a_2) & \text{if } \psi_n \text{ is an odd function in } n \end{cases}$

Equation: $i \, \Phi_{nt} + \Phi_{n+1} + \Phi_{n-1} - 2 \, \Phi_n + a_2 \, (\Phi_{n+1} + \Phi_{n-1}) \, \Phi_n \, \bar{\Phi}_n = 0$

#	Example	Conditions	Name	Equation #
1.	$\Phi(\zeta, t) = A_0 \, \text{sech}[A_1 \, \zeta] \, e^{-i \, [(A_2 + 2) \, (t-t_0) + \phi_0]}$	$A_2 = -2 \, \cosh(A_1), \ a_2 = \dfrac{\sinh^2(A_1)}{A_0^2}, \ \zeta = n - n_0, \ A_0 \neq 0, \ A_0, A_1,$ $n_0, t_0, \text{ and } \phi_0 \text{ are arbitrary real constants}$	Discrete bright soliton	(9.65)
2.	$\Phi(\zeta, t) = A_0 \, \tanh[A_1 \, \zeta] \, e^{-i \, [(A_2 + 2) \, (t-t_0) + \phi_0]}$	$A_2 = -2 \, \text{sech}(A_1), \ a_2 = \dfrac{\tanh^2(A_1)}{A_0^2}, \ \zeta = n - n_0, \ A_0 \neq 0, \ A_0, A_1,$ $n_0, t_0, \text{ and } \phi_0 \text{ are arbitrary real constants}$	Discrete dark soliton	(9.66)

(*Continued*)

Nonlocal Cubic-Quintic Discrete NLSE

Transformation: $\Phi_n(n, t; a_1, a_2, a_3, a_4) = \dfrac{\psi_n(n, t; a_1, a_2, a_3, a_4)}{\psi_n(n, t; a_1, -a_2, -a_3, a_4)}$ if ψ_n is an even function in n,
     if ψ_n is an odd function in n

Equation: $i\,\Phi_{nt} + a_1\,(\Phi_{n+1} + \Phi_{n-1} - 2\,\Phi_n) + a_2\,\Phi_n^2\,\bar{\Phi}_n + (a_3\,\Phi_n\,\bar{\Phi}_n + a_4\,\Phi_n^2\,\bar{\Phi}_n^2)(\Phi_{n+1} + \Phi_{n-1}) = 0$

#	Example	Conditions	Name	Equation #
1.	$\Phi(\zeta, t) = A_0\,\operatorname{sech}[A_1\,\zeta]\,e^{i\,[A_2\,(t-t_0)+\phi_0]}$	$A_0 = \pm\sqrt{\dfrac{a_3\,(a_2^2 - 4\,a_1\,a_4) + \sqrt{a_3^2 - 4\,a_1\,a_4\,(a_2^2 + 4\,a_1\,a_4)}}{2\,a_4\sqrt{2\,a_1}}}$, $A_1 = \operatorname{sech}^{-1}\!\left(\dfrac{-a_3 + \sqrt{a_3^2 - 4\,a_1\,a_4}}{a_2}\right)$, $\zeta = n - n_0$, $A_2 = -2\,a_1 - \dfrac{a_2\,(a_3 + \sqrt{a_3^2 - 4\,a_1\,a_4})}{2\,a_4}$, $a_3^2 - 4\,a_4\,a_1 \geqslant 0$, $a_3\,(a_2^2 - 4\,a_1\,a_4) + \sqrt{a_3^2 - 4\,a_1\,a_4\,(a_2^2 + 4\,a_1\,a_4)} > 0$. $a_1 > 0$, t_0, n_0, and ϕ_0 are arbitrary real constants	Discrete bright soliton	(9.68)
2.	$\Phi(\zeta, t) = (-1)^{\zeta+n_0}\,A_0\,\operatorname{sech}[A_1\,\zeta]\,e^{i\,[A_2\,(t-t_0)+\phi_0]}$	$A_0 = \pm\sqrt{\dfrac{a_3\,(a_2^2 - 4\,a_1\,a_4) + \sqrt{a_3^2 - 4\,a_1\,a_4\,(a_2^2 + 4\,a_1\,a_4)}}{2\,a_4\sqrt{2\,a_1}}}$, $A_1 = \operatorname{sech}^{-1}\!\left(\dfrac{a_3 - \sqrt{a_3^2 - 4\,a_1\,a_4}}{a_2}\right)$, $\zeta = n - n_0$, $A_2 = -2\,a_1 - \dfrac{a_2\,(a_3 + \sqrt{a_3^2 - 4\,a_1\,a_4})}{2\,a_4}$, $a_3^2 - 4\,a_4\,a_1 \geqslant 0$, $a_3\,(a_2^2 - 4\,a_1\,a_4) + \sqrt{a_3^2 - 4\,a_1\,a_4\,(a_2^2 + 4\,a_1\,a_4)} > 0$. $a_1 > 0$, t_0, n_0, and ϕ_0 are arbitrary real constants	Staggered discrete bright soliton	(9.69)

3. $\Phi(\zeta, t) = A_0 \tanh[A_1 \zeta] e^{i[A_2(t-t_0)+\phi_0]}$

Discrete dark soliton (9.70)

$$A_0 = \pm\sqrt{\frac{-a_2 - a_3 + \sqrt{a_3^2 - 4a_1 a_4}}{-2a_4}},$$

$$A_1 = \cosh^{-1}\left(\sqrt{\frac{-a_3 + \sqrt{a_3^2 - 4a_1 a_4}}{a_2}}\right), \zeta = n - n_0,$$

$$A_2 = -2a_1 - \frac{-a_2(-a_3 - \sqrt{a_3^2 - 4a_1 a_4})}{2a_4}, a_3^2 - 4a_4 a_1 \geq 0,$$

$$a_4(-a_2 - a_3 + \sqrt{a_3^2 - 4a_1 a_4}) < 0,$$

$$a_2(-a_3 + \sqrt{a_3^2 - 4a_1 a_4}) > 0, t_0, n_0, \text{ and } \phi_0 \text{ are arbitrary real constants}$$

4. $\Phi(\zeta, t) = (-1)^{\zeta+n_0} A_0 \tanh[A_1 \zeta] e^{i[A_2(t-t_0)+\phi_0]}$

Staggered discrete dark soliton (9.71)

$$A_0 = \pm\sqrt{\frac{-a_2 + a_3 - \sqrt{a_3^2 - 4a_1 a_4}}{2a_4}},$$

$$A_1 = \cosh^{-1}\left(\sqrt{\frac{-a_3 + \sqrt{a_3^2 - 4a_1 a_4}}{-a_2}}\right), \zeta = n - n_0,$$

$$A_2 = -2a_1 - \frac{a_2(a_3 + \sqrt{a_3^2 - 4a_1 a_4})}{2a_4}, a_3^2 - 4a_4 a_1 \geq 0,$$

$$a_4(-a_2 + a_3 - \sqrt{a_3^2 - 4a_1 a_4}) > 0,$$

$$a_2(-a_3 + \sqrt{a_3^2 - 4a_1 a_4}) < 0, t_0, n_0, \text{ and } \phi_0 \text{ are arbitrary real constants}$$

References

[1] Ablowitz M J and Musslimani Z H 2013 Integrable nonlocal nonlinear Schrödinger equation *Phys. Rev. Lett.* **110** 064105

[2] Ablowitz M J and Musslimani Z H 2016 *Integrable Nonlocal Nonlinear equations* **vol 139** (New York: Wiley) pp 7–59 (Studies in Applied Mathematics)

[3] Ablowitz M J, Luo X-D and Musslimani Z H 2020 Discrete nonlocal nonlinear Schrödinger systems: integrability, inverse scattering and solitons *Nonlinearity* **33** 3653

[4] Gadzhimuradov T A and Agalarov A M 2016 Towards a gauge-equivalent magnetic structure of the nonlocal nonlinear Schrödinger equation *Phys. Rev.* A **93** 062124

[5] Yang J 2018 Physically significant nonlocal nonlinear Schrödinger equation and its soliton solutions *Phys. Rev.* E **98** 042202

[6] Santos L, Shlyapnikov G V, Zoller P and Lewenstein M 2000 Bose-Einstein condensation in trapped dipolar gases *Phys. Rev. Lett.* **85** 1791

[7] Elhadj K M, Al Sakkaf L, Al Khawaja U and Boudjemâa A 2020 Singular soliton molecules of the nonlinear Schrödinger equation *Phys. Rev.* E **101** 042221

[8] Khare A and Saxena A 2023 New solutions of nonlocal Nls, Mkdv and Hirota equations *Ann Phys.* **460** 169561

IOP Publishing

Handbook of Exact Solutions to the Nonlinear Schrödinger Equations (Second Edition)

Usama Al Khawaja and Laila Al Sakkaf

Chapter 10

Fractional Nonlinear Schrödinger Equation

A Glance at Chapter 10

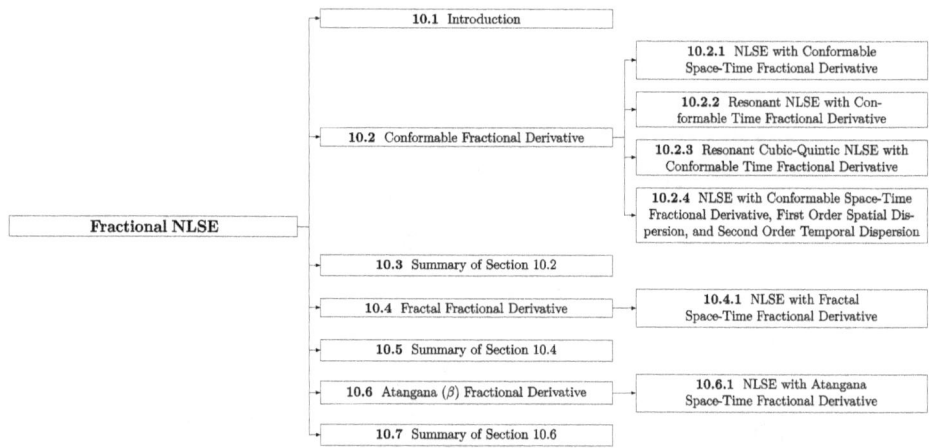

A Statistical View of Chapter 10

	Equation	Solutions
1	$i\,\Phi_t^\alpha + a_1\,\Phi_{xx}^{2\beta} + a_2\,|\Phi|^2\,\Phi = 0$	3
2	$i\,\Phi_t^\alpha + a_1\,\Phi_{xx}^{2\beta} + a_2\,|\Phi|^n\,\Phi = 0$	3
3	$i\,\Phi_t^\alpha + a_1\,\Phi_{xx}^{2\beta} + a_2\,|\Phi|^n\,\Phi + a_3\,|\Phi|^m\,\Phi = 0$	3
4	$i\,\Phi_t^\alpha + a_1\,\Phi_{xx}^{2\beta} + a_2\,|\Phi|^2\,\Phi + i\,a_3\,\Phi_{xxx}^{3\beta} + i\,a_4\,(|\Phi|^2\,\Phi)_x^\beta + i\,a_5\,(|\Phi|^2)_x^\beta\,\Phi = 0$	3
5	$i\,\Phi_t + b_{10}\,\Phi_{xx}^{2\beta} + \frac{a_2\,b_{10}\,c_5}{a_1\,c_5^2\,c_7}\,|\Phi|^2\,\Phi - \frac{3\,c\,c_7}{b_{10}\,c_5}\left(\frac{t^\alpha}{\alpha}\right)\left(\frac{x^\beta}{\beta}\right)\Phi = 0$	3
6	$i\,\Phi_t^\alpha + \alpha_1\,\Phi_{x_1 x_1}^{2\beta} + \alpha_2\,\Phi_{x_2 x_2}^{2\beta} + a_2\,|\Phi|^2\,\Phi = 0$	3
7	$i\,\Phi_{1t}^\alpha + b_0\,\Phi_{1xx}^{2\beta} + (b_1\,|\Phi_1|^2 + b_2\,|\Phi_2|^2)\,\Phi_1 = 0,$ $i\,\Phi_{2t}^\alpha + c_0\,\Phi_{2xx}^{2\beta} + (c_1\,|\Phi_1|^2 + c_2\,|\Phi_2|^2)\,\Phi_2 = 0$	3
8	$i\,\Phi_{nt}^\alpha + \Phi_{n+1} + \Phi_{n-1} - 2\,\Phi_n + \frac{a_2\,|\Phi_n|^2\,\Phi_n}{1+\mu\,|\Phi_n|^2} = 0$	3
9	$i\,\Phi_{nt}^\alpha + \Phi_{n+1} + \Phi_{n-1} - 2\,\Phi_n + \frac{a_2\,\Phi_n^2\,\Phi_n}{1+\mu\,\Phi_n\,\bar{\Phi}_n} = 0$	3
10	$i\,\psi_t^\alpha + a_1\,\psi_{xx} + a_2\,|\psi|^2\,\psi + a_3\,\frac{|\psi|_{xx}}{|\psi|}\,\psi = 0$	5
11	$i\,\psi_t^\alpha + a_1\,\psi_{xx} + a_2\,|\psi|^2\,\psi + a_3\,|\psi|^4\,\psi + a_4\,\frac{|\psi|_{xx}}{|\psi|}\,\psi = 0$	4
12	$i\,\psi_t^\alpha + a_1\,\psi_{xx}^{2\beta} + a_2\,|\psi|^2\,\psi + a_3\,\psi_{tt}^\alpha + i\,a_4\,\psi_x^\beta = 0$	7
Total	**12**	**43**

10.1 Introduction

In 1983, Mandelbrot introduced the term 'fractal' to the domain of physical science [1], bringing with it the concept of fractional calculus. This notion gave rise to the idea of fractional derivatives, which have found numerous applications across various scientific and engineering contexts. These applications demonstrate non-locality, memory-dependent behavior, and multiple time and length scales [2–4], where the boundaries of integer-order derivatives were extended.

Subsequently, Laskin extended the framework by constructing a novel fractional path integral that expanded the Feynman path integral, from the Brownian-like to Lévy-like quantum mechanical paths. This development led to the expansion of the classical Schrödinger equation into a space fractional partial differential equation

$$-i\,\hbar\,\frac{\partial\psi}{\partial t} + D_\alpha(-\hbar^2\Delta)^{\alpha/2}\psi + V(x,t)\,\psi = 0, \tag{10.1}$$

wherein the conventional Laplacian, $\Delta = \partial^2/\partial x^2$, is substituted with the quantum Riesz space-fractional derivative [5, 6]

$$(-\hbar^2 \Delta)^{\alpha/2} \psi = \frac{1}{2\pi\hbar} \int_{-\infty}^{+\infty} e^{i p x/\hbar} |p|^\alpha \phi(p, t) \, dp, \tag{10.2}$$

where $h = 2\pi\hbar$ is the Planck's constant, $V(x, t)$ is an arbitrary potential, $\phi(p, t)$ is the Fourier transformation of the quantum wave function $\psi = \psi(x, t)$, D_α is a physical quantity representing the generalized fractional diffusion coefficient, and $0 < \alpha \leqslant 2$ is the Lévy index. Building on this, Guo and Xu explored practical applications of the space fractional Schrödinger equation [2].

Moreover, the time fractional Schrödinger equation, which features a Caputo fractional time derivative, was explored by Naber based on fractional Brownian motion [7]. Further applications of fractional calculus include fractional geometry, non-Gaussian statistics [8, 9], hydrogeologic behavior of water motion [10, 11], turbulence [12], anomalous diffusion through disordered media [13], and biological systems [14, 15].

These examples emphasize the necessity of employing fractional calculus to accurately describe these physical phenomena. Fractional calculus involves extending the concepts of differentiation and integration from integer orders to fractional orders. It is within this framework that the space–time fractional nonlinear Schrödinger equation (FNLSE) takes the form,

$$i \frac{\partial^\alpha \psi}{\partial t^\alpha} + a_1 \frac{\partial^{2\beta} \psi}{\partial x^{2\beta}} + a_2 |\psi|^2 \psi = 0, \tag{10.3}$$

in which the first order derivative and the second order derivative are replaced with fractional order derivatives, where $0 < \alpha \leqslant 1$ and $0 < \beta \leqslant 1$ denote the fractional order of temporal (t) and spatial (x) dependencies. If $\alpha = 1$, the space-time FNLSE reduces to space FNLSE, if $\beta = 1$, it becomes the time FNLSE. When both α and β equal 1, the equation reverts to the regular integer NLSE.

All variations of the NLSE, including those with higher order terms, $(N + 1)$-dimensional NLSE, discrete NLSE, nonlocal NLSE, and coupled NLSE, can be expressed in terms of their fractional counterparts. In these fractional versions, either the time derivative, the spatial derivative, or both derivatives are converted into fractional derivatives.

Multiple methods exist for extending integer-order derivatives to fractional-order derivatives. Among these, the most explored and studied fractional derivatives of fractional calculus are the Riemann–Liouville, Caputo, and the Grünwald–Letnikov derivatives [16, 17]. The three variants play a crucial role in the investigation of fractional differential equations. The Riemann–Liouville derivative of order α for a function $f(x)$ is defined as

$$f_{RL}^\alpha(x) := \begin{cases} \dfrac{1}{\Gamma(m-\alpha)} \dfrac{d^m}{dx^m} \displaystyle\int_0^x [(x-x_0)^{m-\alpha-1} f(x_0) \, dx_0], & m-1 < \alpha < m, \\ \dfrac{d^m}{dx^m} f(x), & \alpha = m, \end{cases} \tag{10.4}$$

where m is an integer order.

Although the Riemann–Liouville definition is used across various scientific and engineering fields [18], it does not apply for nonzero constant functions. To resolve this, the Caputo definition was introduced, which assigns zero to the derivative of a constant function [18]. The Caputo derivative of order α for a function $f(x)$ is defined as

$$f_C^\alpha(x) := \begin{cases} \dfrac{1}{\Gamma(m-\alpha)} \displaystyle\int_0^x (x-x_0)^{m-\alpha-1} \dfrac{d^m f(x_0)}{dx_0^m} \, dx_0, & m-1 < \alpha < m \\[2ex] \dfrac{d^m}{dx^m} f(x), & \alpha = m, \end{cases} \tag{10.5}$$

where m is an integer order.

The Grünwald–Letnikov derivative of order α for a function $f(x)$ is defined based on discretizing the derivative using a finite difference formula as

$$f_{GL}^\alpha(x) := \lim_{\substack{h \to 0 \\ n \to \infty}} \frac{1}{h^\alpha} \sum_{j=0}^{n} (-1)^j \binom{\alpha}{j} f(x - j\,h), \tag{10.6}$$

where n is the number of grid points, h is the equidistant grid spacing, and

$$\binom{\alpha}{j} = \frac{\Gamma(\alpha+1)}{j!\,\Gamma(\alpha-j+1)} \tag{10.7}$$

is the generalized binomial coefficient.

Several other variations of fractional derivatives have been introduced by extending the concept of fractional integration within the framework of Riemann–Liouville, including the Riesz derivative, Atangana–Baleanu–Caputo, and Caputo–Hadamard definitions [16, 19].

Recently, a new fractional derivative known as the conformable fractional derivative has been introduced. This definition extends the conventional principles of integer-order calculus. In a similar manner, various other fractional derivatives have emerged, such as the M-fractional (truncated M-fractional) derivative, local M-fractional derivative, local fractional derivative, Atangana (β) fractional derivative, and fractal fractional derivative [16, 20–24].

Given the multitude of definitions, a question comes up: what criteria must a differential operator meet to be considered as a fractional operator? According to the wide sense criterion (WSC) proposed in [25], the derivative operator is considered as a fractional derivative if it obeys the following properties:

1. Linearity: the derivative operator is linear.
2. Identity: the zeroth-order derivative of a function yields the function itself.
3. Backward compatibility: for integer orders, fractional derivative and regular derivatives yield equivalent results.
4. Law of exponents: $\frac{d^\alpha}{dx^\alpha} \frac{d^\beta}{dx^\beta} f(x) = \frac{d^{\alpha+\beta}}{dx^{(\alpha+\beta)}} f(x)$, for $\alpha < 0$ and $\beta < 0$.
5. Generalized Leibniz rule: $\frac{d^\alpha}{dx^\alpha}[f(x)g(x)] = \sum_{i=0}^{\infty} \binom{\alpha}{i} \frac{d^i}{dx^i} f(x) \frac{d^{\alpha-i}}{dx^{\alpha-i}} g(x)$.

The inconsistency lies in satisfying these criteria. While fractional derivatives in the Riemann–Liouville and Caputo frameworks fulfill these criteria, other definitions, like the conformable derivative do not. This discrepancy has led to criticism of these criteria [26]. It is important to highlight that fractional derivatives conforming to the aforementioned criteria might still not satisfy certain main characteristics [27]. For example

1. Derivative of a constant c: $\frac{d^\alpha}{dx^\alpha} c = 0$, when α is not an integer.

2. Quotient rule of two functions: $\frac{d^\alpha}{dx^\alpha} \frac{f(x)}{g(x)} = \frac{f(x)\frac{d^\alpha}{dx^\alpha}g(x) - g(x)\frac{d^\alpha}{dx^\alpha}f(x)}{g^2(x)}$.

3. Chain rule: $\frac{d^\alpha}{dx^\alpha}[f(x) \circ g(x)] = \frac{d^\alpha}{dx^\alpha}g(x) \frac{d^\alpha}{dx^\alpha}f(g(x))$, where 'o' is the composition operator.

In this chapter, we consider the fractional NLSE within the framework of the fractional derivative types that admit exact analytical solutions. Specifically, the conformable, fractal, and Atangana types, with their definitions presented subsequently. We refer to the fractional-order derivative as *fractional derivative* and to the integer-order derivative as *regular derivative*.

10.2 Conformable Fractional Derivative

The conformable* fractional derivative of a function $f(x)$ is defined as

$$f_x^\alpha(x) := \lim_{\epsilon \to 0} \left[\frac{f(x + \epsilon\, x^{1-\alpha}) - f(x)}{\epsilon} \right] = f'(x), \qquad (10.8)$$

where α is the fractional derivative order $0 < \alpha \leqslant 1$. The second conformable fractional derivative of order α is then defined as

$$f_{xx}^\alpha(x) := \lim_{\epsilon \to 0} \left[\frac{f'(x + \epsilon\, x^{1-\alpha}) - f'(x)}{\epsilon} \right] = f''(x) \qquad (10.9)$$

and the n conformable fractional derivative of order α is defined as

$$f_{(nx)}^\alpha(x) := \lim_{\epsilon \to 0} \left[\frac{f^{(n-1)}(x + \epsilon\, x^{1-\alpha}) - f^{(n-1)}(x)}{\epsilon} \right] = f^{(n)}(x). \qquad (10.10)$$

*The conformable fractional derivative is equivalent to the *Local M-fractional derivative* defined by

$$f_x^\alpha(x) := \lim_{\epsilon \to 0} \left[\frac{f[x\, E_1(\epsilon\, x^{-\alpha})] - f(x)}{\epsilon} \right] = f'(x), \qquad (10.11)$$

where $E_\gamma(\cdot)$, $\gamma = 1$, is the Mittag–Leffler function.

10.2.1 NLSE with Conformable Space-Time Fractional Derivative

If $\psi(x, t)$ is a solution to any equation from the previous nine chapters or any other versions of the NLSE that involve both regular time and regular space derivatives, then

$$\Phi(x,\, t) = \psi\left(\frac{x^\beta}{\beta},\, \frac{t^\alpha}{\alpha}\right), \tag{10.12}$$

$$\Phi(x,\, t) = \psi\left(\frac{x^\beta}{\beta},\, t\right), \tag{10.13}$$

and

$$\Phi(x,\, t) = \psi\left(x,\, \frac{t^\alpha}{\alpha}\right) \tag{10.14}$$

are solutions to the same equation, where both or any of the regular time and regular space derivatives are replaced by the conformable fractional time and fractional space derivatives, respectively

$$\psi_t \rightarrow \Phi_t^\alpha, \qquad \psi_x \rightarrow \Phi_x^\beta \tag{10.15}$$

and all space- and time-dependent coefficients are transformed as follows

$$g(x,\, t) \rightarrow g\left(\frac{x^\beta}{\beta},\, \frac{t^\alpha}{\alpha}\right), \tag{10.16}$$

where α and β are the time and space fractional derivative of orders $0 < \alpha \leqslant 1$ and $0 < \beta \leqslant 1$, respectively.

***Example* 1:** *Conformable space–time fractional fundamental NLSE*

Given $\psi(x,\, t) = A_0 \sqrt{\frac{2 a_1}{a_2}}\ \mathrm{sech}[A_0\, (x - x_0)]\, e^{i\, [a_1\, A_0^2\, (t - t_0) + \phi_0]}$ is a solution to the fundamental NLSE (2.160), $i\, \psi_t + a_1\, \psi_{xx} + a_2\, |\psi|^2\, \psi = 0$, then

$$\Phi(x,\, t) = A_0 \sqrt{\frac{2 a_1}{a_2}}\ \mathrm{sech}\left[A_0 \left(\frac{x^\beta}{\beta} - x_0\right)\right] e^{i\, \left[a_1\, A_0^2 \left(\frac{t^\alpha}{\alpha} - t_0\right) + \phi_0\right]} \tag{10.17}$$

is a solution to

$$i\, \Phi_t^\alpha + a_1\, \Phi_{xx}^{2\,\beta} + a_2\, |\Phi|^2\, \Phi = 0, \tag{10.18}$$

where $\Phi = \Phi(x,\, t)$ is the complex function profile, x and t are its two independent variables, $a_1\, a_2 > 0$, A_0, x_0, t_0, and ϕ_0 are arbitrary real constants.

***Example* 2:** *Conformable space–time fractional NLSE with power law nonlinearity*

Given $\psi(x,\, t) = \left\{\frac{2 A_0^2\, a_1\, (n + 2)}{a_2\, n^2}\ \mathrm{sech}^2[A_0\, (x - x_0)]\right\}^{\frac{1}{n}} e^{i\, \left[\frac{4 a_1\, A_0^2}{n^2}\, (t - t_0) + \phi_0\right]}$ is a solution to the NLSE with power law nonlinearity (3.9), $i\, \psi_t + a_1\, \psi_{xx} + a_2\, |\psi|^n\, \psi = 0$, then

$$\Phi(x,\, t) = \left\{\frac{2 A_0^2\, a_1\, (n + 2)}{a_2\, n^2}\ \mathrm{sech}^2\left[A_0 \left(\frac{x^\beta}{\beta} - x_0\right)\right]\right\}^{\frac{1}{n}} e^{i\, \left[\frac{4 a_1\, A_0^2}{n^2} \left(\frac{t^\alpha}{\alpha} - t_0\right) + \phi_0\right]} \tag{10.19}$$

is a solution to

$$i\ \Phi_t^\alpha + a_1\ \Phi_{xx}^{2\beta} + a_2\ |\Phi|^n\ \Phi = 0, \tag{10.20}$$

where $\Phi = \Phi(x, t)$ is the complex function profile, x and t are its two independent variables, $a_1\ a_2\ (n + 2) > 0$, n, A_0, x_0, t_0, and ϕ_0 are arbitrary real constants.

Example 3: *Conformable space–time fractional NLSE with dual power law nonlinearity*

Given $\psi(x, t) = \left(\dfrac{A_0\ (n+2)}{a_2 + a_2\ \sqrt{1+\gamma}\ \cosh\{n\ \sqrt{\frac{A_0}{a_1}}\ [(x - x_0)]\}} \right)^{\frac{1}{n}} e^{i\ [A_0\ (t-t_0)+\phi_0]}$ is a solution to the NLSE

with dual power law nonlinearity (3.27), $i\ \psi_t + a_1\ \psi_{xx} + a_2\ |\psi|^n\ \psi + a_3\ |\psi|^m\ \psi = 0$, then

$$\Phi(x, t) = \left(\frac{A_0\ (n+2)}{a_2 + a_2\ \sqrt{1+\gamma}\ \cosh\left\{ n\ \sqrt{\frac{A_0}{a_1}}\ \left[\left(\frac{x^\beta}{\beta} - x_0 \right) \right] \right\}} \right)^{\frac{1}{n}} e^{i\left[A_0\ \left(\frac{t^\alpha}{\alpha} - t_0 \right) + \phi_0 \right]} \tag{10.21}$$

is a solution to

$$i\ \Phi_t^\alpha + a_1\ \Phi_{xx}^{2\beta} + a_2\ |\Phi|^n\ \Phi + a_3\ |\Phi|^m\ \Phi = 0, \tag{10.22}$$

where $\Phi = \Phi(x, t)$ is the complex function profile, x and t are its two independent variables, $m = 2\ n$, $\gamma = a_3/a_{30}$, $a_{30} = a_2^2\ (n + 1)/[A_0\ (n + 2)^2]$, $(1 + \gamma) \geqslant 0$, $A_0\ a_1 > 0$, n, x_0, t_0, and ϕ_0 are arbitrary real constants.

Example 4: *Conformable space–time fractional NLSE with third order dispersion, self-steepening, and self-frequency shift*

Given $\psi(x, t) = \pm\sqrt{-\mu_1}\ \mathrm{sech}\{\sqrt{-\mu_2}\ [x - x_0 + c_1\ (t - t_0)]\} e^{i\ [c_2\ (x-x_0)+c_3\ (t-t_0)+\phi_0]}$ is a solution to the NLSE with third order dispersion, self-steepening, and self-frequency shift (4.2), $i\ \psi_t + a_1\ \psi_{xx} + a_2\ |\psi|^2\ \psi + i\ a_3\ \psi_{xxx} + i\ a_4\ (|\psi|^2\ \psi)_x + i\ a_5\ (|\psi|^2)_x\ \psi = 0$, then

$$\Phi(x, t) = \pm\sqrt{-\mu_1}\ \mathrm{sech}\left\{ \sqrt{-\mu_2}\ \left[\frac{x^\beta}{\beta} - x_0 + c_1\ \left(\frac{t^\alpha}{\alpha} - t_0 \right) \right] \right\} e^{i\left[c_2\left(\frac{x^\beta}{\beta} - x_0 \right) + c_3\left(\frac{t^\alpha}{\alpha} - t_0 \right) + \phi_0 \right]} \tag{10.23}$$

is a solution to

$$i\ \Phi_t^\alpha + a_1\ \Phi_{xx}^{2\beta} + a_2\ |\Phi|^2\ \Phi + i\ a_3\ \Phi_{xxx}^{3\beta} + i\ a_4\ (|\Phi|^2\ \Phi)_x^\beta + i\ a_5\ (|\Phi|^2)_x^\beta\ \Phi = 0, \tag{10.24}$$

where $\Phi = \Phi(x, t)$ is the complex function profile, x and t are its two independent variables, $\mu_1 = \dfrac{6\ (c_1 + 2\ a_1\ c_2 - 3\ a_3\ c_2^2)}{(3\ a_4 + 2\ a_5)} < 0,$ $\mu_2 = \dfrac{c_1 + 2\ a_1\ c_2 - 3\ a_3\ c_2^2}{a_3} < 0,$

$c_2 = \dfrac{-3\ a_2\ a_3 + a_1\ (3\ a_4 + 2\ a_5)}{6\ a_3\ (a_4 + a_5)},$ $c_3 = 8\ a_1\ c_2^2 - 8\ a_3\ c_2^3 - \dfrac{(a_1\ c_1 + 2\ a_1^2\ c_2)}{a_3} + 3\ c_1\ c_2,$

x_0, t_0, c_1, and ϕ_0 are arbitrary real constants.

Example 5: *Conformable space–time fractional NLSE with constant dispersion and linear potential*

Given
$$\psi(x, t) = c_2 A_0 \sqrt{\frac{2 a_1 c_5}{a_2 c_7}} \operatorname{sech}[A_0(c\, t^3 + \tfrac{c_5}{c_7} x - x_0)]$$

$$\times\, e^{i\left\{A_0^2\left[a_1 c_0 + \frac{b_{10} c_5^2}{c_7^2}(t-t_0)\right] - \frac{3\, c\, c_7}{2\, b_{10}\, c_5} t^2 x - \frac{9\, c^2 c_7^2}{20\, b_{10}\, c_5^2} t^5 + c_1 + \phi_0\right\}}$$

is a solution to the NLSE with constant dispersion and linear potential, (5.64) with $g_4(t) = c\, t^3$, $i\, \psi_t + b_{10}\, \psi_{xx} + \frac{a_2 b_{10} c_5}{a_1 c_2^2 c_7} |\psi|^2 \psi - \frac{3\, c\, c_7}{b_{10} c_5} t\, x\, \psi = 0$, then

$$\Phi(x, t) = c_2 A_0 \sqrt{\frac{2\, a_1 c_5}{a_2 c_7}} \operatorname{sech}\left\{A_0\left[c\left(\frac{t^\alpha}{\alpha}\right)^3 + \frac{c_5}{c_7}\frac{x^\beta}{\beta} - x_0\right]\right\} \tag{10.25}$$

$$\times\, e^{i\left\{A_0^2\left[a_1 c_0 + \frac{b_{10} c_5^2}{c_7^2}\left(\frac{t^\alpha}{\alpha}-t_0\right)\right] - \frac{3\, c\, c_7}{2\, b_{10}\, c_5}\left(\frac{t^\alpha}{\alpha}\right)^2 \frac{x^\beta}{\beta} - \frac{9\, c^2 c_7^2}{20\, b_{10}\, c_5^2}\left(\frac{t^\alpha}{\alpha}\right)^5 + c_1 + \phi_0\right\}}$$

is a solution to

$$i\, \Phi_t + b_{10}\, \Phi_{xx}^{2\beta} + \frac{a_2 b_{10} c_5}{a_1 c_2^2 c_7} |\Phi|^2 \Phi - \frac{3\, c\, c_7}{b_{10} c_5}\left(\frac{t^\alpha}{\alpha}\right)\left(\frac{x^\beta}{\beta}\right)\Phi = 0, \tag{10.26}$$

where $\Phi = \Phi(x, t)$ is the complex function profile, x and t are its two independent variables, $(a_1 a_2 c_5 c_7) > 0$, A_0, c, c_2, c_5, c_7, b_{10}, x_0, t_0, and ϕ_0 are arbitrary real constants.

Example 6: *Conformable space–time fractional $(2 + 1)$-dimensional NLSE*

Given $\psi(x_1, x_2, t) = A_0 \sqrt{\frac{2 (c_1^2 \alpha_1 + c_2^2 \alpha_2)}{a_2}} \operatorname{sech}\{A_0 [c_1 (x_1 - x_{01}) + c_2 (x_2 - x_{02})]\}$

$$\times\, e^{i [A_0^2 (c_1^2 \alpha_1 + c_2^2 \alpha_2)(t-t_0)+\phi_0]}$$

is a solution to the $(2 + 1)$-dimensional NLSE (6.9), $i\, \psi_t + \alpha_1 \psi_{x_1 x_1} + \alpha_2 \psi_{x_2 x_2} + u_2 |\psi|^2 \psi = 0$, then

$$\Phi(x_1, x_2, t) = A_0 \sqrt{\frac{2 (c_1^2 \alpha_1 + c_2^2 \alpha_2)}{a_2}} \operatorname{sech}\left\{A_0\left[c_1\left(\frac{x_1^\beta}{\beta} - x_{01}\right) + c_2\left(\frac{x_2^\beta}{\beta} - x_{02}\right)\right]\right\} \tag{10.27}$$

$$\times\, e^{i\left[A_0^2\left(c_1^2 \alpha_1 + c_2^2 \alpha_2\right)\left(\frac{t^\alpha}{\alpha}-t_0\right)+\phi_0\right]}$$

is a solution to

$$i\, \Phi_t^\alpha + \alpha_1 \Phi_{x_1 x_1}^{2\beta} + \alpha_2 \Phi_{x_2 x_2}^{2\beta} + a_2 |\Phi|^2 \Phi = 0, \tag{10.28}$$

where $\Phi = \Phi(x_1, x_2, t)$ is the complex function profile, x_1, x_2, and t are its three independent variables, $a_2 (c_1^2 \alpha_1 + c_2^2 \alpha_2) > 0$, A_0, t_0, x_{01}, x_{02}, c_1, c_2, α_1, α_2, and ϕ_0 are arbitrary real constants.

Example 7: *Conformable space–time fractional coupled NLSE*

Given $\quad \psi_1(x, t) = \dfrac{A_0}{\sqrt{2\,[A_1 + t - t_0]}}\, e^{i\,\phi_1(x,\, t)}, \qquad \psi_2(x, t) = \dfrac{B_0}{\sqrt{2\,[B_1 + t - t_0]}}\, e^{i\,\phi_2(x,\, t)} \qquad$ with

$\phi_1(x, t) = \dfrac{(b_0\, A_2 + x - x_0)^2}{4\, b_0\,(A_1 + t - t_0)} + \dfrac{b_1\, A_0^2}{2} \ln[2(A_1 + t - t_0)] + \dfrac{b_2\, B_0^2}{2} \ln[2\,(B_1 + t - t_0)] + \phi_{01}$,

$\phi_2(x, t) = \dfrac{(c_0\, B_2 + x - x_0)^2}{4\, c_0\,(B_1 + t - t_0)} + \dfrac{c_1\, A_0^2}{2} \ln[2(A_1 + t - t_0)] + \dfrac{c_2\, B_0^2}{2} \ln[2\,(B_1 + t - t_0)] + \phi_{02}$, is a

solution to the CNLSE (7.7), $i\,\psi_{1t} + b_0\,\psi_{1xx} + (b_1\,|\psi_1|^2 + b_2\,|\psi_2|^2)\,\psi_1 = 0$,
$i\,\psi_{2t} + c_0\,\psi_{2xx} + (c_1\,|\psi_1|^2 + c_2\,|\psi_2|^2)\,\psi_2 = 0$, then

$$\Phi_1(x, t) = \frac{A_0}{\sqrt{2\left[A_1 + \dfrac{t^\alpha}{\alpha} - t_0\right]}}\, e^{i\,\phi_1(x,\, t)},$$

$$\Phi_2(x, t) = \frac{B_0}{\sqrt{2\left[B_1 + \dfrac{t^\alpha}{\alpha} - t_0\right]}}\, e^{i\,\phi_2(x,\, t)}$$

$$(10.29)$$

with

$\phi_1(x, t) = \dfrac{(b_0\, A_2 + \frac{x^\beta}{\beta} - x_0)^2}{4\, b_0\,(A_1 + \frac{t^\alpha}{\alpha} - t_0)} + \dfrac{b_1\, A_0^2}{2} \ln[2(A_1 + \frac{t^\alpha}{\alpha} - t_0)] + \dfrac{b_2\, B_0^2}{2} \ln[2\,(B_1 + \frac{t^\alpha}{\alpha} - t_0)]$,

$\quad + \phi_{01}$

$\phi_2(x, t) = \dfrac{(c_0\, B_2 + \frac{x^\beta}{\beta} - x_0)^2}{4\, c_0\,(B_1 + \frac{t^\alpha}{\alpha} - t_0)} + \dfrac{c_1\, A_0^2}{2} \ln[2(A_1 + \frac{t^\alpha}{\alpha} - t_0)] + \dfrac{c_2\, B_0^2}{2} \ln[2\,(B_1 + \frac{t^\alpha}{\alpha} - t_0)]$

$\quad + \phi_{02}$

is a solution to

$$i\,\Phi_{1t}^\alpha + b_0\,\Phi_{1xx}^{2\beta} + (b_1\,|\Phi_1|^2 + b_2\,|\Phi_2|^2)\,\Phi_1 = 0,$$
$$i\,\Phi_{2t}^\alpha + c_0\,\Phi_{2xx}^{2\beta} + (c_1\,|\Phi_1|^2 + c_2\,|\Phi_2|^2)\,\Phi_2 = 0,$$

$$(10.30)$$

where $\Phi = \Phi(x, t)$ is the complex function profile, x and t are its two independent variables, b_0, c_0, b_1, c_1, b_2, c_2, A_0, A_1, A_2, B_0, B_1, B_2, x_0, t_0, ϕ_{01} and ϕ_{02} are arbitrary real constants.

Example 8: *Conformable time fractional discrete NLSE with saturable nonlinearity*

Given $\psi(n, t) = A_0\, \text{sech}[A_1\,(n - n_0)]\, e^{-i\,[A_2\,(t - t_0) + \phi_0]}$ is a solution to the discrete NLSE

with saturable nonlinearity (8.9), $i\,\psi_{nt} + \psi_{n+1} + \psi_{n-1} - 2\,\psi_n + \dfrac{a_2\,|\psi_n|^2\,\psi_n}{1 + \mu\,|\psi_n|^2} = 0$, then

$$\Phi(n, t) = A_0\, \text{sech}[A_1\,(n - n_0)]\, e^{-i\left[A_2\left(\frac{t^\alpha}{\alpha} - t_0\right) + \phi_0\right]} \qquad (10.31)$$

is a solution to

$$i\,\Phi_{nt}^\alpha + \Phi_{n+1} + \Phi_{n-1} - 2\,\Phi_n + \frac{a_2\,|\Phi_n|^2\,\Phi_n}{1 + \mu\,|\Phi_n|^2} = 0, \qquad (10.32)$$

where $\Phi_n = \Phi(n, t)$ is the complex function profile, the integer site index, n, and t are its two independent variables, $A_0 = \frac{\sinh(A_1)}{\sqrt{\mu}}$, $A_2 = \frac{2\mu - a_2}{\mu}$, $\mu = \frac{a_2 \operatorname{sech}(A_1)}{2} > 0$, A_1, t_0, and ϕ_0 are arbitrary real constants, n_0 is an arbitrary real integer.

***Example* 9:** *Conformable time fractional nonlocal discrete NLSE with third saturable nonlinearity*

Given $\psi(\zeta, t) = A_0 \operatorname{sech}[A_1 \zeta] e^{-i[A_2 (t-t_0)+\phi_0]}$ is a solution to the nonlocal discrete NLSE with saturable nonlinearity (9.59), $i\,\psi_{nt} + \psi_{n+1} + \psi_{n-1} - 2\,\psi_n + \frac{a_2 \psi_n^2 \bar{\psi}_n}{1 + \mu \psi_n \bar{\psi}_n} = 0$, then

$$\Phi(\zeta, t) = A_0 \operatorname{sech}[A_1 \zeta] e^{-i\left[A_2 \left(\frac{t^\alpha}{\alpha}-t_0\right)+\phi_0\right]} \tag{10.33}$$

is a solution to

$$i\,\Phi_{nt}^\alpha + \Phi_{n+1} + \Phi_{n-1} - 2\,\Phi_n + \frac{a_2\,\Phi_n^2\,\bar{\Phi}_n}{1 + \mu\,\Phi_n\,\bar{\Phi}_n} = 0, \tag{10.34}$$

where $\Phi_n = \Phi_n(n, t; a_2, \mu)$, $\bar{\Phi}_n = \Phi_n^*(-n, t; a_2, \mu)$, $A_0 = \frac{\sinh(A_1)}{\sqrt{\mu}}$, $A_2 = \frac{2\mu - a_2}{\mu}$, $\mu = \frac{a_2 \operatorname{sech}(A_1)}{2} > 0$, $\zeta = n - n_0$, A_1, t_0, and ϕ_0 are arbitrary real constants, n_0 is an arbitrary real integer.

10.2.2 Resonant NLSE with Conformable Time Fractional Derivative

Equation:

$$i\,\psi_t^\alpha + a_1\,\psi_{xx} + a_2\,|\psi|^2\,\psi + a_3\,\frac{|\psi|_{xx}}{|\psi|}\,\psi = 0, \tag{10.35}$$

where $\psi - \psi(x, t)$ is the complex function profile, x and t are its two independent variables, a_1, a_2, and a_3 are arbitrary real constants, α is the fractional derivative order.

Solutions:

***Solution* 1.** *Bright soliton*

$$\psi(x, t) = \pm\sqrt{\frac{2 A_0 (a_1 + a_3)}{a_2}} \operatorname{sech}\left[\sqrt{A_0}\left(x - 2 a_1 k \frac{t^\alpha}{\alpha} - x_0\right)\right] e^{i\left(k\,x + v\,\frac{t^\alpha}{\alpha}+\phi_0\right)}, \tag{10.36}$$

where $v = A_0(a_1 + a_3) - a_1 k^2$, $A_0 > 0$, $a_2 (a_1 + a_3) > 0$, k, x_0, and ϕ_0 are arbitrary real constants.

- *Reference*: [28], *corrected*.

***Solution* 2.**

$$\psi(x, t) = \pm\sqrt{\frac{-2 A_0 (a_1 + a_3)}{a_2}} \operatorname{csch}\left[\sqrt{A_0}\left(x - 2 a_1 k \frac{t^\alpha}{\alpha} - x_0\right)\right] e^{i\left(k\,x + v\,\frac{t^\alpha}{\alpha}+\phi_0\right)}, \tag{10.37}$$

where $v = A_0(a_1 + a_3) - a_1 k^2$, $A_0 > 0$, $a_2 (a_1 + a_3) < 0$, k, x_0, and ϕ_0 are arbitrary real constants.

- *Reference*: [29], *corrected*.

Solution 3. *Dark soliton*

$$\psi(x, t) = \pm \sqrt{-\frac{2 A_0 (a_1 + a_3)}{a_2}} \tanh\left[\sqrt{A_0}\left(x + 2 a_1 k \frac{t^\alpha}{\alpha} - x_0\right)\right] e^{i\left(-k\, x + v\, \frac{t^\alpha}{\alpha} + \phi_0\right)}, \quad (10.38)$$

where $v = -2 A_0(a_1 + a_3) - a_1 k^2$, $A_0 > 0$, $a_2 (a_1 + a_3) < 0$, k, x_0, and ϕ_0 are arbitrary real constants.

- *Reference*: [28].

Solution 4.

$$\psi(x, t) = \pm \sqrt{\frac{A_0 (a_1 + a_3)}{a_2}} \frac{2}{\sqrt{\sin\left[2\sqrt{-A_0}\left(x - 2 a_1 k \frac{t^\alpha}{\alpha} - x_0\right)\right] + 1}} e^{i\left(k\, x + v\, \frac{t^\alpha}{\alpha} + \phi_0\right)}, \quad (10.39)$$

where $v = A_0(a_1 + a_3) - a_1 k^2$, $A_0 < 0$, $a_2 (a_1 + a_3) < 0$, k, x_0, and ϕ_0 are arbitrary real constants.

- *Reference*: [29], *corrected*.

Solution 5.

$$\psi(x, t) = \pm \sqrt{\frac{A_0 (a_1 + a_3)}{a_2}} \frac{2}{\sqrt{\cos\left[2\sqrt{-A_0}\left(x - 2 a_1 k \frac{t^\alpha}{\alpha} - x_0\right)\right] + 1}} e^{i\left(k\, x + v\, \frac{t^\alpha}{\alpha} + \phi_0\right)}, \quad (10.40)$$

where $v = A_0(a_1 + a_3) - a_1 k^2$, $A_0 < 0$, $a_2 (a_1 + a_3) < 0$, k, x_0, and ϕ_0 are arbitrary real constants.

- *Reference*: [29], *corrected*.

10.2.3 Resonant Cubic-Quintic NLSE with Conformable Time Fractional Derivative

Equation:

$$i\, \psi_t^\alpha + a_1 \psi_{xx} + a_2 |\psi|^2 \psi + a_3 |\psi|^4 \psi + a_4 \frac{|\psi|_{xx}}{|\psi|} \psi = 0, \quad (10.41)$$

where $\psi = \psi(x, t)$ is the complex function profile, x and t are its two independent variables, a_1, a_2, a_3, and a_4 are real constants, α is the fractional derivative order.

Solutions:
Solution 1.

$$\psi(x, t) = \sqrt{\frac{A_0}{1 + A_1^{x + \left(2 a_2 k\right) \frac{t^\alpha}{\alpha} - x_0}}} e^{i\left(-k\, x + v\, \frac{t^\alpha}{\alpha} + \phi_0\right)}, \quad (10.42)$$

where $\quad a_3 = -3\, a_2/(4\, A_0), \qquad a_4 = -[3\, a_1 \ln^2(A_1) + 4\, a_3\, A_0^2]/[3\, \ln^2(A_1)],$
$v = -a_3\, A_0^2/3 - a_1\, k^2$, A_0, A_1, k, x_0, and ϕ_0 are arbitrary real constants.
  • *Reference*: [28].

Solution 2.

$$\psi(x,\, t) = \sqrt{-A_0 + \cfrac{A_0}{1 + A_1^{\, x + \left(2\, a_2\, k\right) \frac{t^\alpha}{\alpha} - x_0}}}\; e^{i\left(-k\, x + v\, \frac{t^\alpha}{\alpha} + \phi_0\right)}, \qquad (10.43)$$

where $\quad a_3 = 3\, a_2/(4\, A_0), \qquad a_4 = -[3\, a_1 \ln^2(A_1) + 4\, a_3\, A_0^2]/[3\, \ln^2(A_1)],$
$v = -a_3\, A_0^2/3 - a_1\, k^2$, A_0, A_1, k, x_0, and ϕ_0 are arbitrary real constants.
  • *Reference*: [28].

Solution 3.

$$\psi(x,\, t) = \sqrt{\frac{3\, a_2}{8\, a_3}}\, \sqrt{\mathrm{sech}\left[\sqrt{A_0}\left(x - 2\, k\, a_1 \frac{t^\alpha}{\alpha} - x_0\right)\right] - 1}\; e^{i\left(k\, x + v\, \frac{t^\alpha}{\alpha} + \phi_0\right)}, \quad (10.44)$$

where $A_0 = 3\, a_2^2/[16\, a_3\, (a_1 + a_4)]$, $v = -15\, a_2^2/(64\, a_3) - a_1\, k^2$, $a_2\, a_3 > 0$, k, x_0,
and ϕ_0 are arbitrary real constants.
  • *Reference*: [29], *corrected.*

Solution 4.

$$\psi(x,\, t) = \sqrt{\frac{3\, a_2}{8\, a_3}}\left[\sqrt{\cfrac{2}{\sqrt{\sin\left(x - 2\, k\, a_1 \frac{t^\alpha}{\alpha} - x_0\right) + 1}}} - 1\right]^{\frac{1}{2}} e^{i\left(k\, x + v\, \frac{t^\alpha}{\alpha} + \phi_0\right)}, \quad (10.45)$$

where $a_3 = -3\, a_2^2/[4\, (a_1 + a_4)]$, $v = -15\, a_2^2/(64\, a_3) - a_1\, k^2$, $a_2\, a_3 > 0$, k, x_0, and
ϕ_0 are arbitrary real constants.
  • *Reference*: [29], *corrected.*

10.2.4 NLSE with Conformable Space–Time Fractional Derivative, First Order Spatial Dispersion, and Second Order Temporal Dispersion

Equation:

$$i\, \psi_t^\alpha + a_1\, \psi_{xx}^{2\beta} + a_2\, |\psi|^2\, \psi + a_3\, \psi_{tt}^\alpha + i\, a_4\, \psi_x^\beta = 0, \qquad (10.46)$$

where $\psi = \psi(x,\, t)$ is the complex function profile, x and t are its two independent
variables, $a_2 = a_4$, a_1, a_3, a_4 are arbitrary real constants, α and β are the time and
space fractional derivative orders, respectively.

Solutions:

Solution 1. *Bright soliton*

$$\psi(x,\, t) = \pm A_0\, \mathrm{sech}\left(\frac{x^\beta}{\beta} + A_1 \frac{t^\alpha}{\alpha}\right) e^{i\left(k\, \frac{x^\beta}{\beta} + v\, \frac{t^\alpha}{\alpha} + \phi_0\right)}, \qquad (10.47)$$

where $k = [-a_2 A_2 + (1 + 2 a_3 v) \sqrt{a_2^2 A_2 + 4 a_1 A_2 (a_1 - v - a_3 v^2)}\,]/(2 a_1 A_2)$,
$A_0 = \sqrt{2 (a_1 + a_2^2 a_3)/(a_2 A_2)}$, $\qquad A_1 = -\sqrt{[4 a_1^2 + a_2^2 - 4 a_1 v (1 + a_3 v)]/A_2}$,
$A_2 = -4 a_1 a_3 + (1 + 2 a_3 v)^2$, $\qquad\qquad [a_2^2 A_2 + 4 a_1 A_2 (a_1 - v - a_3 v^2)] > 0$,
$A_2 [4 a_1^2 + a_2^2 - 4 a_1 v (1 + a_3 v)] > 0$, $(a_1 + a_2^2 a_3)(a_2 A_2) > 0$, v and ϕ_0 are arbitrary real constants.

- *Reference*: [30].

Solution 2.

$$\psi(x, t) = \pm A_0 \operatorname{csch}\!\left(\frac{x^\beta}{\beta} + A_1 \frac{t^\alpha}{\alpha}\right) e^{i\left(k \frac{x^\beta}{\beta} + v \frac{t^\alpha}{\alpha} + \phi_0\right)}, \tag{10.48}$$

where $k = [-a_2 A_2 + (1 + 2 a_3 v) \sqrt{a_2^2 A_2 + 4 a_1 A_2 (a_1 - v - a_3 v^2)}\,]/(2 a_1 A_2)$,
$A_0 = \sqrt{-2 (a_1 + a_2^2 a_3)/(a_2 A_2)}$, $\qquad A_1 = -\sqrt{[4 a_1^2 + a_2^2 - 4 a_1 v (1 + a_3 v)]/A_2}$,
$A_2 = -4 a_1 a_3 + (1 + 2 a_3 v)^2$, $\qquad\qquad [a_2^2 A_2 + 4 a_1 A_2 (a_1 - v - a_3 v^2)] > 0$,
$A_2 [4 a_1^2 + a_2^2 - 4 a_1 v (1 + a_3 v)] > 0$, $(a_1 + a_2^2 a_3)(a_2 A_2) < 0$, v and ϕ_0 are arbitrary real constants.

- *Reference*: [30], *corrected*.

Solution 3. *Dark soliton*

$$\psi(x, t) = \pm A_0 \tanh\!\left(\frac{x^\beta}{\beta} + A_1 \frac{t^\alpha}{\alpha}\right) e^{i\left(k \frac{x^\beta}{\beta} + v \frac{t^\alpha}{\alpha} + \phi_0\right)}, \tag{10.49}$$

where $k = [-a_2 A_2 + (1 + 2 a_3 v) \sqrt{a_2^2 A_2 - 4 a_1 A_2 (2 a_1 + v + a_3 v^2)}\,]/(2 a_1 A_2)$,
$A_0 = \sqrt{-2 (a_1 + a_2^2 a_3)/(a_2 A_2)}$, $\quad A_1 = -\sqrt{[-8 a_1^2 + a_2^2 - 4 a_1 v (1 + a_3 v)]/A_2}$,
$A_2 = 8 a_1 a_3 + (1 + 2 a_3 v)^2$, $\qquad\qquad [a_2^2 A_2 - 4 a_1 A_2 (2 a_1 + v + a_3 v^2)] > 0$,
$A_2 [-8 a_1^2 + a_2^2 - 4 a_1 v (1 + a_3 v)] > 0$, $(a_1 + a_2^2 a_3)(a_2 A_2) < 0$, v and ϕ_0 are arbitrary real constants.

- *Reference*: [30], *corrected*.

Solution 4.

$$\psi(x, t) = \pm A_0 \coth\!\left(\frac{x^\beta}{\beta} + A_1 \frac{t^\alpha}{\alpha}\right) e^{i\left(k \frac{x^\beta}{\beta} + v \frac{t^\alpha}{\alpha} + \phi_0\right)}, \tag{10.50}$$

where $k = [-a_2 A_2 + (1 + 2 a_3 v) \sqrt{a_2^2 A_2 - 4 a_1 A_2 (2 a_1 + v + a_3 v^2)}\,]/(2 a_1 A_2)$,
$A_0 = \sqrt{-2 (a_1 + a_2^2 a_3)/(a_2 A_2)}$, $\quad A_1 = -\sqrt{[-8 a_1^2 + a_2^2 - 4 a_1 v (1 + a_3 v)]/A_2}$,
$A_2 = 8 a_1 a_3 + (1 + 2 a_3 v)^2$, $\qquad\qquad [a_2^2 A_2 - 4 a_1 A_2 (2 a_1 + v + a_3 v^2)] > 0$,
$A_2 [-8 a_1^2 + a_2^2 - 4 a_1 v (1 + a_3 v)] > 0$, $(a_1 + a_2^2 a_3)(a_2 A_2) < 0$, v and ϕ_0 are arbitrary real constants.

- *Reference*: [30], *corrected*.

Solution 5.

$$\psi(x,\, t) = \pm A_0 \coth\left(\frac{x^\beta}{2\,\beta} + A_1\,\frac{t^\alpha}{2\,\alpha}\right) e^{i\left(k\,\frac{x^\beta}{\beta} + v\frac{t^\alpha}{\alpha} + \phi_0\right)}, \qquad (10.51)$$

where

$k = [-2\,a_1\,a_2\,a_3 - a_2\,(1 + 2\,a_3\,v)^2 + (1 + 2\,a_3\,v)$

$\qquad \sqrt{-2\,a_1^2\,A_2 + a_2^2\,A_2 - 4\,a_1\,v\,A_2\,(1 + a_3\,v)}\,]/(2\,a_1\,A_2),$

$A_0 = \sqrt{-(a_1 + a_2^2\,a_3)/(2\,a_2\,A_2)}, \qquad A_1 = -\sqrt{[-2\,a_1^2 + a_2^2 - 4\,a_1\,v\,(1 + a_3\,v)]/A_2},$

$A_2 = 2\,a_1\,a_3 + (1 + 2\,a_3\,v)^2, \qquad [-2\,a_2^2\,A_2 + a_2^2\,A_2 - 4\,a_1\,v\,A_2\,(1 + a_3\,v)] > 0,$

$A_2\,[-2\,a_1^2 + a_2^2 - 4\,a_1\,v\,(1 + a_3\,v)] > 0, \quad (a_1 + a_2^2\,a_3)(a_2\,A_2) < 0, \quad v$ and ϕ_0 are arbitrary real constants.

- *Reference*: [30].

Solution 6.

$$\psi(x,\, t) = \pm A_0 \cot\left(\frac{x^\beta}{2\,\beta} + A_1\,\frac{t^\alpha}{2\,\alpha}\right) e^{i\left(k\,\frac{x^\beta}{\beta} + v\frac{t^\alpha}{\alpha} + \phi_0\right)}, \qquad (10.52)$$

where

$k = [2\,a_1\,a_2\,a_3 - a_2\,(1 + 2\,a_3\,v)^2 - (1 + 2\,a_3\,v)$

$\qquad \sqrt{2\,a_1^2\,A_2 + a_2^2\,A_2 - 4\,a_1\,v\,A_2\,(1 + a_3\,v)}\,]/(2\,a_1\,A_2),$

$A_0 = \sqrt{-(a_1 + a_2^2\,a_3)/(2\,a_2\,A_2)}, \qquad A_1 = \sqrt{[2\,a_1^2 + a_2^2 - 4\,a_1\,v\,(1 + a_3\,v)]/A_2},$

$A_2 = -2\,a_1\,a_3 + (1 + 2\,a_3\,v)^2, \qquad [2\,a_1^2\,A_2 + a_2^2\,A_2 - 4\,a_1\,v\,A_2\,(1 + a_3\,v)] > 0,$

$A_2\,[2\,a_1^2 + a_2^2 - 4\,a_1\,v\,(1 + a_3\,v)] > 0, \quad (a_1 + a_2^2\,a_3)(a_2\,A_2) < 0, v$ and ϕ_0 are arbitrary real constants.

- *Reference*: [30].

Solution 7.

$$\psi(x,\, t) = \pm A_0 \left[\sec\left(\frac{x^\beta}{\beta} + A_1\,\frac{t^\alpha}{\alpha}\right) - \tan\left(\frac{x^\beta}{\beta} + A_1\,\frac{t^\alpha}{\alpha}\right)\right] e^{i\left(k\,\frac{x^\beta}{\beta} + v\frac{t^\alpha}{\alpha} + \phi_0\right)}, \qquad (10.53)$$

where

$k = [2\,a_1\,a_2\,a_3 - a_2\,(1 + 2\,a_3\,v)^2 - (1 + 2\,a_3\,v) \qquad\qquad\qquad ,$

$\qquad \sqrt{2\,a_1^2\,A_2 + a_2^2\,A_2 - 4\,a_1\,v\,A_2\,(1 + a_3\,v)}\,]/(2\,a_1\,A_2)$

$A_0 = \sqrt{-(a_1 + a_2^2\,a_3)/(2\,a_2\,A_2)}, \qquad A_1 = \sqrt{[2\,a_1^2 + a_2^2 - 4\,a_1\,v\,(1 + a_3\,v)]/A_2},$

$A_2 = -2\,a_1\,a_3 + (1 + 2\,a_3\,v)^2, \qquad [2\,a_1^2\,A_2 + a_2^2\,A_2 - 4\,a_1\,v\,A_2\,(1 + a_3\,v)] > 0,$

$A_2\,[2\,a_1^2 + a_2^2 - 4\,a_1\,v\,(1 + a_3\,v)] > 0, \quad (a_1 + a_2^2\,a_3)(a_2\,A_2) < 0, v$ and ϕ_0 are arbitrary real constants.

- *Reference*: [30].

10.3 Summary of Section 10.2

Note: For lengthy conditions, the reader is referred to the solutions in Section 10.2.

NLSE with Conformable Space–Time Fractional Derivative

Transformations:

$$\Phi(x,t) = \psi(\tfrac{x^\beta}{\beta}, \tfrac{t^\alpha}{\alpha}),$$

$$\Phi(x,t) = \psi(\tfrac{x^\beta}{\beta}, t), \quad \psi \text{ is a solution to any equation from the previous nine chapters or any other versions of the NLSE that involve both regular time and regular space derivatives}$$

$$\Phi(x,t) = \psi(x, \tfrac{t^\alpha}{\alpha}),$$

Equation: $i\,\Phi_t^\alpha + a_1\,\Phi_{xx}^{2\beta} + a_2\,|\Phi|^2\,\Phi = 0$

Example	Name	Equation #
$\Phi(x,t) = A_0 \sqrt{\dfrac{2a_1}{a_2}}\ \mathrm{sech}[A_0 (\tfrac{x^\beta}{\beta} - x_0)]\, e^{i\,[a_1 A_0^2 (\frac{t^\alpha}{\alpha} - t_0) + \phi_0]}$	Bright soliton	(10.17)

Conditions: $a_1\, a_2 > 0, A_0, x_0, t_0,$ and ϕ_0 are arbitrary real constants

Equation: $i\,\Phi_t^\alpha + a_1\,\Phi_{xx}^{2\beta} + a_2\,|\Phi|^n\,\Phi = 0$

Example	Name	Equation #
$\Phi(x,t) = \left\{\left[\dfrac{2A_0^2 a_1(n+2)}{a_2 n^2}\ \mathrm{sech}^2\left[A_0\left(\tfrac{x^\beta}{\beta} - x_0\right)\right]\right]^{\frac{1}{n}}\right\}$ $\times\, e^{i\,[\frac{4a_1 A_0^2}{n^2}(\frac{t^\alpha}{\alpha} - t_0) + \phi_0]}$	Bright soliton	(10.19)

Conditions: $a_1\, a_2\,(n+2) > 0, n, A_0, x_0, t_0,$ and ϕ_0 are arbitrary real constants

Equation: $i\,\Phi_t^\alpha + a_1\,\Phi_{xx}^{2\beta} + a_2\,|\Phi|^n\,\Phi + a_3\,|\Phi|^m\,\Phi = 0$

Example	Name	Equation #
$\Phi(x,t) = \left\{\dfrac{A_0(n+2)}{a_2 + a_2\sqrt{1+\gamma}\,\cosh\left[n\sqrt{\frac{A_0}{a_1}}[(\tfrac{x^\beta}{\beta} - x_0)]\right]}\right\}^{\frac{1}{n}}$ $\times\, e^{i\,[A_0(\frac{t^\alpha}{\alpha} - t_0) + \phi_0]}$	Bright soliton	(10.21)

Conditions: $m = 2n, \gamma = a_3/a_{30},$ $a_{30} = a_2^2\,(n+1)/[A_0\,(n+2)^2], A_0\, a_1 > 0,$

(Continued)

$(1 + \gamma) > 0$, n, x_0, t_0, and ϕ_0 are arbitrary real constants

Equation: $i\,\Phi_t^\alpha + a_1\,\Phi_{xx}^{2\beta} + a_2\,|\Phi|^2\,\Phi + i\,a_3\,\Phi_{xxx}^{3\beta} + i\,a_4\,(|\Phi|^2\,\Phi)_x^\beta + i\,a_5\,(|\Phi|^2)_x^\beta\,\Phi = 0$

Example	Name	Conditions	Equation #
$\Phi(x,t) = \pm\sqrt{-\mu_1}\,\mathrm{sech}\left\{\sqrt{-\mu_2}\left[\dfrac{x^\beta}{\beta} - x_0 + c_1\left(\dfrac{t^\alpha}{\alpha} - t_0\right)\right]\right\}$ $\times e^{i\left[c_2\left(\frac{x^\beta}{\beta} - x_0\right) + c_3\left(\frac{t^\alpha}{\alpha} - t_0\right) + \phi_0\right]}$	Bright soliton	$\mu_1 = \dfrac{6(c_1 + 2a_1c_2 - 3a_3c_2^2)}{(3a_4 + 2a_5)} < 0,$ $\mu_2 = \dfrac{c_1 + 2a_1c_2 - 3a_3c_2^2}{a_3} < 0,$ $c_2 = \dfrac{-3a_2a_3 + a_1(3a_4 + 2a_5)}{6a_3(a_4 + a_5)},$ $c_3 = 8a_1c_2^2 - 8a_3c_2^3 - \dfrac{(a_1c_1 + 2a_1^2c_2)}{a_3} + 3c_1c_2,$ x_0, t_0, c_1, and ϕ_0 are arbitrary real constants	(10.23)

Equation: $i\,\Phi_t + b_{10}\,\Phi_{xx}^{2\beta} + \dfrac{a_2\,b_{10}\,c_5}{a_1\,c_2^2\,c_7}\,|\Phi|^2\,\Phi - \dfrac{3\,c\,c_7}{b_{10}\,c_5}\left(\dfrac{t^\alpha}{\alpha}\right)\left(\dfrac{x^\beta}{\beta}\right)\Phi = 0$

Example	Name	Conditions	Equation #
$\Phi(x,t) = c_2\,A_0\sqrt{\dfrac{2a_1c_5}{a_2c_7}}\,\mathrm{sech}\left\{A_0\left[c\left(\dfrac{t^\alpha}{\alpha}\right)^3 + \dfrac{c_5}{c_7}\dfrac{x^\beta}{\beta} - x_0\right]\right\}$ $\times e^{i\left\{A_0^2\left[a_1c_0 + \frac{b_{10}c_5}{c_7^2}\left(\frac{t^\alpha}{\alpha} - t_0\right)\right]\right\}}$ $\times e^{i\left\{-\frac{3cc_7}{2b_{10}c_5}\left(\frac{t^\alpha}{\alpha}\right)^2\frac{x^\beta}{\beta} - \frac{9c^2c_7^2}{20b_{10}c_5}\left(\frac{t^\alpha}{\alpha}\right)^5 + c_1 + \phi_0\right\}}$	Bright soliton	$(a_1\,a_2\,c_5\,c_7) > 0$, A_0, c, c_2, b_{10}, x_0, t_0, and ϕ_0 are arbitrary real constants	(10.25)

Equation: $i\,\Phi_t^\alpha + \alpha_1\,\Phi_{x_1x_1}^{2\beta} + \alpha_2\,\Phi_{x_2x_2}^{2\beta} + a_2\,|\Phi|^2\,\Phi = 0$

Example	Name	Conditions	Equation #
$\Phi(x_1, x_2, t) = A_0 \sqrt{\dfrac{2(c_1^2 \alpha_1 + c_2^2 \alpha_2)}{a_2}}$ $\times \operatorname{sech}\left\{A_0\left[c_1\left(\dfrac{x_1^\beta}{\beta} - x_{01}\right) + c_2\left(\dfrac{x_2^\beta}{\beta} - x_{02}\right)\right]\right\}$ $\times e^{i\left[A_0^2(c_1^2 \alpha_1 + c_2^2 \alpha_2)(\frac{t^\alpha}{\alpha} - t_0) + \phi_0\right]}$	Bright soliton	$a_2(c_1^2 \alpha_1 + c_2^2 \alpha_2) > 0$, A_0, t_0, and ϕ_0 are arbitrary real constants	(10.27)

Equation: $i\Phi_{1t}^\alpha + b_0 \Phi_{1xx}^{2\beta} + (b_1|\Phi_1|^2 + b_2|\Phi_2|^2)\Phi_1 = 0,$
$i\Phi_{2t}^\alpha + c_0 \Phi_{2xx}^{2\beta} + (c_1|\Phi_1|^2 + c_2|\Phi_2|^2)\Phi_2 = 0$

Example	Name	Conditions	Equation #
$\Phi_1(x, t) = \dfrac{A_0}{\sqrt{2[A_1 + \frac{t^\alpha}{\alpha} - t_0]}}\, e^{i\phi_1(x,t)},$ $\Phi_2(x, t) = \dfrac{B_0}{\sqrt{2[B_1 + \frac{t^\alpha}{\alpha} - t_0]}}\, e^{i\phi_2(x,t)}$	Bright soliton	$b_0, c_0, b_1, c_1, b_2, c_2, A_0, A_1, A_2, B_0, B_1, B_2, x_0,$ t_0, ϕ_{01} and ϕ_{02} are arbitrary real constants	(10.29)

Equation: $i\Phi_{nt}^\alpha + \Phi_{n+1} + \Phi_{n-1} - 2\Phi_n + \dfrac{a_2|\Phi_n|^2 \Phi_n}{1+\mu|\Phi_n|^2} = 0$

Example	Name	Conditions	Equation #
$\Phi(n, t) = A_0 \operatorname{sech}[A_1(n - n_0)]\, e^{-i[A_2(\frac{t^\alpha}{\alpha} - t_0) + \phi_{01}]}$	Bright soliton	$A_0 = \dfrac{\sinh(A_1)}{\sqrt{\mu}}$, $A_2 = \dfrac{2\mu - a_2}{\mu}$, $\mu = \dfrac{a_2 \operatorname{sech}(A_1)}{2} > 0$, A_1, t_0, n_0, and ϕ_0 are arbitrary real constants	(10.31)

Equation: $i\Phi_{nt}^\alpha + \Phi_{n+1} + \Phi_{n-1} - 2\Phi_n + \dfrac{a_2 \Phi_n^2 \Phi_n}{1+\mu \Phi_n \Phi_n} = 0$

(Continued)

Example	Conditions	Name	Equation #
$\Phi(\zeta, t) = A_0 \operatorname{sech}[A_1 \zeta] e^{-i[A_2(\frac{t^\alpha}{\alpha} - t_0) + \phi_0]}$	$A_0 = \frac{\sinh(A_1)}{\sqrt{\mu}}, \ A_2 = \frac{2\mu - a_2}{\mu}, \ \mu = \frac{a_2 \operatorname{sech}(A_1)}{2} > 0,$ $\zeta = n - n_0, \ A_1, t_0, n_0,$ and ϕ_0 are arbitrary real constants	Bright soliton	(10.33)

Resonant NLSE with conformable time fractional derivative

Equation: $i\,\psi_t^\alpha + a_1\,\psi_{xx} + a_2\,|\psi|^2\,\psi + a_3\,\frac{|\psi|_{xx}}{|\psi|}\,\psi = 0$

#	Solution	Conditions	Name	Equation #
1.	$\psi(x,t) = \pm\sqrt{\frac{2A_0(a_1+a_3)}{a_2}}\,\operatorname{sech}[\sqrt{A_0}(x - 2a_1 k\,\frac{t^\alpha}{\alpha} - x_0)]$ $\times e^{i(kx + v\frac{t^\alpha}{\alpha} + \phi_0)}$	$v = A_0(a_1 + a_3) - a_1 k^2, \ A_0 > 0,$ $a_2(a_1 + a_3) > 0, k, x_0,$ and ϕ_0 are arbitrary real constants	Bright soliton	(10.36)
2.	$\psi(x,t) = \pm\sqrt{\frac{-2A_0(a_1+a_3)}{a_2}}\,\operatorname{csch}[\sqrt{A_0}(x - 2a_1 k\,\frac{t^\alpha}{\alpha} - x_0)]$ $\times e^{i(kx + v\frac{t^\alpha}{\alpha} + \phi_0)}$	$v = A_0(a_1 + a_3) - a_1 k^2, \ A_0 > 0,$ $a_2(a_1 + a_3) < 0, k, x_0,$ and ϕ_0 are arbitrary real constants	Dark soliton	(10.37)
3.	$\psi(x,t) = \pm\sqrt{\frac{-2A_0(a_1+a_3)}{a_2}}\,\tanh[\sqrt{A_0}(x + 2a_1 k\,\frac{t^\alpha}{\alpha} - x_0)]$ $\times e^{i(-kx + v\frac{t^\alpha}{\alpha} + \phi_0)}$	$v = -2A_0(a_1 + a_3) - a_1 k^2, \ A_0 > 0,$ $a_2(a_1 + a_3) < 0, k, x_0,$ and ϕ_0 are arbitrary real constants	Dark soliton	(10.38)
4.	$\psi(x,t) = \pm\sqrt{\frac{2}{\frac{\sin[2\sqrt{-A_0}(x - 2a_1 k\,\frac{t^\alpha}{\alpha} - x_0)] + 1}{\sqrt{\frac{A_0(a_1+a_3)}{a_2}}}}}$ $\times e^{i(kx + v\frac{t^\alpha}{\alpha} + \phi_0)}$	$v = A_0(a_1 + a_3) - a_1 k^2, \ A_0 < 0,$ $a_2(a_1 + a_3) < 0, k, x_0,$ and ϕ_0 are arbitrary real constants	Dark soliton	(10.39)

5. $\psi(x,t) = \pm \dfrac{2}{\sqrt{\cos[2\sqrt{-A_0}(x - 2a_1 k \frac{t^\alpha}{\alpha} - x_0)] + 1}}$
$\times \sqrt{\dfrac{A_0(a_1+a_3)}{a_2}}\, e^{i(k x + v \frac{t^\alpha}{\alpha} + \phi_0)}$

Dark soliton   (10.40)

$v = A_0(a_1 + a_3) - a_1 k^2$, $A_0 < 0$,
$a_2(a_1 + a_3) < 0$, k, x_0, and ϕ_0 are arbitrary
real constants

Resonant cubic-quintic NLSE with conformable time fractional derivative

Equation: $i\,\psi_t^\alpha + a_1\,\psi_{xx} + a_2\,|\psi|^2\,\psi + a_3\,|\psi|^4\,\psi + a_4\,\dfrac{|\psi|_{xx}}{|\psi|}\,\psi = 0$

#	Solution	Name	Conditions	Equation #
1.	$\psi(x,t) = \sqrt{\dfrac{A_0}{1+A_1^{x+(2a_2 k)\frac{t^\alpha}{\alpha}-x_0}}}\; e^{i(-k x + v\frac{t^\alpha}{\alpha}+\phi_0)}$	Bright soliton	$a_3 = -3\,a_2/(4\,A_0)$, $v = -a_3\,A_0^2/3 - a_1\,k^2$, $a_4 = -[3\,a_1\ln^2(A_1) + 4\,a_3\,A_0^2]/[3\ln^2(A_1)]$, A_0, A_1, k, x_0, and ϕ_0 are arbitrary real constants	(10.42)
2.	$\psi(x,t) = \sqrt{-A_0 + \dfrac{A_0}{1+A_1^{x+(2a_2 k)\frac{t^\alpha}{\alpha}-x_0}}}\; e^{i(-k x + v\frac{t^\alpha}{\alpha}+\phi_0)}$	Dark soliton	$a_3 = 3\,a_2/(4\,A_0)$, $v = -a_3\,A_0^2/3 - a_1\,k^2$, $a_4 = -[3\,a_1\ln^2(A_1) + 4\,a_3\,A_0^2]/[3\ln^2(A_1)]$, A_0, A_1, k, x_0, and ϕ_0 are arbitrary real constants	(10.43)
3.	$\psi(x,t) = \sqrt{\dfrac{3\,a_2}{8\,a_3}}\sqrt{\text{sech}[\sqrt{A_0}(x - 2\,k\,a_1\frac{t^\alpha}{\alpha} - x_0)] - 1}$ $\times\, e^{i(k x + v\frac{t^\alpha}{\alpha}+\phi_0)}$	Dark soliton	$A_0 = 3\,a_2^2/[16\,a_3\,(a_1 + a_4)]$, $v = -15\,a_2^2/(64\,a_3) - a_1\,k^2$, $(a_2\,a_3) > 0$, k, x_0, and ϕ_0 are arbitrary real constants	(10.44)
4.	$\psi(x,t) = \sqrt{\dfrac{3\,a_2}{8\,a_3}}\Big[\dfrac{2}{\sin(x - 2\,k\,a_1\frac{t^\alpha}{\alpha} - x_0) + 1} - 1\Big]^{\frac{1}{2}}$ $\times\, e^{i(k x + v\frac{t^\alpha}{\alpha}+\phi_0)}$	Dark soliton	$a_3 = -3\,a_2^2/[4\,(a_1 + a_4)]$, $v = -15\,a_2^2/(64\,a_3) - a_1\,k^2$, $(a_2\,a_3) > 0$, k, x_0, and ϕ_0 are arbitrary real constants	(10.45)

(Continued)

NLSE with conformable space–time fractional derivative, first order spatial dispersion, and second order temporal dispersion

Equation: $i\,\psi_t^\alpha + a_1\,\psi_{xx}^{2\beta} + a_2\,|\psi|^2\,\psi + a_3\,\psi_{tt}^\alpha + i\,a_4\,\psi_x^\beta = 0$

#	Solution	Conditions	Name	Equation #
1.	$\psi(x,t) = \pm A_0\,\mathrm{sech}\!\left(\dfrac{x^\beta}{\beta} + A_1\dfrac{t^\alpha}{\alpha}\right) e^{\,i\,(k\,\frac{x^\beta}{\beta}+v\frac{t^\alpha}{\alpha}+\phi_0)}$	See text	Bright soliton	(10.47)
2.	$\psi(x,t) = \pm A_0\,\mathrm{csch}\!\left(\dfrac{x^\beta}{\beta} + A_1\dfrac{t^\alpha}{\alpha}\right) e^{\,i\,(k\,\frac{x^\beta}{\beta}+v\frac{t^\alpha}{\alpha}+\phi_0)}$	See text	Dark soliton	(10.48)
3.	$\psi(x,t) = \pm A_0\,\tanh\!\left(\dfrac{x^\beta}{\beta} + A_1\dfrac{t^\alpha}{\alpha}\right) e^{\,i\,(k\,\frac{x^\beta}{\beta}+v\frac{t^\alpha}{\alpha}+\phi_0)}$	See text	Dark soliton	(10.49)
4.	$\psi(x,t) = \pm A_0\,\coth\!\left(\dfrac{x^\beta}{\beta} + A_1\dfrac{t^\alpha}{\alpha}\right) e^{\,i\,(k\,\frac{x^\beta}{\beta}+v\frac{t^\alpha}{\alpha}+\phi_0)}$	See text	Dark soliton	(10.50)
5.	$\psi(x,t) = \pm A_0\,\coth\!\left(\dfrac{x^\beta}{2\beta} + A_1\dfrac{t^\alpha}{2\alpha}\right) e^{\,i\,(k\,\frac{x^\beta}{\beta}+v\frac{t^\alpha}{\alpha}+\phi_0)}$	See text	Dark soliton	(10.51)
6.	$\psi(x,t) = \pm A_0\,\cot\!\left(\dfrac{x^\beta}{2\beta} + A_1\dfrac{t^\alpha}{\alpha}\right) e^{\,i\,(k\,\frac{x^\beta}{\beta}+v\frac{t^\alpha}{\alpha}+\phi_2)}$	See text	Dark soliton	(10.52)
7.	$\psi(x,t) = \pm A_0[\sec\!\left(\dfrac{x^\beta}{\beta} + A_1\dfrac{t^\alpha}{\alpha}\right) - \tan(\dfrac{x^\beta}{\beta} + A_1\dfrac{t^\alpha}{\alpha})]$ $\times\, e^{\,i\,(k\,\frac{x^\beta}{\beta}+v\frac{t^\alpha}{\alpha}+\phi_0)}$	See text	Dark soliton	(10.53)

10.4 Fractal Fractional Derivative

The fractal fractional derivative of a function $f(x)$ is defined as

$$f_x^\alpha(x) := \lim_{\epsilon \to x} \left[\frac{f(\epsilon) - f(x)}{\epsilon^\alpha - x^\alpha} \right] = f'(x), \qquad (10.54)$$

where α is the fractional derivative order $0 < \alpha \leqslant 1$. The second fractal fractional derivative of order α is then defined as

$$f_{xx}^\alpha(x) := \lim_{\epsilon \to x} \left[\frac{f'(\epsilon) - f'(x)}{\epsilon^\alpha - x^\alpha} \right] = f''(x). \qquad (10.55)$$

The n fractal fractional derivative of order α is defined as

$$f_{nx}^\alpha(x) := \lim_{\epsilon \to x} \left[\frac{f^{n-1}(\epsilon) - f^{n-1}(x)}{\epsilon^\alpha - x^\alpha} \right] = f^n(x). \qquad (10.56)$$

10.4.1 NLSE with Fractal Space–Time Fractional Derivative

If $\psi(x, t)$ is a solution to any equation from the previous nine chapters or any other versions of the NLSE that involve both regular time and regular space derivatives, then

$$\Phi(x, t) = \psi(x^\beta, t^\alpha), \qquad (10.57)$$

$$\Phi(x, t) = \psi(x^\beta, t), \qquad (10.58)$$

and

$$\Phi(x, t) = \psi(x, t^\alpha) \qquad (10.59)$$

are solutions to the same equation, where both or any the regular time and regular space derivatives are replaced by the fractal fractional time and fractional space derivatives, respectively

$$\psi_t \to \Phi_t^\alpha, \qquad \psi_x \to \Phi_x^\beta \qquad (10.60)$$

and all space- and time-dependent coefficients are transformed as follows

$$g(x, t) \to g(x^\beta, t^\alpha), \qquad (10.61)$$

where α and β are the time and space fractional derivative of orders $0 < \alpha \leqslant 1$ and $0 < \beta \leqslant 1$, respectively.

***Example* 1:** *Fractal space–time fractional fundamental NLSE*

Given $\psi(x, t) = A_0 \sqrt{\frac{2 a_1}{a_2}} \operatorname{sech}[A_0 (x - x_0)] e^{i [a_1 A_0^2 (t - t_0) + \phi_0]}$ is a solution to the fundamental NLSE (2.160), $i \psi_t + a_1 \psi_{xx} + a_2 |\psi|^2 \psi = 0$, then

$$\Phi(x, t) = A_0 \sqrt{\frac{2\, a_1}{a_2}}\ \text{sech}[A_0\, (x^\beta - x_0)]\ e^{i\left[a_1\, A_0^2\, (t^\alpha - t_0) + \phi_0\right]} \tag{10.62}$$

is a solution to

$$i\, \Phi_t^\alpha + a_1\, \Phi_{xx}^{2\,\beta} + a_2\, |\Phi|^2\, \Phi = 0, \tag{10.63}$$

where $\Phi = \Phi(x, t)$ is the complex function profile, x and t are its two independent variables, $a_1\, a_2 > 0$, A_0, x_0, t_0, and ϕ_0 are arbitrary real constants.

Example 2: *Fractal space–time fractional NLSE with power law nonlinearity*

Given $\psi(x, t) = \left\{ \dfrac{2\, A_0^2\, a_1\, (n+2)}{a_2\, n^2}\ \text{sech}^2[A_0\, (x - x_0)] \right\}^{\frac{1}{n}}\, e^{i\left[\frac{4\, a_1\, A_0^2}{n^2}\, (t - t_0) + \phi_0\right]}$ is a solution to

the NLSE with power law nonlinearity (3.9), $i\, \psi_t + a_1\, \psi_{xx} + a_2\, |\psi|^n\, \psi = 0$, then

$$\Phi(x, t) = \left\{ \frac{2\, A_0^2\, a_1\, (n+2)}{a_2\, n^2}\ \text{sech}^2[A_0\, (x^\beta - x_0)] \right\}^{\frac{1}{n}}\, e^{i\left[\frac{4\, a_1\, A_0^2}{n^2}\, (t^\alpha - t_0) + \phi_0\right]} \tag{10.64}$$

is a solution to

$$i\, \Phi_t^\alpha + a_1\, \Phi_{xx}^{2\,\beta} + a_2\, |\Phi|^n\, \Phi = 0, \tag{10.65}$$

where $\Phi = \Phi(x, t)$ is the complex function profile, x and t are its two independent variables, $a_1\, a_2\, (n+2) > 0$, n, A_0, x_0, t_0, and ϕ_0 are arbitrary real constants.

Example 3: *Fractal space–time fractional NLSE with dual power law nonlinearity*

Given $\psi(x, t) = \left(\dfrac{A_0\, (n+2)}{a_2 + a_2\, \sqrt{1+\gamma}\ \cosh\left[n\, \sqrt{\frac{A_0}{a_1}}\, (x - x_0)\right]} \right)^{\frac{1}{n}}\, e^{i\, [A_0\, (t - t_0) + \phi_0]}$ is a solution to the NLSE

with dual power law nonlinearity (3.27), $i\, \psi_t + a_1\, \psi_{xx} + a_2\, |\psi|^n\, \psi + a_3\, |\psi|^m\, \psi = 0$,

then

$$\Phi(x, t) = \left(\frac{A_0\, (n+2)}{a_2 + a_2\, \sqrt{1+\gamma}\ \cosh\left[n\, \sqrt{\frac{A_0}{a_1}}\, (x^\beta - x_0)\right]} \right)^{\frac{1}{n}}\, e^{i\, [A_0\, (t^\alpha - t_0) + \phi_0]} \tag{10.66}$$

is a solution to

$$i\, \Phi_t^\alpha + a_1\, \Phi_{xx}^{2\,\beta} + a_2\, |\Phi|^n\, \Phi + a_3\, |\Phi|^m\, \Phi = 0, \tag{10.67}$$

where $\Phi = \Phi(x, t)$ is the complex function profile, x and t are its two independent variables, $m = 2\, n$, $\gamma = a_3/a_{30}$, $a_{30} = a_2^2\, (n+1)/[A_0\, (n+2)^2]$, $(1+\gamma) \geq 0$, $A_0\, a_1 > 0$, n, x_0, t_0, and ϕ_0 are arbitrary real constants.

Example 4: *Fractal space–time fractional NLSE with third order dispersion, self-steepening, and self-frequency shift*

Given $\psi(x, t) = \pm\sqrt{-\mu_1} \, \text{sech}\{\sqrt{-\mu_2} \, [x - x_0 + c_1 (t - t_0)]\}e^{i [c_2 (x-x_0)+c_3 (t-t_0)+\phi_0]}$ is a solution to the NLSE with third order dispersion, self-steepening, and self-frequency shift (4.2), $i \, \psi_t + a_1 \, \psi_{xx} + a_2 \, |\psi|^2 \, \psi + i \, a_3 \, \psi_{xxx} + i \, a_4 \, (|\psi|^2 \, \psi)_x + i \, a_5 \, (|\psi|^2)_x \, \psi = 0$, then

$$\Phi(x, t) = \pm\sqrt{-\mu_1} \, \text{sech}\{\sqrt{-\mu_2} \, [x^\beta - x_0 + c_1 (t^\alpha - t_0)]\}e^{i [c_2 (x^\beta - x_0)+c_3 (t^\alpha - t_0)+\phi_0]} \quad (10.68)$$

is a solution to

$$i \, \Phi_t^\alpha + a_1 \, \Phi_{xx}^{2\beta} + a_2 \, |\Phi|^2 \, \Phi + i \, a_3 \, \Phi_{xxx}^{3\beta} + i \, a_4 \left(|\Phi|^2 \, \Phi\right)_x^\beta + i \, a_5 \left(|\Phi|^2\right)_x^\beta \, \Phi = 0, \quad (10.69)$$

where $\Phi = \Phi(x, t)$ is the complex function profile, x and t are its two independent variables, $\mu_1 = \dfrac{6 (c_1 + 2 a_1 c_2 - 3 a_3 c_2^2)}{(3 a_4 + 2 a_5)} < 0$, $\mu_2 = \dfrac{c_1 + 2 a_1 c_2 - 3 a_3 c_2^2}{a_3} < 0$,

$c_2 = \dfrac{-3 a_2 a_3 + a_1 (3 a_4 + 2 a_5)}{6 a_3 (a_4 + a_5)}$, $c_3 = 8 a_1 c_2^2 - 8 a_3 c_2^3 - \dfrac{(a_1 c_1 + 2 a_1^2 c_2)}{a_3} + 3 c_1 c_2$, x_0, t_0, c_1, and ϕ_0 are arbitrary real constants.

Example 5: *Fractal space–time fractional NLSE with constant dispersion and linear potential*
Given

$$\psi(x, t) = c_2 A_0 \sqrt{\frac{2 a_1 c_5}{a_2 c_7}} \, \text{sech}[A_0 (c \, t^3 + \tfrac{c_5}{c_7} x - x_0)]$$

$$\times \, e^{i \, \{A_0^2 \, [a_1 c_0 + \frac{b_{10} c_5^2}{c_7^2} (t-t_0)] - \frac{3 c c_7}{2 b_{10} c_5} t^2 x - \frac{9 c^2 c_7^2}{20 b_{10} c_5^2} t^5 + c_1 + \phi_0\}}$$

is a solution to the NLSE with constant dispersion and linear potential, (5.64) with $g_4(t) = c \, t^3$, $i \, \psi_t + b_{10} \, \psi_{xx} + \dfrac{a_2 b_{10} c_5}{a_1 c_2^2 c_7} \, |\psi|^2 \, \psi - \dfrac{3 c c_7}{b_{10} c_5} \, t \, x \, \psi = 0$, then

$$\Phi(x, t) = c_2 A_0 \sqrt{\frac{2 a_1 c_5}{a_2 c_7}} \, \text{sech}\left[A_0 \left(c \, t^{3\alpha} + \frac{c_5}{c_7} x^\beta - x_0\right)\right]$$

$$\times \, e^{i \, \left\{A_0^2 \left[a_1 c_0 + \frac{b_{10} c_5^2}{c_7^2} (t^\alpha - t_0)\right] - \frac{3 c c_7}{2 b_{10} c_5} t^{2\alpha} x^\beta - \frac{9 c^2 c_7^2}{20 b_{10} c_5^2} t^{5\alpha} + c_1 + \phi_0\right\}} \quad (10.70)$$

is a solution to

$$i \, \Phi_t + b_{10} \, \Phi_{xx}^{2\beta} + \frac{a_2 b_{10} c_5}{a_1 c_2^2 c_7} \, |\Phi|^2 \, \Phi - \frac{3 c c_7}{b_{10} c_5} \, (t^\alpha) \, (x^\beta) \, \Phi = 0, \quad (10.71)$$

where $\Phi = \Phi(x, t)$ is the complex function profile, x and t are its two independent variables, $(a_1 a_2 c_5 c_7) > 0$, A_0, c, c_2, b_{10}, x_0, t_0, and ϕ_0 are arbitrary real constants.

Example 6: *Fractal space–time fractional $(2 + 1)$-dimensional NLSE*
Given

$$\psi(x_1, x_2, t) = A_0 \sqrt{\frac{2 (c_1^2 \, \alpha_1 + c_2^2 \, \alpha_2)}{a_2}} \, \text{sech}\{A_0 \, [c_1 (x_1 - x_{01}) + c_2 (x_2 - x_{02})]\}$$

$$\times \, e^{i \, [A_0^2 \, (c_1^2 \, \alpha_1 + c_2^2 \, \alpha_2) \, (t-t_0)+\phi_0]}$$

is a solution to the $(2 + 1)$-dimensional NLSE (6.9),
$i \, \psi_t + \alpha_1 \, \psi_{x_1 x_1} + \alpha_2 \, \psi_{x_2 x_2} + a_2 \, |\psi|^2 \, \psi = 0$, then

$$\Phi(x_1, x_2, t) = A_0 \sqrt{\frac{2 \, (c_1^2 \, \alpha_1 + c_2^2 \, \alpha_2)}{a_2}} \, \text{sech} \left\{ A_0 \left[c_1 \left(x_1^\beta - x_{01} \right) + c_2 \left(x_2^\beta - x_{02} \right) \right] \right\}$$

$$\times e^{i \left[A_0^2 \left(c_1^2 \, \alpha_1 + c_2^2 \, \alpha_2 \right) (t^\alpha - t_0) + \phi_0 \right]} \quad (10.72)$$

is a solution to

$$i \, \Phi_t^\alpha + \alpha_1 \, \Phi_{x_1 x_1}^{2\beta} + \alpha_2 \, \Phi_{x_2 x_2}^{2\beta} + a_2 \, |\Phi|^2 \, \Phi = 0, \quad (10.73)$$

where $\Phi = \Phi(x_1, x_2, t)$ is the complex function profile, x_1, x_2, and t are its three independent variables, $a_2 \, (c_1^2 \, \alpha_1 + c_2^2 \, \alpha_2) > 0$, A_0, x_{01}, x_{02}, α_1, α_2, c_1, c_2, t_0, and ϕ_0 are arbitrary real constants.

***Example* 7**: *Fractal space–time fractional coupled NLSE*
Given $\quad \psi_1(x, t) = \dfrac{A_0}{\sqrt{2 \, [A_1 + t - t_0]}} \, e^{i \, \phi_1(x, \, t)}, \; \psi_2(x, t) = \dfrac{B_0}{\sqrt{2 \, [B_1 + t - t_0]}} \, e^{i \, \phi_2(x, \, t)} \quad$ with

$\phi_1(x, t) = \dfrac{(b_0 \, A_2 + x - x_0)^2}{4 \, b_0 \, (A_1 + t - t_0)} + \dfrac{b_1 \, A_0^2}{2} \, \ln[2(A_1 + t - t_0)] + \dfrac{b_2 \, B_0^2}{2} \, \ln[2 \, (B_1 + t - t_0)] + \phi_{01}$,

$\phi_2(x, t) = \dfrac{(c_0 \, B_2 + x - x_0)^2}{4 \, c_0 \, (B_1 + t - t_0)} + \dfrac{c_1 \, A_0^2}{2} \, \ln[2(A_1 + t - t_0)] + \dfrac{c_2 \, B_0^2}{2} \, \ln[2 \, (B_1 + t - t_0)] + \phi_{02}$,

is a solution to the CNLSE (7.7), $\quad i \, \psi_{1t} + b_0 \, \psi_{1xx} + (b_1 \, |\psi_1|^2 + b_2 \, |\psi_2|^2)$
$\psi_1 = 0, \, i \, \psi_{2t} + c_0 \, \psi_{2xx} + (c_1 \, |\psi_1|^2 + c_2 \, |\psi_2|^2) \, \psi_2 = 0$, then

$$\Phi_1(x, t) = \frac{A_0}{\sqrt{2 \, [A_1 + t^\alpha - t_0]}} \, e^{i \, \phi_1(x, \, t)},$$

$$\Phi_2(x, t) = \frac{B_0}{\sqrt{2 \, [B_1 + t^\alpha - t_0]}} \, e^{i \, \phi_2(x, \, t)} \quad (10.74)$$

with

$\phi_1(x, t) = \dfrac{(b_0 \, A_2 + x^\beta - x_0)^2}{4 \, b_0 \, (A_1 + t^\alpha - t_0)} + \dfrac{b_1 \, A_0^2}{2} \, \ln[2(A_1 + t^\alpha - t_0)] + \dfrac{b_2 \, B_0^2}{2} \, \ln[2 \, (B_1 + t^\alpha - t_0)],$

$\quad + \phi_{01}$

$\phi_2(x, t) = \dfrac{(c_0 \, B_2 + x^\beta - x_0)^2}{4 \, c_0 \, (B_1 + t^\alpha - t_0)} + \dfrac{c_1 \, A_0^2}{2} \, \ln[2(A_1 + t^\alpha - t_0)] + \dfrac{c_2 \, B_0^2}{2} \, \ln[2 \, (B_1 + t^\alpha - t_0)]$

$\quad + \phi_{02}$
is a solution to

$$i \, \Phi_{1t}^\alpha + b_0 \, \Phi_{1xx}^{2\beta} + (b_1 \, |\Phi_1|^2 + b_2 \, |\Phi_2|^2) \, \Phi_1 = 0,$$
$$i \, \Phi_{2t}^\alpha + c_0 \, \Phi_{2xx}^{2\beta} + (c_1 \, |\Phi_1|^2 + c_2 \, |\Phi_2|^2) \, \Phi_2 = 0, \quad (10.75)$$

where $\Phi = \Phi(x, t)$ is the complex function profile, x and t are its two independent variables, b_0, c_0, b_1, c_1, b_2, c_2, A_0, A_1, A_2, B_0, B_1, B_2, x_0, t_0, ϕ_{01} and ϕ_{02} are arbitrary real constants.

Example 8: *Fractal time fractional discrete NLSE with saturable nonlinearity*

Given $\psi(n, t) = A_0 \operatorname{sech}[A_1 (n - n_0)] \, e^{-i [A_2 (t-t_0)+\phi_0]}$ is a solution to the discrete NLSE with saturable nonlinearity (8.9),

$$i \, \psi_{nt} + \psi_{n+1} + \psi_{n-1} - 2 \, \psi_n + \frac{a_2 \, |\psi_n|^2 \, \psi_n}{1 + \mu \, |\psi_n|^2} = 0, \text{ then}$$

$$\Phi(n, t) = A_0 \operatorname{sech}[A_1 (n - n_0)] \, e^{-i [A_2 (t^\alpha - t_0)+\phi_0]} \qquad (10.76)$$

is a solution to

$$i \, \Phi_{nt}^\alpha + \Phi_{n+1} + \Phi_{n-1} - 2 \, \Phi_n + \frac{a_2 \, |\Phi_n|^2 \, \Phi_n}{1 + \mu \, |\Phi_n|^2} = 0, \qquad (10.77)$$

where $\Phi_n = \Phi(n, t)$ is the complex function profile, the integer site index, n, and t are its two independent variables, $A_0 = \frac{\sinh(A_1)}{\sqrt{\mu}}$, $A_2 = \frac{2\mu - a_2}{\mu}$, $\mu = \frac{a_2 \operatorname{sech}(A_1)}{2} > 0$, A_1, t_0, and ϕ_0 are arbitrary real constants, n_0 is an arbitrary real integer.

Example 9: *Fractal time fractional nonlocal discrete NLSE with saturable nonlinearity*

Given $\psi(\zeta, t) = A_0 \operatorname{sech}[A_1 \zeta] \, e^{-i [A_2 (t-t_0)+\phi_0]}$ is a solution to the nonlocal discrete NLSE with saturable nonlinearity (9.59),

$$i \, \psi_{nt} + \psi_{n+1} + \psi_{n-1} - 2 \, \psi_n + \frac{a_2 \, \psi_n^2 \, \bar{\psi}_n}{1 + \mu \, \psi_n \, \bar{\psi}_n} = 0, \text{ then}$$

$$\Phi(\zeta, t) = A_0 \operatorname{sech}[A_1 \zeta] \, e^{-i [A_2 (t^\alpha - t_0)+\phi_0]} \qquad (10.78)$$

is a solution to

$$i \, \Phi_{nt}^\alpha + \Phi_{n+1} + \Phi_{n-1} - 2 \, \Phi_n + \frac{a_2 \, \Phi_n^2 \, \bar{\Phi}_n}{1 + \mu \, \Phi_n \, \bar{\Phi}_n} = 0, \qquad (10.79)$$

where $\Phi_n = \Phi_n(n, t; a_2, \mu)$, $\bar{\Phi}_n = \Phi_n^*(-n, t; a_2, \mu)$, $A_0 = \frac{\sinh(A_1)}{\sqrt{\mu}}$, $A_2 = \frac{2\mu - a_2}{\mu}$, $\mu = \frac{a_2 \operatorname{sech}(A_1)}{2} > 0$, $\zeta = n - n_0$, A_1, t_0, and ϕ_0 are arbitrary real constants, n_0 is an arbitrary real integer.

10.5 Summary of Section 10.4

NLSE with Fractal Space–Time Fractional Derivative

Transformations: $\Phi(x, t) = \psi(x^\beta, t^\alpha)$, ψ is a solution to any equation from the previous nine chapters or any other versions of
$\Phi(x, t) = \psi(x^\beta, t)$, the NLSE that involve both regular time and regular space derivatives
$\Phi(x, t) = \psi(x, t^\alpha)$,

Equation: $i\,\Phi_t^\alpha + a_1\,\Phi_{xx}^{2\,\beta} + a_2\,|\Phi|^2\,\Phi = 0$

Example	Conditions	Name	Equation #
$\Phi(x, t) = A_0\sqrt{\dfrac{2\,a_1}{a_2}}\,\mathrm{sech}[A_0\,(x^\beta - x_0)]\,e^{i\,[a_1\,A_0^2\,(t^\alpha - t_0) + \phi_0]}$	$a_1\,a_2 > 0$, A_0, x_0, t_0, and ϕ_0 are arbitrary real constants	Bright soliton	(10.62)

Equation: $i\,\Phi_t^\alpha + a_1\,\Phi_{xx}^{2\,\beta} + a_2\,|\Phi|^n\,\Phi = 0$

Example	Conditions	Name	Equation #
$\Phi(x, t) = \left\{ \dfrac{2\,A_0^2\,a_1\,(n+2)}{a_2\,n^2}\,\mathrm{sech}^2[A_0\,(x^\beta - x_0)] \right\}^{\frac{1}{n}}$ $\times\, e^{i\left[\frac{4\,a_1\,A_0^2}{n^2}\,(t^\alpha - t_0) + \phi_0\right]}$	$a_1\,a_2\,(n+2) > 0$, n, A_0, x_0, t_0, and ϕ_0 are arbitrary real constants	Bright soliton	(10.64)

Equation: $i\,\Phi_t^\alpha + a_1\,\Phi_{xx}^{2\,\beta} + a_2\,|\Phi|^n\,\Phi + a_3\,|\Phi|^m\,\Phi = 0$

Example	Conditions	Name	Equation #
$\Phi(x, t) = \left(\dfrac{A_0\,(n+2)}{a_2 + a_2\sqrt{1 + \gamma}\,\cosh[n\sqrt{\frac{A_0}{\eta}}\,(x^\beta - x_0)]} \right)^{\frac{1}{n}}$ $\times\, e^{i\,[A_0\,(t^\alpha - t_0) + \phi_0]}$	$m = 2\,n$, $\gamma = a_3/a_{30}$, $a_{30} = a_2^2\,(n+2)^2/[A_0\,(n+2)^2]$, $A_0\,a_1 > 0$,	Bright soliton	(10.66)

$(1 + \gamma) \geq 0$, n, x_0, t_0, and ϕ_0 are arbitrary real constants

Equation: $i\,\Phi_t^\alpha + a_1\,\Phi_{xx}^{2\beta} + a_2\,|\Phi|^2\,\Phi + i\,a_3\,\Phi_{xxx}^{3\beta} + i\,a_4\,(|\Phi|^2\,\Phi)_x^\beta + i\,a_5\,(|\Phi|^2)_x^\beta\,\Phi = 0$

Example	Name	Equation #
	Bright soliton	(10.68)

$\Phi(x,\,t) = \pm\sqrt{-\mu_1}\ \mathrm{sech}\{\sqrt{-\mu_2}\,[x^\beta - x_0 + c_1\,(t^\alpha - t_0)]\}$
$\times\,e^{i\,[c_2\,(x^\beta - x_0) + c_3\,(t^\alpha - t_0) + \phi_0]}$

Conditions

$\mu_1 = \dfrac{6\,(c_1 + 2\,a_1\,c_2 - 3\,a_3\,c_2^2)}{(3\,a_4 + 2\,a_5)} < 0,$

$\mu_2 = \dfrac{c_1 + 2\,a_1\,c_2 - 3\,a_3\,c_2^2}{a_3} < 0,$

$c_2 = \dfrac{-3\,a_2\,a_3 + a_1\,(3\,a_4 + 2\,a_5)}{6\,a_3\,(a_4 + a_5)},$

$c_3 = 8\,a_1\,c_2^2 - 8\,a_3\,c_2^3 - \dfrac{(a_1\,c_1 + 2\,a_1^2\,c_2)}{a_3} + 3\,c_1\,c_2,$

x_0, t_0, c_1, and ϕ_0 are arbitrary real constants

Equation: $i\,\Phi_t + b_{10}\,\Phi_{xx}^{2\beta} + \dfrac{a_2\,b_{10}\,c_5}{a_1\,c_2^2\,c_7}\,|\Phi|^2\,\Phi - \dfrac{3\,c\,c_7}{b_{10}\,c_5}\,\left(\dfrac{t^\alpha}{\alpha}\right)\left(\dfrac{x^\beta}{\beta}\right)\Phi = 0$

Example	Name	Equation #
	Bright soliton	(10.70)

$\Phi(x,\,t) = c_2\,A_0\,\sqrt{\dfrac{2\,a_1\,c_5}{a_2\,c_7}}\ \mathrm{sech}\left[A_0\left(c\,t^{3\,\alpha} + \dfrac{c_5}{c_7}\,x^\beta - x_0\right)\right]$
$\times\,e^{i\,\left\{A_0^2\left[a_1\,c_0 + \dfrac{b_{10}\,c_5^2}{c_7^2}\,(t^\alpha - t_0)\right]\dfrac{3\,c\,c_7}{2\,b_{10}\,c_5}\,t^{2\,\alpha}\,x^\beta\right\}}$
$\times\,e^{i\,\left\{-\dfrac{9\,c^2\,c_7^2}{20\,b_{10}\,c_5^2}\,t^{5\,\alpha} + c_1 + \phi_0\right\}}$

Conditions

$(a_1\,a_2\,c_5\,c_7) > 0$, A_0, c, c_2, b_{10}, x_0, t_0, and ϕ_0 are arbitrary real constants

Equation: $i\,\Phi_t^\alpha + \alpha_1\,\Phi_{x_1x_1}^{2\beta} + \alpha_2\,\Phi_{x_2x_2}^{2\beta} + a_2\,|\Phi|^2\,\Phi = 0$

Example	Name	Equation #
	Bright soliton	(10.72)

$\Phi(x_1,\,x_2,\,t) = A_0\,\sqrt{\dfrac{2\,(c_1^2\,\alpha_1 + c_2^2\,\alpha_2)}{a_2}}$
$\times\,\mathrm{sech}\{A_0\,[c_1\,(x_1^\beta - x_{01}) + c_2\,(x_2^\beta - x_{02})]\}$
$\times\,e^{i\,[A_0^2\,(c_1^2\,\alpha_1 + c_2^2\,\alpha_2)\,(t^\alpha - t_0) + \phi_0]}$

Conditions

$a_2\,(c_1^2\,\alpha_1 + c_2^2\,\alpha_2) > 0$, A_0, t_0, and ϕ_0 are arbitrary real constants

(*Continued*)

Equation:
$$i\,\Phi_{1t}^{\alpha} + b_0\,\Phi_{1xx}^{2\beta} + (b_1\,|\Phi_1|^2 + b_2\,|\Phi_2|^2)\,\Phi_1 = 0,$$
$$i\,\Phi_{2t}^{\alpha} + c_0\,\Phi_{2xx}^{2\beta} + (c_1\,|\Phi_1|^2 + c_2\,|\Phi_2|^2)\,\Phi_2 = 0$$

Example	Name	Equation #
$\Phi_1(x,t) = \dfrac{A_0}{\sqrt{2[A_1+t^{\alpha}-t_0]}}\,e^{i\,\phi_1(x,t)},$ $\Phi_2(x,t) = \dfrac{B_0}{\sqrt{2[B_1+t^{\alpha}-t_0]}}\,e^{i\,\phi_2(x,t)}$	Bright soliton	(10.74)

Conditions
$b_0, c_0, b_1, c_1, b_2, c_2, A_0, A_1, A_2, B_0, B_1, B_2,$ x_0, t_0, ϕ_{01} and ϕ_{02} are arbitrary real constants

Equation: $i\,\Phi_{nt}^{\alpha} + \Phi_{n+1} + \Phi_{n-1} - 2\,\Phi_n + \dfrac{a_2\,|\Phi_n|^2\,\Phi_n}{1+\mu\,|\Phi_n|^2} = 0$

Example	Name	Equation #
$\Phi(n,t) = A_0\,\mathrm{sech}[A_1\,(n-n_0)]\,e^{-i\,[A_2\,(t^{\alpha}-t_0)+\phi_0]}$	Bright soliton	(10.76)

Conditions
$A_0 = \dfrac{\sinh(A_1)}{\sqrt{\mu}}$, $A_2 = \dfrac{2\mu-a_2}{\mu}$, $\mu = \dfrac{a_2\,\mathrm{sech}(A_1)}{2} > 0$, A_1, $t_0, n_0,$ and ϕ_0 are arbitrary real constants

Equation: $i\,\Phi_{nt}^{\alpha} + \Phi_{n+1} + \Phi_{n-1} - 2\,\Phi_n + \dfrac{a_2\,\Phi_n^2\,\bar{\Phi}_n}{1+\mu\,\Phi_n\,\bar{\Phi}_n} = 0$

Example	Name	Equation #
$\Phi(\zeta,t) = A_0\,\mathrm{sech}[A_1\,\zeta]\,e^{-i\,[A_2\,(t^{\alpha}-t_0)+\phi_0]}$	Bright soliton	(10.78)

Conditions
$A_0 = \dfrac{\sinh(A_1)}{\sqrt{\mu}}$, $A_2 = \dfrac{2\mu-a_2}{\mu}$, $\mu = \dfrac{a_2\,\mathrm{sech}(A_1)}{2} > 0$, $\zeta = n - n_0$, A_1, $t_0, n_0,$ and ϕ_0 are arbitrary real constants

10.6 Atangana (β) Fractional Derivative

The Atangana fractional derivative of a function $f(x)$ is defined as

$$f_x^\alpha(x) := \lim_{\epsilon \to 0} \left[\frac{f\left\{x + \epsilon \left[x + \frac{1}{\Gamma(\alpha)}\right]^{1-\alpha}\right\} - f(x)}{\epsilon} \right] = f'(x), \qquad (10.80)$$

where α is the fractional derivative order $0 < \alpha \leqslant 1$. The second Atangana fractional derivative of order α is then defined as

$$f_{xx}^\alpha(x) := \lim_{\epsilon \to 0} \left[\frac{f'\left\{x + \epsilon \left[x + \frac{1}{\Gamma(\alpha)}\right]^{1-\alpha}\right\} - f'(x)}{\epsilon} \right] = f''(x). \qquad (10.81)$$

The n Atangana fractional derivative of order α is defined as

$$f_{nx}^\alpha(x) := \lim_{\epsilon \to 0} \left[\frac{f^{n-1}\left\{x + \epsilon \left[x + \frac{1}{\Gamma(\alpha)}\right]^{1-\alpha}\right\} - f^{n-1}(x)}{\epsilon} \right] = f^n(x). \qquad (10.82)$$

10.6.1 NLSE with Atangana Space–Time Fractional Derivative

If $\psi(x, t)$ is a solution to any equation from the previous nine chapters or any other versions of the NLSE that involve both regular time and regular space derivatives, then

$$\Phi(x, t) = \psi\left(\frac{1}{\beta}\left[x + \frac{1}{\Gamma(\beta)}\right]^\beta, \frac{1}{\alpha}\left[t + \frac{1}{\Gamma(\alpha)}\right]^\alpha\right), \qquad (10.83)$$

$$\Phi(x, t) = \psi\left(\frac{1}{\beta}\left[x + \frac{1}{\Gamma(\beta)}\right]^\beta, t\right), \qquad (10.84)$$

and

$$\Phi(x, t) = \psi\left(x, \frac{1}{\alpha}\left[t + \frac{1}{\Gamma(\alpha)}\right]^\alpha\right) \qquad (10.85)$$

are solutions to the same equation, where both or any of the regular time and regular space derivatives are replaced by the Atangana fractional time and fractional space derivatives, respectively

$$\psi_t \to \Phi_t^\alpha \qquad \psi_x \to \Phi_x^\beta \tag{10.86}$$

and all space- and time-dependent coefficients are transformed as follows

$$g(x, t) \to g\left(\frac{1}{\beta}\left[x + \frac{1}{\Gamma(\beta)}\right]^\beta, \frac{1}{\alpha}\left[t + \frac{1}{\Gamma(\alpha)}\right]^\alpha\right), \tag{10.87}$$

where α and β are the time and space fractional derivative of orders $0 < \alpha \leqslant 1$ and $0 < \beta \leqslant 1$, respectively.

Example 1: *Atangana space–time fractional fundamental NLSE*

Given $\psi(x, t) = A_0 \sqrt{\frac{2 a_1}{a_2}} \operatorname{sech}[A_0 (x - x_0)] e^{i [a_1 A_0^2 (t - t_0) + \phi_0]}$ is a solution to the fundamental NLSE (2.160), $i \psi_t + a_1 \psi_{xx} + a_2 |\psi|^2 \psi = 0$, then

$$\Phi(x, t) = A_0 \sqrt{\frac{2 a_1}{a_2}} \operatorname{sech}\left(A_0 \left\{\frac{1}{\beta}\left[x + \frac{1}{\Gamma(\beta)}\right]^\beta - x_0\right\}\right) e^{i\left(a_1 A_0^2 \left\{\frac{1}{\alpha}\left[t + \frac{1}{\Gamma(\alpha)}\right]^\alpha - t_0\right\} + \phi_0\right)} \tag{10.88}$$

is a solution to

$$i \Phi_t^\alpha + a_1 \Phi_{xx}^{2\beta} + a_2 |\Phi|^2 \Phi = 0, \tag{10.89}$$

where $\Phi = \Phi(x, t)$ is the complex function profile, x and t are its two independent variables, $a_1 a_2 > 0$, A_0, x_0, t_0, and ϕ_0 are arbitrary real constants.

Example 2: *Atangana space–time fractional NLSE with power law nonlinearity*

Given $\psi(x, t) = \left\{\frac{2 A_0^2 a_1 (n + 2)}{a_2 n^2} \operatorname{sech}^2[A_0 (x - x_0)]\right\}^{\frac{1}{n}} e^{i\left[\frac{4 a_1 A_0^2}{n^2} (t - t_0) + \phi_0\right]}$ is a solution to the NLSE with power law nonlinearity (3.9), $i \psi_t + a_1 \psi_{xx} + a_2 |\psi|^n \psi = 0$, then

$$\Phi(x, t) = \left\{\frac{2 A_0^2 a_1 (n + 2)}{a_2 n^2} \operatorname{sech}^2\left[A_0 \left\{\frac{1}{\beta}\left[x + \frac{1}{\Gamma(\beta)}\right]^\beta - x_0\right\}\right]\right\}^{\frac{1}{n}}$$

$$\times e^{i\left(\frac{4 a_1 A_0^2}{n^2} \left\{\frac{1}{\alpha}\left[t + \frac{1}{\Gamma(\alpha)}\right]^\alpha - t_0\right\} + \phi_0\right)} \tag{10.90}$$

is a solution to

$$i \Phi_t^\alpha + a_1 \Phi_{xx}^{2\beta} + a_2 |\Phi|^n \Phi = 0, \tag{10.91}$$

where $\Phi = \Phi(x, t)$ is the complex function profile, x and t are its two independent variables, $a_1 a_2 (n + 2) > 0$, n, A_0, x_0, t_0, and ϕ_0 are arbitrary real constants.

Example 3: *Atangana space–time fractional NLSE with dual power law nonlinearity*

Given $\psi(x, t) = \left\{ \dfrac{A_0\,(n+2)}{a_2 + a_2\,\sqrt{1+\gamma}\,\cosh\left[n\sqrt{\frac{A_0}{q}}\,(x-x_0)\right]} \right\}^{\frac{1}{n}} e^{i\,[A_0\,(t-t_0)+\phi_0]}$ is a solution to the

NLSE with dual power law nonlinearity (3.27), $i\,\psi_t + a_1\,\psi_{xx} + a_2\,|\psi|^n\,\psi + a_3\,|\psi|^m\,\psi = 0$, then

$$\Phi(x, t) = \left[\dfrac{A_0\,(n+2)}{a_2 + a_2\,\sqrt{1+\gamma}\,\cosh\left(n\sqrt{\frac{A_0}{a_1}}\left\{\frac{1}{\beta}\left[x+\frac{1}{\Gamma(\beta)}\right]^{\beta} - x_0\right\}\right)} \right]^{\frac{1}{n}} e^{i\left(A_0\left\{\frac{1}{\alpha}\left[t+\frac{1}{\Gamma(\alpha)}\right]^{\alpha} - t_0\right\}+\phi_0\right)} \quad (10.92)$$

is a solution to

$$i\,\Phi_t^\alpha + a_1\,\Phi_{xx}^{2\beta} + a_2\,|\Phi|^n\,\Phi + a_3\,|\Phi|^m\,\Phi = 0, \qquad (10.93)$$

where $\Phi = \Phi(x, t)$ is the complex function profile, x and t are its two independent variables, $m = 2\,n$, $\gamma = a_3/a_{30}$, $a_{30} = a_2^2\,(n+1)/[A_0\,(n+2)^2]$, $(1+\gamma) \geqslant 0$, $A_0\,a_1 > 0$, n, x_0, t_0, and ϕ_0 are arbitrary real constants.

Example 4: *Atangana space–time fractional NLSE with third order dispersion, self-steepening, and self-frequency shift*

$\psi(x, t) = \pm\sqrt{-\mu_1}\,\text{sech}\{\sqrt{-\mu_2}\,[x-x_0+c_1\,(t-t_0)]\}e^{i\,[c_2\,(x-x_0)+c_3\,(t-t_0)+\phi_0]}$ is a solution to the NLSE with third order dispersion, self-steepening, and self-frequency shift (4.2), $i\,\psi_t + a_1\,\psi_{xx} + a_2\,|\psi|^2\,\psi + i\,a_3\,\psi_{xxx} + i\,a_4\,(|\psi|^2\,\psi)_x + i\,a_5\,(|\psi|^2)_x\,\psi = 0$, then

$$\Phi(x, t) = \pm\sqrt{-\mu_1}\,\text{sech}\left(\sqrt{-\mu_2}\left\{\frac{1}{\beta}\left[x+\frac{1}{\Gamma(\beta)}\right]^{\beta} - x_0 + c_1\,(t-t_0)\right\}\right)$$

$$\times\, e^{i\left(c_2\left\{\frac{1}{\beta}\left[x+\frac{1}{\Gamma(\beta)}\right]^{\beta}-x_0\right\}+c_3\left\{\frac{1}{\alpha}\left[t+\frac{1}{\Gamma(\alpha)}\right]^{\alpha}-t_0\right\}+\phi_0\right)} \qquad (10.94)$$

is a solution to

$$i\,\Phi_t^\alpha + a_1\,\Phi_{xx}^{2\beta} + a_2\,|\Phi|^2\,\Phi + i\,a_3\,\Phi_{xxx}^{3\beta} + i\,a_4\left(|\Phi|^2\,\Phi\right)_x^{\beta} + i\,a_5\left(|\Phi|^2\right)_x^{\beta}\,\Phi = 0, \quad (10.95)$$

where $\Phi = \Phi(x, t)$ is the complex function profile, x and t are its two independent variables, $\mu_1 = \dfrac{6\,(c_1+2\,a_1\,c_2-3\,a_3\,c_2^2)}{(3\,a_4+2\,a_5)} < 0$, $\mu_2 = \dfrac{c_1+2\,a_1\,c_2-3\,a_3\,c_2^2}{a_3} < 0$, $c_2 = \dfrac{-3\,a_2\,a_3 + a_1\,(3\,a_4+2\,a_5)}{6\,a_3\,(a_4+a_5)}$, $c_3 = 8\,a_1\,c_2^2 - 8\,a_3\,c_2^3 - \dfrac{(a_1\,c_1+2\,a_1^2\,c_2)}{a_3} + 3\,c_1\,c_2$, x_0, t_0, c_1, and ϕ_0 are arbitrary real constants.

Example 5: *Atangana space–time fractional NLSE with constant dispersion and linear potential*

Given

$$\psi(x,\, t) = c_2 \, A_0 \, \sqrt{\frac{2\, a_1\, c_5}{a_2\, c_7}} \; \text{sech}[A_0\,(c\, t^3 + \tfrac{c_5}{c_7}\, x - x_0)]$$

$$\times \, e^{i\,\{A_0^2\,[a_1\, c_0 + \frac{b_{10}\, c_5^2}{c_7^2}\,(t - t_0)] - \frac{3\, c\, c_7}{2\, b_{10}\, c_5}\, t^2\, x - \frac{9\, c^2\, c_7^2}{20\, b_{10}\, c_5^2}\, t^5 + c_1 + \phi_0\}}$$

is a solution to the NLSE with constant dispersion and linear potential, (5.64), with

$$g_4(t) = c\, t^3,\; i\, \psi_t + b_{10}\, \psi_{xx} + \frac{a_2\, b_{10}\, c_5}{a_1\, c_5^2\, c_7}\, |\psi|^2\, \psi - \frac{3\, c\, c_7}{b_{10}\, c_5}\, t\, x\, \psi = 0,\; \text{then}$$

$$\Phi(x,\, t) = c_2\, A_0\, \sqrt{\frac{2\, a_1\, c_5}{a_2\, c_7}}\; \text{sech}\left[A_0\left(c\left\{\frac{1}{\alpha}\left[t + \frac{1}{\Gamma(\alpha)}\right]^{\alpha}\right\}^3 + \frac{c_5}{c_7}\left\{\frac{1}{\beta}\left[x + \frac{1}{\Gamma(\beta)}\right]^{\beta}\right\} - x_0\right)\right]$$

$$\times\, e^{i\left[A_0^2\left(a_1\, c_0 + \frac{b_{10}\, c_5^2}{c_7^2}\left\{\frac{1}{\alpha}\left[t + \frac{1}{\Gamma(\alpha)}\right]^{\alpha} - t_0\right\}\right) - \frac{3\, c\, c_7}{2\, b_{10}\, c_5}\left\{\frac{1}{\alpha}\left[t + \frac{1}{\Gamma(\alpha)}\right]^{\alpha}\right\}^2\left\{\frac{1}{\beta}\left[x + \frac{1}{\Gamma(\beta)}\right]^{\beta}\right\}\right]} \tag{10.96}$$

$$\times\, e^{i\left[-\frac{9\, c^2\, c_7^2}{20\, b_{10}\, c_5^2}\left\{\frac{1}{\alpha}\left[t + \frac{1}{\Gamma(\alpha)}\right]^{\alpha}\right\}^5 + c_1 + \phi_0\right]}$$

is a solution to

$$i\, \Phi_t^{\alpha} + b_{10}\, \Phi_{xx}^{2\,\beta} + \frac{a_2\, b_{10}\, c_5}{a_1\, c_5^2\, c_7}\, |\Phi|^2\, \Phi - \frac{3\, c\, c_7}{b_{10}\, c_5}\left\{\frac{1}{\alpha}\left[t + \frac{1}{\Gamma(\alpha)}\right]^{\alpha}\right\}\left\{\frac{1}{\beta}\left[x + \frac{1}{\Gamma(\beta)}\right]^{\beta}\right\}\Phi = 0,\; (10.97)$$

where $\Phi = \Phi(x,\, t)$ is the complex function profile, x and t are its two independent variables, $(a_1\, a_2\, c_5\, c_7) > 0$, A_0, c, c_2, b_{10}, x_0, t_0, and ϕ_0 are arbitrary real constants.

Example 6: *Atangana space–time fractional (2 + 1)-dimensional NLSE*

Given $\psi(x_1,\, x_2,\, t) = A_0\, \sqrt{\frac{2\,(c_1^2\, \alpha_1 + c_2^2\, \alpha_2)}{a_2}}\; \text{sech}\{A_0\,[c_1\,(x_1 - x_{01}) + c_2\,(x_2 - x_{02})]\}$

$$\times\, e^{i\,[A_0^2\,(c_1^2\, \alpha_1 + c_2^2\, \alpha_2)\,(t - t_0) + \phi_0]}$$

is a solution to the (2 + 1)-dimensional NLSE (6.9),

$i\, \psi_t + \alpha_1\, \psi_{x_1 x_1} + \alpha_2\, \psi_{x_2 x_2} + a_2\, |\psi|^2\, \psi = 0$, then

$$\Phi(x_1,\, x_2,\, t) = A_0\, \sqrt{\frac{2\,(c_1^2\, \alpha_1 + c_2^2\, \alpha_2)}{a_2}}\; \text{sech}\left[A_0\left(c_1\left\{\frac{1}{\beta}\left[x_1 + \frac{1}{\Gamma(\beta)}\right]^{\beta} - x_{01}\right\}\right.\right.$$

$$\left.\left. + c_2\left\{\frac{1}{\beta}\left[x_2 + \frac{1}{\Gamma(\beta)}\right]^{\beta} - x_{02}\right\}\right)\right] \tag{10.98}$$

$$\times\, e^{i\left[A_0^2\left(c_1^2\, \alpha_1 + c_2^2\, \alpha_2\right)\left\{\frac{1}{\alpha}\left[t + \frac{1}{\Gamma(\alpha)}\right]^{\alpha} - t_0\right\} + \phi_0\right]}$$

is a solution to

$$i\, \Phi_t^{\alpha} + \alpha_1\, \Phi_{x_1 x_1}^{2\,\beta} + \alpha_2\, \Phi_{x_2 x_2}^{2\,\beta} + a_2\, |\Phi|^2\, \Phi = 0, \tag{10.99}$$

where $\Phi = \Phi(x_1, x_2, t)$ is the complex function profile, x_1, x_2, and t are its three independent variables, $a_2 (c_1^2 \alpha_1 + c_2^2 \alpha_2) > 0$, A_0, t_0, and ϕ_0 are arbitrary real constants.

Example 7: *Atangana space–time fractional coupled NLSE*

Given $\quad \psi_1(x, t) = \dfrac{A_0}{\sqrt{2\,[A_1 + t - t_0]}}\, e^{i\,\phi_1(x,\, t)}$, $\psi_2(x, t) = \dfrac{B_0}{\sqrt{2\,[B_1 + t - t_0]}}\, e^{i\,\phi_2(x,\, t)} \quad$ with

$\phi_1(x, t) = \dfrac{(b_0 A_2 + x - x_0)^2}{4\, b_0\, (A_1 + t - t_0)} + \dfrac{b_1 A_0^2}{2} \ln[2(A_1 + t - t_0)] + \dfrac{b_2 B_0^2}{2} \ln[2\,(B_1 + t - t_0)] + \phi_{01}$,

$\phi_2(x, t) = \dfrac{(c_0 B_2 + x - x_0)^2}{4\, c_0\, (B_1 + t - t_0)} + \dfrac{c_1 A_0^2}{2} \ln[2(A_1 + t - t_0)] + \dfrac{c_2 B_0^2}{2}$

$\ln[2\,(B_1 + t - t_0)] + \phi_{02}$, $\quad$ is $\quad$ a $\quad$ solution $\quad$ to $\quad$ the $\quad$ CNLSE $\quad$ (7.7),

$i\,\psi_{1t} + b_0\,\psi_{1xx} + (b_1\,|\psi_1|^2 + b_2\,|\psi_2|^2)\,\psi_1 = 0$, $i\,\psi_{2t} + c_0\,\psi_{2xx} + (c_1\,|\psi_1|^2 + c_2\,|\psi_2|^2)\,\psi_2 = 0$, then

$$\Phi_1(x, t) = \dfrac{A_0}{\sqrt{2\left\{ A_1 + \dfrac{1}{\alpha}\left[t + \dfrac{1}{\Gamma(\alpha)} \right]^\alpha - t_0 \right\}}}\, e^{i\,\phi_1(x,\, t)},$$

$$\Phi_2(x, t) = \dfrac{B_0}{\sqrt{2\left\{ B_1 + \dfrac{1}{\alpha}\left[t + \dfrac{1}{\Gamma(\alpha)} \right]^\alpha - t_0 \right\}}}\, e^{i\,\phi_2(x,\, t)}$$

(10.100)

with

$$\phi_1(x, t) = \dfrac{\{b_0 A_2 + \frac{1}{\beta}[x + \frac{1}{\Gamma(\beta)}]^\beta - x_0\}^2}{4\, b_0\, \{A_1 + \frac{1}{\alpha}[t + \frac{1}{\Gamma(\alpha)}]^\alpha - t_0\}} + \dfrac{b_1 A_0^2}{2} \ln(2\,\{A_1 + \frac{1}{\alpha}[t + \frac{1}{\Gamma(\alpha)}]^\alpha - t_0\})$$

$$+ \dfrac{b_2 B_0^2}{2} \ln(2\,\{B_1 + \frac{1}{\alpha}[t + \frac{1}{\Gamma(\alpha)}]^\alpha - t_0\}) + \phi_{01}$$

,

$$\phi_2(x, t) = \dfrac{\{c_0 B_2 + \frac{1}{\beta}[x + \frac{1}{\Gamma(\beta)}]^\beta - x_0\}^2}{4\, c_0\, \{B_1 + \frac{1}{\alpha}[t + \frac{1}{\Gamma(\alpha)}]^\alpha - t_0\}} + \dfrac{c_1 A_0^2}{2} \ln(2\,\{A_1 + \frac{1}{\alpha}[t + \frac{1}{\Gamma(\alpha)}]^\alpha - t_0\}) + \dfrac{c_2 B_0^2}{2}$$

$$\ln(2\,\{B_1 + \frac{1}{\alpha}[t + \frac{1}{\Gamma(\alpha)}]^\alpha - t_0\}) + \phi_{02}$$

is a solution to

$$i\,\Phi_{1t}^\alpha + b_0\,\Phi_{1xx}^{2\beta} + (b_1\,|\Phi_1|^2 + b_2\,|\Phi_2|^2)\,\Phi_1 = 0,$$
$$i\,\Phi_{2t}^\alpha + c_0\,\Phi_{2xx}^{2\beta} + (c_1\,|\Phi_1|^2 + c_2\,|\Phi_2|^2)\,\Phi_2 = 0,$$

(10.101)

where $\Phi = \Phi(x, t)$ is the complex function profile, x and t are its two independent variables, b_0, c_0, b_1, c_1, b_2, c_2, A_0, A_1, A_2, B_0, B_1, B_2, x_0, t_0, ϕ_{01} and ϕ_{02} are arbitrary real constants.

Example 8: *Atangana time fractional discrete NLSE with saturable nonlinearity*

Given $\psi(n, t) = A_0 \,\mathrm{sech}[A_1\,(n - n_0)]\, e^{-i\,[A_2\,(t - t_0) + \phi_0]}$ is a solution to the discrete NLSE

with saturable nonlinearity (8.9), $i\,\psi_{nt} + \psi_{n+1} + \psi_{n-1} - 2\,\psi_n + \dfrac{a_2\,|\psi_n|^2\,\psi_n}{1 + \mu\,|\psi_n|^2} = 0$, then

$$\Phi(n,\, t) = A_0 \, \text{sech}[A_1 \, (n - n_0)] \, e^{-i \left(A_2 \left\{ \frac{1}{\alpha} \left[t + \frac{1}{\Gamma(\alpha)} \right]^{\alpha} - t_0 \right\} + \phi_0 \right)} \qquad (10.102)$$

is a solution to

$$i \, \Phi_{nt}^{\alpha} + \Phi_{n+1} + \Phi_{n-1} - 2 \, \Phi_n + \frac{a_2 \, |\Phi_n|^2 \, \Phi_n}{1 + \mu \, |\Phi_n|^2} = 0, \qquad (10.103)$$

where $\Phi_n = \Phi(n,\, t)$ is the complex function profile, the integer site index, n, and t are its two independent variables, $A_0 = \frac{\sinh(A_1)}{\sqrt{\mu}}$, $A_2 = \frac{2\,\mu - a_2}{\mu}$, $\mu = \frac{a_2 \, \text{sech}(A_1)}{2} > 0$, A_1, t_0 and ϕ_0 are arbitrary real constants, n_0 is an arbitrary real integer.

Example 9: *Atangana time fractional nonlocal discrete NLSE with saturable nonlinearity*

Given $\psi(\zeta,\, t) = A_0 \, \text{sech}[A_1 \, \zeta] \, e^{-i \, [A_2 \, (t - t_0) + \phi_0]}$ is a solution to the nonlocal discrete NLSE with saturable nonlinearity (9.59),

$i \, \psi_{nt} + \psi_{n+1} + \psi_{n-1} - 2 \, \psi_n + \frac{a_2 \, \psi_n^2 \, \bar{\psi}_n}{1 + \mu \, \psi_n \, \bar{\psi}_n} = 0$, then

$$\Phi(\zeta,\, t) = A_0 \, \text{sech}[A_1 \, \zeta] \, e^{-i \left(A_2 \left\{ \frac{1}{\alpha} \left[t + \frac{1}{\Gamma(\alpha)} \right]^{\alpha} - t_0 \right\} + \phi_0 \right)} \qquad (10.104)$$

is a solution to

$$i \, \Phi_{nt}^{\alpha} + \Phi_{n+1} + \Phi_{n-1} - 2 \, \Phi_n + \frac{a_2 \, \Phi_n^2 \, \bar{\Phi}_n}{1 + \mu \, \Phi_n \, \bar{\Phi}_n} = 0, \qquad (10.105)$$

where $\Phi_n = \Phi_n(n,\, t;\, a_2,\, \mu)$, $\bar{\Phi}_n = \Phi_n^*(-n,\, t;\, a_2,\, \mu)$, $A_0 = \frac{\sinh(A_1)}{\sqrt{\mu}}$, $A_2 = \frac{2\,\mu - a_2}{\mu}$, $\mu = \frac{a_2 \, \text{sech}(A_1)}{2} > 0$, $\zeta = n - n_0$, A_1, t_0, and ϕ_0 are arbitrary real constants, n_0 is an arbitrary real integer.

10.7 Summary of Section 10.6

NLSE with Atangana Space–Time Fractional Derivative

$$\Phi(x,t) = \psi\left(\frac{1}{\beta}\left[x + \frac{1}{\Gamma(\beta)}\right]^\beta, \frac{1}{\alpha}\left[t + \frac{1}{\Gamma(\alpha)}\right]^\alpha\right),$$

ψ is a solution to any equation from the previous nine chapters or any other versions of the NLSE that involve both regular time and regular space derivatives

Transformations: $\Phi(x,t) = \psi\left(\frac{1}{\beta}\left[x + \frac{1}{\Gamma(\beta)}\right]^\beta, t\right),$

$$\Phi(x,t) = \psi\left(x, \frac{1}{\alpha}\left[t + \frac{1}{\Gamma(\alpha)}\right]^\alpha\right),$$

Equation: $i\,\Phi_t^\alpha + a_1\,\Phi_{xx}^{2\beta} + a_2\,|\Phi|^2\,\Phi = 0$

Example	Conditions	Name	Equation #
$\Phi(x,t) = A_0\sqrt{\dfrac{2a_1}{a_2}}\,\text{sech}\left(A_0\left\{\dfrac{1}{\beta}\left[x + \dfrac{1}{\Gamma(\beta)}\right]^\beta - x_0\right\}\right)$ $\times e^{\,i\left(a_1 A_0^2\left\{\frac{1}{\alpha}\left[t + \frac{1}{\Gamma(\alpha)}\right]^\alpha - t_0\right\} + \phi_0\right)}$	$a_1\,a_2 > 0, A_0, x_0, t_0,$ and ϕ_0 are arbitrary real constants	Bright soliton	(10.88)

Equation: $i\,\Phi_t^\alpha + a_1\,\Phi_{xx}^{2\beta} + a_2\,|\Phi|^n\,\Phi = 0$

Example	Conditions	Name	Equation #
$\Phi(x,t) = \left\{\dfrac{2A_0^2 a_1(n+2)}{a_2 n^2}\,\text{sech}^2\left[A_0\left\{\dfrac{1}{\beta}\left[x + \dfrac{1}{\Gamma(\beta)}\right]^\beta - x_0\right\}\right]\right\}^{\frac{1}{n}}$ $\times e^{\,i\left(\frac{4a_1 A_0^2}{n^2}\left\{\frac{1}{\alpha}\left[t + \frac{1}{\Gamma(\alpha)}\right]^\alpha - t_0\right\} + \phi_0\right)}$	$a_1\,a_2\,(n+2) > 0, n, A_0, x_0, t_0,$ and ϕ_0 are arbitrary real constants	Bright soliton	(10.90)

Equation: $i\,\Phi_t^\alpha + a_1\,\Phi_{xx}^{2\beta} + a_2\,|\Phi|^n\,\Phi + a_3\,|\Phi|^m\,\Phi = 0$

(Continued)

Example	Name	Conditions	Equation #
$\Phi(x,t) = \left[\dfrac{A_0(n+2)}{a_2 + a_2\sqrt{1+\gamma}\,\cosh\left(n\sqrt{\dfrac{A_0}{q}}\left\{\dfrac{1}{\beta}\left[x+\dfrac{1}{\Gamma(\beta)}\right]^{\beta} - x_0\right\}\right)}\right]^{\frac{1}{n}}$ $\times e^{i\left(A_0\left\{\frac{1}{\alpha}\left[t+\frac{1}{\Gamma(\alpha)}\right]^{\alpha}-t_0\right\}+\phi_0\right)}$	Bright soliton	$m = 2n$, $\gamma = a_3/a_{30}$, $a_{30} = a_2^2(n+1)/[A_0(n+2)^2]$, $A_0\,a_1 > 0$, $(1+\gamma) \geq 0$, n, x_0, t_0, and ϕ_0 are arbitrary real constants	(10.92)

Equation: $i\,\Phi_t^{\alpha} + a_1\,\Phi_{xx}^{2\beta} + a_2\,|\Phi|^2\,\Phi + i\,a_3\,\Phi_{xxx}^{3\beta} + i\,a_4\,(|\Phi|^2\,\Phi)_x^{3\beta} + i\,a_5\,(|\Phi|^2)_x^{\beta}\,\Phi = 0$

Example	Name	Conditions	Equation #
$\Phi(x,t) = \pm\sqrt{-\mu_1}\,\operatorname{sech}\left(\sqrt{-\mu_2}\left\{\dfrac{1}{\beta}\left[x+\dfrac{1}{\Gamma(\beta)}\right]^{\beta} - x_0 + c_1(t-t_0)\right\}\right)$ $\times e^{i\left(c_2\left\{\frac{1}{\beta}\left[x+\frac{1}{\Gamma(\beta)}\right]^{\beta}-x_0\right\}+c_3\left\{\frac{1}{\alpha}\left[t+\frac{1}{\Gamma(\alpha)}\right]^{\alpha}-t_0\right\}+\phi_0\right)}$	Bright soliton	$\mu_1 = \dfrac{6(c_1 + 2a_1c_2 - 3a_3c_2^2)}{(3a_4 + 2a_5)} < 0$, $\mu_2 = \dfrac{c_1 + 2a_1c_2 - 3a_3c_2^2}{c_1 + 2a_1c_2 - 3a_3c_2^2} < 0$, $c_2 = \dfrac{-3a_2a_3 + a_1(3a_4 + 2a_5)}{6a_3(a_4 + a_5)}$, $c_3 = 8a_1c_2^2 - 8a_3c_2^3$ $-\dfrac{a_3}{(a_1c_1 + 2a_1^2c_2)} + 3c_1c_2$, x_0, t_0, c_1, and ϕ_0 are arbitrary real constants	(10.94)

Equation: $i\,\Phi_t + b_{10}\,\Phi_{xx}^{2\beta} + \dfrac{a_2\,b_{10}\,c_5}{a_1\,c_2^2\,c_7}\,|\Phi|^2\,\Phi - \dfrac{3\,c\,c_7}{b_{10}\,c_5}\left(\dfrac{t^{\alpha}}{\alpha}\right)\left(\dfrac{x^{\beta}}{\beta}\right)\Phi = 0$

Example	Name	Conditions	Equation #
(see below)	Bright soliton	$(a_1 a_2 c_5 c_7) > 0, A_0, c, c_2, b_{10}, x_0, t_0,$ and ϕ_0 are arbitrary real constants	(10.96)

$$\Phi(x, t) = c_2 A_0 \sqrt{\frac{2 a_1 c_5}{a_2 c_7}}$$

$$\times \text{sech}\left[A_0 \left(c \left\{ \frac{1}{\alpha}\left[t + \frac{1}{\Gamma(\alpha)}\right]^\alpha \right\}^3 + \frac{c_5}{c_7}\left\{ \frac{1}{\beta}\left[x + \frac{1}{\Gamma(\beta)}\right]^\beta \right\} - x_0 \right) \right]$$

$$\times e^{\,i\left[A_0^2 \left(a_1 c_0 + \frac{b_{10} c_5^2}{c_7^2}\left(\frac{1}{\alpha}\left[t+\frac{1}{\Gamma(\alpha)}\right]^\alpha - t_0\right)\right)\right]}$$

$$\times e^{\,i\left[-\frac{3 c\, c_7}{2 b_{10} c_5}\left\{ \frac{1}{\alpha}\left[t+\frac{1}{\Gamma(\alpha)}\right]^\alpha \right\}^2 \left\{\frac{1}{\beta}\left[x+\frac{1}{\Gamma(\beta)}\right]^\beta\right\}\right]}$$

$$\times e^{\,i\left[-\frac{9 c^2 c_7^2}{20 b_{10} c_5^2}\left\{ \frac{1}{\alpha}\left[t+\frac{1}{\Gamma(\alpha)}\right]^\alpha \right\}^5 + c_1 + \phi_0\right]}$$

Equation: $i\, \Phi_t^\alpha + \alpha_1 \Phi_{x_1 x_1}^{2\beta} + \alpha_2 \Phi_{x_2 x_2}^{2\beta} + a_2 |\Phi|^2 \Phi = 0$

Example	Name	Conditions	Equation #
(see below)	Bright soliton	$a_2 (c_1^2 \alpha_1 + c_2^2 \alpha_2) > 0, A_0, t_0,$ and ϕ_0 are arbitrary real constants	(10.98)

$$\Phi(x_1, x_2, t) = A_0 \sqrt{\frac{2(c_1^2 \alpha_1 + c_2^2 \alpha_2)}{a_2}}$$

$$\times \text{sech}\left[A_0 \left(c_1 \left\{ \frac{1}{\beta}\left[x_1 + \frac{1}{\Gamma(\beta)}\right]^\beta - x_{01} \right\}\right.\right.$$

$$\left.\left. + c_2 \left\{ \frac{1}{\beta}\left[x_2 + \frac{1}{\Gamma(\beta)}\right]^\beta - x_{02} \right\} \right) \right]$$

$$\times e^{\,i\left[A_0^2 (c_1^2 \alpha_1 + c_2^2 \alpha_2)\left\{\frac{1}{\alpha}\left[t+\frac{1}{\Gamma(\alpha)}\right]^\alpha - t_0\right\} + \phi_0\right]}$$

Equation: $i\, \Phi_{1t}^\alpha + b_0 \Phi_{1xx}^{2\beta} + (b_1 |\Phi_1|^2 + b_2 |\Phi_2|^2) \Phi_1 = 0,$
$i\, \Phi_{2t}^\alpha + c_0 \Phi_{2xx}^{2\beta} + (c_1 |\Phi_1|^2 + c_2 |\Phi_2|^2) \Phi_2 = 0$

(Continued)

Example	Name	Conditions	Equation #
$\Phi_1(x,t) = \dfrac{A_0}{\sqrt{2\left\{A_1 + \frac{1}{a}\left[t + \frac{1}{\Gamma(\alpha)}\right]^\alpha - t_0\right\}}}\, e^{i\,\phi_1(x,t)},$ $\Phi_2(x,t) = \dfrac{B_0}{\sqrt{2\left\{B_1 + \frac{1}{a}\left[t + \frac{1}{\Gamma(\alpha)}\right]^\alpha - t_0\right\}}}\, e^{i\,\phi_2(x,t)}$	Bright soliton	$b_0, c_0, b_1, c_1, b_2, c_2, A_0, A_1, A_2, B_0,$ $B_1, B_2, x_0, t_0, \phi_{01}$ and ϕ_{02} are arbitrary real constants	(10.100)

Equation: $i\,\Phi_{nt}^{\alpha} + \Phi_{n+1} + \Phi_{n-1} - 2\,\Phi_n + \dfrac{a_2\,|\Phi_n|^2\,\Phi_n}{1+\mu\,|\Phi_n|^2} = 0$

Example	Name	Conditions	Equation #
$\Phi(n,t) = A_0\,\operatorname{sech}[A_1\,(n-n_0)]$ $\times\, e^{-i\left(A_2\left\{\frac{1}{a}\left[t+\frac{1}{\Gamma(\alpha)}\right]^\alpha - t_0\right\}+\phi_0\right)}$	Bright soliton	$A_0 = \dfrac{\sinh(A_1)}{a_2\,\operatorname{sech}(A_1)},\ A_2 = \dfrac{2\,\mu - a_2}{\mu},$ $\mu = \dfrac{2}{\operatorname{sech}(A_1)} > 0,$ $A_1, t_0, n_0,$ and ϕ_0 are arbitrary real constants	(10.102)

Equation: $i\,\Phi_{nt}^{\alpha} + \Phi_{n+1} + \Phi_{n-1} - 2\,\Phi_n + \dfrac{a_2\,\Phi_n^2\,\bar\Phi_n}{1+\mu\,\Phi_n\,\bar\Phi_n} = 0$

Example	Name	Conditions	Equation #
$\Phi(\zeta,t) = A_0\,\operatorname{sech}[A_1\,\zeta]\, e^{-i\left(A_2\left\{\frac{1}{a}\left[t+\frac{1}{\Gamma(\alpha)}\right]^\alpha - t_0\right\}+\phi_0\right)}$	Bright soliton	$A_0 = \dfrac{\sinh(A_1)}{a_2\,\operatorname{sech}(A_1)},\ A_2 = \dfrac{2\,\mu - a_2}{\mu},$ $\mu = \dfrac{2}{\operatorname{sech}(A_1)} > 0,\ \zeta = n - n_0,\ A_1,$ $t_0, n_0,$ and ϕ_0 are arbitrary real constants	(10.104)

References

[1] Mandelbrot B B 1982 *The Fractal Geometry of Nature* **vol 1** (New York: WH Freeman)

[2] Guo X and Xu M 2006 Some physical applications of fractional Schrödinger equation *J. Math. Phys.* **47** 082104

[3] Atanackovic T M, Pilipovic S, Stankovic B and Zorica D 2014 *Fractional Calculus with Applications in Mechanics: Vibrations and Diffusion Processes* (New York: Wiley)

[4] Herrmann R 2011 *Fractional Calculus: An Introduction for Physicists* (World Scientific)

[5] Laskin N 2000 Fractional quantum mechanics and Lévy path integrals *Phys. Lett.* **A268** 298–305

[6] Laskin N 2002 Fractional Schrödinger equation *Phys. Rev.* **E66** 056108

[7] Naber M 2004 Time fractional Schrödinger equation *J. Math. Phys.* **45** 3339–52

[8] Kilbas A A, Srivastava H M and Trujillo J J 2006 *Theory and Applications of Fractional Differential equations* **vol 204** (Elsevier)

[9] Machado J T, Kiryakova V and Mainardi F 2011 Recent history of fractional calculus *Commun. Nonlinear Sci. Numer. Simul.* **16** 1140–53

[10] Benson D A, Meerschaert M M and Revielle J 2013 Fractional calculus in hydrologic modeling: a numerical perspective *Adv. Water Resour.* **51** 479–97

[11] Schumer R, Benson D A, Meerschaert M M and Baeumer B 2003 Multiscaling fractional advection-dispersion equations and their solutions *Water Resour. Res.* **39** 1022

[12] Kim J and Moin P 1985 Application of a fractional-step method to incompressible Navier-Stokes equations *J. Comput. Phys.* **59** 308–23

[13] Bouchaud J P and Georges A 1990 Anomalous diffusion in disordered media: statistical mechanisms, models and physical applications *Phys. Rep.* **195** 127–293

[14] Lundstrom B N, Higgs M H, Spain W J and Fairhall A L 2008 Fractional differentiation by neocortical pyramidal neurons *Nat. Neurosci.* **11** 1335–42

[15] Anastasio T J 1994 The fractional-order dynamics of brainstem vestibulo-oculomotor neurons *Biol. Cybern.* **72** 69–79

[16] De Oliveira E C and Tenreiro Machado J A 2014 A review of definitions for fractional derivatives and integral *Math. Problems Eng.* **2014** 1–6

[17] Jalalinejad H, Tavakoli A and Zarmehi F 2018 A simple and flexible modification of Grünwald-Letnikov fractional derivative in image processing *Math. Sci.* **12** 205–10

[18] Podlubny I 1998 Fractional *An Introduction to Fractional Derivatives, Fractional, to Methods of Their Solution and Some of Their Applications* (Elsevier)

[19] Atagana A and Baleanu D 2016 New fractional derivative with non-local and non-singular kernel *Therm. Sci.* **20** 757–63

[20] Salahshour S, Ahmadian A, Abbasbandy S and Baleanu D 2018 M-fractional derivative under interval uncertainty: theory, properties and applications *Chaos Solitons Fractals* **117** 84–93

[21] Anastassiou G 2020 About the right fractional local general M-derivative *Analele Univ. Oradea, Fasc. Mate.* **27** 87–94

[22] Chen Y, Yan Y and Zhang K 2010 On the local fractional derivative *J. Math. Anal. Appl.* **362** 17–33

[23] Bas E and Ozarslan R 2018 Real world applications of fractional models by Atangana-Baleanu fractional derivative *Chaos Solitons Fractals* **116** 121–5

[24] Imran M A 2020 Application of fractal fractional derivative of power law kernel $(FFP_0\ D_x^{\alpha,\beta})$ to MHD viscous fluid flow between two plates *Chaos Solitons Fractals* **134** 109691

[25] Ortigueira M D and Machado J T 2015 What is a fractional derivative? *J. Comput. Phys.* **293** 4–13

[26] Katugampola U N 2016 Correction to 'What is a fractional derivative?' by Ortigueira and Machado [Journal of Computational Physics, volume 293, 15 July 2015, pages 4–13. Special issue on Fractional PDEs] *J. Comput. Phys.* **321** 1255–7

[27] Khalil R, Al Horani M, Yousef A and Sababheh M 2014 A new definition of fractional derivative *J. Comput. Appl. Math.* **264** 65–70

[28] Ilie M, Biazar J and Ayati Z 2018 Resonant solitons to the nonlinear Schrödinger equation with different forms of nonlinearities *Optik* **164** 201–9

[29] Hafez M G, Iqbal S A, Akther S and Uddin M F 2019 Oblique plane waves with bifurcation behaviors and chaotic motion for resonant nonlinear Schrödinger equations having fractional temporal evolution *Results Phys.* **15** 102778

[30] Bulut H, Sulaiman T A and Baskonus H M 2018 Dark, bright optical and other solitons with conformable space-time fractional second-order spatiotemporal dispersion *Optik* **163** 1–7

IOP Publishing

Handbook of Exact Solutions to the Nonlinear Schrödinger Equations (Second Edition)

Usama Al Khawaja and Laila Al Sakkaf

Appendix A

Derivation of Some Solutions of Chapters 2 and 3

Remark: Throughout this appendix, x_0, t_0, ϕ_0, A_0, A_1, and A_2 are arbitrary real constants.

A.1 Derivation of Some Solutions of Section 2.2

A.1.1 Schematic Representation

Equation	
$i\,\psi_t + a_1\,\psi_{xx} + a_2\,\lvert\psi\rvert^2\,\psi = 0$	
Solutions with	
real a_1 and a_2	complex a_1 and a_2
Case A1: $\psi(x, t) = A_0\, e^{i\,\phi(t)}$	**Case A1′:** $\psi(x, t) = A_0\, e^{i\,\phi(t)}$
Solution (2.161)	Solution (2.224)
Case A2: $\psi(x, t) = A_0\, e^{i\,\phi(x)}$	**Case A2′:** $\psi(x, t) = A_0\, e^{i\,\phi(x)}$
Solution (2.162)	Solution (2.225)
Case A3: $\psi(x, t) = A_0\, e^{i\,\phi(x, t)}$	**Case A3′:** $\psi(x, t) = A(x)\, e^{i\,\phi_0}$
	Solution (2.231)
Solution (2.163)	Solution (2.232)
	Solution (2.233)

(*Continued*)

Case A4: $\psi(x, t) = A(x) e^{i\,\phi_0}$	**Case A4′:** $\psi(x, t) = A(x) e^{i\,\phi(t)}$
$c \neq 0^*$ $\qquad c = 0$	
Solution (2.187)	
Solution (2.189) $\qquad$ Solution (2.165)	
Solution (2.191)	Solution (2.226)
Case A5: $\psi(x, t) = A(x) e^{i\,\phi(t)}$	**Case A5′:** $\psi(x, t) = A(t) e^{i\,\phi(x, t)}$
$c \neq 0$ $\qquad c = 0$	
Solution (2.169)	Solution (2.227)
Solution (2.186)	
Solution (2.173) $\qquad$ Solution (2.167)	Solution (2.228)
Solution (2.188)	
Solution (2.171)	Solution (2.172)
Solution (2.190)	Solution (2.229)
Case A6: $\psi(x, t) = A(x) e^{i\,\phi(x, t)}$	Solution (2.230)
Solution (2.195)	
Case A7: $\psi(x, t) = A(t) e^{i\,\phi(x, t)}$	
Solution (2.164)	

$^*c$ is an arbitrary constant of integration resulting from integrating the NLSE, as detailed below.

A.1.2 Detailed Derivations

A.1.2.1 Real Coefficients

The general solution to (2.160) can be written in the polar form:

$$\psi(x, t) = Z\, e^{i\,\phi}, \tag{A.1.1}$$

where $Z = Z(x, t)$ and $\phi = \phi(x, t)$ are real functions. Substituting (A.1.1) in the fundamental NLSE (2.160) generates from its real and imaginary parts:

$$\text{Re[equation(2.160)]:} \quad a_2 Z^3 - Z\,\phi_t - a_1 Z\,\phi_x^2 + a_1 Z_{xx} = 0, \tag{A.1.2}$$

and

$$\text{Im[equation(2.160)]:} \quad Z_t + 2 a_1 Z_x\,\phi_x + a_1 Z\,\phi_{xx} = 0, \tag{A.1.3}$$

where the subscripts indicate differentiation with respect to x and t. In the following, we take cases for Z and φ.

Case A1:

$$Z(x, t) = A_0, \tag{A.1.4}$$

$$\phi(x, t) = \phi(t). \tag{A.1.5}$$

Substituting back into (A.1.2) and (A.1.3) leads to the following Re[equation(2.160)] and Im[equation(2.160)]:

$$\text{Re[equation(2.160)]:} \quad A_0^3 a_2 - A_0\,\phi'(t) = 0, \tag{A.1.6}$$

A-2

$$\text{Im}[\text{equation}(2.160)]: \quad 0 = 0. \tag{A.1.7}$$

Solving (A.1.6) for $\phi(t)$ gives

$$\phi(t) = A_0^2 \, a_2 \, (t - t_0) + \phi_0. \tag{A.1.8}$$

Substituting (A.1.4) and (A.1.8) back into (A.1.1) leads to solution (2.161).
   Case A2:

$$Z(x, t) = A_0, \tag{A.1.9}$$

$$\phi(x, t) = \phi(x). \tag{A.1.10}$$

Substituting back into (A.1.2) and (A.1.3) leads to the updated forms of Re[equation(2.160)] and Im[equation(2.160)]:

$$\text{Re}[\text{equation}(2.160)]: \quad A_0^3 \, a_2 - A_0 \, a_1 \, \phi'^2(x) = 0, \tag{A.1.11}$$

$$\text{Im}[\text{equation}(2.160)]: \quad A_0 \, a_1 \, \phi''(x) = 0. \tag{A.1.12}$$

Solving (A.1.12) for $\phi(x)$ gives

$$\phi(x) = A_0 \sqrt{\frac{a_2}{a_1}} \, (x - x_0) + \phi_0. \tag{A.1.13}$$

Substituting (A.1.9) and (A.1.13) back into (A.1.1) leads to solution (2.162).
   Case A3:

$$Z(x, t) = A_0, \tag{A.1.14}$$

$$\phi(x, t) = \phi(x, t), \tag{A.1.15}$$

where A_0 is an arbitrary real constant. Substituting back into (A.1.2) and (A.1.3) leads to the following Re[equation(2.160)] and Im[equation(2.160)]:

$$\text{Re}[\text{equation}(2.160)]: \quad A_0^3 \, a_2 - A_0 \, \phi_t(x, t) - A_0 \, a_1 \, \phi_x^2(x, t) = 0, \tag{A.1.16}$$

$$\text{Im}[\text{equation}(2.160)]: \quad A_0 \, a_1 \, \phi_{xx}(x, t) = 0. \tag{A.1.17}$$

Solving (A.1.17) for $\phi(x, t)$ leads to

$$\phi(x, t) = s_1(t) + (x - x_0) \, s_2(t), \tag{A.1.18}$$

where $s_1(t)$ and $s_2(t)$ are real functions of t. Substituting back into (A.1.16), collecting coefficients of $(x - x_0)^0$ and $(x - x_0)$, separately, and equating to zero, we obtain

$$(x - x_0)^0: \quad A_0^3 \, a_2 - A_0 \, a_1 \, s_2^2(t) - A_0 \, s_1'(t) = 0, \tag{A.1.19}$$

$$(x - x_0): \quad -A_0 \, s_2'(t) = 0. \tag{A.1.20}$$

Solving (A.1.20) for $s_2(t)$ reads

$$s_2(t) = A_1. \tag{A.1.21}$$

Substituting this result in (A.1.19) and solving for $s_1(t)$, we get

$$s_1(t) = (A_0^2 a_2 - a_1 A_1^2)(t - t_0) + \phi_0. \tag{A.1.22}$$

The form of $\phi(x, t)$ in (A.1.18) will then be

$$\phi(x, t) = A_1(x - x_0) + (A_0^2 a_2 - a_1 A_1^2)(t - t_0) + \phi_0. \tag{A.1.23}$$

Using (A.1.23) back into (A.1.1) leads to solution (2.163).

Case A4:

$$Z(x, t) = A(x), \tag{A.1.24}$$

$$\phi(x, t) = \phi_0, \tag{A.1.25}$$

where $A(x)$ is a real function to be determined. Substituting back into (A.1.2) and (A.1.3) leads to the following new forms of Re[equation(2.160)] and Im[equation(2.160)]:

$$\text{Re[equation(2.160)]:} \quad a_2 A^3(x) + a_1 A''(x) = 0, \tag{A.1.26}$$

$$\text{Im[equation(2.160)]:} \quad 0 = 0. \tag{A.1.27}$$

Employing the separation of variables method with using the chain rule $A''(x) = \frac{A'(x) \, d \, A'(x)}{d \, A(x)}$ in (A.1.26), we get the following integral of the independent variable x:

$$x - x_0 = \int \frac{1}{\sqrt{c - \dfrac{a_2}{2 \, a_1} A^4(x)}} \, d \, A(x), \tag{A.1.28}$$

where c is a real constant of integration. In the following, we take two categories of c.

If $c \neq 0$:

The integration above can be written as

$$x - x_0 = \frac{1}{\sqrt{c}} \int \frac{1}{\sqrt{1 - b \, A^2(x)} \, \sqrt{1 + b \, A^2(x)}} \, d \, A(x), \tag{A.1.29}$$

where $b = \sqrt{\dfrac{a_2}{2 \, c \, a_1}}$. In the following, we take two options for $A(x)$.

1. $A(x) = \frac{1}{\sqrt{b}} \sin(\theta)$, $0 > \theta > \frac{\pi}{2}$. The integral in (A.1.29) becomes

$$x - x_0 = \frac{1}{\sqrt{c \, b}} \int \frac{\cos(\theta)}{\sqrt{1 - \sin^2(\theta)} \, \sqrt{1 + \sin^2(\theta)}} \, d \, \theta. \tag{A.1.30}$$

This integration gives

$$x - x_0 = \frac{1}{\sqrt{c\,b}}\, F(\theta, -1), \tag{A.1.31}$$

where F gives the elliptic integral of the first kind. Resubstituting $\theta = \sin^{-1}[\sqrt{b}\, A(x)]$ and $b = \sqrt{\frac{a_2}{2\,c\,a_1}}$ in (A.1.31) and solving for $A(x)$ leads to

$$A(x) = \left(\frac{2\,c\,a_1}{a_2}\right)^{\frac{1}{4}} \mathrm{sn}\left[\left(\frac{c\,a_2}{2\,a_1}\right)^{\frac{1}{4}} (x - x_0), -1\right], \tag{A.1.32}$$

From (A.1.32) and (A.1.25), (A.1.1) will lead to (2.187), where $(\frac{c\,a_2}{2\,a_1})^{\frac{1}{4}} \to A_0$. **2.** $A(x) = \frac{1}{\sqrt{b}}\cos(\theta)$, $0 > \theta > \frac{\pi}{2}$. The integral in (A.1.29) becomes

$$x - x_0 = \frac{1}{\sqrt{c\,b}} \int \frac{-\sin(\theta)}{\sqrt{1 - \cos^2(\theta)}\,\sqrt{1 + \cos^2(\theta)}}\, d\,\theta. \tag{A.1.33}$$

This integration leads to

$$x - x_0 = \frac{-1}{\sqrt{2\,c\,b}}\, F(\theta, \frac{1}{2}), \tag{A.1.34}$$

where F gives the elliptic integral of the first kind. Resubstituting $\theta = \cos^{-1}[\sqrt{b}\, A(x)]$ and $b = \sqrt{\frac{a_2}{2\,c\,a_1}}$ in (A.1.34) and solving for $A(x)$ lead to

$$A(x) = \left(\frac{2\,c\,a_1}{a_2}\right)^{\frac{1}{4}} \mathrm{cn}\left[\left(\frac{2\,c\,a_2}{a_1}\right)^{\frac{1}{4}} (x - x_0), \frac{1}{2}\right], \tag{A.1.35}$$

From (A.1.25) and (A.1.35), (A.1.1) will lead to (2.189), where $\left(\frac{2\,c\,a_2}{a_1}\right)^{\frac{1}{4}} \to A_0$.

3.

$$A(x) = c\,\sqrt{\frac{2\,a_1}{a_2}}\, \mathrm{dn}[c\,(x - x_0), 2], \tag{A.1.36}$$

From (A.1.25) and (A.1.36), (A.1.1) will lead to (2.189), where $c \to A_0$.

The procedure of how we get this solution will be shown in the next case, Case A5, 4.

If $c = 0$:

Solving (A.1.28) for $A(x)$, we get

$$A(x) = \sqrt{\frac{-2\,a_1}{a_2}}\, \frac{1}{x - x_0}. \tag{A.1.37}$$

Substituting (A.1.25) and (A.1.37) back into (A.1.1) leads to solution (2.226).

Case A5:

$$Z(x, t) = A(x), \tag{A.1.38}$$

$$\phi(x, t) = e^{-i[\lambda(t-t_0)+\phi_0]}, \tag{A.1.39}$$

where $A(x)$ is a real function to be determined and λ is an arbitrary real constant. (2.160) takes the form

$$\lambda A(x) + a_1 A''(x) + a_2 A^3(x) = 0. \tag{A.1.40}$$

Using the chain rule $A''(x) = \frac{A'(x) d A'(x)}{d A(x)}$, we get the following integral of the independent variable x:

$$x - x_0 = \int \frac{1}{\sqrt{c - \dfrac{\lambda}{a_1} A^2(x) - \dfrac{a_2}{2 a_1} A^4(x)}} \, d A(x), \tag{A.1.41}$$

where c is a real constant of integration. In the following, we take two categories of c.
  If $c \neq 0$:
  1. From (A.1.41), with $c = \frac{-\lambda^2}{2 a_1 a_2}$, $x - x_0$ will be given by

$$x - x_0 = \sqrt{\frac{-2 a_1}{\lambda}} \, \tan^{-1}[\sqrt{\frac{a_2}{\lambda}} \, A(x)]. \tag{A.1.42}$$

Solving (A.1.42) for $A(x)$, we get

$$A(x) = \sqrt{\frac{\lambda}{a_2}} \, \tan[\sqrt{\frac{-\lambda}{2 a_1}} \, (x - x_0)]. \tag{A.1.43}$$

Substituting (A.1.39) and (A.1.43) back into (A.1.1) leads to (2.169), where $\lambda \to -2 a_1 A_0^2$.
  2. The integral definition of the Jacobi $\text{sn}(x, m)$ elliptic function with modulus m is given by:

$$x - x_0 = \int_0^{\text{sn}(x,m)} \frac{1}{\sqrt{[1 - A^2(x)] [1 - m A^2(x)]}} \, d A(x). \tag{A.1.44}$$

Equating this integration with (A.1.41) gives

$$c - \frac{\lambda}{a_1} A^2(x) - \frac{a_2}{2 a_1} A^4(x) - c [1 - c_1 A^2(x)] [1 - c_1 m A^2(x)] = 0, \tag{A.1.45}$$

where c_1 is a required real constant to be determined. Equating the coefficients of $A^2(x)$ and $A^4(x)$ to zero, separately, and solving for c and c_1, we get

$$c_1 = \frac{-a_2 (1 + m)}{2 m \lambda}, \qquad c = \frac{-2 m \lambda^2}{a_1 a_2 (1 + m)^2}. \tag{A.1.46}$$

Solving

$$x - x_0 = \int_0^{\operatorname{sn}(x,m)} \frac{1}{\sqrt{c\,[1 - c_1\,A^2(x)]\,[1 - c_1\,m\,A^2(x)]}}\, d\,A(x), \qquad (A.1.47)$$

for $A(x)$ with the expressions of c an c_1 in (A.1.46), we get

$$A(x) = \sqrt{\frac{2\,m\,\lambda}{-a_2\,(1 + m)}}\;\operatorname{sn}\!\left[\sqrt{\frac{\lambda}{a_1\,(1 + m)}}\,(x - x_0),\, m\right]. \qquad (A.1.48)$$

Substituting (A.1.39) and (A.1.48) back into (A.1.1) leads to solution (2.186), where $\lambda \to A_0^2$.

If we take the limit of (A.1.48) when $\lambda \to c\,a_1\,(1 + m)$ and replace $m \to -1$, we return back to (A.1.32).

For the special case of $m = 1$, Equation (A.1.48) reads

$$A(x) = \sqrt{\frac{-\lambda}{a_2}}\;\tanh\!\left[\sqrt{\frac{\lambda}{2\,a_1}}\,(x - x_0)\right], \qquad (A.1.49)$$

which leads to solution (2.173).

3. The integral definition of the Jacobi elliptic function, $\operatorname{cn}(x, m)$, with modulus m is given by:

$$x - x_0 = \int_{\operatorname{cn}(x,\, m)}^{1} \frac{1}{\sqrt{[1 - A^2(x)]\,[1 - m\,A^2(x)]}}\, d\,A(x). \qquad (A.1.50)$$

By pulling out a factor of $\frac{1}{1 - m}$ and replacing $m \to \frac{m}{m - 1}$ and $x - x_0 \to \frac{x - x_0}{\sqrt{1 - m}}$, this integral definition can be transformed into

$$x - x_0 = \int_{\operatorname{cn}(x,\, m)}^{1} \frac{1}{\sqrt{[1 - A^2(x)]\,[1 - m + m\,A^2(x)]}}\, d\,A(x). \qquad (A.1.51)$$

Equating the above integral with (A.1.41), we get

$$c - \frac{\lambda}{a_1}\,A^2(x) - \frac{a_2}{2\,a_1}\,A^4(x) - c\,[1 - c_1\,A^2(x)]\,[1 - m + c_1\,m\,A^2(x)] = 0,\,(A.1.52)$$

where c_1 is a real constant to be determined. Equating the coefficients of $A^2(x)$ and $A^4(x)$ to zero, separately, and solving for c and c_1, we obtain

$$c_1 = \frac{a_2\,(1 - 2\,m)}{2\,m\,\lambda}, \qquad c = \frac{2\,m\,\lambda^2}{a_1\,a_2\,(1 - 2\,m)^2}. \qquad (A.1.53)$$

Solving

$$x - x_0 = \int_{\operatorname{cn}(x,\, m)}^{1} \frac{1}{\sqrt{c\,[1 - c_1\,A^2(x)]\,[1 - m + c_1\,m\,A^2(x)]}}\, d\,A(x). \qquad (A.1.54)$$

for $A(x)$ with the help of (A.1.53), we find

$$A(x) = \sqrt{\frac{2\, m\, \lambda}{a_2\, (1 - 2\, m)}}\ \mathrm{cd}[\sqrt{\frac{\lambda\, (1 - m)}{a_1\, (1 - 2\, m)}}\ (x - x_0),\ \frac{m}{m - 1}], \qquad \text{(A.1.55)}$$

where $\mathrm{cd}(x, m)$ is a Jacobi elliptic function with modulus m. This is equivalent to

$$A(x) = \sqrt{\frac{2\, m\, \lambda}{a_2\, (1 - 2\, m)}}\ \mathrm{cn}[\sqrt{\frac{\lambda}{a_1\, (1 - 2\, m)}}\ (x - x_0),\ m]. \qquad \text{(A.1.56)}$$

Substituting (A.1.39) and (A.1.56) back into (A.1.1) leads to the solution in (2.188), where $\lambda \to -A_0^2$.

If we take the limit of (A.1.56) when $\lambda \to c^2\, a_1\, (1 - 2\, m)$ and replace $m \to \frac{1}{2}$, we return back to (A.1.35).

For the special case of $m = 1$, Equation (A.1.56) reads

$$A(x) = \sqrt{\frac{-2\, \lambda}{a_2}}\ \mathrm{sech}[\sqrt{\frac{-\lambda}{a_1}}\ (x - x_0)], \qquad \text{(A.1.57)}$$

which leads to (2.171).

4. The integral definition of the Jacobi elliptic function, $\mathrm{dn}(x, m)$, with modulus m is given by:

$$x - x_0 = \int_{\mathrm{dn}(x,\, m)}^{1} \frac{1}{\sqrt{[1 - A^2(x)]\, [A^2(x) - 1 + m]}}\, d\, A(x). \qquad \text{(A.1.58)}$$

Equating the above integration with the integral in (A.1.41)

$$c - \frac{\lambda}{a_1}\, A^2(x) - \frac{a_2}{2\, a_1}\, A^4(x) - c\, [1 - c_1\, A^2(x)]\, [c_1\, A^2(x) - 1 + m] = 0, \quad \text{(A.1.59)}$$

where c_1 is a real constant to be determined. Equating the coefficients of $A^2(x)$ and $A^4(x)$ to zero, and solving for c and c_1, we get

$$c_1 = \frac{a_2\, (m - 2)}{2\, \lambda}, \qquad c = \frac{2\, \lambda^2}{a_1\, a_2\, (m - 2)^2}. \qquad \text{(A.1.60)}$$

Solving

$$x - x_0 = \int_{\mathrm{dn}(x,\, m)}^{1} \frac{1}{\sqrt{c\, [1 - c_1\, A^2(x)]\, [c_1\, A^2(x) - 1 + m]}}\, d\, A(x). \qquad \text{(A.1.61)}$$

for $A(x)$ with the help of (A.1.60), we obtain

$$A(x) = \sqrt{\frac{2\, \lambda}{a_2\, (m - 2)}}\ \mathrm{cd}[\sqrt{\frac{\lambda\, (m - 1)}{a_1\, (m - 2)}}\ (x - x_0),\ \frac{1}{1 - m}], \qquad \text{(A.1.62)}$$

where $\mathrm{cd}(x, m)$ is a Jacobi elliptic function with modulus m, which is equivalent to

$$A(x) = \sqrt{\frac{2\,\lambda}{a_2\,(m-2)}}\ \mathrm{dn}[\sqrt{\frac{\lambda}{a_1\,(m-2)}}\ (x - x_0), m].\qquad(\text{A.1.63})$$

Substituting (A.1.39) and (A.1.63) back into (A.1.1) leads to (2.190), where $\lambda \to -A_0^2$.

If we take the limit of (A.1.63) when $\lambda \to c^2\,a_1\,(m-2)$ and replace $m \to 2$, we get (A.1.36).

For the special case of $m = 1$, Equation (A.1.63) reads

$$A(x) = \sqrt{\frac{-2\,\lambda}{a_2}}\ \mathrm{sech}[\sqrt{\frac{-\lambda}{a_1}}\ (x - x_0)],\qquad(\text{A.1.64})$$

which again leads to (2.171).

If $c = 0$:

From (A.1.41), $x - x_0$ is given by

$$x - x_0 = \sqrt{\frac{-a_1}{\lambda}}\ (\ln[A(x)] - \ln\{2\,\lambda + \sqrt{2\,\lambda}\ \sqrt{[a_2\,A^2(x) + 2\,\lambda]}\,\}).\qquad(\text{A.1.65})$$

Solving (A.1.65) for $A(x)$, we get

$$A(x) = \frac{-4\,\lambda\ e^{\sqrt{\frac{-\lambda x^2}{a_1}}}}{2\,\lambda\,a_2\ e^{\sqrt{\frac{4\,\lambda x^2}{a_1}}} - 1}.\qquad(\text{A.1.66})$$

In the following, we represent the different possible cases of $A(x)$ depending upon the sign of each of a_1, a_2, and λ.

1. $a_1 > 0$, $a_2 < 0$, and $\lambda > 0$:

$$A(x) = 2\,\lambda\ \sec[\sqrt{\frac{\lambda}{a_1}}\ (x - x_0)].\qquad(\text{A.1.67})$$

Substituting (A.1.39) and (A.1.67) back into (A.1.1) leads to (2.167), where $\lambda \to a_1\,A_0^2$.

2. $a_1 < 0$, $a_2 > 0$, and $\lambda > 0$:

$$A(x) = 2\,\lambda\ \mathrm{csch}[\sqrt{\frac{\lambda}{a_1}}\ (x - x_0)].\qquad(\text{A.1.68})$$

Substituting (A.1.39) and (A.1.68) back into (A.1.1) leads to (2.229), where $\lambda \to -a_1\,A_0^2$. Other cases produce either imaginary or repeated solutions.

Case A6:

$$Z(x, t) = A(x),\qquad(\text{A.1.69})$$

$$\phi(x, t) = \phi(x, t),\qquad(\text{A.1.70})$$

A-9

where $A(x)$ is a real function to be determined. Substituting back into (A.1.2) and (A.1.3) leads to the following forms of Re[equation(2.160)] and Im[equation(2.160)]:

Re[equation(2.160)]: $a_2 A^3(x) + a_1 A''(x) - A(x) [\phi_t(x, t) + a_1 \phi_x^2(x, t)] = 0,$ (A.1.71)

Im[equation(2.160)]: $a_1 [2 A'(x) \phi_x(x, t) + A(x) \phi_{xx}(x, t)] = 0.$ (A.1.72)

Solving (A.1.72) for $\phi(x, t)$ will give

$$\phi(x, t) = \int \frac{s_1(t)}{A^2(x)} d x + s_2(t),$$ (A.1.73)

where $s_1(t)$ and $s_2(t)$ are two real functions of t to be determined. Taking a special case of these two functions, $s_1(t) = \lambda_0$ and $s_2(t) = \lambda_2 (t - t_0)$, Equation (A.1.71) becomes

Re[equation(2.160)]: $a_2 A^3(x) - \dfrac{a_1 \lambda_0^2}{A^3(x)} - \lambda_2 A(x) + a_1 A''(x) = 0,$ (A.1.74)

where λ_0 and λ_2 are arbitrary real constants. Solving (A.1.74) for $A(x)$, we get

$$A(x) = \sqrt{R_3 + m_1 \operatorname{sn}^2[\sqrt{\frac{-a_2 m_2}{2 a_1}} (x - x_0), m]},$$ (A.1.75)

where sn is a Jacobi elliptic function of the modulus $m = \frac{R_2 - R_3}{R_1 - R_3}$, R_j, $j = 1, 2, 3$, are the three roots of $Y(x) = 2 a_1 \lambda_0^2 - 2 a_1 \lambda_1 x - 2 \lambda_2 x^2 + a_2 x^3$, $m_1 = R_1 - R_3$, $m_2 = R_1 - R_3$, and λ_1 is an arbitrary real constant. From (A.1.73), $\phi(x, t)$ takes the form

$$\phi(x, t) = \frac{\sqrt{2} \lambda_0 \Pi\left\{\dfrac{R_3 - R_2}{R_3}, \operatorname{am}[\sqrt{\dfrac{-a_2 m_2}{2 a_1}} (x - x_0), m], m\right\} \operatorname{dn}[\sqrt{\dfrac{-a_2 m_2}{2 a_1}} (x - x_0), m]}{R_3 \sqrt{\dfrac{-a_2 m_2}{a_1}} \sqrt{1 - m \operatorname{sn}^2[\sqrt{\dfrac{-a_2 m_2}{2 a_1}} (x - x_0), m]}} + \lambda_2 (t - t_0),$$ (A.1.76)

where dn is the Jacobi elliptic function of the modulus m, Π is the incomplete elliptic integral, and am is the amplitude for Jacobi elliptic functions. Substituting (A.1.75) and (A.1.76) back into (A.1.1) leads to (2.195).

Case A7:

$$Z(x, t) = A(t),$$ (A.1.77)

$$\phi(x, t) = \phi(x, t),$$ (A.1.78)

where $A(t)$ is a real function to be determined. Substituting back into (A.1.2) and (A.1.3) leads to the following new forms of Re[equation(2.160)] and Im[equation(2.160)]:

Re[equation(2.160)]: $A(t) [a_2 A^2(t) - \phi_t(x, t) - a_1 \phi_x^2(x, t)] = 0,$ (A.1.79)

Im[equation(2.160)]: $A'(t) + a_1 A(t) \phi_{xx}(x, t) = 0.$ (A.1.80)

Solving (A.1.80) for $\phi(x, t)$ gives

$$\phi(x, t) = s_1(t) + s_2(t) (x - x_0) - \frac{(x - x_0)^2 A'(t)}{2 a_1 A(t)},$$ (A.1.81)

where $s_1(t)$ and $s_2(t)$ are real functions to be determined. Using (A.1.81) in (A.1.79), we get

Re[equation(2.160)]: $A(t)[a_2 A^2(t) - a_1 s_2^2(t) - s_1'(t)] + [2 s_2(t) A'(t) - s_2'(t) A(t)] (x - x_0)$

$$+ \left[\frac{A(t) A''(t) - 3 A'^2(t)}{2 a_1 A(t)} \right] (x - x_0)^2 = 0.$$ (A.1.82)

Collecting coefficients of $(x - x_0)^0$, $(x - x_0)$, and $(x - x_0)^2$, separately, and equating to zero, we obtain

$$(x - x_0)^0: \quad a_2 A^2(t) - a_1 s_2^2(t) - s_1'(t) = 0,$$ (A.1.83)

$$(x - x_0)^1: \quad 2 s_2(t) A'(t) - s_2'(t) A(t) = 0,$$ (A.1.84)

$$(x - x_0)^2: \quad \frac{A(t) A''(t) - 3 A'^2(t)}{2 a_1 A(t)} = 0.$$ (A.1.85)

Solving for $A(t)$, $s_1(t)$, and $s_2(t)$, we get

$$A(t) = \frac{A_0}{\sqrt{A_1 + 2 (t - t_0)}},$$ (A.1.86)

$$s_1(t) = \frac{a_1 A_2^2}{2 A_1 + 4 (t - t_0)} + \frac{1}{2} a_2 A_0^2 \ln[A_1 + 2 (t - t_0)] + \phi_0,$$ (A.1.87)

$$s_2(t) = \frac{A_2}{A_1 + 2 (t - t_0)}.$$ (A.1.88)

Substituting (A.1.86), (A.1.87), and (A.1.88) in (A.1.81) and then back into (A.1.1) leads to (2.164).

A.1.2.2 Complex Coefficients

For complex parameters, we define $a_1 = a_{1r} + i\, a_{1i}$ and $a_2 = a_{2r} + i\, a_{2i}$, where $a_{1r}, a_{1i},$ $a_{2r},$ and a_{1i} are real constants. The real and imaginary parts of (2.160) are given by

Re[equation(2.160)]: $a_{2r} Z^3 - Z \phi_t - 2 a_{1i} Z_x \phi_x - a_{1r} Z \phi_x^2 + a_{1r} Z_{xx} - a_{1i} Z \phi_{xx} = 0,$ (A.1.89)

and

Im[equation(2.160)]: $a_{2i} Z^3 + Z_t + 2 a_{1r} Z_x \phi_x - a_{1i} Z \phi_x^2 + a_{1i} Z_{xx} + a_{1r} Z \phi_{xx} = 0,$ (A.1.90) respectively. In the following, we take cases for Z and φ.

Case A1′:

$$Z(x, t) = A_0, \tag{A.1.91}$$

$$\phi(x, t) = \phi(t). \tag{A.1.92}$$

Substituting back into (A.1.89) and (A.1.90) leads to the following Re[equation(2.160)] and Im[equation(2.160)]:

$$\text{Re[equation(2.160)]:} \quad A_0^3 \, a_{2r} - A_0 \, \phi'(t) = 0, \tag{A.1.93}$$

$$\text{Im[equation(2.160)]:} \quad A_0^3 \, a_{2i} = 0. \tag{A.1.94}$$

Solving (A.1.93) for $\phi(t)$ gives

$$\phi(t) = A_0^2 \, a_{2r} \, (t - t_0) + \phi_0. \tag{A.1.95}$$

Using (A.1.91) and (A.1.95) in (A.1.1) leads to solution (2.224).

Case A2′:

$$Z(x, t) = A_0, \tag{A.1.96}$$

$$\phi(x, t) = \phi(x). \tag{A.1.97}$$

Substituting back into (A.1.89) and (A.1.90) leads to the updated forms of Re[equation(2.160)] and Im[equation(2.160)]:

$$\text{Re[equation(2.160)]:} \quad A_0^3 \, a_{2r} - A_0 \, a_{1r} \, \phi'^2(x) - A_0 \, a_{1i} \, \phi''(x) = 0, \tag{A.1.98}$$

$$\text{Im[equation(2.160)]:} \quad A_0^3 \, a_{2i} - A_0 \, a_{1i} \, \phi'^2(x) + A_0 \, a_{1r} \, \phi''(x) = 0. \tag{A.1.99}$$

Multiplying (A.1.98) by a_{1r} and (A.1.99) by a_{1i} and taking the summation of the resulting equations leads to

$$A_0^3 \, (a_{1i} \, a_{2i} + a_{1r} \, a_{2r}) - A_0 \, (a_{1i}^2 + a_{1r}^2) \, \phi'^2(x) = 0. \tag{A.1.100}$$

Solving (A.1.100) for $\phi(x)$ gives

$$\phi(x) = \pm \sqrt{\frac{A_0^2 \, (a_{1i} \, a_{2i} + a_{1r} \, a_{2r})}{a_{1i}^2 + a_{1r}^2}} \, (x - x_0) + \phi_0. \tag{A.1.101}$$

Resubstituting (A.1.101) into (A.1.98) and (A.1.99), equating the two resulting equations, and solving for a_{1i}, we find

$$a_{1i} = \frac{a_{1r} \, a_{2i}}{a_{2r}}. \tag{A.1.102}$$

Substituting (A.1.102) back into (A.1.101) and then, into (A.1.1) with (A.1.96) leads to solution (2.225).

Case A3′:

$$Z(x, t) = A(x), \tag{A.1.103}$$

$$\phi(x, t) = \phi_0, \tag{A.1.104}$$

where $A(x)$ is a real function to be determined. Substituting back into (A.1.89) and (A.1.90) leads to the following Re[equation(2.160)] and Im[equation(2.160)]:

$$\text{Re[equation(2.160)]}: \quad a_{2r}\, A^3(x) + a_{1r}\, A''(x) = 0, \tag{A.1.105}$$

$$\text{Im[equation(2.160)]}: \quad a_{2i}\, A^3(x) + a_{1i}\, A''(x) = 0. \tag{A.1.106}$$

Solving the above two equations separately by following the steps used with real a_1 and a_2 in Case A4, we get two expressions for $A(x)$. For each of the two solutions to satisfy (2.160), a condition arises, namely $a_{1i} = \frac{a_{1r}\, a_{2i}}{a_{2r}}$. The resulting solutions are (2.231), (2.232), and (2.233).

Case A4′:

$$Z(x, t) = A(t), \tag{A.1.107}$$

$$\phi(x, t) = \phi_0, \tag{A.1.108}$$

where $A(t)$ is a real function to be determined. Substituting back into (A.1.89) and (A.1.90) leads to the following new forms of Re[equation(2.160)] and Im[equation(2.160)]:

$$\text{Re[equation(2.160)]}: \quad a_{2r}\, A^3(t) = 0, \tag{A.1.109}$$

$$\text{Im[equation(2.160)]}: \quad a_{2i}\, A^3(t) + A'(t) = 0. \tag{A.1.110}$$

Solving (A.1.110) for $A(t)$

$$A(t) = \frac{1}{\sqrt{2\, a_{2i}\, (t - t_0)}}. \tag{A.1.111}$$

Substituting (A.1.108) and (A.1.111) back into (A.1.1) leads to solution (2.165).

Case A5′:

$$Z(x, t) = A(t), \tag{A.1.112}$$

$$\phi(x, t) = \phi(x, t), \tag{A.1.113}$$

where $A(t)$ is a real function to be determined. Substituting back into (A.1.89) and (A.1.90) leads to the following Re[equation(2.160)] and Im[equation(2.160)]:

$$\text{Re[equation(2.160)]}: \quad A(t)\,[a_{2r}\, A^2(t) - \phi_t(x, t) - a_{1r}\, \phi_x^2(x, t) - a_{1i}\, \phi_{xx}(x, t)] = 0, \tag{A.1.114}$$

Im[equation(2.160)]: $a_{2i} A^3(t) + A'(t) + A(t) [a_{1r} \phi_{xx}(x, t) - a_{1i} \phi_x^2(x, t)] = 0.$ (A.1.115)

Multiplying (A.1.114) by a_{1r} and (A.1.115) by a_{1i} and taking the summation of the resulting equations, we get

$$(a_{1i} a_{2i} + a_{1r} a_{2r}) A^3(t) + a_{1i} A'(t) - A(t) [a_{1r} \phi_t(x, t) + (a_{1i}^2 + a_{1r}^2) \phi_x^2(x, t)] = 0.$$ (A.1.116)

Solving (A.1.116) for $\phi(x, t)$

$$\phi(x, t) = A_0 (x - x_0) + \frac{1}{a_{1r}} \int \left[(a_{1i} a_{2i} + a_{1r} a_{2r}) A^2(t) - A_0^2 (a_{1i}^2 + a_{1r}^2) + \frac{a_{1i} A'(t)}{A(t)} \right] dt + \phi_0$$ (A.1.117)

and resubstituting in both (A.1.114) and (A.1.115), equating the two resulting equations, and solving for a_{1i}, we get

$$a_{1i} = \frac{a_{2i} A^3(t) + A'(t)}{A_0^2 A(t)}.$$ (A.1.118)

In the following, we take four cases for a_{1i} and a_{2i}, simultaneously.

1. $a_{1i} \neq 0$ and $a_{2i} \neq 0$:
Solving (A.1.118) for $A(t)$

$$A(t) = \frac{\sqrt{a_{1i} A_0^2} \; e^{a_{1i} A_0^2 (t-t_0)}}{\sqrt{-1 + a_{2i} \; e^{2 a_{1i} A_0^2 (t-t_0)}}},$$ (A.1.119)

2. $a_{1i} = 0$ and $a_{2i} \neq 0$:
Solving (A.1.118) for $A(t)$

$$A(t) = \frac{1}{\sqrt{2 a_{2i} (t - t_0)}},$$ (A.1.120)

3. $a_{1i} \neq 0$ and $a_{2i} = -a_{1r}$:
Solving (A.1.118) for $A(t)$

$$A(t) = \sqrt{\frac{a_{1r} A_0^2}{-a_{2i} + e^{2 a_{1r} A_0^2 (t-t_0)}}},$$ (A.1.121)

4. $a_{1i} = -a_{1r}$ and $a_{2i} = 0$:
Solving (A.1.118) for $A(t)$

$$A(t) = A_0 \; e^{-a_{1r} A_1^2 (t-t_0)}.$$ (A.1.122)

We then use the resulting four expressions of $A(t)$ individually in (A.1.117) to find the corresponding $\phi(x, t)$ for each case. Substituting $A(t)$ and $\phi(x, t)$ for each case into (A.1.1) leads to solutions (2.172), (2.227), (2.228), and (2.230), respectively.

A.2 Derivation of Some Solutions of Section 3.2

A.2.1 Schematic Representation

Equation
$i\,\psi_t + a_1\,\psi_{xx} + a_2\,|\psi|^n\,\psi = 0$

	Solutions	
	Case B1: $\psi(x,\,t) = A_0\,e^{i\,\phi(t)}$	
Solution (3.14)		
	Case B2: $\psi(x,\,t) = A_0\,e^{i\,\phi(x)}$	
Solution (3.15)		
	Case B3: $\psi(x,\,t) = A_0\,e^{i\,\phi(x,\,t)}$	
Solution (3.16)		
	Case B4: $\psi(x,\,t) = A(x)\,e^{i\,\phi_0}$	
$c \neq 0$		$c = 0$
Solution (3.23)		Solution (3.18)
	Case B5: $\psi(x,\,t) = A(x)\,e^{i\,\phi(t)}$	
$c \neq 0$		$c = 0$
Solution (3.24)		
Solution (3.25)		Solution (3.21)
	Case B6: $\psi(x,\,t) = A(x)\,e^{i\,\phi(x,\,t)}$	
Solution (3.26)		
	Case B7: $\psi(x,\,t) = A(t)\,e^{i\,\phi(x,\,t)}$	
Solution (3.17)		

A.2.2 Detailed Derivations

The general solution to (3.9) can be written in the polar form:

$$\psi(x,\,t) = Z\,e^{i\,\phi}, \tag{A.2.1}$$

where $Z = Z(x,\,t)$ and $\phi = \phi(x,\,t)$ are real functions. Substituting (A.2.1) in (3.9) generates its real and imaginary parts:

$$\text{Re[equation(3.9)]:} \quad a_2\,Z^{n+1} - Z\,\phi_t - a_1\,Z\,\phi_x^2 + a_1\,Z_{xx} = 0, \tag{A.2.2}$$

and

$$\text{Im[equation(3.9)]:} \quad Z_t + 2\,a_1\,Z_x\,\phi_x + a_1\,Z\,\phi_{xx} = 0, \tag{A.2.3}$$

In the following, we take cases for Z and φ.

Case B1:

$$Z(x,\,t) = A_0, \tag{A.2.4}$$

$$\phi(x, t) = \phi(t). \tag{A.2.5}$$

Substituting back into (A.2.2) and (A.2.3) leads to the following Re[*equation*(3.9)] and Im[equation(3.9)]:

$$\text{Re[equation(3.9)]}: \quad A_0 [A_0^n a_2 - \phi'(t)] = 0, \tag{A.2.6}$$

$$\text{Im[equation(3.9)]}: \quad 0 = 0. \tag{A.2.7}$$

Solving (A.2.6) for $\phi(t)$ gives

$$\phi(t) = A_0^n a_2 (t - t_0) + \phi_0. \tag{A.2.8}$$

Substituting (A.2.4) and (A.2.8) back into (A.2.1) leads to solution (3.14).
Case B2:

$$Z(x, t) = A_0, \tag{A.2.9}$$

$$\phi(x, t) = \phi(x). \tag{A.2.10}$$

Substituting back into (A.2.2) and (A.2.3) leads to the following Re[equation(3.9)] and Im[equation(3.9)]:

$$\text{Re[equation(3.9)]}: \quad A_0 [A_0^n a_2 - a_1 \phi'^2(x)] = 0, \tag{A.2.11}$$

$$\text{Im[equation(3.9)]}: \quad A_0 a_1 \phi''(x) = 0. \tag{A.2.12}$$

Solving (A.2.12) for $\phi(x)$ gives

$$\phi(x) = A_0^{n/2} \sqrt{\frac{a_2}{a_1}} (x - x_0) + \phi_0. \tag{A.2.13}$$

Substituting (A.2.9) and (A.2.13) back into (A.2.1) leads to the continuous wave solution (3.15).
Case B3:

$$Z(x, t) = A_0, \tag{A.2.14}$$

$$\phi(x, t) = \phi(x, t). \tag{A.2.15}$$

Substituting back into (A.2.2) and (A.2.3) leads to the following Re[equation(3.9)] and Im[equation(3.9)]:

$$\text{Re[equation(3.9)]}: \quad A_0 [A_0^n a_2 - \phi_t(x, t) - a_1 \phi_x^2(x, t)] = 0, \tag{A.2.16}$$

$$\text{Im[equation(3.9)]}: \quad A_0 a_1 \phi_{xx}(x, t) = 0. \tag{A.2.17}$$

Solving (A.2.17) for $\phi(x, t)$ gives

$$\phi(x, t) = s_1(t) + (x - x_0) s_2(t), \tag{A.2.18}$$

where $s_1(t)$ and $s_2(t)$ are real functions of t. Substituting back into (A.2.16), collecting coefficients of $(x - x_0)^0$ and $(x - x_0)$, separately, and equating to zero, we obtain:

$$(x - x_0)^0: \quad A_0 \left[A_0^n a_2 - a_1 s_2^2(t) - s_1'(t) \right] = 0, \tag{A.2.19}$$

$$(x - x_0): \quad -A_0 s_2'(t) = 0. \tag{A.2.20}$$

Solving (A.2.20) for $s_2(t)$ reads

$$s_2(t) = A_1. \tag{A.2.21}$$

Substituting this result in (A.2.19) and solving for $s_1(t)$, we get

$$s_1(t) = (A_0^n a_2 - A_1^2 a_1)(t - t_0) + \phi_0. \tag{A.2.22}$$

Then, $\phi(x, t)$ in (A.2.18) becomes

$$\phi(x, t) = A_1(x - x_0) + (A_0^n a_2 - A_1^2 a_1)(t - t_0) + \phi_0. \tag{A.2.23}$$

Using (A.2.23) back into (A.2.1) leads to the continuous wave solution (3.16).

Case B4:

$$Z(x, t) = A(x), \tag{A.2.24}$$

$$\phi(x, t) = \phi_0, \tag{A.2.25}$$

where $A(x)$ is a real function to be determined. Substituting back into (A.2.2) and (A.2.3) leads to the following Re[equation(3.9)] and Im[equation(3.9)]:

$$\text{Re[equation(3.9)]:} \quad a_2 A^{1+n}(x) + a_1 A''(x) = 0, \tag{A.2.26}$$

$$\text{Im[equation(3.9)]:} \quad 0 = 0. \tag{A.2.27}$$

By employing the chain rule $A''(x) = \frac{A'(x) \, d \, A'(x)}{d \, A(x)}$ in (A.2.26), we get the following integral of the independent variable x:

$$x - x_0 = \int \frac{1}{\sqrt{c - \dfrac{2 \, a_2}{(n + 2) \, a_1} A^{n+2}(x)}} \, d \, A(x), \tag{A.2.28}$$

where c is the real constant of integration. In the following, we take two categories of c.

If $c \neq 0$:

The integration above reads

$$x - x_0 = \frac{A(x)}{\sqrt{A_0}} \, {}_2F_1 \left[\frac{1}{2}, \frac{1}{n + 2}, \frac{n + 3}{n + 2}, \frac{2 \, a_2 \, A^{n+2}(x)}{a_1 \, A_0 \, (n + 2)} \right]$$

$$= Y[A(x)] \tag{A.2.29}$$

$$\psi(x, t) = A(x) \, e^{i \, \phi_0}, \tag{A.2.30}$$

where

$$Y[A(x)] = \frac{A(x)}{\sqrt{A_0}} \, {}_2F_1\left[\frac{1}{2}, \frac{1}{n+2}, \frac{n+3}{n+2}, \frac{2 \, a_2 \, A^{n+2}(x)}{a_1 \, A_0 \, (n+2)}\right], \tag{A.2.31}$$

which is formally solved as

$$A(x) = Y^{-1}(x - x_0). \tag{A.2.32}$$

Here, ${}_2F_1$ is the hypergeometric function and Y^{-1} indicates the inverse operator of the function $Y[A(x)]$, and hence, we infer the solution expressed in (3.23).

If $c = 0$:
Solving (A.2.28) for $A(x)$

$$A(x) = \left[\frac{1}{\sqrt{\dfrac{-a_2 \, n^2}{2 \, a_1 \, (2+n)}} \, (x - x_0)}\right]^{\frac{2}{n}}. \tag{A.2.33}$$

Substituting (A.2.25) and (A.2.33) back into (A.2.1) leads to (3.18).

Case B5:

$$Z(x, t) = A(x), \tag{A.2.34}$$

$$\phi(x, t) = e^{i \, [\lambda \, (t-t_0) + \phi_0]}, \tag{A.2.35}$$

where $A(x)$ is a real function to be determined and λ, t_0, and ϕ_0 are arbitrary real constants. Equation (3.9) becomes

$$-\lambda \, A(x) + a_1 \, A''(x) + a_2 \, A^{n+1}(x) = 0. \tag{A.2.36}$$

Using the chain rule $A''(x) = \frac{A'(x) \, d \, A'(x)}{d \, A(x)}$, we get the following integral of the independent variable x:

$$x - x_0 = \int \frac{1}{\sqrt{c + \dfrac{\lambda}{a_1} \, A^2(x) - \dfrac{2 \, a_2}{(n+2) \, a_1} \, A^{n+2}(x)}} \, d \, A(x), \tag{A.2.37}$$

where c is real constant of integration. In the following, we take two categories of c.
If $c \neq 0$:
1. From (A.2.37), with $n = 1$, $x - x_0$ reads

$$x - x_0 = \frac{2 \, (R_3 - R_2) \sqrt{\dfrac{3 \, a_1 \, [R_1 - A(x)] \, [R_3 - A(x)] \, [A(x) - R_2]}{(R_1 - R_3) \, (R_2 - R_3)^2}} \, F\left\{\sin^{-1}\left[\sqrt{\dfrac{R_3 - A(x)}{R_3 - R_2}}\right], \dfrac{R_2 - R_3}{R_1 - R_3}\right\}}{\sqrt{-2 \, a_2 \, A^3(x) + 3 \, a_1 \, c + 3 \, \lambda \, A^2(x)}}, \tag{A.2.38}$$

where R_j, $j = 1, 2, 3$, are the three roots of $Y(x) = 3\,c + \frac{3\lambda x}{a_1} - \frac{a_2\,x^3}{a_1}$, and F is the elliptic integral of the first kind. By solving (A.2.38) for $A(x)$, we obtain

$$A(x) = R_3 + (R_2 - R_3)\ \mathrm{sn}^2\left[\sqrt{\frac{-a_2\,(R_1 - R_3)}{6\,a_1}}\,(x - x_0),\ \frac{R_2 - R_3}{R_1 - R_3}\right], \quad \text{(A.2.39)}$$

where sn is the Jacobi elliptic function. Substituting (A.2.35) and (A.2.39) back into (A.2.1), we get (3.24), where $\frac{R_2 - R_3}{R_1 - R_3} \to m$, $c \to A_1$, and $\lambda \to A_0$.

2. From (A.2.37), with $n = 4$, $x - x_0$ is given by

$$x - x_0 = -\sqrt{\frac{9\,a_1}{a_2\,R_2\,(R_1 - R_3)}}\ \mathrm{F}\left\{\sin^{-1}\left[\sqrt{\frac{A^2(x)\,(R_3 - R_1)}{R_3\,[A^2(x) - R_1]}}\right],\ \frac{R_3\,(R_1 - R_2)}{R_2\,(R_1 - R_3)}\right\}, \quad \text{(A.2.40)}$$

where R_j, $j = 1, 2, 3$, are the three roots of $m_0 = 3\,c + \frac{3\lambda x^2}{a_1} - \frac{2\,a_2\,x^3}{a_1}$, and F gives the elliptic integral of the first kind. By solving (A.2.40) for $A(x)$, we obtain

$$A(x) = -\frac{\sqrt{R_1}\ \mathrm{sn}\left[\sqrt{\dfrac{a_2\,R_2\,(R_1 - R - 3)}{3\,a_1}}\,(x - x_0),\ \dfrac{R_3\,(R_1 - R_2)}{R_2\,(R_1 - R_3)}\right]}{\sqrt{\dfrac{R_1 - R_3}{R_3} + \mathrm{sn}^2\left[\sqrt{\dfrac{a_2\,R_2\,(R_1 - R - 3)}{3\,a_1}}\,(x - x_0),\ \dfrac{R_3\,(R_1 - R_2)}{R_2\,(R_1 - R_3)}\right]}}, \quad \text{(A.2.41)}$$

where sn is the Jacobi elliptic function. Substituting (A.2.35) and (A.2.41) back into (A.2.1), we get (3.25), where $\frac{R_3\,(R_1 - R_2)}{R_2\,(R_1 - R_3)} \to m$, $c \to A_1$, and $\lambda \to A_0$.

If $c = 0$:

Solving (A.2.37) for $A(x)$ with $c = 0$, we get

$$A(x) = \left\{\frac{\lambda\,(2 + n)}{2\,a_2}\ \mathrm{sech}^2\left[\sqrt{\frac{n^2\,\lambda}{4\,a_1}}\,(x - x_0)\right]\right\}^{\frac{1}{n}}. \quad \text{(A.2.42)}$$

Substituting (A.2.35) and (A.2.41) back into (A.2.1), we get (3.21), where $\lambda \to \frac{4\,a_1\,A_0^2}{n^2}$.

Case B6:

$$Z(x, t) = A(x), \quad \text{(A.2.43)}$$

$$\phi(x, t) = \phi(x, t), \quad \text{(A.2.44)}$$

where $A(x)$ is a real function to be determined. Substituting back into (A.2.2) and (A.2.3) leads to the following Re[equation(3.9)] and Im[equation(3.9)]:

Re[equation(3.9)]: $a_2\,A^{1+n}(x) + a_1\,A''(x) - A(x)\,[\phi_t(x, t) + a_1\,\phi_x^2(x, t)] = 0$, (A.2.45)

Im[equation(3.9)]: $a_1\,[2\,A'(x)\,\phi_x(x, t) + A(x)\,\phi_{xx}(x, t)] = 0$. (A.2.46)

Solving (A.2.46) for $\phi(x, t)$

$$\phi(x, t) = \int \frac{s_1(t)}{A^2(x)}\, d\,x + s_2(t), \qquad (A.2.47)$$

where $s_1(t)$ and $s_2(t)$ are two real functions to be determined. Taking a special case of these two functions, $s_1(t) = \lambda_0$ and $s_2(t) = \lambda_2\,(t - t_0)$, the Equation (A.2.45) becomes

$$\text{Re[equation(3.9)]:} \quad a_2\,A^{1+n}(x) - \frac{a_1\,\lambda_0^2}{A^3(x)} - \lambda_2\,A(x) + a_1\,A''(x) = 0, \quad (A.2.48)$$

where λ_0 and λ_2 are arbitrary real constants. Using the chain rule $A''(x) = \frac{A'(x)\,d\,A'(x)}{d\,A(x)}$, we get the following integral of the independent variable x:

$$x - x_0 = \int \frac{1}{\sqrt{c + \dfrac{\lambda}{a_1}\,A^2(x) - \dfrac{2\,a_2}{(n + 2)\,a_1}\,A^{n+2}(x) - \dfrac{a_1\,\lambda_0^2}{A^2(x)}}}\, d\,A(x). \qquad (A.2.49)$$

Solving (A.2.49) for $A(x)$ with $n = 4$, we get

$$A(x) = \frac{\sqrt{R_1\,(R_2 - R_4) + R_2\,(R_4 - R_1)\,\text{sn}^2\left[\sqrt{\dfrac{-a_2\,(R_1 - R_3)\,(R_2 - R_4)}{3\,a_1}}\,(x - x_0),\, \dfrac{(R_2 - R_3)\,(R_1 - R_4)}{(R_1 - R_3)\,(R_2 - R_4)}\right]}}{\sqrt{-R_2 + R_4 + (R_1 - R_4)\,\text{sn}^2\left[\sqrt{\dfrac{-a_2\,(R_1 - R_3)\,(R_2 - R_4)}{3\,a_1}}\,(x - x_0),\, \dfrac{(R_2 - R_3)\,(R_1 - R_4)}{(R_1 - R_3)\,(R_2 - R_4)}\right]}}, \qquad (A.2.50)$$

and hence, from (A.2.47), $\phi(x, t)$ will read

$$\phi(x, t) = \frac{A_0\left(\sqrt{\dfrac{-3\,a_1\,(R_1 - R_2)^2}{a_2\,m_2}}\,\Pi\left\{\dfrac{R_2\,(R_1 - R_4)}{R_1\,(R_2 - R_4)},\, \text{am}\left[\sqrt{\dfrac{-a_2\,m_2}{3\,a_1}}\,(x - x_0),\, m\right],\, m\right\}\,\text{dn}\left[\sqrt{\dfrac{-a_2\,m_2}{3\,a_1}}\,(x - x_0),\, m\right]\right)}{R_1\,R_2\,\sqrt{\dfrac{m_2 + (R_3 - R_2)\,(R_1 - R_4)\,\text{sn}^2\left[\sqrt{\dfrac{-a_2\,m_2}{3\,a_1}}\,(x - x_0),\, m\right]}{m_2}}}$$

$$+ \frac{A_0}{R_2}\,(x - x_0) + A_2\,(t - t_0) + \phi_0, \qquad (A.2.51)$$

where $m = \frac{m_1}{m_2}$, R_j, $j = 1, 2, 3, 4$, are the four roots of $Y(x) = 3\,a_1^2\,\lambda_0^2 - 3\,a_1\,c\,x - 3\,\lambda_2\,x^2 + a_2\,x^4$, $m_1 = (R_2 - R_3)\,(R_1 - R_4)$, and $m_2 = (R_1 - R_3)\,(R_2 - R_4)$. Here, sn and dn are Jacobi elliptic functions, Π is the incomplete elliptic integral, am is the amplitude for Jacobi elliptic functions. Substituting (A.2.50) and (A.2.51) back into (A.2.1) leads to (3.26), where $\lambda_0 \to A_0$, $c \to A_1$ and $\lambda_2 \to A_2$.

Case B7:

$$Z(x, t) = A(t), \qquad (A.2.52)$$

$$\phi(x,\,t) = \phi(x,\,t), \tag{A.2.53}$$

where $A(t)$ is a real function to be determined. Substituting back into (A.2.2) and (A.2.3) leads to the following forms of Re[equation(3.9)] and Im[equation(3.9)]:

$$\text{Re[equation(3.9)]:} \quad A(t)\,[a_2\,A''(t) - \phi_t(x,\,t) - a_1\,\phi_x^2(x,\,t)] = 0, \tag{A.2.54}$$

$$\text{Im[equation(3.9)]:} \quad A'(t) + a_1\,A(t)\,\phi_{xx}(x,\,t) = 0. \tag{A.2.55}$$

Solving (A.2.55) for $\phi(x,\,t)$ reads

$$\phi(x,\,t) = s_1(t) + s_2(t)\,(x - x_0) - \frac{(x - x_0)^2\,A'(t)}{2\,a_1\,A(t)}, \tag{A.2.56}$$

where $s_1(t)$ and $s_2(t)$ are real functions to be determined. Using the later expression of $\phi(x,\,t)$ in (A.2.54) gives

$$\text{Re[equation(3.9)]:} \quad A(t)[a_2\,A''(t) - a_1\,s_2^2(t) - s_1'(t)] + [2\,s_2(t)\,A'(t) - s_2'(t)\,A(t)]\,(x - x_0)$$

$$+ \left[\frac{A(t)\,A''(t) - 3\,A'^2(t)}{2\,a_1\,A(t)}\right]\,(x - x_0)^2 = 0. \tag{A.2.57}$$

Collecting coefficients of $(x - x_0)^0$, $(x - x_0)$, and $(x - x_0)^2$, separately, and equating to zero, we obtain:

$$(x - x_0)^0: \quad a_2\,A''(t) - a_1\,s_2^2(t) - s_1'(t) = 0, \tag{A.2.58}$$

$$(x - x_0)^1: \quad 2\,s_2(t)\,A'(t) - s_2'(t)\,A(t) = 0, \tag{A.2.59}$$

$$(x - x_0)^2: \quad \frac{A(t)\,A''(t) - 3\,A'^2(t)}{2\,a_1\,A(t)} = 0. \tag{A.2.60}$$

Solving for $A(t)$, $s_1(t)$, and $s_2(t)$, we get

$$A(t) = \frac{A_0}{\sqrt{A_1 + 2\,(t - t_0)}}, \tag{A.2.61}$$

$$s_1(t) = \frac{a_1\,A_2^2}{2\,A_1 + 4\,(t - t_0)} + \frac{a_2\,(A_1 + 2\,t)}{(2 - n)}\left[\frac{A_0}{\sqrt{A_1 + 2\,(t - t_0)}}\right]^n + \phi_0, \tag{A.2.62}$$

$$s_2(t) = \frac{A_2}{A_1 + 2\,(t - t_0)}. \tag{A.2.63}$$

Substituting (A.2.61), (A.2.62), and (A.2.63) in (A.2.56) and then back into (A.2.1) leads to (3.17).

A.3 Derivation of Some Solutions of Section 3.4

A.3.1 Schematic Representation

Equation
$i\,\psi_t + a_1\,\psi_{xx} + a_2\,|\psi|^n\,\psi + a_3\,|\psi|^m\,\psi = 0$

	Solutions
	Case C1: $\psi(x,\,t) = A_0\,e^{i\,\phi(t)}$
Solution (3.28)	
	Case C2: $\psi(x,\,t) = A_0\,e^{i\,\phi(x)}$
Solution (3.29)	
	Case C3: $\psi(x,\,t) = A_0\,e^{i\,\phi(x,\,t)}$
Solution (3.30)	
	Case C4: $\psi(x,\,t) = A(t)\,e^{i\,\phi(x,\,t)}$
Solution (3.31)	

A.3.2 Detailed Derivations

The general solution to Equation (3.27) can be written in the polar form:

$$\psi(x,\,t) = Z\,e^{i\,\phi}, \tag{A.3.1}$$

where $Z = Z(x,\,t)$ and $\phi = \phi(x,\,t)$ are real functions. Substituting (A.3.1) in (3.27) generates its real and imaginary parts:

Re[equation(3.27)]: $\quad a_3\,Z^{m+1} + a_2\,Z^{n+1} - Z\,\phi_t - a_1\,Z\,\phi_x^2 + a_1\,Z_{xx} = 0,$ (A.3.2)

and

Im[equation(3.27)]: $\quad Z_t + 2\,a_1\,Z_x\,\phi_x + a_1\,Z\,\phi_{xx} = 0.$ (A.3.3)

In the following, we take cases for Z and φ:
Case C1:

$$Z(x,\,t) = A_0, \tag{A.3.4}$$

$$\phi(x,\,t) = \phi(t). \tag{A.3.5}$$

Substituting back into (A.3.2) and (A.3.3) leads to the following Re[equation(3.27)] and Im[equation(3.27)]:

Re[equation(3.27)]: $\quad A_0\,[A_0^n\,a_2 + A_0^m\,a_3 - \phi'(t)] = 0,$ (A.3.6)

Im[equation(3.27)]: $\quad 0 = 0.$ (A.3.7)

Solving (A.3.6) for $\phi(t)$ gives

$$\phi(t) = [A_0^n\, a_2 + A_0^m\, a_3]\,(t - t_0) + \phi_0. \tag{A.3.8}$$

Substituting (A.3.4) and (A.3.8) back into (A.3.1) leads to solution (3.28).
Case C2:

$$Z(x,\, t) = A_0, \tag{A.3.9}$$

$$\phi(x,\, t) = \phi(x). \tag{A.3.10}$$

Substituting (A.3.9) and (A.3.10) back into (A.3.2) and (A.3.3) leads to the following Re[equation(3.27)] and Im[equation(3.27)]:

Re[equation(3.27)]: $A_0\,[A_0^n\, a_2 + A_0^m\, a_3 - a_1\,\phi'^2(x)] = 0, \tag{A.3.11}$

Im[equation(3.27)]: $A_0\, a_1\,\phi''(x) = 0. \tag{A.3.12}$

Solving (A.3.12) for $\phi(x)$ gives

$$\phi(x) = \pm\left(A_0^{n/2}\,\sqrt{\frac{a_2}{a_1}} + A_0^{m/2}\,\sqrt{\frac{a_3}{a_1}}\,\right)(x - x_0) + \phi_0. \tag{A.3.13}$$

Substituting (A.3.9) and (A.3.13) back into (A.3.1) leads to solution (3.29).
Case C3:

$$Z(x,\, t) = A_0, \tag{A.3.14}$$

$$\phi(x,\, t) = \phi(x,\, t). \tag{A.3.15}$$

Substituting back into (A.3.2) and (A.3.3) leads to the following Re[Equation (3.27)] and Im[Equation (3.27)]:

Re[equation(3.27)]: $A_0\,[A_0^n\, a_2 + A_0^m\, a_3 - \phi_t(x,\, t) - a_1\,\phi_x^2(x,\, t)] = 0, \tag{A.3.16}$

Im[equation(3.27)]: $A_0\, a_1\,\phi_{xx}(x,\, t) = 0. \tag{A.3.17}$

Solving (A.3.17) for $\phi(x,\, t)$ gives

$$\phi(x,\, t) = s_1(t) + (x - x_0)\, s_2(t), \tag{A.3.18}$$

where $s_1(t)$ and $s_2(t)$ are real functions of t. Substituting back into (A.3.16), collecting coefficients of $(x - x_0)^0$ and $(x - x_0)$, separately, and equating to zero, we obtain:

$(x - x_0)^0$: $A_0\,[A_0^n\, a_2 + A_0^m\, a_3 - a_1\, s_2^2(t) - s_1'(t)] = 0, \tag{A.3.19}$

$(x - x_0)$: $-A_0\, s_2'(t) = 0. \tag{A.3.20}$

Solving (A.3.20) for $s_2(t)$ gives

$$s_2(t) = A_1. \tag{A.3.21}$$

Substituting this result in (A.3.19) and solving for $s_1(t)$, we get

$$s_1(t) = (A_0^n \, a_2 + A_0^m \, a_3 - A_1^2 \, a_1) \, (t - t_0) + \phi_0. \tag{A.3.22}$$

The final form of $\phi(x, t)$ in (A.3.18) will be

$$\phi(x, t) = A_1 \, (x - x_0) + (A_0^n \, a_2 + A_0^m \, a_3 - a_1 \, A_1^2) \, (t - t_0) + \phi_0. \tag{A.3.23}$$

Using (A.3.23) back into (A.3.1) leads to solution (3.30).

Case C4:

$$Z(x, t) = A(t), \tag{A.3.24}$$

$$\phi(x, t) = \phi(x, t), \tag{A.3.25}$$

where $A(t)$ is a real function to be determined. Substituting back into (A.3.2) and (A.3.3) leads to the following Re[equation(3.27)] and Im[equation(3.27)]:

Re[equation(3.27)]: $\quad A(t) \, [a_2 \, A^n(t) + a_3 \, A^m(t) - \phi_t(x, t) - a_1 \, \phi_x^2(x, t)] = 0 \tag{A.3.26}$

Im[equation(3.27)]: $\quad A'(t) + a_1 \, A(t) \, \phi_{xx}(x, t) = 0. \tag{A.3.27}$

Solving (A.3.27) for $\phi(x, t)$

$$\phi(x, t) = s_1(t) + s_2(t) \, (x - x_0) - \frac{(x - x_0)^2 \, A'(t)}{2 \, a_1 \, A(t)}, \tag{A.3.28}$$

where $s_1(t)$ and $s_2(t)$ are real functions to be determined. Using the expression of $\phi(x, t)$ in (A.3.26), we get

Re[equation(3.27)]: $\quad A(t)[a_2 \, A^n(t) + a_3 \, A^m(t) - a_1 \, s_2^2(t) - s_1'(t)] + [2 \, s_2(t) \, A'(t) - s_2'(t) \, A(t)] \, (x - x_0)$

$$+ \left[\frac{A(t) \, A''(t) - 3 \, A'^2(t)}{2 \, a_1 \, A(t)} \right] (x - x_0)^2 = 0. \tag{A.3.29}$$

Collecting coefficients of $(x - x_0)^0$, $(x - x_0)$, and $(x - x_0)^2$, separately, and equating to zero, we obtain:

$$(x - x_0)^0: \quad a_2 \, A^n(t) + a_3 \, A^m(t) - a_1 \, s_2^2(t) - s_1'(t) = 0, \tag{A.3.30}$$

$$(x - x_0)^1: \quad 2 \, s_2(t) \, A'(t) - s_2'(t) \, A(t) = 0, \tag{A.3.31}$$

$$(x - x_0)^2: \quad \frac{A(t) \, A''(t) - 3 \, A'^2(t)}{2 \, a_1 \, A(t)} = 0. \tag{A.3.32}$$

Solving for $A(t)$, $s_1(t)$, and $s_2(t)$, we get

$$A(t) = \frac{A_0}{\sqrt{A_1 + 2 \, (t - t_0)}}, \tag{A.3.33}$$

$$s_1(t) = \frac{a_1 A_2^2}{2 A_1 + 4 (t - t_0)} + \frac{a_2 [A_1 + 2 (t - t_0)]}{(2 - n)} \left[\frac{A_0}{\sqrt{A_1 + 2 (t - t_0)}} \right]^n \quad \text{(A.3.34)}$$

$$+ \frac{a_3 [A_1 + 2 (t - t_0)]}{(2 - m)} \left[\frac{A_0}{\sqrt{A_1 + 2 (t - t_0)}} \right]^m + \phi_0, \quad \text{(A.3.35)}$$

$$s_2(t) = \frac{A_2}{A_1 + 2 (t - t_0)}. \quad \text{(A.3.36)}$$

Substituting (A.3.33), (A.3.35), and (A.3.36) in (A.3.28) and then back into (A.3.1) leads to (3.31).

IOP Publishing

Handbook of Exact Solutions to the Nonlinear Schrödinger Equations (Second Edition)

Usama Al Khawaja and Laila Al Sakkaf

Appendix B

Darboux Transformation: Single Soliton and Breather Solutions

B.1 Darboux Transformation

The fundamental nonlinear Schrödinger equation (NLSE) to be solved is

$$i\, u_t + \frac{1}{2}u_{xx} + |u|^2\, u = 0,\ u = u(x,\, t). \tag{B.1}$$

Darboux transformation applies only for linear differential equations. Therefore, this equation is associated with a linear system as follows.

Consider the field

$$\Phi = \begin{pmatrix} \psi_1 & \psi_2 \\ \phi_1 & \phi_2 \end{pmatrix}, \tag{B.2}$$

with all components being complex functions of x and t. Consider the linear (Zakharov–Shabat) system of differential equations in this field

$$\Phi_x = U \cdot \Phi + J \cdot \Phi \cdot \Lambda, \tag{B.3}$$

$$\Phi_t = V \cdot \Phi + i\, U \cdot \Phi \cdot \Lambda + i\, J \cdot \Phi \cdot \Lambda^2, \tag{B.4}$$

where

$$U = \begin{pmatrix} 0 & u \\ -u^* & 0 \end{pmatrix}, \tag{B.5}$$

$$V = \frac{i}{2}\begin{pmatrix} |u|^2 & u_x \\ u_x^* & -|u|^2 \end{pmatrix}, \tag{B.6}$$

$$J = \begin{pmatrix} 1 & 0 \\ 0 & -1 \end{pmatrix}, \tag{B.7}$$

$$\Lambda = \begin{pmatrix} \lambda_1 & 0 \\ 0 & \lambda_2 \end{pmatrix}. \tag{B.8}$$

The arbitrary complex constants, $\lambda_{1,2}$, are called the *spectral parameters*. We refer to (B.1.3) and (B.1.4) as the *Lax pair* (LP). It refers also sometimes to the matrices U and V.

Link between NLSE and LP

The *compatibility condition*

$$\Phi_{xt} = \Phi_{tx}, \tag{B.9}$$

requires

$$U_t - V_x + [U, V] = 0, \tag{B.10}$$

where $[U, V]$ is the commutator of U and V. Substituting for U and V from (B.1.5) and (B.1.6), the compatibility condition reads

$$\begin{pmatrix} -i\,\phi_1\left(i\,u_t + \frac{1}{2}u_{xx} + |u|^2\,u\right) & -i\,\phi_2\left(i\,u_t + \frac{1}{2}u_{xx} + |u|^2\,u\right) \\ -i\,\psi_1\left(-i\,u_t^* + \frac{1}{2}u_{xx}^* + |u|^2\,u^*\right) & -i\,\psi_2\left(-i\,u_t^* + \frac{1}{2}u_{xx}^* + |u|^2\,u^*\right) \end{pmatrix} = 0. \tag{B.11}$$

Clearly, the compatibility condition is nothing but the NLSE and its complex conjugate; it requires that u is a solution to the NLSE and u^* is a solution to the complex conjugate of the NLSE. This is the link between the NLSE and the linear system.

Seed solution

For a given (seed) solution of the NLSE, namely u_0, the linear system will have a solution Φ_0.

Darboux transformation

Darboux transformation is defined as

$$\Phi[1] = \Phi \cdot \Lambda - \sigma\,\Phi, \tag{B.12}$$

where

$$\sigma = \Phi_0 \cdot \Lambda \cdot \Phi_0^{-1}. \tag{B.13}$$

Here, Φ_0 is a seed solution of the linear system for a given seed solution of the NLSE, Φ denotes any solution of the linear system, and $\Phi[1]$ is the transformed (new) solution of the linear system.

We request that the LP, (B.1.3) and (B.1.4), is *covariant* under the Darboux transformation

$$\Phi[1]_x = U[1] \cdot \Phi[1] + J \cdot \Phi[1] \cdot \Lambda, \tag{B.14}$$

$$\Phi[1]_t = V[1] \cdot \Phi[1] + i\, U[1] \cdot \Phi[1] \cdot \Lambda + i\, J \cdot \Phi[1] \cdot \Lambda^2. \tag{B.15}$$

By substituting in this system for $\Phi[1]$ from (B.1.12) and using (B.1.5) and (B.1.6), the transformed LP $U[1]$ and $V[1]$, must satisfy

$$U[1] = U_0 + [J, \sigma], \tag{B.16}$$

$$V[1] = V_0 + i\,[U_0, \sigma], \tag{B.17}$$

where U_0 and V_0 are the LP in terms of the seed solution. It should be noted that J and Λ are constant matrices and do not change under the Darboux transformation. The new solution, $u[1]$, is obtained from the last equation by noting that

$$
U[1] = \begin{pmatrix} 0 & u[1] \\ -u[1]^* & 0 \end{pmatrix}
$$
$$
= \begin{pmatrix} 0 & u_0 \\ -u_0^* & 0 \end{pmatrix} + \begin{pmatrix} 0 & \dfrac{2(\lambda_1 - \lambda_2)\psi_1\psi_2}{\varphi_1\psi_2 - \varphi_2\psi_1} \\ \dfrac{2(\lambda_1 - \lambda_2)\varphi_1\varphi_2}{\varphi_1\psi_2 - \varphi_2\psi_1} & 0 \end{pmatrix}, \tag{B.18}
$$

and

$$V[1] = \frac{i}{2}\begin{pmatrix} |u[1]|^2 & u[1]_x \\ u[1]_x^* & -|u[1]|^2 \end{pmatrix}, \tag{B.19}$$

lead to a covariant compatibility condition

$$U[1]_t - V[1]_x + [U[1], V[1]] = 0. \tag{B.20}$$

This means that $u[1]$ is a solution of the NLSE

$$i\, u[1]_t + \frac{1}{2}u[1]_{xx} + |u[1]|^2 = 0. \tag{B.21}$$

Thus, from the solution u to the NLSE, we obtained a new solution $u[1]$ to the same NLSE. The new solution is extracted from (B.1.18) as

$$u[1] = u_0 + \frac{2(\lambda_1 - \lambda_2)\,\psi_1\,\psi_2}{\varphi_1\,\psi_2 - \varphi_2\,\psi_1}, \tag{B.22}$$

together with its complex conjugate

$$u[1]^* = u_0^* - \frac{2(\lambda_1 - \lambda_2)\,\varphi_1\,\varphi_2}{\varphi_1\,\psi_2 - \varphi_2\,\psi_1}.$$

(B.23)

The second term on the right hand side of the last two equations is called the *Darboux dressing*.

Symmetry reduction

For a general seed the linear system reads
  x-**equations:**

$$\psi_{1x} - u\,\phi_1 - \lambda_1\,\psi_1 = 0,$$

(B.24)

$$\psi_{2x} - u\,\phi_2 - \lambda_2\,\psi_2 = 0,$$

(B.25)

$$\phi_{1x} + u^*\,\psi_1 + \lambda_1\,\phi_1 = 0,$$

(B.26)

$$\phi_{2x} + u^*\,\psi_2 + \lambda_2\,\phi_2 = 0.$$

(B.27)

t-**equations:**

$$i\,\psi_{1t} + \left(\frac{1}{2}|u|^2 + \lambda_1^2\right)\psi_1 + \left(\frac{1}{2}u_x + \lambda_1\,u\right)\phi_1 = 0,$$

(B.28)

$$i\,\psi_{2t} + \left(\frac{1}{2}|u|^2 + \lambda_2^2\right)\psi_2 + \left(\frac{1}{2}u_x + \lambda_2\,u\right)\phi_2 = 0,$$

(B.29)

$$i\,\phi_{1t} - \left(\frac{1}{2}|u|^2 + \lambda_1^2\right)\phi_1 + \left(\frac{1}{2}u_x^* - \lambda_1\,u^*\right)\psi_1 = 0,$$

(B.30)

$$i\,\phi_{2t} - \left(\frac{1}{2}|u|^2 + \lambda_2^2\right)\phi_2 + \left(\frac{1}{2}u_x^* - \lambda_2\,u^*\right)\psi_2 = 0.$$

(B.31)

Symmetry reduction:
  With the relations:

$$\phi_2^* = \psi_1,$$

(B.32)

$$\psi_2^* = -\phi_1,$$

(B.33)

$$\lambda_2^* = -\lambda_1,$$

(B.34)

the linear system of eight equations reduces to four equations

$$\psi_{1x} - u\,\phi_1 - \lambda_1\,\psi_1 = 0,$$

(B.35)

$$\phi_{1x} + u^* \psi_1 + \lambda_1 \phi_1 = 0, \tag{B.36}$$

$$i \psi_{1t} + \left(\frac{1}{2}|u|^2 + \lambda_1^2\right) \psi_1 + \left(\frac{1}{2}u_x + \lambda_1 u\right) \phi_1 = 0, \tag{B.37}$$

$$i \phi_{1t} - \left(\frac{1}{2}|u|^2 + \lambda_1^2\right) \phi_1 + \left(\frac{1}{2}u_x^* - \lambda_1 u^*\right) \psi_1 = 0. \tag{B.38}$$

The new solution (B.1.22) then takes the form

$$u[1] = u + \frac{2(\lambda_1 + \lambda_1^*) \psi_1 \phi_1^*}{|\phi_1|^2 + |\psi_1|^2}, \tag{B.39}$$

and

$$u[1]^* = u^* + \frac{2(\lambda_1 + \lambda_1^*) \phi_1 \psi_1^*}{|\phi_1|^2 + |\psi_1|^2}. \tag{B.40}$$

B.1.1 Bright Soliton Solution: Zero Seed

For $u = 0$, the linear system simplifies to

$$\psi_{1x} - \lambda_1 \psi_1 = 0, \tag{B.41}$$

$$\phi_{1x} + \lambda_1 \phi_1 = 0, \tag{B.42}$$

$$i \psi_{1t} + \lambda_1^2 \psi_1 = 0, \tag{B.43}$$

$$i \phi_{1t} - \lambda_1^2 \phi_1 = 0, \tag{B.44}$$

with general solution

$$\psi_1 = c_1 e^{\lambda_1 x + i \lambda_1^2 t}, \tag{B.45}$$

$$\phi_1 = c_2 e^{-\lambda_1 x - i \lambda_1^2 t}. \tag{B.46}$$

and upon employing the symmetry reduction, we get

$$\psi_2 = -c_2^* e^{-\lambda_1^* x + i \lambda_1^{*2} t}, \tag{B.47}$$

$$\phi_2 = c_1^* e^{\lambda_1^* x - i \lambda_1^{*2} t}. \tag{B.48}$$

The new solution is then given by

$$u[1] = \frac{2(\lambda_1 + \lambda_1^*) c_1 e^{\lambda_1 x + i \lambda_1^2 t} c_2^* e^{-\lambda_1^* x + i \lambda_1^{*2} t}}{|c_2|^2 e^{-(\lambda_1^* + \lambda_1) x - i (\lambda_1^2 - \lambda_1^{*2}) t} + |c_1|^2 e^{(\lambda_1^* + \lambda_1) x + i (\lambda_1^2 - \lambda_1^{*2}) t}}. \tag{B.49}$$

which, without loss of generality, simplifies to

$$u[1] = \alpha \operatorname{sech}\left[\alpha\left(x - x_0 - v\,t\right)\right] e^{i\left[\frac{1}{2}(\alpha^2 - v^2)\,t + v\,x + \phi_0\right]}, \tag{B.50}$$

where $\alpha = 2\lambda_{1r}$, $v = 2\lambda_{1i}$, $c_2/c_1 = e^{2\lambda_{1r}\,x_0 + i\,\phi_0}$, and ϕ_0 is an arbitrary real constant.

Remark: This is the bright soliton solution (2.171) characterized by four arbitrary parameters: initial position x_0, initial speed v, amplitude (or inverse width) α, and arbitrary global phase ϕ_0.

B.1.2 Generalized Breather Solution for Focusing and Defocusing Nonlinearity: CW Seed

Here, we derive the generalized breather solution of the fundamental NLSE

$$i\,u_t + \frac{1}{2}u_{xx} - c\,|u|^2\,u = 0, \tag{B.51}$$

where $c = 1(-1)$ corresponds to the defocusing (focusing) case.

The corresponding LP takes the form of (B.1.3) and (B.1.4) where

$$U = \sqrt{-c}\begin{pmatrix} 0 & u \\ -u^* & 0 \end{pmatrix}, \tag{B.52}$$

$$V = \frac{i}{2}\begin{pmatrix} -c\,|u|^2 & \sqrt{-c}\,u_x \\ \sqrt{-c}\,u_x^* & c\,|u|^2 \end{pmatrix}, \tag{B.53}$$

and J and Λ are given by (B.1.7) and (B.1.8). The linear system (B.1.3) and (B.1.4) reads explicitly

x-equations:

$$-i\,\sqrt{c}\,u\,\phi_1 - \lambda_1\,\psi_1 + \psi_{1x} = 0, \tag{B.54}$$

$$-i\,\sqrt{c}\,u\,\phi_2 - \lambda_2\,\psi_2 + \psi_{2x} = 0, \tag{B.55}$$

$$\lambda_1\,\phi_1 + i\,\sqrt{c}\,u^*\,\psi_1 + \phi_{1x} = 0, \tag{B.56}$$

$$\lambda_2\,\phi_2 + i\,\sqrt{c}\,u^*\,\psi_2 + \phi_{2x} = 0, \tag{B.57}$$

t-equations:

$$-i\,\lambda_1^2\,\psi_1 + (\sqrt{c}\,\lambda_1\,\phi_1 + \frac{1}{2}\,i\,c\,u^*\,\psi_1)\,u + \psi_{1t} + \frac{1}{2}\,\sqrt{c}\,\phi_1\,u_x = 0, \tag{B.58}$$

$$-i\,\lambda_2^2\,\psi_1 + (\sqrt{c}\,\lambda_2\,\phi_2 + \frac{1}{2}\,i\,c\,u^*\,\psi_2)\,u + \psi_{2t} + \frac{1}{2}\,\sqrt{c}\,\phi_2\,u_x = 0, \tag{B.59}$$

$$\frac{1}{2}\,i\,(2\lambda_1^2 - c\,|u|^2)\,\phi_1 - \sqrt{c}\,\lambda_1\,u^*\,\psi_1 + \phi_{1t} + \frac{1}{2}\,\sqrt{c}\,\psi_1\,u_x^* = 0, \tag{B.60}$$

$$\frac{1}{2} i \, (2\lambda_2^2 - c \, |u|^2) \, \phi_2 - \sqrt{c} \, \lambda_2 \, u^* \, \psi_2 + \phi_{2t} + \frac{1}{2} \sqrt{c} \, \psi_2 \, u_x^* = 0. \tag{B.61}$$

Symmetry reduction:

Requiring the complex conjugate of (B.1.54) and (B.1.55) to be identical with (B.1.57) and (B.1.56), respectively, is possible with

$$\phi_2 = \psi_1^*, \tag{B.62}$$

$$\psi_2 = c \, \phi_1^*, \tag{B.63}$$

$$\lambda_2 = -\lambda_1^*, \tag{B.64}$$

where the system of eight equations reduces to the four equations

$$-i \, \sqrt{c} \, u \, \phi_1 - \lambda_1 \, \psi_1 + \psi_{1x} = 0, \tag{B.65}$$

$$-i \, \sqrt{c} \, u^* \, \psi_1 - \lambda_1 \, \phi_1 - \phi_{1x} = 0, \tag{B.66}$$

$$-\frac{1}{2} i \, (2\lambda_1^2 - c \, |u|^2) \, \psi_1 + \sqrt{c} \, \lambda_1 \, u \, \phi_1 + \psi_{1t} + \frac{1}{2} \sqrt{c} \, \phi_1 \, u_x = 0, \tag{B.67}$$

$$\frac{1}{2} i \, (2\lambda_1^2 - c \, |u|^2) \, \phi_1 - \sqrt{c} \, \lambda_1 \, u^* \, \psi_1 + \phi_{1t} + \frac{1}{2} \sqrt{c} \, \psi_1 \, u_x^* = 0. \tag{B.68}$$

For the CW seed

$$u_0 = A \, e^{i \, c \, A^2 \, t} \tag{B.69}$$

with arbitrary real constant A, the general solution of the reduced linear system (B.1.65)–(B.1.68) is

$$\psi_1(x, t) = [(A\sqrt{c} \, c_1 \, i + c_2 \lambda_1) \sinh((x + i \, \lambda_1 \, t) \, \omega) + c_2 \, \omega \, \cosh((x + i \, \lambda_1 \, t)\omega)] \frac{1}{\omega} e^{-\frac{1}{2} i \, A^2 \, c \, t}, \tag{B.70}$$

$$\phi_1(x, t) = [-(A\sqrt{c} \, c_2 \, i + c_1 \lambda_1) \sinh((x + i \, \lambda_1 \, t) \, \omega) + c_1 \, \omega \, \cosh((x + i \, \lambda_1 \, t)\omega)] \frac{1}{\omega} e^{-\frac{1}{2} i \, A^2 \, c \, t}, \tag{B.71}$$

where

$$\omega = \sqrt{A^2 \, c + \lambda_1^2}, \tag{B.72}$$

and c_1 and c_2 are arbitrary constants. The seed solution of the linear system, thus reads

$$\Phi_0 = \begin{pmatrix} \psi_1 & c \, \phi_1^* \\ \phi_1 & \psi_1^* \end{pmatrix}, \tag{B.73}$$

where the symmetry reductions (B.1.62), (B.1.63), and (B.1.64) have been taken into account and ψ_1 and ϕ_1 are given by (B.1.70) and (B.1.71), respectively.

The new solution, given formally by (B.1.16) and upon using the symmetry reduction conditions, now reads

$$u[1] = u_0 - 4i\,\sqrt{c}\,\lambda_{1r}\,\frac{\psi_1\,\phi_1^*}{c\,|\phi_1|^2 - |\psi_1|^2}, \tag{B.74}$$

where λ_{1r} is the real part of λ_1.

The breather solution can be viewed at as a combination of two *generalized* solitons where each soliton has a localization component and an oscillatory component, both in x and t. This is verified by rewriting the breather solution as

$$u[1] = \left[1 - i\,\frac{8\lambda_{1r}\,c^{3/2}}{A}\,\frac{\cos(\zeta_1 - i\,\chi_1)\cos(\zeta_2 + i\,\chi_2)}{\cos(2\zeta_1) - c\cos(2\zeta_2) + \cosh(2\chi_1) - c\cosh(2\chi_2)}\right] A\,e^{i\,c\,A^2\,t}, \tag{B.75}$$

where

$$\zeta_{1,2} = \kappa\,X_{1,2} - \Omega\,T_{1,2}, \tag{B.76}$$

$$\chi_{1,2} = \frac{X_{1,2}}{\alpha} - \frac{T_{1,2}}{\tau}, \tag{B.77}$$

$$X_{1,2} = x - x_{01,02}, \tag{B.78}$$

$$T_{1,2} = t - T_{01,02}, \tag{B.79}$$

$$\kappa = \mathrm{Im}[\omega], \tag{B.80}$$

$$\Omega = -\mathrm{Re}[\lambda_1\,\omega], \tag{B.81}$$

$$\alpha = \frac{1}{\mathrm{Re}[\omega]}, \tag{B.82}$$

$$\tau = \frac{1}{\mathrm{Im}[\lambda_1\,\omega]}, \tag{B.83}$$

$$x_{02} = x_{01} + \frac{1}{2\lambda_{1r}}\left(\lambda_{1i}\,\mathrm{Im}\left[\frac{\log q}{\omega}\right] + \lambda_{1r}\,\mathrm{Re}\left[\frac{\log q}{\omega}\right]\right), \tag{B.84}$$

$$t_{02} = t_{01} + \frac{1}{2\lambda_{1r}}\,\mathrm{Im}\left[\frac{\log q}{\omega}\right], \tag{B.85}$$

$$q = \frac{\lambda_1 + \omega}{\lambda_1 - \omega}. \tag{B.86}$$

In addition to the arbitrary CW amplitude, A, there are four arbitrary parameters: x_{01}, t_{01}, λ_{1r}, and λ_{1i}:

x_{01}: sets the reference for x, t_{01}: sets the reference for t, λ_{1r} and λ_{1i} set:

α: width of localization in x, τ: width of localization in t, κ: frequency of oscillation in x, Ω: frequency of oscillation in t.

Note that x_{02} and t_{02}, which correspond to the second generalized soliton, are not arbitrary as they are given in terms of the above four arbitrary parameters. Furthermore, since the four parameters α, τ, κ, and Ω, are given in terms of λ_{1r}, and λ_{1i}, only two out these four parameters are to be considered arbitrary while the other two are not. Any two of the four parameters can be chosen to be the arbitrary ones.

IOP Publishing

Handbook of Exact Solutions to the Nonlinear Schrödinger Equations (Second Edition)

Usama Al Khawaja and Laila Al Sakkaf

Appendix C

Derivation of the Similarity Transformations in Chapter 5

C.1 Function Coefficients

Given the generalized NLSE (5.77)

$$i\,\Phi_t + b_1(x,\,t)\,\Phi_{xx} + b_2(x,\,t)\,|\Phi\,|^2\,\Phi + [b_{3r}(x,\,t) + i\,b_{3i}(x,\,t)]\,\Phi = 0, \quad (C.1)$$

we aim at transforming it to the fundamental NLSE

$$p(x,\,t)(i\,\psi_T + a_1\,\psi_{XX} + a_2\,|\psi\,|^2\,\psi) = 0, \quad (C.2)$$

with the scaling (similarity) transformation

$$\Phi(x,\,t) = A(x,\,t)\,e^{i\,B(x,\,t)}\,\psi[X(x,\,t),\,T(x,\,t)], \quad (C.3)$$

where $a_{1,2}$ are arbitrary real constants and $b_{1,2}(x,\,t)$, $b_{3r,i}(x,\,t)$, and $p(x,\,t)$ are arbitrary real functions. The unknown functions $X(x,\,t)$, $T(x,\,t)$, $A(x,\,t)$, and $B(x,\,t)$ are assumed to be real and need to be determined in terms of the function coefficients of the generalized NLSE and the constant coefficients of the fundamental NLSE. It is essential to have the function $p(x,\,t)$ since otherwise the transformation will restrict the coefficients $b_{1,2}(x,\,t)$ to only time-dependent ones and consequently the potential $b_{3r,i}(x,\,t)$ will be real and quadratic in x.

Substituting (C.1.3) in (C.1.1) and requesting the result to take the form of the fundamental NLSE (C.1.2) gives

$$A\,b_1\,p\,T_x^2 = 0, \text{ (from } \psi_{TT}), \quad (C.4)$$

$$e^{i\,B}p[2b_1\,A_x\,T_x + A\,(i\,T_t + b_1\,(2i\,B_x\,T_x + T_{xx}))] = i, \quad \text{(from } \psi_T), \quad (C.5)$$

$$e^{i\,B}p\,b_1\,A\,X_x^2 = a_1, \quad \text{(from } \psi_{XX}), \quad (C.6)$$

$$e^{i\,B} p\, b_2\, A^3 = a_2, \qquad \text{(from } |\psi|^2\psi), \qquad (C.7)$$

$$e^{i\,B} p\, [2b_1\, A_x\, X_x + A\,(i\,X_t + b_1(2i\,B_x\,X_x + X_{xx}))] = 0, \qquad \text{(from } \psi_X), \qquad (C.8)$$

$$e^{i\,B} p\, [i\,A_t + b_1\,(2i\,A_x\,B_x + A_{xx})$$
$$+ A\,(i\,b_{3i} + b_{3r} - B_t - b_1\,(B_x^2 - i\,B_{xx}))] = 0, \text{(from } \psi). \qquad (C.9)$$

The solution of this system is given by (5.31)–(5.37), and

$$p(r,\, t) = \frac{1}{e^{i\,B}\, A\, g_1'}. \qquad (C.10)$$

As a hint on the procedure of solving this system, one starts with solving (C.1.4) for T, then (C.1.5) for p, then (C.1.6) and (C.1.7) for b_1 and b_2, respectively, then the real part of (C.1.8) gives A while the imaginary part gives B, and finally, the real part of (C.1.9) is solved for b_{3r} and the imaginary part of (C.1.9) is solved for b_{3i}.

C.2 Solution-Dependent Transformation

This is very much similar to the previous case. The only difference is in the procedure of solving (C.1.8) and (C.1.9). While in the previous section we set the coefficient of ψ and ψ_X to zero separately, here we require the sum of the two terms to be zero, namely

$$[2b_1\, A_x\, X_x + A\,(i\,X_t + b_1(2i\,B_x\,X_x + X_{xx}))]\,\psi$$
$$+ \left[i\,A_t + b_1\,(2i\,A_x\,B_x + A_{xx}) + A\,(i\,b_{3i} + b_{3r} - B_t - b_1\,(B_x^2 - i\,B_{xx}))\right]\psi_X = 0. \qquad (C.2.1)$$

The two procedures should be both valid since they both lead to satisfying the generalized and fundamental NLSE. Solving the real part of the last equation for b_{3r} and the imaginary part for b_{3i} gives (5.96) and (5.97), respectively. The solutions of $b_{1,2}$, X, T, A, and B remain the same as in the previous case.

C.3 Similarity Transformation for the NLSE in (N + 1)-Dimensions

Given the generalized NLSE in N-spacial dimensions and polar coordinates

$$i\,\Phi_t + b_1(r,\, t)\,(\Phi_{rr} + \frac{N-1}{r}\,\Phi_r) + b_2(x,\, t)\,|\Phi|^2\,\Phi + [b_{3r}(r,\, t) + i\,b_{3i}(r,\, t)]\,\Phi = 0, \qquad (C.3.1)$$

we aim at transforming it to the fundamental NLSE

$$p(r,\, t)(i\,\psi_T + a_1\,\psi_{RR} + a_2\,|\psi|^2\,\psi) = 0, \qquad (C.3.2)$$

with the similarity transformation

$$\Phi(r,\, t) = A(r,\, t)\,e^{i\,B(r,\, t)}\,\psi[R(r,\, t),\, T(r,\, t)], \qquad (C.3.3)$$

where $a_{1,2}$ are arbitrary real constants and $b_{1,2}(r,\, t)$, $b_{3r}(r,\, t)$, $b_{3i}(r,\, t)$, and $p(r,\, t)$ are arbitrary real functions. The unknown functions $R(r,\, t)$, $T(r,\, t)$, $A(r,\, t)$, and $B(r,\, t)$

are assumed to be real and need to be determined in terms of the function coefficients of the generalized NLSE and the constant coefficients of the fundamental NLSE.

Substituting (C.3.3) in (C.3.1) and requesting the result to take the form of the fundamental NLSE (C.3.2) gives

$$A \, b_1 \, p \, T_r^2 = 0, \qquad \text{(from } \psi_{TT}), \tag{C.3.4}$$

$$\frac{1}{r} \, e^{i \, B} p \, [2r \, b_1 \, A_r \, T_r$$
$$+ A \, (i \, r \, T_t + b_1 \, ((N - 1 + 2i \, r \, B_r) \, T_r + r \, T_{rr}))] = i, \qquad \text{(from } \psi_T), \tag{C.3.5}$$

$$e^{i \, B} p \, b_1 \, A \, R_r^2 = a_1, \qquad \text{(from } \psi_{RR}), \tag{C.3.6}$$

$$e^{i \, B} p \, b_2 \, A^3 = a_2, \qquad \text{(from } |\psi|^2 \psi), \tag{C.3.7}$$

$$\frac{1}{r} \, e^{i \, B} p \, [2r \, b_1 \, A_r \, R_r$$
$$+ A \, (i \, r \, R_t + b_1((N - 1 + 2ir \, B_r) \, R_r + r \, R_{rr}))] = 0, \qquad \text{(from } \psi_R), \tag{C.3.8}$$

$$\frac{1}{r} \, e^{i \, B} p \, [i \, r \, A_t + b_1 \, ((N - 1 + 2i \, r \, B_r) \, A_r$$
$$+ r \, A_{rr}) + i \, A \, (r \, b_{3i} + (N - 1) \, b_1 \, B_r$$
$$+ i \, r \, (-b_{3r} + B_t + b_1 \, B_r^2) + r \, b_1 \, B_{rr})] = 0, \qquad \text{(from } \psi). \tag{C.3.9}$$

The solution of this system is given by (6.48)–(6.55). The procedure of solving the system is similar to that described in Section C.1.

www.ingramcontent.com/pod-product-compliance
Lightning Source LLC
Chambersburg PA
CBHW080902170526
45158CB00008B/1958